Advances in Intelligent Systems and Computing

Volume 1156

The series "Advances in Intelligent Systems and Computing" contains publications on theory, applications, and design methods of Intelligent Systems and Intelligent Computing. Virtually all disciplines such as engineering, natural sciences, computer and information science, ICT, economics, business, e-commerce, environment, healthcare, life science are covered. The list of topics spans all the areas of modern intelligent systems and computing such as: computational intelligence, soft computing including neural networks, fuzzy systems, evolutionary computing and the fusion of these paradigms, social intelligence, ambient intelligence, computational neuroscience, artificial life, virtual worlds and society, cognitive science and systems, Perception and Vision, DNA and immune based systems, self-organizing and adaptive systems, e-Learning and teaching, human-centered and human-centric computing, recommender systems, intelligent control, robotics and mechatronics including human-machine teaming, knowledge-based paradigms, learning paradigms, machine ethics, intelligent data analysis, knowledge management, intelligent agents, intelligent decision making and support, intelligent network security, trust management, interactive entertainment, Web intelligence and multimedia.

The publications within "Advances in Intelligent Systems and Computing" are primarily proceedings of important conferences, symposia and congresses. They cover significant recent developments in the field, both of a foundational and applicable character. An important characteristic feature of the series is the short publication time and world-wide distribution. This permits a rapid and broad dissemination of research results.

**** Indexing: The books of this series are submitted to ISI Proceedings, EI-Compendex, DBLP, SCOPUS, Google Scholar and Springerlink ****

More information about this series at http://www.springer.com/series/11156

Sergey Kovalev · Valery Tarassov ·
Vaclav Snasel · Andrey Sukhanov
Editors

Proceedings of the Fourth International Scientific Conference "Intelligent Information Technologies for Industry" (IITI'19)

Springer

Editors
Sergey Kovalev
Rostovskogo Strelkovogo Polka Narodnogo
Rostov State Transport University
Rostov-on-Don, Russia

Vaclav Snasel
Department of Computer Science
VSB-Technical University of Ostrava
Ostrava-Poruba, Czech Republic

Valery Tarassov
Bauman Moscow State Technical University
Moscow, Russia

Andrey Sukhanov
Rostovskogo Strelkovogo Polka
Rostov State Transport University
Rostov-on-Don, Russia

ISSN 2194-5357 ISSN 2194-5365 (electronic)
Advances in Intelligent Systems and Computing
ISBN 978-3-030-50096-2 ISBN 978-3-030-50097-9 (eBook)
https://doi.org/10.1007/978-3-030-50097-9

This Springer imprint is published by the registered company Springer Nature Switzerland AG
The registered company address is: Gewerbestrasse 11, 6330 Cham, Switzerland

Preface

This volume of Advances in Intelligent Systems and Computing contains papers presented in the main track of IITI 2019, the Fourth International Scientific Conference on Intelligent Information Technologies for Industry held in December 2–7 in Ostrava and Prague, Czech Republic. The conference was jointly co-organized by Rostov State Transport University (Russia) and VŠB – Technical University of Ostrava (Czech Republic) with the participation of Russian Association for Artificial Intelligence (RAAI).

IITI 2019 is devoted to practical models and industrial applications related to intelligent information systems. It is considered as a meeting point for researchers and practitioners to enable the implementation of advanced information technologies into various industries. Nevertheless, some theoretical talks concerning the state of the art in intelligent systems and soft computing were also included into proceedings.

There were 130 paper submissions from 11 countries. Each submission was reviewed by at least three chairs or PC members. We accepted 71 regular papers (55%). Unfortunately, due to limitations of conference topics and edited volumes the program committee was forced to reject some interesting papers, which did not satisfy these topics or publisher requirements. We would like to thank all authors and reviewers for their work and valuable contributions. The friendly and welcoming attitude of conference supporters and contributors made this event a success!

October 2019

Sergey Kovalev
Valery Tarassov
Vaclav Snasel
Andrey Sukhanov

Organization

Organizing Institutes

Rostov State Transport University, Russia
VSB – Technical University of Ostrava, Czech Republic
Russian Association for Artificial Intelligence, Russia

Conference Chairs

Sergey M. Kovalev	Rostov State Transport University, Russia
Václav Snášel	VSB-TU Ostrava, Czech Republic

Organizing Chairs

Alexander N. Guda	Rostov State Transport University, Russia
Pavel Krömer	VSB-TU Ostrava, Czech Republic

Organizing Vice-chair

Andrey V. Sukhanov	Rostov State Transport University, Russia

Conference Organizers

Andrey V. Chernov	Rostov State Transport University, Russia
Anna E. Kolodenkova	Samara State Technical University, Russia
Jan Platoš	VSB-TU Ostrava, Czech Republic
Maria A. Butakova	Rostov State Transport University, Russia
Vitezslav Styskala	VSB-TU Ostrava, Czech Republic
Vladislav S. Kovalev	JSC "NIIAS", Russia

International Program Committee

Aboul Ella Hassanien	Cairo University, Egypt
Alexander I. Dolgiy	JSC "NIIAS", Rostov Branch, Russia
Alexander L. Tulupyev	St. Petersburg Institute for Informatics and Automation of the Russian Academy of Sciences, Russia
Alexander N. Shabelnikov	JSC "NIIAS", Russia
Alexander N. Tselykh	Southern Federal University, Russia
Alexander P. Eremeev	Moscow Power Engineering Institute, Russia
Alexander V. Smirnov	St. Petersburg Institute for Informatics and Automation of the Russian Academy of Sciences, Russia
Alexey B. Petrovsky	Institute for Systems Analysis of Russian Academy of Sciences, Russia
Alexey N. Averkin	Dorodnicyn Computing Centre of Russian Academy of Sciences
Alla V. Zaboleeva-Zotova	Volgograd State Technical University, Russia
Bronislav Firago	Belarusian National Technical University, Belarus
Anton Beláň	Slovak University of Technology in Bratislava, Slovakia
Dusan Husek	Institute of Computer Science, Academy of Sciences of the Czech Republic
Eid Emary	Cairo University, Egypt
Eliska Ochodkova	VSB-Technical University of Ostrava, Czech Republic
František Janíček,	Slovak University of Technology in Bratislava, Slovakia
Gennady S. Osipov	Institute for Systems Analysis of Russian Academy of Sciences, Russia
Georgy B. Burdo	Tver State Technical University, Russia
Habib M. Kammoun	University of Sfax, Tunisia
Hussein Soori	VSB - Technical University of Ostrava, Czech Republic
Igor B. Fominykh	Moscow Power Engineering Institute, Russia
Igor D. Dolgiy	Rostov State Transport University, Russia
Igor N. Rozenberg	JSC "NIIAS", Russia
Igor V. Kotenko	St. Petersburg Institute for Informatics and Automation of the Russian Academy of Sciences, Russia
Ildar Batyrshin	National Polytechnic Institute, Mexico
Ilias K. Savvas	University of Thessaly, Greece
Ivan Zelinka	VSB-Technical University of Ostrava, Czech Republic

Contents

Bayesian Networks and Trust Networks, Fuzzy-Stocastical Modelling

Decision Making Intelligent Systems

Neural Networks

Adaptive Diagnosis Model of Dempster-Shafer Based on Recurrent Neural-Fuzzy Network

Alexaner N. Shabelnikov[1,2], Sergey M. Kovalev[1,2], and Andrey V. Sukhanov[1,2(✉)]

[1] JSC NIIAS, Rostov Branch, Rostov-on-Don, Russia
a.suhanov@rfniias.ru
[2] Rostov State Transport University, Rostov-on-Don, Russia

Abstract. The paper considers a new class of adaptive diagnosis models based on implementation of probability combination scheme of Dempster-Shafer in the form of recurrent neural-fuzzy network. The proposed model is focused on a sequential data processing, where the data about technical state of controlled object is used. The model allows to continuously update its outputs due to the fusion of obtained data with the previously observed one. The continuous clarification of the information and growth of the decision reliability are achieved due to the consequent processing of obtained heterogeneous data, which comprehensively characterizes the controlled object. Architecture of a recurrent neural-fuzzy model is presented and criteria for an adaptive diagnosis model based on the Dempster-Shafer scheme together with the training process for this model are shown. The possible application domains are presented in the paper.

Keywords: Dempster-Shafer theory · Recurrent neural-fuzzy network · Data fusion

1 Introduction

The development of modern systems for the technical diagnosis is based on the hybridization principles with attraction of complex data obtained from various systems for preliminary data collection and processed using several diagnosis models. Here, the key role is played by the dynamical models, which can predict the changes of the technical state for controlled object (CO) based on dynamically changed information continuously observed in the system. Information about the technical state of CO is clarified and reliability of produced decisions is increased, when the new data portion is obtained in the diagnosis model. Some approaches to the construction of similar models can be highlighted [1–3], where time series are predicted with the sequential clarification of data via temporal-difference learning [1]. Nevertheless, the theory of these models is not sufficiently developed.

S. Kovalev et al. (Eds.): IITI 2019, AISC 1156, pp. 3–11, 2020.
https://doi.org/10.1007/978-3-030-50097-9_1

This paper develops a new approach of adaptive network models, which can be used for control over working capacity of technical object and based on analysis of the dynamical information. The proposed model is based on the adaptive form of Dempster-Shafer combination scheme. Recurrent neural-fuzzy network is used as a basis of the model. Because of these properties, the model is called Recurrent Neural-Fuzzy Dempster-Shafer model (RNFDS). The paper is organized as follows. Section 2 provides the typical problem statement of the considered area on the example of railway humping process. Section 3 formalizes the temporal scenario for the humping process. Section 4 shows the criteria for the working capacity assessment. Section 5 presents neural network architecture for Dempster-Shafer scheme implementation. The training process of the neural network is shown in Sect. 5. Section 6 shows the conclusions and future work.

2 Problem Statement

In the presented work, the development of adaptive RNFDS is shown on the example of control over the important item of railway automation, which is car retarder [4].

Car retarder is the executive device, which is used to brake rolling cuts of cars in the railway marshalling process. Several faults can be appeared during the operation of car retarder that reduces its efficiency. Because of this, the most complex and responsible task is the fault prediction and detection during the car retarder operation.

Nowadays, there is a set of methods for fault detection assessing the brake power deviation of car retarder from the passport data. The most common approach [4] is the power assessment via car velocity decrease, when the car properties and its ideal power are known. However, this method has essential disadvantage, which is connected with those fact, that the information about the ideal power is limited, because velocity decrease requires different power over the braking time. As a result, it is difficult to extract really helpful information for the assessment. In addition, because of this fact, prediction models for car retarder control lose their accuracy.

In this work, complex approach to the technical state assessment for car retarder is proposed. The approach is based on the analysis of the whole information about retarder operation, which is consequently obtained at the system during the humping process. The decision making is realized using fusion of data about the car behavior, which is received from various measures obtained during the braking time. Here, the synergy of measures is achieved using the evidence combination for different estimates of the retarder state in the various operation modes. The evidence combination is implemented in the form of Dempster-Shafer scheme.

3 Temporal Scenario of the Considered Dynamical System

Commonly, control process over car retarder can be described as follows (Fig. 1) [5]. Based on the data obtained from measure devices, the control strategy is performed for retarder. This strategy is to define the axles of railway car (active zones), which must be affected during car motion inside braking zone. For each zone, the braking grade and its time interval is computed.

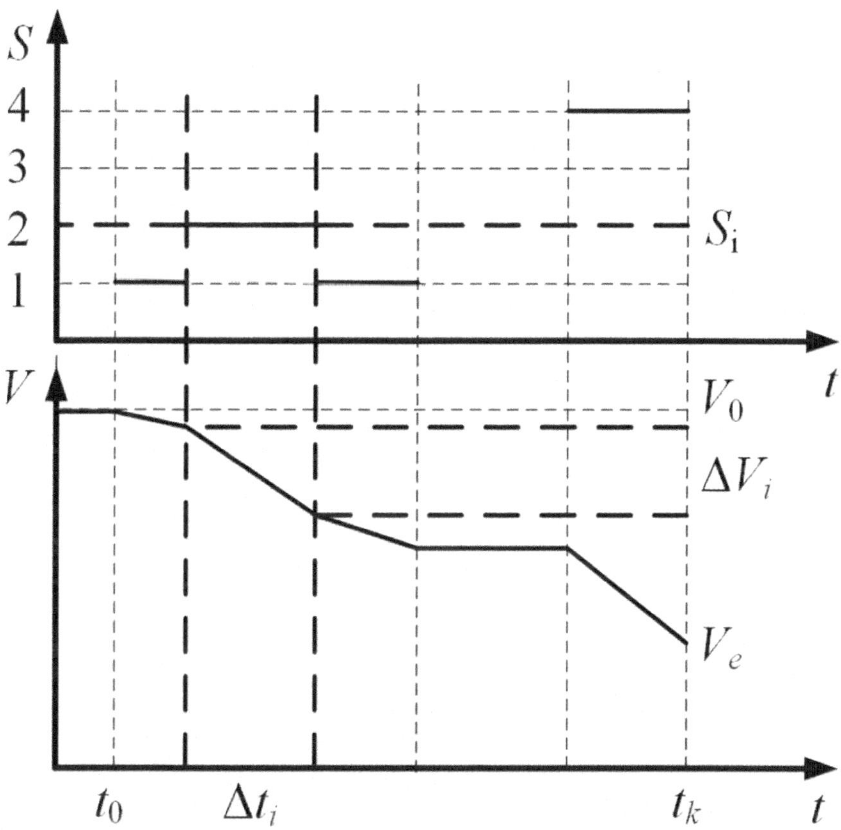

Fig. 1. Schematic illustration of car braking process (the upper figure is the braking grades of retarder, the lower figure is the car velocity change). S is the braking grade, t is the braking time, V is the car velocity.

Figure 1 shows the velocity regulation process, where the braking is performed at the time period of $[t_0, t_k]$ by consequent switching between grades based on the chosen braking strategy. When i-th particular grade S_i is switched on (Fig. 1, upper item), the velocity descent is obtained (Fig. 1, lower item). Changing value ΔV_i depends on the duration of time interval Δt_i, cut properties set P and technical state of the retarder, which defines power characteristics for the chosen grade. In the ideal case (when the retarder is fully serviceable), it is possible to compute velocity decrease via the following dependency:

$$\Delta V_i = f(S_i, P, \Delta t_i) \tag{1}$$

The theoretical braking graph, which corresponds to the chosen braking strategy, can be constructed (Fig. 1, lower item).

When wear or breakage of retarder are obtained, power characteristics are changed. As a result, theoretical value of ΔV_i is markedly different from the real change $\Delta V_i'$. It can be suggested that the difference between ΔV_i and $\Delta V_i'$ reflects the working capacity of car retarder and can be used for the control and prediction of its technical state.

4 Criteria of Working Capacity Based on Evidence Combination

In this paper, the conjunction of the several evidences is used to choose the working capacity criteria. These evidences are made in respect to the working capacity for each braking grade, which is switched on during the braking process.

The process of evidential reasoning can be described on the example of Fig. 1. A cut of railway cars is braked using braking grade S_i with velocity decrease $\Delta V_i'$ on time interval Δt_i. Predicted velocity decrease ΔV_i is computed via Eq. (1). Let δ_i be the following relative estimate, which is used to assess the deviation of ith theoretical decrease from the real one:

$$\delta_i = \frac{\Delta V_i - \Delta V_i'}{\Delta V_i}. \tag{2}$$

δ_i reflects the working capacity of the retarder based on the information about the behavior of a cut on S_i during Δt_i. Particularly, $\delta_i = 0$ means that device is workable. In other cases, the working capacity is decreased.

Let $\{N, A\}$ be the hypotheses set about working capacity of CO characterizing its two global states: N is the normal state, A is the abnormal one. Let linguistic variable *WORKING CAPACITY* characterize these states. *WORKING CAPACITY* depends on δ_i (2) and takes on a linguistic value of *NORMAL* or *ABNORMAL*. Taking into account that working capacity (or probability of normal behavior) depends inversely on δ_i, the membership function (MF) of *NORMAL* can be represented as the Gaussian function decreased in relation to δ:

$$\mu_{NORMAL}(\delta_i) = \exp(-c_1 \cdot \delta_i), \tag{3}$$

where c_1 is the configured parameter.

Analogically, assuming that abnormal state depends directly on δ_i, the membership function of *ABNORMAL* can be represented as the following Gaussian function:

$$\mu_{ABNORMAL}(\delta_i) = 1 - \exp(-c_2 \cdot \delta_i), \tag{4}$$

where c_2 is the configured parameter.

In the present research, functions (3) and (4) are used to compute basic belief assignment (BBA) of the hypotheses from the methodology of Dempster-Shafer for evidence combination [6].

Let R be the complete hypotheses set:

$$R = \{N, A, T\},$$

where N and A are described above, T is the transitional state, which can be interpreted as uncertain, intermediate or pre-failure depending on circumstances. BBAs of N and A are calculated as follows:

$$m(H|\Delta t_i) = \mu_L(\delta_i), \tag{5}$$

where H is the corresponding hypothesis (N or A, respectively) and L is the corresponding linguistic value ($NORMAL$ or $ABNORMAL$, respectively).

BBA of the transitional state can be calculated as follows:

$$m(T|\Delta t_i) = m(N|\Delta t_i) \cdot m(A|\Delta t_i). \tag{6}$$

The following description shows the example of evidential reasoning for consequent time intervals Δt_i and Δt_{i+1} from Fig. 1.

Let two BBA sets $\{m^i(N), m^i(A), m^i(T)\}$ and $\{m^{i+1}(N), m^{i+1}(A), m^{i+1}(T)\}$ characterize the independent evidences about the retarder state, which are based on external information about braked cut behavior at time intervals Δt_i and Δt_{i+1}. These two evidences can be combined using Dempster-Shafer scheme[1]:

$$\begin{array}{c} m^{\Sigma}(N) = m^i(N) \cdot m^{i+1}(N) + m^i(N) \cdot m^{i+1}(T) + m^i(T) \cdot m^{i+1}(N) \\ m^{\Sigma}(A) = m^i(A) \cdot m^{i+1}(A) + m^i(A) \cdot m^{i+1}(T) + m^i(T) \cdot m^{i+1}(A) \\ m^{\Sigma}(T) = m^i(T) \cdot m^{i+1}(T) \end{array} \tag{7}$$

Probabilistic estimations $m^{\Sigma}(N)$, $m^{\Sigma}(A)$ and $m^{\Sigma}(T)$ are the integral BBAs of retarder state's hypothesis, which are obtained for both Δt_i and Δt_{i+1}.

It is obvious, that complete estimation of the working capacity can be made based on the evidence reasoning for the complete time interval $[t_0, t_k]$ including k time subintervals, each of which has separated braking grade. Based on the additivity principle of Dempster-Shafer operations, it can be made by iterative conjunction process for BBAs of all intervals beginning with $i = 1$ and ending with $i = k - 1$. In the following section, this iterative process is performed based on the recurrent neural model.

[1] It should be noted that the used scheme is based on unnormalized conjunction rule of Smets [7]. Here, the probabilistic estimations obtained as a result of conjunction rule may not satisfy to Dempster restriction ($\sum m = 1$). This condition can be eliminated when the used estimations are normalized.

5 Neural Implementation of Dempster-Shafer Scheme

In the considered work, BBA reflects the generalized subjective opinion of an expert about the technical state of car retarder depending on δ (2). This opinion reflects empirical representation of the interaction process between retarder and the axles of railway car taking into account car properties, weather conditions, etc.

Because probabilistic estimations from Eq. (7) mostly depend on MF form, it is required to actualize BBA using statistical information. With this aim, the adaptive diagnosis model based on a recurrent neural-fuzzy network, where the parameters of MF are adjusted to the real statistics of retarder's operation, is developed in this paper.

Initially, adaptive network model of Dempster-Shafer evidence combination was proposed in [8] in the form of neural classifier. The further development of this idea was made in [9,10]. The main shortcoming of this network classifier is a big number of neurons in the hidden layer, which is proportional to the product of number of classes and number of patterns. As a result, this leads to the big number of adjusted parameters.

Presented adaptive recurrent neural-fuzzy model of Dempster-Shafer (RNFDS) is based on the recurrent neural-fuzzy network, where iterative evidence combination mechanism is naturally implemented and fuzzy system of BBA formation is integrated. The number of adjusted parameters is minimized due to the recurrent form of the network. The structure of RNFDS is presented in Fig. 2.

The network consists of 5 layers (L_1–L_5). L_1 is the input layer, L_2 and L_3 are the hidden layers, L_4 is the output layer and L_5 is the delaying layer.

First two layers perform the fuzzy part of RNFDS from Eq. (5) and Eq. (6), were the BBAs are obtained. L_3 and L_4 implement the scheme of Dempster-Shafer from Eq. 7. L_5 provides the delay for the outputs obtained on the previous iterations. According to the structure, accumulated values from $(i-1)$th iteration combined with the ones obtained from ith iteration.

It should be noted, that presented architecture of the network is less complicated and requires less computations per an iteration than one presented in [8]. It is proved in the fact that operation of L_1 has $O(||R||)$ complexity for every input δ ($||\cdot||$ is the power of a set) and combination of basic probabilities in L_2 and L_3 has $O(2||R||)$ complexity. Therefore, the total complexity for every δ is $O(||R||)$ and it has $O(k||R||)$ complexity for k braking actions.

The problem of choice for the resulting hypothesis is the main goal of the decision making based on Dempster-Shaffer theory. To solve this problem, it is required to transform estimations obtained from the scheme (Fig. 2) into the probability function of decision making. In this aim, Smets [7] proposed so-called pignistic transformation. In this transformation, BBA for every hypotheses is distributed equally between all internal elements. Therefore, pignistic probability distribution is defined as:

$$BetP^{\Sigma}(H) = \sum_{h \in H} \Big(\frac{1}{||H||} \cdot \frac{m(h)}{1 - m(\emptyset)} \Big), \forall H \in R \tag{8}$$

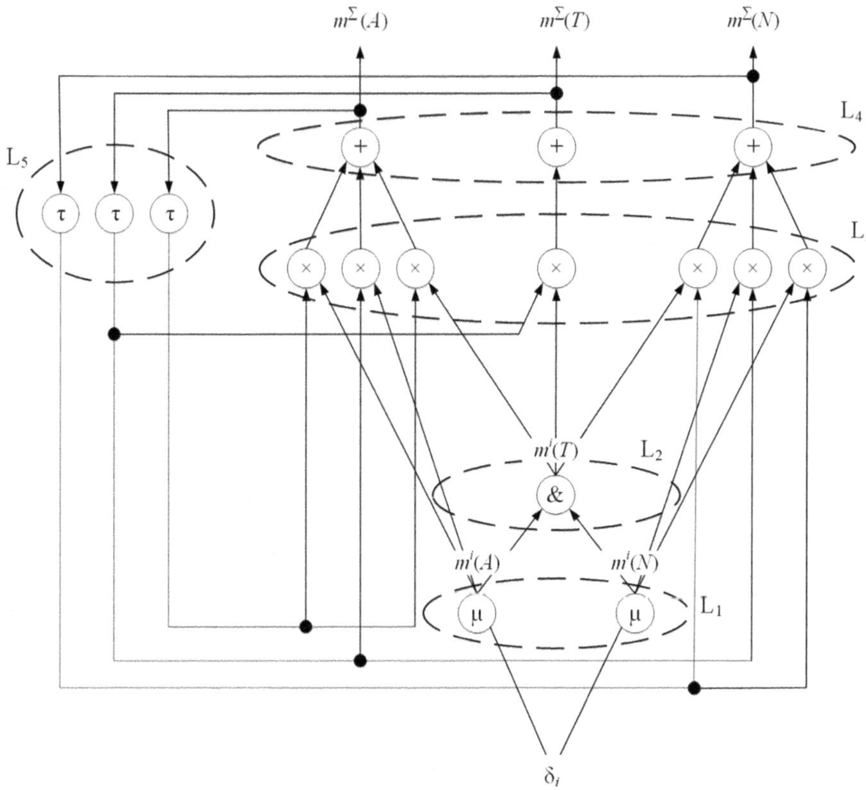

Fig. 2. Structure of RNFDS

The final choice of hypotheses is made according to maximal value of $BetP$ (8).

The procedure of RNFDS training is described below. μ_N and μ_A are adapted parameters. They are computed using conventional backpropagation algorithm. The backpropagation criterion is formed based on the training set using the following procedure.

Training input vector x is the the data obtained during the braking from the velocimeters. When current braking grade is switched on, real velocity descent is compared with the computed one obtained from the passport data. As a result, training vector $x = \{\delta_i | i \in [1, k]\}$ is formed for each braking period having k braking intervals. For each vector x, hypotheses $\omega \in R$ is made by the human expert. In other words, output vector $z(x)$ is formed.

The training error for x is computed as the mean-squared error:

$$E(x) = \frac{1}{2}(m(x) - z(x))^2 = \frac{1}{2}\sum_{i=1}^{k}(z_i - m^{\Sigma}(q_i))^2. \tag{9}$$

The training error for training set X is computed as follows:

$$E^{\Sigma} = \frac{1}{||X||} \sum_{x \in X} E(x) \tag{10}$$

The values obtained from Eq. (9) and (10) are used as the criteria for RNFDS training.

6 Conclusion

The proposed model is oriented on the processing of sequential data about the technical state of the controlled object. The presented model simultaneously implements the following advantages of diagnosis systems:

1. The continuous clarification of information is provided due to the consequent data processing.
2. Network implementation of Dempster-Shafer scheme provides the possibility of adaptation and training of probability model based on the statistical information from the operation logs of CO attracting expert knowledge in form of BBAs.
3. Recurrent form of neural-fuzzy model provides compactness of representation and minimizes the number of adjusted parameters simplifying the training process.

Proposed idea can find an implementation in many tasks, where the input data are obtained as consequent portions, which clarify the previously observed data.

Acknowledgment. This work was supported by RFBR (Grants No. 19-07-00263, No. 19-07-00195, No. 20-07-00100).

References

1. Sukhanov, A., Kovalev, S., Stýskala, V.: Advanced temporal-difference learning for intrusion detection. IFAC-PapersOnLine **48**(4), 43–48 (2015)
2. Shabelnikov, A.N., Kovalev, S.M., Sukhanov, A.V.: Intelligent approach to emergency situations forecasting in the humping processes (intellektualnyy podkhod k prognozirovaniyu neshtatnykh situatsiy v protsesse rasformirovaniya poyezdov na sortirovochnykh gorkakh). Izvestiya YUFU. Tekhnicheskiye nauki 8 (181) (2016). (in Russian)
3. Yeung, D.Y., Ding, Y.: Host-based intrusion detection using dynamic and static behavioral models. Pattern Recognit. **36**(1), 229–243 (2003)
4. Modin, N.K.: Security of hump devices functioning (bezopasnost funktsionirovaniya gorochnykh ustroystv). Transport 173 (1994). (in Russian)
5. Kovalev, S.M., Shabelnikov, A.N.: Hybrid intelligent automation systems for process control at marshalling yards (gibridnyye intellektualnyye sistemy avtomatizatsii upravleniya tekhnologicheskimi protsessami na sortirovochnykh stantsiyakh). Izv. Vuzov Sev. (4), 11 (2002). (in Russian)

6. Shafer, G.: A Mathematical Theory of Evidence, vol. 42. Princeton University Press, Princeton (1976)
7. Smets, P.: The combination of evidence in the transferable belief model. IEEE Trans. Pattern Anal. Mach. Intell. **12**(5), 447–458 (1990)
8. Denoeux, T.: An evidence-theoretic neural network classifier. In: 1995 IEEE International Conference on Systems, Man and Cybernetics. Intelligent Systems for the 21st Century, vol. 1, pp. 712–717. IEEE (1995)
9. Denoeux, T.: A neural network classifier based on dempster-shafer theory. IEEE Trans. Syst. Man Cybern. Part A Syst. Hum. **30**(2), 131–150 (2000)
10. Benmokhtar, R., Huet, B.: Classifier fusion: combination methods for semantic indexing in video content. In: International Conference on Artificial Neural Networks, pp. 65–74. Springer (2006)

The Method of Clearing Printed and Handwritten Texts from Noise

S. Chernenko(✉), S. Lychko, M. Kovalkova, Y. Esina, V. Timofeev, K. Varshamov, A. Karlov, and A. Pozdeev

Moscow Polytechnic University, Moscow, Russia
sergeychrnk@gmail.com

Abstract. The article reviews the existing methods and algorithms for clearing printed and handwritten texts from noise and proposes an alternative approach. Among the solutions analyzed, a group of methods based on adaptive threshold conversion is distinguished. Our method for clearing print and handwritten documents from noise is based on using of a convolutional neural network ensemble with a U-Net architecture and a multi-layer perceptron. Using consequently a convolutional neural network and a multilayer perceptron demonstrates high efficiency in small training sets. As a result of applying our method to the entire test sample, an image cleaning degree of 93% was achieved. In the future, these methods can be introduced in libraries, hospitals, news companies where people work with non-digitized papers and digitization is needed.

Keywords: Computer vision · Artificial intelligence · CNN · Neural networks · Segmentation · OCR · U-net · Backpropagation

1 Introduction

Recognizing noisy text is difficult for most OCR algorithms [16]. Many documents that need to be digitized contain spots, pages with curved corners and many wrinkles - the so-called noise. Often that results in recognition errors. But when an image is cleared, then the accuracy can increase up to 100%. The quality of character recognition varies greatly depending on the text recognition used, filtering and image segmentation algorithm [13].

Currently, there are many different solutions to this problem, but most of them either do not provide satisfactory result, or require hardware resource and are highly time-consuming [3].

This article proposes an effective method for clearing printed and handwritten texts from noise [2], based on the use of a sequential convolutional neural network with a U-Net architecture and a multi-layer perceptron.

Neural network definition lies in the field of artificial intelligence, which based on attempts to reproduce the human nervous system. The main problem is the ability to learn and correct errors.

S. Kovalev et al. (Eds.): IITI 2019, AISC 1156, pp. 12–19, 2020.
https://doi.org/10.1007/978-3-030-50097-9_2

2 The Concept of a Neural Network

Neural network definition lies in the field of artificial intelligence, which based on attempts to reproduce the human nervous system. The main problem is the ability to learn and correct errors.

A mathematical neuron is a computational unit that receives input data, performs calculations, and transmits it further along the network.

Each neuron has inputs ($x_1, x_2, x_3, \ldots, x_n$), they receive data. Each neuron stores the weights of its connections. When a neuron is activated, a nonlinear transformation (activation function) of the weighted sum of the neuron input values is calculated. In other words, the output value of the neuron is calculated as:

$$O = f_a(x_1 w_1 + x_2 w_2 + \cdots + x_n w_n) \tag{1}$$

Where $x_{i \ldots n}$ is the input values of the neuron, $w_{i \ldots n}$ are the weight coefficients of the neuron, f_a is the activation function. The task of training a neuron is to adjust the weights of the neuron using a particular method.

We considered architectures of neural networks that are feedforward networks [11] - sequentially connected layers of mathematical neurons. The input values of each subsequent layer are the output values of the neurons of the previous layer. When the network is activated, values are fed to the input layer, each layer is activated sequentially. As a result of network activation are output values of the last layer.

Direct propagation networks are usually trained using the backpropagation method [12] and their modifications. This method refers to guided training methods and is itself a form of gradient descent. The output values of the network are subtracted from the desired values, then, as a result, an error is generated that propagates through the network in the opposite direction. These weights are adjusted to maximize the output of the network to the desired.

3 Overview of Existing Methods

Until now, classical computer vision algorithms are the most popular in the tasks of clearing images from noise.

One way to eliminate dark spots and other similar artifacts in an image is the adaptive threshold [8]. This operation does not binarize the image by a constant threshold, but takes into account the values of the neighboring pixels, thus the areas of the spots will be eliminated. However, after performing a threshold transformation, noise remains in the image, in the form of a set of small dots in place of spots (some parts of the spot exceed the threshold). Result of adaptive threshold is demonstrated in Fig. 1.

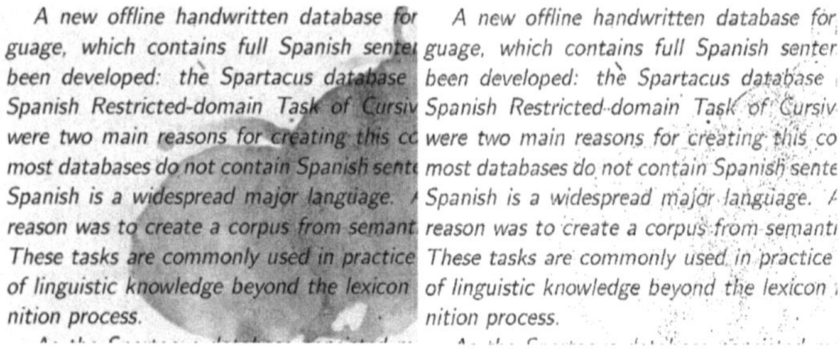

Fig. 1. Result of adaptive threshold

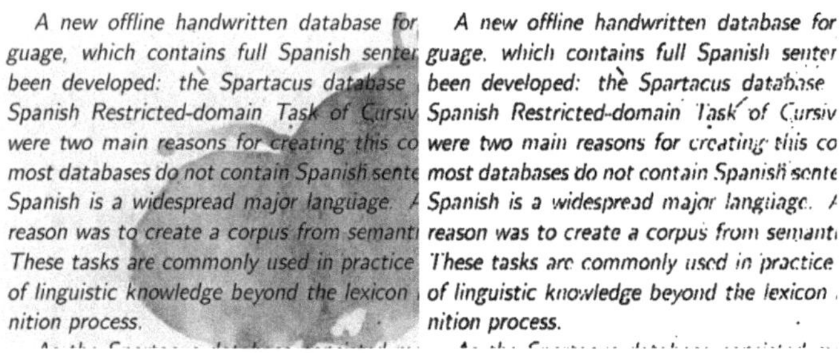

Fig. 2. Erosion and dilatation filters applied

The successive overlay of filters of erosion and dilation [9] helps to get rid of this effect, but such operations can damage the letters of the text. This is shown in Fig. 2.

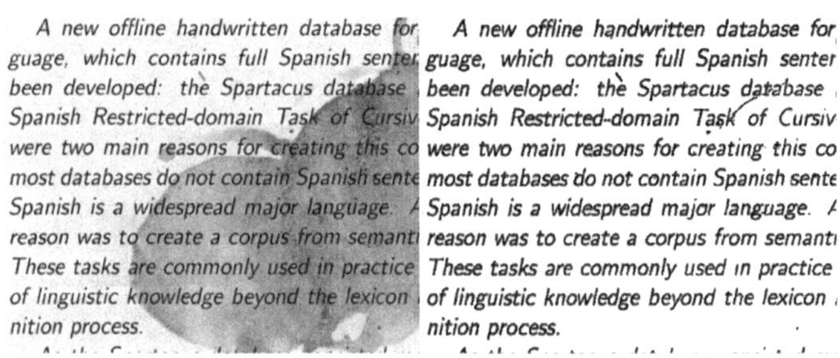

Fig. 3. Non-local means algorithm applied first

This result is not sufficient for recognition, so it can be improved using the non-local means method [10]. This method is applied before the adaptive threshold conversion and allows you to reduce the values of those points where small noise occurs. The result shown in Fig. 3 is much better, but it still shows small artifacts such as lines and points.

Analysis of existing methods has shown that the use of classical computer vision algorithms does not always show a good result and needs to be modernized.

4 Description of the Method Developed for Clearing Print and Handwritten Texts from Noise

4.1 Task Setting

The task of clearing text from noise, recognizing text in an image and converting it into text format consists of a number of subtasks [4, 6]:

1. Select a test image;
2. Converting color images to shades of gray;
3. Scaling and cutting images to a certain size;
4. Clearing the text from noise with the help of a convolutional neural network and a multilayer perceptron.

4.2 Preparation of Images Before Training

After reading, the input image is converted to a single channel (grayscale), where the value of each pixel is calculated as:

$$Y' = 0.299R + 0.587G + 0.114B \tag{2}$$

This equation is used in the OpenCV library and is justified by the characteristics of human color perception [14, 15].

Image sizes can be arbitrary, but too large sizes are undesirable.

For training 144 pictures are used. Since the size of the available training sample was not enough to train the convolutional network, it was decided to divide the images into parts. Since the training sample consists of images of different sizes, each image was scaled to the size of 448 × 448 pixels using the linear interpolation method. After that, they were all cut into non-overlapping windows measuring 112 × 112 pixels. All images were rotated 90, 180 and 270°. Thus, about 9216 images were obtained in the training sample. As a result, an array with the dimension (16,112,112,1) is fed to the input of the network. In the same way, the test sample was processed. The test sample consisted of similar images, the differences were only in the texture of the noise and in the text. We can see process of slicing and resizing of an image in Fig. 4

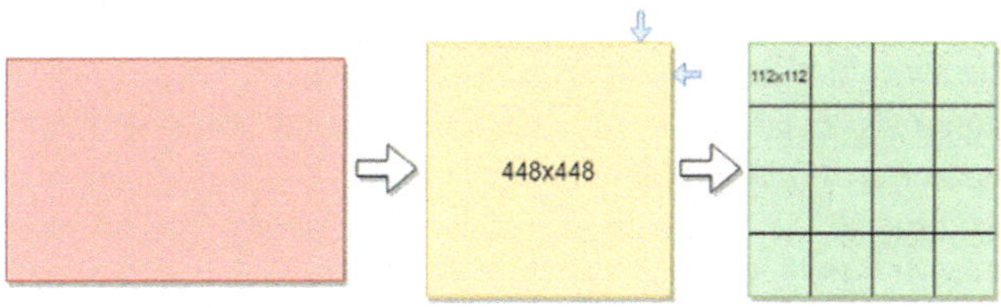

Fig. 4. Process of slicing and resizing of an image

The training sample of a single-layer perceptron is formed as follows [7]:

1. Images of the training set are passed through the pre-trained network of the U-Net architecture. Of the 144 images, only 36 were processed;
2. 28 different filters are superimposed on the resulting images. Thus, from each image we get 29 different, using the initial one;
3. Next, pairs are formed (input vector; resultant vector). The input vector is formed from pixels located at the same place in the 29 resulting images. The resulting vector consists of one corresponding pixel from the cleaned image;
4. Operation (3) is performed for each of the 36 images. As a result, the training sample has a volume of 36 * 448 * 488 elements.

As a result, the training sample has a volume of 36 * 448 * 488 elements.

4.3 Artificial Neural Network Training

In the proposed method, to clean print and handwritten texts from noise, a sequential convolutional neural network with U-Net architecture [17] and a multilayer perceptron are used. An array of non-overlapping areas of the original image measuring 112 × 122 is fed to the input of the network, and the output is a similar array with processed areas.

A smaller version of the U-Net architecture was selected, consisting of only two blocks (the original version of four) (Fig. 5).

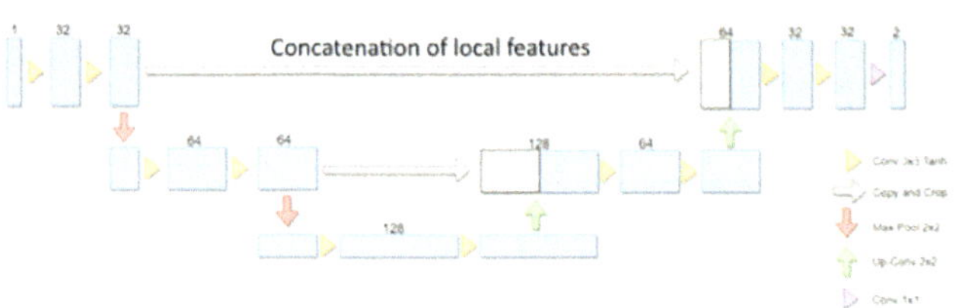

Fig. 5. Abbreviated architecture of U-net

The advantage of this architecture is that a small amount of training data is required for network training. At the same time, the network has a relatively small number of weights due to its convolutional architecture.

The architecture is a sequence of layers of convolution and pooling [18], which reduce the spatial resolution of the image, then increase it by combining the image with the data and passing it through other layers of the convolution.

Despite the fact that the convolutional network coped with the majority of noise, the image became less clear and left artifacts on it. To improve the quality of the text in the image, another artificial neural network is used - the multilayer perceptron.

The output array of the convolutional network is glued together into a single image with dimensions of 448 × 488 pixels, after which it is fed to the input of a multilayer perceptron. The format of the training set was described in Sect. 4.2.

The structure of the multilayer perceptron consists of 3 layers: 29 input neurons, 500 neurons on the hidden layer and one output neuron [1].

4.4 Testing of an Artificial Neural Network

The results of processing the original image using the reduced U-Net architecture are shown in Fig. 6.

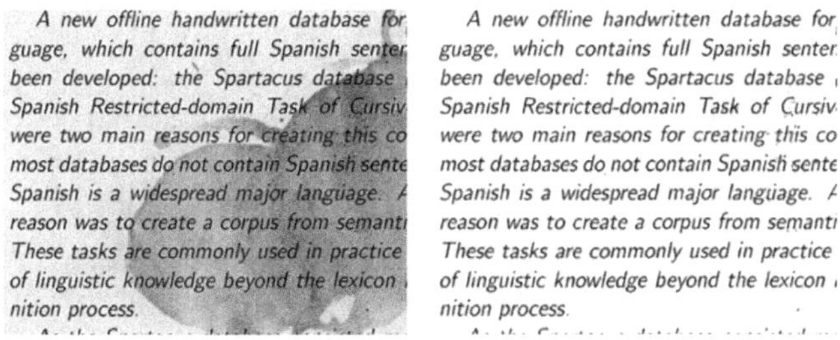
A new offline handwritten database for
guage, which contains full Spanish senter
been developed: the Spartacus database
Spanish Restricted-domain Task of Cursiv
were two main reasons for creating this co
most databases do not contain Spanish sente
Spanish is a widespread major language.
reason was to create a corpus from semanti
These tasks are commonly used in practice
of linguistic knowledge beyond the lexicon
nition process.

Fig. 6. Comparison of the original image and processed using a convolutional neural network.

As a result of the subsequent processing of the obtained image, its accuracy and contrast increased significantly. Small artifacts were also removed. An example is shown in Fig. 7.

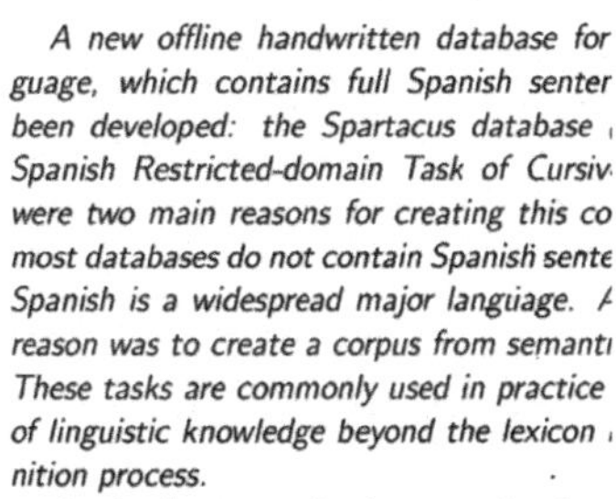
A new offline handwritten database for
guage, which contains full Spanish senter
been developed: the Spartacus database
Spanish Restricted-domain Task of Cursiv
were two main reasons for creating this co
most databases do not contain Spanish sente
Spanish is a widespread major language.
reason was to create a corpus from semanti
These tasks are commonly used in practice
of linguistic knowledge beyond the lexicon
nition process.

Fig. 7. Cleared image

5 Developed Solution

During the study, a software module was developed for digitizing damaged documents using this method. Python 3 was chosen as the development language. The Keras open neural network library, the Numpy library, and the OpenCV computer vision library have been used.

The module also has the ability to recognize text from the processed image using the Tesseract OCR engine [5].

When processing the input data in the form of a noisy image with text using this module, the output data is obtained in the form of text in a format suitable for its processing.

6 Conclusion

We reviewed several existing methods for clearing noisy printed documents, identified their shortcomings and proposed a method that has higher efficiency. The method described in the work requires a small training sample, it works quickly and has an average accuracy of noise removal of 93% [20].

Thus, the image processed by the method described in this article is quite clean, has no significant distortion and is easily recognized by most OCR engines and applications.

In the future, these methods can be used in libraries, hospitals [19], news companies where people work with non-digitized papers and their digitization is needed.

References

1. Khorosheva, T.: Neural network control interface of the speaker dependent computer system «Deep Interactive Voice Assistant DIVA» to help people with speech impairments. In: International Conference on Intelligent Information Technologies for Industry. Springer, Cham (2018)
2. Cai, J., Liu, Z.-Q.: Off-line unconstrained handwritten word recognition. Int. J. Pattern Recognit. Artif. Intell. **14**(03), 259–280 (2000)
3. Fan, K.-C., Wang, L.-S., Tu, Y.-T.: Classification of machine-printed and handwritten texts using character block layout variance. Pattern Recogn. **31**(9), 1275–1284 (1998)
4. Imade, S., Tatsuta, S., Wada, T.: Segmentation and classification for mixed text/image documents using neural network. In: Proceedings of 2nd International Conference on Document Analysis and Recognition (ICDAR 1993). IEEE (1993)
5. Impedovo, S., Ottaviano, L., Occhinegro, S.: Optical character recognition—a survey. Int. J. Pattern Recognit. Artif. Intell. **5**(01n02), 1–24 (1991)
6. Rehman, A., Kurniawan, F., Saba, T.: An automatic approach for line detection and removal without smash-up characters. Imaging Sci. J. **59**(3), 177–182 (2011)
7. Brown, M.K., Ganapathy, S.: Preprocessing techniques for cursive script word recognition. Pattern Recogn. **16**(5), 447–458 (1983)
8. Bradley, D., Roth, G.: Adaptive thresholding using the integral image. J. Graph. Tools **12**(2), 13–21 (2007)
9. Jawas, N., Nanik, S.: Image inpainting using erosion and dilation operation. Int. J. Adv. Sci. Technol. **51**, 127–134 (2013)
10. Buades, A., Coll, B., Morel, J.M.: A non-local algorithm for image denoising. In: 2005 IEEE Computer Society Conference on Computer Vision and Pattern Recognition (CVPR 2005), vol. 2. IEEE (2005)
11. Hornik, K., Maxwell, S., Halbert, W.: Multilayer feedforward networks are universal approximators. Neural Networks **2**(5), 359–366 (1989)
12. Hecht-Nielsen, R.: Theory of the backpropagation neural network. In: Neural Networks for Perception, pp. 65–93. Academic Press, New York (1992)

13. Smith, R.: An overview of the Tesseract OCR engine. In: Ninth International Conference on Document Analysis and Recognition, vol. 2. IEEE (2007)
14. OpenCV: Color Conversions. https://docs.opencv.org/3.4/de/d25/imgproc_color_conversions.html. Accessed 01 May 2019
15. Güneş, A., Habil, K., Efkan, D.: Optimizing the color-to-grayscale conversion for image classification. Signal Image Video Process. **10**(5), 853–860 (2016)
16. Sahare, P., Dhok, S.B.: Review of text extraction algorithms for scene-text and document images. IETE Tech. Rev. **34**(2), 144–164 (2017)
17. Ronneberger, O., Fischer, P., Brox, T.: U-Net: convolutional networks for biomedical image segmentation. In: International Conference on Medical Image Computing and Computer-Assisted Intervention. Springer, Cham (2015)
18. He, K.: Spatial pyramid pooling in deep convolutional networks for visual recognition. IEEE Trans. Pattern Anal. Mach. Intell. **37**(9), 1904–1916 (2015)
19. Adamo, F.: An automatic document processing system for medical data extraction. Measurement **61**, 88–99 (2015)
20. Denoising dirty documents. https://www.kaggle.com/c/denoising-dirty-documents. Accessed 01 July 2019

$$RF = \cup_{i=1}^{m}\{(T_1\backslash\mu_1), (T_2\backslash\mu_2), (T_3\backslash\mu_3)\}_i, \tag{1}$$

where $\mu_{1,\ 2,\ 3}$ are corresponding to attribute values for the terms $T_{1,\ 2,\ 3}$.

The above fuzzification procedure (1) is used in the algorithm for classifying graphs using NLHS [10]. As a result, n classification rules are generated for n training set classes (TrS). The rules reflect the parameters of TrS objects by corresponding separation marks.

Integration of the developed NLHS and the fuzzy inference algorithm makes it possible to find a corresponding class for each new object (with the maximum degree of object belonging). Consequently, there are localized areas that have the same values of fuzzy attributes. This actually leads to the problem of segmentation of two-dimensional graphs into sections, which will close estimates of their structural attributes. This paper proposes a hypothesis on the possibility of using the segmentation procedure for two-dimensional graphs to create a new space based on enlarged attributes. In order to test the proposed hypothesis, there is a new algorithm that extends the capabilities of NLHS.

3 An Algorithm for Enlargement of Interval Signs

Let us analyze the solution of the problem of classifying 2 classes of two-dimensional graphs. Each class has a set of informative attributes M_k (K is a number of classes, k = 1, 2,..., K). For each M_k set, there are determined sections, according to which linguistic variables for attributes take the same values. Such sections of the M_k set are "intervals of constancy":

$$\text{for } \forall M_k \left\{ \begin{array}{c} P_i = P_{i+1}, i = \overline{m, z} \\ \text{for } i = z, P_{z+1} \neq P_z \\ \text{interval } In_{k,m} : (\tilde{P}_m, \tilde{P}_{m+1}, \tilde{P}_z) \end{array} \right\}, \tag{2}$$

where i is a attribute number, $\tilde{P}_i$ is a fuzzy value of the attribute P_i.

Figure 1 shows a description of a graph fragment that is a conjunction of attributes (107∧118∧ … ∧258). For Class 1 objects, the set M_1 is taken from the description (all signs in the conjunction take on the value "HIGH"). Fuzzification characteristics by attributes are considered in detail in [10]. The Class 2 also has an allocated set M_2 (all signs in the conjunction take on the value "LOW"). The sets M_1 and M_2 intersect: $P = M_1 \cap M_2$ = 150, 161, 172, 183, 193, 204, 215, 226 (i.e., Class 1 objects have a high value of the selected attributes, and Class 2 objects have small values). Therefore, the entire allocated interval P (2) is characterized by identical values according to linguistic scales in terms of primary characteristics. Values change only when passing from one class to another. The allocated regularity makes it possible to analyze the entire indicated interval of attributes P as a new attribute $P_{4\text{-}11}$, which characterizes the entire fragment of a two-dimensional characteristic curve.

Procedures for adding new attributes have the following process sequence. The allocated intervals of attributes ($\forall k$) $In_{k,m}$ have intersections of the analyzed classes:

		Number	Name	M_1	M_2		
		0	107	HIGH:17			
		1	118	HIGH:17	MIDDLE:7		
		2	129	HIGH:17			
		3	139	HIGH:17	MIDDLE:7		
		4	150	HIGH:17	LOW:8		
	$In_{1,0}$	5	161	HIGH:17	LOW:8		
		6	172	HIGH:17	LOW:8		
$P\_In_{1,0}$		7	183	HIGH:17	LOW:8		
		8	193	HIGH:17	LOW:8		$P\_In_{2,4}$
		9	204	HIGH:17	LOW:8	$In_{2,4}$	
		10	215	HIGH:17	LOW:8		
		11	226	HIGH:17	LOW:8		
		12	236	MIDDLE:7	LOW:8		
	$In_{1,13}$	13	247	HIGH:17	LOW:8		
		14	258	HIGH:17	LOW:8		

Fig. 1. Distribution of significant attributes for two classes.

1) a description fragment for Class 1: $In_{1,0} = \cup_{i=0}^{11} P_i$; $C_1 = \cup_i^M P_i$;
2) a fragment for Class 2: $In_{2,4} = \cup_{k=4}^{14} P_k$; $C_2 = \cup_k^T P_k$;

where M, T is a number of attributes that define Class 1 and Class 2, consequently: $((\forall i)$ $\{P_i \in In_{1,0} \mid P_i = \text{HIGH}\}$, $(\forall k)$ $\{P_k \in In_{2,4} \mid P_k = \text{LOW}\})$.

A new characteristic ($P_{4\text{-}11}$) is common for $In_{1,0}$ and $In_{2,4}$. For the entire interval, the corresponding subintervals for each class are the following: $P\_In_{1,0} \subseteq In_{1,0}$, $P\_In_{2,4} \subseteq In_{2,4}$, $P\_In_{1,0}$ and $P\_In_{2,4}$, which include the same attributes, but not the same values. To use the subinterval $P\_In_{1,0}$ as an independent feature, it is necessary to apply the union by fuzzy sets.

Let us demonstrate a similar union at the subinterval $P\_In_{1,0}$ (f_1, f_n) that is described by similarly-named terms. It can be used as a value for a new characteristic $P_{4\text{-}11}$ (Fig. 2, a). The value $P_{4\text{-}11}$ for an object l_1 will be a set of points in the interval $(P_3 \leq P_2 \leq P_1)$ :: ΔP. The interval ΔP will be treated as $Supp(P)$, i.e. as a carrier for the fuzzy set P. The middle of the interval ΔP corresponds to the vertex $\bar{P}$.

Consideration of the (f_1, f_n) section for all objects TrS provides a finite number of fuzzy sets l_1, l_2, $l_k = L$ (Fig. 2, b). After combining all the fuzzy sets, the output is a new set that will determine the value of the new attribute: $P_{4\text{-}11} = l_1 \cup l_2 \cup \ldots \cup l_k$.

The basis of $P_{4\text{-}11}$ is the interval [min(P), max(P)]. This procedure allows creating a new linguistic scale: HIGH$\{P_{4\text{-}11}\}$ = HIGH(l_1)$\cup$ HIGH (l_2)$\cup$ … $\cup$ HIGH(l_k). As a result, we can change class descriptions ($m = M \backslash P_{4\text{-}11}$, $m = T \backslash P_{4\text{-}11}$):

1) $C_1 = \bigcup_{i=0}^{m} \tilde{P}_i \cup \left(\tilde{P}_{4-11}\right),\ \tilde{P}_{4-11} = P\_In_{1,0};$

2) $C_2 = \bigcup_{k=0}^{t} \tilde{P}_k \cup \left(\tilde{P}_{4-11}\right),\ \tilde{P}_{4-11} = P\_In_{2,4}.$

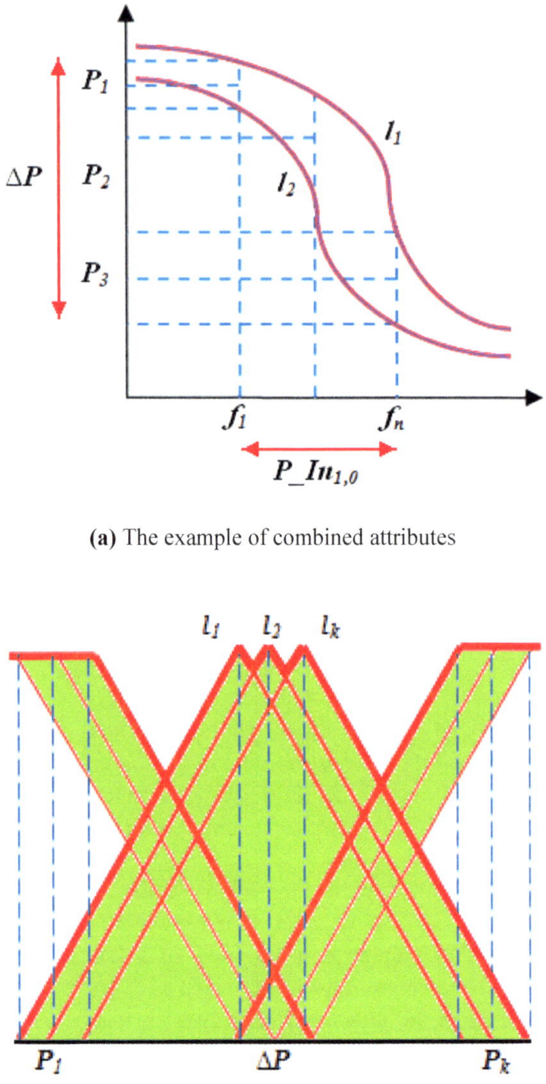

(a) The example of combined attributes

(b) The example of fuzzy attribute sets

Fig. 2. Attribute space characteristics.

Subinterval are removed from outdated attribute spaces $P\_In_{1,0}$, $P\_In_{2,4}$. There follows the procedure of adding a new attribute $P_{4\text{-}11}$ (Fig. 3).

Number	Name	Class 1	Class 2
0	107	HIGH:17	
1	118	HIGH:17	MIDDLE:7
2	129	HIGH:17	
3	139	HIGH:17	MIDDLE:7
4 ($P_{4\text{-}11}$)	150-226	HIGH:17	LOW:8
5	236	MIDDLE:7	
6	247, 258	HIGH:17	

Fig. 3. The example of attribute distribution after NLHS training procedure.

When describing Class 1 (C_1) and Class 2 (C_2) we use the following attribute values: a) $\tilde{C}_1 = (C_1 \backslash P\_In_{1,0}) \cup \tilde{P}_{4-11}$, $\tilde{C}_2 = (C_2 \backslash P\_In_{2,4}) \cup \tilde{P}_{4-11}$; b) the value of the removed feature is equal to the value of new characteristics. If the condition (b) is not fulfilled, i.e. $\cup \tilde{P}_i \neq \tilde{P}_{4-11}$, we adjust the sub-interval boundary (the number of attributes in $P_{4\text{-}11}$ changes).

4 Algorithm Testing

The problem of classification of EMG and EEG, which reflect the variation in the valence of human emotions, is the example for testing the algorithm. The experimental sample of signals includes objects (EMG and EEG patterns) that illustrate changing valence of testees' emotions under audiovisual stimulation. The papers [4, 12, 13] describe the routine of such experiments in detail. Participants: 10 men and 10 women aged 18 to 60 took part in forming the sample.

The face electromyogram (EMG signals) registration was performed on the left side of testee's face. The position of electrodes is related to the work of two muscle groups: "corrugator supercilii" and "zygomaticus major". An EMG signal (fEMG) has been recorded according to the Fridlund and Cacioppo method [14]. The sampling frequency of the received record was 1000 Hz. The duration of the EMG samples (400 patterns) was 6 s.

The sensors (for registration of brain electrical activity signals) were fixed on testee's head according to the international system of leads "10–20" [15]. EEG samples (400 19-channel non-artifactual patterns) with 6 s duration are saved in *.ASCII files with a sampling frequency of 250 Hz.

Table 1 shows compositions of training samples (TrS) and test samples (TsS) (Class 1 is negative emotions, Class 2 is a neutral state, Class 3 is positive emotions).

Table 1. The experimental sample.

Samples		Object classes		
		Class 1	Class 2	Class 3
EMG	**TrS (150)**	50	50	50
	TsS (250)	85	80	85
EEG	**TrS (150)**	50	50	50
	TsS (250)	85	80	85

Here, separating attributes are power spectral density (PSD) readings calculated by the Welch method using the window Fast Fourier Transform (FFT) [16]. Each object is described by the vector $X = \{x_1, x_2, \ldots, x_i, \ldots, x_i\}$, where x_i is the ordinate of the power spectrum at the frequency $f_i = \Delta f \cdot i$; x_i is the value of the i-th attribute; Δf is the frequency step; $\Delta f = f_x / Fw$; f_x is a sampling frequency; Fw is the FFT window width. We used the Hamming transformation window (width 512 for EMG, width 128 for EEG). The frequency range limits for EMG are 0–500 Hz, for EEG they are 0–25 Hz.

The description for the sample object (see Table 1) is represented as follows:

$$X(z)_{EMG} = \{x_1, x_2, \ldots, x_k\}, \tag{3}$$

$$X(s)_{EEG} = < \{x_1, x_2, \ldots, x_r\}_1, \{x_1, x_2, \ldots, x_r\}_2, \ldots, \{x_1, x_2, \ldots, x_r\}_l >, \tag{4}$$

where $X(z)_{\mathrm{EMG}}$, $X(s)_{\mathrm{EEG}}$ are PSD attribute vectors; z is a number of EMG object; s is a number of EEG object; $z, s = 1, 2, \ldots, 400$; l is a number of EEG reading; $l = 1, 2, \ldots, 19$; k is a number of PSD feature for EMG; $k = 1, 2, \ldots, 200$ (PSD calculation step is 2.5 Hz); r is a number of PSD attribute for EEG; $r = 1, 2, \ldots, 40$ (PSD calculation step is 3 Hz).

Investigation of PSD signals using NLHS demonstrated possibilities of localizing intervals with the most informative features that allow providing good levels of generalization and detailing object descriptions in the given classes.

For EMG samples, the most informative PSD attributes are 50–100 Hz range characteristics. For EEG samples, similar characteristics are from 2 leads of the right hemisphere (F4-A2, F8-A2).

Thus, each EMG object $X(z)_{\mathrm{EMG}}$ (3) is described by 40 attributes (two muscle groups: “corrugator supercilii” and “zygomaticus major”), and the object $X(s)_{\mathrm{EEG}}$ (4) is described by 80 attributes (two leads: F4-A2, F8-A2).

Figure 4 shows fragments of PSD attribute space (3 and 4) for experimental sample objects (Fig. 4, a is a frequency range 50–100 Hz, Fig. 4, b is a frequency range 0–120 Hz).

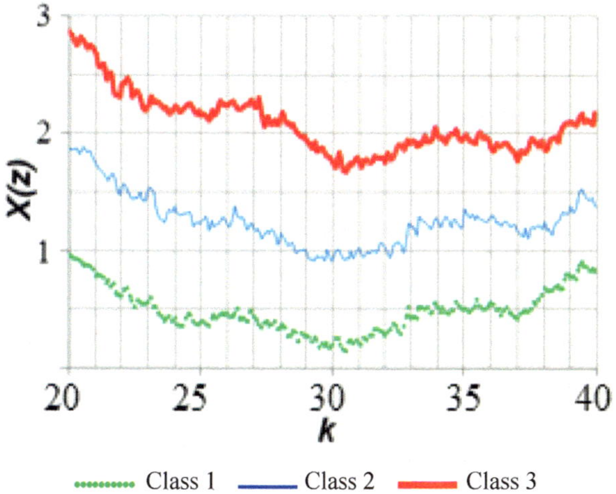

(a) EMG patterns (muscle group "zygomaticus major")

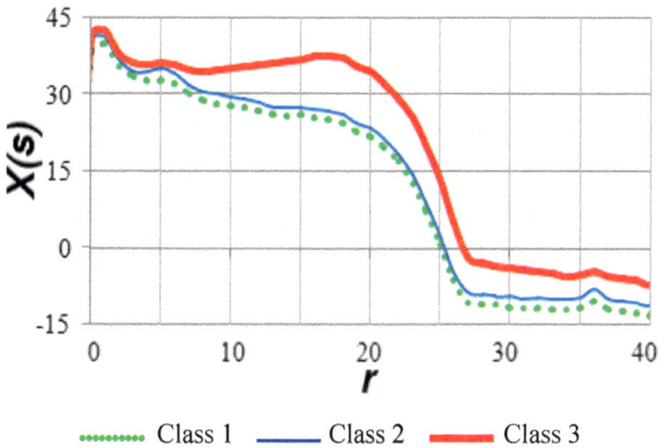

(b) EEG patterns ("F8-A2" lead, "The 10-20 System of Electrode Placement")

Fig. 4. PSD range in the centre classes (an abscissa is the number of a attribute; an ordinate is the number of PSD, c.u.).

There are some rules based on NLHS that describe EMG and EEG objects (see Table 1), their application to TrS and TsS is shown in Table 2.

Table 2. Classification results.

Classification accuracy, %	EMG		EEG	
	TrS	TsS	TrS	TsS
Class 1	100	89	100	91
Class 2	100	90	100	93
Class 3	100	95	100	97
All	100	91	100	94

The work includes the analysis of classifications results of EMG and EEG patterns when applying two NLHS training modes (mode 1 is without attribute integration, mode 2 is with enlarged attributes (see Table 2)).

The transition from mode 1 to mode 2 illustrates 2 trends: 1) a decrease in the total number of attributes (at least by 30–35%), which provide good levels of generalization and refinement of object descriptions in the considered classes; 2) increasing classification accuracy.

5 Conclusion

The NLHS-based classifier with the algorithm of integration of interval attributes makes it possible to generate descriptions of the most informative intervals of the attribute space. Consequently, it reduces the dimension of class descriptions and increases classification accuracy.

The testing results performed at two-dimensional PSD characteristic curves (3 and 4) for EMG and EEG, prove its operability with different types of experimental graphs.

References

1. Gratch, J., Marsella, S.: Evaluating a computational model of emotion. Auton. Agent. Multi-Ag. **11**(1), 23–43 (2005). https://doi.org/10.1007/s10458-005-1081-1
2. Marsella, S., Gratch, J., Petta, P.: Computational models of emotion. In: Scherer, K.R., et al. (eds.) A Blueprint for Affective Computing: A Sourcebook and Manual. Oxford University Press, Oxford (2010)
3. Rabinovich, M.I., Muezzinoglu, M.K.: Nonlinear dynamics of the brain: emotion and cognition. Adv. Phys. Sci. **180**(4), 371–387 (2010). https://doi.org/10.3367/UFNr.0180.201004b.0371. (in Russian Uspekhi Fizicheskikh Nauk)
4. Filatova, N.N., Sidorov, K.V.: Computer models of emotions: construction and methods of research. Tver State Technical Univ. Publ., Tver (2017). (in Russ., Kompyuternye Modeli Emotcy: Postroenie i Metody Issledovaniya)
5. Gladun, V.P.: The growing pyramidal networks. Artif. Intell. News **2**(1), 30–40 (2004). (in Russian Novosti Iskusstvennogo Intellekta)
6. Ifeachor, E.C., Jervis, B.W.: Digital Signal Processing: a Practical Approach, 2nd edn. Pearson Education, New Jersey (2002)
7. Loskutov, A.Yu.: Time series analysis: lectures. Moscow State Univ. Publ. Moscow (2006). (in Russian Analiz Vremennykh Ryadov: Kurs Lektsiy)
8. Rangayyan, R.M.: Biomedical Signal Analysis, 2nd edn. Wiley-IEEE Press, NewYork (2015). https://doi.org/10.1002/9781119068129
9. Sidorov, K.V., Filatova, N.N., Shemaev, P.D.: An interpreter of a human emotional state based on a neural-like hierarchical structure. In: Abraham, A., et al. (eds.) Proceedings of the Third International Scientific Conference "Intelligent Information Technologies for Industry" (IITI 2018), Advances in Intelligent Systems and Computing 874, vol. 1, pp. 483–492. Springer, Switzerland (2019). https://doi.org/10.1007/978-3-030-01818-4_48

10. Sidorov, K.V., Filatova, N.N., Shemaev, P.D.: A neural-like hierarchical structure in the problem of automatic generation of hypotheses of rules for classifying the objects specified by sets of fuzzy features. In: Abraham, A., et al. (eds.) Proceedings of the Third International Scientific Conference "Intelligent Information Technologies for Industry" (IITI 2018), Advances in Intelligent Systems and Computing 874, vol. 1, pp. 511–523. Springer, Switzerland (2019). https://doi.org/10.1007/978-3-030-01818-4_51
11. Lan, Z., Sourina, O., Wang, L., Liu, Y.: Real-Time EEG-based emotion monitoring using stable features. Vis. Comput. **32**(3), 347–358 (2016). https://doi.org/10.1007/s00371-015-1183-y
12. Filatova, N.N., Bodrina, N.I., Sidorov, K.V., Shemaev, P.D.: Organization of information support for a bioengineering system of emotional response research. In: Proceedings of the XX International Conference on Data Analytics and Management in Data Intensive Domains "DAMDID/RCDL 2018", Russia, Moscow, 9–12 October 2018. CEUR-WS 2018. vol. 2277, pp. 90–97. http://ceur-ws.org/Vol-2277/paper18.pdfAccessed 25 May 2019
13. Filatova, N.N., Sidorov, K.V., Terekhin, S.A., Vinogradov, G.P.: The system for the study of the dynamics of human emotional response using fuzzy trends. In: Abraham, A., et al. (eds.) In: Proceedings of the First International Scientific Conference "Intelligent Information Technologies for Industry" (IITI 2016), Advances in Intelligent Systems and Computing 451, vol. 2, part III, pp. 175–184. Springer, Switzerland (2016). https://doi.org/10.1007/978-3-319-33816-3_18
14. Fridlund, A.J., Cacioppo, J.T.: Guidelines for human electromyographic research. Psychophysiology **23**(5), 567–589 (1986). https://doi.org/10.1111/j.1469-8986.1986.tb00676.x
15. Jasper, H.H.: Report of the committee on methods of clinical examination in electroencephalography. Electroencephalogr. Clin. Neurophysiol. **10**(2), 370–375 (1958). https://doi.org/10.1016/0013-4694(58)90053-1
16. Welch, P.D.: The use of fast Fourier transform for the estimation of power spectra: a method based on time averaging over short, modified periodograms. IEEE Trans. Audio Electroacoust. **15**(2), 70–73 (1967). https://doi.org/10.1109/TAU.1967.1161901

Age and Gender Recognition on Imbalanced Dataset of Face Images with Deep Learning

Dmitry Yudin[1(✉)], Maksim Shchendrygin[2], and Alexandr Dolzhenko[2]

[1] Moscow Institute of Physics and Technology (National Research University), Dolgoprudny, Institutskiy per., 9, Moscow 141700, Russia
yudin.da@mipt.ru

[2] Belgorod State Technological University named after V.G. Shukhov, Kostukova str. 46, Belgorod 308012, Russia

Abstract. The paper describes usage of deep neural networks based on ResNet and Xception architectures for recognition of age and gender of imbalanced dataset of face images. Described dataset collection process from open sources. Training sample contains more than 210000 images. Testing sample have more 1700 special selected face images with different ages and genders. Training data has imbalanced number of images per class. Accuracy for gender classification and mean absolute error for age estimation are used to analyze results quality. Age recognition is described as classification task with 101 classes. Gender recognition is solved as classification task with two categories. Paper contains analysis of different approaches to data balancing and their influence to recognition results. The computing experiment was carried out on a graphics processor using NVidia CUDA technology. The average recognition time per image is estimated for different deep neural networks. Obtained results can be used in software for public space monitoring, collection of visiting statistics etc.

Keywords: Age recognition · Gender recognition · Classification · Face image · Imbalanced dataset · Deep neural network

1 Introduction

Recognition of gender and age of people on images has high importance when creating computer vision systems for retail, visitors monitoring in public spaces, etc. Development of reliable and fast face recognition algorithms on data from video cameras in various visibility conditions is still an active research topic.

Proposed in [1] approach is based on two separate deep-convolutional neural networks for age range ((0–2), (4–6), (8–12), (15–20), (25–32), (38–43), (48–53), (60–100)) and gender classification on face images from the Adience benchmark [2].

The modern researchers note the contradiction and significant differences between the concepts of real age and apparent age and attempt to develop datasets for estimating a person's age from an image of a person's face based on the experience of experts, such as LAP [3] and APPA-REAL [4].

S. Kovalev et al. (Eds.): IITI 2019, AISC 1156, pp. 30–40, 2020.
https://doi.org/10.1007/978-3-030-50097-9_4

The age-based deep learning approach based on regression is inferior to the approach based on image classification, when a separate class is assigned to each age. Post-processing of the output of such a network allows to further improve the quality of recognition [5]. There is approach to apparent and real age recognition using VGG deep neural network as classifier with subsequent expectation age calculation which evaluation was done on APPA-REAL database [6].

Many datasets contain real data on the age and gender of a large number of famous people collected on the Internet, for example, the huge IMDB-WIKI database [7], UTKFace dataset [8].

Some datasets contain only annotated images with age, for example FG Net [9], have additional information for personal identification, as in Cross-Age Celebrity Dataset (CACD) [10], information about personality, race, or glasses presence as in MORPH [11] or IBM Research DiF dataset (Diversity in faces) [12].

There are approaches to creation single deep neural network for age and gender recognition, for example, based on InceptionResnetV1 architecture using Tensorflow [13].

Separately, it should be noted a group of approaches to determining the gender and age of people in a video stream, which use additional filtering of the results of the neural networks [14], aggregation and voting procedure [15], expectations calculation or Dempster-Shapher rule usage [16].

The most datasets with labeled data about gender and age have a much smaller number of images corresponding to children and teenagers younger than 18 years old and older people over 60 years old compared to the range of 18–60 years old. We have a significantly imbalanced dataset, which requires the study of deep neural network training on images from it. There are some approaches to dataset balancing:

1) weights balancing using direct procedure in deep learning framework, for example Keras [17], or Focal Loss [18] which down-weights the well-classified examples. It is directed to putting more training emphasis on that data that is hard to classify.
2) data undersampling and oversampling. Undersampling consists in selection only some of the images from the majority class, only using as many examples as the minority class has. Oversampling means that we will create copies or augmented images of our minority class in order to have the same number of examples as the majority class has.

To increase training efficiency different data augmentation techniques are used: mixup approach [19], random erasing [20, 21], affine transformation, noise adding, coarse dropout [22].

One of the effective but resource-intensive oversampling approach is synthetic data generating using Generative Adversarial Nets [23], for example Transparent Latent-space GAN (TL-GAN) [24] or Progressive Growing of GAN (PG-GAN) [25] and attributes to image techniques [26].

2 Task Formulation

In this paper we will solve the task of determining of human gender and age on a single cropped face image.

As input for recognition algorithm we will use grayscale image with equal width and heights sizes from popular datasets IMDB-WIKI, UTK Face and manually annotated dataset with images corresponding to “rare” ages – younger than 20 and older than 70 years. At the same time, it is necessary to check the correctness of the annotations of images contained in open-source datasets. The dataset in question is essentially imbalanced in terms of the distribution of images by age, which requires the study of balancing approaches in preparing the training sample.

Recognition of the gender and age on person’s face image must be carried out simultaneously by one deep neural network. Such approach saves computational resources and speeds up the recognition process, since the layers designed to extract the signs of gender and age in the images belong to the same subject area of facial recognition.

Two output values of the recognition algorithm have to be determined in parallel:

– gender values of one of two classes (0 - woman, 1 - man) with a degree of confidence in each of the classes, and the sum of all degrees of confidence should be equal to one. Predicted gender class g_{out} on base of network output y_g is calculated with formula

$$g_{out} = argmax(y_g);$$

– age values of one of the 101 classes (from 0 to 100), where the class number corresponds to the age, with a degree of confidence in each of the classes, and the sum of all degrees of confidence should be equal to one. Age estimation a_{out} on base of network output y_a is calculated with formula

$$a_{out} = argmax(y_a).$$

Also we should determine the accuracy of gender classification and the mean absolute error (MAE) of age estimation on the training and test sample both with and without training sample balancing.

3 Dataset Preparation

The images from popular datasets IMDB-WIKI, UTK Face and manually annotated data with images corresponding to “rare” ages - younger than 20 years and older than 80 years are combined to form datasets for training of deep neural networks. At the same time, we checked the correctness of the image annotations contained in open-source datasets, which revealed quite a lot of errors in their files with annotations. Full training sample of developed dataset contains 211060 images, testing sample contains 1793 images. The distribution by age in the training and test samples is shown in Fig. 1 and Fig. 2.

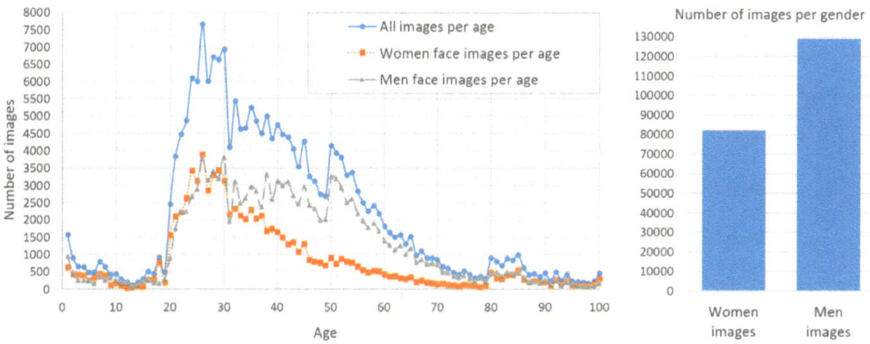

Fig. 1. Full training dataset statistics for age and gender recognition on face images

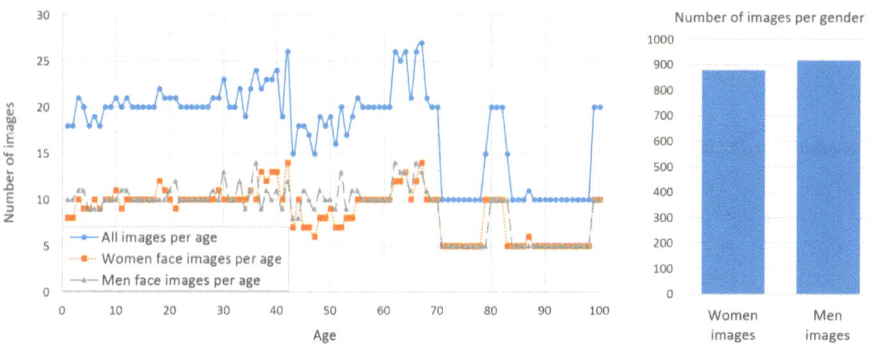

Fig. 2. Testing dataset statistics for age and gender recognition on face images

Formed dataset is significantly imbalanced in terms of the distribution of images by age, which requires the study of balancing approaches in the preparation of the training sample. Undersampling of the most common ages in combination with oversampling for rare ages using the augmentation procedure was chosen as the main approach.

To conduct experiments for learning neural networks, augmentation of images was carried out, and a balanced training sample was formed with 1000 images (500 men and 500 women images) and 600 images (300 men and 300 women image) per class (age).

For image augmentation we have used 4 sequential steps using Imgaug library [22]:

1. Affine transformation – image rotation on random degrees from −15 to 15.
2. Flipping of image along vertical axis with 0.9 probability.
3. Addition Gaussian noise to image with standard deviation of the normal distribution from 0 to 15.
4. Cropping away (cut off) random value of pixels on each side of the image from 0 to 10% of the image height/width.

Results of this augmentation procedure are shown in Fig. 3.

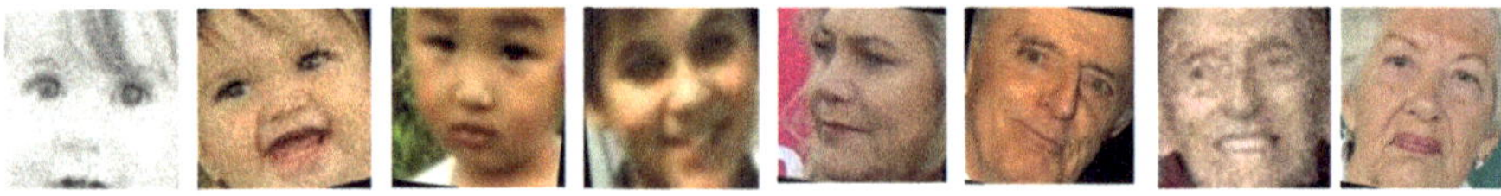

Fig. 3. Examples of augmented images for balanced training sample

4 Training of Deep Convolutional Neural Networks for Age and Gender Recognition

In this paper to solve formulated task we investigate the application of a deep convolutional neural networks of two light and fast architectures: ResNetMAG based on ResNet modules, and modern Xception architecture based on separable convolutions instead of common convolutions. The main criterion for the choice of these architectures was their speed and the expected high recognition quality of images with about 100 × 100 pixels size. Details are listed below:

- ResNetMAG architecture for age and gender classification in one network inspired from ResNet [27] and implemented by authors in previous works [28]. Its structure is shown in Fig. 4 and contains 3 convolutional blocks, 5 identity blocks, 2 max pooling layers, 1 average pooling layer and output gender and age dense layers in parallel. First 11 layers and blocks provide automatic feature extraction and the last fully connected layers allows us to find 2 classes corresponding to gender and 101 classes corresponding to people age;

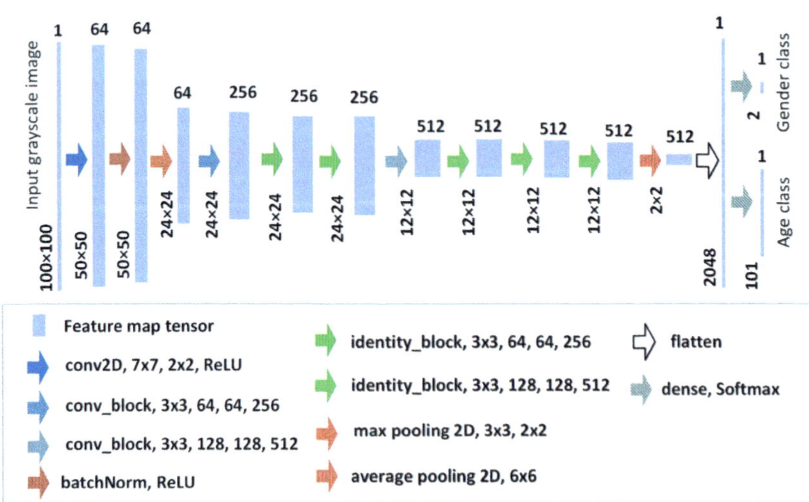

Fig. 4. ResNetMAG architecture.

- XceptionAG architecture [29] with changed input tensor to 114 × 114 × 1 for grayscale images and output parallel gender and age dense layers. This structure is based on prospective Separable convolutional blocks (see Fig. 5) as a development of the Inception concept [30].

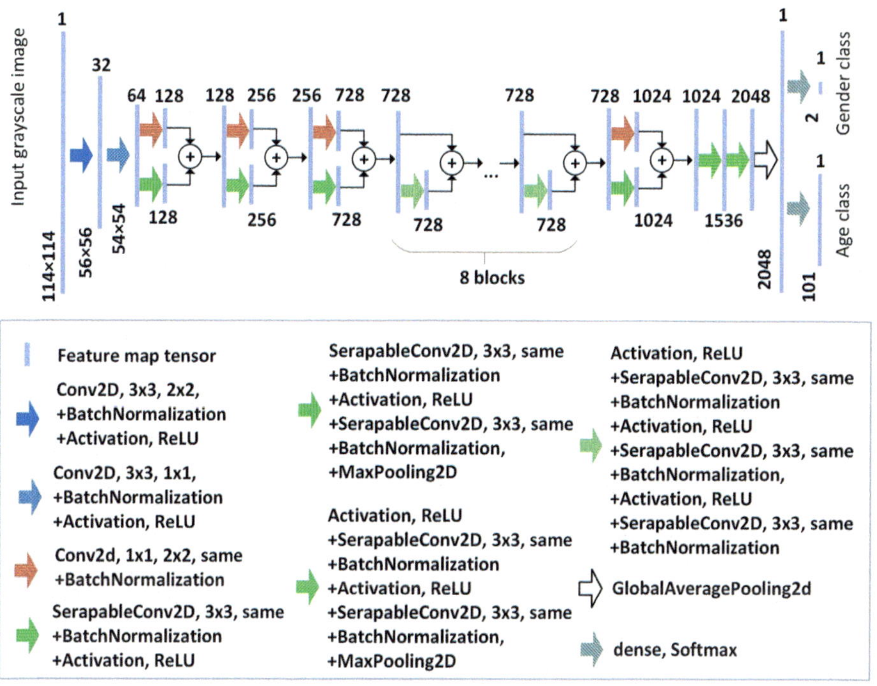

Fig. 5. XceptionAG architecture.

Output layer in all architecture has gender output layer with 2 neurons with "Softmax" activation function and age output layer in parallel with 101 neurons also "Softmax" activation function. All input images are pre-scaled to a size of 100 × 100 pixels for ResNetMAG architecture, 114 × 114 pixels for XceptionAG architecture. Neural networks works with grayscale (one-channel) images.

To train the neural networks we have used "categorical crossentropy" loss function, Stochastic Gradient Descent (SGD) as training method with 0.0005 learning rate. Accuracy is used as classification quality metric during training. The batch size was 16 images.

The training process of deep neural networks is shown in Fig. 6. The training was performed on 50 epochs using our developed software tool implemented on Python 3.5 programming language with Keras and Tensorflow frameworks [31]. We can see that for both ResNetMAG and XceptionAG networks training process for gender recognition is stable. At the same time, for age recognition, there is a rapid increase in the loss function in a validation sample and its decrease in the training sample, which is an undesirable phenomenon. Almost in all the considered cases, training on a balanced sample gave a less effective result than on a full imbalanced training set of data (Fig. 7).

Fig. 6. Training of deep neural network with ResNetMAG architecture.

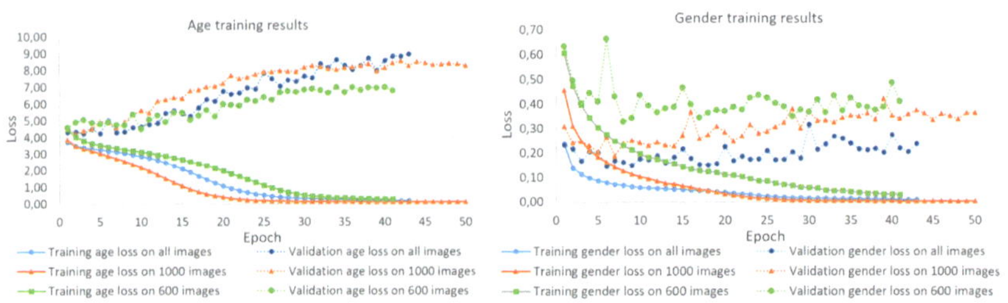

Fig. 7. Training of deep neural network with XceptionAG architecture.

5 Experimental Results

Computational experiments had performed using the NVidia CUDA technology on the graphics processor of the GeForce GTX 1080 graphics card with 8.00 GB, central processor Intel Core i-5-4570, 4 Core with 3.2 GHz and 16 GB RAM.

Table 1 shows the results of the facial expression recognition using imbalanced and balanced training samples using ResNetMAG and XceptionAG architectures.

Analysis of the obtained results shows the highest accuracy on all samples for XceptionAG architecture trained on imbalanced data: gender accuracy on full training sample 99.53% on training sample, 94.76% on testing sample, MAE of age estimation on full training sample is 0.45 years, MAE of age estimation on testing sample is 10.89 years.

ResNetMAG architecture have better MAE of age estimation on testing sample – 10.25 years and 1.6 times faster than XceptionAG network: about 11 ms for processing a single image against 19 ms for XceptionAG. Also, this architecture has satisfactory gender accuracy on test and training sample: 93.31% and 97.89% respectively.

For detail analysis of age recognition quality we had found MAE distribution for different trained ResNetMAG and XceptionAG networks (see Fig. 8 and Fig. 9). It demonstrates mean absolute error of age evaluation per each year. We can see that MAE metric does not exceed 5 years on Training sample for networks trained of full dataset. Also its metric exceed 15 years on Testing sample on difficult range of ages: from 38 years to 70. Some increase in MAE is observed at ages close to 100 years, this is due to the fact that in this case, even for human experts, it is difficult to determine the age of a person from the face image.

Table 1. Quality of gender and age recognition on different samples

Metric	ResNetMAG trained on			XceptionAG trained on		
	Full train sample	1000 images per age	600 images per age	Full train sample	1000 images per age	600 images per age
Gender accuracy on full training sample	0.9789	0.9470	0.9365	**0.9953**	0.9490	0.8134
Gender accuracy on test sample	0.9331	0.9096	0.9018	**0.9476**	0.9119	0.8371
MAE of age estimation on full training sample	3.79	7.72	9.45	**0.45**	11.05	10.65
MAE of age estimation on testing sample	**10.25**	12.90	11.88	10.89	13.64	15.06
Recognition time per image, s	0.0114	**0.0113**	0.0114	0.0181	0.0193	0.0191

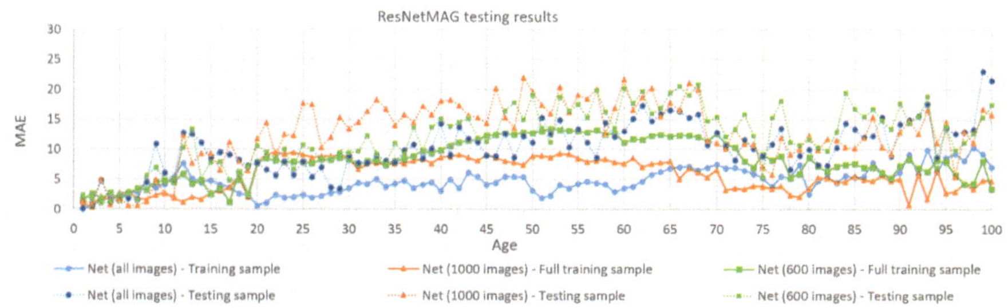

Fig. 8. MAE distribution for different trained ResNetMAG networks.

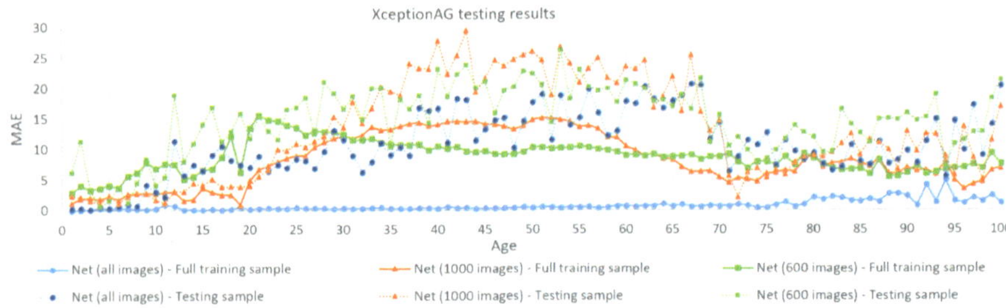

Fig. 9. MAE distribution for different trained XceptionAG networks.

6 Conclusions

It follows from the Table 1, Fig. 8 and Fig. 9 that the most accurate results were obtained for deep neural networks training on imbalanced dataset. Composition of undersampling and oversampling approaches gave worse result. Thus, we confirmed the existing hypothesis that the more images we have, the better the recognition result.

XceptionAG architecture shows comparable results on the quality of age recognition with ResNetMAG, but it has the best indicators on gender recognition. At the same time, the obtained quality and speed of gender and age recognition on cropped face images indicates the possibility to use the trained networks of both the ResNetMAG architecture and XceptionAG for computer vision systems for analyzing people's biometric indicators.

As a topic of future research we consider dataset balancing procedures for ages up to 20 years and over 60 years using a more complex approach based on Generative Adversarial Networks, as well as additional image alignment and the usage of face key points.

Acknowledgment. The research was made possible by Government of the Russian Federation (Agreement № 075-02-2019-967).

References

1. Levi, G., Hassner, T.: Age and gender classification using convolutional neural networks. In: IEEE Workshop on Analysis and Modeling of Faces and Gestures (AMFG), at the IEEE Conference on Computer Vision and Pattern Recognition (CVPR), Boston (2015)
2. Eidinger, E., Enbar, R., Hassner, T.: Age and gender estimation of unfiltered faces. In: Transactions on Information Forensics and Security (IEEE-TIFS), Special Issue on Facial Biometrics in the Wild, vol. 9, no. 12, pp. 2170–2179 (2014)
3. Escalera, S., Fabian, J., Pardo, P., Baro, X., Gonzalez, J., Escalante, H. J., Guyon, I.: Chalearn 2015 apparent age and cultural event recognition: datasets and results. In: ICCV, ChaLearn Looking at People workshop (2015)
4. Agustsson E., Timofte R., Escalera S., Baro X., Guyon I., Rothe R.: Apparent and real age estimation in still images with deep residual regressors on APPA-REAL database. In: Proceedings of FG (2017)

5. Rothe, R., Timofte, R., Gool, L.V.: DEX: Deep EXpectation of apparent age from a single image. In: Proceedings of ICCV (2015)
6. Clapes, A., Bilici, O., Temirova, D., Avots, E., Anbarjafari, G., Escalera, S.: From apparent to real age: gender, age, ethnic, makeup, and expression bias analysis in real age estimation. In: Proceedings of the IEEE Conference on Computer Vision and Pattern Recognition Workshops, pp. 2373–2382 (2018)
7. Rothe, R., Timofte, R., Gool, L.V.: Deep expectation of real and apparent age from a single image without facial landmarks. In: IJCV (2016)
8. Zhang, Z., Song, Y., Qi, H.: Age progression/regression by conditional adversarial autoencoder. In: IEEE Conference on Computer Vision and Pattern Recognition (CVPR) (2017)
9. Panis, G., Lanitis, A., Tsapatsoulis, N., Cootes, T.F.: Overview of research on facial ageing using the FG-net ageing database. IET Biometrics **5**(2), 37–46 (2016)
10. Chen, B.-C., Chen, C.-S., Hsu, W.H.: Face recognition using cross-age reference coding with cross-age celebrity dataset. IEEE Trans. Multimedia **17**, 804–815 (2015)
11. Ricanek, K., Tesafaye, T.: MORPH: a longitudinal image database of normal adult age-progression. In: 7th International Conference on Automatic Face and Gesture Recognition (FGR06) (2006)
12. IBM Research DiF dataset. https://www.research.ibm.com/artificial-intelligence/trusted-ai/diversity-in-faces/#access. Accessed 25 May 2019
13. Jiang, B.: Age and gender estimation based on Convolutional Neural Network and TensorFlow. https://github.com/BoyuanJiang/Age-Gender-Estimate-TF. Accessed 25 May 2019
14. Tommola, J., Ghazi, P., Adhikari, B., Huttunen, H.: Real time system for facial analysis. In: EUVIP 2018 (2018)
15. Becerra-Riera, F., Morales-González, A., Vazquez, H. M.: Exploring local deep representations for facial gender classification in videos. In: Conference: International Workshop on Artificial Intelligence and Pattern Recognition (IWAIPR) (2018)
16. Kharchevnikova, A.S., Savchenko, A.V.: Neural networks in video-based age and gender recognition on mobile platforms. Opt. Memory Neural Netw. **27**(4), 246–259 (2018)
17. Seif, G.: Handling imbalanced datasets in deep learning (2018). https://towardsdatascience.com/handling-imbalanced-datasets-in-deep-learning-f48407a0e758. Accessed 25 May 2019
18. Lin, T.-Y., Goyal, P., Girshick, R., He, K., Doll'ar P.: Focal loss for dense object detection. arXiv:1708.02002v2 (2018)
19. Zhang, H., Cisse, M., Dauphin, Y.N., Lopez-Paz, D.: mixup: beyond empirical risk minimization. arXiv:1710.09412 (2017)
20. Zhong, Z., Zheng, L., Kang, G., Li,, S., Yang, Y.: Random erasing data augmentation. arXiv:1708.04896 (2017)
21. Augmentor library. https://github.com/mdbloice/Augmentor. Accessed 25 May 2019
22. Imgaug library. https://imgaug.readthedocs.io. Accessed 25 May 2019
23. Wang, X., Wang, K., Lian, S.: A survey on face data augmentation. In: CVPR. arXiv:1904.11685v1 (2019)
24. Guan, S.: TL-GAN: transparent latent-space GAN (2018). https://github.com/SummitKwan/transparent_latent_gan. Accessed 25 May 2019
25. Karras, T., Aila, T., Laine, S., Lehtinen, J.: Progressive growing of GANs for improved quality, stability, and variation. In: ICLR 2018. arXiv:1710.10196v3 (2018)
26. Yan, X., Yang, J., Sohn, K., Lee, H.: Attribute2Image: conditional image generation from visual attributes. arXiv:1512.00570v2 (2016)
27. Kaiming, H., Xiangyu, Z., Shaoqing, R., Jian S.: Deep residual learning for image recognition. In: ECCV. arXiv:1512.03385 (2015)

28. Yudin, D., Kapustina, E.: Deep learning in vehicle pose recognition on two-dimensional images. In: Advances in Intelligent Systems and Computing, vol. 874, pp. 434–443 (2019)
29. Chollet, F.: Xception: deep learning with depthwise separable convolutions. In: CVPR 2017. arXiv:1610.02357 (2017)
30. Szegedy, C., Vanhoucke, V., Ioffe, S., Shlens, J., Wojna Z.: Rethinking the inception architecture for computer vision. In: ECCV. arXiv:1512.00567 (2016)
31. Chollet, F.: Keras: deep learning library for Theano and tensorflow. https://keras.io/. Accessed 26 May 2019

A Complex Approach to the Data Labeling Efficiency Improvement

E. V. Melnik[1] and A. B. Klimenko[2(✉)]

[1] Federal Research Centre, Southern Scientific Centre of the Russian Academy of Sciences, 41, Chehova st, 344006 Rostov-on-Don, Russia

[2] Scientific Research Institute of Multiprocessor Computer Systems of Southern Federal University, 2, Chehova st, 347928 Taganrog, Russia

anna_klimenko@mail.ru

Abstract. In this paper the issue of labeling noise is considered. Data labeling is the integral stage of the most part of the machine learning projects, so the problem spotlighted in the paper is quite topical.

According to the labeling noise source classification, a new approach is proposed, affecting both of the labeling noise sources. The first component of the approach is based on the distributed ledger technologies principles, including the automatic consensus between the experts. The second component includes the devices dependability improvement by means of fog- and edge-computing usage. Also some models are developed to estimate the approach and selected results of the simulation are presented and discussed.

Keywords: Machine learning · Dataset forming · Distributed ledger · Labeling noise

1 Introduction

Supervised machine learning is used in various businesses nowadays, improving the operations. The diversity of machine learning projects is considerable, embracing various fields, e.g., AutoKeras [1], AirSim [2], Detectron [3] and others. Yet the integral stage of all ML projects is the training dataset forming, including data labeling [4]. The latter is a serious issue for project teams for its cost, time consumption and lack of guarantee to get the datasets of acceptable quality.

According to the [4], there are different ways to form the labeled dataset for machine learning, including: in-house labeling, outsourcing, crowdsourcing, specialized outsourcing companies, synthetic labeling, data programming.

The first four approaches are oriented to human involvement into the labeling process, others are automated. Yet despite the automatic nature of labeling the experts are involved to correct the collected data and to eliminate mislabeling and label errors.

Label errors, or labeling noise is a serious issue for every ML project.

The labeling noise is anything that obscures the relationship between the features of an instance and its class [5]. The main sources of labeling noise are as follows [6]:

S. Kovalev et al. (Eds.): IITI 2019, AISC 1156, pp. 41–50, 2020.
https://doi.org/10.1007/978-3-030-50097-9_5

- the information which is provided to the expert may be insufficient to provide reliable labeling;
- the errors can occur in the expert's labeling process itself;
- the labeling task is subjective, as, for example, in medical applications and image data analysis;
- the presence of data encoding errors and communication problems, which is caused by insufficient level of dependability.

Generalizing, one can say that there are two main causes for labeling noise: the human-based and the infrastructure-based ones.

At least four major techniques for labeling noise elimination are described in literature:

- label noise-robust methods; [7–9]
- probabilistic label noise-tolerant methods [10, 11];
- data cleansing methods [12, 13];
- model-based label noise-tolerant methods [14–16].

Yet, no one of techniques listed above gives the definite solution to the problem.

In this paper the problem of labeling noise is considered. Partially, this work continues and extends our previous works [17, 18]. A new approach, based on consensus element of distributed ledger and the fog-computing concept, is proposed. The approach handles both of general sources of labeling noise by the distributed ledger technology and fog-computing concept application.

2 Distributed Ledger Elements

A distributed ledger (or distributed ledger technology, DLT) is a consensus of replicated, shared, and synchronized digital data geographically spread across multiple sites, countries, or institutions [19]. A peer-to-peer network is required as well as consensus algorithms to ensure replication across nodes is undertaken.

A considerable number of DLT-based systems are designed and developed till now, e.g., Bitcoin, Etherium, Nano, etc. Some systems use the blockchain data structure to store the transactions [20], and some use the relatively new blocklattice [21] data structure, which can be considered as an extention of the blockchain.

So, the key elements of the DLT are the way of data storing and the consensus method. While the blocklattice is quite new, the blockchain structure is used frequently with the numerous consensus methods.

The comprehensive classification of consensus methods used with blockchain is presented in [22]. According to the analysis conducted, the consensus methods can be divided into the following groups:

- proof-based consensus algorithms (PoW, Prime number finding–based PoW, Generalized PoW, PoS, Hybrid forms of PoW and PoS, Proof of activity, Proof of burn, proof of space, proof of elapced time, proof of luck, multichain);
- vote-based consensus algorithms (PBFT, Symbiont, Iroha with Sumeragi, Ripple, Stellar);

- crash fault-tolerance–based consensus (Quorum with Raft, Chain).

Each of the methods mentioned above has its pros and cons, e.g., PoW can generate forks in the chain, the transactions commit eventually, consumes much energy, but oriented to the usage in large fully distributed systems, PBFT commits transactions definitely, yet possesses poor scalability and generates the redundant network traffic, etc. So, the consensus method must be selected according to the requirements to the system and its field of application.

3 Fog- and Edge-Computing

Fog-computing is quite a new, but extensively growing technological field. Announced in 2012 for the first time, fog-computing aims to support the Internet of Things concept and to deliver facilities of big data processing with communicational environment offloading and the decrease of system latency [23–25].

In contemporary concept descriptions the network is considered as a three-layer structure:

- Edge of the network, which consists of end-point devices, sensors, computers, notepads, etc.;
- Fog layer, which consists of communication facilities, routers, gateways, etc.;
- Cloud, which consists of servers located in a datacenter and, as usual, interconnected by a fast network.

There is a significant difference between edge-computing and fog-computing: the first concept presupposes that the logic of tasks distribution at the edge of the network is implemented at the edge devices, while the fog-computing presupposes that the decision how and where to process the data is made on the fog-layer node. Besides, fog-computing does not exist without a cloud, and the part of the computations has to be done there.

As was mentioned in [24], there are three main computational models in the field of fog-computing:

- the offloading model, when the data generated from edge devices are offloaded to the nearest fog node and then at the Cloud (i.e., up-offloading) and in the reverse order from the Cloud to edge devices (i.e., down-offloading).
- the aggregation model, when data streams generated by multiple edge devices are aggregated and possibly processed at the nearest Fog node before being uploaded to Cloud datacenter.
- the peer-to-peer (P2P) model, Fog nodes, which are at the proximity of edge devices, share their computing and storage capabilities and cooperate in order to offer an abstraction storage and computing layer to edge users.

So, summing up, the fog-computing concept presupposes that the computational workload can be shifted from cloud to the fog- or edge- layer and so to decrease the communicational channels load. Yet there are some studies, where the fog-computing concept application is used for the devices reliability improvement [26–28]. It makes the elements of fog-computing concept to be usable in the approach to the data labeling efficiency improvement.

4 The Technique of Data Labeling

As was mentioned earlier, the causes of labeling noise can be divided into two major classes: human-based causes and infrastructure-based ones.

The approach proposed in this paper aims at decreasing the labeling noise affecting both of the classes in the following way, as is shown in Fig. 1.

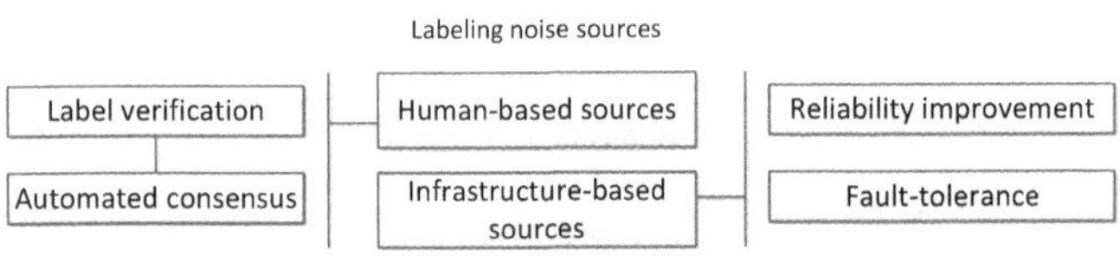

Fig. 1. The approach to the data labeling efficiency improvement.

Consider the generic scheme of data labeling process: a software entity sends the content to be labeled to the expert, expert labels the content and saves it in the data storage.

The procedure of one-piece content labeling is an analogue of the transaction in cryptocurrency system in the context of this paper. So, as it is done in cryptocurrencies, having the piece of content labelled, the next step is to verify its correctness sending it to all participants and reaching the consensus about the correctness of the transaction in automatically manner. Obviously this stage decreases the labeling noise doing this before the labeled data are put into the storage. The well-known methods of data cleansing [29, 30] propose some filters for data which has been put into the storage.

Yet, the automatic consensus (e.g., BFT [31] or Paxos [32]) is possible after all experts received the labeled content from other experts, so there can be a considerable delay. Another way to organize the label verification and the automated consensus is to send to the expert community the same content and to process the overall content to be labeled consequently.

In this case there is no need to wait before the majority of experts label their content, so, such approach can be quite prospective. To estimate the cases described, consider the simple models of the processes, taking into account the dependency between the voting-consensus time and the number of participants.

When all experts receive different pieces of data, the process will be as follows:

1. An expert receives the content to be labeled.
2. The expert labels the content and waits for the content to be verified from other community participants.
3. After all pieces of content are received and verified the consensus takes place.
4. A package of verified content is added to the block.

This procedure can be described by the following expression:

$$t_p = t + \alpha N + (N - 1)t + Nt_{cons},$$

where t is the time needed for the content labeling;
αN is the waiting time for content to be verified;
$(N - 1)t$ is the time of verification of content received from $N-1$ experts, is the time needed to reach a consensus for all labelled data.

Considering $t_{cons} = kN$, where k is the ratio, which describes the dependency between consensus procedure speed and the number of participants, the time of package verification will be as follows:

$$t_p = t + \alpha N + (N - 1)t + kN^2.$$

When the community receives the same data for verification, the time can be modelled in the following expressions.

To verify 1 piece of content, the time needed is as follows:

$$t_p = t + kN;$$

To verify N pieces of content consequently, the time needed is:

$$t_N = Nt + kN^2.$$

Then, if we want to label M pieces of content, the time needed for this will be as follows:

$T = M(t + \alpha + kN)$ for the first case, and $T = M(t + kN)$ for the second case.

The second component of our approach is the data labeling process dependability improvement. It is interconnected with the consensus procedure location in the network and relates to the fog- and edge-computing concepts.

There are two possible ways to organize the consensus procedure among the expert devices:

- the centralized one, when the chosen leader gathers the expert labeled data, analyses it and chooses the label, which has been selected by the majority;
- the distributed one, when every expert device gathers the labeled data pieces from other participants, and each selects the label, which is voted by the majority.

Considering the workload distribution among the network devices as a measure of their reliability [33] we use a simplified network infrastructure model to estimate the workload of the network devices (Fig. 2).

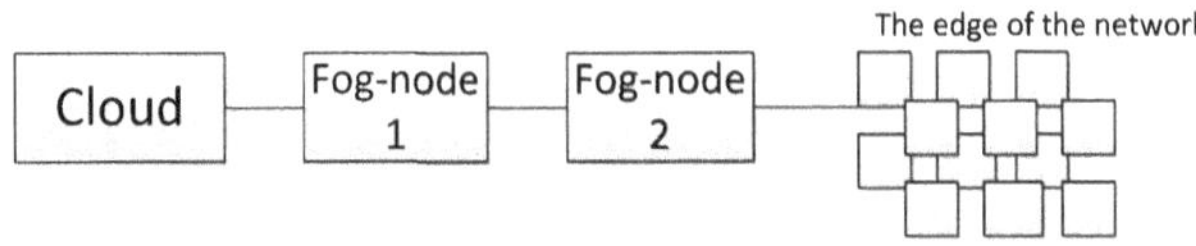

Fig. 2. A simplified network infrastructure model.

The simple procedure of labeling and reaching consensus is as follows:

1. to receive a piece of content;
2. to present a piece of content to expert by means of GUI;
3. to distribute the label to the other participants;
4. to receive the labels from other participants;
5. to send the chosen label, selected by majority, to the central service.

Consider the following model parameters: v is the volume of content to be labeled by the expert, the number of computational operations to transmit the information through the device is $l_{tr} = \xi v$, where ξ - is a ratio between the information volume to be transmitted and the computational complexity of this procedure. Then, the computational complexity of sending and receiving data is as follows: $l_{send} = l_{receive} = \eta v$, where η - is a ratio between the information volume and the computational complexity of sending and receiving data. Also assume that the complexity of choosing the label for the voting is $l_{est} = kN$, where N – the number of voting participants.

To decrease cluttering of the paper, the workload estimation models are presented in the Table 1.

Table 1. The workload estimation models.

	Edge-located consensus	Fog-located consensus	Cloud-located consensus	Distributed edge-located consensus
Cloud workload	$L_c = Nl_{send} + l_{receive}$	$L_c = Nl_{send} + Nl_{receive} + l_{est}$	$L_c = Nl_{send} + Nl_{receive} + l_{est}$	$L_c = Nl_{send} + Nl_{receive}$
Fog workload	$L_{f1} = l_{tr} \cdot N$ $L_{f2} = l_{tr} \cdot N$	$L_{f2} = l_{tr} \cdot N + l_{receive} * N + l_{est}$	$L_{f1} = l_{tr} \cdot N$ $L_{f2} = l_{tr} \cdot N$	$L_{f1} = 2l_{tr} \cdot N$ $L_{f2} = 2l_{tr} \cdot N$
Edge workload	$L_{ei} = l_{receive} + (N-1)l_{send} + (N-1)l_{receive} + l_{est} + l_{send}$ $L_{ei} = l_{receive} + l_{send}$	$L_{ei} = l_{receive} + l_{send}$	$L_{ei} = l_{receive} + l_{send}$	$L_{ei} = l_{receive} + (N-1)l_{send} + (N-1)l_{receive} + l_{est} + l_{send}$

Consider the cloud performance is P_c, fog-node performance – P_f, edge device performance is P_e. Assume that there is a time period T of label verification. Than the cloud, fog and edge nodes workload in percent is as follows:

$$D_c = \frac{L_c}{P_c \cdot T},$$

$$D_{fi} = \frac{L_{fi}}{P_{fi} \cdot T},$$

$$D_{ei} = \frac{L_{ei}}{P_{ei} \cdot T}.$$

Assuming that $\lambda = \lambda_0 \cdot 2^{\Delta T/10}$, and $\Delta T \sim D$, D is a workload percentage, the reliability function of the device is as follows:

$$P(t) = e^{-\lambda t} = e^{-\lambda_0 \cdot 2^{kD/10}}.$$

One can see, that fully distributed consensus located on the edge of the network gives the worst overall system workload distribution, while the centralized solution located on the edge of the network affects only the workload of leader device. Also, one can see that the centralized edge-located consensus procedure is quite poor in terms of fault-tolerance, because of leader "bottleneck" existence. This situation can be improved significantly by the distributed leader approach application. The latter is used in such distributed consensus protocols as ViewStamped Replication and Raft, and consists of leader role applying to the different nodes consequently.

Assume the edge node community is functioning in the course τ, with the labeling of Φ content pieces. For each piece a new leader is elected, e.g., with the simple "round-robin" procedure.

Then, assuming $P(t) = e^{-\int_0^{\tau} \lambda(t)dt}$, and $\Phi = 2$, the computations performed by one edge device are estimated by the following expression:

$$\begin{aligned} L_{ei} &= l_{receive} + (N-1)l_{send} + (N-1)l_{receive} + l_{est} + l_{send} + l_{receive} + l_{send} \\ &= L_{ei\_leader} + L_{ei\_follower} \end{aligned}$$

The average workload for the one piece of content is:

$$L_{ei\_av} = \frac{L_{ei\_leader} + L_{ei\_follower}}{\Phi}.$$

With the following increase of Φ and fixed N, the average device workload will be:

$$L_{ei\_av} = \frac{\left\lceil\frac{\Phi}{N}\right\rceil L_{ei\_leader} + \left(\Phi - \left\lceil\frac{\Phi}{N}\right\rceil\right) L_{ei\_follower}}{\Phi}.$$

With the increase of Φ and the large N, the L_{ei_leader} can be neglected, and with $\Phi \to \infty$ $L_{ei\_av} \to L_{ei\_floower}$. Besides, such distributed leader approach affects the fault-tolerance positively: in case of leader's failure the next leader will be elected.

5 Discussion and Conclusion

In the current paper a new approach to the data labeling efficiency improvement is proposed. It consists of two components: the first one is the application of the distributed ledger elements to the data labeling procedure, and the second one is the application of the edge- and fog-computing concepts to distribute the device workload in an appropriate way and so to increase the reliability and fault-tolerance of the participant nodes. These two components improve the data labeling efficiency, affecting both of the labeling errors causes.

Modeling, estimation and analysis allow to choose the most promising combination of applied techniques: to improve the efficiency of data labeling it is promising to apply the consequent consensus-based labeling among the experts and the implementation of labeling consensus on the edge of the network by means of the distributed leader approach (see Fig. 3).

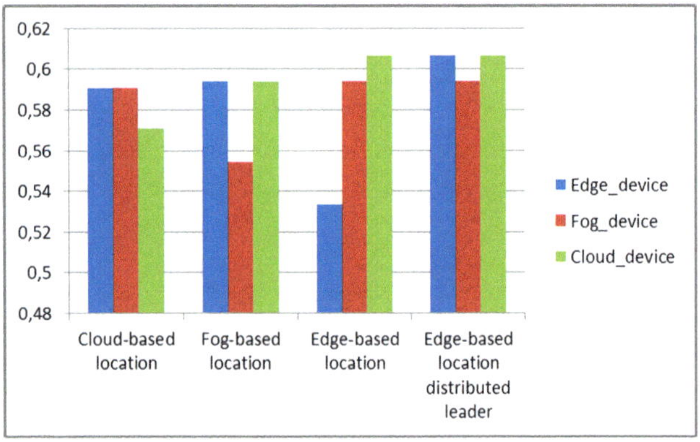

Fig. 3. The comparison of reliability function value for different leader location.

As is shown in the Fig. 3, the application of distributed leader in combination with the edge-located consensus makes it possible to improve the overall system reliability. At the same time, the distributed leader solves the issue of fault-tolerance in centralized systems, and, as a consequence, increases the overall dependability of the system.

Acknowledgements. The paper has been prepared within the RFBR project 18-29-22086 and RAS presidium fundamental research №7 «New designs in the prospective directions of the energetics, mechanics and robotics», № gr.project AAAA-A18-118020190041-1.

References

1. Auto-Keras. https://autokeras.com/. Accessed 19 May 2019
2. Welcome to AirSim. https://github.com/microsoft/AirSim. Accessed 19 May 2019
3. Detectron. https://research.fb.com/downloads/detectron/. Accessed 19 May 2019

4. Machine Learning Project Structure: Stages, Roles, and Tools. https://www.altexsoft.com/blog/datascience/machine-learning-project-structure-stages-roles-and-tools/. Accessed 19 May 2019
5. Hickey, R.J.: Noise modeling and evaluating learning from examples. Artif. Intell. **82**(1–2), 157–179 (1996)
6. Frénay, B., Verleysen, M.: Classification in the presence of label noise: a survey. IEEE Trans. Neural Netw. Learn. Syst. **25**(5), 845–869 (2014)
7. Manwani, N., Sastry, P.S.: Noise tolerance under risk minimization. IEEE Trans. Cybern. **43**(3), 1146–1151 (2013)
8. McDonald, A., Hand, D.J., Eckley, I.A.: An empirical comparison of three boosting algorithms on real data sets with artificial class noise. In: Proceedings 4th International Workshop Multiple Classifier Systems, Guilford, UK, pp. 35–44, June 2003
9. Abellán, J., Masegosa, A.R.: Bagging decision trees on datasets with classification noise. In: Link, S., Prade, H. (eds.) Foundations of Information and Knowledge Systems. FoIKS 2010. Lecture Notes in Computer Science, vol. 5956, pp. 248–265. Springer, Heidelberg (2010)
10. Joseph, L., Gyorkos, T.W., Coupal, L.: Bayesian estimation of disease prevalence and the parameters of diagnostic tests in the absence of a gold standard. Am. J. Epidemiol. **141**(3), 263–272 (1995)
11. Perez, C.J., Giron, F.J., Martin, J., Ruiz, M., Rojano, C.: Misclassified multinomial data: a bayesian approach. Rev. R. Acad. Cien. Serie A. Mat. **101**(1), 71–80 (2007)
12. Brodley, C.E., Friedl, M.A.: Identifying mislabeled training data. J. Artif. Intell. Res. **11**, 131–167 (1999)
13. Gamberger, D., Boskovic, R., Lavrac, N., Groselj, C.: Experiments with noise filtering in a medical domain. In: Proceedings 16th International Conference on Machine Learning, Bled, Slovenia, June 1999, pp. 143–151. Springer, San Francisco (1999)
14. Krauth, W., Mezard, M.: Learning algorithms with optimal stability in neural networks. J. Phys. A: Gen. Phys. **20**(11), L745 (1987)
15. Clark, P., Niblett, T.: The CN2 induction algorithm. Mach. Learn. **3**(4), 261–283 (1989)
16. Cantador, I., Dorronsoro, J.R.: Boosting parallel perceptrons for label noise reduction in classification problems. In: Proceedings of First International Work-Conference on the Interplay Between Natural and Artificial Computation, IWINAC 2005. Lecture Notes in Computer Science, Las Palmas, Canary Islands, Spain, 15–18 June 2005, vol. 3562, pp. 586–593 (2005)
17. Kalyaev, I., Melnik, E., Klimenko, A.: A technique of adaptation of the workload distribution problem model for the fog-computing environment. In: Silhavy, R. (ed.) Cybernetics and Automation Control Theory Methods in Intelligent Algorithms. CSOC 2019. Advances in Intelligent Systems and Computing, vol. 986. Springer, Cham (2019)
18. Melnik, E.V., Klimenko, A.B., Ivanov, D.Y.: The Distributed ledger-based technique of the neuronet training set forming. In: Silhavy, R., Silhavy, P., Prokopova, Z. (eds.) Computational Statistics and Mathematical Modeling Methods in Intelligent Systems. CoMeSySo 2019. Advances in Intelligent Systems and Computing, vol. 1047. Springer, Cham (2019)
19. Distributed ledger technology: beyond blockchain. https://www.gov.uk/government/news/distributed-ledger-technology-beyond-block-chain. Accessed 20 May 2019
20. Wüst, K., Ritzdorf, H., Karame, G.O., Glykantzis, V., Capkun, S., Gervais, A.: On the security and performance of proof of work blockchains. In: Proceedings of the 2016 ACM SIGSAC Conference on Computer and Communications Security, pp. 3–16. ACM, New York (2016)
21. An introduction to the Block-Lattice. https://medium.com/coinmonks/an-introduction-to-the-block-lattice-382071fc34ac. Accessed 20 May 2019

22. Nguyen, G., Kim, K.: A survey about consensus algorithms used in blockchain. J. Inf. Process. Syst. **14**(1), 101–128 (2018)
23. Bonomi, F, Milito, R, Zhu, J., Addepalli, S.: Fog computing and its role in the internet of things. In: Proceedings of the first edition of the MCC Workshop on Mobile Cloud Computing, pp. 13–16. ACM, Mew York (2012)
24. Moysiadis, V., Sarigiannidis, P., Moscholios, I.: Towards distributed data management in fog computing. Wirel. Commun. Mob. Comput. **2018** (2018). article ID 7597686, 14 p
25. Chiang, M., Zhang, T.: Fog and IoT: an overview of research opportunities. IEEE Internet Things J. **3**(6), 854–864 (2016)
26. Melnik, E.V., Klimenko, A.B., Ivanov, D.Y.: Fog-computing concept usage as means to enhance information and control system reliability. J. Phys: Conf. Ser. **1015**(3), 032175 (2018)
27. Melnik, E.V., Klimenko, A.B., Ivanov, D.Y.: Distributed information and control system reliability enhancement by fog-computing concept application. In: IOP Conference Series: Materials Science and Engineering, vol. 327, no. 2 (2018)
28. Melnik, E., Klimenko, A., Ivanov, D.: The model of device community forming problem for the geographically-distributed information and control systems using fog-computing concept. In: IV International research conference Information technologies in Science, Management, Social sphere and Medicine (ITSMSSM 2017), Advances in Computer Science Research, vol. 72, pp. 132–136. Atlantis Press, Amsterdam (2017)
29. Wilson, R., Martinez, T.R.: Instance pruning techniques. In: Proceedings of the 14th International Conference on Machine Learning, Nashville, TN, July 1997, pp. 403–411 (1997)
30. Hart, P.: The condensed nearest neighbor rule. IEEE Trans. Inf. Theory **14**, 515–516 (1968)
31. pBFT—Understanding the Consensus Algorithm. https://medium.com/coinmonks/pbft-understanding-the-algorithm-b7a7869650ae. Accessed 19 May 2019
32. Paxos Made Simple. https://lamport.azurewebsites.net/pubs/paxos-simple.pdf. Accessed 19 May 2019
33. Strogonov, S.A.: Individual reliability forecasting of IC chip with the help of ARIMA models. Mag. Compon. Technol. **10**, 44–49 (2006)

Automation of Musical Compositions Synthesis Process Based on Neural Networks

Nikita Nikitin, Vladimir Rozaliev(✉), Yulia Orlova, and Alla Zaboleeva-Zotova

Volgograd State Technical University, 28 Lenin Avenue, Volgograd 400005, Russia
vladimir.rozaliev@gmail.com

Abstract. This work is devoted to development and approbation of the methods for automated sound generation based on image color spectrum with using the neural networks. The work contains a description of the transition between color and music characteristics, the rationale for choosing and the description of a used neural network. The choice of the neural network implementation technology is described. It also contains the detailed description about the experiments to choose the best neural network parameters.

Keywords: Automated music generation · HSV color space · Newton correlation table · J. Caivano correlation scheme · Image analysis · Sound synthesis · Recurrent neural network

1 Introduction

Nowadays, music is composed by computer has become a mass phenomenon and is turning into an important component of modern musical culture. Computers provide a lot of brand-new opportunities in the development of a musician's professional thinking in all areas of musical creativity, which led to the increasing introduction of music and computer technologies. This allows to significantly complement and even change the very nature of the work of a musicologist, composer, performer, a significant impact on the process of teaching and learning music.

Initially, the term "computer music" was used by specialists to denote the field of engineering development related to digital synthesis of musical sounds, digital processing of audio signals, digital recording of various sound structures, etc. Currently, the definition of "computer music" is often used by many musicians and listeners in relation to any music that is created using one or another music and computer technology.

Since the music began to be recorded on paper in the form of musical notation, the original "ways" of its composition began to appear. One of the first methods of

This work was partially supported by RFBR and administration of Volgograd region (grants 17-07-01601, 18-07-00220, 18-47-342002, 19-47-343001, 19-47-340003, 19-47-340009).

S. Kovalev et al. (Eds.): IITI 2019, AISC 1156, pp. 51–59, 2020.
https://doi.org/10.1007/978-3-030-50097-9_6

algorithmic composition was the method of composing music invented by Mozart - "The Musical Game of the Dice". The first computer musical composition - "Illiac Suite for String Quartet" - was created in 1956 by the pioneers of using computers in music - Lejaren Hiller and Leonard Isaacson [1].

The development of computer music, including the sound generation by image, in the last century was severely limited by computing resources - only large universities and laboratories could afford to buy and hold powerful computers, and the first personal computers lacked computing power. However, in the 21st century, almost everyone can study computer music.

Now, computer music can be used in many industries: creating music for computer games, advertising and films. Now, to create background music compositions in computer games and advertising, companies hire professional composers or buy rights to already written musical works. However, in this genre the requirements for musical composition are low, which means that this process can be automated, which will allow companies to reduce the cost of composing songs. Also, the generation of sounds based on image can be applied in the educational process [2]. The development of musical perception in preschool children can be in the form of integrated educational activities, which is based on combinations of similar elements in music and arts (the similarity of their mood, style and genre) [3].

The greatest success of the theory of automation of the process of writing and creating music made up relatively recently (at the end of XX century), but mostly associated with the study and repetition of different musical styles [4].

Since the process of creating music is difficult to formalize, artificial neural networks are best suited for automated sound generation – they allow identifying connections that people do not see [5]. In addition, to reduce the user role in the generation of music, it was decided to take some of the musical characteristics from the image. Thus, the purpose of this work is to increase the harmony and melodicity of sound generation based on image colour spectrum through the use of neural networks.

To achieve this purpose the following tasks were identified:

- Determine the correlation scheme between colour and musical characteristics.
- Review the types of neural networks and choose the most suitable type for generating musical compositions.
- Describe the neural network used to generate music compositions by image.
- Choose neural network implementation technology.
- Choose a method for sounds synthesizing.
- Design and develop a program for sound generation using neural networks.
- Make an experiment to assess the harmony and melody of the output musical composition.

2 Image Analysis

The algorithm of image analysis in this work allows to get musical characteristics from image – to determine the nature of the result musical composition. To do this, firstly, it is necessary to transform original image to HSV color space. This transformation allows to easily get general characteristic of each pixel from image – hue, saturation and brightness. The second part of image analysis is determining the predominant color of image. The predominant color in this work will allow to get the tonality of the result musical composition. For this task, we use the K-means clusterization algorithm, since it has the following advantages:

- relatively high efficiency with ease of implementation;
- high quality clustering;
- the possibility of parallelization;
- existence of many modifications.

K-Means Clustering is an unsupervised learning method. If there are labels in the sample datasets, it is necessary to use a controlled method, but in the real world, as a rule, we have no labels, and therefore we prefer clustering methods that are known as uncontrolled methods. The purpose of this algorithm is to find groups in the data with the number of groups represented by the variable K. The algorithm works iteratively to assign each data point to one of the K groups based on the provided functions. Data points are grouped based on feature similarity. Results of the K-means clustering algorithm:

- centroids of clusters K, which can be used to label new data;
- labels for training data (each data point is assigned to one cluster).

First, it is necessary to read the image data using the cv2.imread function from OpenCV. After the image is read using cv, our color image channel becomes us Blue-Green-Red. But we want to use red-green-blue as the color channel of the image, so then, the algorithm converts image to the desired channel using the cv2.cvtcolor function. Now there are three-dimensional parameters in the image data: row number X, column number X and the color channel number. But it is not necessary to separate information about rows and columns. In addition, it is difficult to deal with the 3D matrix, so the method changes the image and make it with the data of the 2D matrix.

Since K-Means has been imported, the program can easily use it by specifying only n_clusters, which initially represents the cluster number. After that, the fit function is used to apply the K-Means clustering algorithm to pre-processed image data, and the result will return to the clt objects. Then, the find_histogram function is used to limit the number of histograms to the desired number of clusters. Since there is no need to find a histogram for all pixels and the entire color palette, it is necessary to limit it to the required number of clusters.

3 Synthesis of the Musical Materials

To reduce the user role in the generation of music, some of the musical characteristics are obtained by analysing the colour scale of the image. Thus, the character of the output musical composition will correspond to the input image. This feature makes possible to use this approach for creating background music in computer games, advertising and films.

The key characteristics of a musical work are its tonality and tempo. These parameters are determined by analysing the colour scheme of the image. To begin with, we determine the ratio of colour and musical characteristics [6] (Table 1).

Table 1. Correlation between color and musical characteristics

Colour characteristics	Musical characteristics
Hue (red, blue, yellow…)	Pitch (c, c#, d, d#, …)
Colour group (warm/cold)	Musical mode (major/minor)
Brightness	Octave
Saturation	Duration

Then, it is necessary to define the correlation scheme between the color and pitch name. At the moment, there are a large number of such schemes, but in this work, was chosen the Newton scheme (Table 2).

Table 2. Correlation scheme between color and pitch names

Colour name	Pitch
Red	C
Red – orange	C#
Orange	D
Yellow – orange	D#
Yellow	E
Green	F
Green – blue	F#
Blue	G
Blue – violet	G#
Violet	A
Yellow – green	A#
Pink	H

As can be seen from Table 1, the tonality of a composition is determined by two colour characteristics - a hue and a colour group, the tempo by brightness and saturation. The algorithm for determining the tonality relies on image analysis and Table 1. It consists of 3 steps and described below.

Step 1. Converting the input image from RGB to HSV colour space. This step allows transforming the image to a more convenient form, because HSV space already contains the necessary characteristics - the name of the colour (determined by the parameter hue), saturation and brightness (value parameter).
Step 2. Analysing the whole image and determining the predominant colour.
Step 3. Determining the name and colour group of predominant colour.

Step 4. According to Table 1 and Table 2 define the tone of the musical composition (pitch and the musical mood).

To determine the tempo of composition, it`s necessary to get the brightness and saturation of predominant color, and calculate the tempo, according to these parameters.

4 Neural Network

In this work, the neural network is supposed to be used to generate musical material, since neural networks can find such dependencies in a pre-existing corpus of musical compositions that cannot be seen by humans.

Recurrent neural network (RNN) has recurrent connections which allow the network to store information related to the inputs. These relationships can be considered similar to memory. Exactly this feature of RNN makes them particularly well applicable for generating musical material. RNN is especially useful for the study of sequential data, such as music. However, the great complexity of RNN networks is the problem of explosive gradient, which consists in the rapid loss of information over time. Of course, this only affects the weights, not the states of the neurons, but it is in them information accumulates. Networks with long-short term memory (LSTM) try to solve this problem through the using filters and an explicitly specified memory cell. Each neuron has a memory cell and three filters: input, output and forgetting. The purpose of these filters is to protect information. The input filter determines how much information from the previous layer will be stored in the cell. The output filter determines how much information the following layers will receive. Such networks are able to learn how to create complex structures, for example, compose texts in the style of a certain author or compose simple music, but they consume a large amount of resources [7].

Thus, to automate the process of sound generation, recurrent neural networks with long short-term memory (RNN LSTM) area used. This kind of neural networks is used to generate musical compositions in various programs such as Magenta. Magenta is an open source music project from Google. Also, RNN LSTM is used in BachBot. This is the program that creates the musical composition in the Bach style. And this kind of neural network is used in DeepJaz - the system that allows to generate jazz compositions based on analysis of midi files [8].

For implementation of neural network Keras library was used as the high-level library of machine learning methods, since this library allows to work based on Theano and TensorFlow, taking advantage of them, while the process of developing neural networks using this library is simple and fast, which allows to create prototypes for rapid experimentation.

5 The Experiments

The neural network was created using the Keras library and was trained on classic Bach compositions. A neural network consists of one or more LSTM levels, each of which is accompanied by an exclusion layer that helps streamline training and prevent retraining of the network. To evaluate the neural network training, it was decided to use the loss function of binary cross-entropy instead of the classical mean-square error. This is due to the nature of the problem under consideration, which is close to the classification problem by several labels, and since binary cross-entropy gives a multivalued classification and also gives an estimate of losses, which more accurately reflects the accuracy of the model [9]. To assess network performance, the original dataset was divided into 75% training data and 25% verification data.

The F-measure is then used, which is the harmonic mean value of accuracy and completeness, to calculate the accuracy of the network [10].

5.1 Methodology for Evaluating the Results of Experiments

Accuracy (precision) and completeness (recall) are metrics that are used in evaluating information extraction algorithms. Sometimes they are used on their own, sometimes as a basis for derived metrics such as F-measure or R-Precision.

The accuracy of a system within a class is the proportion of documents actually belonging to a given class with respect to all documents that the system has assigned to this class. The completeness of the system is the proportion of documents found by the classifier belonging to a class relative to all documents of this class in the test set [11].

These values are easily calculated on the basis of a contingency table, which is compiled for each class separately (Table 3).

Table 3. Contingency table

Category		Expert assessment	
		Positive	Negative
System assessment	Positive	TP	FP
	Negative	FN	TN

The table contains information on how many times the system made the right and how many times the wrong decision on the documents of a given class. Namely [11]:

- TP is a true positive decision;
- TN is a true negative solution;
- FP - false positive solution;
- FN - false negative solution.

Then, accuracy and completeness are defined as follows [11]:

$$Precision = \frac{TP}{TP + FP} \quad (1)$$

$$Recall = \frac{TP}{TP + FN} \quad (2)$$

The higher accuracy and completeness, the better the system is trained, but in real life maximum accuracy and completeness are not achievable at the same time and it is necessary to find a balance. Such a metric is the F-measure.

F-measure is the harmonic average between accuracy and completeness. It tends to zero if precision or completeness tends to zero.

$$F = 2 * \frac{Precision * Recall}{Precision + Recall} \quad (3)$$

This formula gives the same weight of accuracy and completeness, so the F-measure will fall equally with decreasing accuracy and completeness.

5.2 The Count of LSTM Layers

During the learning process, it has been found that having multiple LSTM layers makes the learning process slower and less accurate than with one or two layers with a sufficient number of neurons. This may be due to the problem of a fading (or explosive) gradient, when the depth of the network interferes with the layers near the entrance to effectively update weights. As a result, it can be concluded that the use of LSTM layers in the range from 2 to 4 allows to obtain the most acceptable results.

5.3 The Size of LSTM Layers

The size of the LSTM layers has a great influence on the quality of the resulting musical composition. The larger the size, the faster the network converges; however, this leads to retraining. This may be due to the fact that more neurons allow the network to store more data from the training set in the weights, instead of optimizing the way in which the music template is summarized as a whole. It has been found, that 512 size of each LSTM layers give the better results for sound generation.

5.4 The Size of the Corpus for Network Training

Training data should be divided into small parts, which is exactly the neural network that is being trained. After each such batch, the LSTM network status is reset, and the weights are updated. Different batch size values were tested, and it was found that smaller batch sizes make the model more efficient. However, the presence of small batch sizes makes learning very long. It was found that a batch size of 512 gives the best result while maintaining relatively fast learning.

6 Conclusion

In this work, the ways to automate the process of sound generation have been analyzed and described. There are three part of general process:

- Image analysis;
- Synthesis of musical material;
- Sound synthesis.

For image analysis, the method of getting the color characteristic has been described. This method bases on the light-music theories and color transformation methods, allows to determine musical characteristics from image. Also, K-means clustering algorithm has been used to determine predominant color of given image, and then transform this color to tonality of the result musical composition.

For synthesis of musical material, corresponding method has been described. This method bases on the Caivano music-color theory and Newton color-pitch correlation table, describe the method to generate musical material. Also, for generation of musical material the core method is neural networks, which can produce some pitches in midi format, based on data, determined by previous algorithm. Exactly this method of joint usage of the light music theory and the neural network that is of particular interests in this work.

Finally, three experiments have been carried out:

- The count of LSTM layers;
- The size of LSTM layers;
- The size of the corpus for network training;

References

1. Ariza, C.: Two pioneering projects from the early history of computer-aided algorithmic composition. Comput. Music J. **35**(3), 40–56 (2012)
2. Chereshniuk, I.: Algorithmic composition and its role in modern musical education. Art Educ. (3), 65–68 (2015)
3. Acar, I.H.: Early childhood development and education through nature-child interactions: a conceptual paper. Int. J. Educ. Researchers **4**(2), 1–10 (2013)
4. Koops, H.V., Magalhaes, P., Bas de Haas, W.: A functional approach to automatic melody harmonisation. In: Proceedings of the First ACM SIGPLAN Workshop on Functional Art, Music, Modeling & Design, FARM 2013, pp. 47–58. ACM (2013)
5. Mazurowski, L.: Computer models for algorithmic music composition. In: Proceedings of the Federated Conference on Computer Science and Information Systems, pp. 733–737 (2012)
6. Palmer, E., Schloss, K., Xu, Z., Prado-León, L.: Music–color associations are mediated by emotion. Duke-National University of Singapore Graduate Medical School (2013)
7. Doornbusch, P.: Gerhard Nierhaus: algorithmic composition: paradigms of automated music generation. Comput. Music J. **34**(3) (2014)
8. Brinkkemper, F.: Analyzing Six Deep Learning Tools for Music Generation. http://www.asimovinstitute.org/analyzing-deep-learning-tools-music/. Accessed 04 May 2019

9. Kline, D.M.: Revisiting squared-error and cross-entropy functions for training neural network classifiers. https://link.springer.com/article/10.1007/s00521-005-0467-y/. Accessed 04 May 2019
10. Fernández, J.D., Vico, F.: AI methods in algorithmic composition: a comprehensive survey. J. Artif. Intell. Res. **48**, 513–582 (2013)
11. Waite, E., Eck, D., Roberts, A., Abolafia, D.: Project magenta: generating long-term structure in songs and stories (2016)

Convolutional Neural Network Application for Analysis of Fundus Images

Nataly Yu. Ilyasova[1,2(✉)], Aleksandr S. Shirokanev[1,2], Ilya Klimov[2], and Rustam A. Paringer[1,2]

[1] IPSI RAS - Branch of the FSRC «Crystallography and Photonics» RAS, Samara, Russia
ilyasova.nata@gmail.com
[2] Samara National Research University, Samara, Russia

Abstract. The paper proposes the apply of convolutional neural networks (CNN) for segmentation of fundus images. A new neural network architecture was made to classify three classes of images, which are made up of exudates, blood vessels and healthy areas. CNN architecture has been constructed, allowing a testing error of no more than 10% to be attained. Based on a 3×3 convolution kernel, CNN training was conducted on 12×12 images, thus enabling the best result of CNN testing to be achieved. CNN-aided segmentation of the input image conducted in this work has shown the CNN to be capable of identifying all training dataset classes with high accuracy. The segmentation error was calculated on the exudates class, which is key for laser coagulation surgery. The segmentation error on the exudates class was 5%. In the paper we utilized the HSL color model because it renders color characteristics of eye blood vessels and exudates most adequately. We have demonstrated the H channel to be most informative.

Keywords: Eye fundus image · Image processing · Convolution neural networks · Diabetic retinopathy

1 Introduction

Convolutional neural networks are recommended for classification [1]. This fact has been confirmed by research in the field of medical image analysis. Neural networks are commonly used for the intelligent analysis and segmentation of medical images. The first detailed review about deep learning using for medical image analysis was published in 2017 [2]. Nowadays digital medicine is actively developing.

In Ref. [3], a classification model based on a convolutional neural network was used for diagnosing the H. Pylori infection. The special architecture was used in this work. The authors came to the conclusion that the particular disease was possible to diagnose based on endoscopic images obtained using CNN. In Ref. [4], diagnosing an early-stage hypertension retinopathy was discussed. The blood hypertension resulted in

This work was partly funded by the Russian Foundation for Basic Research under grants # 19-29-01135, # 19-31-90160 and the RF Ministry of Science and Higher Education within the government contract of the FSRC "Crystallography and Photonics", RAS.

S. Kovalev et al. (Eds.): IITI 2019, AISC 1156, pp. 60–67, 2020.
https://doi.org/10.1007/978-3-030-50097-9_7

disease in the eye retina. In Ref. [5], a toolkit was developed for the automated analysis of psoriasis-affected skin biopsy images. This work is of considerable significance in clinical treatment. The study resulted in a practical system based on machine analysis. Approaches based on CNN outperform the traditional hand-crafted feature-based classification approaches. In work [6] the relative location prediction in computed tomography (CT) scan images is considered. To solve this problem different machine learning methods were applied. Unfortunately, the accuracy and speed of these methods don't provide the required accuracy and speed. In this paper the regression model based on one-dimensional convolutional neural networks (CNN) was proposed. This allowed determining the relative location of a CT scan image both quickly and precisely [6]. This CNN model consists primarily of three one-dimensional convolutional layers, three dropout layers and two fully-connected layers with appropriate loss functions. In this work, the proposed CNN method can contribute to a quick and accurate relative location prediction in CT scan images. As a result, the efficiency of the medical images archiving and communication system is improving.

2 Application of Convolution Neural Networks in Eye Fundus Image Analysis

A severe consequence of diabetic retinopathy (DRP) is vision loss. DRP affects all parts of the retina, leading to macular edema, which in turn causes fast worsening of eyesight [7]. The accurate and early diagnosis alongside an adequate treatment can prevent the vision loss in more than 50% of cases [8, 9]. There are a number of approaches to treating DRP, one of which involves laser photocoagulation [10]. During this procedure, a number of retina areas where edema occurs are cauterized with a laser (Fig. 1). The procedure is conducted via coagulating near-edema zones. The development of diagnostic systems enabling an automatic identification of the edema zone is currently a relevant task [2]. For the laser coagulation procedure to be automated [11, 12], objects in the eye fundus image need to be classified [13–16], which can be done in a number of ways [17, 18].

In this work, we study a class of eye fundus images with pathological changes that can be found at different stages of the disease. Manifestations of the diabetic retinopathy include exudates, which cause the retina thickening (Fig. 1).

In general, the image of an eye fundus with pathology contains three classes of objects, such as blood vessels, healthy areas and exudate zones.

In the [20] article, this problem was solved using a convolutional neural network trained on a small amount of data, and therefore an objective assessment cannot be made. The initial data for the analysis contained a training sample. For this experiment, the new data set is balanced and in total contained the 52680 images of dimension 12×12. For the present work, CNN training was conducted on the three classes of fundus images described above.

The original data set consisted of 75% of the training images and 25% of the test images. A control dataset was also used to prevent overtraining. The 3×3 convolution core was chosen because it is optimal for 12×12 images. The accuracy of training on the architecture used in [20] reached 82%. This accuracy is insufficient for effective results. The recognition accuracy of 93.1% was achieved with architecture in Table 1, which is the best recognition result for the three above-mentioned image classes.

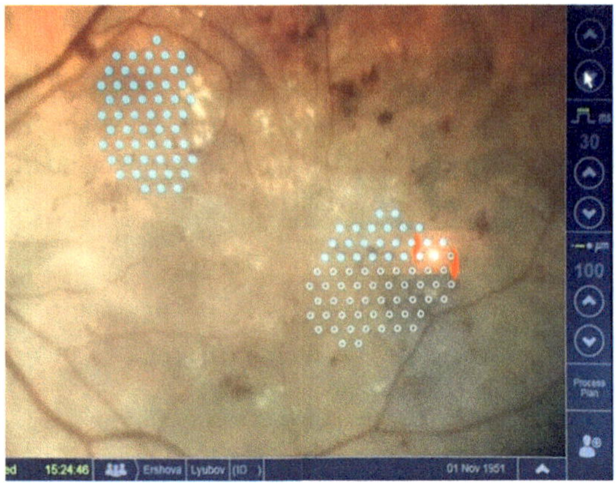

Fig. 1. Laser coagulation for the treatment of diabetic retinopathy of a patient using the NAVILAS system [19].

Table 1. The new architecture of the convolutional neural network for fundus images analysis.

Layer number	Layers	Parameters
1	Convolutional	144 neurons
1	Activation	Function: RELU
2	Convolutional	72 neurons
2	Activation	Function: RELU
2	MaxPooling	Size: 2 × 2
3	Convolutional	62 neurons
3	Activation	Function: RELU
3	Dropout	0.5
3	MaxPooling	Size: 2 × 2
4	Convolutional	150 neurons
4	Activation	Function: RELU
5	Convolutional	150 neurons
5	Activation	Function RELU
5	MaxPooling	Size: 2 × 2
5	Dropout	0.5
6	Fully-connected	3
6	Activation	Softmax

Figure 2 shows the dependence of the training accuracy on the number of eras. To achieve 90% recognition accuracy, CNN completed 125 training runs in the original image. Figure 3 shows the average learning outcome. The results in Fig. 3 show the dependence of learning error on the number of eras.

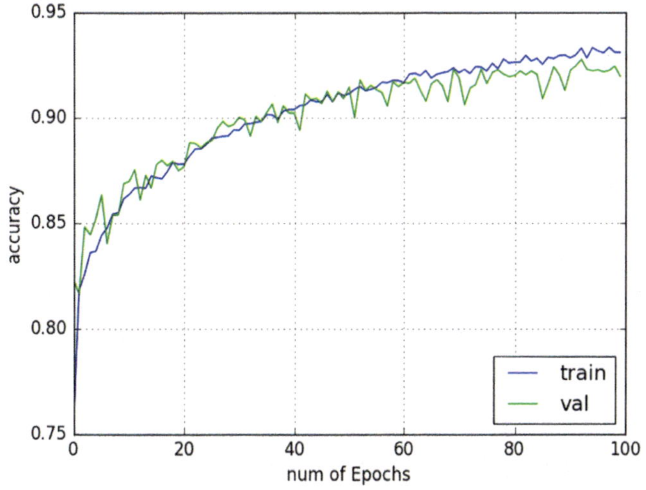

Fig. 2. Graph of the training accuracy on the number of eras.

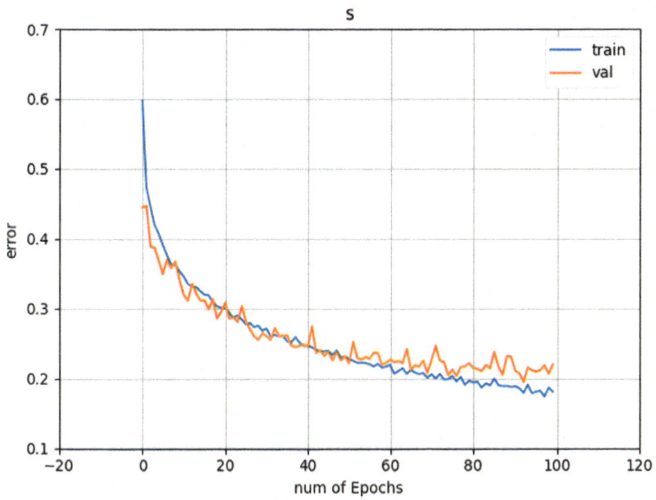

Fig. 3. Graph of the dependence of training errors on the number of eras.

3 Experimental Segmentation Study

Data sets were created containing the three above-described 12×12 image classes. In this study, fundus images were segmented through deep learning. Figure 4b shows the result of image segmentation using CNN on the old architecture and 4c shows the result of image segmentation on the new architecture. To evaluate segmentation error using CNN, an experienced ophthalmologist introduced manual segmentation as a reference image. The study was conducted in the class of exudates, which was highlighted in a separate image. The segmentation error of these exudate areas using CNN was calculated relative to expert judgment.

The segmentation error using CNN for exudates was defined as and amounted to 8% (where N × M is the image size, k is the number of expert-highlighted pixels that CNN failed to recognize as exudates, t is the number of exudate pixels recognized by CNN but missing from the expert's image). The error of the first kind, defined as, where l is the number of falsely recognized exudates classes and F is the total number of exudate-containing pixels in the expert's image, amounted to 7%.

In this paper, the segmentation error is calculated in the color space HSL. Chromaticity plays an important role in determining the exudate zone. Figure 5 shows the pathological zones on channel L of the color space HSL, highlighted by an expert.

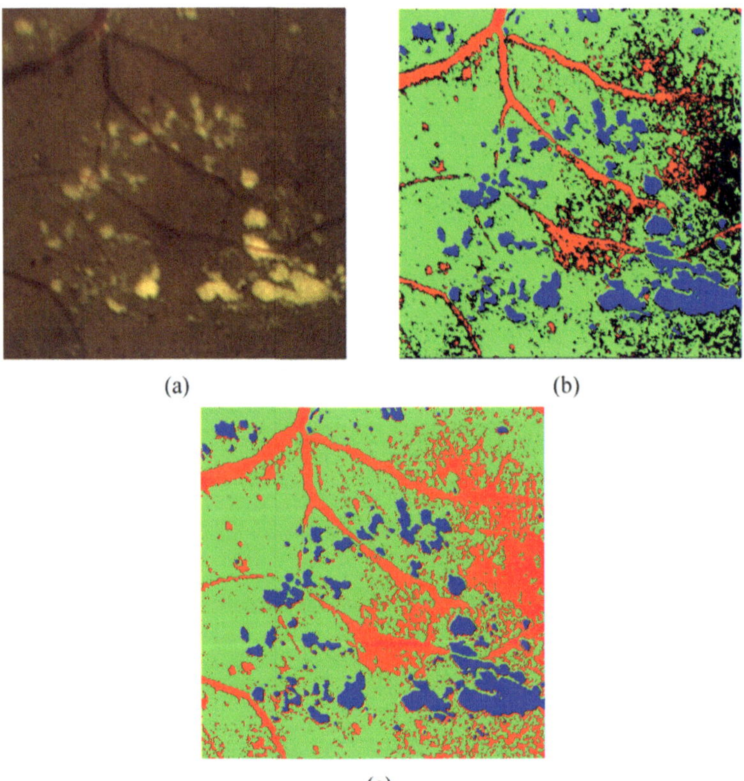

Fig. 4. (a) The original eye fundus image; (b) classes of objects highlighted in the image using CNN; (c) three classes of objects highlighted in the image using CNN (new model).

The convolutional neural network was able to identify 93% of the area of the exudate zone in the image of the expert. 2.8% of the neural network mistook a healthy area for a diseased area. 4.2% neural network mistook exudate zone for the healthy area.

The reliability of exudate extraction using CNN was confirmed by comparing histograms of images obtained using CNN and expert images, which were superimposed for each corresponding channel of the HSL color system, and expert histograms are marked with green bars. and CNN-based histograms are marked in red (Fig. 6). Expert histograms determine the range of values for the affected areas of the fundus. From the histograms, the interval between exudate areas using CNN is considered narrower than the interval obtained on the basis of expert estimates. The areas of the histogram that correspond to the false classification using CNN are with-in the intervals shown by tangled paths (Fig. 6). Table 2 shows the segmentation errors calculated for each channel in the HSL color model.

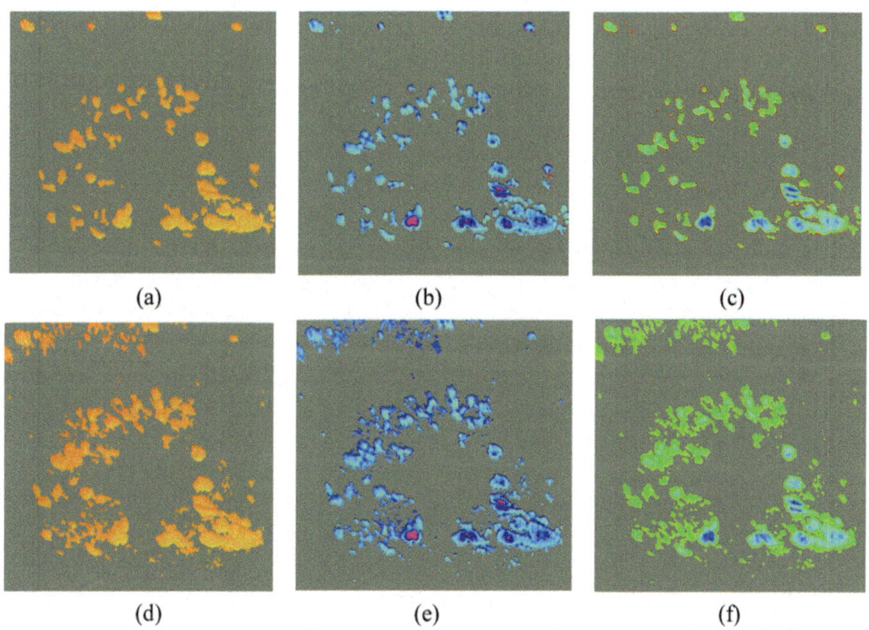

Fig. 5. Expert-highlighted affected fundus areas for the HSL color model for the channel (a) H, (b) S, and (c) L; CNN-highlighted affected fundus areas for the HSL color model for the channel (d) H, (e) S, and (f) L.

Table 2. Segmentation error for each channel.

Channel	H	S	L
Segmentation error (the exudates), %	5	8,1	8,4

Table 2 suggests that the H channel is the most informative channel with the least error of the segmentation.

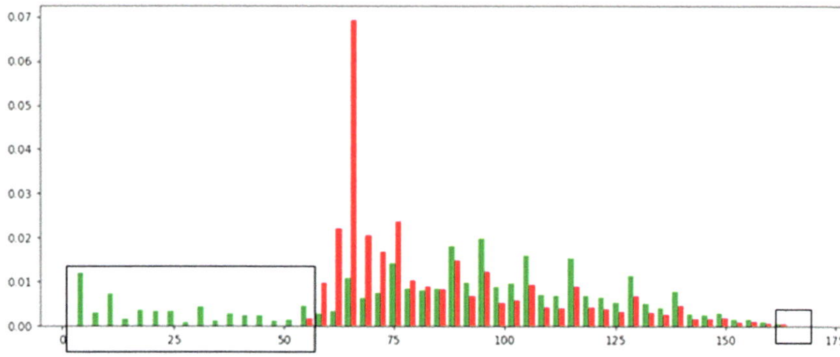

Fig. 6. An example of the exudate fundus histogram (L channel).

4 Conclusion

In this work, a convolutional neural network (CNN) was used to segment the fundus images. A convolutional network with a new architecture was applied, which allowed achieving testing errors of no more than 10. CNN-aided segmentation of the input image conducted in this work has shown CNN to be capable of identifying all training dataset classes with high accuracy. The segmentation error was calculated on the exudates class, which is key for laser coagulation surgery. The segmentation error on the exudates class was 8%, with the error of the first kind being 7%. In the study, we utilized the HSL color model because it renders color characteristics of eye blood vessels and exudates most adequately. We have demonstrated the H channel to be most informative, with the segmentation error amounting to 5%.

References

1. Guido, S., Andreas, C.: Introduction to Machine Learning with Python. O'Reilly Media, Sebastopol, p. 392 (2017)
2. Shichijo, S., et al.: Application of convolutional neural networks in the diagnosis of helicobacter pylori infection based on endoscopic images. Lancet **25**, 106–111 (2017)
3. Litjens, G.A., Litjens, G., Kooi, T., Babak, E.B.: Survey on deep learning in medical image analysis. Med. Image Anal. **42**, 60–88 (2017)
4. Bambang, K. T.: The classification of hypertensive retinopathy using convolutional neural network. In: ICCSCI, pp. 166–173 (2017)
5. Anabik, P.: Psoriasis skin biopsy image segmentation using Deep Convolutional Neural Network. Comput. Methods Programs Biomed. **159**, 59–69 (2018)
6. Jiajia, G., Hongwei, D., Jianyue, Z.: Relative location prediction in CT scan images using convolutional neural networks. Comput. Methods Programs Biomed. **159**, 43–49 (2018)
7. Shadrichev, F.: Diabetic retinopathy. Mod. Optom. **36**(4), 8–11 (2008)

8. Ilyasova, N.: Evaluation of geometric features of the spatial structure of blood vessels. Comput. Opt. **38**(3), 529–538 (2014)
9. Khorin, P.A., Ilyasova, N.Yu., Paringer, R.A.: Informative feature selection based on the Zqrnike polynomial coefficients for various pathologies of the human eye cornea. Comput. Opt. **42**(1), 159–166 (2018)
10. Astakhov, Y.S., Shadrichev, F.E., Krasavira, M.I., Grigotyeva, N.N.: Modern approaches to the treatment of diabetic macular edema. Ophthalmol. Sheets **4**, 59–69 (2009)
11. Ilyasova, N., Kirsh, D., Paringer, R., Kupriyanov, A., Shirokanev, A.: Coagulate map formation algorithms for laser eye treatment. IEEE Xplore, pp. 1–5 (2017)
12. Shirokanev, A.S., Kirsh, D.V., Ilyasova, NYu., Kupriyanov, A.V.: Investigation of algorithms for coagulate arrangement in fundus images. Comput. Opt. **42**(4), 712–721 (2018)
13. Ilyasova, N., Kirsh, D., Paringer, R., Kupriyanov, A.: Intelligent feature selection technique for segmentation of fundus images. In: lyasova, N., Kirsh, D., Paringer, R., Kupriyanov, A. A. (eds.) Proceedings of the Seventh International Conference on Innovative Computing Technology (INTECH) 2017, IEEE Xplore, pp. 138–143 (2017)
14. Shirokanev, A.S., Ilyasova, NYu., Paringer, R.A.: A smart feature selection technique for object localization in ocular fundus images with the aid of color subspaces. Procedia Eng. **201**, 736–745 (2017)
15. Ilyasova, N., Paringer, R., Kupriyanov, A.: Regions of interest in a fundus image selection technique using the discriminative analysis methods. In: Lecture Notes in Computer Science (Including Subseries Lecture Notes in Artificial Intelligence and Lecture Notes in Bioinformatics), vol. 9972, pp. 408–417. Springer, Cham (2016)
16. Ilyasova, N.Yu., Kupriyanov, A.V., Paringer, R.A.: Formation of features for improving the quality of medical diagnosis based on descriminant analysis methods. Comput. Opt. **38**(4), 851–855 (2017)
17. Ilyasova, NYu.: Methods for digital analysis of human vascular system. Literature review. Comput. Opt. **37**(4), 517–541 (2013)
18. Nikitaev, V.G.: Experimental study of color models in automated image analysis tasks. Sci. Session MIFI **1**, 253–254 (2004)
19. Kernt, M., Cheuteu, R., Liegl, R.: Navigated focal retinal laser therapy using the NAVILAS® system for diabetic macula edema. Ophthalmologe **109**, 692–700 (2012)
20. Ilyasova, N., Shirokanev, A., Demin, N.: Analysis of convolutional neural network for fundus image segmentation. J. Phys: Conf. Ser. **1438**, 1–7 (2019)

Approximation Methods for Monte Carlo Tree Search

Kirill Aksenov[1] and Aleksandr I. Panov[2,3](✉)

[1] National Research University Higher School of Economics, Moscow, Russia
kaaksyonov@edu.hse.ru
[2] Artificial Intelligence Research Institute, Federal Research Center "Computer Science and Control" of the Russian Academy of Sciences, Moscow, Russia
[3] Moscow Institute of Physics and Technology, Moscow, Russia
panov.ai@mipt.ru

Abstract. Today planning algorithms are among the most sought after. One of the main such algorithms is Monte Carlo Tree Search. However, this architecture is complex in terms of parallelization and development. We presented possible approximations for the MCTS algorithm, which allowed us to significantly increase the learning speed of the agent.

Keywords: Reinforcement learning · Monte-Carlo tree search · Neural network approximation · Deep learning

1 Introduction

Today, reinforcement learning is one of the most promising areas on the road to creating artificial intelligence. This is largely due to the major victories of an agent trained in this way over a human. Planning is an integral and important component of such algorithms. Perhaps due to the flexibility of these algorithms, i.e. they can be well adapted to solve a specific problem. For example, the performance can be increased by modifying the rules that select the trajectory to traverse, the principle of tree extension, the evaluation function and the backup rule by which those evaluations are propagated up the search tree.

One of the strongest and most important search algorithms is Monte Carlo Tree Search, which is the main one in recent AlphaZero, AlphaStar and OpenAI Five. The MCTS consists of four main parts that we will discuss, and we will also present possible approximations of some phases of the MCTS algorithm. To begin with, at the moment of trajectory simulation, when an unknown state is found, the MCTS algorithm does not initialize q-value. We use a neural network for this procedure, which greatly accelerates learning. Also, instead of using a random policy, we will use a neural network trained for this, since we have approximation of q-values for each state.

In Sect. 3.1 we will discuss the classic MCTS algorithm. In the next paragraph 3.2, we will offer some approximations for it, and also describe the main neural network structures that we used. At the end, in Sect. 4, we will compare the

S. Kovalev et al. (Eds.): IITI 2019, AISC 1156, pp. 68–74, 2020.
https://doi.org/10.1007/978-3-030-50097-9_8

classical MCTS with the MCTS, in which we present the various approximations we proposed. And also compare which neural networks have greater success in the environment of Sokoban.

2 Related Work

A lot of work in the field of reinforcement learning is associated with various applications of search algorithms. In most of these algorithms can be divided into two large groups: Dyna [1] and Monte Carlo Tree Search [2].

The popularity of the planning algorithm based tree search has been found for a long time, however, the main peak falls on the present, or more precisely, after the victory of the AlphaGo algorithm over a professional go player Lee Sedol [3]. There are many variations of this type of search algorithm, such as UCT [4], AMAF [5], open loop execution [6], but most of them use a similar architecture. Only type of selection function of the next state and the simulation strategy change.

Currently, there is a variety of planning architectures based on tree searches and using neural networks. The main work is to call the work [7], it offers a neural network architecture that repeats the Monte Carlo Tree Search. However, authors of this work did not provide the source code, without which it was impossible to repeat the experiments performed. Also worth noting is the I2A algorithm [8], which aggregates the results of several simulations into its network computation.

The main advantage of using neural networks is the ability to parallelize the algorithm, which is very important, since these algorithms work in huge spaces. Various options for parallelization are described in the articles [9,10].

In our article, we will offer a possible version of the approximation of some parts of MCTS algorithm by neural networks, by analogy with the article [7]. And also we will provide experiments on a rather complex environment Sokoban.

3 Algorithm

3.1 MCTS Algorithm

The goal of planning is to find the optimal strategy that maximises the total reward in an environment defined by a deterministic transition model $s' = T(s, a)$, mapping each state and action to a successor state s', and a reward model $r(s, a)$, describing the goodness of each transition.

Consider the classic Monte Carlo Tree Search algorithm proposed in [2]. The main algorithm consists of 4 main parts:

- **Selection.** We consider each position as a task of a multi-armed bandit. Nodes at each stage are selected according to the adopted tree policy (greedy for MCTS). This phase is valid until a node is found in which not all subsidiaries have statistics.

- **Expansion.** When a strategy cannot be applied due to the fact that there are uncharted nodes, the tree is expanded by adding one or more child nodes reached from the selected node through uncharted actions.
- **Simulation.** On the selected node or on one of its recently added child nodes (if any), a full episode is simulated with actions selected by a default policy (random for MCTS). The result is a Monte Carlo test with actions chosen first in accordance with the tree policy, and outside the tree, default policy.
- **Backpropagation.** At this stage, information about the game played is distributed up the tree, updating the information in each of the previously passed nodes. Values for states and actions that are visited by a default policy are not saved. At the end of the algorithm, the node visited the most times is selected (Fig. 1).

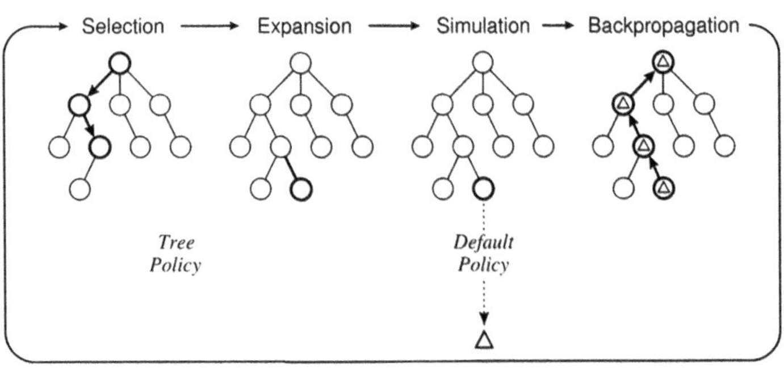

Fig. 1. MCTS algorithm

3.2 Approximation

Now we will present the possible approximations of some parts of the MCTS algorithm. First of all, let's pay attention to the initialization of the vector of statistics when getting into a new state. Having a good initialization, we can significantly improve the default policy, which will significantly speed up convergence.

Initialization of Statistics. Initialization of statistics occurs at the time of the meeting of a new state. This happens at the time of walking through the tree, at the time of the execution of a new action for the previous state. In this situation, we cannot know the statistics vector, therefore, it is usually not initialized or initialized by infinity, so that in the future any unknown action would be more priority than the one already tested. However, for some games in which the state presents the location of objects, it is possible to perform initialization using neural networks. So, as an entrance we have maps with the location of all objects.

For this task, MLP was originally chosen, but since the input data is a map of locations, it is much more correct to use a convolutional neural network. After a large number of tests, it was decided to stop at the network with four convolutional layers with batch normalisation and ReLU.

Simulation Policy. In the original work on MCTS, it was proposed to choose an action randomly at the time of the simulation, since we did not know the statistics. But now, using embedding net, we have initialized values, which allows us to choose actions for simulation much better.

An input in this subtask is a statistics vector for the current state, at the output we have the probabilities of choosing an action from the current state. In its basic, unstructured form, the simulation policy network is a simple MLP mapping the statistics to the logits.

4 Experiments

To confirm the work we conducted experiments in the environment of Sokoban, which is a classic Japanese puzzle. There is a 10 by 10 field, on which the play area is veiled, there are also boxes, places where they need to be installed and the player. The player can not pull the box behind him, and can only push, so most actions are irreversible and one wrong action can lead to the insolubility of the whole puzzle. In our case, the environment is procedurally generated. All enviroment was taken from [11] (Figs. 2 and 3).

Fig. 2. Example of Sokoban-env

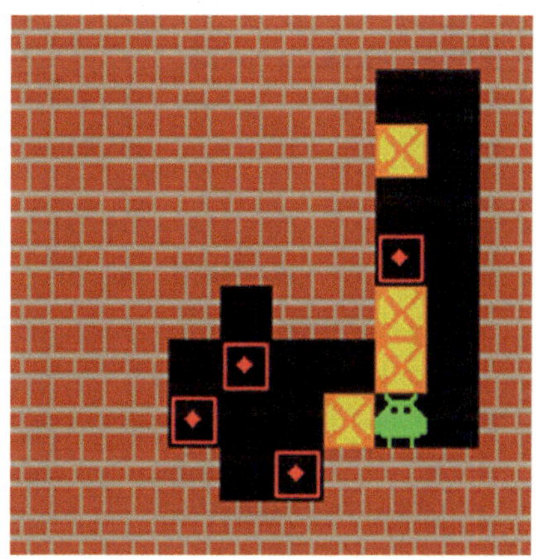

Fig. 3. Another example

4.1 Main Result

To begin with, we will compare the classic MCTS with the MCTS version with the approximations proposed by us. As far as can be seen on the Fig. 4, using approximations, the optimal policy found much faster, given that both algorithms carry out the same number of simulations.

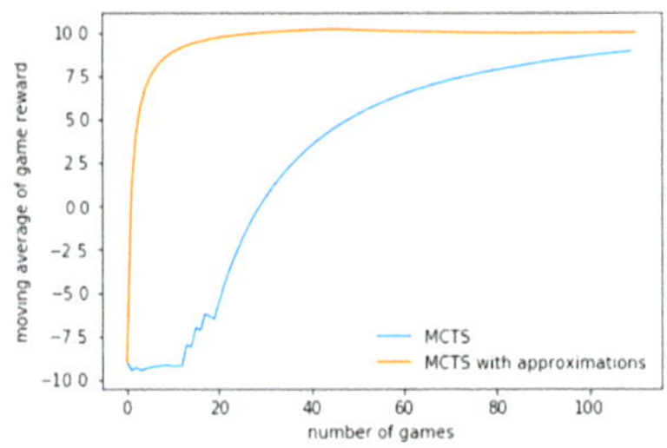

Fig. 4. The blue shows the learning function for the UCT algorithm with 100 simulations, while the orange shows the MCTS algorithm with an approximation of the statistics initialization at the tree node with 4-layer CNN with ReLU activation function, for which the number of simulations is also 100.

Both algorithms will sooner or later converge to the same policy, but our version will require much fewer simulations for this, as well as the possibility of optimizing NN on the GPU.

4.2 Neural Network Architecture

During work on the initialization of statistics in a tree node, a number of different architectures of approximating neural net were surveyed; we can distinguish the main ones and make a comparison between them:

- MLP
- Simple CNN
- Deep CNN with batch normalization

On the Fig. 5 you can see how much the perceptron loses to convolution networks. However, it is also worth noting that the correct architecture greatly speeds up the learning process, which shows the difference between two CNN.

4.3 Default Policy

The classic default policy for the MCTS algorithm is a random policy, since the MCTS does not provide for the initialization of statistics. However, we have the ability to use a small MLP to select an action based on the initialized statics of a tree node during the simulation.

Let's compare how much this affects the speed of learning. As you can see on the Fig. 6, learning really happens faster, but you cannot say that this is a significant change.

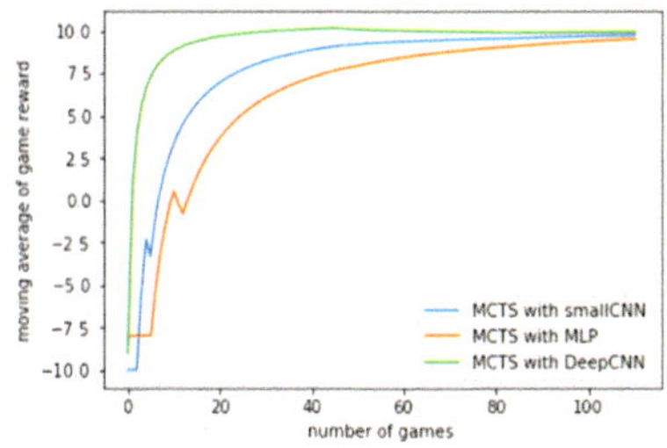

Fig. 5. On this figure, the green MCTS algorithm is depicted in which Deep CNN is used as an approximator, the blue is the same algorithm, but with a three-layer CNN. The orange is MLP.

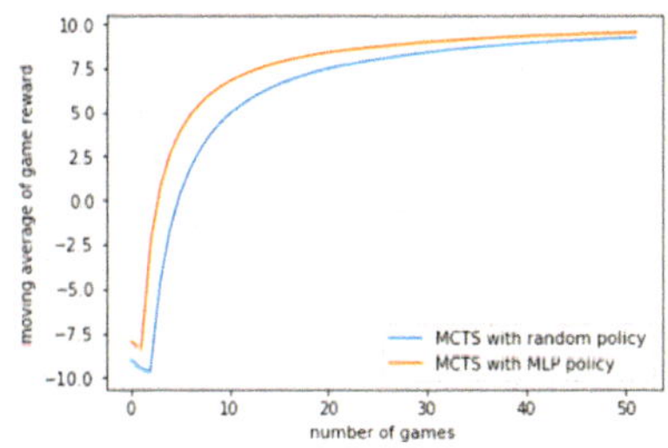

Fig. 6. On this figure, the orange is MCTS algorithm in which the MLP is set as the default policy, the blue one is the same algorithm, but the random policy is used as the default policy.

5 Conclusion

We have shown that some parts of the MCTS algorithm can be approximated by neural networks, which leads to:

- Increase learning speed
- Ability to calculate by gradient methods

These approximation methods make it possible to use the conditions of the problem itself, i.e. adjust neural networks under the state view, as well as use additional conditions, for example, symmetry in game of go. Also, due to the use of a trained neural network instead of a random default policy during simulation, the reliability to each individual simulation is significantly increased.

In the future, we plan to transfer the entire MCTS to neural networks in order to significantly speed up the planning process by optimizing the gradient. That today it is possible to solve a huge number of tasks using planning methods. Some really applied problems can be solved in this way, for example, the logistics tasks described in [12].

Acknowledgements. The reported study was supported by RFBR, research Projects No. 17-29-07079 and No. 18-29-22047.

References

1. Sutton, R.S.: Dyna, an integrated architecture for learning, planning, and reacting. SIGART Bull. **2**(4), 160–163 (1991)
2. Coulom, R.: Efficient selectivity and backup operators in monte-carlo tree search. In: International Conference on Computers and Games, pp. 72–83. Springer (2006)
3. Silver, D., Schrittwieser, J., Simonyan, K., Antonoglou, I., Huang, A., Guez, A., Hubert, T., Baker, L., Lai, M., Bolton, A., et al.: Mastering the game of go without human knowledge. Nature **550**(7676), 354 (2017)
4. Kocsis, L., Szepesvári, C.: Bandit based monte-carlo planning. In: European Conference on Machine Learning, pp. 282–293. Springer (2006)
5. Gelly, S., Silver, D.: Monte-carlo tree search and rapid action value estimation in computer go. Artif. Intell. **175**(11), 1856–1875 (2011)
6. Lecarpentier, E., Infantes, G., Lesire, C., Rachelson, E.: Open loop execution of tree-search algorithms. arXiv preprint arXiv:1805.01367 (2018)
7. Guez, A., Weber, T., Antonoglou, I., Simonyan, K., Vinyals, O., Wierstra, D., Munos, R., Silver, D.: Learning to search with MCTSNets. arXiv preprint arXiv:1802.04697 (2018)
8. Racanière, S., Weber, T., Reichert, D., Buesing, L., Guez, A., Rezende, D.J., Badia, A.P., Vinyals, O., Heess, N., Li, Y., et al.: Imagination-augmented agents for deep reinforcement learning. In: Advances in Neural Information Processing Systems, pp. 5690–5701 (2017)
9. Chaslot, G.M.J.-B., Winands, M.H.M., van Den Herik, H.J.: Parallel monte-carlo tree search. In: International Conference on Computers and Games, pp. 60–71. Springer (2008)
10. Ali Mirsoleimani, S., Plaat, A., van den Herik, J., Vermaseren, J.: A new method for parallel Monte Carlo tree search. arXiv preprint arXiv:1605.04447 (2016)
11. Schrader, M.-P.B.: gym-sokoban (2018). https://github.com/mpSchrader/gym-sokoban
12. Edelkamp, S., Gath, M., Greulich, C., Humann, M., Herzog, O., Lawo, M.: Monte-Carlo tree search for logistics. In: Commercial Transport, pp. 427–440. Springer (2016)

Labor Intensity Evaluation Technique in Software Development Process Based on Neural Networks

Pavel V. Dudarin(✉), Vadim G. Tronin, Kirill V. Svatov, Vladimir A. Belov, and Roman A. Shakurov

Ulyanovsk State Technical University, Ulyanovsk, Russia
p.dudarin@ulstu.ru
http://ulstu.ru

Abstract. Software development process is actively studied by experts from different spheres of science and different view points. However the degree of success of projects in the development of software intensive systems (Software Intensive Systems, SIS) has changed insignificantly, remaining at the level of 50% inconsistency with the initial requirements (finance, time and functionality) for medium-sized projects. The annual financial losses in the world because of the total failures are counted by hundreds of billions of dollars. Its high complexity leads to constant mistakes in labor intensity evaluation, and even new agile development paradigm does not solve this problem. In this paper an approach to retrospective labor intensity evaluation in software development process based on neural networks is proposed. This technique provides impersonal expertise of program code developed during the sprint in agile paradigm. Experiments performed on the real life software project show the effectiveness of proposed technique.

Keywords: Software development process · Neural network · Data augmentation · Complexity evaluation metrics · Halstead metrics · Cyclomatic metrics

1 Introduction

Software development process is actively studied by experts from different spheres of science and different view points [2,12]. And all the spheres of our live use software products [9]. However, according to the CHAOS report from Standish Group during 1992–2017, the degree of success of projects in the development of software intensive systems (Software Intensive Systems, SIS) has changed insignificantly, remaining at the level of 50% inconsistency with the initial requirements (finance, time and functionality) for medium-sized projects. The annual financial losses in the world because of the total failures are in the order of hundreds of billions of dollars.

S. Kovalev et al. (Eds.): IITI 2019, AISC 1156, pp. 75–84, 2020.
https://doi.org/10.1007/978-3-030-50097-9_9

Due to high complexity of software development process the standard development paradigm, which is know as 'waterfall', nowadays is crowded out by different agile techniques like 'scrum' and 'kanban'. These techniques allow to develop huge and complex systems step by step. Each step has the full development cycle: planning, coding, testing, analysis. This paper is focused on code evaluation aspect, as a key factor of success. In agile techniques during the planning phase developers evaluate tasks in a labor intensity units called story-points. This expert evaluation depends on level of expertise, team coherence, motivation, money and other factors. Regular mistakes could lead to team ineffectiveness due its overworking or under-working. In both cases the project loses money, reputation and stuff. Team of 10 or more people practically does not able to cope with thorough retrospective of each task. The automated instrument of task retrospective evaluation could be really helpful. In this paper such an instrument is described. For developers this instrument works as a trainer in future task evaluation. It could reveal tasks that are underestimated or overestimated. And should be discussed on the retrospective meeting in details. For team-lead this instrument works as a tool of individual assessment. Some people tend to overestimate their task some underestimate. For product manager this tools works as a tool of whole team productivity assessment instrument without the necessity of diving deep into the project context. Different teams on the same projects could be compared based in their productivity.

The rest of this paper is organized as follows. The Sect. 2 shows the overview of existing techniques. In Sect. 3 the detailed technique description is presented. Section 4 shows an experimental results. And Sect. 5 concludes the paper.

2 Evaluation Approaches Review

One of the most standard and common approaches to determine the timing and labor intensity of development task is expert review. This approach suggests one or several experts are putting forward their assumptions about labor intensity of the development, based on their knowledge and experience. Casually this job is made by leaders of the development team (employees with extensive experience in this field).

However, the assessment obtained is subjective, even if it is obtained as a result of discussion by a group of specialists. So if the assessment was carried out on some other day or other specialists took part in it, the result of the assessment would differ. At the same time, such an assessment costs a lot of money for the project, because instead of performing the current tasks, a group of leading experts will spend their time on retrospective evaluation.

Another approach suggests automated evaluation based on code analysis. Developed program code can be assessed by formal features - code metrics [4]. It allows to determine the quality, complexity, labor intensity of the written code. The following is a set of the most common classes of metrics and their representatives [3,6].

Quantitative metrics are metrics that count number of different program components. Since these metrics are the easiest to implement, they were among the first to be used to calculate labor intensity of software projects. Quantitative metrics include:

- SLOC - the number of lines of code
- Number of comments
- Percentage of comments - the ratio of the number of comments to the entire amount of code
- Average number of lines in methods and classes
- Number of methods and class fields

The main drawback of these metrics that the same business function can be implemented in with a completely different amount of code lines. So the correlation among code complexity and task complexity is weak.

The quantitative metrics also include *Halstead metrics*, which are calculated based on the number of operators and operands in the source code. Based on this metric, the following estimates can be made:

- Halstead Vocabulary - dictionary of the program
- Halstead Length - the length of the program
- Halstead Volume - program code volume
- Halstead Difficulty - program code complexity
- Halstead Effort - developer's mental expenses for code creation
- Halstead Operators Total/Distinct - the number of total/unique operators
- Halstead Operands Total/Distinct - the number of total/unique operands

The main advantage of Halstead metrics among the other quantitative metrics is a relative independence from coding style. But all the operators are treated equally, and complexity of control flow remains not rated.

Metrics of control flow of a program - class of metrics are based on the analysis of the program control graph. One of the most common estimates of this graph is the cyclomatic complexity of the program (Cyclomatic [5]). This metric calculates the number of independent routes that can be drawn in a graph constructed from the program code. The number of routes increases with each conditional statement or cycle. The main flaw of this assessment is the absence of differences between conditional constructions and cycles. Also, the metric does not take into account the number and complexity of condition predicates.

Data management flow complexity metrics - this class of metrics are based on the analysis of the program inputoutput stream. This class includes the Chepin metric where the informational strength of a program module or class is evaluated by analyzing the nature of the use of list variables. The list of variables is divided into the following sets:

- P - input output variables required for calculations
- M - variables modified or created inside the code.
- C - variables that control the state of the program module
- T - unused variables.

The final result of the metric estimated by the following formula: $Q = A1 * P + A2 * M + A3 * C + A4 * T$, where $A1, A2, A3, A4$ - weight coefficients. According to the author, these coefficients are equal: $1, 2, 3, 0.5$, respectively.

Object-Oriented Metrics - class of metrics are used in projects that use an object-oriented approach.

- WMC (Weighted Methods Per Class) - this metric assumes class complexity based on the cyclomatic complexity of its methods. If class contains more methods and the complexity of their implementation, than higher value of the metric for the class.
- WMC2 (Weighted methods per class) - class complexity is calculated depending on the number of methods in the class, as well as the number of arguments in the method.
- DIT (Depth of Inheritance Tree) - metric counts the levels of inheritance from top of the hierarchy.
- NOC (Number of children) - metric determines the number of classes that are the heirs of the current class.
- CBO (Coupling between objects) - metric determines the number of classes that use or are used by the current class.
- RFC (Response For Class) - metric determines the number of methods that can be called by a class object.

Presented metrics perfectly reflect the complexity of implementing an object-oriented application. However, they are completely inapplicable for other programming approaches.

All these metrics have their advantages and disadvantages. Some researches show that linear combination of metrics could work even better. In this paper non linear approach is proposed based on neural network.

There are some studies dedicated to neural network metrics [11] and [7]. These studies used on small or artificially generated peaces of code, not from the real commercial projects. Moreover this researches focused on re-creating existing metrics on neural network, not on training neural network to assess labor intensity.

As long as the researched project used as data source for the analysis of labor intensity, is implemented in TypeScript language using the library for the JavaScript language - ReactJS, object-oriented metrics are not suitable. Since this project implements the client part, it does not include complex calculations, data conversion and analysis, but serves to query and display them. In accordance with the arguments presented, it was decided to avoid the data management complexity metrics too. To calculate the input vector of the neural network, the following metrics were selected: Cyclomatic, Cyclomatic Density, Param Count, Halstead Bugs, Halstead Difficulty, Halstead Effort, Halstead Length, Halstead Time, Halstead Vocabulary, Halstead Volume, Halstead Operators Total, Halstead Operators Distinct, Halstead Operands Total, Halstead Operands Distinct, Sloc Logical, Sloc Physical.

3 Labor Intensity Evaluation Technique

As it was shown above different classes of metrics take into account different aspects of programming code. A nonlinear function of different sets of formal features, taking into account the different aspects of the features of the code, will give better results compared to the expert and formal approach. However, the right choice of function type and its coefficient is not trivial task as long as it depends on many factors: programming language, code convention accepted on the project, project specific and team members' specific. This leads to necessity of neural network training for each project. To accomplish this following steps need to be done:

- extract completed task history form the project task tracking system with labor intensity evaluation
- extract related programming code parts
- calculate vectors of metrics for all the code parts related to the task
- choose the architecture of a neural network and construct a model
- train the neural network and analyze obtained results

All the tasks, statements, assessments of tasks of a software project are usually stores in the system of task tracking and project management - JIRA. Evaluation of tasks in this system rated in a format of story points - a relative assessment of the amount of work in the user story.

For getting task numbers, their description, ratings, task performer, the integration service with Jira was implemented. The service was implemented in Java using the jira-rest-java-client library[1]. Further it is required to receive the source code of implemented tasks.

The source code for the project usually is stored in the Git version control system (for example, hosted on Bitbucket). The developed service was supplemented with the integration functionality with bitbucket repositories. Thus all the hashes of commits were loaded with their description from the repository. The development team adhered to the following model of the description of the commit- "Task # number # - what has been done". Commit hashes and tasks from Jira were mapped by task number. This functionality was implemented using regular expressions and search by substring. Given the hash of the commit the source code itself was obtained from Git repository.

For each task all the related files were scanned and evaluated using the complexity assessment service written in Node JS using the typhonjs-escomplex library [8]. Allocation of this service from the general application was necessary, both from architectural and in terms of performance. In the future, this approach will expand the list of supported programming languages without major changes in the main application. In addition, it will allow the use of more appropriate libraries and implementations written for a particular language, taking into account its features.

[1] All the codes are accessible on https://gitlab.com/wertqed/labor-intensity-evaluation.

To assess partial changes in the code firstly the complexity of the origin code is estimated (before the changes were made) in the commit. Then the complexity is analyzed after the changes made. Such actions are performed for all commits within each task. The accumulated increase of complexity is the final result of the evaluation of the program code of a specific task.

The average software developer solve 1–3 tasks a week, which means that standard team of 5–7 people even after a year of work will have only few amount of tasks appropriate for automated evaluation. Thus a data augmentation technique should be used. Another reason for data augmentation is a low quality of expert assessments, because the evaluation is made during the working process when expert has a lot of other things to do. Additional code samples with more precise evaluation could be generated as following: firstly, experts evaluate some code samples 10–20; then tasks are created as a random sets from evaluated code samples and total evaluated calculated as a sum. This approach allows to generate as many samples as it needed without going into artificial code generation.

Experiments show that task with a large amount of story points could not be evaluated correctly, so such tasks should be excluded from evaluation and avoided in development process. Task with a large amount of story points should be split into a set of small ones.

Mathematically, evaluation task could be treated as prediction one. Thus from all the variety of neural network architectures like RNN, CNN, GAN and others, the architecture provided by fastai framework [1] for grocery sales prediction [10] has been chosen. In Fig. 1 neural network architecture could bee seen. This neural network model consists of embedded layers for categorical variables, followed by Dropout and BatchNorm for continuous variables. The results are combined and followed by the BatchNorm, Dropout, Linear and ReLU blocks (the first block is missing BatchNorm and Dropout, the last block is missing the ReLU). Used sigmoid activation function. The network includes 200 neurons on the first and 100 neurons on the second inner layer, respectively.

4 Experiment Results

A sample of one commercial project was used as experimental data. In this project, one story point is approximately equal to 3 h of developer's work. Using developed auxiliary service 1177 tasks were downloaded, but only 704 of them were rated by story points.

During the comparison process of JIRA and BitBucket data, the sample was reduced to 303 tasks, that are assessed by an expert group and having corresponding commits in the repository. In this sample, the assessment of tasks was carried out in story points in the range from 0.5 to 46 units. There were some incorrectly estimated tasks that had to be eliminated in order to improve the quality of the model. It was decided to exclude tasks with a rate above 20 units, since their number was too large, and the difference in metrics between objects of similar evaluation varied in a fairly large range. Also, tasks with rate equals to zero were excluded. After filtering the sample, its size was reduced to only 281 objects. In Fig. 2 some feature vectors are presented.

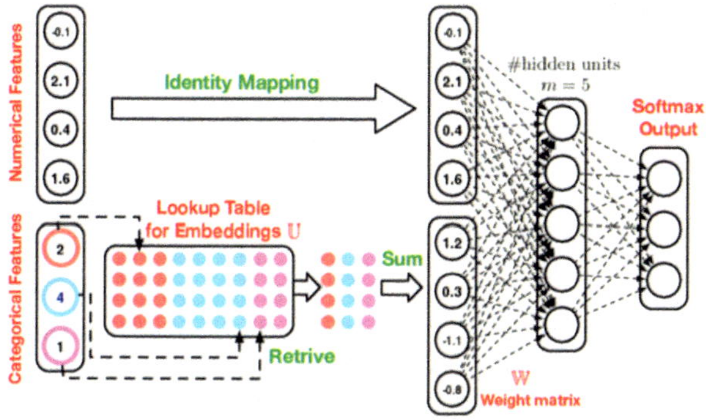


```
(0): ModuleList()
(1): Dropout(p=0.0)
(2): BatchNorm1d(16, eps=1e-05, momentum=0.1, affine=True, track_running_stats=True)
(3): Linear(in_features=16, out_features=200, bias=True)
(4): ReLU(inplace)
(5): BatchNorm1d(200, eps=1e-05, momentum=0.1, affine=True, track_running_stats=True)
(6): Linear(in_features=200, out_features=100, bias=True)
(7): ReLU(inplace)
(8): BatchNorm1d(100, eps=1e-05, momentum=0.1, affine=True, track_running_stats=True)
(9): Linear(in_features=100, out_features=1, bias=True)
```

Fig. 1. Predictive neural network architecture

	Story Points	Cyclomatic	CyclomaticDensity	ParamCount	Halstead Bugs	Halstead Difficulty	Halstead Effort	Halstead Length	Halstead Time	Halstead Vocabulary	Halstead Volume	Halstead Operators Total	Halstead Operators Distinct	Halstead Operands Total	Halstead Operands Distinct	Sloc Logical	Sloc Physical
0	4.0	50.0	270.172	10.0	1.113	34.055	95682.391	534.0	5315.690	143.0	3346.906	251.0	29.0	283.0	114.0	115.0	211.0
1	7.0	39.0	141.757	14.0	1.280	26.750	177074.152	485.0	9837.453	96.0	3827.512	245.0	12.0	241.0	84.0	119.0	166.0
2	5.0	51.0	38.722	19.0	1.274	20.870	325664.820	494.0	18092.470	55.0	3630.548	239.0	2.0	258.0	57.0	62.0	111.0
3	5.0	64.0	141.766	24.0	2.225	74.940	173208.372	1005.0	9622.687	273.0	6666.752	463.0	53.0	542.0	220.0	230.0	433.0
4	8.0	128.0	873.672	32.0	6.448	303.902	275398.876	3058.0	15299.940	1245.0	19327.079	1408.0	284.0	1650.0	961.0	640.0	1761.0
5	6.0	34.0	176.749	6.0	0.779	16.410	104577.748	321.0	5809.875	60.0	2320.053	160.0	9.0	163.0	51.0	77.0	143.0
6	5.0	11.0	37.078	1.0	0.253	6.471	13765.356	121.0	764.742	31.0	760.680	52.0	7.0	71.0	26.0	30.0	65.0
7	3.0	3.0	0.543	1.0	0.242	3.459	150954.510	82.0	8386.363	4.0	725.102	45.0	1.0	37.0	3.0	21.0	21.0
8	6.0	26.0	16.564	8.0	1.565	38.295	169573.202	709.0	9420.732	116.0	4692.480	349.0	19.0	360.0	97.0	150.0	176.0
9	3.0	0.0	0.589	0.0	0.076	5.664	32519.150	20.0	1806.619	12.0	226.303	12.0	4.0	8.0	8.0	4.0	5.0
10	4.0	2.0	3.256	0.0	0.093	1.050	3449.117	30.0	191.616	11.0	280.180	12.0	0.0	18.0	11.0	3.0	17.0
11	1.0	0.0	0.000	0.0	0.001	0.120	27.017	0.0	1.501	2.0	4.144	0.0	0.0	0.0	2.0	0.0	2.0
12	4.0	6.0	4.314	0.0	0.419	1.881	28799.068	151.0	1599.947	45.0	1256.851	62.0	0.0	89.0	45.0	19.0	48.0
13	3.0	2.0	0.015	0.0	0.125	1.801	74935.776	36.0	4163.099	6.0	374.506	20.0	1.0	16.0	5.0	6.0	21.0
14	5.0	172.0	273.890	0.0	7.014	60.837	510752.219	2933.0	28375.122	546.0	21015.523	1651.0	26.0	1282.0	520.0	306.0	498.0
15	5.0	4.0	5.660	1.0	0.206	3.465	25092.071	75.0	1394.003	23.0	611.601	34.0	3.0	41.0	20.0	9.0	23.0
16	3.0	2.0	1.856	2.0	0.128	3.566	27472.995	46.0	1526.279	9.0	386.034	20.0	0.0	26.0	9.0	10.0	14.0
17	1.0	12.0	186.205	1.0	0.489	27.321	15944.175	244.0	885.784	103.0	1378.919	109.0	27.0	135.0	76.0	28.0	132.0
18	13.0	51.0	256.241	22.0	3.041	87.927	305671.904	1306.0	16981.770	384.0	9118.725	577.0	74.0	729.0	312.0	268.0	474.0
19	5.0	2.0	5.068	0.0	0.064	1.123	2897.415	30.0	160.968	7.0	192.274	19.0	1.0	11.0	6.0	8.0	11.0
20	3.0	0.0	0.278	0.0	0.010	0.268	28.564	3.0	1.586	2.0	30.431	1.0	0.0	2.0	2.0	1.0	5.0

Fig. 2. Filtered data sample

To augment data the proposed approach was used, and 1000 of samples were generated.

The neural network was trained using the Google Colab service. As a result of neural network training, the loss-function (the sum of square errors) on the test part of the sample reached values less than 5 units. Figure 3 shows the training epochs.

epoch	train_loss	valid_loss	time
0	45.247143	26.154900	00:00
1	42.605099	22.374788	00:00
2	40.612766	17.865685	00:00
3	37.896023	12.954546	00:00
4	35.434860	9.990437	00:00
5	32.487514	8.639280	00:00
6	29.802570	6.890329	00:00
7	27.047462	7.728254	00:00
8	24.737619	11.083056	00:00
9	22.838421	11.333658	00:00
10	21.319412	8.851075	00:00
11	20.144808	8.983871	00:00
12	19.075329	7.845511	00:00
13	18.146326	6.960416	00:00
14	17.273411	6.572458	00:00
15	16.644100	6.538685	00:00
25	12.417384	7.146481	00:00
26	12.173486	6.835426	00:00
27	11.935629	7.749971	00:00
28	11.729676	5.519372	00:00
29	11.420516	4.679532	00:00
30	11.226760	4.719009	00:00
31	11.083948	4.998270	00:00
32	11.068912	5.917696	00:00
33	10.913160	7.892910	00:00
34	10.802450	8.165093	00:00
35	10.614226	6.211092	00:00
36	10.525489	5.501328	00:00
37	10.413771	5.010170	00:00
38	10.300986	5.977239	00:00
39	10.200490	7.082429	00:00
40	10.131946	7.225183	00:00

Fig. 3. Neural network training epochs

As a demonstrative comparison to the results of the neural network, prediction based Cyclomatic metrics are provided in Fig. 4. Square error of this prediction equals to 9, 5. The chart steady increase to the value of the 12 units of Story Points, then it alternately changes from growth to fall and vice versa.

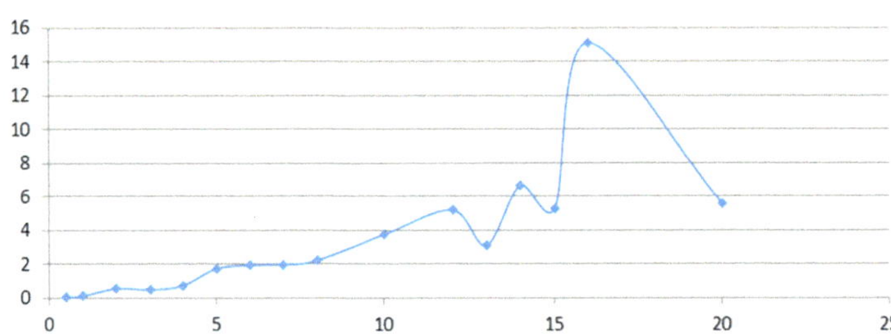

Fig. 4. Predictions based on Cyclomatic metric

Figure 5 shows a plot of predicted values from Story Points by neural network, which is quite more accurate. In the rang of [0, 6] the chart shows a line $y = x$, which means an ideal prediction. After the mark of 12 story points chart become unstable like in Fig. 4. The reason for this phenomenon are errors in the work of the expert group in evaluating large and complex tasks. Experts of the observed team tend to overestimate large tasks and underestimate huge ones.

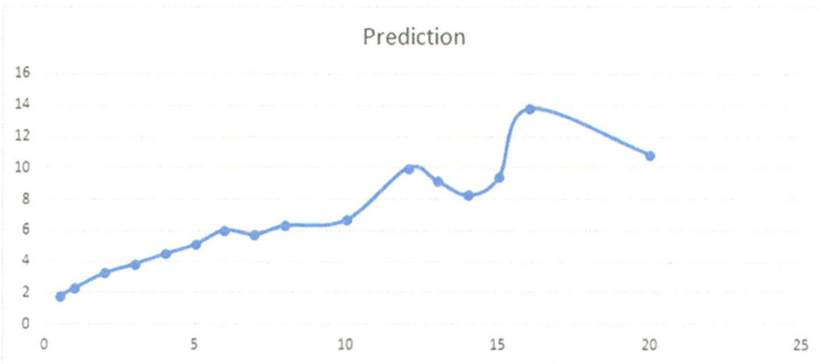

Fig. 5. Neural network prediction results

5 Conclusion

In this paper an approach to software development tasks retrospective evaluation was presented. Each project and each team need to train network on their data sources this needs some history of tasks. The lack of history could be leveled by data augmentation technique which is also proposed in this paper. For developers this instrument works as a trainer in future task evaluation. It could reveal tasks that are underestimated or overestimated. And should be discussed on the retrospective meeting in details. For team-lead this instrument works as a tool of individual assessment. Some people tend to overestimate their task some underestimate. For product manager this tools works as a tool of whole team productivity assessment instrument without the necessity of diving deep into the project context. Different teams on the same projects could be compared based in their productivity.

Moreover this approach helps to reveal a boundary of evaluation capability of the team or expert. Tasks with evaluations greater than this boundary almost always will not be done on time.

Further works will study the aspect of changes during the time. Team of the project evolve and along this evolve the code: it's style, technique, maturity. Quality of code evaluation could characterize the team in aspects of it's balance, expertise and so on.

Acknowledgements. The reported study was funded by RFBR and the government of Ulyanovsk region according to the research project Num. 18-47-732005 and by RFBR and the government of Ulyanovsk region according to the research project Num. 18-47-732004.

References

1. FastaAI Neural Network Framework. https://www.fast.ai/about/
2. Nadezhda, Y., Gleb, G., Pavel, D., Vladimir, S.: An approach to similar software projects searching and architecture analysis based on artificial intelligence methods. In: Abraham, A., Kovalev, S., Tarassov, V., Snasel, V., Sukhanov, A. (eds.) Proceedings of the Third International Scientific Conference "Intelligent Information Technologies for Industry" (IITI 2018), IITI 2018. Advances in Intelligent Systems and Computing, vol. 874. Springer, Cham (2019)
3. Isola, E., Olabiyisi, S., Omidiora, E., Ganiyu, R.: Performance evaluation of improved cognitive complexity metric and other code based complexity metrics. Int. J. Sci. Eng. Res. **3**(9) 2012
4. Kaur, K., Minhas, K., Mehan, N., Kakkar, N.: Static and dynamic complexity analysis of software metrics. World Acad. Sci. Eng. Technol. Int. J. Comput. Electr. Autom. Control Inf. Eng. **3**(8), 1936–1938 (2009)
5. Kemerer, F., Gill, K.: Cyclomatic complexity density and software maintenance productivity. IEEE Trans. Software Eng. **17**(12), 1284–1288 (1992). https://doi.org/10.1109/32.106988
6. Khan, A.A., Mahmood, A., Amralla, S.M., Mirza, T.H.: Comparison of software complexity metrics. Int. J. Comput. Netw. Technol. **4**(1) (2016). ISSN 2210-1519
7. NASA: A neural net-based Approach to Software Metrics. https://ntrs.nasa.gov/archive/nasa/casi.ntrs.nasa.gov/19930003204.pdf
8. Coleman, D., Ash, D., Lowther, B., Oman, P.: Using metrics to evaluate software system maintainability. Computer **27**(8), 44–49 (1994). https://doi.org/10.1109/2.303623
9. Dudarin, P., Pinkov, A., Yarushkina, N.: Methodology and the algorithm for clustering economic analytics object. Autom. Control Process. **47**(1), 85–93 (2017)
10. Rossmann Store Sales - Kaggle contest. https://www.kaggle.com/c/rossmann-store-sales
11. Senan, S., Sevgen, S.: Measuring software complexity using neural networks. J. Electr. Electron. Eng. **17**(2), 3503–3508 (2017)
12. Vlasov, I., Dudarin, P., Yusupov, A.: An experience of software development for information system integration in production process of aeronautical operator. News Samara Sci. Center Russ. Acad. Sci. **14**(2), 577–582 (2012)

An Analysis of Convolutional Neural Network for Fashion Images Classification (Fashion-MNIST)

Khatereh Meshkini[1], Jan Platos[2(✉)], and Hassan Ghassemain[3]

[1] Department of Information Engineering and Computer Science, Trento University, Trento, Italy
[2] Faculty of Electrical Engineering and Computer Science, VSB Technical University of Ostrva, Ostrava, Czechia
jan.platos@vsb.cz
[3] Image Processing and Information Analysis Lab, Tarbiat Modares University Tehran, Tehran, Iran

Abstract. Recently, Convolutional Neural Networks (CNN) has been used in variety of domains, including fashion classification. Social media, e-commerce, and criminal law are extensively applicable in this field. CNNs are efficient to train and found to give the most accurate results in solving real world problems. In this paper, we use Fashion MNIST dataset for evaluating the performance of convolutional neural network based deep learning architectures. We compare most common deep learning architectures such as AlexNet, GoogleNet, VGG, ResNet, DenseNet and SqueezeNet to find the best performance. We additionally propose a simple modification to the architecture to improve and accelerate learning process. We report accuracy measurements (93.43%) and the value of loss function (0.19) using our proposed method and show its significant improvements over other architectures.

Keywords: Deep learning · Convolutional neural networks · Fashion MNIST · Squeeze network

1 Introduction

Image classification is one of the significant problems in the field computer vision. There are a remarkable number of researchers who have a special interest to describe an image and study different ways of image classification [1]. Obviously, it is easy for humanbeings to identify and classify different images based on their features, but a computer needs complicated algorithms to successfully recognize various image. Nowadays, Convolutional Neural Network (CNN) plays an important role in classification of images [2] and image segmentation [3] due to its great performance in different problems. There are numerous image datasets which are introduced to simplify the procedure of classification for investigators. The handwritten digits (MNIST) dataset, firstly introduced by LeCun et al. in 1998 [4], contains 70000 grayscale images in the size of 28 * 28. MNIST is an appropriate dataset for deep learning researchers to

S. Kovalev et al. (Eds.): IITI 2019, AISC 1156, pp. 85–95, 2020.
https://doi.org/10.1007/978-3-030-50097-9_10

quickly test their CNN algorithms and compare their results with the vast range of experiments which were done by other people.

The growing number of internet businesses pave the ways for people to buy their commodities through internet websites faster and easier. There are considerable number of fashion websites which have a great number of cloths in any types for men and women. Introducing an effective method for retrieving relevant images (with the desired characteristics) from the users' textual queries is decisive [5]. Moreover, considering the continuous stream of new products being added to the catalogues, the automatic generation of image labels can alleviate the workload of human annotators, improve quick access to relevant products, and help generate textual product descriptions. Making some experiments on Fashion-MNIST would be an interesting idea to improve the previous techniques and get a better results with minimum error rates.

In this paper, we focus on different convolutional neural networks and implement our experiments on Fashion-MNIST dataset with 60000 training images and 10000 testing images in the size of 28 * 28 [6]. Each image is associated with a label from10 classes as shown in the Fig. 1.

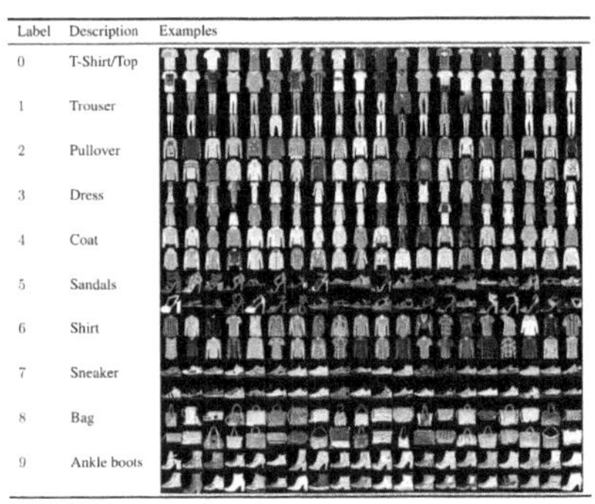

Fig. 1. Fashion-MNIST dataset [8]

Since the invention of CNN, a remarkable number of researches have been introduced each year with major advances and new methods of computer vision solving problems. Basically, convolutional network receives a normal image as a rectangular box and takes square patches of pixels and passes them through a filter. It is also called a kernel [7]. Each convolutional layers apply a convolution operation to the input and passing the result to the next layer. Convulsion neural networks are largely similar to artificial neural networks, these types of networks consist of weightless neurons and readable bias. Each neuron receives the input value, and then the product obtained from the multiplication in the input values is computed and finally yields the result by using an informal transform. The entire network still provides a differentiable rating function, on the one hand, the raw image pixels of the input image and on the other side the scores for each category. This kind of network is still a loss function (like SVM, Softmax) in the last layer (all connected or fully connected), and all the points in the normal neural networks are true here [7].

This paper is arranged as follow. First, we introduce some related prior works following with definitions of the methods and materials which are employed in our assessment. The experiments and the results are presented in the next part. At last, we discuss the results and draw the conclusion.

2 Related Works

Since the creation of The Convolutional Neural Networks (CNN) and promotion of multilayer neural networks by LeCun in 1998 [4], various neural models have been introduced. Generally, there has been continuous improvement in networks with the innovation of new layers and involvement of different computer vision techniques [8].

The invention of GPGPUs pave the ways for researchers to go deeper in the field of CNN. They gained remarkable improvement in the results of their studies [9]. Moreover, training larger and more complex models became possible by introduction of large scale datasets. The CNNs is not restricted to image classification, there are several experiments have been employed in object detection like faces [10] and pedestrian [11].

In 2018 an end-to-end architecture based on the hierarchical nature of the annotations was introduced for classification of fashion images [12]. They defined category and subcategory levels in their experiments where the category level influences the subcategory. An analysis of convolutional neural network was proposed to compare the performance of different popular models of image classification such as AlexNet, GoogleNet and ResNet50 [13]. Some researchers focused on combination of convolutional neural networks and other frequent image classification techniques like support vector machine (SVM) [14]. The applications of CNN in face recognition, scene labelling, image classification, action recognition, human pose estimation and document analysis were considered by majority of scientist [7, 15]. Introducing new convolutional neural network architectures by adding new layers or changing the parameters is another challenging competition between researchers. Some of them tried to use residual skip connections [2] or normalized pre-processed datasets [16] to win the contest and gain the best result.

3 Models

In this section, the methods used for analysis and the developed deep learning frameworks will be discussed.

3.1 Alex Network

AlexNet designed by Alex Krizhevsky in 2012. It's the winner of the ImageNet Large Scale Visual Recognition Challenge which achieved a top-5 error of 15.3%, more than 10.8 percentage points lower than that of the runner up. The architecture of AlexNet was more complicated than prior models and had several filters per layer with stacked convolutional layers. The size of convolution blocks were 11 * 11, 5 * 5 and 3 * 3 followed by max pooling, dropout, data augmentation, ReLU activations, SGD with momentum [17].

3.2 Google Network/Inception V1

The size of filters is a significant factor which have direct effect on the performance of the model. The idea of GoogleNet was introducing filters with multiple sizes operate on the same level. It performs convolution on an input, with 3 different sizes of filters (1×1, 3×3, 5×5). Moreover, GoogleNet won the ILSVRC 2014 competition with top-5 error rate of 6.67% and achieved 92.7% top-5 test accuracy. This was very close to human level performance which the organizers of the challenge were now forced to evaluate [18].

3.3 VGG Network

This neural network is simple and practical architecture with 7.3% error rate. VGG is the runner-up in the 2014 ImageNet challenge and currently the most preferred choice in the community for extracting features from images. What makes this network really elegant is the depth of the architecture. It replaced large kernel-sized filters (11 and 5 in the first and second AlexNet convolutional layer, respectively) with multiple 3×3 kernel-sized filters one after another. As a result, more convolutional layers are used with about 138 million parameters which is the key design of the VGG [19]. In Fig. 2 an image is the input of multiple convolutional and pooling layers and

3.4 Residual Network

In ResNet features are combined through summation before passing into the next layer. Furthermore, the filters are less than VGGNet and the architecture is simpler. ResNets help to eliminate the varnishing and exploding gradient problems by some skip connections or shortcuts. It follows two simple design rules: for the same output feature map size, the layers have the same number of filters and if the feature map size is halved, the number of filters is doubled so as to preserve the time complexity per layer. The winner of the ILSVRC 2015 classification achieved 3.57% error on the ImageNet test set [20].

3.5 Dense Network

DenseNet introduces connection from one layer to all its subsequent layer in a feed forward fashion. This connection is done using concatenation not through summation. DenseNet has very narrow layers, it means that each block will only produce k features. And these k features will be concatenated with previous layers features and will be given as input to the next layer. Another significant feature of this architecture is that it alleviates the vanishing-gradient problem [21].

3.6 Squeeze Network

The idea of developing SqueezeNet by researchers originate from the desire of creating smaller neural networks with fewer parameters that can more easily fit into computer memory and can more easily be transmitted over a computer network. SqueezeNet achieved AlexNet-level accuracy on ImageNet with 50x fewer parameters [22]. For achieving their goals they employed 3 strategies: Replacing 3×3 filters with 1×1 filters, Decreasing the number of input channels to 3×3 filters and Down sampling late in the network so that convolution layers have large activation maps. By replacing the filters and reducing the number of channels, the quantity of parameters will decrease while the accuracy level preserves. Furthermore, this methodology maximize the accuracy level using limited amount of parameters.

The Fire module was introduced by scientists for designing building block of the CNN architectures and implementing mentioned strategies. For achieving first strategy, a 1×1 squeeze convolution layer leading to 1×1 and 3×3 expand convolution filters was defined. Three tunable dimensions ($S1 \times 1$, $E1 \times 1$, $E3 \times 3$) express the number of filters in squeeze layers, 1×1 and 3×3 filters in the expand layer respectively. The second strategy is attained by putting $S1 \times 1$ to be less than $E1 \times 1 + E3 \times 3$ [22].

SqueezeNet CNN architecture is illustrated in Table 4. It comprises two convolutional layers at the first and final layer (conv1, conv10), followed by 8 Fire modules and three max-poolings with a stride of 2. The number of filters increases in each layer steadily. To fulfill the third strategy, mentioned max-poolings are placed after three or four Fire module layers. The full SqueezeNet architecture is shown in Table 4.

Table 1. Macroarchitectural view of our SqueezeNet architecture

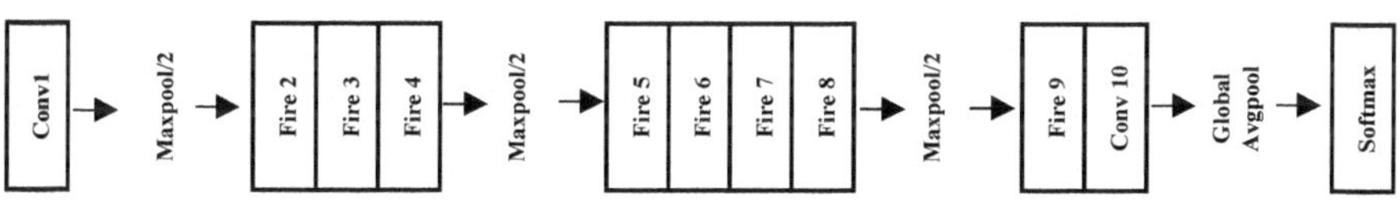

Moreover, some scientists produced a compressed set of SqueezeNet parameters that was 1.2 percentage-points and also an uncompressed set of SqueezeNet parameters that was 4.3 percentage-points more accurate than AlexNet [23]. In this approach, a dense-sparse-dense training flow is used to regularize the SqueezeNet through sparsity-constrained optimization. Furthermore, recovering and retraining on pruned weights increases the capacity of architecture and maintains dimension of original model.

Table 2. The architecture of Squeeze Network

Layer	Output size	Filter size/Stride	Depth	$S_{1\times1}$ (Squeeze)	$E_{1\times1}$ (Expand)	$E_{3\times3}$ (Expand)	$S_{1\times1}$ Sparsity	$E_{1\times1}$ Sparsity	$E_{3\times3}$ Sparsity	Parameters before pruning	Parameters before pruning
Input image	224 × 224 × 3										
Conv1	111 × 111 × 96	7 × 7/2	1				100%(7 × 7)			14,128	14,128
Maxpool1	55 × 55 × 96	3 × 3/2	0								
Fire 2	55 × 55 × 128		2	16	64	64	100%	100%	33%	11,920	5,748
Fire3	55 × 55 × 128		2	16	64	64	100%	100%	33%	12,432	6,256
Fire 4	55 × 55 × 256		2	32	128	128	100%	100%	33%	45,344	20,646
Maxpool4	27 × 27 × 256	3 × 3/2	0								
Fire 5	27 × 27 × 256		2	32	128	128	100%	100%	33%	49,440	24,742
Fire 6	27 × 27 × 384		2	48	192	192	100%	50%	33%	104,880	44,700
Fire 7	27 × 27 × 384		2	48	192	192	50%	100%	33%	111,024	46,236
Fire 8	27 × 27 × 512		2	64	256	256	100%	50%	33%	118,992	77,581
Maxpool8	13 × 13 × 512	3 × 3/2	0								
Fire 9	13 × 13 × 512		2	64	256	256	50%	100%	30%	197,184	77,581
Conv 10	13 × 13 × 1000	1 × 1/1	1				20%(3 × 3)			513,000	103,400
Avgpool 10	1 × 1×1000	13 × 13/1	0								
										1,248,424	421,098

4 Experiments

All experiments in this study were conducted on a laptop computer with Intel core i7-7500U, up to 3.58 Hz, 12 GB of Memory, and NVIDIA GeForce 940MX.

The performance analysis of CNN's is done by testing each of the networks on Fashion-MNIST dataset with 60000 training images and 10000 testing images in the size of 28 * 28. All models are trained for 40 epochs and report the loss function, the rate of accuracy and the confusion matrix. All networks have been tested independently and use the implementation provided by Caffe, a Deep Learning framework. Furthermore, results can be computed extremely fast in a graphical processing unit (GPU).

In all training process we used 'ReLU' activation function in the hidden layer and a softmax activation function in the final layer. Normalized exponential function or softmax activation function, transforms a K-dimensional vector to be between 0 and 1. Its output is equivalent to a categorical probability distribution and will add up to 1. The softmax function is shown below mathematically:

$$\varphi(z_j) = \frac{e^{z_j}}{\sum_{k=1}^{k} e^{z_k}} \, for\, j = 1, \ldots, k \tag{1}$$

We optimized the parameters of the network so that the loss is gradually minimized. We have used the 'Adam' optimizer [24] for optimization of the loss function.

To get better results in SqueezeNet, we proposed batch normalization and compared the result of our models with and without batch normalization. Actually, batch normalization normalizes the input of each layer to overcome covariance shift problem. In addition, it has trainable layers and properties which makes the model to be able to learn generalized features more accurately. Batch normalization improves the stability of a neural network by scaling the activations [25]. As a result, we add this fruitful methodology before every convolutional layer to enhance the training speed and accuracy of the model. The Caffe framework does not natively support a convolution layer that contains multiple filter resolutions. To get around this, we implement expand layers of SqueezeNet with two separate convolution layers: a layer with 1×1 filters, and a layer with 3×3 filters.

The architecture of all SqueezeNet and its building block is shown in Table 1 and 2. Furthermore, the results of the training and testing process of each models are depicted in Figs. 2 to 3 and Table 3.

5 Results and Discussion

According to our experiments and the results shown in Fig. 2 to 3, the AlexNet had the lowest accuracy level and the SqueezeNet with batch normalization took the first place of the accuracy contest. By observing the Fig. 2(a), AlexNet, GoogleNet and VGG followed same trend of penalizing the false classifications. They reached the average loss after 5000 iterations. Along these three networks VGG had better loss and accuracy value. The ResNets train with some fluctuations until about 30000 iterations and then the loss function drops to around 0.3. Figure 2(b) compares the testing loss of ResNet44, DenseNet, SqueezeNet and BN-SqueezeNet respectively and indicates that SqueezeNet with batch normalization has won the loss contest. Obviously, BN-SqueezeNet had better performance in modeling the given data and its epoch loss reaches to about 0.2 in 40000th iteration.

Generally, classification accuracy is the fraction of predictions our model got right and it's the other metric which we used to be sure about the performance of our models. Looking at Fig. 3(a), with the same learning strategy, ResNet20 and ResNet32 had better accuracy percentage between the rest of the models (AlexNet, GoogleNet and VGG). But, our best models were ResNet44 and BN-SqueezeNet shown in Fig. 3(b). Although SqueezeNet hadn't superior performance in comparison with other models shown in Fig. 3(b), the batch normalization had great impact and turned it to the winner of the competition. Moreover, the output shows significant difference by addition of batch normalization.

In Table 3 we displayed best testing accuracy and testing loss after 40000 iterations to figure out the exact amount of the metrics evaluated in our machine learning algorithms. After 40000 iterations, AlexNet, GoogleNet, VGG, ResNet20, ResNet32, ResNet44, DenseNet, SqueezeNet and SqueezeNet with batch normalization gives an accuracy of 88.50%, 88.75%, 90.28%, 90.62%, 90.80%, 91.40%, 90.37%, 88.93% and 92.56% respectively. We restrict the number of epochs to only 40000 and use the same learning strategy described for all the architectures.

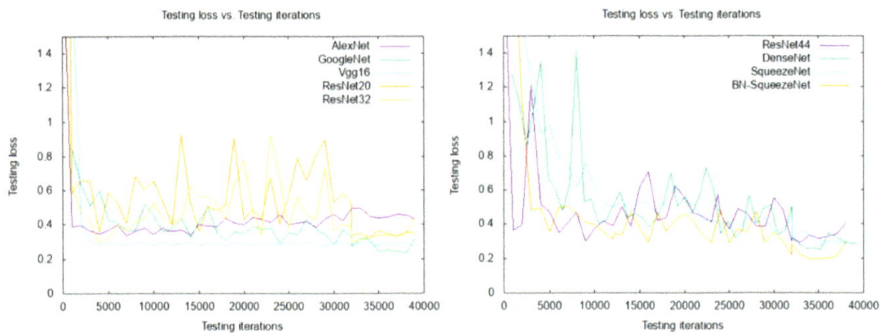

(a) (b)

Fig. 2. Comparison of testing loss of (a) AlexNet, GoogleNet, VGG, ResNet20, ResNet32 and (b) ResNet44, DenseNet, SqueezeNet and BN-SqueezeNet

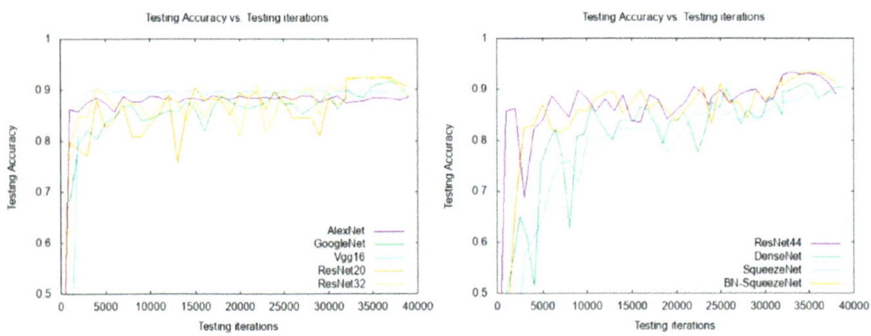

(a) (b)

Fig. 3. Comparison of testing accuracy of (a) AlexNet, GoogleNet, VGG, ResNet20, ResNet32 and (b) ResNet44, DenseNet, SqueezeNet and BN-SqueezeNet

Table 3. Best performance of each architectures in 40000 iterations

Model	AlexNet	GoogleNet	VGG	ResNet20	ResNet32	ResNet44	DenseNet	SqueezeNet	BN-SqueezeNet
Loss	0.40	0.23	0.27	0.32	0.29	0.29	0.25	0.26	0.19
Accuracy %	88.82	91.75	90.28	92.59	92.56	93.39	90.75	89.56	93.43

6 Error Analysis

The Confusion matrixes of the models determine the number of samples and the accuracy of each class. As we defined before, we had 10 classes with 6400 test samples in our experiment. The confusion matrix considers each class separately and shows how many samples are categorize correctly. By observing all our models, the SqueezeNet with batch normalization had less mistakes in classification process. The confusion matrix of BN-SqueezeNet is illustrated in Table 4. Figuring out Table 4, the class 1 and 6 (T-Shirts and Shirts) are the most trouble makers among the classes and the lowest accuracy percentage in all experimented models belongs to Shirt. About 17 images of Shirts are wrongly predicted as T-Shirts and 11 images are misclassified as coat. Moreover, 13 images of T-Shirts are mistaken as Shirt. Pullover is another category with high misclassification. As you can see in Table 4, 10 images of pullovers are predicted as Shirt and 7 images are classified as Dress.

Table 4. The confusion matrix of BN-SqueezeNet

Confusion matrix											Accuracy
T-Shirt	**155**	0	2	2	1	0	13	0	1	0	89.08%
Trouser	0	**173**	0	1	0	0	0	0	0	0	99.43%
Pullover	4	0	**115**	0	7	0	10	0	0	0	84.56%
Dress	1	0	1	**143**	2	0	3	0	0	0	95.33%
Coat	0	1	5	1	**144**	0	4	0	0	0	92.90%
Sandal	0	0	0	0	0	**150**	0	2	0	0	98.68%
Shirt	17	0	4	3	11	0	**129**	0	0	0	78.66%
Sneaker	0	0	0	0	0	1	0	**184**	0	1	98.92%
Bag	0	0	0	0	0	0	0	0	**148**	0	100%
Ankle Boats	0	0	0	0	0	1	0	6	0	**154**	95.65%

In addition, there are three most wrongly classified images which are shown in Fig. 3. The image 3(a) and (b) actually belong to shirt and T-shirt in order, but are classified reversely. Pullover shown in Fig. 3 looks like a shirt to computer and is classified as a shirt. These misclassifications are because of some factors that affect the classification process. One of the reasons is that 28 $\times$ 28 pixel size fashion images do not satisfy initial values which algorithms require for classification. As a result, classes seem similar in such small images (Fig. 4).

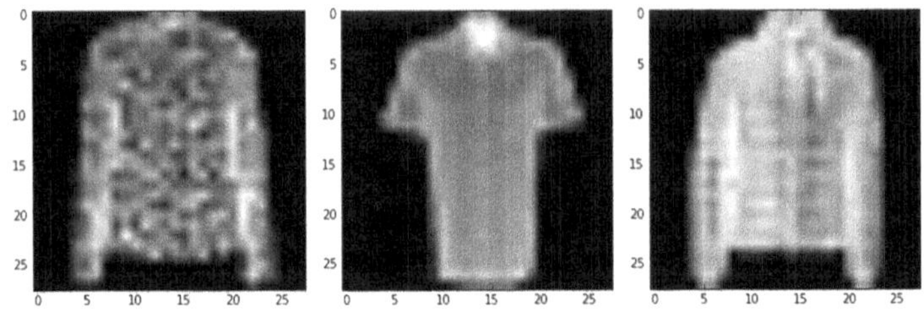

Fig. 4. Most wrongly classified images (Shirt, T-shirt, Pullover)

7 Conclusion

In this study, we tried to analyze some well-known machine learning architectures and compare their results on fashion MNIST dataset. In addition, we considered the Squeeze network as one of the new and practical models of image classification and made an attempt to increase its accuracy by using batch normalization between the convolutional layers and activation layers. Although Batch normalization is an optional layer in machine learning architectures and a few studies have been used its feature in their experiments, we showed it is inevitable in deep networks. Squeeze network without batch normalization performed weakly and failed to learn anything or learn initially. But, batch normalization paves the ways for architecture to learn well and improved both loss and accuracy. At last, we obtain the highest percentage of accuracy (around 93.50%) by using batch normalization. Moreover, in this work, we implement nine famous deep learning models on fashion MNIST and became aware of their strength and blind points.

References

1. Meshkini, K.H., Ghassemian, H.: Texture classification using shearlet transform and GLCM. In: Iranian International Conference of Electrical Engineering (2017)
2. Bhatnagar, S.H., Ghosal, D., Kolekar, M.H.: Classification of fashion article images using convolutional neural networks. In: Fourth International Conference on Image Information Processing (ICIIP) (2017)
3. Lin, K., et al.: Deep learning-based segmentation and quantification of cucumber powdery mildew using convolutional neural network. Frontiers Plant Sci. (2019)
4. LeCun, Y., et al.: Gradient-based learning applied to document recognition. In: Processing of the IEEE (1998)
5. Madhavi, K.V., Tamikodi, R., Sudha, K.J.: An innovative method for retrieving relevant images by getting the top-ranked images first using interactive genetic algorithm. In: 7th International Conference on Communication, Computing and Virtualization (2016)
6. Xiao, H., Rasul, K., Vollgraf, R.: Fashion-MNIST: a novel image dataset for benchmarking machine learning algorithms. arXiv preprint arXiv:1708.07747
7. Yamashita, R., et al.: Convolutional neural networks: an overview and application in radiology. Insights Imaging **9**(4), 611–629 (2018)

8. Zhao, Z.-Q., et al.: Object detection with deep learning: a review. J. Latex Class Files **14**(8) (2017)
9. Hinton, G.E., Osindero, S., Teh, Y.W.: A fast learning algorithm for deep belief nets. J. Neural Comput. **18**, 1527–1554 (2006)
10. Rowley, H.A., Baluja, S.H., Kanade, T.: Neural network-based face detection. Comput. Vis. Pattern Recogn. (1996)
11. Sermanet, P., et al.: Pedestrian detection with unsupervised multi-stage feature learning. arXiv preprint arXiv:1212.0142
12. Ferreira, B.Q., Faria, J., Baia, L., Sousa, R.G.: A unified model with structured output for fashion images classification. arXiv preprint arXiv:1806.09445v1
13. Sharma, N., Jain, V., Mishra, A.: An analysis of convolutional neural networks for image classification. In: International Conference on Computational Intelligence and Data Science (ICCIDS) (2018)
14. Agarap, A.M.: An architecture combining convolutional neural network (CNN) and support vector machine (SVM) for image classification. arXiv preprint arXiv:1712.03541v1
15. Bhandre, A., et al.: Applications of convolutional neural networks. Int. J. Comput. Sci. Inf. Technol. **7**(5), 2206–2215 (2016)
16. Shamsuddin, M.R., et al.: Exploratory analysis of MNIST handwritten digit for machine learning modelling. In: International Conference on Soft Computing in Data Science (2018)
17. Krizhevsky, A., Sutskever, I., Hinton, G.E.: ImageNet classification with deep convolutional neural networks. Commun. ACM **60** (2017)
18. Szegedy, C.H., et al.: Going deeper with convolutions. arXiv preprint arXiv:1409.4842
19. Simonyan, K., Zisserman, A.: Very deep convolutional networks for large-scale image recognition. arXiv preprint arXiv:1409.1556
20. He, K., Zhang, X., Ren, S.H., Sun, J.: Deep residual learning for image recognition. arXiv preprint arXiv:1512.03385
21. Huang, G., Liu, Z.H., Maaten, L., Weinberger, K.Q.: Densely connected convolutional networks. arXiv preprint arXiv:1608.06993v5
22. Iandola, F.N., et al.: SqueezeNet: AlexNet-level accuracy with 50x fewer parameters and < 0.5 MB model size. arXiv preprint arXiv:1602.07360v4
23. Han, S., et al.: Regularizing deep neural networks with dense-sparse-dense training flow. arXiv:1607.04381 (2016)
24. Kingma, D.P., Lei Ba, J.: Adam: a method for stochastic optimization. arXiv preprint arXiv: 1412.6980v9
25. Bjorck, J., et al.: Understanding batch normalization. arXiv preprint arXiv:1806.02375v4

Multiagent Systems

Implementation of the Real-Time Intelligent System Based on the Integration Approach

A. P. Eremeev(✉), A. A. Kozhukhov, and A. E. Gerasimova

National Research University "MPEI", Krasnokazarmennaya Street, 14, Moscow 111250, Russia
eremeev@appmat.ru, saaanchezzz@yandex.ru, leksiel50@mail.ru

Abstract. The paper describes the possibilities and basic means of constructing intelligent systems of real-time based on an integrated approach. A multi-agent approach, flexible decision search algorithms, forecasting algorithms based on reinforced learning are used. The architecture of forecasting module, module of deep reinforced learning and the architecture of the forecasting subsystem are given. The results of computer simulation of reinforcement learning algorithms based on temporal differences are presented and the corresponding recommendations for their use in multi-agent systems are given.

Keywords: Artificial intelligence · Intelligent system · Real time · Reinforcement learning · Deep learning · Forecasting · Decision support · Anytime algorithm

1 Introduction

Reinforcement learning (RL) methods [1] are based on the using large amount of information for learning in arbitrary environment. They are the most rapidly developing areas of artificial intelligence, related with the development of advanced intelligent systems of real-time (IS RT). Typical example is an intelligent decision support system of real-time (IDSS RT) [2]. An overview and design principles of IDSS RT for nuclear power facilities are given in [3]. In addition, deep reinforcement learning (DRL) approach allows RL-algorithms to work effectively in situations where the data that the RL-agent operates with is unstable, frequently updated, and highly correlated [4–6].

One of the most promising in terms of use in IDSS RT and central in RL is Temporal Difference (TD) learning [1, 6, 7]. TD-learning process is based directly on experience with TD-error, bootstrapping, in a model-free, online, and fully incremental way. Therefore, process does not require knowledge of the environment model with its rewards and the probability distribution of the next states. The fact that TD-methods are adaptive is very important for the IS RT of semiotic type able to adapt to changes in the controlled object and environment.

The work was supported by the Russian Foundation for Basic Research, projects №№ 18-01- 00201 a, 18-01-00459 a, 18-51-00007 Bel-a, 18-29-03088 MK.

S. Kovalev et al. (Eds.): IITI 2019, AISC 1156, pp. 99–108, 2020.
https://doi.org/10.1007/978-3-030-50097-9_11

The multi-agent approach contains of groups of autonomous interacting agents having a common integration environment and capable to receive, store, process and transmit information in order to address their own and corporate (common to the group of agents) analysis tasks and synthesis information. It is the fastest growing and promising approach for dynamic distributed control systems and data mining systems, including IDSS RT. Multi-agent systems could be characterized by the possibility of parallel computing, exchange of experience between the agents, resiliency, scalability, etc. [8].

In the perspective IDSS RT, which operates under strict time constraints and with noisy data, it is necessary to have tools for predicting the development of situations on the object and the consequences of decisions made, as well as machine learning tools [9].

2 Implementation of Intelligent Forecasting Subsystem of Real-Time

The Forecasting Module

On the basis of statistical and expert methods of forecasting combined (integrated) prediction method was suggested [10]. Method contains of an averaging the results obtained on the basis of the moving average method and the Bayesian approach, based on weighting coefficients. Then, resulting prediction is corrected by values of series obtained by the method of exponential smoothing. After that, forecast is adjusted by results of the expert methods: ranking and direct evaluations. The probability of each outcome acquired by statistical methods, increased or decreased depending on the expert assessment values for this outcome. Because of more time is required to obtain expert assessments, the system allows not to take them into account at certain steps, and use them only if available. The forecasting module is based on the methods and algorithms described above.

The module has two paths: the main and the alternative. Under normal system conditions, sequential results obtained by each algorithm are collected and calculated. After that, the analysis and the formation of the final result of the calculation are performed. Throughout the entire calculation process, the status of the system, the available resources and the presence of time constraints are monitored, during which it is necessary to form the result immediately, using the anytime algorithm. In such situations, the process proceeds along an alternative path when the calculations are transferred to the background (if possible) or stopped and a quick result is generated based on the performed calculations by parts of the system at the current time. The architecture of the forecasting module is presented in Fig. 1.

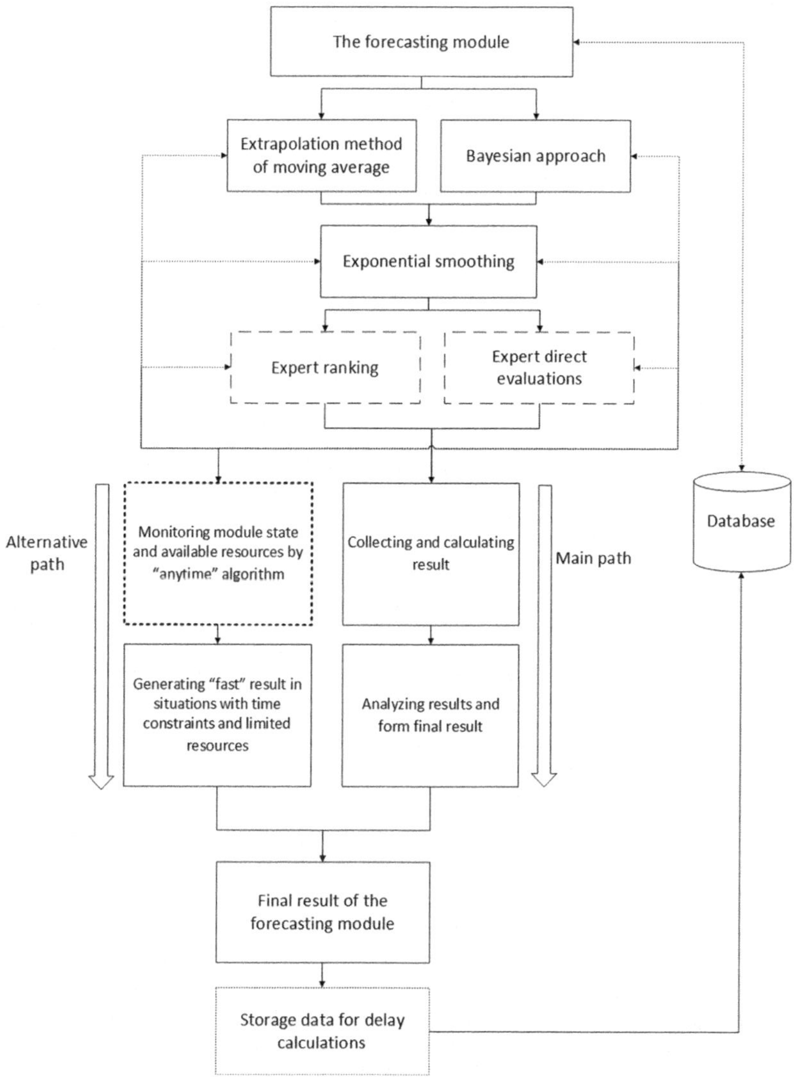

Fig. 1. The architecture of the forecasting module

As a result, each of the ways forms the final result: "full" in the case of normal operation of the system and "approximate" in the presence of instabilities. Finally, the results are stored in a database for use in the next iterations of the subsystem.

The Deep Reinforcement Learning Module

DRL-module consists of the group of independent agents that learning based on developed TD-methods (TD (0), TD (λ), SARSA, and Q-learning), focused on IS RT (IDSS RT) [1, 9]. Next, consider some of these methods.

SARSA – on-policy TD control method. For an on-policy method we must estimate $Q^{\pi}(s, a)$ for the current behavior policy π and for all states s and actions a:

$$Q(s_t, a_t) \leftarrow Q(s_t, a_t) + \alpha[\mathrm{r}_{t+1} + \gamma Q(s_{t+1}, a_{t+1}) - Q(s_t, a_t)],$$

where α - constant step length; γ - the valuation of the terminal state; r_t – reward on the step t. This update is done after every transition from a nonterminal state s_t.

The algorithm converges to the optimal strategy and the value function of the action, provided that all the "state-action" pairs have been visited infinite number of times, and the strategy converges to the greedy strategy.

Q-learning - off-policy TD control method that finds the optimal value of function Q to select the follow-up actions and at the same time determines the optimal strategy. Similarly, to method TD (0) in each iteration there is only knowledge of the two states: s and one of its predecessors. Thus, the values of Q allow to get a glimpse of the future actions quality in the previous states and to make the decision task easier. One-step Q-learning is characterized by the following relationship:

$$Q(s_t, a_t) \leftarrow Q(s_t, a_t) + \alpha[r_{t+1} + \gamma \, max \, \alpha \, Q(s_{t+1}, a) - Q(s_t, a_t)],$$

where α - constant step length; γ - the valuation of the terminal state; r_t – reward on the step t; *max* - maximizing over all those actions possible in the next state. In this case, the desired function Q directly approximates Q^{π} - optimum function values of action, regardless to the applied strategies.

To ensure convergence, it is necessary that all pairs continue to be adjusted. This is the minimum requirement in the sense that each method, which is guaranteed to find the optimal course of conduct, generally satisfies this condition. For a sequence of step length values, it is shown that the function Q_t converges to Q^*.

Sub-module represents of a multi-agent network that learning in parallel by various algorithms are divided into two networks also learning in parallel - one determines the behavior and second the objective function. Each agent consist of several additional intermediate hidden learning layers created between the input and output layer. The initial implementation of the module and its subsequent modernization was initially presented in [6, 19, 20]. At each layer except input layer, we compute the input to each unit, as the weighted sum of units from the previous layer. Then we are using activation function to obtain a new representation of the input from previous layer. We have weights on links between units from layer to layer. After computations flow forward from input to output, at output layer and each hidden layer, we can compute error derivatives backward and back propagate gradients towards the input layer, so that weights can be updated to optimize some loss function. In addition, each agent storing separate data in experience replay memory where the data can be batched or randomly sampled from different time-steps [11, 12].

The architecture of the DLR-module is presented in Fig. 2.

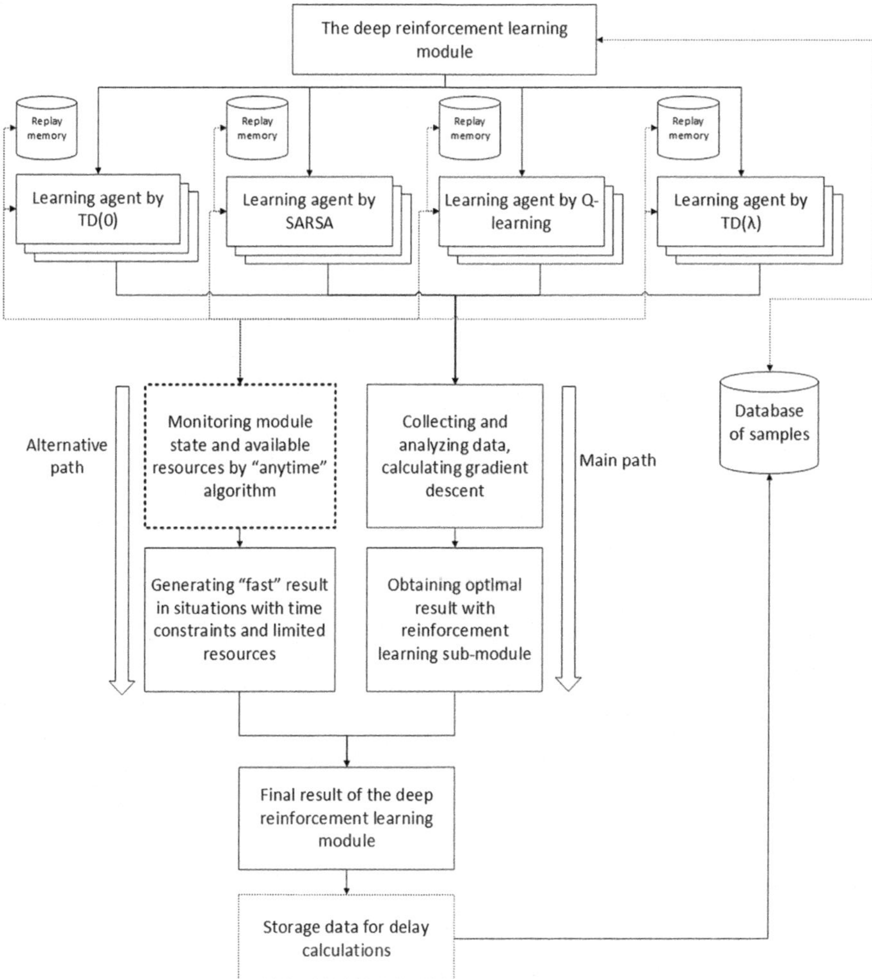

Fig. 2. The architecture of DLR-module

The DRL-module also has two paths: the main and the alternative. Under normal system conditions, agents are learning in parallel. After the end of episode data is collecting and analyzing, the gradient descent is calculating. Network become completely updated, formation of the final result of the calculation is performed.

Under conditions of severe time or resources constraints, system switches to alternative path: the milestone method is applied to the system [13, 14]. That allows making a tradeoff between computation time and solution quality, making it possible to compute approximate solutions to complex problems under different constraints. In this method algorithm chooses which of the paths is the most promising, relative to the accuracy of the forecast and the execution time, and calculates the result only by methods capable of obtaining the necessary optimal results at the current moment. In this case, all other steps can also be executed in the background and could be included in the analysis in the next steps. As a result, each of the way forms the final result: "full" in the case of normal operation of the system with completely updated network and "approximate" in the presence of instabilities. Finally, the results are stored in a database for use in the next episodes of the subsystem. The architecture of the forecasting subsystem is presented in Fig. 3.

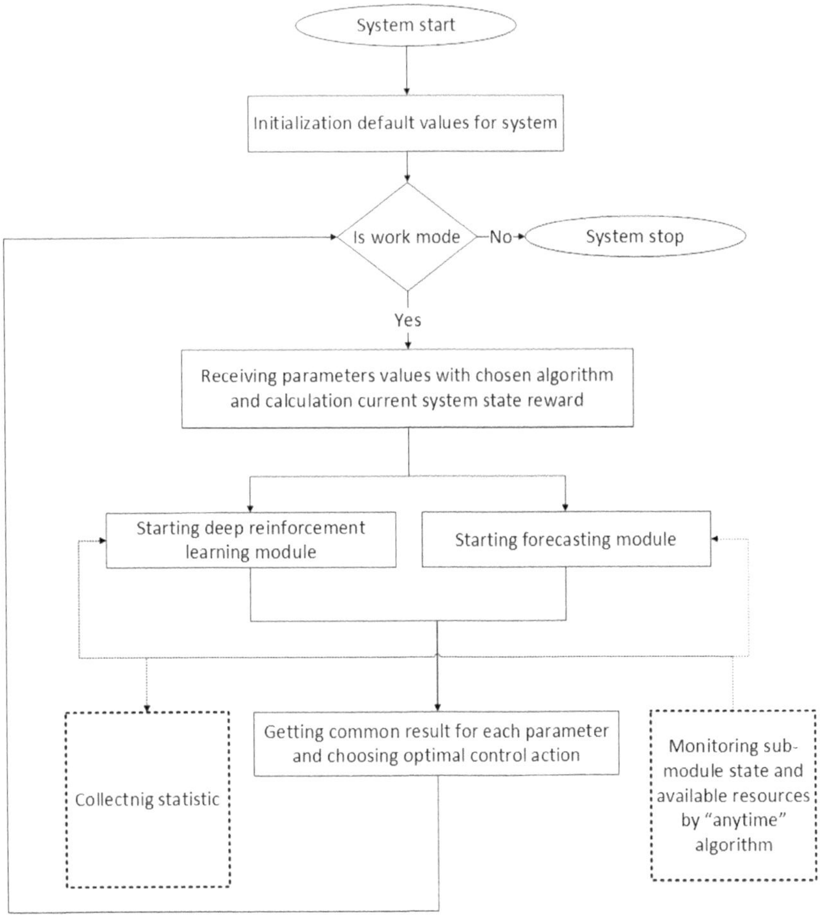

Fig. 3. The architecture of the forecasting subsystem

The Forecasting Subsystem

Proposed architecture of the forecasting subsystem includes:

- the emulator, which simulates the state of the environment with using of various system parameters change algorithms (linear and random) in the online database. Emulator capable simulate different constraints for the system such as time and resource constraints;
- the forecasting module based on statistical methods (extrapolation method of moving average, exponential smoothing and the Bayesian approach) and forecasting expert methods (ranking and direct evaluation). This module contains (as sub-modules) the monitoring module, that capable of generating fast results, and the analyzing module;
- the multi-agent DRL-module consists of the group of independent agents that learning on the basis of a developed TD-methods (TD (0), TD (λ), SARSA, Q-learning) divided into two networks learning in parallel. This module also contains (as sub-modules) the monitoring module, that of generating fast results, and the analyzing sub-module;
- the decision-making module designed for the data analysis coming from the forecasting module and multi-agent DRL-module, making decisions on follow-up actions and adjusting management strategies; the item module that collecting and analyzing statistic for valuation the effectiveness and performance of the system;
- the monitoring module based on milestone anytime algorithm, that analyze system state and could initialize getting fast result from all other modules.

3 Implementation of Algorithms and Results

For the testing and comparative analysis of the algorithms, a ε-greedy strategy was chosen, which allows achieving maximum remuneration and the feature of which is that the agent can combine training and solving the problem.

Next, we present the results obtained for algorithms based on Q-learning methods and SARSA. To test the TD algorithms, the problem was solved with different values of ε: $\varepsilon = 0.1$, $\varepsilon = 0.5$ and $\varepsilon = 0.9$. With $\varepsilon = 0.1$, the agent will more often perform "greedy" actions, and with $\varepsilon = 0.9$ - "research" actions. The parameter α influences the "trust" of the agent of new information - the higher it is, the more new information influences the perception of the agent. The parameter γ shows how much the agent thinks about the future benefits that the chosen action will bring him. Both of these parameters are assumed to be 0.5 to ensure even distribution of parameters. The task of a multi-armed bandit was chosen for the test, where agent at each step of the decision chooses which hand (machine handle) to pull in order to get the maximum reward. We considered this task in the work [15].

In Fig. 4 and Fig. 5, the diagram represents the results of working of RL-methods.

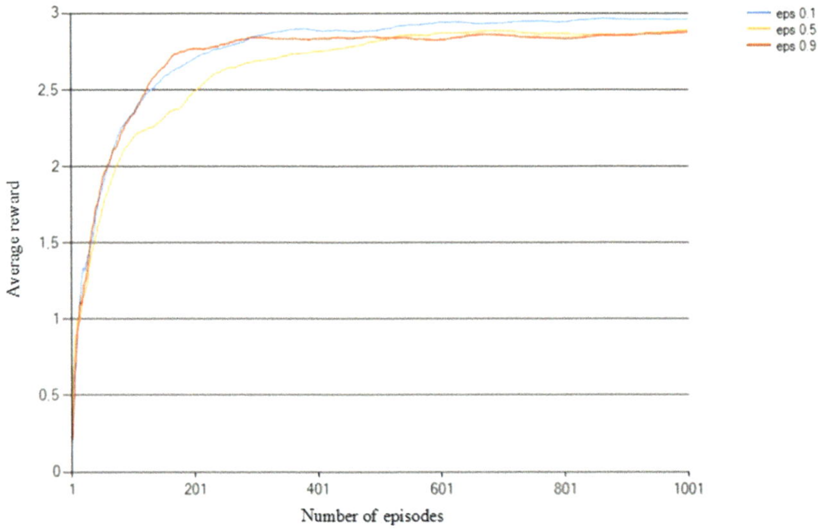

Fig. 4. Average episode reward for Q-learning method

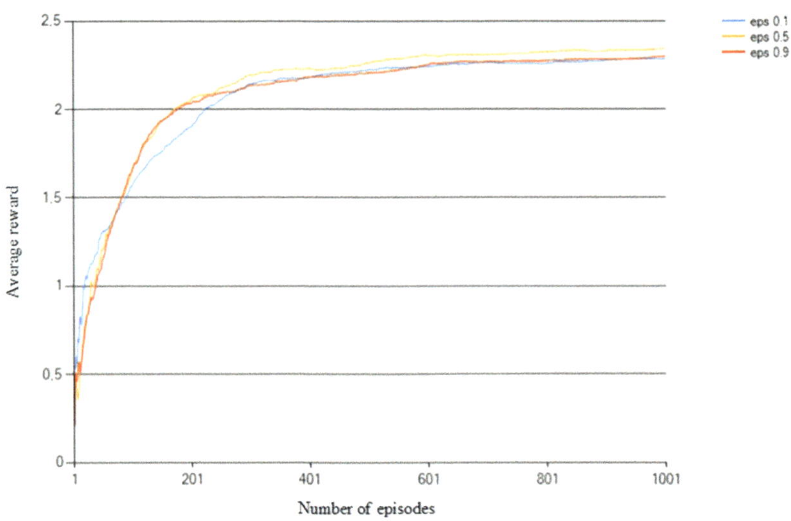

Fig. 5. Average episode reward for SARSA method

As a result of tests and analysis carried out, it was found that the best solution (the greatest reward) cannot always be achieved. As a rule, the greater the value of ε, the greater the chance that the agent will choose the action that gives the greatest reward (as the number of random actions increases with increasing ε, therefore, the agent will not "loop" on the close, but not the best solution). However, at small values of ε, the average remuneration may be greater, since the agent more often makes the decision that seems best to him (gives a greater reward).

One of the problems for IDSS RT is the problem of finding solutions in systems with noisy or distorted information, as well as in systems, information about which is unknown. TD-methods are well suited for solving problems in such systems, due to the ability of these methods to learn through trial and error. They make it possible to identify the existing patterns by analyzing the history of the process, thereby reducing the influence of random phenomena and improving the quality of the solutions found the effectiveness of the IDSS RT [16, 17].

The monitoring module based on milestone anytime algorithm, could obtain approximate fast results in the situations with time and resource constraints [18]. A description of the implementation of anytime algorithm designed to find a solution with time constraints is given in author's work [19, 20].

4 Conclusion

In the paper, architecture of intelligent forecasting system of real-time, consist of the forecasting module, the DRL-module and the main decision-making and monitoring module were proposed. The forecasting module contains of statistical, expert methods and monitoring modules (as submodules). At the output, module acquires predicted parameter changes based on statistics.

The multi-agent DRL-module, contains of a group of independent agents, each of that is learning in parallel on the basis of one of the developed TD-methods (TD (0), TD (λ), SARSA, Q-learning) as well as used for the accumulation of knowledge about the environment and capable of adaptation, modification and accumulation of knowledge. In addition, each agent has hidden layers that created between the input and output layer, separate agent's storing data in an experience replay memory and after the end of the episode, the gradient descent is calculated.

The decision-making module is designed to analyze the data coming from the forecasting and DRL-modules, making decisions on follow-up actions and methods to adjust management strategies.

In situations where the agent is forced to act in conditions of distorted or noisy information, as well as in dynamic environments that change over time, it is necessary to perform research actions often enough to get a complete picture of the environment. In this case, it is advisable to apply the ε-greedy learning strategy, allowing you to find a compromise between studying the environment and finding a solution. The obtained test results showed that the Q-learning method allows getting more reward than SARSA.

References

1. Sutton, R.S., Barto, A.G.: Reinforcement Learning. The MIT Press, London (2012)
2. Vagin, V.N., Eremeev, A.P.: Some basic principles of design of intelligent systems for supporting real-time decision making. J. Comput. Syst. Sci. Int. **6**, 953–961 (2001)
3. Bashlykov, A.A., Eremeev, A.P.: Fundamentals of Design of Intelligent Decision Support Systems in Nuclear Power Engineering: Textbook. INFRA-M, Moscow (2018). (in Russian)

4. Mnih, V., Badia, A.P., Mirza, M., Graves, A., Harley, T., et al.: Asynchronous methods for deep reinforcement learning. In: Proceedings of the 33rd International Conference on Machine Learning (PMLR 48), pp. 1928–1937 (2016)
5. Nikolenko, S., Kadurin, A., Archangelskaya, E.: Deep learning. In: Immersion in the World of Neural Networks. PITER, St. Petersburg (2017). (in Russian)
6. Eremeev, A.P., Kozhukhov, A.A., Guliakina, N.A.: Implementation of intelligent forecasting subsystem of real-time. In: Proceedings of the International Conference on Open Semantic Technologies for Intelligent Systems (OSTIS-2019), Minsk, pp. 201–204 (2019)
7. Alekhin, R., Varshavsky, P., Eremeev, A., Kozhevnikov, A.: Application of the case-based reasoning approach for identification of acoustic-emission control signals of complex technical objects. In: 2018 3rd Russian-Pacific Conference on Computer Technology and Applications (RPC), pp. 28–31 (2018)
8. Busoniu, L., Babuska, R., De Schutter, B.: Multi-agent reinforcement learning: an overview. In: Innovations in Multi-agent Systems and Applications, vol. 310, pp. 183–221. Springer, Heidelberg (2010)
9. Eremeev, A.P., Kozhukhov, A.A.: About implementation of machine learning tools in real-time intelligent systems. J. Softw. Syst. **2**, 239–245 (2018). (in Russian)
10. Sort, J., Singh, S., Lewis, R.L.: Variance-based rewards for approximate Bayesian reinforcement learning. In: Proceedings of Uncertainty in Artificial Intelligence, pp. 564–571 (2010)
11. Mnih, V., Kavukcuoglu, K., Silver, D., Rusu, A.A., Veness, J., Bellemare, M.G., et al.: Human-level control through deep reinforcement learning. Nature **518**, 529–533 (2015)
12. Li, Y.: Deep reinforcement learning: an overview, arXiv (2017). http://arxiv.org/abs/1701.07274
13. Hansen, E.A., Zilberstein, S.: Monitoring and control of anytime algorithms: a dynamic programming approach. J. Artif. Intell. **126**, 139–157 (2001)
14. Mangharam, R., Saba, A.: Anytime algorithms for GPU architectures. In: IEEE Real-Time Systems Symposium (2011)
15. Eremeev, A.P., Gerasimova, A.T., Kozhukhov, A.A.: Comparative analysis of machine reinforcement learning methods applied to real time systems. In: Proceedings of the International Conference on Intelligent Systems and Information Technologies (IS&IT 2019), Taganrog, vol. 1, pp. 213–222 (2019)
16. Golenkov, V.V., Gulyakina, N.A., Grakova, N.V., Nikulenka, V.Y., Eremeev, A.P., Tarasov, V.B.: From training intelligent systems to training their development means: In: Proceedings of the International Conference on Open Semantic Technologies for Intelligent Systems (OSTIS-2018), Minsk, vol. 2, no. 8, pp. 81–99 (2018)
17. Eremeev, A.P., Kozhukhov, A.A., Golenkov, V.V., Guliakina, N.A.: On the implementation of the machine learning tools in intelligent systems of real-time. J. Softw. Syst. **31**(2), 81–99 (2018). (in Russian)
18. Likhachev, M., Ferguson, D., Gordon, G., Stentz, A., Thrun, S.: Anytime dynamic A*: an anytime, replanning algorithm. In: Proceedings of the International Conference on Automated Planning and Scheduling (ICAPS), pp. 262–271 (2005)
19. Eremeev, A.P., Kozhukhov, A.A.: About the integration of learning and decision-making models in intelligent systems of real-time. In: Proceedings of the Second International Scientific Conference "Intelligent Information Technologies for Industry" (IITI 2018), vol. 2, pp. 181–189. Springer (2018)
20. Eremeev, A.P., Kozhukhov, A.A.: Methods and program tools based on prediction and reinforcement learning for the intelligent decision support systems of real-time. In: Proceedings of the Second International Scientific Conference "Intelligent Information Technologies for Industry" (IITI 2017), vol. 1, pp. 74–83. Springer (2017)

Agent-Based Situational Modeling and Identification Technological Systems in Conditions of Uncertainty

Marina Nikitina[1(✉)] and Yuri Ivashkin[2]

[1] V.M. Gorbatov Federal Research Center for Food Systems of Russian Academy of Sciences, 26, Talalikhina Str., 109316 Moscow, Russia
nikitinama@yandex.ru

[2] Moscow Technical University Communication and Informatics, 8a, Aviamotornaya Str., 111024 Moscow, Russia

Abstract. A structural-parametric situational model for identifying the state of a complex technological system under conditions of uncertainty, implemented by a trained intelligent agent, is proposed. The algorithm of neural network learning of an intelligent agent for recognizing difficult situations with finding the dividing surface between them upon presentation of a training sample of the parametric vector of the current state of the system is formalized. The methodology of developing a self-educating intelligent agent capable of identifying the current situation with incomplete and fuzzy information and making adequate decisions on their normalization in real time in the management of the technological system of the processing enterprise is described. The main stages of the software implementation of a trained intelligent agent in identifying and predicting the anomalous state of technological systems are formulated. For software implementation of a self-learning intelligent agent, a universal simulation system Simplex 3 with a specialized object-oriented language Simplex 3-MDL (Model Description Language) is proposed for describing system-dynamic, discrete-event and multi-agent models. The procedure for training an intelligent agent in the dynamics of its behavior is based on a multilayered neural network with pairs of interconnected input and output vectors and recurrent tuning of synaptic links by similarity measures.

Keywords: Intelligent agent · Incomplete and fuzzy information · Neural network · Situational analysis · Structural-parametric system model · Technological system

1 Introduction

Information technology of situational analysis and identification of the state of complex chemical-technological and biotechnological systems based on processing large volumes of statistical data in the case of dynamically changing,

S. Kovalev et al. (Eds.): IITI 2019, AISC 1156, pp. 109–119, 2020.
https://doi.org/10.1007/978-3-030-50097-9_12

incomplete and insufficient information about the controlled object require the development of intelligent modules for adapting and learning to recognize complex situations for computer support to make adequate decisions in the face of uncertainty and risk [1–3].

Uncertainty in this case may be associated with noise distortion of incoming signals, disturbance and cycling of known causal relationship between controlled variables, drift of extremum of the target function of the system and boundaries of acceptable solutions, manifestation of previously unaccounted factors and random influences. Then the identification of the state of the system is reduced to determining by methods of artificial intelligence the closest to it a priori known anomalous situation contained in the knowledge base of the computer decision support system [4–8].

2 Structural-Parametric Situational Model of the Technological System in Certainty Conditions

The existing direction of situational analysis of data associated with their structurization according to a functional or objective principle and the definition of links between state and goal parameters of a technological system [9] is reduced to the formation of a **cognitive matrix** $\|C_{ij}\|^n$ dimensionless characteristics of relations between the i-th and j-th parameters, on the basis of which, **the structural-parametric situational matrix system model** (SPSM) is built $C_{ij} \cdot \Delta x_j, \quad i,j = \overline{1,n}$, by multiplying it by the diagonal matrix of the vector of normalized deviations $\Delta x_1, ..., \Delta x_n$,

$$\begin{Vmatrix} 1 & c_{12} & \dots & c_{1n} \\ c_{21} & 1 & \dots & c_{2n} \\ \dots & \dots & \dots & \dots \\ c_{n1} & c_{n2} & \dots & 1 \end{Vmatrix} \cdot \begin{Vmatrix} \Delta x_1 & & & \\ & \Delta x_2 & & \\ \dots & \dots & \dots & \dots \\ & & & \Delta x_n \end{Vmatrix} = \begin{Vmatrix} \Delta x_1 & c_{12}\Delta x_2 & \dots & c_{1n}\Delta x_n \\ c_{21}\Delta x_1 & \Delta x_2 & \dots & c_{2n}\Delta x_n \\ \dots & \dots & \dots & \dots \\ c_{n1}\Delta x_1 & c_{n2}\Delta x_2 & \dots & \Delta x_n \end{Vmatrix} \quad (1)$$

where $\Delta x_i = \frac{|x_i - x_i{}^0|}{\Delta x_i{}^0}$, $i = \overline{1,n}$ – normalized deviations of state parameters from the range of tolerances Δx_i^0.

As a result, the elements of the main diagonal of the situational matrix reflect the normalized deviations x_i^0 of controlled factors from the given values x_i^0, and the nondiagonal $C_{ij} \cdot \Delta x_j; \quad i,j = \overline{1,n} \quad (j \neq i)$ – contributions of the deviations $\Delta x_j, \quad j = \overline{1,n}$ to the deviation $\Delta x_i, \quad i = \overline{1,n}$ in accordance with the system of equations

$$\Delta x_i = \sum_{j}^{N} C_{ij} \cdot \Delta x_j; \; i = \overline{1,n}; \; i \neq j \quad (2)$$

ordering the rows of all a priori known reasons for the deviation of Δx_i, and for the columns - possible investigative effects of the deviation of Δx_i on other parameters.

In the general case, a situational matrix $\|c_{ij} \cdot \Delta x_j\|^n$ with a set of functional elements $\{x_1 ... x_n\}$ and relations between them $\|c_{ij}\|^n$ describes the dynamic state of the system in a discrete-event sequence of temporary situations combining an a priori knowledge base about the structure of cause-effect relationships with current information Δx.

The formalized algorithm for **identifying an anomalous situation** in a technological system is as follows.

In the line of the maximum diagonal element corresponding to the maximum deviation from the norm – Δx_i^0 in the observed set of state parameters, the maximum non-diagonal element corresponding to the main reason that caused this deviation is searched.

The found column is followed by a transition to a new element of the main diagonal, after which the next main cause of the anomaly is in its row. The search continues until a diagonal deviation is found, in the row of which all non-diagonal elements will be zero, which means finding the original cause of the abnormal situation.

The registration of each current situation in real time supplements the initial database with the subsequent recalculation of the regression coefficients.

3 Agent-Oriented Situational Model of the Technological System in Conditions of Uncertainty

Identification of the state of a complex technological system in real time under the conditions of uncertainty and risk of decision making, with insufficient and fuzzy information, requires further intellectualization of the SPM based on artificial intelligence methods using agent and neural network technologies.

At the same time, the intellectual function of the learning intellectual agent (IA) and the neural network module is to clarify and correct the initially established characteristics of the relationships between the monitored state and target parameters, as well as to recognize and classify anomalous situations in the system with an accumulation of reliable quantitative and qualitative ratings, which is formed by the **situational classifier** and system knowledge base.

Learning and self-learning **intellectual agents** are able to accumulate knowledge based on current data and ontology of events in the process of interaction with other agents and the environment, adapt to the situation, choose a strategy to achieve the selected goal and evaluate the degree of its achievement.

The general algorithm of the behavior of an intelligent agent [10–12] includes the identification of a situation, the assessment of one's own state and the correction of a target, followed by a reflexive reaction or meaningful (intellectual) decision making in the direction of achieving the goal.

The parametric description of the agent includes a knowledge base and a set of goals in a certain area, the vectors of its state variables; a description of relations between agents and the influence of the environment, a bank of models and strategies of agent behavior.

An identification algorithm with a small training sample size with a declared set of monitored parameters describes the agent's recognition of the current state of the technological system in real time with preliminary classification of situations based on presenting the current system states as belonging to certain decision-making clusters.

For this, a variety of Hamming's artificial neural network architecture (ANN) with a multilayer recurrent structure, the operation of which consists in minimizing the Hamming distance between the input vector (system parameters) and the vectors of training samples encoded in the network structure is proposed [13].

4 Algorithm of Neural Network Learning Agent to Recognize Situations

The task of teaching an agent to recognize situations on the basis of a training sample with the presentation of input vectors of state variables of the system (images) in situations related to certain clusters is connected with the construction of the separating function $y = f(x_1, \ldots, x_n)$, where $\{x_1, \ldots, x_n\}$. Is the n-dimensional parametric vector of the situation, and y is the value of belonging to a specific cluster.

Training samples are composed of parametric vectors of anomalous situations based on a structural-parametric situational model of the system (SPSM) and an algorithm for causal identification under certainty conditions.

The neural network structure of training an intelligent agent in a simplified version with input, one hidden and output layers is shown in Fig. 1.

The inputs of the network receive the values of n components of the current situation vector $x_1, \ldots, x_n$ and the network task is to iteratively select the weights $w_{1j}^{(1)}$ and $w_{1j}^{(2)}$ for the elements of all network layers so that for a given input vector of parameters $\{x_1, \ldots, x_n\}$ the signal at the network output coincides with acceptable accuracy test vector y according to the criterion of least squares of the differences between the input and reference vectors of training samples.

In the agent's perception, the output signal of the i-th neuron of the hidden layer in general form is:

$$y_i^{(2)} = f\left(\sum_{j=1}^{N} w_{ij}^{(1)} x_j\right); \quad i = \overline{1, k} \tag{3}$$

where $w_{ij}^{(1)}$ is the weight of connection of the j-th neuron of the first layer with the input of the i-th neuron of the hidden layer.

Then for the output layer, where the image of the situation is formed, the following expression is valid:

$$Y = f\left(\sum_{i=1}^{k} w_i^{(2)} y_i\right) \tag{4}$$

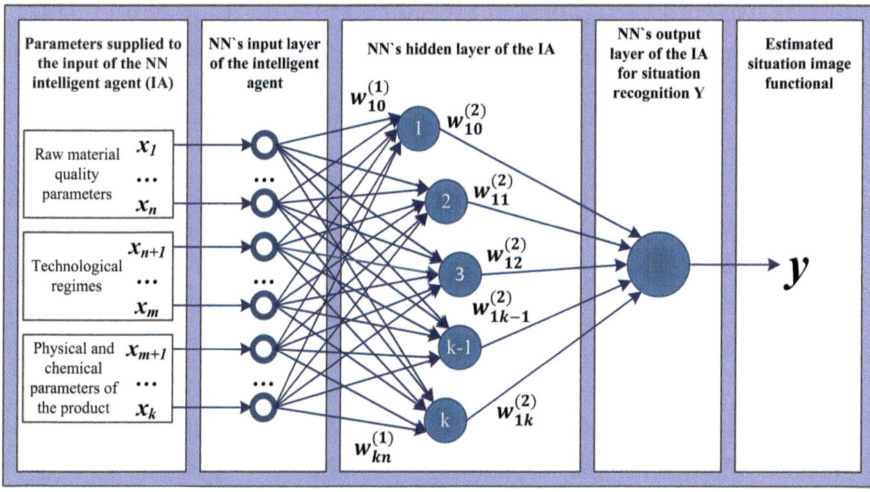

Fig. 1. Neural network structure of intellectual agent training with one hidden layer

or

$$Y = f\left(\sum_{i=1}^{k} w_i^{(2)} y_i \, f\left(\sum_{j=1}^{n_i} w_{ij}^{(1)} x_j\right)\right) \tag{5}$$

With a sigmoidal neuron activation function, the signal at the output of the i-th neuron of the hidden layer will be:

$$y_i = \frac{1}{1 + e^{-\sum\limits_{j=1}^{n} w_{ij}^{(1)} x_j}} \tag{6}$$

and the signal at the output of the network will be expressed by the formula

$$Y = \frac{1}{1 + \exp(-\sum\limits_{i=1}^{k} w_i^{(2)} \frac{1}{1+\exp(-\sum\limits_{j}^{n} w_j^{(1)} x_j)} y_i)} \tag{7}$$

As an agent learning algorithm, the gradient method of back propagation of an error is chosen, the essence of which is to select such values of weights $w_{1j}^{(1)}$ and $w_{1j}^{(2)}$ for all network layer elements so that for a given input vector of parameters $\{x_1, \ldots, x_n\}$ with acceptable accuracy to get at the output, coincidence with the test reference vector of the training sample.

This minimizes the sum of squared differences between the actual and expected values of the output signals in the form:

$$Q = \sum_{l=1}^{p} (f_l(\sum_{i=1}^{k} w_i^{(2)} y_i) - d_l)^2 \tag{8}$$

where p is the number of training samples; d_l is the expected (test) vector at the output for a given vector of input parameters in the l-th representation (l-th learning step).

The full algorithm of neural network learning of an intelligent agent consists of the following steps:

1. The agent in the observation mode perceives the vector of signals of the current situation in one of the possible images and calculates the values of the i-th output of the n-th layer by the formula:

$$y_i^{(n)} = \sum_{j=1}^{m_i} y_j^{(n-1)} w_{ij}^{(n)} \tag{9}$$

   where $w_{ij}^{(n)}$ is the weight of the connection of the i-th neuron of the n-th layer with the j-th input;

   m_i is the number of neurons of the $(n-1)$-th layer associated with the i-th neuron of the n-th layer;

   with sigmoidal neuron activation function formulas (6), (7) of the n-th layer as $y_i^{(n)} = F(y_i^{(n)})$.
2. Calculation of the error $\delta^{(n)}$ in the assessment of the image of the current k-th situation Y in the output, N-th layer of its neuron layer by the formula:

$$\delta_l^{(n)} = (Y - d_k)\frac{dY}{dy_i^{(n-1)}} \tag{10}$$

   with the correction of the weights of its inputs $w^{(N-1)}$ as:

$$w_i^{(N-1)} = w_i^{(N-1)} + \Delta w_i^{(N-1)} \tag{11}$$

   where $\Delta w_{1i}^{(N-1)} = -\eta\delta_l^{(2)} y_i$; η – coefficient of learning speed, $0 < \eta < 1$.
3. Calculation of the error $\delta^{(n)}$ and $\Delta w^{(n)}$ for hidden layers $n = \overline{N-1, \ldots, 1}$ with the correction of the weights $w_{ij}^{(n)}$ of their inputs by the formulas (10), (11).
4. If the error is significant, you should return to the first step, otherwise, the agent can be considered trained.

Before starting operation, the trained network is checked for the quality of training with a check for permissible errors when the sample is fed to the network input with known values of the output parameters, but different from the training sample.

The training of an intelligent agent based on the work of the INS makes it possible to ensure the necessary accuracy in assessing the influence of each parameter on the state of the technological system and making decisions in real time.

5 Agent Technology of Situational Modeling of Technological Systems in Conditions of Uncertainty

The practical implementation of agent technologies is associated with the description of the agent's state variables and the dynamics of its behavior in a specialized agent-oriented language in a simulation system with an experimentation environment, a set of simulation programs and software for carrying out the experiment [14,15].

The general algorithm of the behavior of an intelligent agent (Fig. 2) includes identifying the situation, assessing one's own state and correcting the target, followed by reflexion of reactions or meaningful (intellectual) decision making in the direction of achieving the target. The criterion of the agent's intelligence is the degree of completeness and depth of a priori knowledge, learning strategies and decision-making algorithms under conditions of uncertainty, risk and conflict.

Each operation has its own algorithmic and software module that provides:

- perception of information and the accumulation of knowledge about the environment and the environment of interaction or conflict (sensory module);
- the mechanism of interaction and data processing from counterparties;
- analysis of the own state and the state of counterparties with the selection or correction of the objective functions (intellectual module);
- making autonomous decisions and choosing strategies.

The behavior of the agent can be represented by some recursive form that describes the finding and selection at the next step of the transition function from the initial state to the new state in the direction of improving the objective function.

In some cases, such a problem can be solved by mathematical programming with the correction of the objective function and individual constraints at the next step of changing the agent's state depending on the situation and approaching the goal under conditions of uncertainty and fuzzy information.

The parametric description of an intelligent agent includes many goals and a knowledge base in a certain area, a vector of characteristics of its state; a bank of models and behavioral strategies, descriptions of external relations with agents and the environment.

The main stages of modeling a learning agent are as follows:

1. Decomposition of the technological system into a set of functional divisions and production processes with the formalization of a variety of state parameters and factors of influence and control;
2. Parametric description of the functional units of the technological system in the form of a set of vectors of input and output variables, state parameters and the objective function;
3. Development of a mathematical model of agent training and an algorithm for its behavior with training procedures and identification of current situations and decision making;

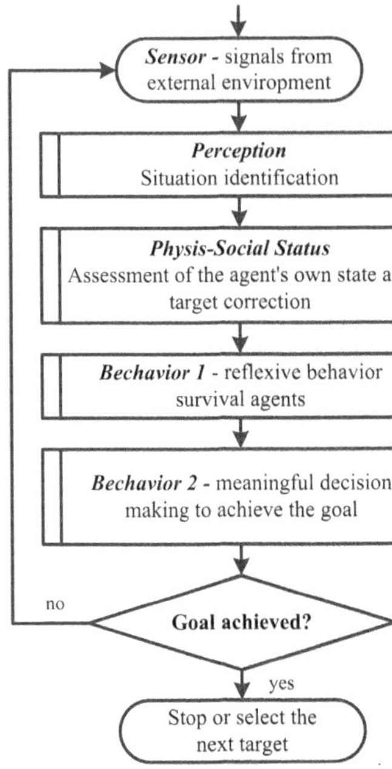

Fig. 2. General algorithm for the behavior of an intelligent agent

4. Description of an autonomous intelligent agent with a variety of state variables, input sensory variables that communicate with other agents and the environment, as well as discrete-event dynamics of agent behavior in a specialized agent-oriented modeling language or high-level algorithmic language [14,15];
5. Compilation of an agent-oriented situational model of a technological system in real time, with algorithms for identification and prediction of its state under conditions of uncertainty and risk;
6. Software implementation of a multi-agent simulation model in a universal simulation system using special software packages.

For software implementation, it is proposed to use the universal simulation system Simplex 3, developed at the universities of Nuremberg-Erlangen, Passau and Magdeburg (Germany). Simplex 3 has its own programming language Simpkex3-MDL (Model Description Language) describing system-dynamic, discrete-event and multi-agent models [16,17].

The basic component includes sections describing the state variables of the intelligent agent DECLARATION OF ELEMENTS and the dynamics of its state change (behavior) DYNAMIC BEHAVIOUR with analytical or discrete-event description. The syntactic form of the basic component is as follows:

```
BASIC COMPONENT < name >
 [mobile_subclass_declaration - mobile components]
 [subunit_declaration - base units]
 [local_definitions - arrays, functions, distribution
    laws]
 DECLARATION OF ELEMENTS
  [list_of_constants - constants]
  [list_of_state_variables - state variables]
  [list_of_dependent_variables - calculated variables]
  [list_of_sensor_variables - touch Variables]
  [list_of_random_variables - random variables]
  [list_of_transition_indicators - transition
     indicators]
  [list_of_sensor_indicators - touch indicators]
  [list_of_locations - cumulative arrays]
  [list_of_sensor_locations - touch drives]
 DYNAMIC BEHAVIOUR
  algebraic_equation - algebraic equations
  |differential_equations - differential equations
  |region_defining_statement - areas of certain states
  |event_defining_statement - events
 END
END OF < name >
```

6 Conclusion

Agent technologies with neural network behavior algorithms of self-learning intelligent agents with recognition of current situations in conditions of fuzzy information, uncertainty and risk opens up a new direction of intellectualization of expert computer systems for decision making in control systems of complex technological processes, as well as in virtual research of the influence of various technological factors on abnormal system states.

For the software implementation of a self-learning intelligent agent, a universal simulation system Simplex 3 with a specialized object-oriented language Simplex 3-MDL (Model Description Language) describing system-dynamic, discrete-event and multi-agent models is proposed. The procedure of learning an intelligent agent in the dynamics of its behavior is based on a multilayered neural network with pairs of interconnected input and output vectors and recurrent tuning of synaptic connections by measures of similarity (for example, Hamming distance). The vectors of the training sample are formed using a structural-parametric situational model and an algorithm for the causal identification of the technological system under certainty conditions [18].

The proposed direction of the intellectualization of situational modeling systems is the basis for building intelligent computer decision support systems and the operational management of the quality of food products at processing enterprises of the Agro-industrial complex.

References

1. Pulido, B., Zamarreno, J.M., Merino, A., Bregon, A.: State space neural networks and model-decomposition methods for fault diagnosis of complex industrial systems. Eng. Appl. Artif. Intell. **79**, 67–86 (2019). https://doi.org/10.1016/j.engappai.2018.12.007
2. Bobka, P., Heyn, J., Henningson, J.O., Romer, M.: Development of an automated assembly process supported with an artificial neural network. J. Mach. Eng. **18**, 28–41 (2018). https://doi.org/10.5604/01.3001.0012.4605
3. Rymarczyk, T., Klosowski, G., Cieplak, T., Kozlowski, E.: Industrial processes control with the use of a neural tomographic algorithm. Przeglad elektrotechniczny **95**(2), 96–99 (2019). https://doi.org/10.15199/48.2019.02.22
4. Ovsyanikova, I., Tarapanov, A.: Neural network management of technological systems at the finish operations. In: International Conference on Modern Trends in Manufacturing Technologies and Equipment (ICMTMTE), vol. 179, no. 01025 (2017). https://doi.org/10.1051/matecconf/201712901025
5. Rojek, I., Kujawinska, A., Hamrol, A., Rogalewicz, M.: Artificial neural networks as a means for making process control charts user friendly. In: International Conference on Intelligent Systems in Production Engineering and Maintenance, vol. 637, pp. 168–178 (2018). https://doi.org/10.1007/978-3-319-64465-3_17
6. Almassri, A.M.M., Hasan, W.Z.W., Ahmad, S.A., Shafie, S., Wada, C., Horio, K.: Self-calibration algorithm for a pressure sensor with a real-time approach based on an artificial neural network **18**(8), 2561 (2018). https://doi.org/10.3390/s18082561
7. Gomez-Espinosa, A., Sundin, R.C., Eguren, I.L., Cuan-Urquizo, E., Trevino-Quintanilla, C.D.: Neural network direct control with online learning for shape memory alloy manipulators **19**(11), 2576 (2019). https://doi.org/10.3390/s19112576
8. Jain, L.C., Seera, M., Lim, C.P., Balasubramaniam, P.: A review of online learning in supervised neural networks. Neural Comput. Appl. **25**(3–4), 491–509 (2014)
9. Ivashkin, Yu.I.: Structural-parametric modeling and identification of anomalous situations in complex technological systems. Control Probl. **3**, 39-43 (2004)
10. Kennedy, W.G.: Modeling human behavior in agent-based models. In: Agent-Based Models of Geographical Systems, pp. 167–179. Springer, New York (2011)
11. Schmidt, B.: The modeling of human behavior: the PECS reference model. In: Proceedings 14th European Simulation Symposium, Dresden, Germany, 23–26 October 2002 (2002)
12. Stanilov, K.: Space in agent-based models. In: Agent-Based Models of Geographical Systems, pp. 253–271. Springer, New York (2012)
13. Eremin, D.M., Gartsev, I.B.: Artificial neural networks in intelligent control systems: a monograph. MIREA, Moscow (2004)
14. Schmidt, B.: The Art of Modeling and Simulation. SCS-Europe BVBA, Ghent, Belgium (2001)
15. Karpov, Yu.G.: Simulation systems. In: Introduction to modeling with AnyLogic 5. BHV-S.-Petersburg, Sankt-Petersburg (2005)
16. Simplex3: Simulation komplexer systeme. http://www.simplex3.net

17. Eschenbacher, P.: Die Modellschreibungssprache Simplex-MDL. In: Operations Research Proceedings, pp. 119–125 (1998)
18. Ivashkin, Y.A., Blagoveschensky, I.G., Nikitina, M.A.: Neural networks and agent technologies in the structural-parametric modelling of technological systems. In: CEUR Workshop Proceedings OPTA-CSL-2018-Proceedings of the School-Seminar on Optimizations Problems and Their Applications, pp. 169–180 (2018)

Features of Data Warehouse Support Based on a Search Agent and an Evolutionary Model for Innovation Information Selection

Vladimir K. Ivanov[1(✉)], Boris V. Palyukh[1], and Alexander N. Sotnikov[2]

[1] Tver State Technical University, 22, Quay A. Nikitin, Tver 170026, Russia
{mtivk,pboris}@tstu.tver.ru

[2] Joint Supercomputer Centre of RAS, 32a, Leninskiy Av., Moscow 119991, Russia
asotnikov@jscc.ru

Abstract. Innovations are the key factor of the competitiveness of any modern business. This paper gives the systematized results of investigations on the data warehouse technology with an automatic data-replenishment from heterogeneous sources. The data warehouse is suggested to contain information about objects having a significant innovative potential. The selection mechanism for such information is based on quantitative evaluation of the objects innovativeness, in particular their technological novelty and relevance for them. The article presents the general architecture of the data warehouse, describes innovativeness indicators, considers Theory of Evidence application for processing incomplete and fuzzy information, defines basic ideas of measurement processing procedure to compute probabilistic values of innovativeness components, summarizes using evolutional approach in forming the linguistic model of object archetype, gives information about an experimental check if the model developed is adequate. The results of these investigations can be used for business planning, forecasting technological development, investment project expertise.

Keywords: Data warehouse · Intelligent agent · Subject search · Genetic algorithm · Innovativeness · Novelty · Relevance

1 Introduction

The basis model of R&D management involves the competitive analysis and forecasting of the technological development based on scientometric analytical services and semantic systems for searching commercially valuable information. This field features the obvious world-wide trend of using the global and already

S. Kovalev et al. (Eds.): IITI 2019, AISC 1156, pp. 120–130, 2020.
https://doi.org/10.1007/978-3-030-50097-9_13

existing innovative potential. Innovations are the key factor of the competitiveness of any modern business, both in competitive and monopoly markets.

This paper is the first one to give the systematized review of the results of the works performed within the project of *Data Warehouse Support on the Base Intellectual Web Crawler and Evolutionary Model for Target Information Selection*. The objective of the project is to develop the theoretical basis for and pilot implementation of data warehouse technology with an automated data supply from sources belonging to different subject segments. The data warehouse is suggested to contain technical, economic, social and other characteristics of objects having a significant innovative potential [1]. We suppose that the algorithms and technologies developed would be used for the expert system operation to control the evolution of multistage production processes [2] and [3]. Further, the article considers the tasks and basic results of the project, the architectural solutions of the data warehouse, objects innovativeness indicators, application of the theory of evidence, procedure for objects innovativeness calculation, evolutional approach to forming the linguistic model of the object, and some experimental results.

2 Project Tasks

The project results are supposed to be used in decision support systems (DSS) and expert systems to support solving the following applied tasks: *(a)* determine the characteristics of new domains in business planning; *(b)* forecast the technological development of the business; *(c)* provide an information support for expert groups and individual experts.

The researches focused on the development and experimental testing the model of the evolutional process of query generation and search results filtration. Further, some substantial aspects of the researches done will be detailed.

3 Data Warehouse Support System Architecture

According to [4], the software architecture includes layers of presentation, services, business logic, data access, as well as through functionality to provide the interaction of users and external systems with data sources. Figure 1 shows the general architecture of the system. Below, there is a brief description of the composition and purpose of the basic components (circles):

1. *User's Applications.* Specialized applied systems for the information support of innovation implementation.
2. *Decision Support Systems. Expert Systems.*
3. *Data Presentation.* Visualization of innovation data warehouse composition, search patterns and results of the latter, objects innovativeness indicators, sets of associated objects.
4. *Services.* Software interfaces for interaction with DSS and expert systems, as well as presentation layer components.

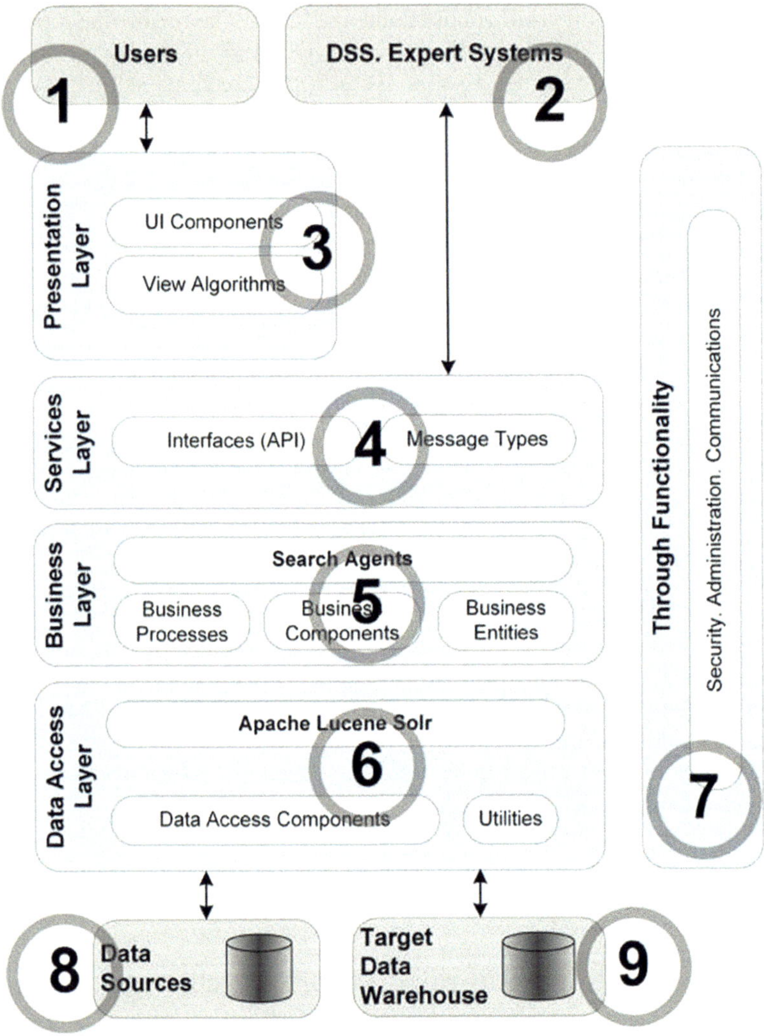

Fig. 1. The general architecture of the data warehouse support system

5. *Search Agents.* Innovation solutions information retrieval.
   (a) *Business Processes.* Evolutional generation of the linguistic model of the object archetype and effective multiset of queries. Raw data measurements processing to compute the probabilistic values of the objects innovativeness indicators. Processing of measurements obtained from several sources. Applying the Dempster-Shafer Theory of Evidence. Arranging intelligent search agents interactions.
   (b) *Business Components.* A genetic algorithm to produce the effective multiset of queries. An algorithm to filtrate the search results. A model to

calculate the objects innovativeness indicators. An algorithm for a group processing of the object innovativeness level measurements.

(c) *Business Entities.* A concepts basis. The linguistic model of the sought-for object archetype. Search pattern. Objects innovativeness indicators. Fuzzy indicators of the probability of specified innovativeness properties. Limitations on the reference information-model in the specified domain.

6. *Apache Lucene Solr* (http://lucene.apache.org). A software implementation of the model of the vector space of the documents: an object model, a software library for document data warehouse access, data indexer, data storage.
7. *Through Functionality.* Security. Administration. Network communications.
8. *Data Sources.* Internet resources. Specialized data warehouses and databases.
9. *Target Data Warehouse.* A register of innovative solutions.

4 Objects Innovativeness Indicators

We introduced concepts of technological novelty, relevance and implementability as the components of the object innovativeness criterion. *The novelty* means significant improvements, a new way of object usage or presentation (novelty subjects are potential users or a producer oneself). *The relevance* is a potential producer's recognized need for an object formed as the demand. And *the implementability* determines the technological validity, physical feasibility and integrability of an object into the system in order to obtain the effects desired.

The linguistic model of the sought-for-object archetype has been proposed. The model terms are classified as key properties describing the object structure, application conditions or functional results. A marker determines the archetype definition domain. The queries are constructed as terms-and-marker combinations. The genetic algorithm of queries generation and results filtration is used to obtain a quasi-optimal queries set.

The following expressions to compute the values of the innovativeness indicators have been proposed:

$$Nov = 1 - \frac{1}{S}\sum_{k=1}^{S}[1 - exp(1 - \frac{R_k}{R_{min}})] \tag{1}$$

where Nov is the object novelty (the value is normalized for range [0;1]), S is the total number of executed queries; R_k is the number of documents found in the database as a result of the k-th query; R_{min} is the minimum number of the documents found among all queries.

$$Rel = \frac{1}{S}\sum_{k=1}^{S}[1 - exp(1 - \frac{F_k}{F_{min}})] \tag{2}$$

where Rel is the object relevance (the value is normalized for range [0; 1]), S is the total number of executed queries; F_k is the frequency of users' executed queries similar to the k-th query; F_{min} is the minimum frequency of the query execution among all query.

A hypothesis of the adequate representation of real processes in the information space. The object novelty evaluation is based on the normalized integral evaluation of the number of results of the object information search in the heterogeneous databases. It is suggested that the number of the search results which are relevant to the search pattern would be less for new objects rather than for long-time existing and well-known ones. The object relevance estimation is based on the normalized integral estimation of the frequency of users' executed queries similar to the queries generated from the search pattern. Taking into account the direct quantitative evaluation of the innovativeness, we suppose that this approach can be complementary to the conventional ones [5] and [6].

5 Applying the Theory of Evidence

Since an obviously incomplete and inaccurate object information is expected to be obtained from different sources, fuzzy indicators for the probability of the technological novelty and relevance for the object have been introduced. To calculate the above probabilities, application of the Dempster-Shafer Theory of Evidence is validated [7–9]. So, the basic probability m of the fact that the object innovativeness indicator (Nov or Rel) measurements belong to the interval A can be evaluated from the following:

$$m : P(\Omega) \rightarrow [0,1], m(\emptyset) = 0, \sum_{A \in P(\Omega)} m(A) = 1 \tag{3}$$

where Ω is the indicator measurements set, $P(\Omega)$ is the set of all Ω subsets. Further, the belief

$$Bel = \sum_{A_k : A_k \subseteq A} m(A_k) \tag{4}$$

and plausibility

$$Pl = \sum_{A_k : A_k \cap A \neq \emptyset} m(A_k) \tag{5}$$

for specified k intervals are calculated. These functions determine the upper and lower boundaries for the probability that the object has the property specified. This is the way to estimate the values of indicators Nov or Rel in the conditions of the incomplete and inaccurate information about the objects. We also studied the applicability of Theory of Evidence for solving tasks in complex technical systems diagnostics and optimum control over the evolution of multistage processes in the fuzzy dynamic medium [3] and [10]. The objective of these investigations is to suggest a new architecture (for interactions among the intelligent search agents in the system of the heterogeneous data warehouses) based on the concept of an "abnormal" agent. The abnormal state (AS) of a search agent can be interpreted as the presence of a challenger for innovation as a result of the search by this agent. AS can be diagnosed as an exit of the objects innovativeness indicators beyond their characteristical values. The following necessary and sufficient condition can be used to indicate the search agent AS, $s \in S$:

$$(\forall s \in S)(F = 0) \leftrightarrow P^* \neq 0 \tag{6}$$

where F is the indicator function, P^* is the set of registered AS.

Taking into account the potentially great number of information sources, testing diagnostic hypotheses provides enhanced efforts to avoid missing a true innovation and indicating a false one. The indicator function allows the quantitative evaluation how "reasonable" or "useful" is the search agent.

6 Evolutional Approach

The evolutional approach is validated for and applied to solving the task of forming the linguistic model of the object archetype. The main idea is to use a special genetic algorithm (GA) to arrange an evolution process generating the stable and effective set of queries to obtain the most relevant results. The basic concepts of the approach used in the development of the GA proposed by the authors are described in [11]. Thus, the initial interpretations are as follows: *a query* is *an individual*, *an encoded set of query terms* represents *a genotype*, the replacement of a query term with another term is defined as *crossover*, and the replacement of a query term with its synonym is *mutation*. The procedure of a fitness function calculation consists in executing a query by a search engine and getting the set of relevant documents found – *a phenotype*.

The GA search pattern K is a set of terms related to a certain subject area. Each search query is represented by a vector $\overline{q} = (c_1, c_2, \dots c_n, \dots c_m)$, where $c_n = \{k_n, w_n, z_n\}$, $k_n \in K$ is a term, w_n is a term weight, z_n is a set of term synonyms k_n, m is the number of terms in a query. The result of the query is a set of documents R, $|R| = D$. The initial population of S queries is a set Q_0, where $|Q_0| = S$, $S < |K|/2$, $q \in Q_0$. The fitness function for the query population is calculated as follows:

$$\overline{W}(Q) = \frac{1}{S}\sum_{j=1}^{S}\frac{1}{R}\sum_{i=1}^{R} w_{ji}(g, f, s) \tag{7}$$

where $Q = (q_1, q_2, \dots q_S)$ is the population of S queries; w_{ji} is the fitness function of i-th results of j-th query. Here, w_i depends on position g in the search engine results list, frequency f of this search result in all S query result lists, similarity measure s of the short result text and search pattern K.

To understand how GA works, Holland's Schema Theorem plays a key role. It was stated for the canonical GA and proves its convergence. It is obviously reasonable to check if the theorem of schemes works for any modifications of the canonical GA. Our investigations specify conditions for the correct check of the theorem of schemes. So, a new encoding method (geometric coding) has been proposed. To code individuals, we suggest using distance $Dist(\overline{q}_i, \overline{q}_0)$ between vector $\overline{q}_i$ and initial vector $\overline{q}_0$. In the case of a cosine measure we have:

$$Dist(\overline{q}_i, \overline{q}_0) = \frac{\overline{q}_i * \overline{q}_0}{\parallel \overline{q}_i \parallel \cdot \parallel \overline{q}_0 \parallel} \tag{8}$$

Encoding method applicability criterion based on the uniform continuity of the fitness function has been suggested too. Fitness function $\overline{w}(q_j)$ is called uniformly continuous on the set Q, if $\forall \epsilon > 0 \ \exists \lambda > 0$, such that $\forall q', q'' \in Q$ satisfying the condition $|q'' - q'| < \lambda$, the inequality $|\overline{w}(q'') - \overline{w}(q')| < \epsilon$ is valid. It means that small changes of individual code q_j lead to small changes of fitness function $\overline{w}(q_j)$. Also, it means that the value of λ limiting the deviation of individual code q_j only depends on the value ϵ of the deviation of fitness function $\overline{w}(q_j)$ and does not depend on the value of individual code q_j, i.e. it is constant on the whole domain of the function.

7 Method for Objects Innovativeness Calculation

In the framework of the project, a version of a method for raw indicators measurements processing has been processed to calculate the probabilistic values of the objects innovativeness indicators. The main steps of the method are as follows:

1. Execute the specified number of quasi-optimal queries generated by the GA from the search pattern. From the viewpoint of the Theory of Evidence, such queries are observed subsets or focal elements. For all retrieved documents R the number of group intervals is determined as $I = S^{1/2}$. In terms of measuring Nov the mentioned intervals correspond to the nominal scale "It is novel","It is evidently novel","It is evidently not novel","It is not novel".
2. Compute the basic probability $m(A_k)$ of the appearance of innovativeness indicators according to (3) for each of the subsets observed. Note, $m(A_k)$ can be estimated as follows:

$$m(A_k) = q_k/S, \quad \sum q_k = S \tag{9}$$

   where q_k is a number of observed subsets (queries).
3. Compute the belief Bel and plausibility Pl for each A_k according to (4) and (5).
4. Processing the measurement results retrieved from different search engines. The combined base probability m_{12} for two search engines:

$$m_{12}(A_k) = \frac{1}{1-K} \sum_{A_i^{(1)} \cap A_j^{(2)} = A} m_1(A_i^{(1)}) m_2(A_j^{(2)}) \tag{10}$$

$$K = \sum_{A_i^{(1)} \cap A_j^{(2)} =} m_1(A_i^{(1)}) m_2(A_j^{(2)}) \tag{11}$$

   where K – the conflict factor.

5. Evaluating source credibility. It can be considered with the introduction of discount factor α for base probability $m(A)$. Discounted base probabilities are estimated as follows:

$$m^{\alpha}(A) = (1 - \alpha)m(A) \tag{12}$$

An own algorithm for the group processing of the objects innovativeness level measurements has been developed. The combining is executed recursively, from couples of sources: two evidence sources form a single conditional one, the evidences of which are combined with the next actual source.

8 Data Warehouse Support System Functioning

Figure 2 shows the chart of UML sequences that describes the general functioning of the Data Storage Support System. The figures correspond to the system components described in Sect. 3 above. It shows the sequence of communications among the interacting objects (components and actors). Note two important activity periods: data presentation (component 3) and search agents functioning (component 5) including obtaining variants of an innovation solution and its associated objects. The intermediate steps reflect the algorithmic aspects of the interactions among the system components.

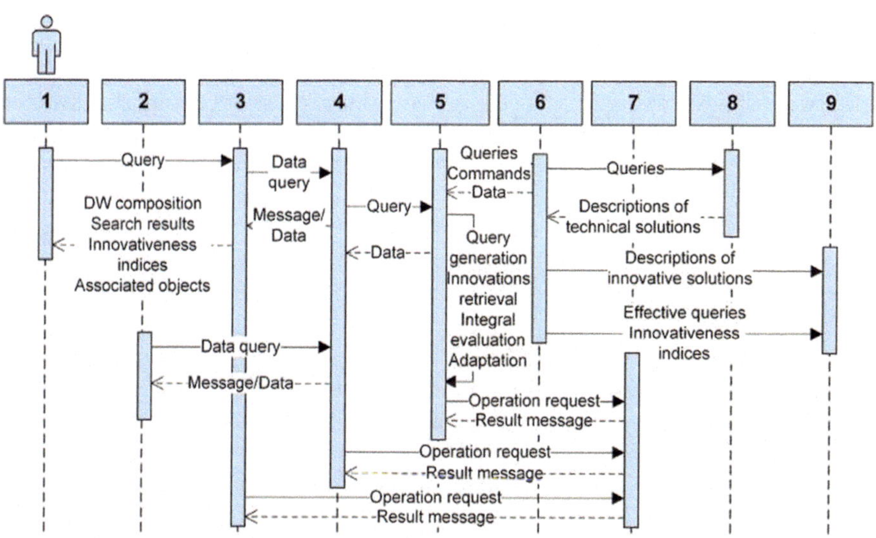

Fig. 2. Data warehouse support system components functioning

9 Experimental Investigations

To experimentally check if the computational model developed is adequate, we stated the following tasks:

1. Approve the computation procedure for objects innovativeness indicators.
2. Compare the computed values of the innovativeness indicators to the expert-estimated ones.
3. Compare the computed values of the innovativeness indicators obtained after data processing from different search engines.
4. Evaluate the dynamics of the object innovativeness indicators in time.
5. Validate the feasibility of the measured innovativeness indicators for further processing.

The following search engines were selected as objects information sources: http://new.fips.ru, https://elibrary.ru, https://rosrid.ru, https://yandex.ru, https://wordstat.yandex.ru, https://google.com, https://adwords.google.com, https://patents.google.com, https://scholar.google.ru.

The objects to analyze were ten top inventions made in 2017 and selected by Rospatent (Russian patent authority) experts, and ten random inventions made in 2017. The search patterns were prepared by experts. The document bodies to analyze were formed.

Our experiments proved the validity of the methods of the Theory of Evidence for processing of the measured innovativeness indicators. Despite anticipated differences in the absolute values of the measurements from different data sources, our model can adequately evaluate relative changes in the values of the object novelty and relevancy indicators (combined values of the indicators show similar results). The general conclusion is: the average novelties of the objects estimated as the best objects by the experts are greater than the average novelties of the random objects.

Our experiments involved an analysis of the objects novelty evaluated for twenty years. Figure 3 give as examples Optic Nerve Electric Stimulator object archetype novelty and relevancy plots (it is specified by the corresponding linguistic model).

Approximation of the values obtained (solid trend lines) validates the hypothesis that the object novelty lowers in time. However, the object relevancy increase in time. It is clear that the object becomes more and more popular among the users, so it enjoys a growing potential interest to it.

Our experiments validated the well-known cyclic regularities (revealed in the analysis of correlation between the innovations and economic growth). In a quite long interval, the values of the innovativeness indicators show cyclic changes (dotted trend lines on the plots). Though the computation results as a whole are ambiguous, we do identify the cycles which require testing the hypothesis of the innovation cycles in the particular usage domain.

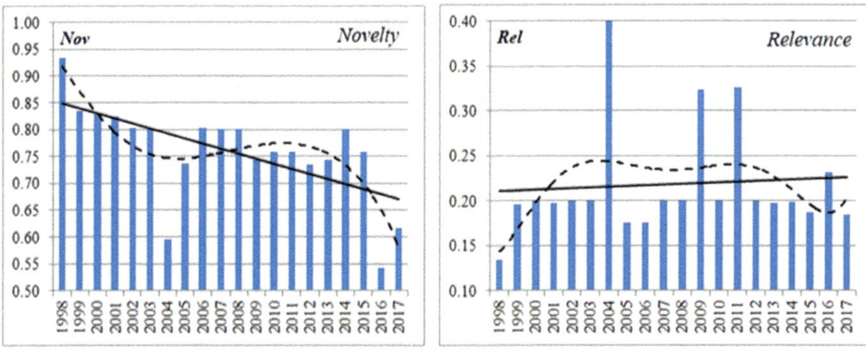

Fig. 3. Object innovativeness indicators behavior (example)

10 Conclusion

The works on the project directions discussed herein are finished. Further we are planning to carry out investigations on the following:

- Formalize the description of the linguistic model of the object archetype, including aiming the search pattern at the innovativeness of the sought-for objects and specifying limitations on the reference information model.
- Develop a behavior model for the intelligent search agent working with a data source in a multiagent system with the heterogeneous data warehouses.

Acknowledgements. This work was done at the Tver State Technical University with supporting of the Russian Foundation of Basic Research (projects No. 18-07-00358 and No. 17-07-01339) and at the Joint Supercomputer Center of the Russian Academy of Sciences – Branch of NIISI RAS within the framework of the State assignment (research topic 065-2019-0014).

References

1. Ivanov, V.K.: Computational model to quantify object innovativeness. In: CEUR Workshop Proceedings, vol. 2258, pp. 249-258 (2019). http://ceur-ws.org/Vol-2258/paper31.pdf
2. Palyukh, B.V., Vinogradov, G.P., Egereva, I.A.: Managing the evolution of a chemical engineering system. Theor. Found. Chem. Eng. **48**(3), 325–331 (2014)
3. Palyukh, B.V., Vetrov, A.N., Egereva, I.A.: Architecture of an intelligent optimal control system for multi-stage processes evolution in a fuzzy dynamic environment. Softw. Syst. **4**, 619–624 (2017)
4. Microsoft Application Architecture Guide, p. 529, 2nd edn., October 2009. www.microsoft.com/architectureguide)
5. Tucker, R.B.: Driving Growth Through Innovation: How Leading Firms Are Transforming Their Futures, 2nd edn. Berrett-Koehler Publishers, San Francisco (2008)

6. Oslo Manual: Guidelines for Collecting and Interpreting Innovation Data, 3rd edn. The Measurement of Scientific and Technological Activities (2005). https://www.oecd-ilibrary.org/science-and-technology/oslo-manual/9789264013100-en
7. Shafer, G.: A Mathematical Theory of Evidence. Princeton University Press, Princeton (1976)
8. Yager, R., Liu, L.: Classic Works of the Dempster-Shafer Theory of Belief Functions. Springer, London (2010)
9. Ivanov, V.K., Vinigradova, N.V., Palyukh, B.V., Sotnikov, A.N.: Modern directions of development and application areas of Dempster-Schafer theory (review). Artif. Intell. Decis. Making **4**, 2–42 (2018)
10. Palyukh, B., Ivanov, V., Sotnikov, A.: Evidence theory for complex engineering system analyses. In: 3rd International Scientific Conference on Intelligent Information Technologies for Industry, IITI 2018. Advances in Intelligent Systems and Computing, vol. 874, pp. 70–79 (2019)
11. Ivanov, V.K., Palyukh, B.V., Sotnikov, A.N.: Efficiency of genetic algorithm for subject search queries. Lobachevskii J. Math. **37**(3), 244–254 (2016). https://doi.org/10.1134/S1995080216030124

Multi-Agent System of Knowledge Representation and Processing

Evgeniy I. Zaytsev(✉), Rustam F. Khalabiya, Irina V. Stepanova, and Lyudmila V. Bunina

MIREA - Russian Technological University, Moscow, Russia
zajcev@mirea.ru

Abstract. The article reviews the functional and structural organization of the Multi-Agent System of Knowledge Representation and Processing (MASKRP). The architecture of the MASKRP, the models of reactive and cognitive software agents are described. The examples of the interaction, states and transitions diagrams of the software agents, the systems of rules and the queries of the problem-oriented multi-agent solver are given. A classical logic solver and the multi-agent solver, which is developed on the basis of the software agents models described in this article, are compared. The results of fuzzy queries to the knowledge base, which were realized according to the sets of fuzzy rules and membership functions specified in the example, are shown. The results demonstrate the practical viability and efficiency of the presented approach to implementation of a multi-agent system based on fuzzy knowledge. At the heart of design decisions at creation of the MASKRP the requirement of support of users of different competence level is put. To support the process of knowledge acquisition and user's comfort setting up mechanisms of automated reasoning developed the problem-oriented toolkit of visual design.

Keywords: Multi-agent systems · Knowledge-base systems · Intelligent software agents · Distributed systems · Fuzzy systems

1 Introduction

Agent-oriented technology provides developers of artificial intelligence systems with convenient high-level abstractions (Cognitive Data Structures, CDS), such as goals, desires, intentions, knowledge, and beliefs of an agents, which are used in the design and implementation of a new class of distributed intelligent systems - Multi-Agent Systems of Knowledge Representation and Processing (MASKRP).

The goals of MASKRP software agents are represented by a certain kit of intermediate and final states, the achievement of which realizes the intentions of the agents. Knowledge of software agents that change over time is considered as agent beliefs. That beliefs became knowledge (true predicates) their confirmation by the new facts or explanations is necessary. The desires of software agents are considered states and situations to which they aspire at a certain stage of the solution of a task. To achieve the target state, software agents go through a chain of desired states [17, 19, 20].

S. Kovalev et al. (Eds.): IITI 2019, AISC 1156, pp. 131–141, 2020.
https://doi.org/10.1007/978-3-030-50097-9_14

Cognitive data structures, the features of which are determined by the rules of logical inference, methods of machine self-learning and adaptation, allow software agents to act rationally. Even if the actions of a separate software agent are quite simple, thanks to self-organization, the MASKRP's complex behavior is realized, which makes it possible to find a solution to a partly formalizable problem, in which qualitative and quantitative dependencies and parameters can be combined. Self-organization is possible due to the ability of software agents to formulate their own goals and implement individual plans [2, 10, 14].

In MASKRP self-organization is combined with an organization that, at a certain hierarchical level, enter a single plan and common goals for the agent community, expressed primarily in the roles played by agents. The distribution of roles in the design of MASKRP is carried out in accordance with the stages of solving the problem. The task of identifying these stages and abstractions associated with these stages is specific to the problem area and depends on the level of detail of the process of knowledge processing. The organization's characteristic use of a common plan for all agents, in which each agent has a certain role, allows to take into account all the circumstances of achieving the goal. However, because of impact on agents of random factors, self-organization is the main form in the MASKRP.

Classical knowledge processing systems use a single intelligent solver that implements logical inference based on a complete and consistent knowledge base. The intelligent solver of this type is designed for SMP (Shared Memory Processors) systems, in which a automated reasoning is possible on the basis of the Resolution Principle. To speed up the generation of an empty resolvent, a Linear resolution with Selection function for Definition clauses (SLD-Resolution) is usually used. Effective parallel algorithms implementing the Resolution Principle have been developed for SMP-computers. However, the implementation of such algorithms in NORMA (No Remote Memory Access) systems requires the use of additional mechanisms such as DSM (Distributed Shared Memory) or RPC (Remote Procedure Call). These mechanisms are provided by the Operation System and middleware.

Traditional OS hide information about hardware resources behind universal high-level abstractions, making it impossible for a MASKRP developer to implement problem-dependent optimization, since new abstractions can be described only by inconvenient emulation over existing ones. Traditional OS architectures have not yet exhausted their capabilities, and the rapid increase in their resource consumption is successfully compensated by the cheapening and development of hardware. However, experts think about the most effective use of available resources [1, 3, 4, 6, 9, 11, 16].

An effective MASKRP implementation is possible on the basis of an Library Operating System (LibOS) and event-driven software components that can directly interact with the low-level access interface to the computational resources. Specialized software module for Virtual Memory Management (VMM) and Inter-Process Communication mechanism (IPC) defined in the LibOS are significantly faster than primitives that perform the same work in the OS with monolithic or microkernel architecture. Agent-oriented technology integrated with specialized system modules for the planning and management of computational resources is becoming the primary to create high-performance knowledge processing systems.

2 Architecture and Structural Organization of MSKRP

In classical logical systems of knowledge representation and processing the solution of the problem is considered as a search (logical inference) on the reduction graph. In the MASKRP, the problem is solved by several loosely coupled solvers (program agents) that are able to take into account the so-called "NON-knowledge factors" (inaccuracy, incompleteness, uncertainty, fuzziness of knowledge) [18]. After the initial problem is decomposed into separate subtasks, these tasks are distributed among the software agents and roles are assigned to each of the agents. Roles are defined by the interaction protocols of software agents. By setting up protocols and establishing connections, the cognitive agent, which coordinates the work of reactive agents, can choose different strategies. For example, it may assume that the knowledge source associated with the first agent has the highest priority. The priorities of other knowledge sources can be set, for example, according to the principle of decreasing the priorities with the increase of sequence numbers of the software agents. This approach is used in the problem-oriented multi-agent solver, the structure of which is presented in Fig. 1, where the first software agent is associated with the most priority knowledge base tables: TableU_1 and TableSovU_1.

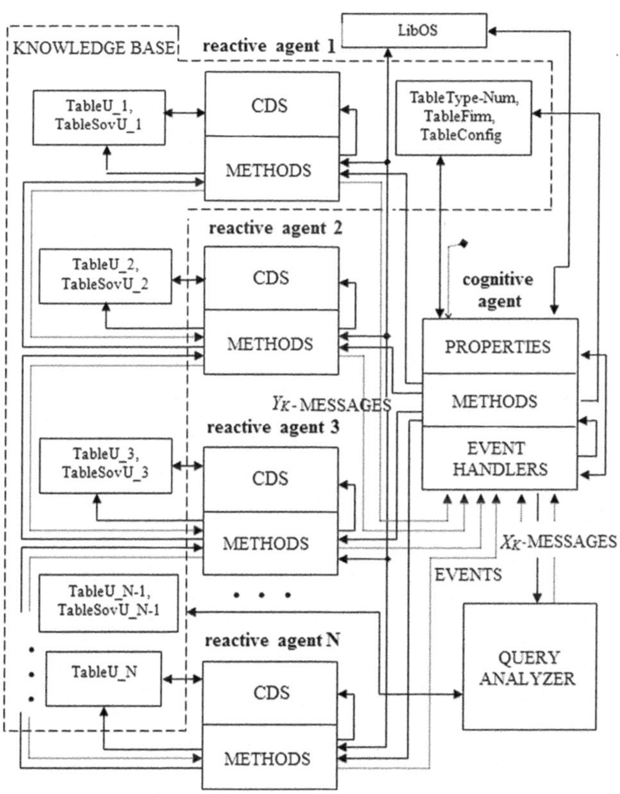

Fig. 1. Structure of the two-level specialized multi-agent solver.

The presented multi-agent solver has a two-level architecture. At the upper level, the cognitive software agent, using knowledge about the situations and states of the problem, coordinates the actions of reactive software agents, which are knowledge sources of the lower level. In Fig. 2 is presented the States and Transitions diagram of the cognitive software agent. From the diagram it is clear that this agent has five basic states: Initialization, Selection, Coordination, Solution and No Solution.

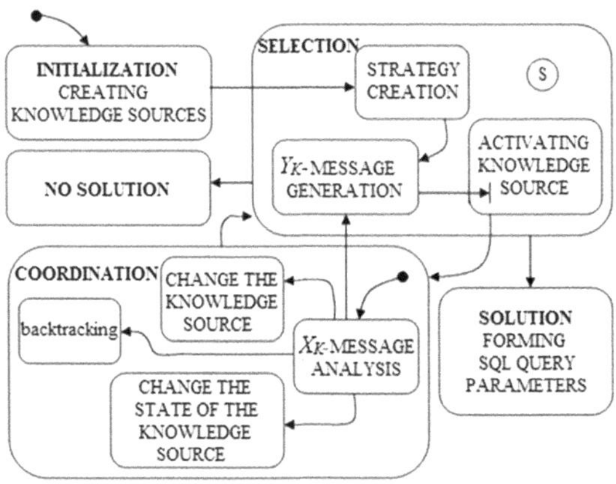

Fig. 2. State and transition diagram.

After initialization and establishment of connections between reactive agents, the transition to the "Selection" state occurs, in which the cognitive software agent selects the necessary knowledge source, taking into account the informative signals from the reactive agents. The cognitive agent then goes into a "Coordination" state in which it coordinates the actions of the lower-level software agents. If at the next step, the reactive agents do not find a coordinated solution, multi-agent solver backtracking to the previous partial solution. The interaction diagram of the first level agents, each of which is associated with only one neighbor, is presented in Fig. 3.

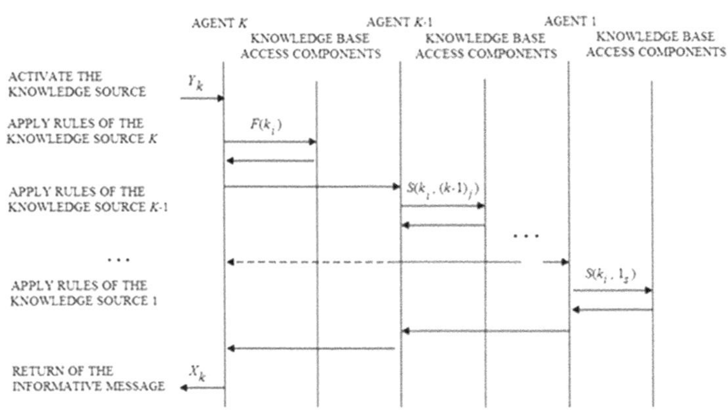

Fig. 3. Interaction diagram of the reactive agents.

In the process of a tasks solution, software agents form fuzzy queries to the knowledge base, using the values of linguistic variables. Examples of the membership functions used in the MASKRP and the result of a made on their basis fuzzy query (with use of two linguistic variables) to the knowledge base table, which includes five records, are presented in Fig. 4.

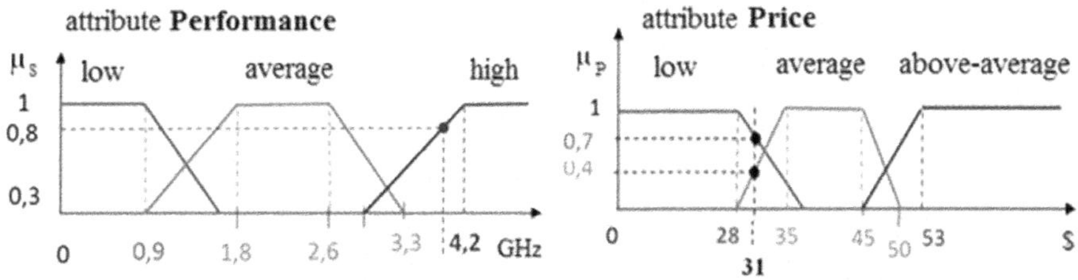

ID	Denomination	Price	Performance
1	N1	47	3,2
2	N2	25	1,5
3	N3	26	3,9
4	N4	31	4,8
5	N5	23	2,5

SELECT * from TableU_1 where (Price = " low" AND Performance = "high")

ID	Denomination	Price	Performance	FP
3	N3	26	3,9	0,8
4	N4	31	4,8	0,7

Fig. 4. Examples of the membership functions and the result of a fuzzy query.

Fuzzy queries allow to find records in the knowledge base that satisfy the query to a certain extent. For realization of fuzzy queries in MASKRP, various types of membership functions $\mu_S(x)$ are used, which determine the degree of membership of the element x of the data domain to the fuzzy set S, which is the term of the linguistic attribute of the software agent [5, 7, 8]. In order to select the membership functions that are most suitable for a task solution, the knowledge expert required to make a series of computational experiments.

3 The Models of Software Agents

In MASKRP two types of program agents are used: cognitive and reactive. A cognitive software agent is used to establish connections between reactive software agents, develop a common strategy for finding a solution, activate the necessary sources of knowledge, coordinate the work of reactive agents, taking into account incoming X_K–messages [19]. A mathematical model of a cognitive software agent is formally defined as follows:

$$A_K = (Q_K, X_K, Y_K, PS, PL),$$

where Q_K – is a set of states of a cognitive agent, corresponding to a set of situations that depend on the desires, intentions and beliefs of software agents, as well as on events occurring in the system; X_K – is a set of input (informative) messages; Y_K – is a set of output (control, information) messages; PS – is a production system that determines agent transitions from one state to another and generates output messages; $PL(W, q_0, GS)$ – is a planning system in the subtask space for the problem domain W with the initial state q_0 and the dynamic set of goals GS.

The reactive software agent model can be defined as follows:

$$A_R = (X_R,\ Y_R,\ MS(X_R, Q_R), Q_R(RL, Atr(V))),$$

where X_R – is a set of input messages; Y_R – is a set of output messages; MS – is a set of methods that determine the reaction of the agent to the input messages X_R; Q_R – is a set of states determined by the values of the selected kit of software agent attributes:

$$intensional\, RL_i = \{\ldots [Atr_j, D(Atr_j)], \ldots\};$$

$$extensional\, RL_i = \{B_1, \ldots B_p\};$$

$$B_k = \{Atr_1(V_1), \ldots Atr_s(V_n)\},$$

where RL – is a set of relations stored in the knowledge base of a software agent; $V = \{V_1, \ldots V_n\}$ – is a set of values of the attribute kit Atr_j.

Concrete values for the relationship attributes are selected from the domain D. The information that characterizes the semantics of the problem domain is contained in the *intensional* part of the knowledge base. In the *extensional* part of the knowledge base, the relationships and states of the software agents are stored. As noted above, reactive agents act as sources of knowledge, forming and implementing in the course of work fuzzy queries to a distributed dynamic knowledge base.

The processing of knowledge about the conceptual components of the MASKRP is carried out at receipt of requests for which it is necessary to calculate relations on the set of entities, or to specify the characteristics of the entities. Requests are processed using special methods: comparisons (*M_CMP*), associations (*M_ASS*) and analysis (*M_ANS*), which calculate various relations on the set of events and their subjects, and also the specification method (*M_VAL*).

The specification method (*M_VAL*) uses fuzzy queries to the knowledge base, which allow to operate quality characteristics and fuzzy conditions. The linguistic variables used in fuzzy queries are defined by a set of properties, including the variable name, terms (qualitative characteristics that define the variable definition domain), the associated area of quantitative values, syntactic and semantic rules. The *M_VAL* method implements fuzzing actual data, activating rules for which the degree of truth of the precondition is more than zero, aggregating and accumulating input variables and conditional parts of the rules, forming a ranked list of effective records that are output as a result of the query if the threshold value of the membership function is exceeded.

The *M_CMP*-method is called when comparing events ($Event_i$ и $Event_j$) or objects ($Objekt_s$ и $Objekt_k$); The *M_AS*-method is used to receive responses to requests for relationships between objects and events; *M_ANS*-method realizes the logical analysis of events.

For processing knowledge of the problematic components of the ISAS, special cognitive data structures (frame-scripts) are used that describe the plan for solving the target task, define the rules for reducing the task to the subtasks, and establish links between the software agents responsible for solving these subtasks.

When processing requests that require the synthesis of action plans, the following rules are applied:

$$TrgtTsk \rightarrow Sbtsk_1, Sbtsk_2, \ldots, Sbtsk_m,$$

where *TrgtTsk* – is a target task; the symbol $'' \rightarrow ''$ denotes the reduction; the symbol "," means the logical "AND"; $Sbtsk_1$, $Sbtsk_2$,... $Sbtsk_m$ – is a set of subtasks, to which the solution of *TrgtTsk* task is reduced.

For the synthesis of an action plan, special planning functions are used (*M_SLV*-method), which decompose subtasks $Sbtsk_1$, $Sbtsk_2$,..., $Sbtsk_m$ to the level of elementary (*SmplTsk*) tasks and represent the synthesized plan as an ordered composition of subtasks that are distributed between program agents selectable by specific criteria. For selected software agents, interaction protocols are defined that describe the roles of agents in solving a common task, as well as the rules for synthesizing the overall result from particular solutions.

So, the agent software agent $Agent_i$ can independently solve the subtask $Sbtsk_i$ or perform its decomposition by requesting assistance from other software agents. The software agent that initiated the task distribution should then collect the results of the work of other software agents and synthesize the overall result. The initiating agent is usually located at a higher level of management and can coordinate the actions of a group of agents established to solve a specific task.

4 The Examples of the System of Rules

Let the solution of tasks $Sbtsk_1$, $Sbtsk_2$, $Sbtsk_3$ be carried out by software agents $Agent_1$, $Agent_2$, $Agent_3$. The simplest scenario may require agents to sequentially solve particular tasks and then transmit the results to their nearest neighbors. After solving the $Sbtsk_i$ task, the agent $Agent_i$ sends the X_j message (results obtained at this stage) to the agent $Agent_j$ and proceeds to search for another possible solution to the $Sbtsk_i$ task. In distributed systems, the interaction of agents can be organized on the basis of synchronous or asynchronous messages. In the former, the agent that sent the message is blocked until it receives a confirmation message that the sent message has been received. This explicit synchronization is suitable not for all types of interaction. In an asynchronous transfer, control is returned to the sending agent immediately after the operating system determines the operative memory buffer in which to place the message. Thanks to buffering, the agent sender can do the work in parallel with sending the message.

Let tasks $Sbtsk_1$, $Sbtsk_2$, $Sbtsk_3$ be described by the following rule system:

$$Sbtsk_1 \rightarrow Sbtsk_{11}, Sbtsk_{12}, Sbtsk_{13}$$

$$Sbtsk_2 \rightarrow Sbtsk_{21}, Sbtsk_{12}, Sbtsk_{13}$$

$$Sbtsk_3 \rightarrow Sbtsk_{21}, Sbtsk_{12}, Sbtsk_{33}$$

$$Sbtsk_{11} \rightarrow Event_1, Event_2$$

$$Sbtsk_{12} \rightarrow Event_3$$

$$Sbtsk_{13} \rightarrow Event_4, Event_5$$

$$Sbtsk_{21} \rightarrow Sbtsk_{11} \uparrow \{Event_2 \sim > Event_2, Event_6\}$$

$$Sbtsk_{33} \rightarrow Sbtsk_{13} \uparrow \{Event_4 \sim > Event_4, Event_7\}$$

where the "↑" symbol indicates the need to transform plans for solving $Sbtsk_{11}$ and $Sbtsk_{13}$ tasks using transformation rules that are defined in curly braces. $Sbtsk_{lk}$ tasks can also be distributed among software agents.

The synthesis of action plan for solving the task $Sbtsk_1$ following:

$$Sbtsk_1 \rightarrow (Sbtsk_{11}, Sbtsk_{12}, Sbtsk_{13}) \rightarrow (Event_1, Event_2, Sbtsk_{12}, Sbtsk_{13})$$

$$\rightarrow (Event_1, Event_2, Event_3, Sbtsk_{13}) \rightarrow (Event_1, Event_2 Event_3 Event_4, Event_5).$$

The synthesis of action plan for solving the task $Sbtsk_2$ following:

$$Sbtsk_2 \rightarrow (Sbtsk_{21}, Sbtsk_{12}, Sbtsk_{13}) \rightarrow (Sbtsk_{11} \uparrow, Sbtsk_{12}, Sbtsk_{13})$$

$$\rightarrow (Sbtsk_{11} \uparrow, Event_3, Sbtsk_{13}) \rightarrow (Sbtsk_{11} \uparrow, Event_3, Event_4, Event_5)$$

$$\rightarrow (Event_1, Event_2, Event_6, Event_3, Event_4, Event_5).$$

The chain of conclusion, performed in order to synthesize the action plan for solving the task $Sbtsk_3$, will look like this:

$$Sbtsk_3 \rightarrow (Sbtsk_{21}, Sbtsk_{12}, Sbtsk_{33}) \rightarrow (Sbtsk_{11} \uparrow, Sbtsk_{12}, Sbtsk_{33})$$

$$\rightarrow (Sbtsk_{11} \uparrow, Event_3, Sbtsk_{33}) \rightarrow (Sbtsk_{11} \uparrow, Event_3, Sbtsk_{13} \uparrow)$$

$$\rightarrow (Event_1, Event_2, Event_6, Event_3, Sbtsk_{13} \uparrow)$$

$$\rightarrow (Event_1, Event_2, Event_6, Event_3, Event_4, Event_7, Event_5).$$

Obtained in finite expressions, the elements $Event_s$ are the ordered sets of events included in the synthesized action plans.

Let the $Agent_1$ software agent be responsible for solving the subtasks $Sbtsk_{11}$, $Sbtsk_{12}$, $Sbtsk_{13}$, for which it is required to calculate relations using the *M_CMP*-method. The same method is used by the software agents $Agent_2$ (for solving tasks $Sbtsk_{21}$, $Sbtsk_{22}$) and $Agent_3$ (for solving task $Sbtsk_{31}$). Let it is required to find the relation between two types of entities:

$$\begin{aligned} Sbtsk_{11} &: M\_CMP_1(A, B), \\ Sbtsk_{12} &: M\_CMP_1(A, C), \\ Sbtsk_{13} &: M\_CMP_1(A, D), \\ Sbtsk_{21} &: M\_CMP_2(B, C), \\ Sbtsk_{22} &: M\_CMP_2(B, D), \\ Sbtsk_{31} &: M\_CMP_3(C, D). \end{aligned}$$

Let that the agents $Agent_1$, $Agent_2$, $Agent_3$ implement the processing of knowledge about conceptual components using the specification method:

$$\begin{aligned} Agent_1 &: M\_VAL_1(A), \\ Agent_2 &: M\_VAL_2(B), \\ Agent_3 &: M\_VAL_3(C). \end{aligned}$$

To solve the *TrgtTsk* target task, software agents need to interact according to a frame-script. Let the following links are established in this frame- script:

$$\begin{aligned} Agent_3 &\leftrightarrow Agent_2, \\ Agent_2 &\leftrightarrow Agent_1, \end{aligned}$$

where agent $Agent_3$ interacts with its nearest neighbor – software agent $Agent_2$, and that, in turn, is connected with the software agent $Agent_1$. Let the software agents use the d_s value obtained from the message of the X_q component responsible for interaction with the end-user.

The following example, in which the symbol "|" means a logical "OR", demonstrates the logical rules according to which the task presented as a predicate *SBTSK (A, B, C, D)* is reduced:

$$SBTSK(A, B, C, D) \sim > SBTSK(A, B, C, d_1) | SBTSK(A, B, C, d_2) | \ldots | SBTSK(A, B, C, d_n);$$
$$\begin{aligned} SBTSK(A, B, C, d_1) \sim > & SBTSK_{11}(A, B), SBTSK_{12}(A, C), SBTSK_{13}(A, d_1), SBTSK_{21}(B, C), \\ & SBTSK_{22}(B, d_1), SBTSK_3(c, d_1), M\_VAL_1(A), M\_VAL_2(B), M\_VAL_3(C); \end{aligned}$$

… …

$$\begin{aligned} SBTSK(A, B, C, d_s) \sim > & SBTSK_{11}(A, B), SBTSK_{12}(A, C), SBTSK_{13}(A, d_s), SBTSK_{21}(B, C), \\ & SBTSK_{22}(B, d_s), SBTSK_3(C, d_s), M\_VAL_1(A), M\_VAL_2(B), M\_VAL_3(C); \end{aligned}$$

… …

$$SBTSK(A, B, C, d_n) \sim > SBTSK_{11}(A, B), SBTSK_{12}(A, C), SBTSK_{13}(A, d_n), SBTSK_{21}(B, C), SBTSK_{22}(B, d_n), SBTSK_3(C, d_n), M\_VAL_1(A), M\_VAL_2(B), M\_VAL_3(C);$$

$\ldots\ \ldots$

Software agents can use Semantic model, which can be built into MASKRP in the form of relationships based on services built on SQL-queries [12, 15].

5 Conclusion

The architecture of the MASKRP described in the article uses two types of software agents. The models of reactive and cognitive software agents combine precise formal methods and ways to solve problems and fuzzy methods and models. Structural and functional organization of MASKRP is based on key ideas about systematic reuse of components and model-driven development. To support the development of the MASKRP based on the agent models discussed in the article, the Multi-Agent-KB5 toolkit is used. This toolkit include interactive wizards and property panels that simplify and optimize the design of reactive and cognitive software agents for problem-oriented MASKRP. The Multi-Agent-KB5 toolkit gives knowledge experts the opportunity to focus their efforts on solving domain - specific problems and create applications based on high-level domain-specific abstractions and special system libraries. The use of special agent-oriented system libraries that implement algorithms for planning and management of computing resources, taking into account the specifics of the inter-process communication, increases the performance of MASKRP.

References

1. Aly, S., Badoor, H.: Performance evaluation of a multi-agent system using fuzzy model. In: 1st International Workshop on Deep and Representation Learning (IWDRL), Cairo, pp. 175–189 (2018)
2. Baranauskas, R., Janaviciute, A., Jasinevicius, R., Jukavicius, V.: On multi-agent systems intellectics. Inf. Technol. Control **1**, 112–121 (2015)
3. Batouma, N., Sourrouille, J.: Dynamic adaption of resource aware distributed applications. Int. J. Grid Distrib. Comput. **4**(2), 25–42 (2011)
4. Cardoso, R., Hübner, J., Bordini, R.: Benchmarking Communication in Actor-and Agent-based Languages. Engineering Multi-agent System, pp. 58–77. Springer, Heidelberg (2013)
5. Chen, J., Li, J., Duan, R.: T-S fuzzy model-based adaptive repetitive consensus control for second-order multi-agent systems with imprecise communication topology structure. Neurocomputing **331**, 176–188 (2019)
6. Darweesh, S.; Shehata, H.: Performance evaluation of a multi-agent system using fuzzy model. In: 1st International Workshop on Deep and Representation Learning (IWDRL), pp. 7–12 (2018)
7. Er, M.J., Deng, C., Wang, N.: A novel fuzzy logic control method for multi-agent systems with actuator faults. In: IEEE International Conference on Fuzzy Systems (FUZZ-IEEE), 08–13 July 2018

8. Guarracino, M., Jasinevicius, R., Krusinskiene, R., Petrauskas, V.: Fuzzy hyperinference-based pattern recognition, In: Towards Advanced Data Analysis by Combining Soft Computing and Statistics, pp. 223–239 (2013)
9. Hadzibeganovic, T., Cui, P., Wu, Z.: Nonconformity of cooperators promotes the emergence of pure altruism in tag-based multi-agent networked systems. Knowl.-Based Syst. **171**, 1–24 (2019)
10. Houhamdi, Z., Athamena, B., Abuzaineddin, R., Muhairat, M.: A multi-agent system for course timetable generation. TEM J. **8**, 211–221 (2019)
11. Jurasovic, K., Jezic, G., Kusek, M.: Performance analysis of multi-agent systems. Int. Trans. Syst. Sci. Appl. **4**, 601–608 (2006)
12. Khalabiya, R.F.: Organization and structure of dynamic distributed database. Inf. Technol. **3**, 54–56 (2011)
13. Langbort, C., Gupta, V.: Minimal interconnection topology in distributed control design. SIAM J. Control Optim. **48**(1), 397–413 (2009)
14. Lihtenshtejn, V.E., Konyavskij, V.A., Ross, G.V., Los', V.P.: Mul'tiagentnye sistemy: samoorganizaciya i razvitie. Finansy i statistika, Moscow (2018)
15. Nurmatova, E., Shablovsky A.: The development of cross-platform applications semi-structured data. Herald of MSTU MIREA **3** (2015)
16. Sethia P.: High performance multi-agent System based Simulations. Center for Data Engineering International Institute of Information Technology, India (2011)
17. Tarasov, V.B.: Ot mnogoagentnykh sistem k intellektual'nym organizatsiyam. - M.: Editorial URSS (2002)
18. Narin'yani, A.S.: NE-faktory i inzheneriya znanij: ot naivnoj formalizacii k estestvennoj pragmatike. Nauchnye trudy Nacional'noj konferencii po iskusstvennomu intellektu. T.1. - Tver': AII, pp. 9–18 (1994)
19. Zaytsev, E.I.: Method of date representation and processing in the distributed intelligence information systems. Autom. Modern Technol. **1**, 29–34 (2008)
20. Wooldridge, M.: An Introduction to Multi-Agent Systems. Willey, Chichester (2008)

The Technique of Data Analysis Tasks Distribution in the Fog-Computing Environment

E. V. Melnik[1], V. V. Klimenko[2], A. B. Klimenko[2(✉)], and V. V. Korobkin[2]

[1] Federal Research Centre, Southern Scientific Centre of the Russian Academy of Sciences, 41, Chehova Street, 344006 Rostov-on-Don, Russia
[2] Scientific Research Institute of Multiprocessor Computer Systems of Southern Federal University, 2, Chehova Street, 347928 Taganrog, Russia
anna_klimenko@mail.ru

Abstract. Cognitive assistants are the promising and intensively growing field nowadays. Some of them provide powerful mechanisms of monitoring and control of individual's health, and mental health, in particular. Yet, some aspects of mental health and individual's safety are still uncovered: the problem has not been considered in terms of Social Media effect on the individual's life and self-estimation level. In the current paper the problem of individual's safety in the Internet is proposed to be solved with the cognitive assistant, based on the authorship identification techniques. The search of fake accounts is a time-consuming procedure, so it is highly desirable to decrease the time of data processing. Taking into account the peculiarities of the fog-computing concept, the technique of data analysis tasks distribution is proposed. It is based on the appropriate location of the search agents nearby the data source, where the information stores.

Keywords: Cognitive assistant · Fake account detection · Authorship identification · Multiagent systems · Tasks location

1 Introduction

Cognitive assistants are quite new and intensively growing trend in the IT. The general definitions of the term "cognitive assistant" are [1]:

- it is a software agent that "augments human intelligence";
- it performs tasks and offer services (assists human in decision making and taking actions);
- it complements human by offering capabilities that is beyond the ordinary power and reach of human (intelligence amplification);
- cognitive assistant offers computational capabilities typically based on Natural Language Processing (NLP), Machine Learning (ML), and reasoning chains, on large amount of data, which provides cognition powers that augment and scale human intelligence.

S. Kovalev et al. (Eds.): IITI 2019, AISC 1156, pp. 142–151, 2020.
https://doi.org/10.1007/978-3-030-50097-9_15

Cognitive assistants are used in a wide range of application fields, e.g., healthcare [2, 3], mental health [4, 5], emergency and safety [6, 7], monitoring and control [8] and many others.

The application of the cognitive assistants to the mental health monitoring seems to be successful: Patterns in our speech and writing analyzed by new cognitive systems will provide tell-tale signs of early-stage developmental disorders, mental illness and degenerative neurological diseases that can help doctors and patients better predict, monitor and track these conditions [9].

Besides, there is another side of the mental health monitoring and control: Social Media and the relations in the virtual space affect the individual's real life. Chatting with strangers, sharing the individual information, carelessness, or, in contrary, aggressive behavior, lead to mental health damage, self-estimation decrease and causes the real crimes up to suicide. The latter is topical for children, in particular, as far as they are involved easily to different communities and trends.

The classical, automated parental control facilities are not efficient frequently because of numerous and uncontrolled Internet access points. At the same time, a few Social Media gives the information about the access points and the content published from those access points. Moreover, some communities, which publish the hazardous content can use different cyphers for participants and hardly be discovered by the Social Media security [10].

As was mentioned frequently in studies [11, 12], the existence of fake accounts can be an important indicator of the potentially hazardous activity of the individual in the Internet. The problem of fake account detection is solved by facilities of Social Media sometimes, yet the Internet space is not limited by the Social Media, so the fake account detection is much more comprehensive and complex than the detection within the social network. The application of the cognitive assistant-based parental control to the fake accounts detection brings the possibility of self-learning and self-adaptation, which, finally, enhance the efficiency of the parental control.

In the previous study [13] the authorship identification-based fake account detection was proposed. Actually, the authorship identification is quite promising in the similar field of fake texts authorship identification, so we believe that it is applicable to the fake accounts identification [14–16].

Yet there is problem of a large data analysis in conditions of real time system functioning. Conducting the procedure of a fake account detection, the cognitive assistant service has to process large amounts of data and do it as quick as possible. The search within the Social Media is a time consuming procedure, so, the objective of the current research is to minimize the time of the stylometry-based fake identification by the application of the fog-computing concept.

The following sections of this paper contain:

- the design, architecture and the general algorithmic description of the fake account detection cognitive assistant (FADCA);
- the analysis of cases of data analysis tasks distribution;
- the technique of data analysis tasks distribution;
- conclusion.

2 Functional and Algorithmic Structure of the FADCA

Consider the general functional of the FADCA (see Fig. 1).

Parents, who are interested in their children fake account detection, connect to the FADCA service and form the text feature vector by the loading of text samples.

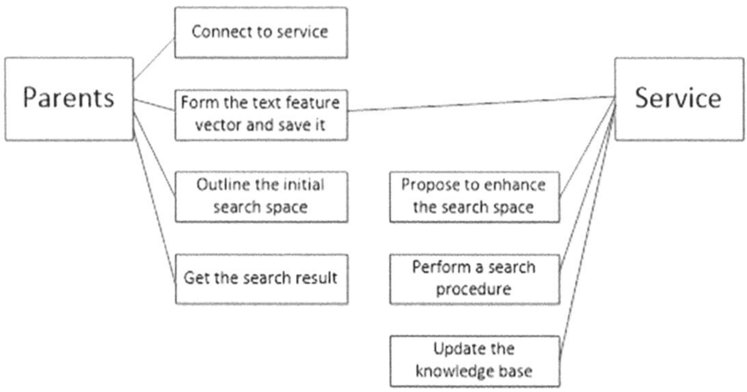

Fig. 1. The general functional scheme of the FADCA.

Then, the initial search space is given as a list of the Internet resources, on which fake accounts can be found. At this stage, the FADCA conducts the analysis in its own kernel database, which is, in fact, the adaptive classifier of any kind. The classification task is to find other Internet resources, on which fake accounts can be according the learning history. For example, consider the initial search space as "resource A, resource B". The input considered generates the outputs as "resource A, resource C"; "resource B, resource D". It means that the most individuals, which have fake accounts on resource A, possible have fakes on resource C, and so on. This is the adaptive mechanism of the FADCA, which enhances the search space with the usage of self-learning. Than enhanced search space is proposed to the users, and the search is conducted. If there are accounts found, which are supposed to be fakes of the individual characterized by the given text samples, the results are return to the users and are used for classifier adaptation. Such adaptive learning enhances the search, reduces the time of search and improves the results of fake accounts detecting.

The generic algorithmic structure of the FADCA is presented below (see Fig. 2).

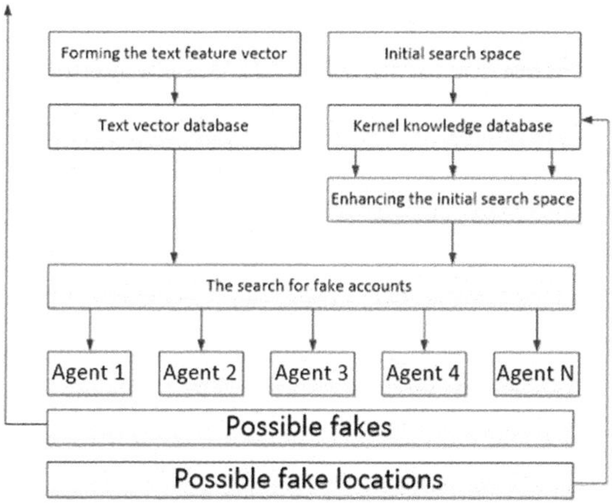

Fig. 2. The generic algorithmic structure of the FADCA.

One can see, that the search procedure is distributed. It is quite natural for the search in the Internet and improves the scalability and the performance of the procedure. Yet there is a question of the quantity of search agents, their location and the maintenance of the agents community integrity in case of failures.

3 The Fog-Computing Concept and the Distributed Data Analysis Procedure

In this section the location of the FADCA search agents is considered in terms of time needed for the processing of information volume *V*. The simple models are developed to estimate tendencies of the FADCA time consumption in various cases of agents distribution through the network and locations of the information processing. So the strategy of the search agents distribution through the network can be chosen according to the simulation results.

Consider the following network structure, as is shown in Fig. 3.

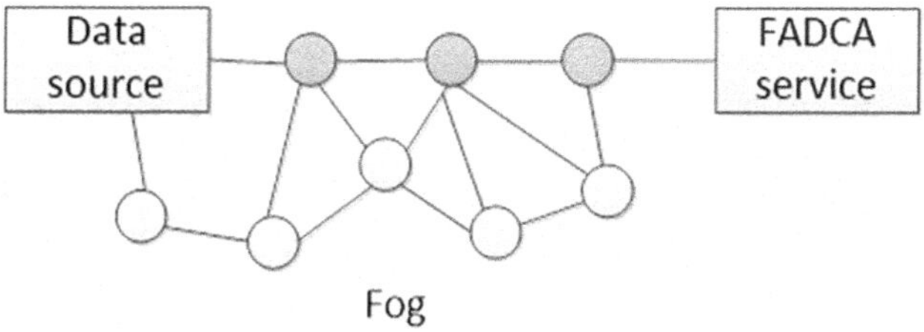

Fig. 3. The scheme of a network structure.

In the Fig. 3, FADCA service is located on the computational node in a datacenter, while the data source, where the information to be processed presents, is located in the other datacenter.

A fog-computing concept, emerged in the 2012, enhances the possibilities of software elements location. We do not consider fog-computing precisely, as the concept is examined and overviewed widely in the literature [17–19]. Yet, in the current paper, we will explore the possibilities of software elements placement on the basis of fog-computing concept.

Consider the following cases of search agents location:

- service agent is located at the same computational node as FADCA service;
- service agent is located at the same computational node as Data Source and preprocess the information;
- service agent is located in the fog-layer of the network near the Data Source node and preprocess the information retrieved.

To estimate roughly the cases in terms of time of information processing, consider the following model parameters:

- n_1 is the distance between FADCA service and the data source in network hops;
- n_2 is the distance between the data source node and the fog nodes on which the search agent is located (in network hops);
- n_3 is the distance between the search agent and the FADCA service computational node;
- γ is the percentage of data processing volume of the overall data volume V;
- v is the speed of information processing/analysis of the agent;
- V_{req} is the volume of request for information block from a search agent to the data source;
- ξ is a ratio between time and number of network hops.

With the simple models the cases of search agents location can be presented in following way.

$$T_{agent\_in\_house} = \xi n_1 V_{req} + \xi n_1 V + \frac{V}{v},$$

where $T_{agent\_in\_house}$ is the time, which is needed for search agent on the FADCA node to process the information with the volume V.

$$T_{agent\_in\_place} = \xi n_1 (1 - \gamma) V + \gamma \frac{V}{v},$$

where $T_{agent\_in\_place}$ is the time needed to process the information, when the search agent is located at the same node with the data source, assuming the possible information preprocessing.

$$T_{agent\_in\_fog} = \xi n_2 V_{req} + \gamma \frac{V}{v} + \xi n_3 (1 - \gamma) V,$$

where $T_{agent\_in\_fog}$ is the time needed to process the information, when the search agent is located on the fog-node with the partial information preprocessing.

In the Fig. 4 one can see that with the fixed distance between the data source and FADCA the best time is got when the search agent is located on the same node with the data source.

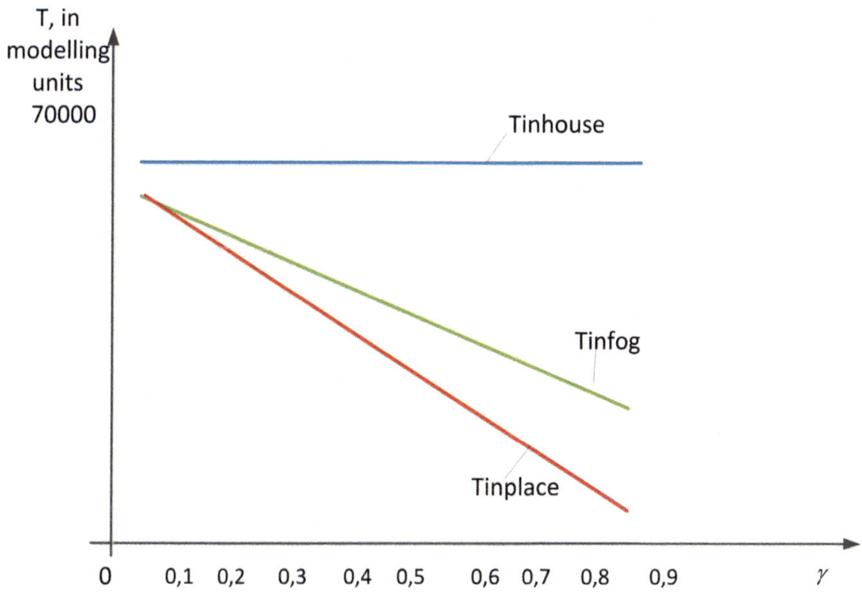

Fig. 4. Simulation results with the variables values: n1 = 10; n2 = 1; n3 = 9; $\gamma = 0..0.9$.

With the increase of distance between data source and the search agent the time continues to decrease, up to the time equal to that of an agent located on the FADCA node (Fig. 5).

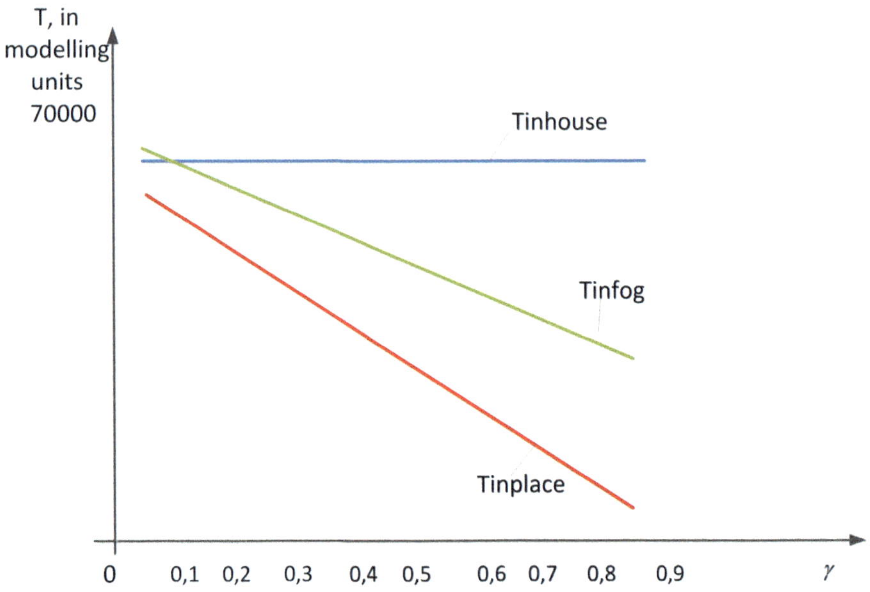

Fig. 5. Simulation results with the variables values: n1 = 10; n2 = 3; n3 = 7; $\gamma = 0..0.9$.

4 The Technique of Data Analysis Tasks Distribution

As is shown in the previous section, it is expedient to locate the search agents as near to the data source as possible, or as near to data source as it is enough to meet the time requirements of the FADCA functioning.

To distribute the search agents around the data source, an LDG-based workload distribution technique can be used. This technique is described in details in [], here we present its brief description. Local device group is a set of computational nodes interconnected by communication channels without transitional nodes, or participating in a solving of common computational task.

1. Leader election is conducted among the nodes, which perform the overall computational task;
2. The leader asks its local group for information about resources to place the additional workload;
3. If the answer is positive, the set of nodes is fixed (and the boundaries of the search space a set). Then the leader models the distribution of the computational subtasks to be relocated through the nodes selected, solving the scheduling optimization problem.
4. If the answer is negative (i.e., there are no nodes in the local group with sufficient resources), then the local group is extended in the following way: each node of the local group retransmits the request of the initial node to its local group. This procedure is iterative and repeats until the nodes with resources are found.

5. If the problem has been solved successfully, and the result meets all problem boundaries, the computational subtasks are binded to the fixed nodes set. If there is no acceptable solution, the procedure repeats from the step 4 with the extention of the local group.

In the conditions of the search agents distribution the issue is to find the initial node to be a leader. Then, this leader begins to examine the nodes of its local group and, if it is necessary, extends the local group. The initial node can be found on the basis of route tables analysis. Such an analysis makes it possible to choose some nodes not far from the data source.

Practically, the contemporary utilities such as MTR or Traceroute present the partial information of the route from node A to node B taking into account the transitory nodes. The example of MTR usage is presented below:

```
1.|-- ***.**.**.*
   2.|-- ???
   3.|-- ???
   4.|-- ???
   5.|-- ???
   6.|-- ???
   7.|-- 100.65.13.177
   8.|-- 52.93.29.43
   9.|-- ???
  10.|-- ash-b1-link.telia.net (213.248.92.170)
  11.|-- ash-bb4-link.telia.net (62.115.143.120)
  12.|-- prs-bb3-link.telia.net (62.115.112.243)
  13.|-- ffm-bb3-link.telia.net (62.115.123.12)
  14.|-- ffm-b4-link.telia.net (62.115.120.0)
  15.|-- vkontakte-ic-338045-ffm-b4.c.telia.net
(213.248.97.63)
  16.|-- srv214-190-240-87.vk.com (87.240.190.214)
  17.|-- ???
  18.|-- 93.186.225.197
```

So, the possibility to find the nodes nearby the source of information is quite real.

After the initial nodes are found, the LDG-based technique continues the procedure of search agents location.

There is one more quite practical way to locate the search agents as near the information source as possible, and nowadays this procedure is rather organizational:

1. As the initial search address is given by the users of FADCA, the IP-addresses can be retrieved.
2. Then, the geographical locations of the retrieved IP-addresses are found.
3. The servers are rented in the same datacenter as the information sources are located.
4. The search agents are distributed through the rented servers and operates from them, minimizing the distance to information sources.

5 Conclusion

In this paper the problem of data analysis tasks location is considered for the fog-computing environments. The issue of big data processing and analysis relates to the problem of the cognitive assistant functioning: the challenge is to search the fake accounts by the authorship identification methods. The search is quite time- and resource- consuming, besides, a predefined number of the Internet resources must be examined with the authorship attribution techniques. To improve the speed of search procedure the following is proposed: firstly, the search must be distributed and conducted by a set of search agents, secondly, they should be distributed in the neighborhood of the information source and preprocess the information. With the simple models we showed that the location of the search agents does affect the overall search time. So, it is possible to improve the performance of the cognitive assistant service by the particular placement of the search agents.

Acknowledgements. The paper has been prepared within the RFBR projects 18-29-22093, 18-29-22046 and RAS presidium fundamental research №7 «New designs in the prospective directions of the energetics, mechanics and robotics» , № gr. project AAAA-A18-118011290099-9.

References

1. https://www.slideshare.net/hrmn/cognitive-assistants-opportunities-and-challenges-slides. Accessed 15 May 2019
2. Costa, A., Heras, S., Palanca, J., Jordán, J., Novais, P., Julián, V.: Argumentation schemes for events suggestion in an e-Health platform. In: de Vries, P., Oinas-Kukkonen, H., Siemons, L., Beerlage-de Jong, N., van Gemert-Pijnen, L. (eds.) Persuasive Technology: Development and Implementation of Personalized Technologies to Change Attitudes and Behaviors. PERSUASIVE 2017. Lecture Notes in Computer Science, vol. 10171, pp. 17–30. Springer, Cham (2017)
3. Rincon, J.A., Costa, A., Novais, P., Julián, V., Carrascosa, C.: A new emotional robot assistant that facilitates human interaction and persuasion. Knowl. Inf. Syst., 1–21 (2018)
4. OCIO Connect (2019). https://ocioconnect2018.sched.com/event/FcTr/watson-cognitive-assistant-for-mental-health. Accessed 15 May 2019
5. Using Artificial Intelligence for Mental Health. https://www.verywellmind.com/using-artificial-intelligence-for-mental-health-4144239. Accessed 15 May 2019
6. Cognitive Assistant Systems for Emergency Response. https://www.nist.gov/ctl/pscr/cognitive-assistant-systems-emergency-response. Accessed 15 May 2019
7. Preum, S.M., Shu, S., Ting, J., Lin, V., Williams, R., Stankovic, J., Alemzadeh, H.: Towards a cognitive assistant system for emergency response. In: Bilof, R. (ed.) 2018 ACM/IEEE 9th International Conference on Cyber-Physical Systems (ICCPS), pp. 347–348. IEEE Press, Piscataway (2018)
8. Tokadlı, G., Dorneich, M.C.: Development of design requirements for a cognitive assistant in space missions beyond low earth orbit. J. Cogn. Eng. Decis. Making **12**(2), 131–152 (2017)
9. With AI, our words will be a window into our mental health. https://www.research.ibm.com/5-in-5/mental-health/. Accessed 15 May 2019

10. Babutskiy, V., Sidorov, I.: A novel approach to the potentially hazardous text identification under theme uncertainty based on intelligent data analysis. In: Silhavy, R., Silhavy, P., Prokopova, Z. (eds.) Computational and Statistical Methods in Intelligent Systems, CoMeSySo 2018. Advances in Intelligent Systems and Computing, vol. 859, pp. 32–38. Springer, Cham (2019)
11. Man Faces Cyber-Bullying Charge in Ex-Girlfriend's Fake Adult-Date Profile. https://5newsonline.com/2013/11/07/man-faces-cyber-bullying-charge-in-ex-girlfriends-fake-adult-date-profile/. Accessed 15 May 2019
12. Facebook has deleted 1.3 BILLION fake accounts. https://www.tweaktown.com/news/63113/facebook-deleted-1-3-billion-fake-accounts/index.html. Accessed 15 May 2019
13. Melnik, E., Korovin, I., Klimenko, A.: A cognitive assistant functional model and architecture for the social media victim behavior prevention. In: Silhavy, R. (ed.) Artificial Intelligence Methods in Intelligent Algorithms, CSOC 2019. Advances in Intelligent Systems and Computing, vol. 985, pp. 51–61. Springer, Cham (2019)
14. Afroz, S., Brennan, M., Greenstadt, R.: Detecting hoaxes, frauds, and deception in writing style online. In: Kellenberger, P. (ed.) 2012 IEEE Symposium on Security and Privacy, pp. 461–475. IEEE Press, Piscataway (2012)
15. Argamon-Engelson, S., Koppel, M., Avneri, G.: Style-based text categorization: what newspaper am I reading? In: Proceedings of the AAAI Workshop on Learning for Text Categorization, pp. 1–4 (1998)
16. Rubin, V., Conroy, N., Chen, Y.: Towards news verification: deception detection methods for news discourse. In: Proceedings of the Hawaii International Conference on System Sciences (HICSS48) Symposium on Rapid Screening Technologies, Deception Detection and Credibility Assessment Symposium, Kauai, Hawaii, USA (2015)
17. Wang, Y., Uehara, T., Sasaki, R.: Fog computing: issues and challenges in security and forensics. In: Ahamed, S.I., Chang, C.K. Chu, W., Crnkovic, I., Hsiung, P.-A., Huang, G., Yang, J. (eds.) 2015 IEEE 39th Annual Computer Software and Applications Conference Proceedings, vol. 3, pp 53–59. IEEE Press, Piscataway (2015)
18. Yi, S., Li, C., Li, Q.: A survey of fog computing: concepts, applications and issues. In: Proceedings of the 2015 Workshop on Mobile Big Data, pp. 37–42. ACM, New York (2015)
19. Bonomi, F., Milito, R., Zhu, J., Addepalli, S.: Fog computing and its role in the internet of things. In: Proceedings of the First Edition of the MCC Workshop on Mobile Cloud Computing, pp. 13–16. ACM, New York (2012)

Non-classical Logic

Model of the Operating Device with a Tunable Structure for the Implementation of the Accelerated Deductive Inference Method

Vasily Yu. Meltsov(✉), Dmitry A. Strabykin, and Alexey S. Kuvaev

Vyatka State University, Kirov, Russia
meltsov69@mail.ru

Abstract. It is proposed to use the first-order predicate logic to represent knowledge in the selected subject area. The accelerated parallel inference method based on the disjuncts division operation is taken as a method for processing knowledge. To analyze the functioning of the abstract inference engine operating device, a model of logical-flow computing is used. Software implementation model of this device allows you to explore the possibilities of improving the performance of inference mechanisms and evaluate the effectiveness of various configurations of the executive part. Based on the analysis of the conducted experiments, formulas and recommendations for users on the choice of the operating device optimal structure are proposed, taking into account the existing features of specific applied tasks.

Keywords: Knowledge processing · Inference method · Inference engine · Operating device · Unification unit · Tunable structure

1 Introduction

In the early eighties in a number of leading countries of the world, the start of the development of computers of the fifth generation was announced. The main point in the transition to a new generation of computers was the transition from a data-oriented architecture to a knowledge-oriented architecture [6, 7, 11, 16].

One of the main components of the developed knowledge processing systems was the inference engine (IE). At present, a sufficiently large number of such machines (solvers) are known, differing in the knowledge models that are embedded in them, types, methods and techniques of implementing logical inference [7–9, 12, 14–17].

2 Analysis of Problem Area

The main problem of the fifth generation systems being created was the problem of choosing a basic programming language and its hardware support either on serial produced processors or on new specialized symbol processing units [7]. In the Japanese project, the most widely known, the Prolog language was chosen as the main development environment [1, 13]. Its implementation on serial processors encountered the

S. Kovalev et al. (Eds.): IITI 2019, AISC 1156, pp. 155–164, 2020.
https://doi.org/10.1007/978-3-030-50097-9_16

serious problem of a "semantic gap" between the high-level programming language constructions and the simplest machine-language instructions.

The specialized Prolog processor designed a bit later, unfortunately, did not show the required speed due to the limitations of parallel processing of information both in the programming language (it is based on a consistent SLD-resolution procedure) and in the processor architecture itself [7].

To improve the speed of programs written in this Prolog programming language, attempts were made to modify it in parallel versions [2]. Despite this, the expected effect was not achieved. In this regard, the relevance of the development of parallel methods of logical inference and specialized processors, which must implement high-speed processing of information stored in the knowledge base, is increasing [7, 14].

3 Task Statement

As the mathematical basis of the projected system of artificial intelligence, it is proposed to take a high-performance accelerated deductive inference method (ADIM) based on the disjuncts division operation [14]. In general, the problem of the derivation of conclusions in the first-order predicate calculus can be formulated as follows. It is necessary to find out whether there is a logical conclusion for a certain set of assumptions that are presented in the form of expressions of the first-order predicate calculus, which is also presented in the form of an expression of the calculus of the first-order predicates. If the problem is solved positively, then from a semantic point of view, this suggests that, with the truth of the assumptions, the conclusion will be true.

Example. For the well-known test logical task [1, 13] many rules are as follows.

Source facts and rules:

1) leads (Peter, John):	$1 \rightarrow P(c, b);$
2) leads (John, Anna):	$1 \rightarrow P(b, a);$
3) leads (Anna, Fred):	$1 \rightarrow P(a, e);$
4) leads (X, Y) → reports (Y, X):	$1 \rightarrow \overline{P}(X, Y) \vee O(Y, X);$
5) leads (Z, U) & reports (S, U) → reports (S, Z):	$1 \rightarrow \overline{P}(Z, U) \vee \overline{O}(S, U) \vee O(S, Z);$

Inferred conclusion:

6) reports (Fred, Peter):	$1 \rightarrow O(e, c)$

Taking into account the formal description of the source data in the form of sequences-disjuncts, it is necessary to develop a logical computations model using the accelerated method, the structure of the abstract machine and the functional organization of the devices for the most effective implementation of this method.

4 Model of Logical-Streaming Computing

ADIM is based on the procedure for the formation of finite remainders and the inference procedure itself [7, 14]. The method is in their multiple use and consists of a number of steps. At each of them, the inference procedure is applied to the original and inferencing clauses, thus forming new output clauses that will be used in the new step. The process ends when, at the next step, a clause is found that cannot be output, or signs of successful completion of the output will be generated for all clauses that were inferenced at that step.

This method is applicable if the inferenced clauses are not tautologies, and their conjunction is a contradiction. The method of parallel-accelerated inference is simple, but extremely effective [7]. For the above example, the inference will be successfully completed in the third step, while even for the parallel version of Prolog (resolution method) at least five steps will be required [1, 13].

In paper [7] a formal system of predicate logic and the basic procedures of accelerated inference were defined. This is the inference procedure itself (V-procedure), the procedure for the formation of finite remainders (N-procedure), the procedure for filling and analyzing the matrix of partial derivatives (M-procedure), the procedure for unifying literals (U-procedure) and a special procedure for completing a logical inference (ANSWER-procedure).

To describe the accelerated method at the operational level, it is most convenient to use a tree view of the model. With this approach, the model of logical-flow computing can be represented as a set of abstract objects (actors) [10] connected by means of messages (Fig. 1).

The elementary actor defines a process which is called for execution by its name. Actors have internal discrete states that change when receiving messages from other actors. It is necessary to associate each unique logic with each procedure of logical-stream computing. The inference process ends when all activated V-tops send positive solutions to the input of the ANSWER top, which gives the answer. If at least one of the triggered V-tops reported failure (F = 1), then the ANSWER top stops the calculations and reports the output process failed. It means that it is impossible to prove the deductibility of the r rule from the set of source assumptions (M_p, M_f), where M_p is the set of source rules, and M_f is the set of facts.

The analysis scheme shows that all V-tops with $<M_p, M_f, r>$ input data can be processed independently of each other (except for parent-child relationships). It does not require any additional clock signals since the activation of the top occurs immediately after receiving data on the input port.

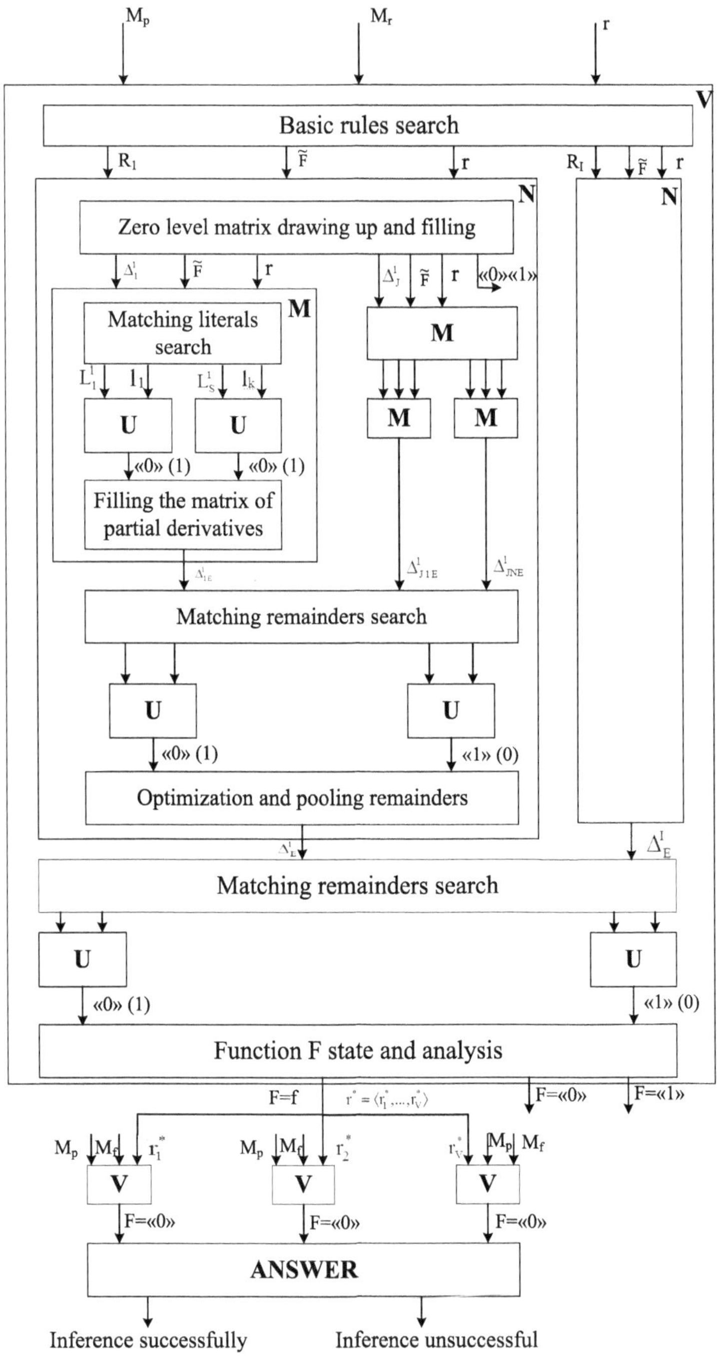

Fig. 1. Diagram of logical computing flows

5 Functional Organization of the Operating Device

To implement logical-stream computing, an abstract inference engine was developed [7]. Its main elements are: Control Device (CD), Operating Device (OD) and main memory module (MM) for storing operands and service information. The formal description of the task (source data) is loaded in the MM associated with the Operating Device. The generalized structure of the inference engine is shown in Fig. 2.

The Control Device fills in the frames of commands created in the dynamic memory of packages. Commands ready for execution are sent to the Operating Device. For interaction with OD, a queue of command packets is introduced into the Control Device, it functions according to the principle FIFO [18].

Operating Device is the most important component of the inference engine, since its speed largely determines the speed of the whole system [4, 7]. The OD processes the process in accordance with the algorithm provided for this type of process. The result of the calculation is stored in memory, and the address of this cell is sent to the Control Unit.

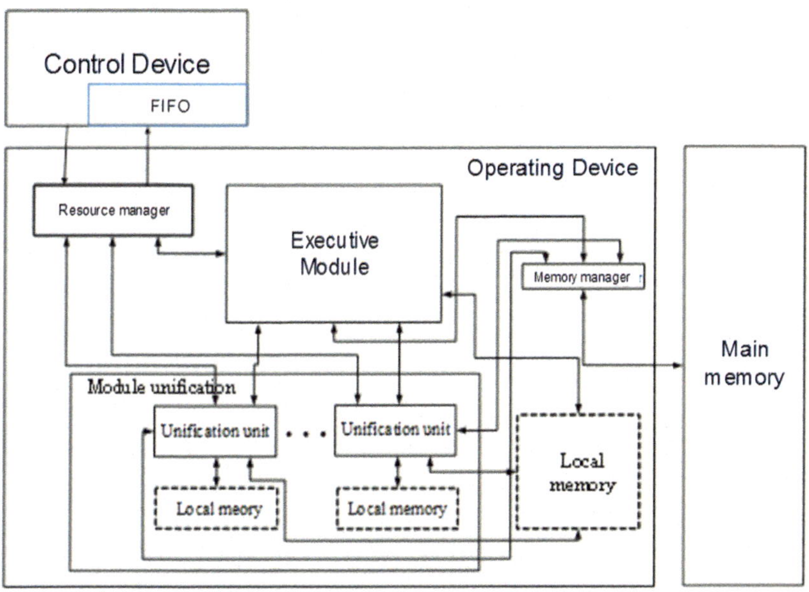

Fig. 2. Generalized structure of the inference engine

The analysis of the functioning of the flow model [7] showed that unification commands (U-processes) will be executed much more frequently on the Operating Device. That is why it is necessary that each OD could have several parallel unification units (UUs) operating in parallel. A resource manager has been introduced to work with the command queue and accounting for the using of various units of the OD. Executive module (EM) performs all ADIM model procedures, with the exception of the unification procedure (U-top).

To improve the performance of the inference engine, it was decided to analyze the possibility of connecting several ODs working in parallel to one Control Device [4].

All calls to the common main memory for the execution modules with a lack of data in local storages are performed through the memory manager. Thus, the developed structure satisfies the requirement of both parallel execution of unification commands and parallel work of several EMs as part of the IE.

The general algorithm of functioning of the Operating Device is as follows. The OD operation begins with initialization, then it checks there are commands ready for execution in the command queue. After the appearance of the command in the queue, it is read, decrypted and set the busy OD flag. Depending on the operation code, it is executed either on EM or on UU. Based on the results of the operation, a special message packet is generated to be sent to the Control Device.

6 Experiments

Taking into account the analysis of architectural and structural solutions of the inference engine and a detailed algorithm for the functioning of the execution module, the computer model OD was written. The efficiency of the model was tested on well-known test problems of inference, as well as on synthetic examples. In all cases, the simulation result coincided with the "manual" step-by-step solution of test problems, which indicates a correct simulation of both ADIM itself and the developed Operational Device.

Some results of experiments without preliminary optimization of the program code are given in Table 1.

Table 1. Experimental results

Experiment	Number of V-processes	Number of N-processes	Number of M-processes	Number of U-processes	1 OD, time (ms)	4 OD, time (ms)	32 OD, time (ms)	128 OD, time (ms)
No. 1	3	45	1236	31321	199	132	110	105
No. 2	3	48	1864	68360	300	199	160	154
No. 3	3	72	4776	129444	1029	718	414	395
No. 4	3	48	1664	43344	254	169	142	133
No. 5	257	4112	20512	606680	29814	28511	26846	7083

The following set of rules was used as the source assumptions in experiment No. 1:

- 18 one-literal facts, in each predicate there are two term arguments (terms);
- 8 rules with two predicates, each predicate has two arguments;
- 8 rules with three predicates, each predicate has two arguments;
- 1 single-literal rule-conclusion, the predicate has two arguments.

In each subsequent experiment, the number of facts, rules, predicates in conclusion and arguments in predicates increased by 50%.

An analysis of the experiments showed that with a small amount of calculations (experiments No. 1 to No. 4) an increase in the number of Operating Device does not give a special advantage, which is explained by the downtime of the added OD. With a large amount of computation (experiment No. 5), increasing the number of OD to 4–32 entails some increase in the speed of the OD. The insignificance of the acceleration is due to the increase in the total overhead of the organization of queues and the waiting processes of free OD. As the number of ODs increases to 128, a significant acceleration of the calculations is observed, since the queue of processes for each OD in this case will be much less.

In addition, experiments have shown that the number of unification units included in the operating device has a significant impact on the performance of the OD and the engine as a whole. As mentioned above, in the OD model, it is possible to vary the number of concurrently operating UUs. To evaluate the effectiveness of various configurations of IE in solving specific problems of inference and the influence of the above factors (including various modifications of the working memory subsystem), a software statistics module was developed. To determine runtime of various operations in the inference engine, it was decided to take the duration of the operation of the unification of two constant terms for the duration of one clock cycle [7]. Such an operation is a matching of two arguments: if they are equal - the unification is successful, otherwise it is not. The statistics module aggregates information during the execution of logical inference from all nodes and blocks of IE: operation time (depending on the duration of the cycle), the number of created processes of all types during operation, the queue length at each device operation step, as well as the level of load of execution units and blocks unifications in them. In addition, you can get information about all the created processes - on which execution unit it was running, the time it takes to receive resources, the time it takes, the time it takes to wait for results from child processes (Fig. 3).

Of course, we cannot talk about a strict unambiguous dependence of the initial parameters of the problem and the structure of the machine. It is extremely difficult to predict how many V-processes will be created for the whole process of logical inference. It is difficult to predetermine the generation of new M-processes after the execution of previous parent processes. In addition, the choice of the number of unification blocks is significantly influenced by the presence of bound variables in the initial rules-rules. However, the information collected by the statistics module allows you to generate valuable recommendations to the user and even design a semi-automatic static configuration module IE. For example, the number of initiated N-processes almost linearly depends on the number of rules recorded in the knowledge base. The number of M-processes correlates with the number of potentially unified pairs of predicates, i.e. predicates with the same name (and the number of arguments) from the base source rule and the derivation rule.

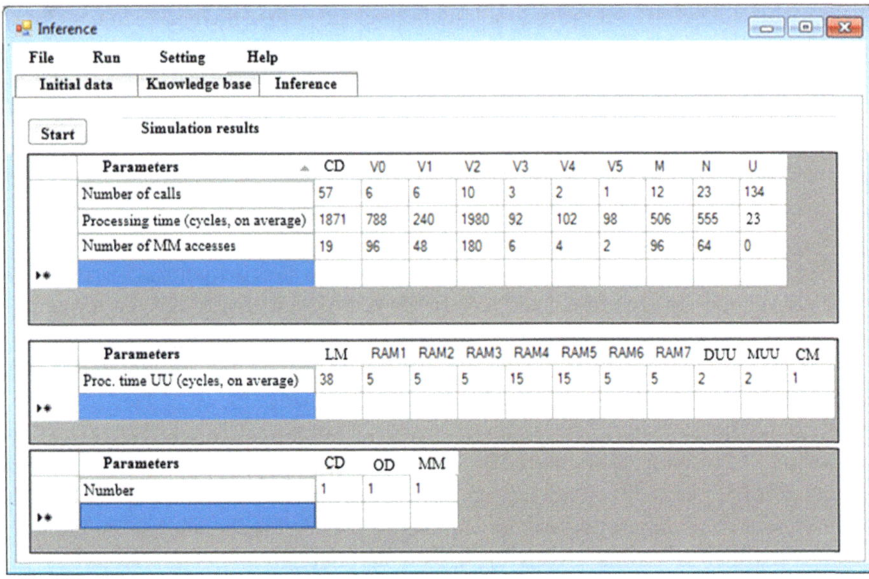

Fig. 3. Results of logical inference

Based on the above, by determining the number of source assumptions and the number of pairs of predicates in the rules, it is possible to make recommendations for the efficient connection of parallel operating units.

$$N_{OD} = \sum\nolimits_{i=0}^{N} C_i + N, \tag{1}$$

here: N_{OD} – Approximate number of parallel working OD, C_i – number of pairs of predicates in the i-th rule, N – number of source rules.

As for the unification blocks, U-processes are executed on them, the number of which at the same time being created will depend on the maximum number of matched pairs of predicates.

$$N_{UU} = max(P_{N_i} * P_F), \tag{2}$$

here: N_{uu} – approximate number of concurrently used CU, P_{Ni} – number of predicates in the i-th rule, P_F – number of fact predicates.

All information collected by the statistics module allows you to identify the bottlenecks in the selected IE structure [5]. After analyzing these values, an experienced user can choose the most optimal configuration of both the execution unit and IE as a whole, depending on the features of the task being solved.

7 Conclusion

Summarizing all the above, let us try to classify the developed model of calculations for the accelerated method of logical inference:

1. The model uses tops of 5 types: V, N, M, U and ANSWER. The number of heirs is not limited. A top receives data, triggers, and returns a result without remembering its internal state.
2. The model is dynamic by deploying a calculation graph from one initial top.
3. The model supports all 4 types of parallelism inherent in the accelerated LI method: V-parallelism (at the system level), N-parallelism (at the rule level), M-parallelism (at the level of partial derivatives), U-parallelism (at the predicate level).
4. The model uses an effective "first in breadth" strategy using heuristic estimates at each processing level.
5. Data processing is based on a principle called "dataflow control", which says: "Operations are activated only when all the operands (data) necessary for their execution are available" (**dataflow processing**). The difference of this model is a purely streaming one-pass mode of organization of logical calculations.
6. Triggering a top that receives all the necessary operands can occur regardless of the state of other tops, i.e. it is possible to simultaneously perform multiple operations (**parallel processing**).
7. The exchange of data between operations is clearly defined, so the dependency relationship between operations is easily detected (**functional processing**).
8. Since operations are controlled by transferring data between them, there is no need to determine the sequence of their execution and, moreover, there is no need for centralized management (**distributed processing**).

Features of this model allow us to estimate the computational and communication requirements for an operating device when solving specific applied tasks of a user [3, 7]. These characteristics are essential for building a specialized inference processor based on the accelerated ADIM method. The tunability of the operating device structure ensures maximum system performance (with existing resource constraints, for example, taking into account the FPGA used) due to its adaptation to computational processes and the composition of tasks [3, 5].

At present, the question of only a static change in the structure of the operational device has been considered, taking into account the preliminary analysis of the main parameters of the problem being solved. Further research should focus on the ability to dynamically reconfigure the structure, and possibly the architecture of the inference engine as a whole, using a special reconfiguration unit. This approach has good perspectives in view of the hardware implementation of the proposed logical inference processor on the FPGA [5].

References

1. Bratko, I.: Prolog Programming for Artificial Intelligence. Addison-Wesley Longman Ltd., Boston (2001)
2. Endriss, U.: An Introduction to Prolog Programming. Lecture Notes. University of Amsterdam, Amsterdam (2018)
3. Korat, U., Alimohammad, A.: A reconfigurable hardware architecture for principal component analysis. Circuits Syst. Signal Process. **38**(5), 2097–2113 (2019)
4. Kuvaev, A., Meltsov, V., Lesnikov, V.: Features of the design operating unit inference engine and its implementation on FPGA. In: Proceedings of the 2019 IEEE Conference of Russian Young Researchers in Electrical and Electronic Engineering, ElConRus 2019, pp. 110–115. Saint Petersburg, Russian Federation (2019)
5. Levin, I., Dordopulo, A., Fedorov, A., Kalyaev, I.: Reconfigurable computer systems: from the first FPGAs towards liquid cooling systems. Supercomput. Front. Innov. **3–1**, 22–40 (2016)
6. Levin, M.: Modular Systems Design and Evaluation. Springer, Cham (2016)
7. Meltsov, V.: High-Performance Systems of Deductive Inference: Monograph. Science Book Publishing House, Yelm (2014)
8. Meltsov, V., Lesnikov, V., Dolzhenkova, M.: Intelligent system of knowledge control with the natural language user interface. In: Proceedings of the 2017 International Conference "Quality Management, Transport and Information Security, Information Technologies", IT and QM and IS 2017, pp. 671–675. St. Petersburg, Russian Federation (2017)
9. Norvig, P., Russell, S.: Artificial Intelligence: A Modern Approach. Pearson Education Limited, Edinburgh (2011)
10. Osipov, G.S., Panov, A.I.: Relationships and operations in a sign-based world model of the actor. Sci. Tech. Inf. Process. **45**(5), 317–330 (2018)
11. Pospelov, D.: Modeling of deeds in artificial intelligence systems. Appl. Artif. Intell. **7**, 15–27 (1993)
12. Rahman, S.A., Haron, H., Nordin, S., Bakar, A.A., Rahmad, F., Amin, Z.M., Seman, M.R.: The decision processes of deductive inference. Adv. Sci. Lett. **23**(1), 532–536 (2017)
13. Sterling, L., Shapiro, E.: The Art of Prolog, 2nd edn. The MIT Press, Cambridge (1994)
14. Strabykin, D.: Logical method for predicting situation development based on abductive inference. J. Comput. Syst. Sci. Int. **52**(5), 759–763 (2013)
15. Strabykin, D., Meltsov, V., Dolzhenkova, M.,Chistyakov, G., Kuvaev, A.: Formal verification and accelerated inference. In: 5th Computer Science Online Conference, CSOC 2016. Advances in Intelligent Systems and Computing, vol. 464, pp. 203–211, Prague, Czech Republic (2016)
16. Vagin, V., Derevyanko, A., Kutepov, V.: Parallel-inference algorithms and research of their efficiency on computer systems. Sci. Tech. Inf. Process. **45**(5), 368–373 (2018)
17. Vagin, V., Antipov, S., Fomina, M., Morosin, O.: Application of intelligent data analysis methods for information security problems. In: 2nd International Conference on Intelligent Information Technologies for Industry, IITI 2017. Advances in Intelligent Systems and Computing, vol. 679, pp. 16–25 (2018)
18. Waibel, P.: Architecture-Driven Design and Configuration of Messaging Systems. Technischen Universität Wien, Vienna (2015)

A Model Checking Based Approach for Verification of Attribute-Based Access Control Policies in Cloud Infrastructures

Igor Kotenko[1,2(✉)], Igor Saenko[1,2], and Dmitry Levshun[1,2]

[1] St. Petersburg Institute for Informatics and Automation of the Russian Academy of Sciences (SPIIRAS), 14-th Liniya, 39, Saint-Petersburg 199178, Russia
{ivkote,ibsaen,levshun}@comsec.spb.ru

[2] St. Petersburg National Research University of Information Technologies, Mechanics and Optics (ITMO University), 49, Kronverkskiy Prospekt, Saint-Petersburg 197101, Russia
http://www.comsec.spb.ru/

Abstract. Attribute-Based Access Control (ABAC) model is a perspective access control model for cloud infrastructures used for automation of industrial, transport and energy systems as they include large number of users, resources and dynamical changed permissions. The paper considers the features of ABAC model and the theoretical background for verification of the ABAC policies based on the model checking. The possibility of applying the model checking is justified on the example of the ABAC policy. Implementation of the proposed approach was made using the UPPAAL verification tool. Experimental assessment shows high acceptability of the model checking not only for finding anomalies in ABAC policies but for finding decisions to eliminate them.

Keywords: Access control · Model checking · Temporal logics · ABAC · Cloud infrastructure

1 Introduction

Access control plays an important role in providing of computer and network security in cloud infrastructures. In clouds users should have different rights on execution of different actions over variable information resources [9,21]. Nowadays, cloud infrastructures are the basis of big corporative information systems and also for cyber-physical systems (smart city, smart house, automated production, robotics, etc.) [17]. Several access control models are considered as traditional ones and developed for solving of access control issues in these systems. Such models are, for example, discretionary access control (DAC), mandatory access control (MAC) and role-based access control (RBAC). However an experience of using traditional access control models shows that in the conditions of high dynamics when required permissions, resources or environment are changing, traditional models become inefficient. Thus, there is a need for using of new, more flexible access control models.

S. Kovalev et al. (Eds.): IITI 2019, AISC 1156, pp. 165–175, 2020.
https://doi.org/10.1007/978-3-030-50097-9_17

One of such sufficiently flexible access control models appeared rather recently is the Attribute-Based Access Control (ABAC) model [6]. This model can successfully replace the traditional access control models [13]. A permission to execute actions over resources (objects) in this model is given on the basis of checking of the correctness of the execution of a set of logical conditions (rules) which are called access control policies. Rules are formed as logical expressions in which values of attributes are used. All attributes can be divided in three groups: attributes of users (subjects), attributes of resources (objects) and attributes of the computer environment. Time belongs to the last group. For this reason the ABAC model is more flexible and capable to react quickly to changes in required permissions.

However, unlike traditional DAC, MAC and RBAC models, the ABAC model in many respects is still at the research level. Development and application of ABAC-based policies are not up to the end investigated. Therefore developers of security systems do not pass to widespread introduction of ABAC in their products yet. One of the main issues is the verification of ABAC policies. The issue is in finding contradictions in access control rules and the way to eliminate them.

The paper investigates a possibility to verify the correctness of ABAC-based policies using model checking. Model checking uses temporal logic and is oriented to the analysis of the set of possible states of the analyzed system. The set of software (UPPAAL, SPIN, DREAM, BLAST, etc.) is developed for its implementation. However, for ABAC policies the verification based on model checking was not investigated yet. The *theoretical contribution of the paper* is defined by it. The *novelty of the received results* consists in the following: (1) the theoretical background for application of the model-checking to ABAC policies verification is considered, (2) the implementation of the model checking for a fragment of ABAC policy is developed, (3) the identification and elimination of contradictions in ABAC policies is shown.

The paper is organized as follows. The review of related work is provided in Sect. 2. Section 3 considers the theoretical background. Implementation issues are discussed in Sect. 4. Section 5 describes experimental results. Section 6 contains main results and directions of the further research.

2 Related Work

In [7] valuable information for developing of ABAC to improve information sharing within organizations is provided. Authors took into account the planning, design, implementation, and operation issues. They explained related standards for ABAC, applications, as well as deployment challenges and some issues of ABAC model verification. However, the issue of automatic verification was not considered.

In [18] the issue of optimization of ABAC model structure is considered. As a solution of this issue it is suggested to use the deep learning methods. However, this work assumes that a priori access control model does not contain anomalies.

Therefore the verification is not taken into account and belongs to the further research.

The analysis of the current issues in the field of ABAC modeling is presented in [20]. In this paper the issue of formal security analysis of the ABAC model was considered. Verification of ABAC rules is a part of this issue. It is emphasized that many investigations, for example [4,11,16], are focused on analyzing the access control policies independently from the formal access control model. It means that while different approaches can be applied to the policies supported by ABAC models, they alone cannot provide a full security analysis of the given ABAC model. It is important to take into account the properties of the underlying model and the way in which policies are combined and enforced. In this regard the issue of verification of the rules of the ABAC model gains rather great value. However, this issue is solved now, generally, by a trusted third party [15].

In [3] the approach for verification of access control policies is presented. This approach is based on pre-designed templates. It simplifies security system design. However, it does not suit for run-time security analysis and verification.

In [8] the graphical constraint expression model to simplify constraints specification and make safety verification practical is presented. Nowadays, visualization is rather perspective direction of the security analysis. It leads to emergence of new access control models using graphic elements, for example, the access control visualization model based on triangular matrices [10]. However, visualization models can not ensure required speed of verification.

To achieve the required quality of verification of ABAC policies it is necessary to use automatic verification methods. The model checking is rather widespread among verification methods and well developed. In [1] an application of the model checking for verification of authorization rules for mobile systems is considered. In [12] the model checking based approach for detection of anomalies in security policy filtering rules is presented. In [19] the model checking based approach for formal modeling and analysis of different attacks on computer networks is considered. These papers show rather high efficiency of application of the model checking for the analysis and verification of different security systems. This gives us the grounds to believe that this method can be successfully applied to ABAC model verification.

3 Background

3.1 Background of the Attributive-Based Access Control Model

Unlike RBAC, in ABAC the users' access control is provided based on attributes, but not roles. The attributes can be divided in three categories: attributes of access subjects, attributes of information resources and environment attributes. Values of the attributes are participating in forming of access rules. These rules are the basis of the decision on access permission or prohibition. Access operations are related to information resources and their attributes. As a result, ABAC allows one to build more flexible access schemes than RBAC. The difference is in ability to adapt well to high dynamics of change of the security policy inherent to modern large-scale information systems.

Both in the RBAC model and in the ABAC model there is an issue of forming of the access control scheme. In the RBAC model this issue is called Role Mining Problem. In the ABAC model some researchers suggest to call this problem ABAC Policies Mining Problem (APMP). Let us describe it in more detail.

Let the large number of users (U), resources (R) and operations (O) over resources are given. Attributes are divided into two types - for users (A_u) and resources (A_r). The attribute a of the user u or the resource r can take empty value or value from the domain D_a. This value is denoted by relations $a(u)$ or $a(r)$. The policy rule in the ABAC model is defined by expression $p =< e; o >$, where (e) - is the conditions of applicability and (o) - is the performed operation. Let us show it on the following example.

Let the user attribute A_u have the name "Department" (A_u = "Department") and the resource attribute A_r has the name "Owner" (A_r = "Owner"). The user u can perform over a resource r an operation $o = read$ if it is carried out one or the other conditions: (1) the user u works in "Management" department, or (2) he/she is the owner of the resource r. The formal representation of this rule has the following appearance:

$$p = \langle A_u(u) = \text{``}Management\text{''} \text{ OR } A_r(r) = u; read(u, r)\rangle, \quad (1)$$

where $read(u, r)$ is the *read* operation carried out by the user u over the resource r.

The APMP problem is formulated as follows. Given log L consisting of the records of a form $< u, r, o, t >$ - the user u performs the operation o over the resource r in the time point t. It is required to find such a policy which maximizes its quality indicator.

3.2 Background of Model Checking

The model checking based verification of access control policies regarding rule anomalies comes down to the following actions. In the beginning the model of *an information system* in which security policies are applied is created. Then the specification of this system by means of a temporal logic is set.

The model of an information system is intended for representation of interrelations of users, resources and actions, their attributes and the involved information processes. It includes two basic components: the system configuration and the access control policies. The system configuration is represented by a large number of users with the logical connection established between them and a set of information resources.

Verification of the access control policies includes the following stages: (1) creation of the model of the information system in an internal format of the verification system; (2) creation of the specification of the system, which is setting properties of the correctness (i.e. lack of anomalies) on the temporal logic language; (3) checking of the model in an appropriate software tool (verifier);

(4) analysis of the verification results and of the examples, where system passed into an incorrect state; (5) comparison and assessment of the verification results according to the requirements to their efficiency.

To create the model of an information system in the model checking it is accepted to use the *Kripke structure* [2]. It consists of the set of states, the set of transitions between states and the function which marks each state with properties, which are true in this state. The Kripke structure M_N over a set of atomic expressions AP can be presented as $M_N = (S_0, S, R, L)$, where S is a finite set of states, $S_0 \subseteq S$ is a set of initial states, $R \subseteq S \times S$ is the relation of transitions where for each state there has to be such state $s' \in S$ that it carried out $R(s, s'), L : S \rightarrow 2^{AP}$ - the function which marks each state with properties, which are true in this state.

The Kripke structure is based on the first order formulas and the following rules: (1) the set of states S is the set of all estimates over a set of variables V; (2) for any pair of states s and s', the relation $R(s, s')$ is observed in that and only in that case when the formula is *True* after for each variable $v \in V$ the value $s(v)$ is assigned and for each variable $v' \in V'$ the value $s'(v')$ is assigned.

Each atomic expression represents an assignment to variables from a set of the V values from domain D. Let us give an example of formalization of variables for ABAC policy: $V = (u, p)$, where u is the access control rule, p is a possible access for the user to a resource. The set of states will be defined as $S = E \times U \times AF$, where E is the set of the possible accesses, U is the set of the rules used in the model and AF is the set of the applied rules.

To create the system of transitions between states the following operations are performed: (1) for the access p the possibility of application of the rule u is checked until all rules are not analyzed (if the rule can be applied to this access, then the couple (p, u) is added to an array of couples including the access and the policy rule); (2) when the access p is processed, the following access p' is analyzed, and all control rules begin to be applied to it; (3) the pair (p, u) is checked, including the access and the rule applied to it. If all sets of the rules, which can be applied to the one access p, are analyzed then they are removed.

4 Implementation of the Verification Method

Let us consider a small organization which contains five employers: two bosses of departments and three workers (Fig. 1).

Bosses are working only in their departments; workers can be transferred from one department to another. Each employee can create and work with his own files. Additionally, bosses can work with any files from their department. Each employee has three attributes: *personal_id* - unique identifier in organization (helps to distinguish one employee from another); *role_id* - identifier of employee role (*Worker* or *Boss* in our example) and *department_id* - identifier of department in which employee works (*department_1* or *department_2* in our example).

A file when created and signed by an employee has two attributes: *owner_id* - an identifier of the employee (his or her *personal_id*) and *department_id* - an

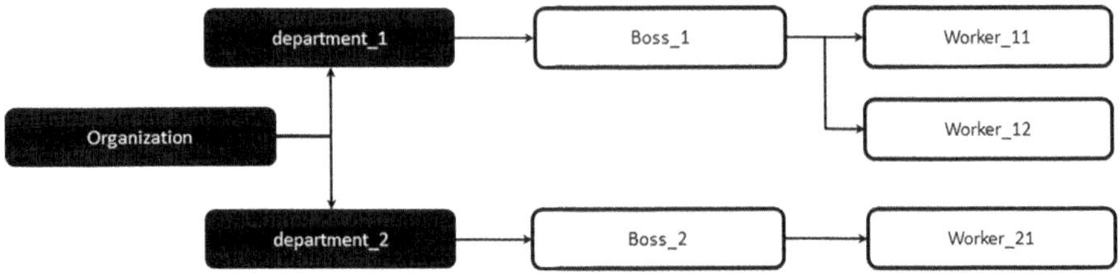

Fig. 1. Hierarchy of employers in organization.

identifier of the department in which the employee worked when file was created (1 or 2 in our example). According to the policy, the access to the file for employee with the *Boss* role is granted if the employee *department_id* and the file *department_id* are equal. For the employee with the *Worker* role the access to the file is granted if the employee *personal_id* and the file *owner_id* are equal. In all other situations the access will be denied. Such policy will prevent situations when employees from the *department_2* will be able to work with the files from *department_1* and in the opposite direction. Moreover, if *Worker_11* from *department_1* will create *File* and then *Worker_11* will be transferred to *department_2*, *Boss_1* will still have access to *File* (*department_id* attribute). But the issue is that according to our policy, *Worker_11* will still have access to *File* too (*owner_id* attribute). To prevent such situations the policy should be adjusted: for employee with *Worker* role the access to the file is granted if the employee *personal_id* and the file *owner_id* are equal and the employee *department_id* and the file *department_id* are equal.

The complexity of checking access policies grows with each new rule and their manual checking becomes a time consuming task. To automate the verification process the model checking based approaches are used. In the next Section an example from this Section will be modelled and verified in UPPAAL.

5 Experimental Results

UPPAAL is an integrated tool environment for modeling, validation and verification of real-time systems modeled as networks of timed automata [14]. The Sect. 4 example was modelled as automatons (Fig. 2).

The employee automaton contains five states initial state, department_1, department_2, work_1, work_2, and three parameters p_id (*personal_id*), r_id (*role_id*) and d_id (textitdepartment_id). State transitions are represented with edges which contain a guard parameter (an access rule) and synchronization events (work with file in our example, *open ! close !*).

The file automaton contains four states created (initial state), signed, opened and closed. State transitions are represented with edges which contain synchronization events (work with the file in our example, *open ? close ?*). File declarations reflect that the file was created by *Worker_11* while he was working in *department_1*.

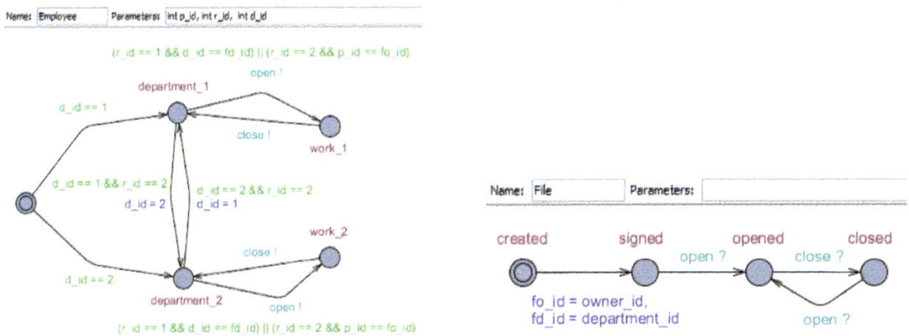

(a) employee (b) file

Fig. 2. UPPAAL automatons

For modeling and verification, we declared a system, which contains five employees and one file (Fig. 3). Declaration follows the example from Sect. 4 (see Listing).

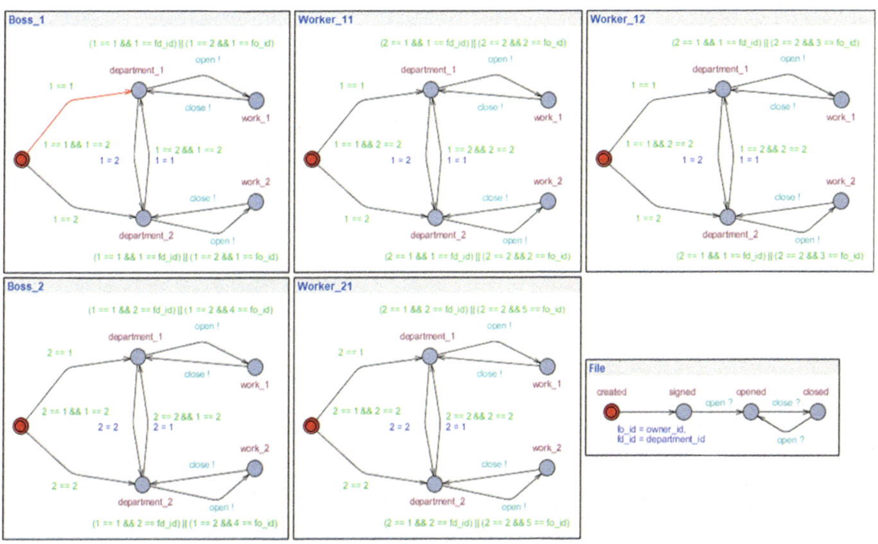

Fig. 3. UPPAAL system simulation (initial state).

Policy rule 1. "access to the file for the employee with *Boss* role is granted if the employee *department_id* and the file *department_id* are equal" is represented with the "r_id == 1 && d_id == fd_id" guard parameter of the *Employee* automaton edge from the *department_1* to *work_1* and from *department_2* to *work_2* states.

```
// Place template instantiations here.
Boss_1 = Employee(1, 1, 1);
Worker_11 = Employee(2, 2, 1);
Worker_12 = Employee(3, 2, 1);

Boss_2 = Employee(4, 1, 2);
Worker_21 = Employee(5, 2, 2);

// List one or more processes to be composed into a system.
system Boss_1, Worker_11, Worker_12, Boss_2, Worker_21, File;
```

Listing: source code for the example.

Policy rule 2. "access to the file for the employee with Worker role is granted if the employee *personal_id* and the file *owner_id* are equal" is represented with the "r_id == 2 && p_id == fo_id" guard parameter of the *Employee* automaton edge from *department_1* to *work_1* and from *department_2* to *work_2*.

For verification of the policy the next properties were model checked: transfer of the employees from department to department and possibility to open the file (Fig. 4).

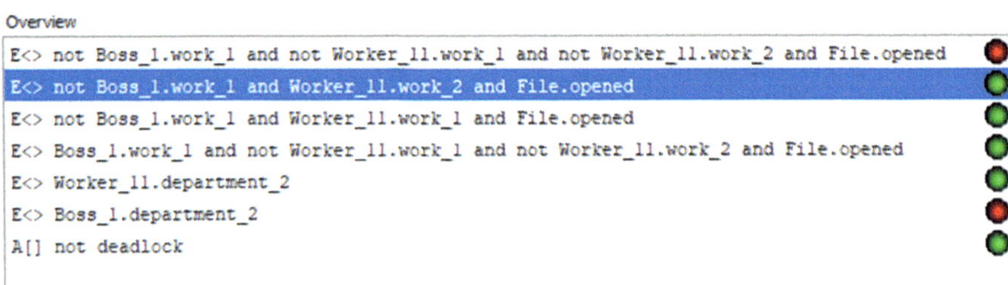

Fig. 4. Verification of the policy (with issue).

According to the results of model checking of "E <> not Boss_1.work_1 and not Worker_11.work_1 and not Worker_11.work_2 and File.opened" we know that the access to *File* can be granted to *Boss_1* (employer of the *department_1*) and *Worker_11* (owner of the file) only.

According to "E <> not Boss_1.work_1 and Worker_11.work_2 and File.opened" we know that the access to *File* can be granted to *Worker_11* even if he or she is working in *department_2* now (issue form Sect. 4).

To prevent such situations, the policy rule 2 should be adjusted to "for employee with Worker role the access to the file is granted if employee *personal_id* and the file *owner_id* are equal and the employee *department_id* and the file *department_id* are equal" which can be represented as the "r_id == 2 && p_id == fo_id && d_id == fd_id" guard parameter of the *Employee* automaton edge from *department_1* to *work_1* and from *department_2* to *work_2*.

Let us verify the adjusted policy with model checking again (see Fig. 5).

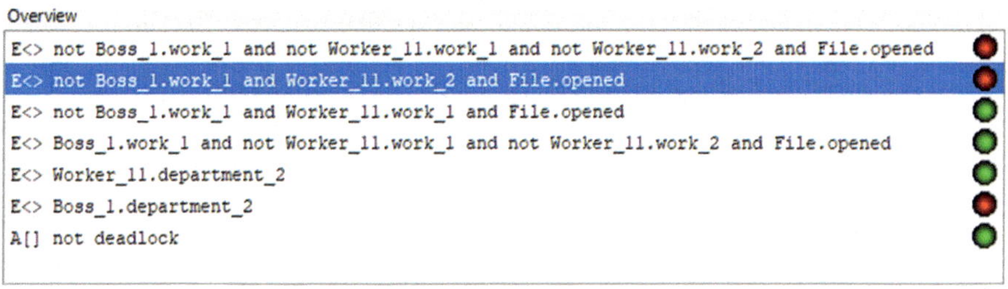

Fig. 5. Verification of the policy (without issue).

With the adjusted policy "E <> not Boss_1.work_1 and Worker_11.work_2 and File.opened" is not satisfied while all other checks are staying the same. It means that our adjustment from one hand solved the issue and from the other hand not created new one.

For the convenience of working with the presented experiment, UPPAAL files of the developed models are available for download on GitHub repository https://github.com/levshun/IITI-UPPAAL.

6 Conclusions

The paper presented a new approach to verification of ABAC policies in cloud infrastructures using Model checking. The mathematical background for ABAC modeling and model checking was considered. It allowed one to propose the verification system for ABAC policies using UPPAAL tool. In this paper the example of the ABAC policy in UPPAAL was modeled and verified to show the applicability of the suggested approach. During the verification the developed UPPAAL code helped us to find anomalies in the ABAC policy and then to prove that after adjustments of the policy rules, the anomalies were solved and new ones were not created. It is important to note that while the UPPAAL automatons are great for graphical representation of small access policies, this tool is not suitable for larger ones. In future research we plan to use the model checking tools which are focused on verifying of the correctness of concurrent models (like SPIN [5]).

Acknowledgements. This work was partially supported by grants of RFBR (projects No. 16-29-09482, 18-07-01369, 18-07-01488, 18-37-20047, 18-29-22034, 18-37-20047, 19-07-00953, 19-07-01246), by the budget (the project No. 0073-2019-0002), and by Government of Russian Federation (Grant 08-08).

References

1. Braghin, C., Sharygina, N., Barone-Adesi, K.: A model checking-based approach for security policy verification of mobile systems. Formal Aspects Comput. **23**(5), 627–648 (2011)
2. Clarke, E.M., Grumberg, O., Peled, D.: Model checking (2000)
3. Deng, Y., Wang, J., Tsai, J.J., Beznosov, K.: An approach for modeling and analysis of security system architectures. IEEE Trans. Knowl. Data Eng. **15**(5), 1099–1119 (2003)
4. Fisler, K., Krishnamurthi, S., Meyerovich, L.A., Tschantz, M.C.: Verification and change-impact analysis of access-control policies. In: Proceedings of the 27th International Conference on Software Engineering, pp. 196–205. ACM (2005)
5. Holzmann, G.J.: The model checker spin. IEEE Trans. Software Eng. **23**(5), 279–295 (1997)
6. Hu, V.: Attribute based access control (ABAC) definition and considerations. Tech. rep, National Institute of Standards and Technology (2014)
7. Hu, V.C., Kuhn, D.R., Ferraiolo, D.F., Voas, J.: Attribute-based access control. Computer **48**(2), 85–88 (2015)
8. Jaeger, T., Tidswell, J.E.: Practical safety in flexible access control models. ACM Trans. Inf. Syst. Secur. (TISSEC) **4**(2), 158–190 (2001)
9. Karataş, G., Akbulut, A.: Survey on access control mechanisms in cloud computing. J. Cyber Secur. Mobility **7**(3), 1–36 (2018)
10. Kolomeets, M., Chechulin, A., Kotenko, I., Saenko, I.: Access control visualization using triangular matrices. In: 2019 27th Euromicro International Conference on Parallel, Distributed and Network-Based Processing (PDP), pp. 348–355. IEEE (2019)
11. Kolovski, V., Hendler, J., Parsia, B.: Analyzing web access control policies. In: Proceedings of the 16th International Conference on World Wide Web, pp. 677–686. ACM (2007)
12. Kotenko, I., Polubelova, O.: Verification of security policy filtering rules by model checking. In: Proceedings of the 6th IEEE International Conference on Intelligent Data Acquisition and Advanced Computing Systems, vol. 2, pp. 706–710. IEEE (2011)
13. Kuhn, D.R., Coyne, E.J., Weil, T.R.: Adding attributes to role-based access control. Computer **43**(6), 79–81 (2010)
14. Larsen, K.G., Pettersson, P., Yi, W.: Uppaal in a nutshell. Int. J. Softw. Tools Technol. Transfer (STTT) **1**(1), 134–152 (1997)
15. Lee, A.J.: Credential-based access control. In: Encyclopedia of Cryptography and Security, pp. 271–272 (2011)
16. Lin, D., Rao, P., Bertino, E., Li, N., Lobo, J.: EXAM: a comprehensive environment for the analysis of access control policies. Int. J. Inf. Secur. **9**(4), 253–273 (2010)
17. Lopez, J., Rubio, J.E.: Access control for cyber-physical systems interconnected to the cloud. Comput. Netw. **134**, 46–54 (2018)
18. Mocanu, D., Turkmen, F., Liotta, A., et al.: Towards ABAC policy mining from logs with deep learning. In: Proceedings of the 18th International Multiconference, IS2015, pp. 124–128 (2015)
19. Rothmaier, G., Kneiphoff, T., Krumm, H.: Using spin and eclipse for optimized high-level modeling and analysis of computer network attack models. In: International SPIN Workshop on Model Checking of Software, pp. 236–250. Springer (2005)

20. Servos, D., Osborn, S.L.: Current research and open problems in attribute-based access control. ACM Comput. Surv. (CSUR) **49**(4), 65 (2017)
21. Subashini, S., Kavitha, V.: A survey on security issues in service delivery models of cloud computing. J. Netw. Comput. Appl. **34**(1), 1–11 (2011)

Detection of Anomalous Situations in an Unforeseen Increase in the Duration of Inference Step of the Agent in Hard Real Time

Michael Vinkov[1], Igor Fominykh[2(✉)], and Nikolay Alekseev[2]

[1] Bauman Moscow State Technical University, Moscow, Russia
vinkovmm@mail.ru
[2] Moscow Power Engineering Institute, Moscow, Russia
igborfomin@mail.ru, alekseevnp@list.ru

Abstract. The paper considers an approach to improving the stability of a cognitive agent functioning in hard real time in the event of anomalous situations (anomalies) associated with its functioning. The approach is based on the concept of metacognition (metareasoning) and is implemented by means of Active Logic. The principle of metacognition is based on a metacognitive cycle that includes stages of self-observation (introspection), self-evaluation and self-improvement. Introduces the concept of multiple granulation of time, when each inference step of the agent corresponds to the individual temporal granules and these granules are different for the various inference steps. It is shown that in some cases multiple granulation of time contributes to the timely detection of anomalies.

Keywords: Duration of inference step · Metacognition · Metacognitive cycle · Active logic · Hard real time · Anomaly · Granulation of time

1 Introduction

One of the most important and complex problems of the theory of intelligent multi-agent systems is to provide the stability ("invulnerability") of cognitive agents to unforeseen situations ("anomalies" [1]). Anomalies occur due to the imperfection of the agent's knowledge of the environment and its own cognitive processes and adversely affect its functioning. Especially serious consequences of anomalies can be in intelligent systems of hard real time. In work [2] the approach to modeling of the behavior of the intellectual (cognitive) agent directed on the increase of resistance to anomalies in the conditions of hard time restrictions was offered. This approach is based on the use of the concept of metacognition, implemented by means of a family of active logics specially developed

This work was supported by the Russian Foundation for Basic Research (RFBR) (projects 19-07-00439, 19-07-00-123, 17-07-00696, 18-07-00213, 18-51-00007, 18-29-03088, 17-07-01374).

S. Kovalev et al. (Eds.): IITI 2019, AISC 1156, pp. 176–183, 2020.
https://doi.org/10.1007/978-3-030-50097-9_18

for this purpose. However, there is an important class of anomalies, discussed below, in the event of which this approach is ineffective. In this paper, we propose a way to overcome this problem, which is based on a specialized version of time granulation.

2 The Concept of Metacognition

Real-time systems can be divided into two classes – soft and hard real time systems. Soft real-time systems are characterized by a smooth deterioration in the quality of their work with increasing response time to changes in the external environment. The situation is different for hard real time systems, which are characterized by the existence of a critical time threshold (deadline) solution to the problem facing the system. Exceeding this threshold is fraught with disastrous consequences. A typical example of an anomaly for intelligent systems of this kind is the situation when the event expected at the appointed time, however, has not come and we can talk about the threat of exceeding the time allowed for the execution of the cognitive process, the purpose of which is to solve the problem, i.e., a catastrophic deterioration in the quality of functioning of the agent.

Since for a cognitive agent operating in hard real time, the time resource is allocated primarily to the execution of a certain cognitive process, then the anomalies that such an agent may encounter are manifested in the fact that the cognitive process begins to go "somehow wrong" as the agent expects. It is clear that in order to timely detect anomalous during the cognitive process the agent is required, in a sense, metacognitive activities.

The term "metacognition" was introduced by Flavell [3, 4] and defined it as the awareness of the individual about their cognitive processes and related policies, or, in his words, as "knowledge and cognition regarding the cognitive phenomena". In other sources metacognition is often defined simply as thinking about thinking (e.g. [5]), referring to the "cognition of the second order". In the future, the term "thinking" will be used instead of the term "reasoning" more familiar to use in relation to the problems of artificial intelligence systems.

It should be noted the difference between cognitive and metacognitive strategies. The former helps the individual to achieve a specific cognitive goal (e.g., to understand the text) and the latter are used to control the achievement of this goal (e.g., self-questioning for understanding this text). Metacognitive components are usually activated when cognition fails (in this case, it may be a misunderstanding of the text from the first reading). This failure activates metacognitive processes that allow the individual to correct the situation. Thus, metacognition is responsible for active monitoring and consistent regulation of cognitive processes.

In [6], the concept of "metacognitive cycle" was proposed in the context of using the principles of metacognition to improve the resistance to anomalies of the agent with a limited time resource. It is defined as the cyclic execution of the following three steps: self-observation (monitoring); self-evaluation (analysis of the identified anomaly); self-improvement (regulation of the cognitive process).

It should be noted that in other works devoted to metareasoning or, more generally, metacognition in multi-agent systems [7–9], the metacognitive cycle of this type is also implied. Common in all these works is the approach based on the fact that the stage of self-observation, which reveals the presence of anomalies of the cognitive process, is built with reference to the possible actions of the agent, affecting the external environment, to the expected consequences of these actions. A sign of an anomaly is the mismatch of the agent's expectations with the incoming information about the external environment. Note that the implementation of the stages of the metacognitive cycle, in any case, does not involve any sophisticated thoughts, so deep that the agent can "get stuck" in them. The agent's metareasoning shouldn't be like that. At the stage of self-observation, they are reduced to checking for the presence in the agent's reasoning of formal signs of the presence of anomalies in the process of reasoning of the agent solving some problem. Due to the fact that, as mentioned above, in the systems of hard real time anomalies are mainly associated with the delay in the appearance of the expected results of cognitive activity of the agent, it is this kind of situation that should be detected in the monitoring process in the first place. At the stage of self-observation, the degree of threat to the quality of the agent's functioning, which is fraught with the identified anomaly, is established, and at the stage of self-improvement, if the threat is real, there is a choice of a new strategy for solving the task facing the agent. A typical way out of this kind of situation is the transition to a new strategy that requires a smaller time resource for its implementation, but provides, although acceptable, but less than the "old" strategy, the level of quality of the solution of the task facing the agent.

Thus, a logical system that formalizes the reasoning of an agent with strictly limited time resources should give him the opportunity to evaluate the time resource available to the agent at any time in such a way that, depending on the results of the evaluation, the agent can change the course of his reasoning (temporal sensitivity [10]). A necessary condition is also the ability of the agent to assess at any time the completeness of his knowledge and to realize not only what he knows, but also what he does not know at the moment. Both of these possibilities are directly related to the phase of self-observation, which will be the main focus of the remainder of this work.

3 Reasoning in Time and Metareasoning Based on Active Logic

In order to make it possible to observe the process of reasoning of the agent during its implementation, a logical system was proposed, called "Step Logic", [10] and became historically the first implementation of a more General concept, called "Active Logic" [6, 11]. While maintaining the ability to reason about agents, as if "looking at them from the outside", at the same time Active Logic allows the agent to relate the process of his reasoning with the events occurring in the external environment as a result of his activities or in addition to it. As a model of deduction, Active Logic is characterized by language, a set of deductive rules, and a set of "observations". If we assume that the agent thinks while in a static environment, a set of observations can be considered as part of the initial knowledge base of the deductive system, i.e. as a set of statements that support the deductive process, as a result of which new knowledge is generated.

However, using the monitoring function allows you to simulate a dynamic environment, information about which comes to the agent as changes occur in this environment.

Reasoning in time is characterized by the execution of deduction cycles called inference steps. Since the Active Logic is based on a discrete time model, the inference steps play the role of a time standard – the time is measured in them. Agent knowledge is associated with the index of the inference step at which it was first obtained.

The principal difference between Active Logic and other temporal epistemic logics is that "temporal arguments are introduced into the language of agents 'own theories" [12]. Thus, the time parameter is associated not only with each statement (formula) that the agent explicitly knows, but also with deductive inference rules. What the agent learned in inference step t (knowledge t) is used to inference new knowledge in step (t + 1). Deductive inference rules in Active Logic have the following form:

$$\frac{t : now(t)}{t+1 : now(t+1)} \text{time counting}$$

$$\frac{t : \phi, \psi}{t+1 : \phi \wedge \psi} \text{conjunction,}$$

$$\frac{t : \phi \wedge \psi}{t+1 : \phi} \text{detachment,}$$

$$\frac{t : \phi}{t+1 : \phi} \text{inheritance,}$$

$$\frac{t : \phi, \phi \rightarrow \psi}{t+1 : \psi} \text{modus ponens,}$$

$$\frac{t\colon \varphi, sub(\ulcorner\phi\urcorner, \ulcorner\varphi\urcorner), [\phi].}{t+1\colon \neg K(t+1, \phi)} \text{ self – observation (negative introspection)}$$

Where ϕ – any formula unknown to agent i at step t, but which is a sub-formula of some formula φ known to it, i.e. conscious by the agent, sub (. , .) – double metapredicate expressing the relation "to be a sub-formula », $\ulcorner\phi\urcorner, \ulcorner\varphi\urcorner$ – constants denoting the names of the formulas ϕ and φ, $[\phi]$ – a notation indicating that formula ϕ is missing from the agent's current knowledge in step t, K(. , .) – double metapredicate used to express the fact that an agent knows some formula at some point in time.

$$\frac{t\colon \phi, \neg\phi}{t+1\colon contra(t, \ulcorner\phi\urcorner, \ulcorner\neg\phi\urcorner)} \text{ identification of "direct" contradictions}$$

where contra (. , .) is the double metapredicate used to express the fact that in the current knowledge of the agent there is a formula ϕ and its negation $\neg\phi$ at some point in time t.

Observations can be made at any step of the deductive process. The result of observation is a formula expressing some statement and associated with the corresponding step. To illustrate the step-by-step reasoning process, assume that the agent initially knows (in step t) that $\phi \rightarrow \psi$ and $\psi \rightarrow \chi$ and at the step (t + 1) agent observes ϕ. The following shows what new knowledge is available at each step when the agent uses deductive inheritance rules and modus ponens.

$$t : \phi \rightarrow \psi, \psi \rightarrow \chi$$

$$t+1 : \phi$$

$$t+2 : \psi$$

$$t+3 : \chi$$

4 Multiple Granulation of Time

As noted above, reasoning in time is carried out through the implementation of cycles of deduction, called the inference steps. Note that such a cyclic character are the arguments in most systems, knowledge-based, i.e., where knowledge is represented explicitly. These steps play the role of a time reference – time in Active Logic is measured in inference steps. It is assumed that the reasoning steps have approximately the same duration: if the duration of the different inference steps is different, then this difference can be ignored. However, in practice this is not always the case – the duration of deductive cycles can be influenced by various factors, sometimes not accounted for. Unforeseen delays may be caused by technical reasons, for example, disruptions in the supply of electricity and associated with a variety of software bugs, in particular, the operating system. An unforeseen delay in the execution of the deductive cycle can also occur when an unusually large amount of information is received by the agent as a result of observation. It is easy to see that in all such cases the property of temporal sensitivity of the agent is lost: the measurement of time in deductive steps is too inaccurate. This means that anomalies of this type may not be recognized by the agent whose behavior is described by means of Active Logic.

It seems that this problem can be overcome if each particular step of the conclusion is compared with its individual duration, while abandoning the postulate of the indistinguishability of the various steps of the conclusion on the duration of their execution. If when using the duration of the inference step as a time reference (as is customary in Active Logic) we have an example of time information granulation [13], then the transition to the individual duration of each step is naturally called multiple time granulation: now each inference step corresponds to an individual time granule and these granules are different for different inference steps. The following approach to solving the problem of unexpected increase in the duration of inference steps is based on the constructions given in [14] for the logic of reasoning planned in time (TRL) and is based on the following principles.

Time is considered as an infinite sequence of natural numbers from the set N. We denote it Gck (global clock). However, in this case it is taken into account that the main purpose of such logical systems is to model the behavior of the cognitive agent in different conditions (= runs). Therefore, each such run is put in accordance with the so-called run clock model Ck, reflecting its specificity (the principle of granulation of time). The model run clock is a finite or infinite strictly increasing subsequence of the global clock whose members are interpreted as the time points (on the global clock) of the completion of the cycle inference steps, for example, $< 3, 5, 7, 10, \ldots >$. The set of all such moments of time will be denoted by Ck*.

Each "tick" of the model run clock, as well as the "tick" of the virtual internal clock considered above, corresponds to one execution of a specific inference step. In this case, the sequence number of this step does not coincide with the time of its completion (as is the case in Active Logic), but only with the sequence number of this time on the model run clock. This makes it possible, by changing the running hours of the model, to simulate different operating conditions of a multi-agent system and better reflect, for example, such features as the increase in the duration of deductive cycles of the agent as the amount of information known to it increases. In addition, different agents can be assigned different local clocks, thus simulating, for example, their different "intelligence" (performance) or the fact that they are activated at different times. In the future, however, for simplicity, we consider the case when the agent is only one and he is assigned the same model run hours. Then, in the conditions of time granulation deductive rules of inference will change as follows (for example, the rule of reference time):

$$\frac{t : now(t)}{next(t) : now(next(t))}$$

5 Example of Detection of Anomalies of the Cognitive Process

Below is an example of the process of metareasoning, when the fact that contrary to the expectations of the agent, some event A did not become known to him in time (time 2), is manifested in the form of a direct contradiction.

Example 1.

Let Ck = (0, 1, 13, 15, 17, ...)

0: $now(0), (now(t) \wedge (\mathrm{t} \geq 10) \wedge \neg K(\mathrm{t}, A)) \rightarrow \text{ANOMALY}, \ldots$

1: $now(1), (now(0) \wedge (0 \geq 10) \wedge \neg K(0, A)) \rightarrow \text{ANOMALY}, \ldots$

13: $now(13), (now(1) \wedge (1 \geq 10) \wedge \neg K(1, A)) \rightarrow \text{ANOMALY}, \neg K(1, A), \ldots$

15: $now(15), (now(13) \wedge (13 \geq 10) \wedge \neg K(13, A)) \rightarrow$
$\text{ANOMALY}, \neg K(1, A), \neg K(13, A), \ldots$

17: $now(17) \rightarrow \text{ANOMALY}$

The rule $now(0), (now(t) \wedge (\mathrm{t} \geq 10) \wedge \neg K(\mathrm{t}, A)) \rightarrow$ ANOMALY is shown only at time 0, although it is inherited at all other times 1, 13, 15, 17... This rule expresses the expectation that formula A will become known earlier than 10 units of time starting from 0, otherwise we can talk about an anomaly in the cognitive process. From this rule, at the moments of time 1, 13, 15, its instances are obtained by substituting the values of the moments of time, respectively, 0, 1, 13, instead of the variable t. As a result of the rule of self-knowledge, which when using multiple granulation of time takes the form:

$$\frac{t: \varphi, sub(\lceil\phi\rceil, \lceil\varphi\rceil), [\phi].}{next(t): \neg K(next(t), \phi)}$$

the following formulas were derived $\neg K(1, A)$, $\neg K(13, A)$ at the time moments 13 and 15 respectively. At the moment 15 worked rule $(now(13) \wedge (13 \geq 10) \wedge \neg K(13, A)) \rightarrow$ ANOMALY, being an instance of a rule $(now(t) \wedge (\mathrm{t} \geq 10) \wedge \neg K(\mathrm{t}, A)) \rightarrow$ ANOMALY and as a result, at the time 17 it was recorded the presence of anomalies in the cognitive process.

Note that without multiple granulation of time, i.e. with Ck = (0, 1, 2, 3, 4, 5, ...), this anomaly would have been discovered much later.

6 Conclusion

The multiple time granulation considered in this paper is oriented for application in multi-agent systems of hard real time. Practical experience metareasoning strategies in combination with multiple granularity of time substantially extend the class of contingencies that can be successfully resolved when modeling the behavior of intelligent agents. Nevertheless, the considered approach should not be interpreted as a panacea that solves all the problems associated with ensuring resistance to anomalies of intelligent systems. There is also no doubt that a metacognitive approach is a necessary link in addressing these problems.

References

1. Anderson, M.L., Perlis, D.: Logic, self-awareness and self-improvement: the metacognitive loop and the problem of brittleness. J. Logic Comput. **15**(1), 21–40 (2005)
2. Vinkov, M.M., Fominykh, I.B.: Increased resistance to anomalies of intelligent agent with a limited time resource: metacognitive approach. In: Proceedings works of the Twelfth National Conference on Artificial Intelligence "CAI-2010", Fizmatlit 2010, vol. 3 (in Russian)
3. Flavell, J.H.: Speculations about the nature and development of metacognition. In: Weinert, F., Kluwe, R. (eds.) Metacognition and Motivation. Lawrence Erlbaum Associates, Hillsdale (1979)
4. Flavell, J.H.: Metacognition and cognitive monitoring: a new era in cognitive- developmental inquiry. Am. Psychol. **34**(10), 906–911 (1987)
5. Metcalfe, J., Shimamura, A.P.: Metacognition: Knowing about Knowing. MIT Press, Cambridge (1994)

6. Anderson, M.L., Oates, T., Chong, W., Perlis, D.: The metacognitive loop I: enhancing reinforcement learning with metacognitive monitoring and control for improved perturbation tolerance. J. Exp. Theor. Artif. Intell. **18**(3), 387–411 (2006)
7. Brown, A.: Metacognition, executive control, self control, and other mysterious mechanisms. In: Weinert, F., Kluwe, R. (eds.) Metacognition, Motivation, and Understanding, pp. 65–116. Erlbaum, Hillsdale (1987)
8. Cox, R.: Metareasoning: Manifesto, in BBN Technical Memo TM-2028 (2007)
9. Raja, L.: A framework for meta-level control in multi-agent systems. Autonom. Agents Multi-Agent Syst. **15**(2), 147–196 (2007)
10. Elgot-Drapkin, J.: Step logic: reasoning situated in time. Ph.D. thesis. Department of computer science, University of Maryland, Colledge-Park, Maryland (1988)
11. Anderson, M.L., Lee, B.: Empirical results for the use of meta-language in dialog management. In: Proceedings of the 26th Annual Conference of the Cognitive Science Society (2004)
12. Purang, K., Purushothaman, D., Traum, D., Andersen, C., Traum, D., Perlis, D.: Practical reasoning and plan executing with active logic. In: Proceedings of the IJCAI 1999 Workshop on Practical Reasoning and Rationality (1999)
13. Tarasov, V.B.: Information Granulation, non-standard and hybrid fuzzy sets. In: Proceedings of 6th International Conference "Integrated Models and Soft Computing in Artificial Intelligence". MSCAI-2011, Fizmatlit 2011, pp. 35–49. (in Russian)
14. Vinkov, M.M.: Time as an external entity in modeling reasoning of rational agent with limited resources. In: Proceedings of 11th National Conference on AI, CAI 2008. Fizmatlit. Publication (2008). (in Russian)

Bayesian Networks and Trust Networks, Fuzzy-Stocastical Modelling

Protection System for a Group of Robots Based on the Detection of Anomalous Behavior

Alexander Basan, Elena Basan(✉), and Oleg Makarevich

Southern Federal University, Chekov Street 2, 347922 Taganrog, Russian Federation
ebasan@sfedu.ru

Abstract. This article is devoted to the development of a security system for a group of mobile robots. A feature of this system is the combination of various methods of security, such as detecting attacks and establishing trusted devices in one system. Thanks to the developed data analysis method, it is possible to use the same set of parameters for different security purposes. This article describes the modular system architecture. In particular, much attention is paid to the module for detecting anomalous activity. This module conducts an initial assessment of the data and identifies the fact of abnormal behavior. The module converts statistical data into a certain type of probability distribution, which is suitable for the parameter and identifies deviations from standard indicators. If an anomaly is detected, then it gives information for further study to other modules. The network simulation results and the initial parameter estimation confirm the effectiveness of the proposed methods.

Keywords: Mobile robots group · Abnormal behavior · Trust · Attack · Group management system · Method · Algorithm · Simulation model

1 Introduction

Today, the popularity of mobile robotic systems is growing. Mobile robots can be divided into the following four groups: autonomous ground vehicles; unmanned aerial vehicles (UAVs); Autonomous marine vehicles; Autonomous underwater vehicles (APA) [1]. Mobile robots are usually controlled by the operator using a wireless channel. They are often located outside the controlled area, have limited processing power and a small-capacity battery. These features cause attacks [2]. The wireless network that is used to control the robot, as well as to exchange data between robots, is one of the main sources of threats from the attacker. There are a large number of attacks aimed at capturing a mobile robot and system crash. Such attacks can be implemented on the channel and on the physical layer of the network by an external attacker [3]. These attacks can cause significant damage to the network. Their goals: complete destruction of the system, disruption of the network, penetration into the mobile robotic system (MRS). If an intruder's task is to infiltrate an IFA, gain access to all devices, change the logic of the system, and obtain confidential information, then an attacker can carry out internal active attacks on MRS. To develop an effective system for detecting and blocking network attacks, it is necessary to understand which components of the

S. Kovalev et al. (Eds.): IITI 2019, AISC 1156, pp. 187–197, 2020.
https://doi.org/10.1007/978-3-030-50097-9_19

system are most susceptible to attacks, as well as which indicators affect the attack. At the same time, during network operation, it is important not only to detect abnormal behavior in time, but also to maintain trusted relations between nodes. Nodes need to be sure that the information they receive is accurate and came from a trusted sender. In this case, tools such as digital signature, authentication, hashing, or encryption may not always be effective. This is due to the physical insecurity of nodes and communication channels. If a node exhibits abnormal behavior, then other nodes cannot trust it fully. Since it can be either supported by the attack and distort the transmitted data, or be an attacker and also affect the accuracy of the information. Therefore, the detection of anomalies is directly related to building trust relationships and calculating the level of trust. Thus, it is possible to combine these two subsystems of protection in one.

2 Related Works

There are several works about the development of intrusion detection systems for the mobile robots, cyber-physical control systems. In cyber-physical control systems, for example, the main attention is paid to the analysis of physical parameters of the technological process, such as: readings of sensors, status of actuators, and mathematical relations between them, controlled by the corresponding programmable logic controllers to detect anomalies. Currently, most anomaly detection methods for cyber-physical systems are based on a forecasting model, such as an autoregressive model [4], a linear dynamic model of the state space [5, 6] and regression models based on a neural network [7], where analyzed and predicted sensor measurements. There are different sources of noise in the processes, a strict threshold between normal and abnormal sensor measurements is usually not determined. Attackers can use this ambiguous decision boundary to introduce malicious measurements that achieve the goals and bypass detection systems that are based on existing methods [8, 9, 10]. The proposed alternative approach to the use of residual errors is an invariant method based on the rules [11, 12]. Rule-based invariant methods analyze physical conditions that are a priority for all states of cyber-physical systems, mobile robotic systems. Any observed values of physical processes that violate these rules are classified as anomalies. There are many hidden invariant rules that are extremely difficult to detect for people, especially those that are in several subsystems. As a result, the effectiveness of existing methods for detecting anomalies based on invariant rules is often limited by both inaccuracy and inadequacy of design-based rules. The authors of [13] propose an approach to creating an intrusion detection system (IDS) based on the analysis of the rules of behavior and the use of monitoring nodes. The authors suggest that the monitor node performs intrusion detection on a trusted neighboring node. One possible design is for the sensor (actuator) to control another sensor (actuator, respectively) within the same unmanned aerial vehicle (UAV). However, this project requires that each sensor (actuator) perform several different measurement functions using a sensor system. Another project that the authors propose is to use a neighboring UAV or a remote HMI monitor for a trusted UAV. At the same time, the authors offer the following indicators of behavior that help detect anomalies: The first indicator is that the UAV turns on its weapon when it is out of the battlefield; The second is that the readings of the built-in sensor of the trusted node are different from the readings of the built-in monitor sensor; The third is that the control UAV gives poor recommendations regarding a reliable UAV that is behaving well; The fourth is that the UAV deploys the chassis outside its internal air base; The fifth is that the node sends data to unauthorized devices; The sixth is that the UAV uses countermeasures without detecting a threat; The seventh is that an unoccupied UAV uses the disparate power necessary to maintain altitude. Renjian Feng et al. [14] describe their approach based on reputation and trust. To efficiently identify selfish and malicious nodes and solve security problems this article proposes a confidence assessment algorithm based on a theory of behavior based on D-S belief theory. The main stages of the algorithm (NBBTE) include the following parts: Confidence factors are determined taking into account the practical environment of the network. Then a quantitative and qualitative analysis is carried out to calculate the direct and indirect significance of trust. The degree of confidence of the nodes in the neighbors is calculated using the theory of fuzzy sets and, accordingly, form the main input vector of the theory of evidence. At the end of the simulation, it is shown that the circuit can effectively evaluate the reliability of nodes, taking into account the subjectivity, uncertainty and fuzziness of the confidence assessment.

3 Simulating Model

The network of mobile robots was implemented to collect statistical information for analysis. In this case, the AODV protocol was used for routing, the UDP protocol was used to transfer data packets with a variable bit rate. Nodes sent packets of 512 mb in size. Not all nodes could send packets directly some acted as transmitters. Next to the network was carried out two types of attacks. The first is a denial of service attack. The attack was carried out in a low-intensity mode, i.e. the attacker sent packets with an intensity four times greater than normal nodes. This attack did not lead to a destructive impact on the network. One measure of the success of a denial of service attack may be a reduction in throughput and an increase in power consumption by the nodes. Such an attack is difficult to fix and rather easily confused with normal behavior. This may cause errors of the first and second kind when the attack detection system is in operation. The second is the Black-Hole Attack. An attacker node is located between trusted nodes and listens for all traffic. An intruder node drops all packets that pass through it.

4 Modular Architecture of the Security System for a Mobile Robots Group

In general, the modular structure of the developed protection system can be represented in Fig. 1.

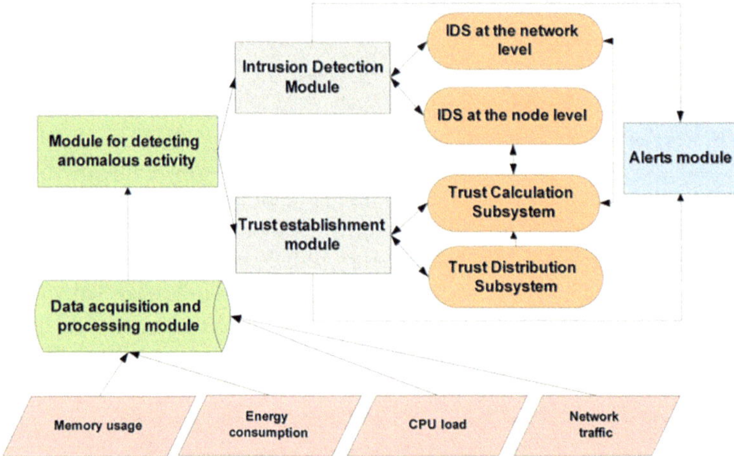

Fig. 1. Architecture protection system for a group of mobile robots

4.1 Data Acquisition and Processing Module

The parameters to be analyzed were obtained on the basis of a previously conducted experimental study [15]. Data acquisition and processing modulecollects statistical information at the current time and then, using the methods of mathematical statistics, performs the primary processing of information. That is, a set of information on the change of fixed parameters will be presented in the form of metrics that will allow us to further assess the presence of signs of abnormal activity. The collected information is divided into three large categories: parameters related to network traffic analysis; parameters associated with the change in the power supply plan of the device; parameters associated with changes in system characteristics, reflecting the load on the processor and the device's RAM.

Each of the parameters is influenced by certain sets of attributes, and then the attributes are analyzed and formalized. One of the main stages of the implementation of this module was to determine the set of parameters that must be analyzed for the effective detection of anomalies. To select the parameters, scenarios of attacks on the full-scale and simulation models of robots were developed and implemented. More broadly, the results of attack modeling and parameter selection results are presented in the works of the authors [16, 17]. A simulation of attacks by external and internal attackers was carried out such as denial of service and man-in-the-middle attacks and the Black Hole. Table 1 presents the simulation results. Not all of the parameters presented in the table are used further for analysis in the protection system. This is due to the fact that some parameters have not changed throughout the entire time of the attack, regardless of the degree of its intensity.

Table 1. The parameters selected for further analysis and development of the protection system

Parameter	Attack						
	Auth-Dos	DeAuth	DoS – Fragment.	SYN-flood	ACK, FIN-flood	Sybil	BlackHole
CPU load (%)	✓	X	✓	✓	X	–	X
RAM usage (Mb)	X	X	X	X	X	–	X
CPU temperature (°C)	X	X	X	X	X	–	X
Energy consumption	✓	X	✓	✓	✓	X	X
Network traffic	X	X	X	✓	✓	✓	X
Channel availability	✓	✓	X	✓	X	X	X
Discard Packets	X	X	X	X	X	X	✓

For example, the CPU temperature parameter did not change, as well as the memory usage parameter. At the same time, when such attacks as SYN-flood, Auth-DoS, DoS – Fragmentation were carried out, then on the attacker's node significant changes were observed in the memory parameter. Therefore, this parameter will be further evaluated. Below is a description of the attacks presented in the table. Auth-DoS attack [18]. Clients trying to connect to a wireless point find an access point in their range and request a connection. Then an authentication process occurs that continues until it joins the network. The attack consists of sending thousands of fake client requests for authentication from random MAC addresses. DeAuth Attack - Deauthentication. After de-authorization (deauthentication) attacks, all clients are disconnected from the wireless network. Disconnect packets are sent to one or all data network clients. Typically, client devices immediately try to connect again, and at this point the attacker has the opportunity to intercept the client's handshake. This attack can cause a denial of service when infinitely sending disconnect packets. The DoS – Fragmentation attack is performed by sending a large number of fragmented RTS frames, as a result of which a nonstandard situation arises and the system reboots in most cases. RTS (Request to Send frame) - frame, request to send. Mobile devices send an RTS frame to another device, as the first phase in a two-step process required before sending the data frame. SYN-flood attack is to send a large number of requests to establish a connection using the TCP protocol to an open port. The essence of the attack lies in the fact that the victim is trying to hold a large number of half-open connections with the malicious node, and ultimately the channel is overloaded and the victim's resources are exhausted. The Sibyl attack is that the attacker appears to be several nodes (entities) on the network and tries to redirect most of the traffic to himself. In this article, issues of processing and obtaining information about changes in parameters will not be considered. These issues will be addressed in subsequent studies.

4.2 Module for Detecting Anomalous Activity

This module is intended for primary normalization and parameter processing. The change of each parameter is reduced to the form of a probability distribution. Multiple data analysis options may be used. Data can be analyzed in real time continuously. Or the data can be analyzed discretely, i.e., for the same time intervals. When analyzing the network traffic parameter, it means that all packets passing through the node will be counted. If this parameter is taken into account continuously, then the collected data correspond to the normal distribution. The correspondence of the distribution to the network traffic parameter to the normal distribution is proved by the plotting construction of the quantile-quantile diagram. With regard to energy consumption, the situation is similar to network traffic. Change this parameter also corresponds to the normal distribution. In addition to building probability distributions, this module also allows for the primary detection of anomalous activity in network traffic. The idea of developed the method of detecting anomalies in network traffic is as follows. The nodes of the robotic network act on a predetermined algorithm. Either they have a predefined set of actions or patterns of behavior depending on the situation. In essence, the functioning of a group of robots is reduced to the following main stages. Strategic management level - when the group of robots, the distribution of goals, the choice of the group leader occurs. Tactical level of management - coordination of robots, coordination of actions and tasks. And the level of autonomous control - when the robot directly reaches the goal. Thus, the interaction of robots within the framework of solving problems at each of the routes is inherently static and predictable. Accordingly, statistical analysis of network traffic can detect abnormal activity.

1. Determination of the ratio of sent and received packets of the network node for each separate period of time and for the entire time of the network operation $Ratio_{r,s}$. As well as an assessment of the ratio taken to the sent packets $Ratio_{s,r}$.

$$Ratio_{r,s}(\Delta t) = r_{i,n}(\Delta t)/s_{i,n}(\Delta t), \quad (1)$$

$$Ratio_{s,r}(\Delta t) = s_{i,n,data}(\Delta t)/r_{i,n,data}(\Delta t). \quad (2)$$

where $r_{i,n,data}(\Delta t)$ - these are packets received by the node n for a set time interval, $s_{i,n}(\Delta t)$ - these are packets sent by the node n for a set time interval.

At first glance it may seem that the formulas overlap. When the network is working normally, and the attack is not carried out, the impact on the network is not observed, and then all these values should strive for one. At the same time, only in some cases deviations could be observed for some nodes in single time intervals. These deviations did not exceed the difference by more than three times from the conventionally normal deviation. Increasing or decreasing the difference between sent or received packets, can be observed if a node changes its role in the network, for example, it becomes a beating node. This situation may occur when the two nodes are far from each other and begin to transmit packets through an intermediary. At the same time, no matter what the situation occurs, according to the results of the simulation it can be said that an increase in the value of more than three times the normal was not observed. Such peak values could be fixed at one or several (no more than three) adjacent intervals. Figure 2 shows a histogram of changes in the ratio of the results of the simulation of the normal operation of the network sent to the received packets for three network nodes. In the future, these nodes will perform the role of the victim of an attack, an attacking node and a node that demonstrates the nominal work. From Fig. 2 it can be seen that at some intervals the nodes have peak values, but as a whole everything is within the normal range.

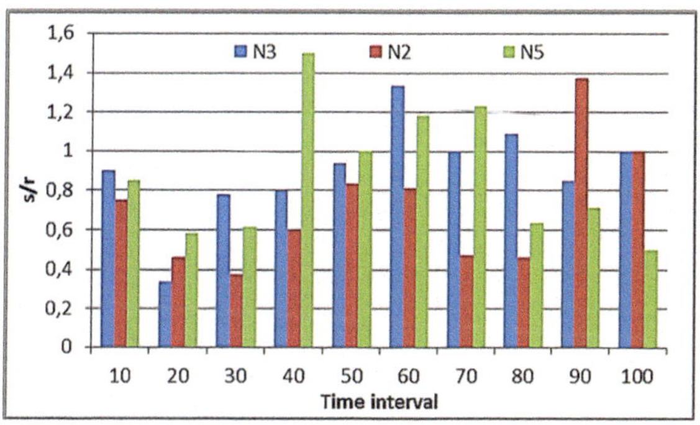

Fig. 2. Measurement of the ratio of sent to received packets for three nodes in the absence of anomalous activity in the network.

Figure 3 shows the evaluation results obtained during the simulation of a denial of service attack. From the Fig. 3 (a) it can be seen that the malefactor's node has readings much higher than the other nodes. And these exceedances can no longer be called peak. The attacker readings obtained for several consecutive intervals above the maximum allowable. Figure 3 (b) shows the ratio of sent to receive packets observed in the victim. It is also more than the minimum allowable and is observed throughout the entire simulation. Next was the Black hole attack. The number of packets received by the node is reduced, since the nodes discard them. In particular, the victim loses connection with some nodes and a significant decrease in received packets is observed, zero values are observed at some time intervals.

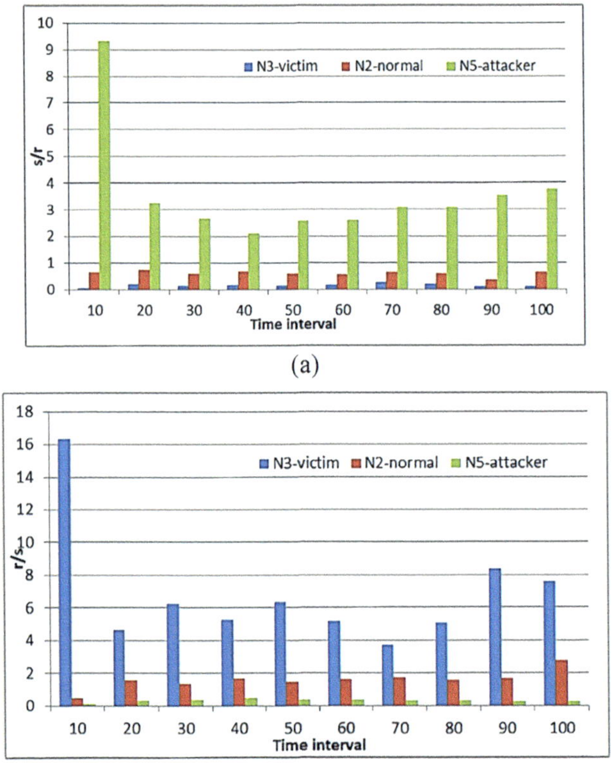

(b)

Fig. 3. The results of evaluating the ratio of (a) sent and received (b) received and sent packets when modeling a denial of service attack

From Fig. 4 the values in the intervals are not exceeded, but the victim has a lot of zero values. At the same time, when the ratio of sent to received packets was calculated, the victim, significantly exceeded the values at some intervals, but not all, for

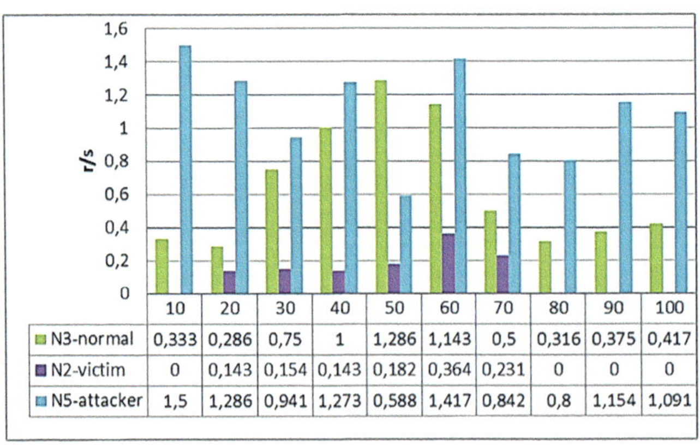

	10	20	30	40	50	60	70	80	90	100
N3-normal	0,333	0,286	0,75	1	1,286	1,143	0,5	0,316	0,375	0,417
N2-victim	0	0,143	0,154	0,143	0,182	0,364	0,231	0	0	0
N5-attacker	1,5	1,286	0,941	1,273	0,588	1,417	0,842	0,8	1,154	1,091

Fig. 4. Result of calculating the ratio of received packets to the sent for the Black Hole attack

some intervals where the value of the received packets was equal to zero. Thus, it turns out that the trusted node that is the victim of the Black Hole attack can be falsely perceived as an attacker implementing the DoS attack.

2. To distinguish between these attacks, we introduce a complementary metric—the variance estimates for the network traffic parameter V_{NT}. Evaluation of this parameter will make it clear how much the level of network traffic on different nodes in one and the same time interval differs from each other.

$$V_{NT}(\Delta t) = \left(\sum_{i=1}^{N} (NT_i(\Delta t) - \overline{NT}_g(\Delta t))^2 \right) / n, \tag{3}$$

where NT- is the total number of packets elapsed in a discrete time interval, n is the number of nodes.

From Fig. 5 it can be seen that the largest variation in values is observed in the attack denial of service, the smallest of the normal operation of the network. For the Black Hole attack, the spread of values gets not so large. This assessment will distinguish the denial of service attack from the Black Hole attack.

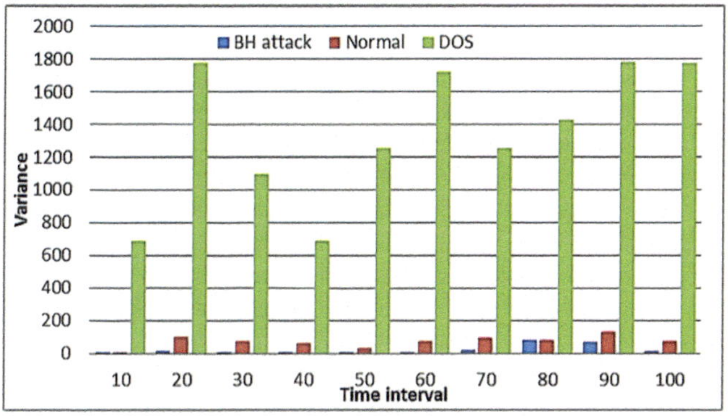

Fig. 5. The result of calculating the variance for all nodes when conducting different types of attacks

5 Conclusion

In conclusion, I would like to note that such a preliminary assessment of anomalous activity has great prospects. Unfortunately, within this article it was not possible to fully disclose all possible metrics. But even those metrics that were presented give an understanding of how to break off anomalous activity and how to distinguish one attack from another. If you select the right conditions, the attack will be detected quite quickly and there is no need to carry out complex calculations. With the limited resources of the network nodes, this is very important.

This work was partially supported by the Ministry of Education and Science of Russian Federation, Initiative research projects No. 2.6244.2017/8.9.

References

1. Vasiliev, D.S., Abilov, A.V.: Routing protocols in MANET. Telecommunication. **11**, 52–54 (2014)
2. Kirichek, R.V., Kucheryavy, A.E.: Flying sensor networks. Problems of technology and telecommunications technology PTiTT-2016. In: The First Scientific Forum "Telecommunications: Theory and Technology" 3T-2016: A Collection, pp. 343–344 (2016)
3. Pshihopov, V.Kh.: Control of moving objects in a priori non-formalized environments. News SFU. Technical Science, no. 3, pp. 6–19. TTI SFU Publishing House, Taganrog (2009)
4. Mitchell, R., Chen, I.-R.: Adaptive intrusion detection of malicious unmanned air vehicles using behavior rule specifications. IEEE Trans. Syst. Man Cybern. Syst. **44**(5), 2168–2216 (2014)
5. Hadžiosmanovi, D., Sommer, R., Zambon, E., Hartel, P.H.: Through the eye of the PLC: semantic security monitoring for industrial processes. In: Proceedings of the 30th Annual Computer Security Applications Conference, pp. 126–135. ACM (2014)
6. Abur, A., Exposito, A.G.: Power System State Estimation: Theory and Implementation. CRC Press, Boca Raton (2004)
7. Urbina, D.I., Giraldo, J.A., Cardenas, A.A. Tippenhauer, N.O., Valente, J., Faisal, M., Ruths, J., Candell, R., Sandberg, H.: Limiting the impact of stealthy attacks on industrial control systems. In: Proceedings of the 2016 ACM SIGSAC Conference on Computer and Communications Security, pp. 1092–1105. ACM (2016)
8. Goh, J., Adepu, S., Tan, M., Lee, Z.S.: Anomaly detection in cyber physical systems using recurrent neural networks. In: IEEE18th International Symposium on High Assurance Systems Engineering (HASE 2017), pp. 140–145. IEEE (2017)
9. Dan, G., Sandberg, H.: Stealth attacks and protection schemes for state estimators in power systems. In: First IEEE International Conference on Smart Grid Communications (Smart Grid Comm. 2010), pp. 214–219. IEEE (2010)
10. Liu, Y., Ning, P., Reiter, M.K.: False data injection attacks against state estimation in electric power grids. ACM Trans. Inf. Syst. Secur. (TISSEC) **14**(1), 13 (2011)
11. Feng, C., Li, T., Zhu, Z., Chana, D.: A deep learning-based framework for conducting stealthy attacks in industrial control systems. arXiv preprint arXiv:1709.06397 (2017)
12. Adepu, S., Mathur, A.: Using process invariants to detect cyberattacks on a water treatment system. In: IFIP International Information Security and Privacy Conference, pp. 91–104. Springer (2016)
13. Adepu, S., Mathur, A.: From design to invariants: detecting attacks on cyber physical systems. In: IEEE International Conference on Software Quality, Reliability and Security Companion (QRS-C 2017), pp. 533–540. IEEE (2017)
14. Keribin, C.: Consistent estimation of the order of mixture models. Indian J. Stat. Ser. A, 49–66 (2000)
15. Feng, C., Palleti, V.R., Chana, D.: A systematic framework to generate invariants for anomaly detection in industrial control systems. Published in NDSS 2019. https://doi.org/10.14722/ndss.2019.23265

16. Basan, E., Medvedev, M., Teterevyatnikov, S.: Analysis of the impact of denial of service attacks on the group of robots. In: Proceedings CyberC 2018: The 10th International Conference on Cyber-Enabled Distributed Computing and Knowledge Discovery, pp. 63–71 (2018). https://doi.org/10.1109/cyberc.2018.00023. 978-0-7695-6584-2/18/
17. Basan, E., Basan, A., Makarevich, O.: Evaluating and detecting internal attacks in a mobile robotic network. In: Proceedings CyberC 2018: The 10th International Conference on Cyber-Enabled Distributed Computing and Knowledge Discovery, pp. 1–9 (2018). 978-0-7695-6584-2/18/
18. Hafiz, M.M., Ali, F.H.-M.: Profiling and mitigating brute force attack in home wireless LAN. In: 2014 International Conference on Computational Science and Technology (ICCST), pp. 1–6 (2014)
19. Basan, E., Makarevich, O., Stepenkin, A.: Development of the methodology for testing the security of group management system for mobile robots. In: 2018 ACM International Conference Proceeding Series. Proceeding SIN 2018 Proceedings of the 11th International Conference on Security of Information and Networks (2018)

Employees' Social Graph Analysis: A Model of Detection the Most Criticality Trajectories of the Social Engineering Attack's Spread

A. Khlobystova[1(✉)], M. Abramov[1,2], and A. Tulupyev[1,2]

[1] Laboratory of Theoretical and Interdisciplinary Problems of Informatics, St. Petersburg Institute for Informatics and Automation of the Russian Academy of Sciences, 14-th Linia, VI, № 39, St. Petersburg 199178, Russia
{aok, mva, alt}@dscs.pro

[2] Mathematics and Mechanics Faculty, St. Petersburg State University, Universitetskaya Emb., 7-9, St. Petersburg 199034, Russia
http://dscs.pro

Abstract. In this research we present the hybrid model of finding the most critical distribution trajectories of multipath Social engineering attacks, passing through which by the malefactor on a global basis has the topmost degree of probability and will bring the greatest loss to the company. The solution of search problem concerning the most critical trajectories rests upon the assumption that the estimated probabilities of the direct Social engineering attack on user, degree evaluation of documents' criticality, the estimated probabilities of Social engineering attack's distribution from user to user are premised on linguistic indistinct variables are already calculated. The described model finds its application at creation when constructing the estimates of information systems users' safety against Social engineering attacks and promotes well-timed informing of decision-makers on the vulnerabilities which being available in system.

Keywords: Social engineering · Multiway social engineering attacks · Hybrid model of linguistic fuzzy variable · Analysis of social graph of company employees · Propagation of the multiway social engineering attack · Finding of the most criticality trajectory of the spread multiway social engineering attack

1 Introduction

1.1 Prerequisites for Research

Despite extensiveness and depth of investigations devoted to protection of information systems against hardware and technical attacks [9, 10, 16], statistics gives evidence of growth of organizations' losses from unauthorized access of malefactors to documents [4]. As usual the principle information disclosure and also the weakest element in

The research was carried out in the framework of the project on state assignment SPIIRAS № 0073-2019-0003, with the financial support of the RFBR (project №18-01-00626, № 18-37-00323).

S. Kovalev et al. (Eds.): IITI 2019, AISC 1156, pp. 198–205, 2020.
https://doi.org/10.1007/978-3-030-50097-9_20

information security of the organization is the individual. Certainly for the longest time one of the most serious risk to the organization's safety is represented by Social engineering attacks [4, 5, 12]. According to data of the known companies on protection of information systems against the unauthorized attacks [5, 12] in 2018 in comparison with 2017 the number of Social engineering attacks increased by 3 times. It is also noted [5], that the most active application of Social engineering attacks was recorded relating to governmental and financial organizations, educational institutions and individuals. According to the data of the news agency BBC [3] the increasing number of people is inclined to excessive exchange of private information (dates and places of birth, names of pets, personal photos, etc.) that can lead to serious losses. Thus, the common goal of line of the research is cushioning of risks, connected with Social engineering attacks' realization.

Usually Social engineering attacks are carried out through a chain of users, that is the malefactor-social engineer attacks one of the company's staff and then he tries to make an attack on his colleagues for the purpose of obtaining bigger volume of information or access to information resources of the organization in which he is interested in. Such attacks are called multipath [1]. Formally, multipath Social engineering attacks are Social engineering attacks which include the attack of more than one employee in such a manner that successfully attacked employees are directly involved in crack of the subsequent victims [1].

It is convenient to conduct the analysis of information systems' protection against multipath Social engineering attacks on the social graph of the organization's staff [7]. As it was shown in [8] earlier, it is possible to model and investigate various trajectories of multipath Social engineering attacks' development on the social graph. Acquaintance of trajectories, which penetration probability is the highest, and trajectories passing upon which will bring the greatest damage to the organization allows authorized persons making time-bound measures reducing the risk of successful influence. Pursuant thereto the problem of searching the most critical trajectories concerning multipath Social engineering attacks' distribution that depend both on probability of this trajectory's realization, and on the expected damage from a successful discredit seems to be currently central. Earlier metrics of assessment related to criticality of trajectories were offered [7]. The objectives achieved in this article are the offer of identification methods concerning such trajectories.

1.2 Related Work

Most researches connected with Social engineering attacks are directed to protection of information systems against phishing attacks [11, 13, 17, 18]. For example, in [11, 18] the influence of following characteristics on susceptibility to phishing letter is considered: argument quality, source credibility, genre conformity, pretexting, authoritativeness of the sender, urgency, within a letter there is an aid proposal directed to the insider, someone has already written back, the potential award is offered or ignoring will cause damage and they came to the conclusion that the strongest influence factor is psychological abuse which can be reached through urgency of the letter, authoritativeness of the sender and possible damage. Results of such researches redound to signification of the new vulnerabilities of users, however they do not give a probability to investigate the information system upon all types of Social engineering attacks, and as a consequence they will not give an aggregate picture concerning vulnerabilities that are available in this system.

The problem of automated information extraction from published sources is described in work [6], the automatic vulnerability scanner is based on obtaining corporate employee's accounts and extension of the known information on these employees due to the comparison of user's accounts on various social networking sites and extraction available information from them is used for this purpose. The article is also concerned with different strategies promoting increase in level of users' protection level from Social engineering attacks, for example, it is offered to conduct trainings, re-examine the policy, introduce restrictions for social networks, remove vulnerable data from the organization's website of and test the system for susceptibility to Social engineering attacks from time to time. In [2] detection of abnormal behavior of users on the basis of the two-stage strategy of detection is offered: Markov chains training based on users with usual behavior pattern is offered, and the use of detection system connected with the activity of abnormal behavior among users is offered too. Results of this research are useful and interesting, but their use in the organization requires installation of the additional software.

The design work for this research is presented by the following articles [1, 7, 8]. In [1] the concept of critical documents was entered, an approach to receiving estimates concerning multipath Social engineering attacks was presented and also the project model of the company staff's social graph whose knots are presented by corporate employees and the edges—by communications between them. In [8] formalization of the most probable trajectories of multipath Social engineering attacks' distribution was given and the algorithm of their finding was presented too. This article serves as an extension to the researchers conducted in [7] in which various distribution models of allocation of rights to critical documents were considered and the metrics for assessment of criticality of a trajectory from the point of view of expected damage was entered. However, in these researches models and algorithms using which it would be possible to find such trajectories were not presented.

Also the researches [14, 15, 18] in which indistinct linguistic variables are applied were considered, however in the area of Social engineering attacks they were applied for the first time.

2 Problem Statement

In [1, 8] there are results of researches allowing finding the most probable trajectories of distribution concerning multipath Social engineering attacks however besides the probability of realization of influence for decision-makers, the caused damage is also important. Thus, there is a question of searching the most critical trajectories of distribution concerning multipath Social engineering attacks, from determination of anticipated or actual damage expected at the successful attack of the critical document also considering the probability of passing on this trajectory by malefactor.

The purpose of article is proposition methods of finding the most critical trajectories of distribution as relating to multipath Social engineering attacks, passing through which by the malefactor on a global basis has the topmost degree of probability and will bring the greatest loss to the company. The solution of search problem concerning the most critical trajectories rests upon the assumption that the estimated probabilities of the direct Social engineering attack on user, degree evaluation of documents' criticality, the estimated probabilities of Social engineering attack's distribution from user to user are premised on linguistic indistinct variables are already calculated.

Getting all this information is subject to separate consideration and it was described in detail in preexisting researches [1, 8]. Also it is supposed that in the information system right of users' access to critical documents are distributed in such a manner that documents of only single critical level can be available to each user.

3 Theoretical Part

The social graph of the company's employees understood as a directed weighted graph $G = (U, E)$, where $U = \{\mathrm{User}_i\}_{i=1}^{n}$ is a multiplicity of vertexes (users), $E = \{(\mathrm{User}_i, \mathrm{User}_j, p_{i,j})\}_{1 \le i,j \le n, i \ne j}$ is a multiplicity of orderly threesomes with specified assessment of probability of attack's dissemination from User_i to User_j. Let we are given social graph of organization employees associated with the critical documents and the levels of access to them, that is each user is assigned a set of documents to which they have access. The example of such graph is represented on (Fig. 1).

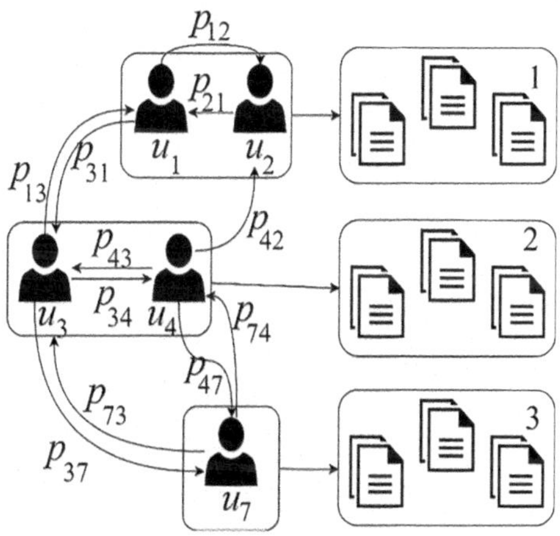

Fig. 1. The example the social graph associated with the critical documents and the levels of access to them.

That is, we consider the object: $G' = (U, E, D, A)$, где $U = \{\mathrm{User}_i\}_{i=1}^{n}$ is a multiplicity of vertexes (users), $E = \{(\mathrm{User}_i, \mathrm{User}_j, p_{i,j})\}_{1 \le i,j \le n, i \ne j}$ is a multiplicity of orderly threesomes with specified assessment of probability of attack's dissemination from User_i to User_j $p_{i,j}$, $D = \{(\mathrm{D}_j, \mathrm{CL}_i)\}_{1 \le j \le m, 1 \le i \le r}$ is a set of critical documents with preset criticality levels, $A = \left\{ (\mathrm{User}_i, \mathrm{D}_j) \right\}_{1 \le i \le n, 1 \le j \le m}$ is a set of pairs corresponding of information system's users and documents, of which they have access.

As a first step, all $\mathrm{D}_i : \mathrm{CL}_i = \underset{1 \le j \le m}{\mathrm{Max}} (\mathrm{CL}_j)$ are found. We introduce the set of all users who have direct access to such documents: $\mathrm{FV} = \left\{ \mathrm{User}_i : \ (\mathrm{User}_i, \mathrm{D}_j) \in A \right\}$.

Let us saying, that the implementation probability of the trajectory Tr more then Tr′, if probabilistic estimates, obtained according to (1), is as follows: $p_{Tr} > p_{Tr'}$.

$$p_{\mathrm{Tr}} = \frac{1}{p_i} + \sum_{l=i}^{j-1} \ln \frac{1}{p_{l,l+1}}, \tag{1}$$

where $\mathrm{Tr} = (\mathrm{User}_i, E_i, \ldots, E_{j-1}, \mathrm{User}_j)$, p_i—probability of success of a Social engineering attack on a user i (considered according to [1]), a $p_{l,l+1}$—assessment of probability of attack's dissemination from User_l to User_{l+1} (more in [8]).

$\forall \mathrm{User}_i \in \mathrm{FV}$ find a set of trajectories ordered by decreasing probability of attack realization and entered to this vertex, wherein if the probability of the calculated trajectories becomes less than a given threshold we have to exclude current trajectory from consideration. Notice, because $p_{l,l+1}$ (those that are part (1)) is the probability estimates, associated with linguistic variables, finding such trajectories is a hybrid model of a linguistic fuzzy variable.

Thus, we get a set:

$$\left\{ \left(\mathrm{D}_i, \left\{ \mathrm{Tr}_j^{(i)} \right\}_j \right) \right\}_i,$$

where $\mathrm{Tr}_j^{(i)} = \left\{ (\mathrm{User}_{k_1}^{(j)}, E_{k_1}^{(j)}, \ldots, E_{k_{l-1}}^{(j)}, \mathrm{User}_{k_l}^{(j)}) : \mathrm{User}_{k_l}^{(j)} \in \mathrm{FV} \right\}$.

4 Practical Significance

The described theoretical part finds its application in the software package, which produces automated analysis of information systems in order to identify vulnerabilities to social engineering attacks. Namely, the theoretical results allows to explore the information system for susceptibility to multipath Social engineering attacks and promotes well-timed informing of decision-makers on the vulnerabilities which being available in system. The algorithm for finding critical trajectories is represented on Fig. 2. He was implemented in the software package.

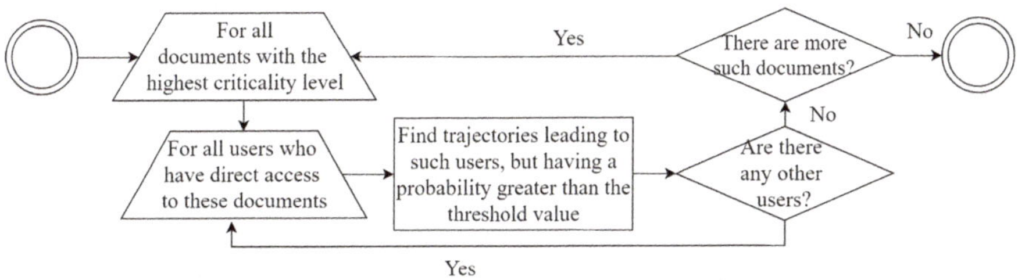

Fig. 2. The algorithm for finding critical trajectories

Screenshot of the program's work is present to (Fig. 3). It shows the result of finding critical paths in the social graph of organization, which consists of 56 employees. The program has revealed 6 trajectories to pose a threat to information security of this company. Also, it should be noted that the all found trajectories consist of only two or three vertexes. This exists because with increasing number of involved users the implementation probability of the trajectory has been decreasing.

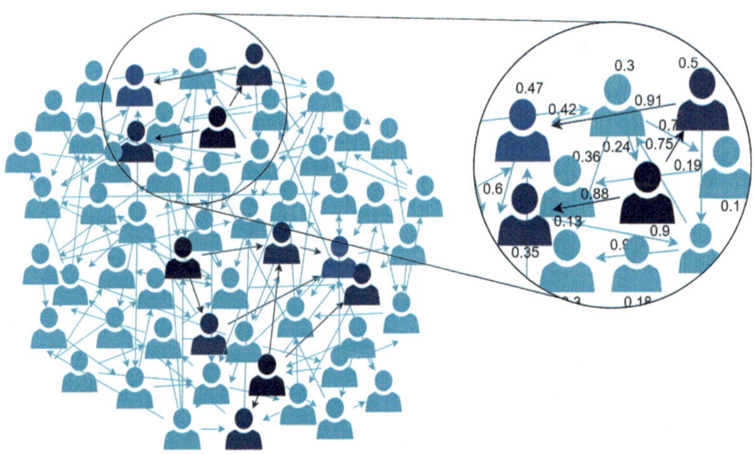

Fig. 3. Screenshot of the program: finding of critical paths in the social graph of the organization's employees

The practical significance of the obtained results consists in the formation of a tool for decision makers, giving the opportunity to reduce the risks of implementation of social engineering attacks.

5 Conclusion

Thus, the hybrid model of the linguistic fuzzy variable is proposed in the article, which allows to find the most critical distribution trajectories of multipath Social engineering attacks, passing through which by the malefactor on a global basis has the topmost degree of probability and will bring the greatest loss to the company. The described model finds its application at creation when constructing the estimates of information systems users' safety against Social engineering attacks. In the future, it is planned to consider information systems with different distribution among users of access rights to the critical documents.

References

1. Abramov, M., Tulupyeva, T., Tulupyev, A.: Social Engineering Attacks: social networks and user security estimates. SUAI, St. Petersburg (2018), 266 p.
2. Amato, F., Castiglione, A., De Santo, A., Moscato, V., Picariello, A., Persia, F., Sperlí, G.: Recognizing human behaviours in online social networks. Comput. Secur. **74**, 355–370 (2018)
3. Coughlan, S.: 'Sharenting' puts young at risk of online fraud. https://www.bbc.com/news/education-44153754. Accessed 03 Apr 2019
4. Cyber security facts and statistics for 2018 fraud. https://us.norton.com/internetsecurity-emerging-threats-10-facts-about-todays-cybersecurity-landscape-that-you-should-know.html . Accessed 11 Apr 2019
5. Cybersecurity threatscape 2018: trends and forecasts. https://www.ptsecurity.com/ww-en/analytics/cybersecurity-threatscape-2018/. Accessed 28 Mar 2019
6. Edwards, M., Larson, R., Green, B., Rashid, A., Baron, A.: Panning for gold: automatically analysing online social engineering attack surfaces. Comput. Secur. **69**, 18–34 (2017)
7. Khlobystova, A., Abramov, M., Tulupyev, A.: An approach to estimating of criticality of social engineering attacks traces. Studies in Systems. Decision and Control, pp. 446–456 (2019)
8. Khlobystova, A., Abramov, M., Tulupyev, A.: Identifying the most critical trajectory of the spread of a social engineering attack between two users. In: The Second International Scientific and Practical Conference "Fuzzy Technologies in the Industry – FTI 2018". CEUR Workshop Proceedings, pp. 38–43 (2018)
9. Li, J., Zhang, Y., Chen, X., Xiang, Y.: Secure attribute-based data sharing for resource-limited users in cloud computing. Comput. Secur. **72**, 1–12 (2018)
10. Muhammad, K., Sajjad, M., Mehmood, I., Rho, S., Baik, S.W.: Image steganography using uncorrelated color space and its application for security of visual contents in online social networks. Future Gener. Comput. Syst. **86**, 951–960 (2018)
11. Musuva, P.M.W., Getao, K.W., Chepken, C.K.: A new approach to modelling the effects of cognitive processing and threat detection on phishing susceptibility. Comput. Hum. Behav. **94**, 154–175 (2019)
12. Protecting People: A Quarterly Analysis of Highly Targeted Cyber Attacks. https://www.proofpoint.com/us/resources/threat-reports/quarterly-threat-analysis. Accessed 20 Jan 2019
13. Sahingoz, O.K., Buber, E., Demir, O., Diri, B.: Machine learning based phishing detection from URLs. Expert Syst. Appl. **117**, 345–357 (2019)

14. Tang, J., Meng, F., Zhang, S., An, Q.: Group decision making with interval linguistic hesitant fuzzy preference relations. Expert Syst. Appl. **119**, 231–246 (2019)
15. Tian, Z.P., Wang, J., Wang, J.Q., Chen, X.H.: Multicriteria decision-making approach based on gray linguistic weighted Bonferroni mean operator. Int. Trans. Oper. Res. **25**(5), 1635–1658 (2018)
16. Vance, A., Lowry, P.B., Eggett, D.L.: Increasing accountability through the user interface design artifacts: a new approach to addressing the problem of access-policy violations. MIS Q. **39**(2), 345–366 (2015)
17. Vishwanath, A., Harrison, B., Ng, Y.J.: Suspicion, cognition, and automaticity model of phishing susceptibility. Commun. Res. **45**(8), 1146–1166 (2018)
18. Williams, E.J., Hinds, J., Joinson, A.N.: Exploring susceptibility to phishing in the workplace. Int. J. Hum Comput Stud. **120**, 1–13 (2018)

An Approach to Quantification of Relationship Types Between Users Based on the Frequency of Combinations of Non-numeric Evaluations

A. Khlobystova[1,2(✉)], A. Korepanova[1,2], A. Maksimov[1,2], and T. Tulupyeva[1,2]

[1] Laboratory of Theoretical and Interdisciplinary Problems of Informatics, St. Petersburg Institute for Informatics and Automation of the Russian Academy of Sciences, 14-th Linia, VI, № 39,, St. Petersburg 199178, Russia
{aok, aak, agm, tvt}@dscs.pro

[2] Mathematics and Mechanics Faculty, St. Petersburg State University, Universitetskaya Emb., 7-9, St. Petersburg 199034, Russia

Abstract. The goal of this article is to propose an approach to linguistic values quantification and to consider an example of its application to the relationship types between users in the popular social network in Russia "VK". To achieve this aim, we used the results of a sociological survey, by which were found the frequency of the order, then the probability theory apparatus was used. This research can be useful in studying of the influence of the types of users' relationships on the execution of requests, also finds its use in building social graph of the organization's employees and indirectly in obtaining estimates of the success of multi-pass Social engineering attacks propagation.

Keywords: Social engineering · Multi-pass social engineering attacks · Linguistic variables · Linguistic values · Quantification · Analysis of social graph of company employees · Frequencies

1 Introduction

1.1 Prerequisites for Research

There is a tendency these days for the number and the quality of cybercrimes committed with social engineering methods to increase [16]. This is confirmed both by reports on major incidents associated with security breaches, private data compromises and fraudulent schemes [5, 6] and by large companies' reports on the current threats [7, 16]. For example, according to the US Federal Trade Commission [13], one of the most popular attack types is romance scams. In the United States alone in 2018 total loses from them are $143 million. All of the above determines the relevance and the necessity of the researches focused on developing automated methods reducing the vulnerability of information system users to Social engineering attacks.

The research was carried out in the framework of the project on state assignment SPIIRAS № 0073-2019-0003, with the financial support of the RFBR (project №18-01-00626, № 18-37-00323).

S. Kovalev et al. (Eds.): IITI 2019, AISC 1156, pp. 206–213, 2020.
https://doi.org/10.1007/978-3-030-50097-9_21

The social graphs of a company's employees are convenient to use in modelling and researching Social engineering attacks (in particular, multi-pass ones [9]) [14]. A social graph is a weighted directed graph which nodes represent information system users; edges represent the connections between them [1]. For building a social graph the information that is gathered automatically from social networks is often used. However, there are difficulties in matching the types of relationships between users (linguistic values—non-numeric data) to the weights of the edges (numeric values). The purpose of this article is to propose an approach to non-numeric data quantification and to consider an example of its application to the relationship types between users in the most popular social network in Russia—"VK" [11].

1.2 Related Work

[12] raised a question about the factors that affect user interaction in social network "Facebook". During this study the hypothesis that there is a direct link between the density of a social graph, the interactions between users of a social network and the clustering of the graph (segregation of the densest parts of the network) has been confirmed. These results can be applied in the further studies focused on the analysis of social graphs of employees in the context of Social engineering attacks, but the article doesn't consider the way to obtain the quantitative estimates of user interaction.

An approach presented in [3] can be also useful for the further studies. It is based on the correlation of the three factors derived from social networks (interaction on the wall, the number of common friends, the number of common photos) and the degree of trust between users. This research revealed that for users with a wide range of friends only common photos affect the degree of trust, for ones with a small range of photos all of the factors matters.

Authors of [4] describe the relationship of psychological traits to the success of phishing attacks. This study revealed that high narcissism and psychopathy are related to greater susceptibility to phishing emails (high probability of the response). The results are interesting, but there is a question of their applicability to other types of Social engineering attacks.

One of the aims of the study [17] is to investigate the connections between the behavior of people that use the application "Ant Forest" and the theories of conviction and motivation. The study revealed that the key factors for conviction are the existence of a main purpose and the interaction with other users. This information may be useful for modelling Social engineering attacks taking into account the competencies of the attacker [2].

Articles [1, 9, 14] provided a basis for the present study. They suggested a model of a social graph of a company's employees, and approaches to analyzing the graph. They also provided approaches to constructing estimates of security of users, represented by the nodes of this graph. However, the quantification of relationship types being derived from social network and matched to the weights of the edges hasn't been considered.

2 Problem Statement

Social graphs of the organization's employees are used for the automated construction of assessments of the level of user protection from social engineering impact [2, 14]. Such graphs can be constructed on the basis of information aggregated from social networks. This article considers the social network "VK" as the basis for analyzing user interaction, where there are several possible designations for relationship types (types of social connections) between two users. Knowledge of such relationships is used in assessing the criticality of multi-pass Social engineering attacks [9] and requires their presentation in the form of probabilistic values.

The purpose of the article is to propose a quantification approach for the values of the linguistic variable "type of relationship", which can be obtained from the social network and which characterize 1-to-1 relation between two users of this network. The quantification method for non-numerical characteristics proposed in [10] is used to achieve this goal.

3 Theoretical Part

Suppose we have the results of a survey in which respondents were asked to match non-numeric values of a certain linguistic variable to numerical values. We count the number of respondents who have arranged the linguistic values in the same order (more precisely, implicitly assigned to them the same rank in the evaluation). Such a number is called the frequency of the order.

Consider the example of its finding. Let the $q_1 - q_5$ correspond to the numerical values that the user with id = 84 has associated with the proposed linguistic variables: $q_1 = 52,\ q_2 = 68,\ q_3 = 84,\ q_4 = 35,\ q_5 = 68$. The variables will be ordered in increasing order of the numeric values associated with them and the order will be fixed: 41[25]3, square brackets denote that the values of the variables enclosed in them coincide. A program in C# has been developed to automate the process of obtaining frequencies, the input is a Microsoft Excel document with the results of the survey, and the output is the same document with added pages. Each of these pages contains the generated order of variables and the number of times the respondents have ordered linguistic variables this way. An example of the output of the program is shown in Table 1.

Then we apply the method of N. V. Khovanov to the received orders and frequencie [10]. To do this, we introduce the scale of possible probabilistic values as a discrete set with a given step N: $\left\{p_0 = 0, p_1 = \frac{1}{N}, \ldots, p_{N-1} = \frac{N-1}{N}, p_N = 1\right\}$. According to [10], we find for all such orders the expected value for Q (the value distribution for orders will be the introduces probability scale) and the expected value of the expected value (in the calculation, obtained frequencies divided by the total number of respondents are used as the distribution).

Table 1. An example of a part of the processed survey results.

Order	Frequency	Deciphering of the order				
[12345]	21	q_1	q_2	q_3	q_4	q_5
[1345]2	7	q_1	q_3	q_4	q_5	q_2
[51]342	6	q_5	q_1	q_3	q_4	q_2
15342	5	q_1	q_5	q_3	q_4	q_2
1[345]2	5	q_1	q_3	q_4	q_5	q_2
13452	5	q_1	q_3	q_4	q_5	q_2
14352	4	q_1	q_4	q_3	q_5	q_2
13[45]2	4	q_1	q_3	q_4	q_5	q_2
15432	4	q_1	q_5	q_4	q_3	q_2
[45]132	4	q_4	q_5	q_1	q_3	q_2
51432	4	q_5	q_1	q_4	q_3	q_2
...						

Let Q_i—random variable, taking some probabilistic value, k—number of the order, r_k—frequency corresponding to this order k. Let's consider the example of the application of this method. Let $k = 5$ corresponds to row 5 of Table 1, so $r_5 = 5$ respondents arranged linguistic variables as follows: $\left\langle Q_1^5, \left[Q_3^5, Q_4^5, Q_5^5\right], Q_2^5\right\rangle$. Q_i^5, $i = 1..5$ may take values from the set: $\left\{0, \frac{1}{4}, \frac{1}{2}, \frac{3}{4}, 1\right\}$. Then may be 10 different variants of the random variables' distribution Q_i^5, $i = 1..5$ (Table 2), and the probability that random variables will be distributed according to the j-th line is 1/10.

Table 2. An example of the random variables' with a fixed order probability distribution.

	0	1/4	1/2	3/4	1
1	Q_1^5	$[Q_3^5, Q_4^5, Q_5^5]$	Q_2^5		
2	Q_1^5	$[Q_3^5, Q_4^5, Q_5^5]$		Q_2^5	
3	Q_1^5	$[Q_3^5, Q_4^5, Q_5^5]$			Q_2^5
4	Q_1^5	$[Q_3^5, Q_4^5, Q_5^5]$	Q_2^5		
5	Q_1^5		$[Q_3^5, Q_4^5, Q_5^5]$	Q_2^5	
6	Q_1^5		$[Q_3^5, Q_4^5, Q_5^5]$		Q_2^5
7		Q_1^5	$[Q_3^5, Q_4^5, Q_5^5]$	Q_2^5	
8		Q_1^5	$[Q_3^5, Q_4^5, Q_5^5]$		Q_2^5
9		Q_1^5		$[Q_3^5, Q_4^5, Q_5^5]$	Q_2^5
10			Q_1^5	$[Q_3^5, Q_4^5, Q_5^5]$	Q_2^5

Therefore, the expected value for Q_1^5 can be calculated as follows: $\mathrm{E}[Q_1^5] = 0 \cdot 6/10 + 1/4 \cdot 3/10 + 1/2 \cdot 1/10 = 1/8$. Similarly: $\mathrm{E}[Q_3^5] = \mathrm{E}[Q_4^5] = \mathrm{E}[Q_5^5] = 9/20$, $\mathrm{E}[Q_2^5] = 33/40$.

Calculate the expected value for each of the presented orders and find:

$$\forall i \ \ \mathrm{EE}\left[Q_i^k\right] = \sum_{k=1}^{m} \frac{r_k}{n} \cdot \mathrm{E}\left[Q_i^k\right], \tag{1}$$

where n—is the number of all respondents, m—is the number of different orders.

4 Practical Significance

The input data for the practical application of the proposed approach were the results of a psychological survey, the question of which was formulated as follows: "Imagine the following situation: You have been invited to join the group "VK". Please, estimate the probability that you would respond to the request, if you received an invitation from a person who is marked in your account "VK" as:". "Type of relationship" between users was seen as linguistic variable. This variable can have values represented in the social network "VK". Examples of such values are: "Friends", "Best friends", "Colleagues", "School friends", "University friends", etc. A total of 17 such values were considered. These values were divided (including the social network itself) into three categories: "Friend lists", "Relationship" and "Relatives". Respondents were asked to set the slider on a scale from 0 to 100, where a value of 100 corresponds to the maximum probability of performing some action if asked by a user who has the specified characteristics. 0 corresponds to the fact that the request received from the user with these characteristics will not be responded under any conditions.

Then, frequencies were calculated according with obtained results. An example of a part of the calculated results is presented in the Table 1, where q_1 corresponds the linguistic value "Friend", q_2—"Best friend", q_3—"Colleagues", q_4—"School friends" and q_5—"University friends". Then in each of the received orders for each random variable its expectation was found. An example of a part of the obtained values is presented in the Table 3.

Table 3. An example of the calculated expectation for a part of the results in the category "Friend lists".

Order	Frequency	$E[Q_1^5]$	$E[Q_2^5]$	$E[Q_3^5]$	$E[Q_4^5]$	$E[Q_5^5]$
[12345]	21	1/2	1/2	1/2	1/2	1/2
[1345]2	7	1/4	3/4	1/4	1/4	1/4
[51]342	6	1/20	19/20	7/20	13/20	1/20
15342	5	0	1	1/2	3/4	1/4
13452	5	0	1	1/4	1/2	3/4
1[345]2	5	1/8	9/20	9/20	9/20	33/40
14352	4	0	1	1/2	1/4	3/4

(continued)

Table 3. (*continued*)

Order	Frequency	$E[Q_1^5]$	$E[Q_2^5]$	$E[Q_3^5]$	$E[Q_4^5]$	$E[Q_5^5]$
13[45]2	4	1/20	19/20	7/20	13/20	13/20
15432	4	0	1	1/2	3/4	1/4
[45]132	4	7/20	19/20	13/20	1/20	1/20
51432	4	1/4	1	3/4	1/2	0
53412	3	3/4	1	1/4	1/2	0
13542	3	0	1	1/4	3/4	1/2
[34]512	3	13/20	19/20	1/20	1/20	7/20
1[54]23	2	1/20	13/20	19/20	7/20	7/20
14532	2	0	1	1/2	1/4	3/4
134[25]	2	1/20	19/20	7/20	13/20	19/20
1[53]42	2	1/20	19/20	7/20	13/20	7/20
[45]312	2	13/20	19/20	7/20	1/20	1/20
[13]542	2	1/20	19/20	1/20	13/20	7/20
[34][15] 2	2	9/20	33/40	1/8	1/8	9/20
[345]21	2	33/40	9/20	1/8	1/8	1/8
54312	2	3/4	1	1/2	1/4	0
…						

Then, according to the formula (1), for all $i = 1..17$ $\mathrm{EE}\left[Q_i^k\right]$ was calculated. For example, $\mathrm{EE}\left[Q_1^5\right] = \frac{1}{2}\cdot\frac{21}{145} + \frac{1}{4}\cdot\frac{7}{145} + \frac{1}{20}\cdot\frac{6}{145} + \ldots + \frac{9}{20}\cdot\frac{1}{145} = \frac{213}{725} \approx 0.2938$. The results of the linguistic values quantification are presented in the Table 4.

Table 4. The results of the quantification of the linguistic values variable "type of relationship"

Types of relationships	Mapped estimate of probability
Friends	$213/725 \approx 0.2938$
Best friends	$29/37 \approx 0.7838$
Colleagues	$11/27 \approx 0.4074$
School friends	$367/826 \approx 0.4443$
University friends	$195/529 \approx 0.3686$
Family	$312/857 \approx 0.3641$
Grandparents	$187/746 \approx 0.2507$
Parents	$13/38 \approx 0.3421$
Siblings	$230/523 \approx 0.4398$
Children	$41/100 \approx 0.41$
Grandchildren	$74/213 \approx 0.3474$
In a relationship	$234/761 \approx 0.3075$
Engaged	$206/663 \approx 0.3107$
Married	$223/588 \approx 0.3793$
In a civil union	$39/121 \approx 0.3223$
In love	$111/265 \approx 0.4189$
It's complicated	$173/900 \approx 0.1922$

5 Conclusion

Thus, in the article, an approach to quantify of the values of linguistic variables was proposed. Also the practical application of the approach to the linguistic variable “type of relationships” was considered, the values of which can be extracted from the social network “VK”. This research can be useful in studying the influence of the types of users’ relationships on the response to request, also finds its use in building social graph of the organization’s employees and indirectly in obtaining estimates of the success of multi-pass Social engineering attacks propagation. Prospects for further research are to consider applying the theory of Bayesian networks to Social engineering attacks [8, 15], in particular, for solving backtracking tasks during multi-pass attacks.

References

1. Abramov, M., Tulupyeva, T., Tulupyev, A.: Social Engineering Attacks: social networks and user security estimates. SUAI, St. Petersburg (2018), 266 p.
2. Azarov, A.A., Abramov, M.V., Tulupyeva, T.V., Tulupyev, A.L.: Users’ of Information System Protection Analysis from Malefactor’s Social Engineering Attacks Taking into Account Malefactor’s Competence Profile, Biologically Inspired Cognitive Architectures (BICA) for Young Scientists, pp. 25–30 (2016)
3. Bapna, R., Gupta, A., Rice, S., Sundararajan, A.: Trust and the strength of ties in online social networks: an exploratory field experiment. MIS Q. **41**(1), 115–130 (2017)
4. Curtis, S.R., Rajivan, P., Jones, D.N., Gonzalez, C.: Phishing attempts among the dark triad: patterns of attack and vulnerability. Comput. Hum. Behav. **87**, 174–182 (2018)
5. FBI Warns of Chinese Law Enforcement Impersonation Scam. https://www.fbi.gov/contact-us/field-offices/seattle/news/press-releases/fbi-warns-of-chinese-law-enforcement-impersonation-scam. Accessed 12 Apr 2019
6. Fifth Bronx Man Pleads Guilty In Multimillion-Dollar Ghana-Based Fraud Scheme Involving Business Email Compromises And Romance Scams Targeting Elderly. https://www.justice.gov/usao-sdny/pr/fifth-bronx-man-pleads-guilty-multimillion-dollar-ghana-based-fraud-scheme-involving. Accessed 15 Apr 2019
7. Is social engineering the biggest threat to your organization? https://www.microsoft.com/security/blog/2017/04/19/is-social-engineering-the-biggest-threat-to-your-organization/. Accessed 12 Dec 2018
8. Kharitonov, N.A., Maximov, A.G., Tulupyev, A.L.: Algebraic Bayesian networks: the use of parallel computing while maintaining various degrees of consistency. Studies in Systems, Decision and Control, vol. 199, pp. 696–704. Springer (2019)
9. Khlobystova, A.O., Abramov, M.V., Tulupyev, A.L., Zolotin, A.A.: Identifying the most critical trajectory of the spread of a social engineering attack between two users. Informatsionno-upravliaiushchie sistemy (Inf. Control Syst.) **6**, 74–81 (2018)
10. Khovanov, N.V.: Measurement of a discrete indicator utilizing nonnumerical, inaccurate, and incomplete information. Meas. Tech. **46**(9), 834–838 (2003)
11. Leading active social media platforms in Russia in 2018. https://www.statista.com/statistics/867549/top-active-social-media-platforms-in-russia/. Accessed 10 Apr 2019
12. Maiz, A., Arranz, N., Fdez. de Arroyabe, J.C.: Factors affecting social interaction on social network sites: the Facebook case. J. Enterp. Inf. Manag. **29**(5), 630–649 (2016)

13. Schifferle. L.W.: Romance scams will cost you. https://www.consumer.ftc.gov/blog/2019/02/romance-scams-will-cost-you. Accessed 04 Apr 2019
14. Suleimanov, A., Abramov, M., Tulupyev, A.: Modelling of the social engineering attacks based on social graph of employees communications analysis. In: Proceedings of 2018 IEEE Industrial Cyber-Physical Systems (ICPS), St.-Petersburg, pp. 801–805 (2018)
15. Tulupyev, A., Kharitonov, N., Zolotin, A.: Algebraic Bayesian networks: consistent fusion of partially intersected knowledge systems. In: The Second International Scientific and Practical Conference "Fuzzy Technologies in the Industry – FTI 2018". CEUR Workshop Proceedings, pp. 109–115 (2018)
16. Weekly Threat Report 10th August 2018. https://www.ncsc.gov.uk/report/weekly-threat-report-10th-august-2018. Accessed 19 Feb 2019
17. Yang, Z., Kong, X., Sun, J., Zhang, Y.: Switching to green lifestyles: behavior change of ant forest users. Int. J. Environ. Res. Public Health **15**(9), 1819 (2018)

Algebraic Bayesian Networks: Parallel Algorithms for Maintaining Local Consistency

Nikita A. Kharitonov[1], Anatolii G. Maksimov[1,2(✉)], and Alexander L. Tulupyev[1,2]

[1] St. Petersburg Institute for Informatics and Automation of the Russian Academy of Sciences, 39, 14-th Line Vasilyevsky Ostrov, St. Petersburg 199178, Russia
{nak,agm}@dscs.pro
[2] St. Petersburg State University, 7-9, University Embankment, St. Petersburg 199034, Russia
alt@dscs.pro

Abstract. Algebraic Bayesian networks belong to the class of machine-learning probabilistic graphical models. One of the main tasks during researching machine learning models is the optimization of their time of work. This paper presents approaches to parallelizing algorithms for maintaining local consistency in algebraic Bayesian networks as one of the ways to optimize their time of work. An experiment provided to compare the time of parallel and nonparallel realizations of algorithms for maintaining local consistency.

Keywords: Algebraic Bayesian networks · Probabilistic graphic models · Consistency · Parallel computing · Knowledge pattern · Machine learning · Bayesian networks · Probabilistic-logical inference

1 Intro

It is often necessary to work with imperfect data [1,2] in machine learning, or in a decisions making, or in a number of other areas. This can happen if the data are incomplete or, for example, when generating data from expert assessments [3]. One of the models that can be learned on such data are algebraic Bayesian networks [4], belonging to the class of probabilistic graphical models [5,6].

One of the most important natural problems arising from the study of algebraic Bayesian networks is the problem of optimizing and shorting the time of work with the model, including the processes of checking and maintaining consistency, that is, the correctness of the estimates of the elements probability presented in the network. This paper addresses the issue of optimizing these processes through their parallelization.

The purpose of this work is the reduction of time costs while maintaining local consistency through the use of parallel algorithms.

S. Kovalev et al. (Eds.): IITI 2019, AISC 1156, pp. 214–222, 2020.
https://doi.org/10.1007/978-3-030-50097-9_22

2 Relevant Papers

As mentioned above, algebraic Bayesian networks [4] belong to the class of probabilistic graphical models (PGM) [7,8]. PGM also include Bayesian belief networks [9,10], Markov chains [11], etc. All these models are able to learn using machine learning, and the process of their learning is different from the process of learning artificial neural networks [12–14].

An algebraic Bayesian network is a collection of knowledge patterns which are represented with conjunct ideals with given scalar or interval estimates of probabilities [4] (Fig. 1).

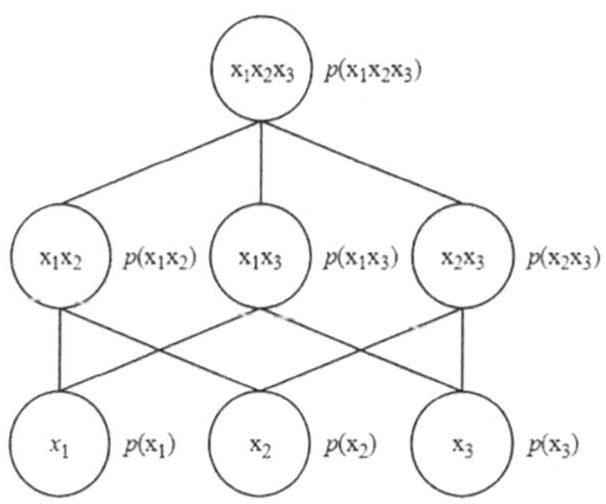

Fig. 1. An example of a knowledge pattern

It is necessary to check the coherence of the estimates of probabilities, that is, to maintain consistency, when creating or changing a network. In the works [4, 15], four types of consistency were outlined:

- local—the consistency of each knowledge pattern;
- external—local consistency and coincidence of assessments of conjuncts at the intersections of knowledge patterns;
- internal—local consistency and the possibility of choosing consistent scalar estimates across the entire network for an arbitrarily chosen estimate of an arbitrary conjunct;
- global—the ability to immerse the entire network in a single consistent knowledge pattern, without changing the estimates.

Sequential algorithms for maintaining local consistency, that is, consistency in a single knowledge pattern, were proposed in [16]. Parallel algorithms for maintaining external and internal consistency already were considered in [17].

Similar issues in Bayesian belief networks were considered in [18].

So, to the author's knowledge, the research about parallelizing processes of local consistency maintaining in knowledge pattern of algebraic Bayesian network has never been published before.

3 Local Consistency in the Case of Scalar Estimates

Let consider a knowledge pattern with scalar estimates of the probability of truths of conjuncts and deal with the task of parallelizing the algorithm for maintaining local consistency in it. Let us denote the number of atoms in a knowledge pattern by n.

The algorithm for maintaining consistency in a knowledge pattern with scalar estimates was described in [16]. To verify consistency with scalar estimates it is necessary and sufficient to check the following condition [16]:

$$\mathbf{I}_n \times \mathbf{P}^{(n)} \geqslant \mathbf{0}^{\langle n \rangle}, \tag{1}$$

when $\mathbf{I}_n$ – defined recursively special type matrix and $\mathbf{0}^{\langle n \rangle}$ – column vector of zeros of height n. The rules for constructing matrices $\mathbf{I}_n$ based on the Kronecker matrix product are given below.

$$\mathbf{I}_1 = \begin{bmatrix} 1 & -1 \\ 0 & 1 \end{bmatrix}, \quad \mathbf{I}_n = \begin{bmatrix} \mathbf{I}_{n-1} & -\mathbf{I}_{n-1} \\ \mathbf{0} & \mathbf{I}_{n-1} \end{bmatrix} = \mathbf{I}_1 \otimes \mathbf{I}_{n-1};$$

$$\mathbf{P}^{(2)} = \begin{Bmatrix} p(e_\wedge) \\ p(x_1) \\ p(x_2) \\ p(x_1x_2) \end{Bmatrix}, \quad \mathbf{I}_2 \times \mathbf{P}^{(2)} = \begin{Bmatrix} p(x_1x_2) \\ p(x_1) - p(x_1x_2) \\ p(x_2) - p(x_1x_2) \\ 1 - p(x_1) - p(x_2) + p(x_1x_2) \end{Bmatrix}.$$

The fulfillment of conditions (1) ensures that the estimates in the knowledge pattern do not contradict the axioms of the theory of probability. In this case, the verification of one inequality begins when the verification of the previous one ends. Note also that the probability values do not change in any way during the operation of the algorithm. All this means that we can check the implementation of inequalities in different processes in parallel. Listing is set out below.

Algorithm 1. Parallel algorithm for maintaining local consistency in a knowledge pattern with scalar estimates

Require: $\mathbf{P}^{(n)}$
Ensure: $\mathbb{T}/\mathbb{F}$
  $\mathbf{I}_n = \mathbf{1}^{(0)}$
  for $i = 1..n$ **do**
    $\mathbf{I}_n = \mathbf{I}_1 \otimes \mathbf{I}_n$
  end for
  $\mathcal{E}^{(n)} := \mathbf{I}_n \times \mathbf{P}^{(n)} \geqslant \mathbf{o}^{(n)}$
  for all $e \in \mathcal{E}^{(n)}$ **do in parallel**
    if $\overline{e}$ **then**
      return $\mathbb{F}$
    end if
  end for
  return $\mathbb{T}$

4 Consistency of a Knowledge Pattern with Interval Estimates

We now consider a knowledge pattern with interval estimates of the probability of truth of conjuncts, and we will maintain local consistency. While maintaining consistency for a knowledge pattern, two sets of constraints are introduced $\mathcal{R}^{(n)} = \mathcal{E}^{(n)} \cup \mathcal{D}^{(n)}$, when $\mathcal{E}^{(n)}$ – constraints due to axiomatic probability theory and $\mathcal{D}^{(n)}$ – constraints due to subject area. The set of constraints $\mathcal{D}^{(n)}$ can be presented as $\mathbf{P}_0^{-,(n)} \leqslant \mathbf{P}^{(n)} \leqslant \mathbf{P}_0^{-,(n)}$, when $\mathbf{P}_0^{+,(n)}$ – vector of lower bounds of probability, and $\mathbf{P}_0^{+,(n)}$ – vector of upper bounds of probability. Example of set $\mathcal{D}^{(n)}$ is set out below. Set of restrictions $\mathcal{E}^{(n)}$, as mentioned above, can be represented as $\mathbf{I}_n \times \mathbf{P}^{(n)} \geqslant \mathbf{0}^{\langle n \rangle}$. Examples of $\mathcal{E}^{(n)}$ is set out below.

$$\mathcal{D}^{(2)} = \left\{ \begin{array}{l} 1 \leqslant 1 \leqslant 1 \\ p_0^-(x_1) \leqslant p(x_1) \leqslant p_0^+(x_1) \\ p_0^-(x_2) \leqslant p(x_2) \leqslant p_0^+(x_2) \\ p_0^-(x_1x_2) \leqslant p(x_1x_2) \leqslant p_0^+(x_1x_2) \end{array} \right\}$$

An example of restrictions imposed by a subject area for a fragment of knowledge with two atoms

$$\mathcal{E}^{(2)} = \left\{ \begin{array}{l} p(x_1x_2) \geqslant 0 \\ p(x_1) - p(x_1x_2) \geqslant 0 \\ p(x_2) - p(x_1x_2) \geqslant 0 \\ 1 - p(x_1) - p(x_2) + p(x_1x_2) \geqslant 0 \end{array} \right\}$$

An example of the limitations imposed by the axiomatic theory of probability for a knowledge pattern with three atoms.

Then for each element f from a knowledge pattern, linear programming problems are solved of the form

$$p^-(f) = \min_{\mathcal{R}^{(n)}} \{p(f)\}$$

$$p^+(f) = \max_{\mathcal{R}^{(n)}} \{p(f)\}.$$

If the linear programming problem has no solution, the conclusion is that the knowledge pattern is contradictory. Otherwise, we obtain the vectors $\mathbf{P}^{-,(n)}$ and $\mathbf{P}^{+,(n)}$ with updated estimates of the probability [16].

In addition, each linear programming problem is solved separately, which makes the process of maintaining consistency rather time consuming. At the same time, the results of each of the linear programming problems are not involved in any way in solving subsequent ones; therefore, this process can be performed not sequentially, but in parallel. Pseudo code is shown in Listing 2.

Algorithm 2. Parallel algorithm for maintaining local consistency in a knowledge pattern with interval estimates

Require: $\mathbf{P}_0^{-,(n)}, \mathbf{P}_0^{-,(n)}$
Ensure: $\mathbf{P}^{-,(n)}, \mathbf{P}^{-,(n)}$ or $\mathbb{F}$
  $\mathbf{I}_n = \mathbf{1}^{(0)}$
  for $i = 1..n$ **do**
    $\mathbf{I}_n = \mathbf{I}_1 \otimes \mathbf{I}_n$
  end for
  $\mathcal{E}^{(n)} := \mathbf{I}_n \times \mathbf{P}^{(n)} \geqslant \mathbf{0}^{(n)}$
  $\mathcal{D}^{(n)} := \mathbf{P}_0^{-,(n)} \leqslant \mathbf{P}^{(n)} \leqslant \mathbf{P}_0^{-,(n)}$
  $\mathbb{R}^{(n)} := \mathbb{E}^{(n)} \cup \mathbb{D}^{(n)}$
  for all f **do in parallel**
    $p^+(f) = \max_{\mathcal{R}^{(n)}}\{p(f)\}$
    $p^-(f) = \min_{\mathcal{R}^{(n)}}\{p(f)\}$
    if no solutions **then**
      return $\mathbb{F}$
    end if
  end for
  return $\mathbf{P}^{-,(n)}, \mathbf{P}^{+,(n)}$

5 Experiments

The proposed approach to the parallel maintenance of consistency was implemented in the framework of the program complex in C# [19,20].

Knowledge patterns containing from 2 to 8 atoms were considered. For each number of atoms, 15 knowledge patterns were constructed with different estimates of the probabilities.

For each knowledge pattern, several maintenance of consistency was carried out with a return to the initial estimates and the time of work was measured. A total of 300 measurements were carried out for each number of atoms in the knowledge pattern (Fig. 2).

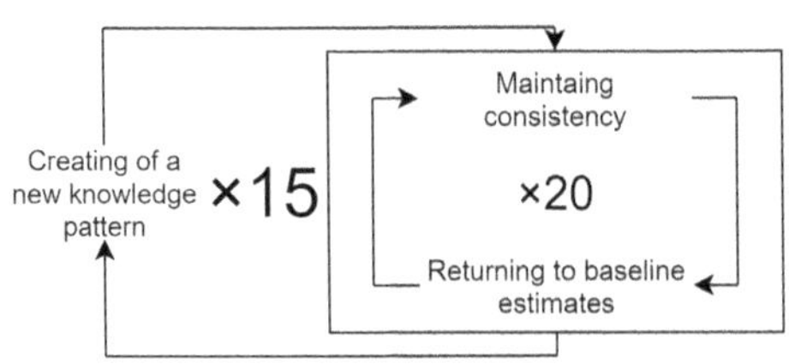

Fig. 2. Design of the experiment for fixed number of atoms in knowledge pattern

Then the time for each knowledge pattern with fixed estimates was averaged. The graphs (Figs. 3 and 4) show the mean (*mean*), median (*median*), maximum (*max*), minimum (*min*), first and third quartiles ($q1$ and $q3$ respectively) and first and ninth deciles ($d1$ and $d9$ respectively) for a set of 15 means for each n. In addition, for clarity, the graph (Fig. 5) shows the average operating time for the sequential (meanunp) and parallel (meanp) algorithms.

The results of experiments confirm that the parallel algorithm does give a gain in efficiency compared to the sequential one, as can be seen from the graphs.

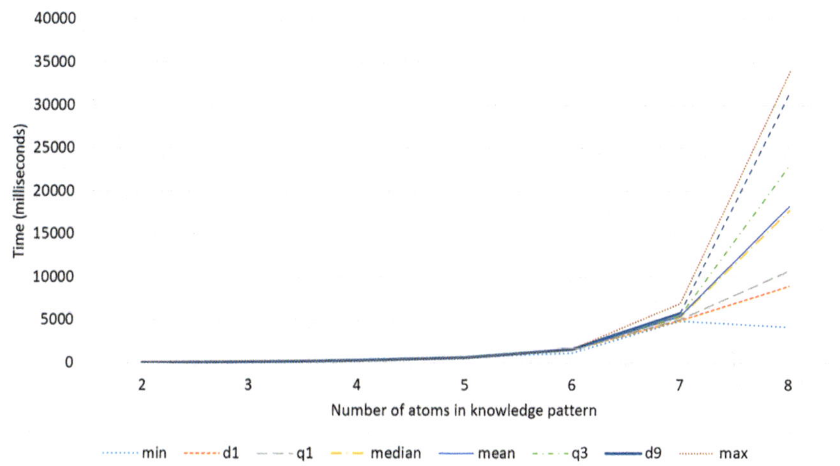

Fig. 3. Experimental results for a sequential algorithm

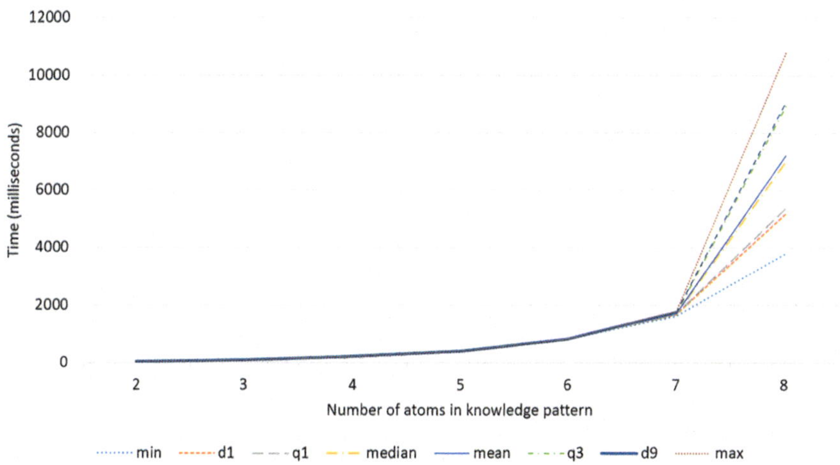

Fig. 4. Experimental results for parallel algorithm

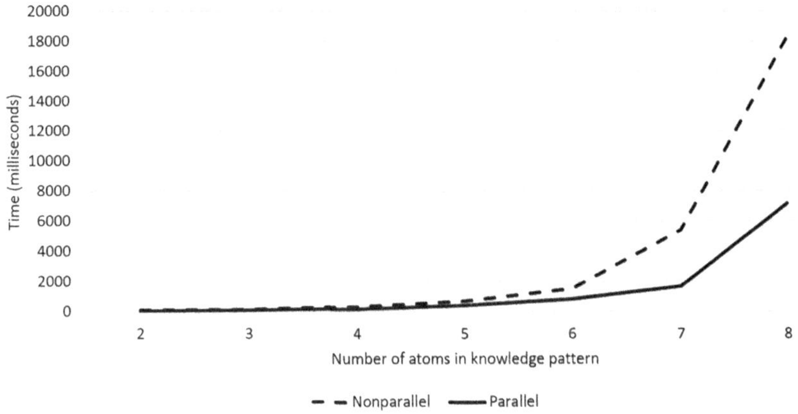

Fig. 5. Matching of average times of algorithms work

6 Conclusion

Parallel algorithms were proposed and implemented to test and maintain consistency in knowledge patterns with scalar and interval probability estimates. Practical measurements of the operating time were carried out, which confirmed the effectiveness of the proposed algorithms. The results of the study, as well as further studies, are planned to be used, for example, in studies of social engineering attacks [21–23].

Acknowledgments. The research was carried out in the framework of the project on SPIIRAS governmental assignment No. 0073-2019-0003, with the financial support of the RFBR (project No. 18-01-00626: Methods of representation, synthesis of truth estimates and machine learning in algebraic Bayesian networks and related knowledge models with uncertainty: the logic-probability approach and graph systems).

References

1. Pelissari, R., Oliveira, M.C., Ben Amor, S., Abackerli, A.J.: A new FlowSort-based method to deal with information imperfections in sorting decision-making problems. Eur. J. Oper. Res. **276**(10), 235–246 (2019). https://doi.org/10.1016/j.ejor.2019.01.006
2. Saha, I., Sarkar, J.P., Maulik, U.: Integrated rough fuzzy clustering for categorical data analysis. Fuzzy Sets Syst. **361**, 1–32 (2019). https://doi.org/10.1016/j.fss.2018.02.007
3. Cheng, J., Wang, J.: An association-based evolutionary ensemble method of variable selection. Expert Syst. Appl. **124**, 143–155 (2019). https://doi.org/10.1016/j.eswa.2019.01.039
4. Tulupyev, A.L., Nikolenko, S.I., Sirotkin, A.V.: Fundamentals of the Theory of Bayesian Networks: Textbook. St.-Petersburg University, Saint-Petersburg (2019). (in Russian)

5. Ye, J., Li, J., Newman, M.G., Adams, R.B., Wang, J.Z.: Probabilistic multigraph modeling for improving the quality of crowdsourced affective data. IEEE Trans. Affect. Comput. **10**(1), 115–128 (2019). https://doi.org/10.1109/TAFFC.2017.2678472
6. Qiang, Y.-T., Fu, Y.-W., Yu, X., Guo, Y.-W., Zhou, Z.-H., Sigal, L.: Learning to generate posters of scientific papers by probabilistic graphical models. J. Comput. Sci. Technol. **34**(1), 155–169 (2019). https://doi.org/10.1007/s11390-019-1904-1
7. Vogel, K., Weise, L., Schroter, K., Thieken, A.H.: Identifying driving factors in flood-damaging processes using graphical models. Water Resour. Res. **54**(11), 8864–8889 (2018). https://doi.org/10.1029/2018WR022858
8. Buscombe, D., Grams, P.E.: Probabilistic substrate classification with multispectral acoustic backscatter: a comparison of discriminative and generative models. Geosciences **8**(11) (2018). Article no. UNSP395. https://doi.org/10.3390/geosciences8110395
9. Huang, Z.M., Yang, L., Jiang, W.: Uncertainty measurement with belief entropy on the interference effect in the quantum-like Bayesian Network. Appl. Math. Comput. **347**, 417–428 (2019). https://doi.org/10.1016/j.amc.2018.11.036
10. Marella, D., Vicard, P.: Toward an integrated Bayesian network approach to measurement error detection and correction. Commun. Stat.-Simul. Comput. **48**(2), 544–555 (2019). https://doi.org/10.1080/03610918.2017.1387664
11. Suwanwimolkul, S., Zhang, L., Gong, D., Zhang, Z., Chen, C., Ranasinghe, D.C., Shi, J.Q.: An adaptive Markov random field for structured compressive sensing. IEEE Trans. Image Process. **28**(3), 1556–1570 (2019). https://doi.org/10.1109/TIP.2018.2878294
12. Dolgiy, A.I., Kovalev, S.M., Kolodenkova, A.E.: Processing heterogeneous diagnostic information on the basis of a hybrid neural model of Dempster-Shafer. Commun. Comput. Inf. Sci. **934**, 79–90 (2018). https://doi.org/10.1007/978-3-030-00617-4_8
13. Ojha, V.K., Abraham, A., Snášel, V.: Metaheuristic design of feedforward neural networks: a review of two decades of research. Eng. Appl. Artif. Intell. **60**, 97–116 (2017). https://doi.org/10.1016/j.engappai.2017.01.013
14. Tai, W.P., Teng, Q.Y., Zhou, Y.M., Zhou, J.P., Wang, Z.: Chaos synchronization of stochastic reaction-diffusion time-delay neural networks via non-fragile output-feedback control. Appl. Math. Comput. **354**, 115–127 (2019). https://doi.org/10.1016/j.amc.2019.02.028
15. Tulupyev, A.L.: Algebraic Bayesian Networks: Global Logical and Probabilistic Inference in Joint Trees: A Tutorial, 2nd edn. SPb: VVM, Saint-Petersburg (2019). (in Russian)
16. Tulupyev, A.L.: Algebraic Bayesian Networks: Local Logical and Probabilistic Inference: A Tutorial, 2nd edn. SPb: VVM, Saint-Petersburg (2019). (in Russian)
17. Kharitonov, N.A., Maximov, A.G., Tulupyev, A.L.: Algebraic Bayesian networks: the use of parallel computing while maintaining various degrees of consistency. Studies in Systems, Decision and Control, vol. 199, pp. 696–704 (2019). https://doi.org/10.1007/978-3-030-12072-6_56
18. Zhao, L., Zhou, Y.H., Lu, H.P., Fujita, H.: Parallel computing method of deep belief networks and its application to traffic flow prediction. Knowl.-Based Syst. **163**, 972–987 (2019). https://doi.org/10.1016/j.knosys.2018.10.025
19. Kharitonov N., Tulupyev A., Zolotin A.: Software implementation of reconciliation algorithms in algebraic Bayesian networks. In: Proceedings of 2017 XX IEEE International Conference on Soft Computing and Measurements (SCM), pp. 8–10 (2017). https://doi.org/10.1109/SCM.2017.7970479

20. Mal'chevskaya, E.A., Berezin, A.I., Zolotin, A.A., Tulupyev, A.L.: Algebraic Bayesian networks: local probabilistic-logic inference machine architecture and set of minimal joint graphs. Advances in Intelligent Systems and Computing, vol. 451, pp. 69–79 (2016)
21. Abramov, M.V., Azarov, A.A.: Identifying user's of social networks psychological features on the basis of their musical preferences. In: Proceedings of 2017 XX IEEE International Conference on Soft Computing and Measurements (SCM), pp. 90–92 (2017). https://doi.org/10.1109/SCM.2017.7970504
22. Bagretsov, G.I., Shindarev, N.A., Abramov, M.V., Tulupyeva, T.V.: Approaches to development of models for text analysis of information in social network profiles in order to evaluate user's vulnerabilities profile. In: Proceedings of 2017 XX IEEE International Conference on Soft Computing and Measurements (SCM), pp. 93–95 (2017). https://doi.org/10.1109/SCM.2017.7970505
23. Shindarev, N., Bagretsov, G., Abramov, M., Tulupyeva, T., Suvorova, A.: Approach to identifying of employees profiles in websites of social networks aimed to analyze social engineering vulnerabilities. Advances in Intelligent Systems and Computing, vol. 679, pp. 441–447 (2018)

Decision Making Intelligent Systems

Development of an Intelligent Decision Support System for Electrical Equipment Diagnostics at Industrial Facilities

Anna E. Kolodenkova[1](✉), Svetlana S. Vereshchagina[1], and Evgenia R. Muntyan[2]

[1] Samara State Technical University, Samara, Russia
anna82_42@mail.ru
[2] Southern Federal University, Taganrog, Russia

Abstract. This paper presents the description of the basic principles of designing an Intelligent Decision Support System (IDSS) for diagnosing electrical equipment (EE) of industrial facilities while in operation based on the data received from the measurement technology using soft computing methods and their combinations, as well as fuzzy cognitive modeling. Since the development of an IDSS for diagnosing EE is a complex task that requires the study of a large number of interconnected modules, the work includes detailed information on the IDSS architecture, IDSS operating principles and basic capabilities of the system. By way of example, some objective-settings solved by the system, as well as fragments of screen forms of the developed system have been shown. The proposed IDSS will make it possible not only to assess the EE condition at a given time under conditions of a wide range of monitored parameters, but also to predict their values under conditions of statistical and fuzzy data. That will help to identify EE defects and failures at an early stage of their development; to prevent emergencies and reduce the risk of man-made disasters; to increase the validity of making decisions on EE faults and the equipment as a whole, as well as to give troubleshooting recommendations.

Keywords: Electrical equipment · Database knowledge · Soft computing methods fuzzy cognitive modeling

1 Introduction

Modern electrical equipment at industrial facilities is characterized by technological complexity, a large number of parameters, complex structural and functional relationships between parameters and equipment, the presence of restrictions on changes to these parameters, operation under external impacts, etc.

All of this may cause the accumulation of defects, an early ultimate state, EE reduced lifetime, equipment failure, a disruption of the production process, power supply interruption and problems in safety of industrial facilities, as well as power failure with large-scale consequences [1–3].

The work was supported by RFBR grants No. 19-07-00195, No. 19-08-00152.

S. Kovalev et al. (Eds.): IITI 2019, AISC 1156, pp. 225–233, 2020.
https://doi.org/10.1007/978-3-030-50097-9_23

Currently, an urgent task is to maintain a good level of reliable operation and a high level of equipment fault tolerance, it is necessary to ensure its high-quality maintenance, as well as the timely diagnostics of possible defects and failures.

The problems of diagnosing the technical conditions of EE at industrial facilities in real time and predicting its efficiency can not be solved without the use of IDSS.

This is due to the fact that these systems can use unstructured data; operate in situations with a high degree of uncertainty; use heterogeneous information, experts' knowledge and accumulated professional experience; adapt easily to changing requirements; improve the efficiency of the decision-making process and generate management recommendations in real time of diagnosing and predicting the equipment.

In view of the above, this paper proposes the IDSS development that provides continuous monitoring and analysis of EE technical conditions, decisions making on the necessary maintenance, repair, and messages on divergence of parameters in real time based on the analysis and continuously available or discrete information on its technical condition, as well as preventing emergencies at an early stage.

2 Analysis of Existing DSS for Electrical Equipment Diagnostics

Currently, there are quite a large number of domestic and foreign DSS for continuous monitoring of EE based on fuzzy logic systems, methods of heterogeneous data clustering and probabilistic-statistical methods [4–8]. However, the problem of EE diagnosing during the operational phase still remains open even now, since the main disadvantages of most such systems are [3, 9–14] as follows:

1. impossibility to fill the system with new knowledge about EE, the external environment and information about previous EE diagnostics;
2. low level of automation of diagnostic heterogeneous information processing;
3. impossibility to use the most of the systems for different types of EE;
4. limited use of artificial intelligence technologies;
5. impossibility to configure the system for application in all industries (oil, nuclear, chemical, etc.);
6. similarity of data structures of systems for EE diagnostics.

The most well known DSS for EE diagnostics used at Russian enterprises and abroad are as follows:

1. The *Diagnostics +* system for certification, technical condition assessment and prediction of electrical equipment operation (transformers, autotransformers, shunt reactors) of stations and substations [12]. This system was developed at the Centre for design and reliability of electrical equipment at Ivanovo State Power Engineering University. The *Diagnostics +* system is notable for the principles of software implementation, a scope of equipment for diagnostics, and a knowledge base (KB) in the form of a shell which allows users to create, modify procedural and declarative knowledge of the system. The proposed system addresses such

objectives as maintenance of technical data sheets of the equipment, viewing the specific objects data obtained by monitoring systems, logging of failures and defects, assessment of the condition of one or several pieces of equipment, establishment of new rules into the KB and generation of the protocol with the diagnostic test results.
2. The expert-diagnostic system *EDIS Albatross* developed by joint efforts of specialists of the Ural Federal University [13]. This system has a KB in the form of a shell but it is enable to change only declarative knowledge (maximum permissible values of monitored parameters). *The Albatross* system in contrast to other systems has such features as ranking of equipment by technical condition, identification of the nature and stage of the defect, as well as recommendations to the staff about steps to be taken. This system makes it possible to analyze the condition of power and instrument transformers, bushings, cables and other electrical equipment.
3. The *HELMOS* expert system for troubleshooting and detailed monitoring of generators and distribution substations at energy plants [14]. This system is based on knowledge obtained earlier and on current values (signals) coming from transformers that are used to assess the condition of the generator and the substation to detect any defect that could indicate an emergency at any time. The developed system allows the staff (even not well-trained) to recognize incoming signals.

All presented DSS have, in varying degrees, disadvantages discussed above which prevent them from the widespread use in practice.

Therefore, to ensure such characteristics as versatility, adaptability, scalability, the DSS data should be applicable not only to certain types of EE but also to the entire EE stock. In addition, it should be possible to be integrated with the monitoring systems of the EE in real time, to use retrospective, expert information when building diagnostic models for increasing the reliability of estimation of EE condition and to include effective methods for processing the data obtained [15, 16].

To address these challenges we developed the IDSS aimed at continuous monitoring of the EE efficiency, as well as making scientifically based decisions on the equipment under the conditions of large flows of raw heterogeneous data.

3 Architecture of IDSS

The IDSS considered is a software package for automated collection of initial information from the EE, processing and visualization of various diagnostic heterogeneous information, as well as generating output information in the form of the current operating condition of the equipment (EE is normal; there are deviations in EE operation).

The *raw data collection and storage module* transmits the values of monitored parameters to the *diagnostics and prediction module* which makes it possible to detect the process of untoward conditions of EE at the operational stage.

In cases EE is normal, the current values of the parameters (omitting the processing module) are sent to the diagnostic subsystem (search for divergence of parameters from permissible norms) that operates in accordance with the specified search algorithm including information from the database (technical data sheets), KB (duty staff) and then are transferred to the *recommendation and decision making module.*

The Chief Engineer is provided with a decision-making recommendation based on yes-or-no principle (for example, whether to diagnose EE and predict the values of its parameters) resulting in the introduction of appropriate changes in the modes of IDSS operation. Note that for further diagnosis of EE and the use of the prediction subsystem, the current values of the monitored parameters can immediately go to the diagnostic subsystem, as well as through the *raw data processing module*. In addition, the diagnostic subsystem functions using not only the database (DB), the KB but the database of methods too.

In cases there are deviations in EE operation, first, the current values of the parameters go to the diagnostic subsystem (search for divergences of parameters from permissible norms), and then to the *chief engineer interface* where abnormal parameters (for example, low insulation resistance) are displayed. Then, the main operator is provided with recommendations to make a decision (the operator either accepts or rejects the recommendations).

The key features of the developed IDSS are as follows:

1. design, editing and analyzing the fuzzy and fuzzy-functional cognitive models for EE diagnostics;
2. design and editing a fuzzy-production model for making final diagnostic decisions;
3. design of predictive models in the form of functional dependencies of any monitored EE parameters and calculation of predicted values of monitored parameters;
4. search for deviations of the values of EE parameters from the permissible norms using DB and KB;
5. modeling and forecasting the development of untoward conditions during the operational phase of EE;
6. creating the KB that contains the expert knowledge, as well as fuzzy and fuzzy-functional cognitive models and fuzzy-production rules.

Figure 1 shows the structure of the IDSS for EE diagnosing. It includes four main modules: raw data collection and storage module; raw data and knowledge processing module; diagnostics and prediction module; recommendations and decision making module.

Next, we consider the diagnostics and prediction module which is the key module in the system.

The *raw data collection and storage module* receives information from the equipment (coefficient of temporary supervoltage, current harmonics, thermal imaging, and others), the external environment (temperature drops, thunderstorms, etc.). It also store this information in *Excel* format (**.xls*) for further processing as well.

The DB is filled with data coming from the equipment and the duty staff; it contains a data archive (daily files with time indications), real-time measurement data, regulatory documentation, state standards, technical data sheets, as well as restrictions on the values of various parameters of EE. As a result, all the data received on EE are in one place and brought to a common format.

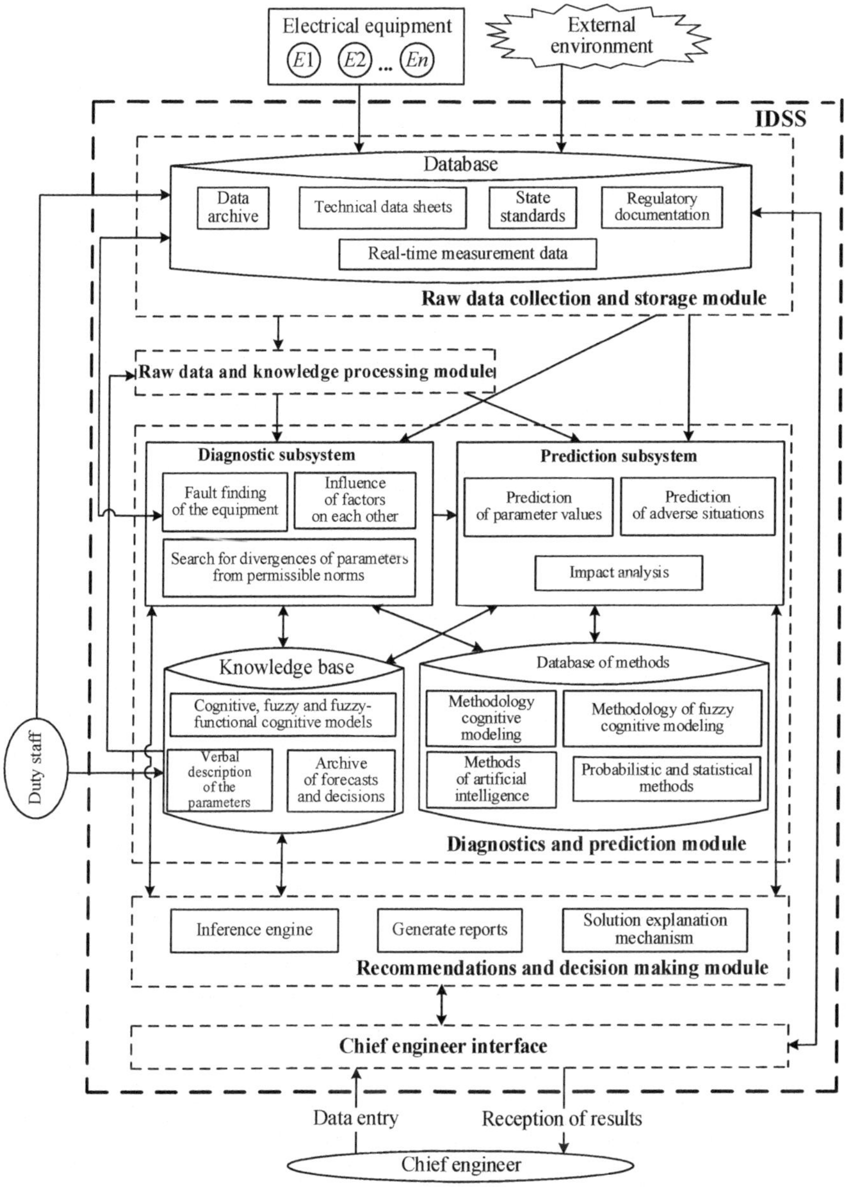

Fig. 1. The structure of the IDSS for electrical equipment diagnostics.

The *raw data processing module* processes data coming from the DB and the KB in order to apply the diagnostics and prediction module.

Processing means the partitioning of knowledge and the normalization of data presented in the form of clear data and fuzzy numbers in order to apply the methodology and fuzzy cognitive modeling and methods of artificial intelligence [17].

Note that for the application of the proposed probabilistic-statistical methods, the data processing is not required.

The *diagnostics and prediction module* carries out the diagnostics of electrical equipment and the prediction of parameter values, the development of processes in case of emergency using the KB and bases of methods.

The knowledge base is one of the key components of the proposed IDSS which is updated through the knowledge of the duty staff, as well as the adaptation of the KB to the operation conditions (replacement of rules or facts in the KB). The proposed KB contains various fuzzy and fuzzy-functional cognitive models for EE diagnostics, scenarios for the process of untoward conditions, an archive of predictive models, as well as recommendations and management decisions [18, 19]. The diagnostics and prediction results come to the *recommendation and decision making module*.

A *Fuzzy Cognitive Model* (FCM) is a weighted directed graph in which vertices are factors, and edges are fuzzy cause-effect relationships between factors:

$$G_{cog} = \langle V, W \rangle,$$

where $V = \{v_i\}$ – the set of vertices, $v_i \in V$, $i = \overline{1,n}$; W – fuzzy cause-effect relationships between the vertices. Elements $w_{i,j}$, $w_{i,j} \in W$ characterize the direction and force of the influence between the vertices v_i and v_j.

Fuzzy cognitive models make it possible to identify which of the factors have the greatest influence on EE (and vice versa), to search the best values of factors reflecting the EE operability under the condition of uncertain raw data.

A *Fuzzy-Functional Model* (FFM) is a functional graph in which the vertices are factors, and the edges are functional dependencies and/or fuzzy cause-effect relationships between factors (Fig. 2).

$$G_{fun} = \langle \langle V, W \rangle, F \rangle,$$

where $G_{fun} = \langle V, W \rangle$ – a weighted directed graph; F – connecting function between the vertices, $f_{i,j}$ ($f_{i,j} \in F$) – functional dependence of the vertices parameters which is assigned to each edge between the vertices v_i and v_j. In addition, this dependence can be not only functional and fuzzy, but also stochastic and can be built from the statistical data of the object.

The design of FCM and FFM is carried out until a fixed point state is reached, at which the vertex does not generate new vertices (there is no data replenishment with new values of the EE parameters).

Note that when we do not work with FCM and FFM as a mathematical model, we use the term *factor*. While working with FCM and FFM, we use the term *vertex*.

Fuzzy-functional models, being added to the FCM, provide additional information about the parameters of the equipment and make it possible to search for the best values of factors reflecting the serviceability of the equipment in the presence of restrictions prescribed in the regulatory documentation, State Standards and technical data sheets.

The design of predictive models (prediction of parameter values) is carried out using the methods of soft calculations.

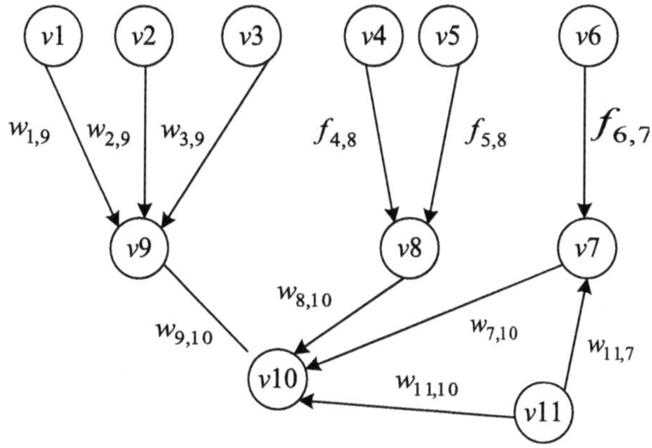

Fig. 2. Fuzzy-functional model of EE diagnostics.

To predict the values of EE parameters under conditions of statistical and fuzzy raw data, it is proposed to use the fuzzy-plural method [20]. The principle of the method is to build up functional dependencies in the form of clear mappings $P_i = \varphi_{i-1,i}(P_{i-1})$ (P_i – parameter, $i = \overline{1;n}$; $P_i^{(\min)} \leq P_i \leq P_i^{(\max)}$ – restrictions placed on the parameters of the standard deviations) of the set *frequency of an emergency for the equipment* with the possibility of subsequent calculation of the predicted parameter value.

The membership functions $\mu_{\widetilde{A}}(P_{i-1})$ and $\mu_{\widetilde{B}}(P_i)$ of fuzzy sets $\widetilde{A}$ – the frequency of the emergency of parameter P_{i-1} and $\widetilde{B}$ – the frequency of the emergency of parameter P_i for the equipment are based on the collected statistical data [9, 10] and regulatory documentation. In this case, the membership functions of the values of the linguistic variable *frequency of an emergency for the equipment* can be represented in the form of triangular and trapezoid fuzzy numbers. The predicted values of the parameters can be used in FCM and FFM.

The duty staff assessing the frequency of an emergency for each specific piece of equipment relies on their own experience and evaluates it using the words *rarely, medium, often.*

The *recommendation and decision making module* is responsible for making a decision for a Chief engineer taking into account the detected faults of the electrical equipment in a convenient and visual form (a report in a textual format).

The structure of this module includes an inferential mechanism designed to obtain new facts based on the comparison of raw data and knowledge from the KB, as well as a report generation for requesting an explanation of the decision's ways in the process of the problem or based on the solution of the problem (explains how it is received, what knowledge at the same time were used).

The *chief engineer interface* allows the chief engineer to run computational experiments and view the results.

4 Screen Forms of the Developed Diagnostics and Prediction Module for Electrical Equipment

The diagnostics and prediction module of IDSS is implemented using probabilistic-statistical methods, soft computing methods, fuzzy cognitive modeling methodologies based on the *Java*, *C++*, *Matlab* and in *MS Excel*.

Figure 3a) and 3b) present fragments of screen forms of the developed software of the diagnostics and prediction module of IDSS.

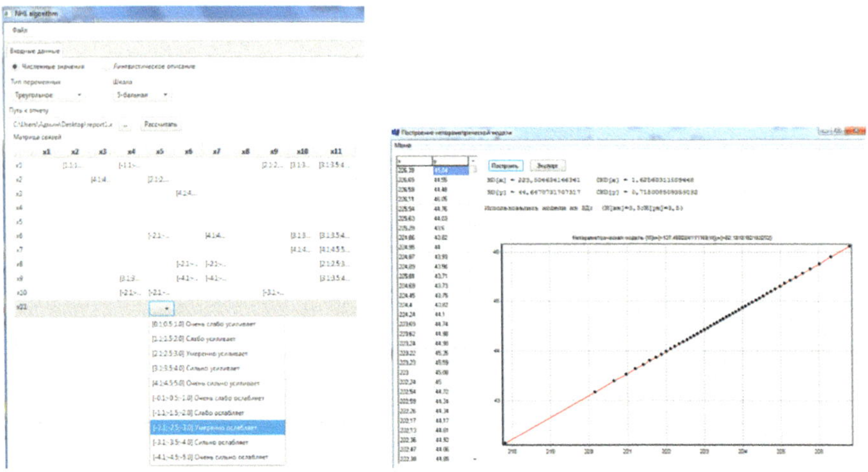

a) *b)*

Fig. 3. Software screen forms for diagnostics and prediction module of IDSS: a) system indicators calculations (determination of the factors influence on each other); b) design of functional dependencies based on statistical data.

The developed software helps to reduce the time spent on EE diagnostics and predicting by 60%. In addition, the proposed IDSS of EE diagnostics requires the duty staff to have only system knowledge and understanding the equipment.

5 Conclusion

This paper demonstrates a fundamentally new IDSS for EE diagnostics under conditions of uncertainty. The proposed system in real time allows one to solve problems not only for control, but also for predicting the EE conditions; to analyze not only the actual values of the monitored equipment parameters, but also their dynamics. All this provides identifying EE faults and failures at an early stage, preventing emergency situations; developing and taking measures to prevent them under conditions of large volumes of heterogeneous raw data. In addition, the developed IDSS is universal, since there is the possibility of its application in all industries (chemical, petroleum, metallurgical, nuclear energy, etc.).

References

1. Mareček, O.: Monitoring and diagnostic system of power plant electrical equipment. In: Conference on Diagnostics in Electrical Engineering (Diagnostika), Pilsen, pp. 1–4 (2016)
2. Dmitriev, S.A., Manusov, V.Z., Ahyoev, J.S.: Diagnosing of the current technical condition of electric equipment on the basis of expert models with fuzzy logic. In: 57th International Scientific Conference on Power and Electrical Engineering of Riga Technical University (RTUCON), Riga, pp. 1–4 (2016)
3. Decision support system in the problems of diagnostics of submersible electrical equipment. https://elibrary.ru/item.asp?id=19141782. Accessed 31 May 2019
4. Yang, P., Liu, S.: Fault diagnosis system for turbo-generator set based on fuzzy neural network. Int. J. Inf. Technol. **11**(12), 76–85 (2005)
5. Yuan, B., Li, W., Wu, Y.: Fault diagnosis on cabin electric power equipments. Adv. Knowl. Syst. Res. (AISR) **145**, 287–289 (2017)
6. Zou, F.H., Zhang, H.: The design of electrical equipment fault diagnosis system based on fuzzy inference. Adv. Mater. Res. **1061**, 716–719 (2015)
7. Intelligent lifecycle management system of power grid equipment. http://lib.omgtu.ru. Accessed 31 May 2019
8. Expert system for diagnosing power transformers of power supply systems. http://cyberleninka.ru. Accessed 31 May 2019
9. Eltyshev, D.K.: On the development of intelligent expert diagnostic system for assessing the conditions of electrical equipment. Syst. Methods Technol. **3**(35), 57–63 (2017)
10. Khoroshev, N.I.: Intellectual decision support in the operation of power equipment based on adaptive cluster analysis. Syst. Methods Technol. **3**, 123–128 (2016)
11. Some aspects of the software implementation of the decision support system for the accounting and diagnostics of electrical equipment. http://iii03.pfo-perm.ru/Data/petrohe2/petrohe2.htm. Accessed 31 May 2019
12. The system for assessing the condition of electrical equipment "DIAGNOSTICS +". http://www.transform.ru/diagnostika.shtml. Accessed 31 May 2019
13. EDIS Albatross system. http://www.edis.guru. Accessed 31 May 2019
14. Architecture of a Fault Diagnosis Expert System for Power Plants Protection. http://masters.donntu.org/2007/eltf/pastuhova/library/st12.htm. Accessed 31 May 2019
15. Kychkin, A.V.: Software and hardware network energy-accounting complex. Sens. Syst. **7**(205), 24–32 (2016)
16. Eltyshev, D.K., Boyarshinova, V.V.: Knowledge decision support in the electrical equipment diagnostics. In: Proceedings of the 19th International Conference on Soft Computing and Measurements, Saint Petersburg, pp. 157–160 (2016)
17. Kolodenkova, A.E.: The process modeling of project feasibility for information management systems using the fuzzy cognitive models. Herald Comput. Inf. Technol. **6**(144), 10–17 (2016)
18. Kolodenkova, A.E., Korobkin, V.V.: Diagnosis in SEMS based on cognitive models: group interaction. Stud. Syst. Decis. Control **174**, 275–284 (2019)
19. Kolodenkova, A.E., Muntyan, E.R., Korobkin, V.V.: Modern approaches to modeling of risk situations during creation complex technical systems. Adv. Intell. Syst. Comput. **875**, 209–217 (2019)
20. Kolodenkova, A.E., Werechagina, S.S.: Knowledge method for forecasting of electrical equipment technical condition in conditions of fuzzy initial data. Vestnik RGUPS **1**(73), 76–81 (2019)

Methodology and Technologies of the Complex Objects Proactive Intellectual Situational Management and Control in Emergencies

B. Sokolov[1], A. Pavlov[2], S. Potriasaev[1], and V. Zakharov[1](✉)

[1] St. Petersburg Institute for Informatics and Automation of the RAS, St. Petersburg, Russia
valeriov@yandex.ru

[2] Mozhaisky Military Aerospace Academy, St. Petersburg, Russia

Abstract. This article discusses the results of solving the problems of complex objects (CO) recovery programs, which are based on the structure dynamics management and control of these objects. The main advantage is the combined use of different models, methods and algorithms (analytical, simulation, logical-algebraic and their combinations) which allows to compensate their objectively existing shortcomings and limitations while enhancing their positive qualities during planning and scheduling for CO proactive intellectual situational management and control in emergencies. Proposed integration of CO management and control methods allows to link at the constructive level the substantive and formal aspects of the solving problem. Within the framework of the presented methodology, it is possible to rely on the fundamental scientific results obtained to date in the modern control theory of complex dynamic systems and knowledge engineering with a tunable structure during the situational management of complex objects recovery. It allows to determine the sequence of tasks and operations (to synthesize CO disaster recovery plans and schedules), to find and reasonably choose compromise solutions in the presence of several options for the sequence of operations. The developed CO disaster recovery plans and schedules are evaluated by involving experts and identifying their implicit expertise with the help of multicriteria approach. The problem statement of recovery is given. The new method of multicriteria evaluation of the CO disaster recovery plans and schedules, based on a combination of theory of experiments design models and models of fuzzy rule-based language is given.

Keywords: Proactive control · Intelligent systems · Decision making systems · Multicriteria approach

1 Introduction

One of the most important problems in the XXI century is the widespread occurrence of crisis, accidents and catastrophes having natural-ecological, technical-production or anthropogenic-social reasons. At the same time the range of threats to economic, physical, information security, and the list of vulnerabilities of the technical infrastructure of a complex object (CO), and in particular, information systems, is constantly

S. Kovalev et al. (Eds.): IITI 2019, AISC 1156, pp. 234–243, 2020.
https://doi.org/10.1007/978-3-030-50097-9_24

growing. The main feature of CO management in emergencies is that information about the main factors and conditions affecting the successful solution of the management tasks of the relevant objects has a different degree of confidence and certainty [1]. The main difficulty is associated with an uncertain state of modeled object. The lack of analytical dependence between these factors and the conditions that determine (describe) the control technology significantly complicates the process of CO control. This is primarily concerned with the factors and conditions that make it difficult to fulfill the objectives of the CO (factors of environmental resistance).

2 Related Works

Nowadays there are many possibilities of structural dynamic control: alteration of functioning means and objectives; alteration of the order of observation tasks and control tasks solving; redistribution of functions, problems and control algorithms between levels; reserve resources control; control of motion of elements and subsystems; reconfiguration of structures [2]. Different types of models (deterministic, logical algebraic, logical-linguistic models) are used for modeling of functioning and CO control [3].

There are a number of approaches that take into account uncertainty factors (fuzzy, stochastic mathematical programming models). However, to solve the control problems within the framework of these models, rather "rigid" assumptions about the parameters of the distribution laws of random (fuzzy) quantities should be made, with the help of which a description of the process of influence of the external environment on the elements and sub-systems of the CO is carried out.

3 Approach Requirements

The peculiarity of the considered automatic control systems is that they should be focused on the application in the conditions of faults, accidents and even catastrophes and, consequently, are endowed with the property of resilience (disaster resistance) [4]. The most important indicator of the CO effectiveness is the quality of functioning under the influence of the catastrophic environment. As a rule, deterministic approach, methods of reliability theory and simulation modeling are used to assess the resilience of CO in case of predictable disruptions [5]. The statistical analysis of previous objects is used for designing new objects. This approach is adequate when creating similar or "invariant" in time objects, but if the created object is significantly different from the previous one, it is different from the "reference" characteristics and the use of the above description method becomes unacceptable. Within this framework, the models, methods and algorithms used in the theory of reliability and simulation become inapplicable to ensure the required level of model's adequacy to the real world.

3.1 Features

The problems of CO structural dynamics control in their content belong to the class of structural-functional synthesis problems of CO and control program synthesis. The main feature of the solution of the problems of the class under consideration is the definition of optimal control programs for the main elements and sub-systems of CO. They can be executed only after the list of functions and algorithms of information processing and management, which has to be realized in the specified elements and subsystems, becomes known. In turn, the distribution of functions and algorithms by elements and subsystems of the CO depends on the structure and parameters of the laws of controlling these elements and subsystems. The difficulty of resolving this controversial situation is compounded by the fact that the composition and structure of CO at different stages of the life cycle changes over time due to various reasons [6].

To ensure the required degree of autonomy, quality and efficiency of COs control in emergencies, it is necessary to develop conceptually new methods of formalization and solving structure-dynamics control problems of CO. In addition, we should take into account all aspects of the problems describing the factors of uncertainty and from the standpoint of a systematic approach to solve the following problems.

- Synthesis of the final image of recoverable CO (to solve the problem of structure-functional synthesis)
- Defining the period of time by which it is necessary to complete.
- Determine recovery technology of the CO (in which at some point in time will operate simultaneously the elements and subsystems of the "old" and "new" CO to ensure the resilience of the corresponding target processes)
- Determine the program of transition from the "old" to the "new" state
- Determine programs of rescheduling or correction of the original recovery program of the CO, if there are uncertainty disturbing effects

It is necessary to construct proactive methods of self-organization allowing through a purposeful managed expansion of the variety of control actions to provide the required level of resilience (disaster resistance) [7]. Increase in CO resilience in emergencies is operational design of procedures.

- Information gaining.
- Analytics.
- Design computing environment.

The main goal is to create the environment in which the detection, localization and elimination of failures and disruption of elements and subsystems of these objects will occur much earlier than the possible consequences of these faults.

4 Formal Problem Statement

The presented requirements for solving the problems of plan synthesis of CO restoration, logical-dynamic models found and the results obtained thanks to them were proved in [8, 19]. The considered approach is based on th dynamic interpretation of CO recovery as processes of multicriteria structural and functional synthesis of the recovered CO image as well as the simultaneous synthesis of its recovery technology and plans (programs) of recovery. This interpretation is developed by the authors in the framework of their theory of proactive management of structural dynamics of CO. In this case, the problems can be described at the set-theoretic level as follows. It is necessary to develop principles, approaches, models, methods, algorithms that allow to find such $\left\langle U_*^t\, S_\delta^{tf} \right\rangle$ under which the following conditions are met:

$$
\begin{aligned}
& J_\theta(X_\chi^t, \Gamma_\chi^t, Z_\chi^t, F^t_{<\chi,\chi'>}, \prod\nolimits^t_{<\tilde{\delta},\tilde{\tilde{\delta}}>}, t \in (t_0, t_1]) \rightarrow \underset{<U^t, S_\delta^{*tf}> \in \Delta g}{extr} \\
& \Delta g \left\{ <U^t, S_\delta^{tf}> | R_\beta \left(X_\chi^t, \Gamma_\chi^t, Z_\chi^t, F^t_{<\chi,\chi'>}, \prod\nolimits^t_{<\tilde{\delta},\tilde{\tilde{\delta}}>} \right) \leq \tilde{R}_\beta; , \right. \\
& \left. U^t = \prod\nolimits^{t_1}_{<\delta_1,\delta_2>} \circ \prod\nolimits^{t_2}_{<\delta_2,\delta_3>} \circ \ldots \circ \prod\nolimits^{t}_{<\tilde{\delta},\tilde{\tilde{\delta}}>} ; \beta \in \mathbf{B} \right\}
\end{aligned}
\tag{1}
$$

where χ is the index characterizing various types of structures of the CO recovery, $\chi \in \{\text{Top, Fun, Tech, OS}\}$ is the set of indices corresponding to the topological, functional and technical structures, organizational structure. $t \in T$ is the set of moments of time. $X_\chi^t = \{x_{\chi l}^t, l \in L_\chi\}$ is the set of elements that makes up the structure of the dynamic alternative system graph (DASG). G_χ^t (vertex set of DASG) with the help of which the controlled structural dynamics is set at the time t. $\Gamma_\chi^t, = \{\gamma^t_{<\chi,1,1'>}, 1, 1' \in \mathrm{L}_\chi\}$ is the set of GASG arch type G_χ^t reflecting the relationship between its elements at time *t*. $Z_\chi^t = \{z^t_{<\chi,1,1'>}, 1, 1' \in \mathrm{L}_\chi\}$ is the set of parameter values that quantitatively characterizes the relationship of the DASG corresponding elements. It is constants. $F^t_{<\chi,\chi'>}$ are mapping of different CO structures to each other at a time *t*. $\Pi^t_{<\tilde{\delta},\tilde{\tilde{\delta}}>}$ is an operation of composition of multistructural macrostates with numbers at time *t*. U^t is the control (program and real-time) effects, allowing to synthesize the structure of the CO recovery. J_θ are the cost, time, resource indicators characterizing quality of functioning of CO. Δ_g is a set of dynamic alternatives (a set of structures and parameters of CO, a set of programs of their functioning). B is the set of numbers of spatio-temporal, technical and technological restrictions that determines the process of implementation of programs of CO recovery for different scenarios of disturbances. $\tilde{\mathrm{R}}_\beta$ - are predatum. $T = (t_0, t_f]$ is the time interval at which the CO is synthesized and recovered.

The problem of search for optimal program controls of restoration is realized by the combined method combining the method of branches and borders and the method of successive approximations [9]. As a result of the solution of this problem, a vector of control actions changing in time is formed, which determines both the control technology and the recovery plan of the CO. In the process of selecting the recovery program (plan) at each iteration, it is necessary to reasonably determine in which place and at what time this or that operation should be performed. In addition, there should be an assessment of how affect changes of structures in different subsystems on the process of restoring the functioning of the CO. To do this, a system of indicators of efficiency and quality of co recovery should be used. These include next indicators:

- Adaptability of structures
- Complexity of structures
- Recovery time
- Period of useful operation-life
- Time in service
- Effectiveness of recovery management and decisions validity
- Reliability and resilience both in nominal conditions of operation and in the event of predictable and unpredictable disruptions
- Operating cost

The approach makes it possible to simultaneously solve the problems of synthesis of the technology of proactive control of CO recovery and the problem of scheduling, flows and resources of the specified system, based on a single polymodel logical-dynamic description of the CO. Combined methods and algorithms for the synthesis of control technologies and recovery plans allow to obtain optimal solutions based on the objectives and requirements for the indicators of CO recovery quality.

5 Expert Evaluation of Recovery Plans

Nowadays, methods based on precedents are widely used, but in the conditions of constant changes and complexity of the elements and subsystems of CO, as well as in the absence of a base of precedents, it is impossible to find a close "reference" solution [10–12]. In the absence of a relevant base of precedents, it is advisable to consult experts, but in emergencies where the speed of decision-making is extremely important, experts may not be available due to circumstances or their assessment in a critical situation will be erroneous.

To eliminate limitation of method it is proposed to use an automated multi-criteria analysis of the synthesized recovery plan. The main idea of this approach is to identify implicit knowledge of experts by conducting a survey and processing responses using the method presented below. It is also necessary to take into account the non-linear nature of the impact of these private indicators of the management areas on the effectiveness of the generalized index of the CO recovery plan, as well as a fuzzy-possibilistic approach of private indicator. We propose the construction of a generalized index using fuzzy-possibilistic convolution. Fuzzy-possibilistic convolution is based on fuzzy measures and fuzzy integrals, and allows us the flexibility to take into account the non-linear nature of the influence of private indicators [13, 14].

Set of indicators of the recovery plan of a complex object be estimated by a set of quality indicators $F = \{F_1, F_2,\ldots, F_m\}$, each of which is a linguistic variable. The linguistic variable F_i = "economic efficiency of CO recovery" can take values from a set of simple and compound terms $T(F_i)$ = {low, below average, average, above average, high}. For qualitative evaluation of the resulting indicator, we will use the linguistic variable "target efficiency of CO recovery", which can take the values $T(F_{res})$ = {"bad", "below average", "above average", "good"}. In general, the knowledge of decision makers about connection with particular quality indicators $F = \{F_1, F_2,\ldots, F_m\}$ with the resulting F_{res} indicators can be represented by production models of the following kind:

$$P_j: \text{"IF } F_1 = A_{1j} \text{ and } \ldots \text{ and } F_m = A_{mj}, \text{ THEN } F_{res} = A_{jres}\text{"}, A_{ij} \in T(F_i), A_{jres} \in T(F_{res}), \quad (2)$$

where A_{ij}, A_{jres} are terms of the corresponding linguistic variables. The bipolar scale $[-1, 0, +1]$ is used as a general scale in relation to all values of indicators, and the terms can be set using fuzzy numbers (L–R) of the type (Fig. 1).

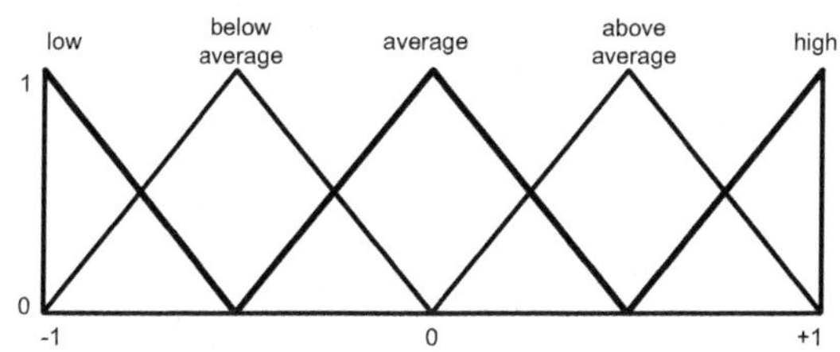

Fig. 1. Terms of the linguistic variable in the scale [−1, +1]

In accordance with the method of solving the problem of multi-criteria evaluation proposed in [15], the extreme ("minimum" and "maximum") values of the linguistic variable F_i scales are labeled "−1" and "+1", and to build the resulting indicator F_{res}, according to the provisions of the theory of planning the experiment [16, 17], form an orthogonal design of the expert survey, the elements of which are the extreme marked values of the partial performance indicators $F = \{F_1, F_2, \ldots, F_m\}$. An example of orthogonal design of the expert survey for the three partial indicators of efficiency are shown in Table 1.

Table 1. Orthogonal design of the expert survey

F_0	F_1	F_2	F_3	F_1F_2	F_1F_3	F_2F_3	$F_1F_2F_3$	F_{res}
1	−1	−1	−1	1	1	1	−1	A_{1res}
1	1	−1	−1	−1	−1	1	1	A_{2res}
1	−1	1	−1	−1	1	−1	1	A_{1res}
1	1	1	−1	1	−1	−1	−1	A_{3res}
1	−1	−1	1	1	−1	−1	1	A_{2res}
1	1	−1	1	−1	1	−1	−1	A_{4res}
1	−1	1	1	−1	−1	1	−1	A_{3res}
1	1	1	1	1	1	1	1	A_{5res}
λ_0	λ_1	λ_2	λ_3	λ_{12}	λ_{13}	λ_{23}	λ_{123}	

In Table 1 the values of the terms of the linguistic variable F_{res} of the resulting efficiency index can be represented by fuzzy triangular numbers (Fig. 2). Then, for example, in the second row of the table, the following expert judgment is presented: "if F_1 is "high", F_2 is "low", F_3 is "low", then the resulting F_{res} indicator is estimated as "below average"." Moreover, the very production rule is considered as the reference situation, when carrying out according to expert survey.

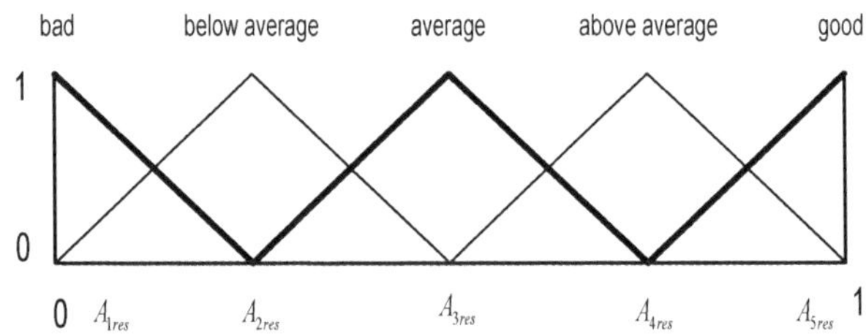

Fig. 2. Scale of the resulting indicator

Calculation of coefficients of the resulting (generalized) indicator:

$$F_{res} = \lambda_0 + \sum_{i=1}^{m} \lambda_i F_i + \sum_{i=1}^{m} \sum_{\substack{j=1 \\ j \neq i}}^{m} \lambda_{ij} F_i F_j + \ldots + \lambda_{12\ldots m} F_1 F_2 \ldots F_m, \tag{3}$$

The equation takes into account the influence of both private indicators and sets of two, three and so on indicators, is carried out according to the rules adopted in the theory of planning the experiment. To do this, the averaged scalar products of the corresponding columns of the orthogonal matrix are calculated (Table 1) on the vector of values of the resulting performance indicator. For example, the value of the coefficient λ_2 is calculated as follows:

$$\lambda_2 = \frac{-A_{1res} - A_{2res} + A_{1res} + A_{3res} - A_{2res} - A_{4res} + A_{3res} + A_{5res}}{8}. \tag{4}$$

The proposed method of multicriteria decision-making consists of the following steps.

Step 1. Define a set linguistic scale for each of the particular indicators and the resulting indicator of the quality of comparable CO recovery plans. Converting of particular indicators into the scale [−1, +1].
Step 2. Construction of an orthogonal design of expert survey and expert survey (answers to questions of production rules).
Step 3. Calculation of the resulting indicator of the quality of comparable plans for CO recovery plans. To assess the quality of the results obtained, the authors propose using the following quality indicators: target, technical, economic efficiency of CO recovery. The resulting indicator is the overall efficiency of CO recovery management.

6 Algorithm of Proactive Intellectual Situational Management and Control in Emergencies

Algorithm of proactive intellectual management and control in emergencies is following:

Step 1. Define a possible CO multi-structure macro-states or the carrying out of structural-functional synthesis of a new image. At this stage, on the basis of multi-dimensional orthogonal design for the set of achievability, formed as a result of integrated modeling of structural dynamics of CO, the set by which the requirements for particular performance indicators of CO recovery (new appearance of CO) are set, the formation of a set of non-final solutions (set of non-dominant, effective alternatives, set of V. Pareto)

Step 2. The first step is repeated, but the target quality indicators change. The result of this phase is an updated recovery plan that meets other requirements (quality criteria), such as maximum recovery rate, etc.

Step 3. CO recovery plans are evaluated using the above methodology and a recovery plan is selected.

Step 4. Implementation of the option selected in the previous step (multi-structural macro-state of CO) with simultaneous synthesis of adaptive plans and control programs for the transition from the current to the selected macro-state.

Step 5. Formation of the program of rescheduling or correction of the current CO recovery program in case of uncertainty disturbing effects based on information received from sensors in real time [18].

Plans must ensure that the restoration of the CO, in which, along with the implementation of the programs of the relevant macro-states provides both the implementation of plans for sustainable recovery control in the intermediate microstates.

7 Conclusion

Within the framework of the proposed approach to the formalization and solution of one of the central problems of CO recovery management associated with situational structural and functional synthesis of CO and the corresponding control actions, it is possible to solve not only the five main tasks of complex recovery planning, but also to carry out an intelligent multicriteria evaluation of the synthesized plans for CO recovery from a single methodological position. Firstly, it is possible to form the final shape of the corresponding recoverable CO. Secondly, it is possible to determine the period by which the recovery should be completed. Thirdly, it is possible to identify technologies for the recovery of the CO (in which "old" and "new" subsystems and components of CO will operate at the same time over a certain period, in order to ensure the sustainability of the operation and CO recovery in emergencies). Fourthly, it is possible to determine the program of transition to a new macrostructural state. Fifthly, it is possible to form a program of rescheduling or correction of the original program of recovery of CO if there are uncertainty disturbing effects.

The joint use of different models will make it possible to compensate objectively existing shortcomings and limitations while improving their positive qualities in planning and scheduling for joint proactive intellectual and situational management and emergency management. Integration of management and control methods connects meaningful and formal aspects of problem solved at the constructive level. The methodology allows us to rely on the fundamental scientific results obtained to date in the modern theory of control of complex dynamic systems and knowledge engineering with a tunable structure in case of situational management of CO recovery.

The research described in this paper is partially supported by the Russian Foundation for Basic Research (grants 16-29-09482-ofi-m, 17-08-00797, 17-06-00108, 17-01-00139, 17-20-01214, 17-29-07073-ofi-i, 18-07-01272, 18-08-01505, 19–08–00989), state order of the Ministry of Education and Science of the Russian Federation №2.3135.2017/4.6, state research 0073–2019–0004 and International project ERASMUS+ , Capacity building in higher education, # 73751-EPP-1-2016-1-DE-EPPKA2-CBHE-JP.

References

1. Zelentsov, V.A., Sokolov, B.V., Sivirko, E.G.: Variants of accounting uncertainty factors in models of disaster-resistant information systems. In: 1th Proceedings of Reliability and Quality, pp. 165–166 (2012)
2. Ohtilev, M.Y., Sokolov, B.V., Yusupov, R.M.: Intellectual technologies for monitoring and control of structure-dynamics of complex technical objects. Nauka, Moscow (2006). (in Russian)
3. Fleming, W.H., Richel, R.W.: Deterministic and Stochastic Optimal Control. Springer, Berlin (1975)
4. Ivanov, D., Sokolov, B.: Control and system-theoretic identification of the supply chain dynamics domain for planning, analysis, and adaptation of performance under uncertainty. Eur. J. Oper. Res. **224**(2), 313–323 (2013)
5. Kim, Y., Chen, Y.S., Linderman, K.: Supply network disruption and resilience: a network structural perspective. J. Oper. Manag. **33–34**, 43–59 (2015)
6. Skurihin, V.I., Zabrodsky, V.A., Kopeychenko, Y.V.: Adaptive Control Systems in Machine-Building Industry. Mashinostroenie, Moscow (1989). (in Russian)
7. Munoz, A., Dunbar, M.: On the quantification of operational supply chain resilience. Int. J. Prod. Res. **53**(22), 6736–6751 (2015)
8. Mikoni, S., Sokolov, B., Yusupov, R.: Qualimetry of Models and Polymodel Complexes, 1st edn. RAS, St. Petersburg (2018). (in Russian)
9. Sokolov, B.V., Kalinin, V.N.: Autom. Rem. Control **5**, 106–114 (1985). (in Russian)
10. Petrosjan, L.A., Zenkevich, N.A.: Game Theory. World Scientific Publication, Singapure (1996)
11. Larichev, O.I., Moshkovich, E.M.: Qualitative Methods of Decision-Making. Fizmatlit, Moscow (1996). (in Russian)
12. Bakhmut, A.D., Krylov, A.V., Krylova, M.A., Okhtilev, M.Y., Okhtilev, P.A., Sokolov, B.V.: Proactive management of complex objects using precedent methodology. In: Silhavy, R. (eds.) Artificial Intelligence and Algorithms in Intelligent Systems, CSOC2018 2018, Advances in Intelligent Systems and Computing, vol. 764. Springer, Cham (2018)
13. Ehrgott, M.: Multicriteria Optimization, 2nd edn, p. 323. Springer, Heidelberg (2005)

14. Grabisch, M., Murofushi, T.: Fuzzy Measures and Integrals: Theory and Applications. Physica-Verlag, Germany (2000)
15. Pavlov, A.N., Pavlov, D.A., Pavlov, A.A., Slin'ko, A.A.: The technique of multicriteria decision-making in the study of semi-structured problems. In: Cybernetics and Mathematics Application in Intelligent Systems: Proceedings of the 6th Computer Science On-line Conference 2017 (CSOC2017), vol. 2. Series Advances in Intelligent Systems and Computing, vol. 574, pp. 131–140. Springer, Heidelberg (2017)
16. Petrovsky, A.B.: Decision Theory. Akademia, Moscow (2009). (in Russian)
17. Petrovsky, A.: Group verbal decision analysis. In: Adam, F., Humphreys, P. (eds.) Encyclopedia of Decision Making and Decision Support Technologies, vol. 1, pp. 418–425. IGI Global, Hershey (2008)
18. Gilchrist, A.: Industry 4.0: The Industrial Internet of Things. Apress, Berkeley (2016). Distributed to the book trade worldwide by Springer
19. LITSAM. http://litsam.ru. Accessed 21 Aug 2019

Rules for the Selection of Descriptions of Components of a Digital Passport Similar to the Production Objects

Julia V. Donetskaya(✉) and Yuriy A. Gatchin

ITMO University, Saint-Petersburg, Russia
donetskaya_julia@mail.ru

Abstract. In this paper, it is shown the function of technologies, which ensure the support of different stages of a product life cycle and the presentation of production objects in digital environment. For this purpose on enterprises can be created a digital passport, based on requirements and experience examination of similar industrial enterprises. Whereas a digital passport contains significant data amount, it implies the application opportunity in order to take timely project-based and managerial decisions at any moment. Thus, it is necessary to ensure the content selection (also known as component description) of a digital passport in accordance with parameters of analysed life cycle stage that is verbally defined.

The problem statement is formulated, according to which it is realized in the form of a selection of descriptions of a digital passport components given by parameters and in the form of a selection of descriptions specified by the parameters of the life cycle stage. Since the passport contains a greater number of parameters than that which characterizes the analyzed stage of the life cycle, their analysis is performed first. Further solution of the problem is performed in accordance with the elements of the theory of fuzzy sets. Initially, matrices of pairwise comparisons are formed, allowing to determine how much one or another parameter of a component or stage of the life cycle corresponds to the analysed component of a digital passport. Further, the eigenvalues and vectors of matrices are determined. The obtained values of the eigenvectors are the values of the membership functions characterizing the elements of fuzzy sets. This will allow using the obtained results to select component descriptions when generating a design solution by performing simple arithmetic operations. Formed solutions will allow to ensure the similarity of the production objects in the digital environment at the greatest extent.

Keywords: Selection of digital passport components · Similarity of the production object · Digital environment

1 Introduction

– The digital environment of an enterprise is intended for storing, analysing and using data of a product at the stages of its life cycle, for which ERP, PDM, MES and/or EAM systems are being introduced [1]. Their functionality allows to manage, among other things:

S. Kovalev et al. (Eds.): IITI 2019, AISC 1156, pp. 244–251, 2020.
https://doi.org/10.1007/978-3-030-50097-9_25

- at the conclusion of the contract - data on customers and contracts concluded with them, data on co-contractors and contracts concluded with them, information on additional agreements to contracts;
- in the course of development, by order numbers, operational work schedules, data on purchased component parts, product design data, product program data;
- when preparing for production - order numbers, technological data about the product, remarks on the design documentation and decisions made on them;
- during production - identification codes of product copies (including purchased components), the results of their checks by representatives of technical control, remarks on design, software and technological documentation, as well as decisions made on them, permission card numbers and defective statements;
- during maintenance and repair, with data on the shipment of products to the customer, numbers of accompanying documents, numbers of reclamation acts and results of failures research, data on service requests.

This information is formed and modified by the means of the integration mechanisms of the previously mentioned systems as a result of automating the management processes of the relevant product data [2]. As a result, a unified digital passport is created, which ensures the adoption of timely design and management decisions based on events initiated by the customer - submitting an application for the development or supply of products, a complaint act or an application for service. Any of them is characterized by the formation of lists of processed data about the electronic product and the sequence of the procedures, defined by the work schedules and stages specified in the customer's contract [3, 4].

At the same time, an employee of an enterprise has the opportunity to receive any information he needs, regardless of where, when and by whom it was uploaded to the digital environment. This reduces the development time of the product and reduces the number of gross errors in the documentation.

However, there are other problems associated with the complexity of technological processes of production and high requirements for accuracy and reliability of products. In order to solve them, some serious measures should be taken, including the automation of production processes, ensuring the assessment and analysis of the quality of manufactured products not only during production and operation, but also in the process of their development and production preparation.

Here, it is logical to argue about the adequacy of the content of a digital passport for automating the formation of design solutions that ensure the production facility in a digital environment. Specifically:

- sequences of product data management processes at the stages of its life cycle;
- lists and structures of electronic products and their components;
- sequences of manufacturing operations for the manufacture of electronic products and their components;
- sequences of procedures for refining data on an electronic product at the stages of development and production preparation.

Automation is performed based on the selection of descriptions of the components of a digital passport, the development of which is the subject of the present work.

2 The Task of Presenting an Object of Production in a Digital Environment

The similarity of an object of the production in the digital environment is achieved by integrating ERP, PDM, MES and/or EAM systems when generating product data and implementing their management processes at a particular enterprise. This is ensured by meeting the invariant requirements imposed on the content of the passport.

$$RS = (RS_1 \ldots RS_S). \quad (1)$$

Furthermore, it is necessary to ensure compliance with variant parameters of the current stage of the product life cycle.

$$RP = (RP_1 \ldots RP_P). \quad (2)$$

In accordance with this, it is logical to state that the listed groups of parameters initiate interaction between the stages of the product life cycle, similar to those initiated by applications or complaint reports. This allows us to represent the production object based on the results of the analysis of the content of a digital passport, which has all the properties of complex systems. Consequently, it is possible to represent it as a set of mathematical models: component models $R(C)$, description models $G(C)$, substitution models $Z(C)$, interaction models $X(C)$ and parameter models U [5].

Then the task of representing production objects in a digital environment is to form a matrix $X(C)$, each element of which satisfies the condition:

$$X_{ij}(C_i, C_j) = \begin{cases} 0, & out(G_i(C_i) \neq in(G_j(C_j)) \\ 1, & out(G_i(C_i) = in(G_j(C_j)) \end{cases},$$

where $i,j = (1\ldots k), k$ – total number of digital passport components; $G_i(C_i)$ and $(G_j(C_j)$ calculated on the basis of the component model and model parameters set verbally, according to the requirements of RS and RP respectively. In order to solve this problem it is necessary:

1. To analyse the components of the digital passport and the parameters describing them.
2. Select the descriptions of the components of the digital passport specified by the parameters.
3. Select the descriptions of the components of the digital passport specified by the parameters of the life cycle stage.

3 Analysis of Digital Passport Components and Parameters Describing Them

The practice of implementing the digital environment in enterprises shows that the decision on its formation is a collective decision of the working group members. This means an assessment of the compliance of the components of a digital passport with the requirements (1) imposed on them, performed by each of the h experts. As a result, a matrix is formed

$$A_h^{RS} = \left|\frac{1}{a_{ij}}\right|, a_{ij} = \frac{a_j}{a_i}, i,j = \overline{1,k},$$

where the values of the elements a_{ij} are determined on a nine-point Saaty scale:

1 – in the absence of the advantage of the element a_j over the element a_i;
3 – with a weak advantage of the element a_j over the element a_i;
5 – with a significant element advantage a_j over the element a_i;
7 – with a clear advantage of the element a_j over the element a_i;
9 – with absolute element advantage a_j over the element a_i.

Perform the valuation of each of the matrices

$$B_h^{RS} = \left|b_i^{RS}\right|, b_i^{RS} = \max\left(max\frac{a_{ij}}{\sum_{j=1}^{k} a_{ij}}\right), i = \overline{1,k}.$$

The results obtained allow to form a set of

$$\overrightarrow{C} = \left(\overrightarrow{C_1}, \overrightarrow{C_2} \ldots \overrightarrow{C_k}\right), \tag{3}$$

elements of which are unique.

As for the parameters describing the components of a digital passport, their list is in a variety $U = (v_1, v_2 \ldots v_a)$ that may exceed a similar list (2), corresponding to a certain stage of the life cycle. Then the formation of a set of parameters is performed by the rule: $\overrightarrow{v_i} \in \overrightarrow{U}$, если $\overrightarrow{v_i} = v_i$ и $\overrightarrow{v_i} \in RP$, $i = \overline{1,P}$.

4 Selection of Descriptions of the Digital Passport Components, Given by Mentioned Parameters

Since the sets $\overrightarrow{C}$ and $\overrightarrow{U}$ are given in verbal form, the descriptions of the components of a digital passport given by the parameters to them are also represented by fuzzy sets. Characteristics of their elements are the values of membership functions. We calculate them by the method proposed by Chernov V. [6], which consists of several steps.

Construction of matrices of pairwise comparisons of elements of the set $\overrightarrow{U}$ for each element of the set (3)

$$M\left(\overrightarrow{C}\right) = \left|\frac{1}{m_{ij}}\right|, m_{ij} = \frac{m_j}{m_i}, i,j = \overline{1,P}, \tag{4}$$

Where the values of the elements m_{ij}, are determined on a nine-point Saaty scale:

1 – in the absence of the advantage of the parameter *j* over the parameter *i* of the set $\overrightarrow{U}$ for the analysed component of a digital passport;

3 – with a weak advantage of the parameter *j* over the parameter *i* of the set $\overrightarrow{U}$ for the analysed component of a digital passport;

5 – with a significant parameter advantage *j* over the parameter *i* of the set $\overrightarrow{U}$ for the analysed component of a digital passport;

7 – with a clear advantage of the parameter *j* over the parameter *i* of the set $\overrightarrow{U}$ for the analysed component of a digital passport;

9 – with absolute parameter advantage *j* over the parameter *i* of the set $\overrightarrow{U}$ for the analysed component of a digital passport.

Here it is necessary to consider that the elements of the matrix (4):

- $m_{ij} = 1$ provided that $i = j$;
- $m_{ij} = 1$ in the lower triangles, since matrix elements *i* have no advantages over elements *j*.

1. Calculation of the eigenvalues of the matrix *M*, written in the equations of the following form:

$$M * \overrightarrow{\omega} = \lambda * \overrightarrow{\omega}; \tag{5}$$

where $\omega = \left(\overrightarrow{\omega_1}, \overrightarrow{\omega_2} \ldots \overrightarrow{\omega_P}\right)$ – matrix eigenvector *M*, is the value of membership function; λ – eigenvalue of the matrix *M*.

The eigenvalues of the matrix (4) are calculated in accordance with the properties of the determinants [7]:

$$\Delta M = \begin{vmatrix} (1-\lambda) & \ldots & m_{1P} \\ \ldots & \ldots & \ldots \\ 1 & \ldots & (1-\lambda) \end{vmatrix} = (1-\lambda) * \prod\left((1-\lambda) - m_{ij}\right),$$

$$i = \overline{1,(P-1)};\ j = \overline{2,P};$$

The result is several values λ, from which the maximum value is chosen - λ_{max}.

2. Calculation of the eigenvector of the matrix *M*

Substitute the calculated value $\lambda = \lambda_{max}$ in the (5):

$$\begin{vmatrix} (1-\lambda) & \ldots & m_{1P} \\ \ldots & \ldots & \ldots \\ 1 & \ldots & (1-\lambda) \end{vmatrix} * \begin{vmatrix} \overrightarrow{\omega_1} \\ \ldots \\ \overrightarrow{\omega_P} \end{vmatrix} = 0.$$

Obtain the corresponding system of equations of the following form:

$$\begin{cases} (1-\lambda) * \overrightarrow{\omega_1} + m_{12} * \overrightarrow{\omega_2} + \ldots + m_{1P} * \overrightarrow{\omega_P} = 0 \\ \ldots \\ \overrightarrow{\omega_1} + \overrightarrow{\omega_2} + \ldots + (1-\lambda) * \overrightarrow{\omega_P} = 0 \end{cases}. \tag{6}$$

To solve it, we introduce the normalization condition of the form $\overrightarrow{\omega_1} + \overrightarrow{\omega_2} + \cdots + \overrightarrow{\omega_P} = 1$. According to Chernov V. [6] this condition can be replaced by any equation in the system (6). A further solution is performed in any known way, to the results of which we obtain the values $\omega = \left(\overrightarrow{\omega_1}, \overrightarrow{\omega_2} \ldots \overrightarrow{\omega_P}\right)$, determining the parameter's belonging to the component of a digital passport.

Steps 1 through 3 are performed until the membership of parameters for all components of the digital passport is determined.

5 Selection of Descriptions of the Digital Passport Components, Given the Parameters of the Life Cycle Stage

The sets $\overrightarrow{C}$ and *RP* are also given in verbal form and make possible to describe the components of a digital passport given by the parameters of the life cycle stage in the form of fuzzy sets. The characteristic of their elements are the values of the membership functions, for the calculation of which the sequence of steps described earlier can be used.

1. The construction of matrices of pairwise comparisons of the elements of the set *RP* for each element of the set (3)

$$N\left(\overrightarrow{C}\right) = \left|\frac{1}{n_{ij}}\right|, n_{ij} = \frac{n_j}{n}, i,j = \overline{1,P}, \tag{7}$$

Where the values of the elements n_{ij}, are determined on a nine-point Saaty scale:

1 – in the absence of the advantage of the parameter *j* over the parameter *i* of the set *RP* for the analysed component of the digital passport;
3 – with a weak advantage of the parameter *j* over the parameter *i* of the set *RP* for the analysed component of the digital passport;
5 – with a significant parameter advantage *j* over the parameter *i* of the set *RP* for the analysed component of the digital passport;
7 – with a clear advantage of the parameter *j* over the parameter *i* of the set *RP* for the analysed component of the digital passport;

9 – with absolute parameter advantage j over the parameter i of the set RP for the analysed component of the digital passport.

It is necessary to consider that the elements of the matrix (7):

- $n_{ij} = 1$ provided that $i = j$;
- $n_{ij} = 1$ in the lower triangles, since matrix elements i have no advantages over elements j.

2. Calculation of the eigenvalues of the matrix N, written in the equations of the following form:

$$N * \overrightarrow{\omega^*} = \lambda * \overrightarrow{\omega^*}; \tag{8}$$

where $\omega^* = \left(\overrightarrow{\omega_1^*}, \overrightarrow{\omega_2^*} \ldots \overrightarrow{\omega_P^*}\right)$ – matrix eigenvector N, is the value of membership function; λ – eigenvalue of the matrix N.

The eigenvalues of the matrix (7) are calculated in accordance with the properties of the determinants [7]:

$$\Delta N = \begin{vmatrix} (1-\lambda) & \ldots & n_{1P} \\ \ldots & \ldots & \ldots \\ 1 & \ldots & (1-\lambda) \end{vmatrix} = (1-\lambda) * \prod((1-\lambda) - n_{ij}),$$

$$i = \overline{1, (P-1)}; j = \overline{2, P};$$

The result is several values λ, from which the maximum value is selected – λ_{max}.

3. Calculation of the eigenvector of the matrix N

Substitute the calculated value $\lambda = \lambda_{max}$ in the (8):

$$\begin{vmatrix} (1-\lambda) & \ldots & n_{1P} \\ \ldots & \ldots & \ldots \\ 1 & \ldots & (1-\lambda) \end{vmatrix} * \begin{vmatrix} \overrightarrow{\omega_1^*} \\ \ldots \\ \overrightarrow{\omega_P^*} \end{vmatrix} = 0.$$

Obtaining the corresponding system of equations of the following form:

$$\begin{cases} (1-\lambda) * \overrightarrow{\omega_1^*} + n_{12} * \overrightarrow{\omega_2^*} + \ldots + n_{1P} * \overrightarrow{\omega_P^*} = 0 \\ \ldots \\ \overrightarrow{\omega_1^*} + \overrightarrow{\omega_2^*} + \ldots + (1-\lambda) * \overrightarrow{\omega_P^*} = 0 \end{cases}. \tag{9}$$

To solve it, we introduce the normalization condition of the form $\overrightarrow{\omega_1^*} + \overrightarrow{\omega_2^*} + \ldots + \overrightarrow{\omega_P^*} = 1$. According to Chernov V. [6] this condition can be replaced by any equation in the system (9). A further solution is performed in any known way, to the results of which we obtain the values $\omega^* = \left(\overrightarrow{\omega_1^*}, \overrightarrow{\omega_2^*} \ldots \overrightarrow{\omega_P^*}\right)$, determining the parameter of the life cycle stage to the component of the digital passport.

Steps 1 through 3 are performed until the membership of parameters for all components of the digital passport is determined.

6 Conclusion

Thus, in the paper are presented the rules for calculating the descriptions of the components of a digital passport, given by the parameters describing them and the parameters of the life cycle stage. The first variant of the parameters corresponds to the parameters defined at the enterprise, directly during the creation of a digital passport. The second option is for parameters where values are fixed at different stages of the life cycle of an electronic product.

This will make possible further development of rules for the selection of descriptions of the components of a digital passport for generating design solutions by performing simple arithmetic operations. Formed solutions will allow to ensure the similarity of the production objects in the digital environment at the greatest extent.

References

1. Petrov, R.: Technology for developing integrated control systems for technical equipment based on standard instruments, standard documentation and software using design automation tools. Inf. Manage. Process. Syst. **1**(32), 80–87 (2016)
2. Shakhovtsev, E., Balandin, A., Shvidko, A.: The interaction of systems of automated pre-production and product lifecycle management: promising options. In: Proceedings of the Conference "Information Technologies in Management (ITU 2014)", pp. 431–439 (2014)
3. Donetskaya, J., Gatchin, Yu.: Integrated management environment for development and production work. Autom. Ind. **8**, 24–30 (2018)
4. Donetskaya, J., Sharygin, B., Butsyk, A.: Electronic technological passport of the product. Instrum. Eng. **3**(60), 280–286 (2017). News from universities
5. Donetskaya, J., Kuznetsova, O., Kuznetsov, A., Tushkanov, E., Gatchin, Yu.: Formation and analysis of engineering alternatives for an integrated electronic product description system. In: 2017 IEEE II International Conference on Control in Technical Systems (CTS), pp. 397–400 (2017)
6. Chernov, V.: Fundamentals of the Theory of Fuzzy Sets: A Textbook. Publishing House of Vladimir State University, Vladimir (2010)
7. Belousov, A., Tkachev, S.: Discrete Mathematics: A Textbook for Universities, 5th edn. Publishing Moscow State Technical University named after Bauman N., Moscow (2015)

High Performance Clustering Techniques: A Survey

Ilias K. Savvas[1(✉)], Christos Michos[1], Andrey Chernov[2], and Maria Butakova[2]

[1] University of Thessaly, Larissa, Greece
isavvas@uth.gr, cs7416012@teilar.gr
[2] Rostov State Transport University, Rostov-on-Don, Russia
{avcher, butakova}@rgups.ru

Abstract. We are living in a world of heavy data bombing and the term Big Data is a key issue these days. The variety of applications, where huge amounts of data are produced (can be expressed in PBs and more), is great in many areas such as: Biology, Medicine, Astronomy, Geology, Geography, to name just a few. This trend is steadily increasing. Data Mining is the process for extracting useful information from large data-sets. There are different approaches to discovering properties of datasets. Machine Learning is one of them. In Machine Learning, unsupervised learning deals with unlabeled datasets. One of the primary approaches to unsupervised learning is clustering which is the process of grouping similar entities together. Therefore, it is a challenge to improve the performance of such techniques, especially when we are dealing with huge amounts of data. In this work, we present a survey of techniques which increase the efficiency of two well-known clustering algorithms, k-means and DBSCAN.

Keywords: Clustering · High performance computing · DBSCAN · K-means

1 Introduction

Clustering is a very useful technique to extract information of data sets but is a task involving high time complexity especially when dealing with large and high dimensional data. The clustering algorithms run a lot of repetitions before finally converging to the desired result that is the clustered data. Because of this fact, it's a challenge to discover techniques that improve already existing clustering algorithms like the K-means [1] and DBSCAN [2] etc. (and their variants). Algorithms are generally serial by nature. Therefore, the parallelization of clustering algorithms, which they are computationally expensive, is very practical approach specially when they deal with large data sets. Hence, processing large data sets, demands the use of parallel and distributed calculation methods for faster results.

The main paradigm for parallel processing are High-Performance Computing (HPC) based on Message Passing Interface (MPI) [3] or Open Multi-Processing (OpenMP) [4]. Also, General Purpose Graphics Processing Units (GPGPUs) have hundreds of processor cores and can be programmed using programming models such as CUDA [5]. Combining MPI, OpenMP, and CUDA can exploit of all available hardware

S. Kovalev et al. (Eds.): IITI 2019, AISC 1156, pp. 252–259, 2020.
https://doi.org/10.1007/978-3-030-50097-9_26

resources of a computational cluster, adding value to the perspectives of clustering techniques. Another newer paradigm for parallel processing is the MapReduce-based Apache Hadoop. MapReduce [6], which has become the most prevailing parallel processing paradigm, it was first introduced in 2003.

This paper gives a survey of parallel implementation of various clustering algorithms and how particular steps of the algorithm can be computed using specific parallel programming paradigms. The rest of the paper is organized as follows. In Sect. 2 Parallel Programming paradigms and software platforms/libraries are presented. In Sect. 3, k-means and its parallel versions are described while the same approach for DBSCAN is given in Sect. 4. Finally, Sect. 5 summarizes this work and future directions are given.

2 Parallel Programming Frameworks

Increasing the performance of existing algorithms and techniques is a big challenge nowadays with the rapidly growth amount of data. Adapting parallelism is one very promising strategy to do it and there are several reasons to use it mainly, since the most of the times there is no need to discover new techniques but parallelizing the existing ones. To achieve this, fortunately, there are many Application Programming Interfaces (APIs), frameworks and libraries which support parallelism [7]. Some representative models are Computational Clusters using Message Passing Interface or the Hadoop framework [6, 8], single computational devices with many physical cores can use them to exploit parallelism or even to utilize their Graphics Processing Unit (GPU) which is consisting of hundreds of cores can use libraries like OpenMP or CUDA

OpenMP [4, 7]. OpenMP provides a set of compiler directives (pragmas) to create threads, synchronize the operations, and manage the shared memory on top of threads. Each thread has its own local memory. Therefore, OpenMP requires specialized compiler support to understand and process these directives. Due to the fact that threads share the same memory space the communications between threads can be very efficient. Typical programming languages that support OpenMP are C/C++ and Fortran and not just them. The main common use of the OpenMP is parallelizing loops.

MPI [3, 7]. MPI is a message passing library specification which defines an extended message passing model for parallel, distributed programming on a distributed computing environment. It is not actually a specific implementation of the parallel programming environment and its several implementations have been made, most notably OpenMPI and MPICH. MPI provides various library functions to provide point to point, collective, one-sided, and parallel I/O communication models. Like OpenMP typical programming languages that support MPI are C/C++ and Fortran.

CUDA [5, 7]. CUDA (Compute Unified Device Architecture) can be considered as a heterogeneous parallel programming model. It is a parallel computing platform created by NVIDIA in order to use the GPU for general purpose computing. For CUDA, a parallel system consists of a host (CPU) and a device (GPU). CUDA provides a software environment that allows developers to use C (with small variations), as high-level programming language. Since CUDA is specialized in matrix operations makes it ideal for numerical analysis and clustering algorithms, increasing their performance.

3 K-Means Clustering Algorithm

Cluster analysis is a well-known technique useful in many cases like machine learning, statistics, anomaly detections, and so on [9]. The most representative techniques are k-means, and DBSCAN. The computational complexity of both of algorithms makes them inefficient to be used especially when applications need real time response.

K-means. Given a data set, *D*, containing *n* objects in space, partitioning methods distribute the objects in *D* into k clusters, $C_1,\ldots,C_{k,}$ that is, $Ci \subset D$ and $Ci \cap Cj = \emptyset$ for $(1 \leq i, j \leq k)$. K-means requires as input the number of clusters (k) while its output is the k-clusters and their centroids. The similarity between objects/data point can be measured using metrics like the Euclidean or Manhattan distance [9]. K-means clustering is a greedy algorithm which is guaranteed to converge to a local minimum but the minimization of its score function is known to be NP-Hard [10]. The selection of initial centroids in K-means is extremely important and can affect the performance and its quality. The time complexity of K-means is O(NKdT), where T is the number of iterations. Since K, d, and T are usually much less than N, the time complexity of K-means is approximately linear [11].

3.1 Algorithms for K-Means Parallelization

In 2008, Farivar et al. [13] implemented a parallel version of K-means based on CUDA technology. As we have said above, K-means in each iteration during the phase in which it assigns the points to the clusters, requires nkd operations. It's obvious that this stage, is the candidate for parallelism. In their algorithm, the n points are partitioned in each GPU processor. In this way the computational cost for this particular stage from nkd decreases to nkd/p assuming there are p GPU processors. They tested their algorithm to a data set that consists of one-dimensional array of one million elements. The test, on an NVIDIA GeForce 8600 with 256 MB of dynamic RAM and 32 streaming processors, showed a 13× speed improvement compared to the serial implementation. Similar techniques were used by other many researchers who used the multi-core power of Graphical Processing Units and CUDA API [16].

In 2009, Zhao et al. [14] implemented an algorithm (**PKMeans**), which is a parallel version of K-means based on MapReduce. The algorithm consists essentially of 3 functions: a mapper, a combiner and a reducer. Firstly, the data-set is split and globally broadcast to all mappers. Then the distance computations are parallel executed. Next, the combiner, on the partial data-set, computes the number and the sum of samples that's belong to the same cluster for all clusters. Since the intermediate data is stored in local disk of the node, the communication cost is reduced. Finally, the reducer can sum all the samples and compute the total number of samples assigned to the same cluster. Therefore, we can get the new centers which are used for next iteration. The process is repeated until a convergence criterion is met. Since mappers, combiners and reducers can be located on several nodes the whole process can speed up very efficiently. The algorithm is tested in datasets of different size (from 1 GB to 8 GB) and the number of computational nodes varied from 1 to 4 giving very promising results [14]. Trying to improve PKMeans performance, Jin et al. [17] introduced **IPKEANS** which using k-d tree try to reduce the MapReduce process to just one job.

In 2012, Savvas et al. [15] presented the **MR** algorithm based on MapReduce. It contains a mapper and a reducer function. Contrary to the "PKMeans" the mapper sends all the data objects and the centroids they are assigned to, to the reducer. Then, for each centroid, the reducer calculates a new value based on the objects assigned to it in that iteration. This new centroid list is sent back to the start-up program. In order to reduce the time overhead between the mappers and the reducer, MR was modified in **MRC** by adding a combiner, thus arriving at a solution similar to PKMeans. In their tests MR outperforms the serial version only if the number of participating nodes in the clusters is large. As for MRC, its efficiency is in any case similar to PKMeans.

In 2016, Shahrivari et al. [18] proposed an algorithm based on MapReduce. The interest fact is that the algorithm run in a single pass. More precisely the dataset is partitioned among the mappers and in each mapper is applied the K-Means++. After, the mappers dispatch the intermediate centroids with the number of points corresponding to them, back to the reducer. In this intermediate dataset (weighting data points) is applied a modified K-Means++ with support of weighting data points.

In [19] the authors implemented a parallel K-Means algorithm that combine MPI and OpenMP. The algorithm performs better than a single MPI or OpenMP solution.

A novel algorithm was implemented in [20] in which the K-means is applied to each dimension of the dataset separately. After, the centroids of each dimension is combined to produce the final centroids. The K-means is implemented with MPI.

At [21] the **K-MeansII** algorithm was implemented. It comes from K-Means++ and has been appropriately modified, so it can be run in parallel. The authors claim that even the serial version of K-MeansII is better than K-Means++. The whole implementation was done with the MapReduce.

4 Density-Based Clustering Algorithm

In the contrary with k-means, DBSCAN is a density oriented spatial clustering algorithm. The main advantages over k-means is that distinguish the data points to core, border, and noise or outliers. This feature is very useful when trying to identify anomalies to a data set (outliers). In addition, clusters can be of any shape thus, centroids are meaningful here [9].

DBSCAN [2]. DBSCAN requires two input parameters; *eps* (distance measurement named as ε) and MinPts (minimum number of points in order to form a cluster). **ε-neighborhood** of a point p is defined as the space within the radius ε centered at p. The set of objects within the ε-neighborhood is $N_\varepsilon(p)$ and if $|N_\varepsilon(p)| \geq$ MinPts then p is called a **core point**. If $|N_\varepsilon(p)| <$ MinPts and in its ε-neighborhood contains at least one core point, is called **border point**. If $|N_\varepsilon(p)| <$ MinPts and in its ε-neighborhood contains no core point, is named as **noise point**. A point q is **directly density-reachable** from a point p, if q belongs to ε-neighborhood of p and p is a core point. A point q is **density-reachable** from p if there is a sequence of directly density-reachable points connecting point q with point p. A point p is **density-connected** to a point q if they are both density-reachable from a point o [22]. The set of points which is density-connected, form a cluster.

4.1 Algorithms for DBSCAN Parallelization

At [23] (1999), the parallel **PDBSCAN** clustering algorithm for mining in large spatial databases is presented. PDBSCAN is implemented using the PVM Platform [24], which is similar to MPI. The method consists of three main steps: The first step is to divide the input into several partitions, and to distribute these partitions to the available computers. The second step is to cluster partitions concurrently using DBSCAN. The third step is to combine or merge the clusterings of the partitions into a clustering of the whole database. To Identify core points which is located in boundaries of the partitions, a structure R*tree[25] is used, which also support efficient access to the distributed data in order to reduce the run-time to $O(|S_i|\log|S_i|)$, for each partition S_i. It was proven that the speedup of the algorithm is almost linear. Also it is evident that the algorithm scales well.

A very interesting work in [26], the authors introduce **DSDBSCAN** where they combine MPI with OpenMP in order to take advantage of both features of distributed and multi-core programming using the master/worker model [23]. To achieve this, they split the data set in order to increase the parallelism of their technique.

Similar technique but using only the existing cores of a computational device use in a shared memory environment, authors in [12] developed **PDSDBSCAN-S**. Like the previous mentioned case, the data set is divided in disjoint subsets depending on the number of cores/threads. The result is for each thread to create its local clusters which in turn after merging using locking techniques lead to global solution.

Another technique (**GSCAN**), that improves CUDA-DClust is found in [28] (based in CUDA). In GSCAN, the entire dataset is evenly divided into gridSize data subsets and each subset (a grid) is assigned to a kernel block. GSCAN reduces the number of unnecessary distance computations, by performing them in the space of a grid cell and possibly in its neighbors.

CUDA-DClust outperforms DBSCAN (not indexed), by a large factor (10x and above). CUDA-DClust* outperforms indexed DBSCAN by a large factor, that is proportional to the size of the data set(at least 10x if we have 250 k points). GSCAN outperformed CUDA-DClust and DBSCAN by up to 13.9 and 32.6 times, respectively.

Finally, at [29] we have an implementation in MPI in which each node performs sequential DBSCAN and with the help of the local centroids with their radiuses, the final clusters can be formed while at [30] the **RP-DBSCAN** is implemented, in which is proposed a cell-based data partitioning scheme, pseudo random partitioning, that randomly distributes small cells rather than the points themselves.

As a resume, the importance of DBSCAN can be easily proved by the number of researchers who tried to increase its performance parallelizing it using relatively new platforms and APIs as MapReduce [8], or CUDA [27].

5 Conclusion and Future Work

In this work, we surveyed about the most important clustering algorithms and presented ways to improve their performance. Such improvement was achieved using mainly OpenMP, MPI, MapReduce and CUDA, which are the most widely recognized parallel or distributed programming frameworks. We noticed that in any model, the data set is divided into as many as the computational resources available. Since tasks in each resource are usually executed serially, the total time of the algorithm equals the latest time that has been achieved individually by the resources. We also noticed that especially on the MPI and MapReduce platforms it is important to minimize the communication overhead (by using for example R-trees), in order to minimize network communication (which is time-consuming).

As a final conclusion we can say that MPI is suitable when the data size is moderate and the problem is computation-intensive. When the data size is large and jobs do not require repeat processing, MapReduce can be an excellent framework. In OpenMP, resources are specific and therefore can ideally handle small data, unlike MPI and MapReduce models that can scale up. OpenMP can easily be combined with MPI and perhaps more difficult to do with MapReduce. CUDA is ideal for performing very fast numerical calculations in parallel, but it does not escalate. In our opinion it could be part of the MPI master node or Reducer node at MapReduce.

The future work includes the extension of this survey to other clustering techniques such as kernel k-means, OPTICS, and on algorithms which try to find out outliers like the Local Outlier Factor (LOF).

Acknowledgments. The reported study was funded by RFBR according to the research project 19-01-246-a, 19-07-00329-a, 18-01-00402-a, 18-08-00549-a.

References

1. Lloyd, S.P.: Least squares quantization in PCM. IEEE Trans. Inf. Theory **28**(2), 129–136 (1982)
2. Ester, M., Kriegel, H.-P., Sander, J., Xu, X.: A density-based algorithm for discovering clusters in large spatial databases with noise. In: Proceedings of 2nd International Conference on Knowledge Discovery, pp. 226–231 (1996)
3. MPICH, Message Passing Interface. http://www.mpich.org/. Accessed 21 Apr 2019
4. OpenMP, Open Multi-Processing. http://www.openmp.org/. Accessed 21 Apr 2019
5. CUDA Zone: NVDIA Accelerated Computing. https://developer.nvidia.com/cudazone. Accessed 21 Apr 2019
6. Dean, J., Ghemawat, S.: MapReduce: simplified data processing on large clusters. Commun. ACM **51**, 137–150 (2004)

7. Kang, S.J., Lee, S.-H., Lee, K.-M.: Performance comparison of OpenMP, MPI, and MapReduce in practical problems. In: Advances in MM, pp. 575687:1–575687:9 (2015)
8. He, Y., Tan, H., Luo, W., Feng, S., Fan, J.: MR-DBSCAN: a scalable MapReduce-based DBSCAN algorithm for heavily skewed data. Front. Comput. Sci. **8**(1), 83–99 (2014)
9. Han, J., Kamber, M., Pei, J.: Data Mining: Concepts and Techniques. Morgan Kaufmann Publishers Inc., San Francisco (2011)
10. Manning, C.D., Raghavan, P., Schütze, H.: Introduction to Information Retrieval. Cambridge University Press, Cambridge (2008)
11. Xu, R., Wunsch, D.: Clustering. Wiley-IEEE Press, Hoboken (2008)
12. Kadam, P., Jadhav, S., Kulkarni, A., Kulkarni, S.: Survey of parallel implementations of clustering algorithms. Int. J. Adv. Res. **6**(10) (2017)
13. Farivar, R., Rebolledo, D., Chan, E., Campbell, R.H.: A parallel implementation of K-Means clustering on GPUs. In: Arabnia, H.R., Mun, Y. (eds.) PDPTA, pp. 340–345. CSREA Press (2009)
14. Zhao, W., Ma, H., He, Q.: Parallel K-means clustering based on MapReduce. In: Jaatun, M. G., et al. (eds.) CloudCom, pp. 674–679. Springer, Heidelberg (2009)
15. Savvas, I.K., Kechadi, M.T.: Mining on the cloud - K-means with MapReduce. In: Leymann, F., et al. (eds.) CLOSER, pp. 413–418. SciTePress (2012)
16. Yang, L., Chiu, S.C., Liao, W.K., Thomas, M.A.: High performance data clustering: a comparative analysis of performance for GPU, RASC, MPI, and OpenMP implementations. J. Supercomput. **70**(1), 284–300 (2014)
17. Jin, S., Cui, Y., Yu, C.: A new parallelization method for k-means. CoRR. Abs/1608.06347 (2016)
18. Shahrivari, S., Jalili, S.: Single-pass and linear-time K-means clustering based on MapReduce. Inf. Syst. **60**, 1–12 (2016)
19. Savvas, I.K., Tselios, D.C.: Combining distributed and multi-core programming techniques to increase the performance of K-Means algorithm. In: Reddy, S., et al. (eds.) WETICE, pp. 95–100. IEEE Computer Society (2017)
20. Savvas, I.K., Sofianidou, G.N.: A novel near-parallel version of k-means algorithm for n-dimensional data objects using MPI. IJGUC **7**(2), 80–91 (2016)
21. Bahmani, B., Moseley, B., Vattani, A., Kumar, R., Vassilvitskii, S.: Scalable K-Means++. CoRR. Abs/1203.6402 (2012)
22. Wowczko, I.A.: Density-based clustering with DBSCAN and OPTICS. Business Intelligence and Data Mining, (2013)
23. Arlia, D., Coppola, M.: Experiments in parallel clustering with DBSCAN. In: Sakellariou, R. et al. (eds.) Euro-Par, pp. 326–331. Springer, Heidelberg (2001)
24. Geist, A., Beguelin, A., Dongarra, J., Jiang, W., Manchek, R., Sunderam, V.: PVM: Parallel Virtual Machine—A Users' Guide and Tutorial for Networked Parallel Computing. MIT Press, Cambridge (1994)
25. Beckmann, N., Kriegel, H., Schneider, R., Seeger, B.: The R*-tree: an efficient and robust access method for points and rectangles. In: Proceedings of ACM-SIGMOD International Conference on Management of Data, Atlantic City, NJ, pp. 322–331 (1990)
26. Patwary, M.M.A., Palsetia, D., Agrawal, A., Liao, K.W., Manne, F., Choudhary, A.N.: A new scalable parallel DBSCAN algorithm using the disjoint-set data structure. In: Hollingsworth, J.K. (ed.) SC, p. 62. IEEE/ACM (2012)
27. Böhm, C., Noll, R., Plant, C., Wackersreuther, B.: Density-based clustering using graphics processors. In: Cheung, D.W.-L., et al. (eds.) CIKM, pp. 661–670. ACM (2009)
28. Loh, W.-K., Moon, Y.-S., Park, Y.-H.: Fast density-based clustering using graphics processing units. IEICE Trans. Inf. Syst. **97**(7), 1947–1951 (2014)

29. Savvas, I.K., Tselios, D.C.: Parallelizing DBSCaN algorithm using MPI. In: Reddy, S., Gaaloul, W. (eds.) WETICE, pp. 77–82. IEEE Computer Society (2016)
30. Song, H., Lee, J.-G.: RP-DBSCAN: a superfast parallel DBSCAN algorithm based on random partitioning. In: Das, G., et al. (eds.) SIGMOD Conference, pp. 1173–1187. ACM (2018)

An Intelligent Data Warehouse Approach for Handling Shape-Shifting Constructions

Georgia Garani[1(✉)], Ilias K. Savvas[1], Andrey V. Chernov[2], and Maria A. Butakova[2]

[1] University of Thessaly, Geopolis, 41110 Larisa, Greece
{garani, isavvas}@uth.gr
[2] Rostov State Transport University, Rostov-on-Don 344038, Russia
{avcher, butakova}@rgups.ru

Abstract. A growing interest has been shown recently, concerning buildings as well as different constructions that use transformative and mobile attributes for adapting their shape, size and position in response to different environmental factors, such as humidity, temperature, wind and sunlight. Responsive architecture as it is called, can exploit climatic conditions and changes for making the most of them for the economy of energy, heating, lighting and much more. In this paper, a data warehouse has been developed for supporting and managing spatiotemporal objects such as shape-shifting constructions. Spatiotemporal data collected from these transformations are good candidates for analysis by data warehouses for decision making and business intelligence. The approach proposed in this research work is based on the integration of space and time dimensions for the management of these kinds of data. A case study is presented where a shape-shifting buildings data warehouse is developed and implemented. A number of spatiotemporal queries have been executed and their run times were compared and evaluated. The results prove the suitability of the proposed approach.

Keywords: Data warehouse · Shape-shifting construction · Spatiotemporal object

1 Introduction

The need to analyze large volume of data has emerged the last decades for extracting meaningful information. Towards this direction, Data Warehousing has been introduced in the late 1980s by two IBM workers, Paul Murphy and Barry Devlin, who developed the Business Data Warehouse for addressing the problem of redundant data acquired from operational systems to support decision support environments. Since then, Data Warehousing consists one of the main techniques for supporting the decision making process within an organization by collecting and managing data from varied and distributed sources to provide meaningful business insights. The surging of the internet during the 1990s resulted in the use of advanced application systems, networking and globalization followed by the need of true Data Warehousing for business intelligence (BI) and strategic decision making.

S. Kovalev et al. (Eds.): IITI 2019, AISC 1156, pp. 260–269, 2020.
https://doi.org/10.1007/978-3-030-50097-9_27

The real concept of a Data Warehouse (DW) was given by Bill Inmon. He has published many books and articles about Data Warehousing and data management. His approach is characterized as a top-down approach compared to the bottom-up approach designed by Ralph Kimball, another pioneer in this field. The top-down approach starts with a normalized data model followed by the complete DW and finally, the dimensional data marts are created, in contrast with the bottom-up design where the data marts are created first and then, these are combined into a large DW.

With the rise of the new millennium it was obvious that business systems depend upon a unified approach for organizing and representing their disparate, heterogeneous and usually, fragmented data. In some way, the data needed to be integrated to provide support to their strategic decision-making efforts. To this perspective, a DW consists the core of the BI system which is built for data consolidation, analysis and reporting at different aggregate levels. It operates as a central repository where information arrives from diverse data sources.

DWs have applications to many different sectors, i.e., healthcare, telecommunication, banking, industry. Civil engineering and environmental sciences are two fields which can also benefit from DWs. Modern civilization is characterized by buildings of different shape, material, size, and format. Nevertheless, all of them are considered to be stationary and rigid constructions losing thus, in flexibility and adaptability. Recently, promising research explores the benefits and advantages of shape-shifting architecture which transforms in response to changing environmental conditions. Global climate is changing and observations have shown that the impact is multi-faceted. Climate changes are seriously affecting humans, animals and plants. By taking advantage of specific conditions referring to temperature, humidity, light and wind, buildings can become smarter and responsive to their external environment.

In this paper a DW is designed for supporting shape-shifting buildings. The DW uses spatiotemporal objects for supporting constructions that change occasionally over time. The analysis of data collected from the modification of shape and position over time of different sides and facades of buildings, can give useful information about how materials are responding according to environmental and climatic conditions.

The paper is structured as follows. Section 2 discusses related work. Section 3 is devoted to presenting motivating example scenarios regarding the topic that handles this paper. The basic idea of spatiotemporal objects is explained in Sect. 4 as well as some fundamental design choices that need to be made. The case study is presented in Sect. 5 where the logical modeling is designed followed by a number of queries tested and executed with their experimental results demonstrated. Finally, Sect. 6 offers conclusions and future work ideas.

2 Current Status and Research Issues

Related research work is presented briefly below in chronological order. In [3] moving points and moving regions are modeled as abstract data types integrated as base attribute data types into DBMS data models. SOLAP (Spatial OLAP) tools are presented and defined in [12] for effective spatiotemporal analysis. A series of essential features, as well as desirable characteristics, are then described. Finally, some technological considerations are considered and an example of spatiotemporal analysis is presented. In [11] several issues are studied about trajectory DWs and in particular, storing aggregate measures. The focus is on the measure presence which is discussed thoroughly. Presence returns the number of distinct trajectories situated at a spatial region during a given temporal interval. A new method of how to compute such a measure is proposed and some experiments have conducted which were implemented on an Oracle prototype DW.

An extension of the Web Ontology Language (OWL) is proposed in [14] enhanced with spatial and temporal features for defining a spatiotemporal DW. [15] presents a conceptual framework for defining spatiotemporal DWs using an extensible data type system. A taxonomy for spatiotemporal OLAP is given, based on temporal dimensions, OLAP, GIS and moving data types. Specifically, Spatio-Temporal Data derives from GIS and moving data types, TOLAP from Temporal OLAP, SOLAP from Spatial OLAP, S-TOLAP from Spatial TOLAP, ST-OLAP from Spatio-Temporal OLAP and ST-TOLAP from Spatio-Temporal TOLAP. Based on this taxonomy, different kinds of queries are provided. Finally, a spatiotemporal calculus for supporting moving data types is defined.

In [7], a survey of spatiotemporal data warehousing, OLAP and mining is given, followed by a comparison of two proposals, the Piet data model with the Hermes system. A main difference is that Piet model considers geometric objects as static and thus, does not provide temporal support for them, but only for traceable objects that move through the geographic space, while Hermes supports moving objects that change their shape or position over time. The GeoCube model is proposed in [2] for integrating geographic information and spatial analysis into multidimensional models. The semantic component of geographic information is considered for the definition of geographic dimensions and geographic measures. Five new operators are provided for extending SOPAL operators. These are called Classify, Specialize, Permute, OLAP-Buffer and OLAP-Overlay.

The notion of a Spatiotemporal DW is discussed in [17]. This work deals with how to store and query data with spatial and temporal features into spatiotemporal DWs. An extended classical relational calculus is used for querying the spatiotemporal DW enhanced with spatial types and moving types. [4] presents a spatiotemporal DW for efficiently querying location information of moving objects. It is based on a star schema dimensional model supporting spatial and temporal dimensions. The proposed schema introduces new direction-based measures, i.e. direction majority which computes the average motion direction among all segments that exist in a spatial region at a given time interval. A number of SQL queries are given for implementing the proposed measures.

A DW supporting spatial and mobility data is presented in [16]. The Mobility DW is developed based on MobilityDB, a moving object database that extends the PostgresSQL database with temporal data types; thus, relational warehouse data is integrated with moving object data. However, the model supports only moving points and not moving lines and moving regions.

3 Motivating Example Scenario

Responsive architecture is an evolving architectural field lately. Buildings can change their shape, form or even location with the integration of computing power into built spaces and structures by measuring actual environmental conditions via sensors or control systems [8]. Rotating constructions, sliding surfaces and kinetic facades are a number of innovative conceptions which have already appeared in different constructions and present a promising research field for a wide range of professionals, such as civil engineers, architects, building designers, computer scientists, mechanical engineers and environmental scientists. The result is to design buildings that reflect the technological and cultural circumstances of current time and take advantage of the climatic conditions that occur during the day and season. Therefore, these building are called shape-shifting structures. A number of new buildings could function like living systems, altering their shapes in response to changing weather conditions or the way people use them. These would truly be smart buildings.

A number of such examples are presented below. Sharifi-ha House is located in Tehran, Iran. The main characteristic of this building is that it opens up to rooms and terraces in summer by rotating 90° three turning pods. During winter these pods can be turned inward. In that way, the residence can adjust to Iran's changeable temperatures [10]. Another house on the outskirts of Warsaw, Poland transforms from a day villa to a night fortress. This is achieved by covering the entire facade and windows with wall panels for minimizing the possibility of breaking in. The Dynamic D* Haus is a promising shape-shifting construction; however, it is not yet implemented. It responds to weather and time conditions. It is based on a mathematical formula for transforming an equilateral triangle into a square which splits the building into four separate modules. Eight different shapes can be derived from the original shape by rotation and inclination.

The optimization of living conditions, by self-adapting buildings in accordance with light, temperature, wind, rain, and other climatic conditions, can be succeed by designing buildings that include moving walls, rotating rooms and sliding surfaces. In particular, several research works have been published recently about adapting building facades from different perspectives, such as physics, chemistry, engineering and architecture. Temperature-responsive systems are discussed in [1] where different materials are examined that can be used in the field of adaptive facades. In [9] properties, characteristics and applications of responsive materials for designing adaptive facades are investigated. A brief overview is presented in [13] about current approaches in this field and the term 'adaptive' is explored under the prism of interaction between technological systems and the environment. The above mentioned technology can also be applied to several other usages for materials that can change shape, size or position including agriculture, biology and medicine.

Inspired by this highly environmental expectations research field, the approach proposed in what follows, tries to encounter the problem of storing spatiotemporal data to DWs for analysis purposes. However, due to their nature, i.e. at least three-dimensional (two-dimensional+time), special treatment is required as presented in the next sections.

4 Spatiotemporal Object

For designing a spatiotemporal DW, a spatiotemporal object is introduced for supporting spatial and temporal data. A spatiotemporal object is considered a geometry that changes over time. It is in fact, a moving region that captures moving as well as growing or shrinking regions. For representing a spatiotemporal object, space and time are integrated and time is not considered an attribute of space. It is a three (2D+time) or higher dimensional entity. A spatiotemporal object is based on the starnest schema approach [6] where dimension tables are nested containing subattributes. The nested feature of the starnest schema is used for defining the geometry of the spatiotemporal object as a number of nested subtuples. For each time interval denoted by the specific Start and Stop attributes, a spatial object is defined by a number of subtuples which represents the number of points that specify it, i.e. for a triangle, three points are needed, for a quadrangle four points and for a polygon many points are required. The spatiotemporal object is defined below as an extension of the spatial object defined in [5].

Definition 1 (Spatiotemporal object, three-dimensional): STO(STOId, d_{a1}, …, d_{ak}, $STDim_i(f_i, SDim_i(X, Y), Start, Stop, f_{a1}, \ldots, f_{aj}))$, where STOId stands for the spatiotemporal object identifier, f_i a unique identifier for each different $STDim_i$ ($i \geq 1$), X, Y are the spatial two-dimensional spatial coordinates of the spatiotemporal object, Start and Stop are the start and stop time points of the corresponding time interval of the form [Start, Stop) where Start < Stop, d_{ak} stands for description attribute k ($k \geq 1$) and f_{aj} for feature attribute j ($j \geq 1$).

Definition 1 can be extended to support four-dimensional spatiotemporal objects, by increasing the degree of SDim by 1 when three-dimensional spatial coordinates are supported.

BuildingID	BuildingName	BuildingFeatures						
		BF	BuildingShape		Start	End	Temperature	Humidity
			X	Y				
B1	Altura	1	x_{11}	y_{11}	$t1_{START}$	$t1_{STOP}$	T1	H1
			x_{12}	y_{12}				
			x_{13}	y_{13}				
			x_{14}	y_{14}				
		2	x_{21}	y_{21}	$t2_{START}$	$t2_{STOP}$	T2	H2
			x_{22}	y_{22}				
			x_{23}	y_{23}				
			x_{24}	y_{24}				
			x_{25}	y_{25}				
		3	x_{31}	y_{31}	$t3_{START}$	$t3_{STOP}$	T3	H3
			x_{32}	y_{32}				
			x_{33}	y_{33}				
			x_{34}	y_{34}				

Fig. 1. Spatiotemporal object building instance

In Fig. 1 an example of a spatiotemporal object is shown where Altura building initially is a quadrangle, then a polygon and at a later time, a quadrangle again. Other features can also describe the spatiotemporal object, such as temperature and humidity as it is shown in Fig. 1.

5 Case Study

5.1 Logical Modeling

The logical design of the spatiotemporal DW is based on spatiotemporal objects. The implemented DW concerns shape-shifting buildings, their usages and residents. The shape-shifting buildings DW consists of one fact table, Usage, and two dimension tables, Building and Resident. Usage fact table consists of the primary key UsageID, two foreign keys, ResidentID and Building ID for establishing links between the fact and the two dimension tables of the DW and four other attributes, including two time attributes, UsageStartDate and UsageStopDate. Building table is a spatiotemporal table consisting of two nested subtables, one inside the other, for defining spatiotemporal objects as explained in Sect. 4. It consists of the two-dimensional spatial coordinates for defining the shape and position along with the time duration by the Start and Stop attributes for each building instance. Resident table is a conventional dimension table consisting of the primary key ResidentID and four more dimensional attributes.

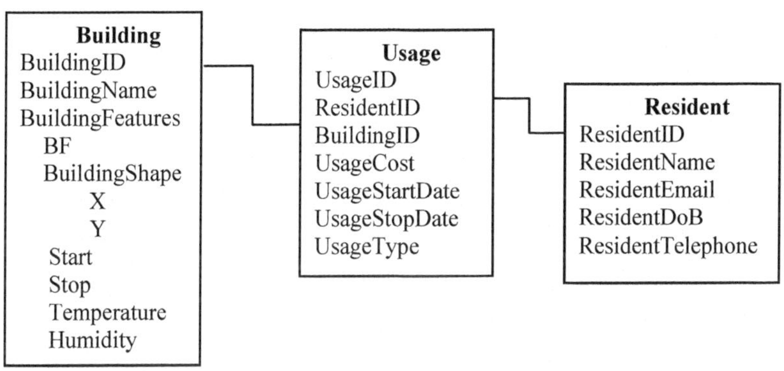

Fig. 2. Shape-shifting buildings DW

5.2 Querying

In this subsection a number of queries are presented in standard SQL concerning spatiotemporal data for the shape-shifting buildings DW presented in Fig. 2.

Query 1: How many times Altura building changed its shape during the year 2018?

```
SELECT COUNT(Building.BuildingFeatures.BuildingShape)
FROM Building
WHERE BuildingName='Altura'
AND Building.BuildingFeatures.Start>='1/1/2018'
AND Building.BuildingFeatures.Stop<='31/12/2018'
```

Query 2: Which building changed its shape the maximum number of times?

```
SELECT MAX(T.NUM)
FROM (SELECT COUNT(Building.BuildingFeatures.BuildingShape) AS NUM
      FROM Building
      GROUP BY BuildingID) T
```

Query 3: What is the total number of days for each building that stayed in one specific shape when the temperature was greater than 30 °C and the humidity less than 50%?

```
SELECT BuildingID, Building.BuildingFeatures.BF,
    SUM(DATEDIFF(Building.BuildingFeatures.Stop-Building.BuildingFeatures.Start))
FROM Building
WHERE Building.BuildingFeatures.Temperature>30
AND Building.BuildingFeatures.Humidity<50
GROUP BY BuildingID, Building.BuildingFeatures.BF
```

Query 4: Which building changed its shape although temperature and humidity remained the same?

```
SELECT B1.BuildingID
FROM Building B1, Building B2
WHERE B1.BuildingID=B2.BuildingID
AND B1.Building.BuildingFeatures.Temperature=B2.Building.BuildingFeatures.Temperature
AND B1.Building.BuildingFeatures.Humidity=B2.Building.BuildingFeatures.Humidity
AND B1.Building.BuildingFeatures.Start<>B2.Building.BuildingFeatures.Start
```

Query 5: Who was the resident of building with name Altura and what was the shape (the coordinates) of this building on 1/1/2019?

```
SELECT ResidentName, Building.BuildingFeatures.BuildingShape
FROM Resident, Usage, Building
WHERE Resident.ResidentID=Usage.ResidentID
AND Building.BuildingID=Usage.BuildingID
AND BuildingName='Altura'
AND Building.BuildingFeatures.Start<='1/1/2019'
AND Building.BuildingFeatures.Stop>='1/1/2019'
```

5.3 Implementation

An Intel(R) Core(TM) i5-3210 M processor CPU with SSD was used for the implementation running at 2.5 GHz with 16 GB ram memory, under Windows 10 (64bit) OS. The proposed spatiotemporal schema for the shape-shifting buildings DW was built in Oracle Data Warehouse builder 11 g Release 2 (11.2.0.1.0) and Oracle SQL Developer 18.3.0.277 was used.

Table 1. Shape-shifting buildings DW features.

Table name	Size (KB)	Number of rows
Building	128	32
Usage	640	10.000
Resident	192	1.000

The total size for the schema is 1.375 MB and the size of the DW (space the files physically consume on disk) is 1.13 GB. Shape-shifting buildings DW features are shown in Table 1.

5.4 Experimental Results

The queries presented in Subsect. 5.2 were executed 10 times each. Table 2 shows queries execution times and Fig. 3 the corresponding average execution times chart.

Table 2. Queries execution times.

	Query1	Query2	Query3	Query4	Query5
T1	0,002	0,009	0,007	0,007	0,013
T2	0,002	0,002	0,005	0,003	0,002
T3	0,002	0,002	0,003	0,003	0,003
T4	0,002	0,003	0,002	0,003	0,002
T5	0,002	0,003	0,003	0,002	0,003
T6	0,001	0,002	0,002	0,002	0,002
T7	0,003	0,002	0,002	0,003	0,002
T8	0,001	0,003	0,002	0,002	0,002
T9	0,002	0,003	0,002	0,003	0,002
T10	0,001	0,002	0,002	0,003	0,003
secs	0,0018	0,0031	0,003	0,0031	0,0034
msecs	1,8	3,1	3	3,1	3,4

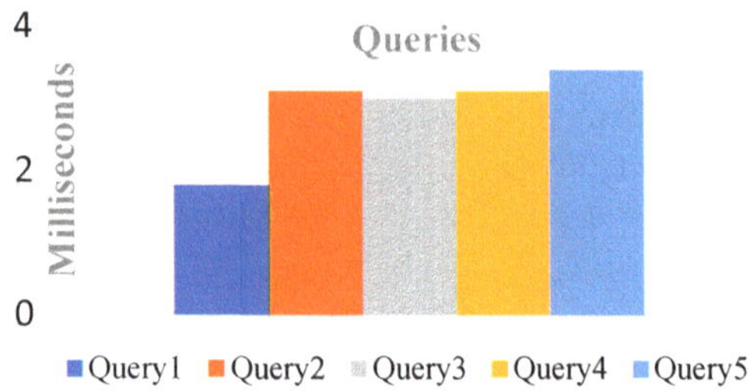

Fig. 3. Queries average execution times chart

All queries have similar execution times apart from Query 1. Query 1 is faster compared to all other queries. This can be justified by the fact that it is the simplest query given that only one aggregate function is used, i.e. count and only one table. Query 2 uses two aggregate functions, max and count, and one subquery expression in the FROM clause that acts as a derived table. Query 3 uses one aggregate function, sum, and one date function, datediff, for calculating number of days. Query 4 uses two copies of the Building table. Query 5 is the slowest, since it involves all tables from the DW, Usage, Building and Resident and therefore, two joins are required.

6 Conclusion

Adaptive buildings responding to environmental and climatic conditions, such as light, temperature, wind and humidity, have started to be designed and built in recent years and a growing interest is observed for their increased production and improved productivity. The analysis of spatiotemporal data concerning the transformation of shape and changing of position for shape-shifting buildings is studied in this research work. Spatiotemporal DWs consist the right solution for collecting and managing such spatial and temporal features. Hence, this paper proposes a spatiotemporal DW for supporting spatiotemporal objects. A case study is presented and implemented for a shape-shifting buildings DW. A number of queries have been tested and their execution times were compared. Results are very promising for analyzing efficiently spatiotemporal data.

Future work involves comparisons of the proposed model with already existed solutions such as Oracle Spatial. In addition, since nowadays, the majority of cloud providers offer DW services like the Apache Hive DW, Azure SQL DW and Amazon DW over Amazon Redshift Cluster, the proposed model will be applied to such an approach and their performances will be compared.

Acknowledgments. The reported study was funded by RFBR according to the research project 19-01-246-a, 19-07-00329-a, 18-01-00402-a, 18-08-00549-a. The authors would like to thank Christos Siopis, undergraduate student from the Department of Computer Science and Engineering of the University of Thessaly, who helped with the implementation of this research work during his diploma thesis.

References

1. Aresta, C.: Temperature-responsive systems: passive strategies for building envelopes. In: FAÇADE 2018, Final Conference of COST TU1403 "Adaptive Facades Network", Lucerne, Switzerland (2018)
2. Bimonte, S., Tchounikine, A., Miquel, M., Pinet, F.: When spatial analysis meets OLAP: multidimensional model and operators. Int. J. Data Warehouse. Min. **6**(4), 33–60 (2010)
3. Erwig, M., Güting, R.H., Schneider, M., Vazirgiannis, M.: Spatio-temporal data types: an approach to modeling and querying moving objects in databases. GeoInformatica **3**(3), 269–296 (1999)
4. Esheiba, L., Mokhtar, H.M.O., El-Sharkawi, M.: Spatio-temporal queries for moving objects data warehousing. Int. J. Database Manag. Syst. **5**(3), 1–13 (2013)
5. Garani, G.: Representing spatial objects in data warehouses: a logical solution. Int. J. Spat. Temporal Multimedia Inf. Syst. **1**, 232–252 (2019)
6. Garani, G., Helmer, S.: Integrating star and snowflake schemas in data warehouses. Int. J. Data Warehouse. Min. **8**(4), 22–40 (2012)
7. Gσmez, L., Kuijpers, B., Moelans, B., Vaisman, A.: A survey of spatio-temporal data warehousing. Int. J. Data Warehouse. Min. **5**(3), 28–55 (2009)
8. Kormaníková, L., Kormaníková, E., Katunský, D.: Shape design and analysis of adaptive structures. Procedia Eng. **190**, 7–14 (2017)
9. Miren, J., Aurora, M.B., Ulrich, K., Tomas, G.A.: Smart and multifunctional materials and their possible application in façade systems. J. Facade Des. Eng. **6**(3), 19–33 (2018)
10. Orhon, A.V.: Adaptive building shells. In: Developments in Science and Engineering, pp. 555–567. St. Kliment Ohridski University Press, Sofia (2016)
11. Orlando, S., Orsini, R., Raffaetà, A., Roncato, A.: Trajectory data warehouses: design and implementation issues. J. Comput. Sci. Eng. **1**(2), 211–232 (2007)
12. Rivest, S., Bédard, Y., Marchand, P.: Toward better support for spatial decision making: defining the characteristics of spatial on-line analytical processing (SOLAP). Geomatica **55**(4), 539–555 (2001)
13. Romano, R., Aelenei, L., Aelenei, D., Mazzucchelli, E.S.: What is an adaptive façade? Analysis of recent terms and definitions from an international perspective. J. Facade Des. Eng. **6**(3), 65–76 (2018)
14. Salguero, A., Araque, F., Delgado, C.: Spatio-temporal ontology based model for data warehousing. In: 7th WSEAS International Conference on Telecommunications and Informatics, TELE-INFO 2008, Istanbul, Turkey, pp. 125–130 (2008)
15. Vaisman, A., Zimanyi, E.: What is spatio-temporal data warehousing? In: Pedersen, T.B., Mohania, M.K., Tjoa, A.M. (eds.) International Conference on Data Warehousing and Knowledge Discovery (DaWaK) 2009. LNCS, vol. 5691, pp. 9–23. Springer, Heidelberg (2009)
16. Vaisman, A., Zimányi, E.: Mobility data warehouses. Int. J. Geo-Inf. **8**(4), 170, 1–22 (2019)
17. Zimanyi, E.: Spatio-temporal data warehouses and mobility data: current status and research issues. In: 19th International Symposium on Temporal Representation and Reasoning (TIME), Leicester, UK (2012)

Effect of Resonance in the Effective Control Model Based on the Spread of Influence on Directed Weighted Signed Graphs

Alexander Tselykh[1], Vladislav Vasilev[1], and Larisa Tselykh[2(✉)]

[1] SFEDU, Taganrog, Russia
[2] RSUE, Taganrog, Russia
l.tselykh58@gmail.com

Abstract. This article proposes a method of selecting the damping factor (DF) based on finding the boundaries of the range of the domain of the DF for a model of effective controls, reflecting the spread of influence on the directed weighted signed graphs. The graph model is a fuzzy cognitive map (FCM). Two ranges of the subset values of the domain of the DF are proposed: (i) the admissible range, within which the co-directionality of the response and impact vectors is guaranteed to prevail, and (ii) the resonance range, within which there is a rapid rearrangement and reversal of the ranks of the vertices. An interpretation of the resonance surge for this model of the spread of influence is proposed.

Keywords: Effective controls · Damping factor · Influence · Directed weighted signed graphs · Fuzzy cognitive map

1 Introduction

This study addresses the problem of choosing the damping factor (DF) in the model of effective controls based on maximizing the influence transfer onto directed weighted signed graphs. The problem of spreading influence in networks has been extensively studied in recent years. Nets can be represented by undirected weighted signed graphs and directed weighted signed graphs. Undirected weighted signed graphs represent social networks, where the weights of edges encode frequency of contacts between vertices. The most well-known algorithm for ranking nodes in networks is the PageRank algorithm [1]. The issues of selecting the DF are addressed for undirected weighted signed graphs, the adjacency matrix of which is a real and symmetric matrix [2].

Fuzzy cognitive map (FCM) is a directed weighted signed graph, where the weights of edges encode a compulsory (causal) nature of relationships between vertices. Their adjacency matrix is a real and asymmetric matrix. The classical Kosko model [3] assumes stability in terms of technical systems and requires finding the state of fixed point equilibrium. Vertices are ranked based on the achieved values of the vertex states in this fixed point [4]. Some papers mention the use of the DF to accelerate

S. Kovalev et al. (Eds.): IITI 2019, AISC 1156, pp. 270–280, 2020.
https://doi.org/10.1007/978-3-030-50097-9_28

convergence [5]. However, the rationale for its choice has not been considered. To our knowledge, the problems of ranking vertices in the models of the spread of influence transfer onto a directed weighted signed graph with causal relationships on edges have not been considered in literature. Therefore, the question of how the DF variations can influence the ranking of influential vertices, induced by the effective control algorithm [6, 7], in the case of a real and asymmetric adjacency matrix, requires appropriate research.

2 Related Work

An analysis of publications using models with the DF showed that the putting of the DF into the model can be in additive (8), [9] or multiplicative form [10–14]. The additive form is expressed in the adjustment of the model parameters by adding or subtracting certain corrections. The multiplicative form is expressed by putting a multiplier in the model. A typical example of the additive form of the DF is the approach offered in the article [8]. By putting three additive dumping factors in the model, the growth and delay of network growth are simulated. Putting the additive DF in the model is manifested in the individual adjustment of each model indicator. In the case of using a FCM, such adjustment of individual parameters of the model is not allowed, but control over the comparative scale of direct and indirect influence is permissible. The multiplicity of the DF consists in weighing direct and indirect influences for the system as a whole. Characteristic representatives of the additive form can be found in the studies of α-centrality in [13] and [14], and the PageRank algorithms [10–12]. Here, the weighing of direct and indirect influences is expressed by putting a multiplier into the model of the adjacency matrix as a whole. The question of choosing a DF value is directly addressed in the article [12], in which the dependence of node ranking from the variation of the DF is noted. The authors in [12] pay attention to the fact that a small change in DF can lead to significant changes in the ranks of the vertices. Bressan and Peserico give an example of a graph when a change in the DF of even 0.001 “can cause a complete reversal of the ranking”. This indicates the importance and necessity of choosing the value of the DF in the influence propagation model. However, the mechanism of using the DF in these studies is applicable to networks with a probabilistic matrix and the assumed uniform propagation of transitions along the outgoing arcs of the graph.

In article [15] a relaxation parameter is introduced and optimized to accelerate convergence. In this case, this parameter is related not to the model, but to the computational method for finding a solution. An analogy can be drawn between this relaxation parameter and the Tikhonov regularization parameter [16].

3 Method

3.1 Algorithm of Effective Controls and Damping Factor

The algorithm of effective controls proposed in articles [6, 7] is considered. The method used in the algorithm is based on solving the optimization problem of finding the vector (vectors) of external impacts maximizing the accumulated growth of the increments of the vertex indicators for the system represented by a directed weighted signed graph with causal relationships on the edges. The graph is a fuzzy cognitive map (FCM) of the system under study. A control model called the growth model is used to spread the influence

$$x_j = u_j + \delta \sum_{i=1}^{n} a_{ij} x_i,$$

where x_j is the value of the index increment of vertex v_j; u_j there is a control impact on vertex v_j; δ is the DF; a_{ij} is the weight of the arc connecting vertex v_i with v_j, $-1 \leq a_{ij} \leq 1$; x_i is the value of the index increment of vertex v_i. The optimal control impact is **u**, maximizing ratio of the square of the norm of the index vector **x** to the square of the norm of the control vector **u**:

$$(\mathbf{x}, \mathbf{x})/(\mathbf{u}, \mathbf{u}) \to max.$$

The result of the algorithm is two sets of graph nodes ranked in order of importance. The importance of a node is determined by the value (modulo) of the components of the impact and influence vectors. One set of graph nodes represents the component values of the impact control vector; the other represents the component values of the influence vector for the corresponding vertices of the graph. The selection of the most influential and influencing nodes (effective controls) is made on the basis of cumulative effect (components of the vector) expressing the potential of the system as a whole. The method is based on the spectral properties of the adjacency matrix of the graph and the determination of its resonance properties.

As demonstrated in Sect. 2, in many models the DF was used to accelerate convergence. It is assumed that the presence of any solution of the system of equations with achieved convergence provides the correct solution. However, in [15] it was noted that there were instances of obtaining contradictory results. In this case, the structure of the model was analyzed and modified through adjusting the weights and factors to obtain more reasonable results. Such cases are found in most FCM. However, analysis shows that this cannot always be explained by the incorrect structure of the model. Then changing the weights, connections and factors can, in fact, become fudging the solution to match the expected result.

Let us consider a "flip the tables" example. Consider the FCM for clinical risk of the drug administration from [17]. FCM contains 30 vertices-factors, 119 edges with a total weight of 84.2. We will analyze the quality of the results: (i) for the ranks assigned to vertices of a FCM; (ii) for compliance of impacts and responses assigned to the vertices under different values of the DF. Analysis of the ranks assigned to the FCM vertices showed that for this FCM, a change in the DF significantly affects the result obtained. Figure 1 shows the changes in the rank of influential vertices (Fig. 1a) and the vertices that received the response of the system (Fig. 1b) with changes in the DF. Thus, for example, vertex 5 moved from the 9th place to the 1st, and vertex 1 moved from the 1st place to the 20th. That means that if the system operates at full capacity in the way in which the expert set the vertices-concepts and the weight of the network connections, then the vertex that received the maximum response is vertex 5 (under $\delta = 1.0$), but not vertex 1 (under $\delta = 0.1$). Accordingly, the most influential vertex, which took the 1st place under $\delta = 1.0$, is vertex with number 23, and vertex number 30 moved from the 1st place under $\delta = 0.1$ to the 4th place under $\delta = 1.0$. Vertex 24 moved from 11th to the 3rd place, correspondingly.

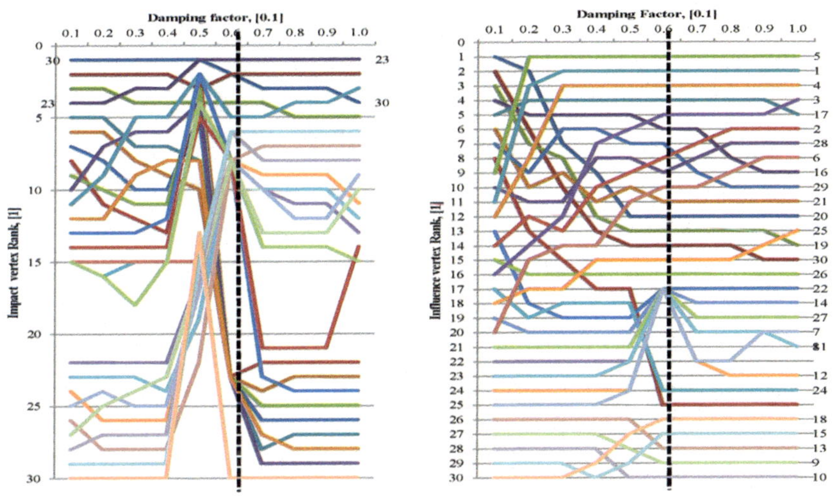

Fig. 1. Changes in the ranks of (a) influential vertices and (b) vertices that received a system response, under $0 < \delta \leq 1.0$.

Analysis of compliance of impacts and responses that vertices received under different values of the DF, will be conducted from the point of view of the semantic content of the system factors. For example, an increase in the factors of fatigue, haste, stress, accidents, and workload (respectively, vertices 1–6) obviously increases the errors made by staff (respectively, vertices 19, 20, 23, 24, 30). Such vertices have "conjugate signs" of changes in their characteristics. In this case, the sign of all values of the vector components must be positive. However, the solution obtained on the interval $0.6 \leq \delta \leq 1$ does not satisfy the consistency requirements of the result (Table 1). That is, the sign of the vector components of the vertices in the second group changed to the opposite.

Table 1. The values of the components of responses and impacts for the conjugate signs of the index increment of the vertices.

Vertex number	The values of the components of the vector under the values of the DF				
	0.6	0.7	0.8	0.9	1.0
1	-0.3937	-0.3998	-0.4029	-0.4011	-0.3928
2	-0.165	-0.1603	-0.1562	-0.1556	-0.1619
3	-0.2083	-0.2138	-0.2191	-0.2288	-0.2492
4	-0.3156	-0.3079	-0.2996	-0.2897	-0.2769
5	-0.6325	-0.668	-0.6973	-0.7209	-0.7386
6	-0.1348	-0.1335	-0.1286	-0.1206	-0.1095
19	0.001	0.00504	0.00605	0.0058	0.00526
20	0.00136	0.00711	0.00896	0.00881	0.00758
23	0.00153	0.00842	0.01078	0.01108	0.01062
24	0.00077	0.00461	0.00648	0.00709	0.00693
30	0.00136	0.00686	0.00833	0.00793	0.00671

For our model, at different values of the DF, $0 < \delta \leq 1$, the resulting solutions display the scaled result of the strength of the impact of all the intra-factor connections of the system. The value of the DF in (1) can be understood as follows. If $\delta = 0.1$, then the influence of indirect impacts of the system as a whole is manifested only by 0.1 of the declared power of the system as a whole; if $\delta = 1$, then the inter-factor connections work to the full; the indirect influence is transmitted to the maximum within the system. Note that there are poor quality results for $0.6 \leq \delta \leq 1$, that is, the tables flip. This means that the power of indirect influence is such that it has enough power to flip the tables. In this case, the solution convergence is achieved for all values of the DF. Thus, the presence of convergence does not guarantee the correctness of the result. Consequently, it is required to find such ranges of acceptable values for the obtained solutions for which (i) the condition of consistency of the result is satisfied; and (ii) the change in the ranks of the vertices is slow. By consistency of the result, we understand the satisfaction of the rules of the system as a whole. These rules can be expressed in imposing restrictions on the status of the vertices and on the sign of impacts and responses.

3.2 Domain of the DF

The constraints to the input data are divided into 3 types: (1) a component (impact, response) equals to zero. $x_i = 0$, $u_i = 0$; (2) a component (impact, response) is non-negative, $x_i \geq 0$, $u_i \geq 0$; (3) A component (impact, response) is non-positive, $x_i \leq 0$, $u_i \leq 0$. The first group of constraints represents model nodes that are non-controllable factors (for example, environmental factors that cannot be directly impacted). The second and third groups determine the factors whose characteristics can have only positive or only negative changes in characteristics, including conjugate ones. Since all constraints are introduced into the system of equations, the solution will also satisfy the required compliance of signs of impacts and responses.

Consider the graph $G = V, E$ for the socio-economic system, where $V = \{V_1, V_2, \ldots, V_N\}$ is a set of vertices and $E \in V \times V$ is a set of arcs. The graph is determined by the adjacency matrix $\mathbf{A}^T = \|a_{i,j}\|_n$, where $a_{i,j}$ is the weight of the arc from vertex V_i to vertex V_j on $-1 \leq a_{i,j} \leq 1$. Consider the control problem under the growth model with the simple relationship of impact and influence (response):

$$\left(\mathbf{I} - \delta A^{\mathrm{T}}\right)\mathbf{x} = \mathbf{u} \tag{1}$$

where $\mathbf{I}$ is the identity matrix; $\mathbf{x} = (\mathrm{x}_1\mathrm{x}_2\ldots\mathrm{x}_\mathrm{n})^{\mathrm{T}}$ is the vector of the influences, called the system response; $\mathbf{u} = (\mathrm{u}_1\mathrm{u}_2\ldots\mathrm{u}_\mathrm{n})^{\mathrm{T}}$ is the vector of the control impacts and δ is the DF $0 < \delta \leq 1$. This problem can be solved either with respect to impacts or with respect to influences (responses). The solution of the problem with respect to influences is considered. The maximization task

$$\mathbf{x}^{\mathrm{T}}\mathbf{x}/\mathbf{u}^{\mathrm{T}}\mathbf{u} \to \max$$

under the constraint $\mathbf{x}^{\mathrm{T}}\mathbf{x} = \mathbf{1}$ (or $\mathbf{u}^{\mathrm{T}}\mathbf{u} = \mathbf{1}$) is equivalent to

$$\mathbf{x}^T\mathbf{B}\mathbf{x} \to min \tag{2}$$

where $\mathbf{B} = (\mathbf{I} - \delta A)\left(\mathbf{I} - \delta A^{\mathrm{T}}\right)$. For the case $\mathbf{x} = \mathbf{u}$ and $\mathbf{u}^{\mathrm{T}}\mathbf{x} > 0$ at $\delta = 0$, the response is co-directional and identical to the impact. Suppose

$$\mathbf{A} = \mathbf{A}_s + \mathbf{A}_a = \frac{1}{2}\left(\mathbf{A} + A^{\mathrm{T}}\right) + \frac{1}{2}\left(\mathbf{A} - A^{\mathrm{T}}\right)$$

where $\mathbf{A}_s$ is a symmetric matrix, $\mathbf{A}_a$ is skew-symmetric matrix. From

$$\mathrm{proj}_{\vec{u}}\vec{x} = \frac{(\vec{u}, \vec{x})}{|\vec{u}|} = \frac{\mathbf{u}^{\mathrm{T}}\mathbf{x}}{\sqrt{\mathbf{u}^{\mathrm{T}}\mathbf{u}}}$$

it follows that the sign of $\mathbf{u}^{\mathrm{T}}\mathbf{x}$ determines the sign of the projection. In general case we have

$$\mathbf{u}^{\mathrm{T}}\mathbf{x} = \mathbf{x}^{\mathrm{T}}(\mathbf{I} - \delta A)\mathbf{x} = \mathbf{x}^{\mathrm{T}}\mathbf{x} - \delta\mathbf{x}^{\mathrm{T}}\mathbf{A}\mathbf{x} = \mathbf{x}^{\mathrm{T}}\mathbf{x} - \frac{1}{2}\delta\mathbf{x}^{\mathrm{T}}\left(\mathbf{A} + A^{\mathrm{T}}\right)\mathbf{x}$$

Since all eigenvalues of the symmetric matrix $\mathbf{A}_s$ are real, we obtain

$$\lambda_{min}\mathbf{x}^{\mathrm{T}}\mathbf{x} \leq \frac{1}{2}\mathbf{x}^{\mathrm{T}}\left(\mathbf{A} + A^{\mathrm{T}}\right)\mathbf{x} \leq \lambda_{max}\,\mathbf{x}^{\mathrm{T}}\mathbf{x}$$

where λ_{min} and λ_{max} are the least and largest eigenvalues of the matrix $\mathbf{A}_s$. All the eigenvalues of sign-defined symmetric matrix have the same sign. If $\lambda_{max} > 0$, then the signs of $\mathbf{u}^{\mathrm{T}}\mathbf{x}$ are constant and $\mathbf{u}^{\mathrm{T}}\mathbf{x} > 0$ for $\delta < \lambda_{max}^{-1}$, and the sign switches from $\mathbf{u}^{\mathrm{T}}\mathbf{x} > 0$ to $\mathbf{u}^{\mathrm{T}}\mathbf{x} < 0$ when $\lambda_{max}^{-1} < \delta < \lambda_{min}^{-1}$. This means that the co-directional components of the impact and response become oppositely directed. Then the sign of $\mathbf{u}^{\mathrm{T}}\mathbf{x}$ is a switching indicator.

Definition 1. The DF (δ) – is a coefficient expressing the level of indirect impact compared to the level of direct impact.

Definition 2. There is an admissible range of DF domain $Im(DF) = [0, \delta_{adm}]$, over which the co-direction is guaranteed to predominate in the components of the response and impact vectors, where $\delta_{adm} = 1/\lambda_{max_{As}}$.

Definition 3. There is a resonance range of DF domain $Im(DF) = [\delta_{adm}, 1]$, over which there is a rapid restructuring of the ranks of the vertices by the components of the response and impact vectors, and ranking reversal.

Definition 4. There is a DF value, called resonance DF (δ_{res}), when a resonant surge in the value of the objective function of the problem of maximizing the influence occurs. It is at the point of the resonant surge that the sign switches from $\mathbf{u}^{\mathrm{T}}\mathbf{x} > 0$ to $\mathbf{u}^{\mathrm{T}}\mathbf{x} < 0$.

4 Experiment

The feasibility of the proposed approach, which determines the domain of the DF, was tested on a test example discussed in Sect. 3.1. Determining the domain of the DF solutions for which the condition of consistency of the model is met, is considered under the following system of constraints to the input data: (1) impact component $x_i = 0$ equals to zero for $i = 1, \ldots, 6$ (uncontrollable factors); (2) response component $u_i \geq 0$ is non-negative for $i = 7, \ldots, 13, 15, 18, 27$; (3) response component $u_i \leq 0$ is non-positive for $i = 1, \ldots, 6, 14, 16, 17, 19, \ldots, 26, 28, \ldots, 30$. The first group of constraints represents model nodes that are non-controllable factors (for example, environmental factors that cannot be directly controlled). The second and third groups determine factors whose characteristics can have only positive or only negative changes in the characteristics, including conjugate ones. Since all restrictions are put into the system of equations, the solution will also satisfy the required compliance of signs of impacts and responses.

The process of identifying effective controls with regard to admissible range of DF domain includes the following steps:

Step 1. Determining the domain of the DF by the calculating the largest eigenvalue of the adjacency matrix.

Step 2. Formalization of the operating conditions of the model in the form of a system of restrictions.

Step 3. Verification and validation of the model.

Step 4. Computation of the components of response and impact on the domain of the DF.

Step 5. Selection of the most influential vertices (impact vertices) and the vertices that received the greatest influence (response vertices).

As shown in Table 2, the values of $\mathbf{u}^T\mathbf{x}$, calculated in step 3, switch the sign from "plus" to "minus" for $\delta_{res} = 0.585$. The corresponding value of the boundary of the admissible range of the DF is $\delta_{adm} = 1/\lambda_{max_{As}} = 0.2550648$. Note that $\delta_{res} > \delta_{adm}$, which is consistent with Definition 2 and Definition 3.

Table 2. Values of $\mathbf{u}^T\mathbf{x}$ in increments of 0.1

Model	Values of $\mathbf{u}^T\mathbf{x}$ for the DF					
	0.1	0.2	0.3	0.4	0.5	0.6
Unconstrainted SLE	0.618237	0.306535	0.116972	0.032865	0.005236	-0.00075
SLE with constraints	0.620998	0.281798	0.100844	0.028337	0.004801	-0.000825
Model	Values $\mathbf{u}^T\mathbf{x}$ for DF					
	0.7	0.8	0.9	1.0	-	-
Unconstrainted SLE	-0.00066	0.000278	0.000864	0.001045	-	-
SLE with constraints	0.044707	0.027549	0.017148	0.010729	-	-

5 Discussion of the Results

Evaluation of the proposed solution is made on the following issues:

1. To what extent does the ranking resulting from the operation of the algorithm of effective controls depend on the variations in the DF?
2. What characteristic does the DF express?
3. How does one determine the region of admissible DF values?
4. Which characteristic determines the choice of the DF on the domain of the DF?

The rank stability test showed that there is a significant change in the ranks of the nodes for any discrete variations of the DF A series of experiments showed that the range of values is divided into 3 types of intervals:

- Type 1 – intervals with slow rearrangement of vertex ranks (or retention of ranks) under consistency of results;
- Type 2 – intervals with rapid rearrangement of vertex ranks; and
- Type 3 – intervals with slow rearrangement of vertex ranks (or retention of ranks) under the inconsistency of results.

It is obvious that the intervals of the 2nd and 3rd types should be excluded from consideration. Finding the boundaries of intervals of the 1st type is prescribed by Theorem 1 and is reduced to finding the maximum eigenvalue of matrix A_s, a very simple task with hardness $O(n^3)$. Then this border will show the interval without rapid rearrangements in ranks and with non-contradictory results, which is sufficient for the decision maker. For the example considered here, this value is $\delta = 1/\lambda_{max_{As}} = 0.2550648$. Intervals of the 2nd and 3rd types are detected by simple algorithms with clearly formalized criteria. Exclusion of type 2 intervals is based on finding DF values at the actual switch of sign from $\mathbf{u}^{\mathrm{T}}\mathbf{x} > 0$ to $\mathbf{u}^{\mathrm{T}}\mathbf{x} < 0$. The solution is reduced to solving the problem of maximizing the influence for some characteristic values of the DF using the dichotomy method. Intervals of the 3rd type immediately follow the intervals of the 2nd, and do not represent interest for the decision maker. The actual sign switch from $\mathbf{u}^{\mathrm{T}}\mathbf{x} > 0$ to $\mathbf{u}^{\mathrm{T}}\mathbf{x} < 0$ is the resonance point of the system.

For the decision maker, resonance of the system means that the response of the system is determined by the internal properties of the system and ceases to depend on the control action. It can be said that the decision maker is deprived of the ability to control the system, and receives only the possibility of exciting its resonant response, which can become explosive. Then the value of the resonant DF is a reasonable boundary by which it is possible to determine the stability limit of the system under study. For the considered example, this value is $\delta = 0.585$. In this case, we can say that the safety margin of such system allows to hope that the system will operate steadily, and will not go into a tailspin until the strength of the impact of all intra-factor connections of reaches $\delta = 0.585$. Then the decision maker must understand that with continued impact on the system (after reaching this boundary), the system will enter into resonance and may collapse. Therefore, other actions are required that will change the system itself and create the possibility of excluding the adverse effects of resonance. For example, the creation of other factors and relationships in the system. When low DF values are obtained for the system, it can be concluded that the system has a small margin of safety and taking urgent actions by the decision maker is needed.

The obtained result of finding resonance boundary is similar to the results obtained in studies of the critical point and its nature [2, 9, 18]. At this critical point, also referred to in various sources as an "attenuation" factor [13], the epidemiological spread of influences is delocalized in a network represented by a real and symmetric matrix. In these studies, models are described in which the main role is given to the principal eigenvector and the critical point is $1/\lambda_{max}$ (α-centrality, PageRank, etc.). However, this approach cannot be used if at least one eigenvalue is complex-valued.

Thus, our contribution for the considered model of effective controls on directed weighted signed graphs is as follows:

1. It was determined that the choice of DF affects not only the value of the objective function of the impact maximization task, but also the vector of effective control, on which this solution is achieved.
2. In the DF domain there is a range of rapid change in the objective function and in the solution, which manifests itself in the inversion of the vertex ranks.
3. If the eigenvalue of the adjacency matrix, which determines this DF value is real, then the resonant surge has a form of discontinuity of the second kind. If this number is complex, then the objective function will have a finite maximum.
4. For real-world graphs, the choice of the DF up to its furthest boundary can be based on the realistic value of the objective function, showing the ratio of the impact and the response of the system.
5. The value of the resonant DF can be interpreted as the limit of the possible controllability of the system, i.e. the limit of the potential impact on the system without destroying it.

6 Conclusion

In this study, we show that for the considered model of effective controls on directed weighted signed graphs (i) the DF expresses the scaling of indirect influence transmitted along the arcs of the graph; (ii) there is a boundary of the domain of the DF, within which a rapid rearrangement of the vertex ranks and a reversal of the ranking are guaranteed not to occur; (iii) the DF value at the resonance point is a reasonable boundary during decision making and represents a kind of "safety margin" of the system; (iv) the selection of the DF on the admissible range can be based on the realism of the value of the objective function.

Acknowledgements. This work was supported by the Russian Foundation for Basic Research [grant number № 19-01-00109].

References

1. Page, L., Brin, S., Motwany, R., Winograd, T.: The PageRank citation ranking: bringing order to the web. In: 7th International World Wide Web Conference, pp. 161–172 (1998)
2. Goltsev, A.V., Dorogovtsev, S.N., Oliveira, J.G., Mendes, J.F.F.: Localization and spreading of diseases in complex networks. Phys. Rev. Lett. **109**(12), 128702 (2012)
3. Kosko, B.: Fuzzy cognitive maps. Int. J. Man Mach. Stud. **24**(1), 65–75 (1986)
4. Knight, C.J.K., Lloyd, D.J.B., Penn, A.S.: Linear and sigmoidal fuzzy cognitive maps: an analysis of fixed points. Appl. Soft Comput. **15**, 193–202 (2014)
5. Hobbs, B.F., Ludsin, S.A., Knight, R.L., Ryan, P.A., Biberhofer, J., Ciborowski, J.J.H.: Fuzzy cognitive mapping as a tool to define management objectives for complex ecosystems. Ecol. Appl. **12**(5), 1548–1565 (2002)
6. Tselykh, A.N., Vasilev, V.S., Tselykh, L.A., Barkovskii, S.A.: Method maximizing the spread of influence in directed signed weighted graphs. Adv. Electr. Electron. Eng. **15**(2), 203–214 (2017)
7. Tselykh, A., Vasilev, V., Tselykh, L.: Management of control impacts based on maximizing the spread of influence. Int. J. Autom. Comput. **16**, 341–353 (2019)
8. Cai, G., Wang, R., Qiang, B.: Online social network evolving model based on damping factor. Procedia Comput. Sci. **9**, 1338–1344 (2012)
9. de Arruda, G.F., Rodrigues, F.A., Moreno, Y.: Fundamentals of spreading processes in single and multilayer complex networks. Phys. Rep. **756**, 1–59 (2018)
10. Son, S.-W., Christensen, C., Grassberger, P., Paczuski, M.: PageRank and rank-reversal dependence on the damping factor. Phys. Rev. E **86**(6), 066104 (2012)
11. Boldi, P., Santini, M., Vigna, S.: PageRank as a function of the damping factor. In: Proceedings of the 14th International Conference on World Wide Web - WWW 2005, p. 557 (2005)
12. Bressan, M., Peserico, E.: Choose the damping, choose the ranking? J. Discrete Algorithms **8**(2), 199–213 (2010)
13. Bonacich, P., Lloyd, P.: Eigenvector-like measures of centrality for asymmetric relations. Soc. Netw. **23**(3), 191–201 (2001)
14. Katz, L.: A new status index derived from sociometric analysis. Psychometrika **18**(1), 39–43 (1953)

15. Hobbs, B.F., Ludsin, S.A., Knight, R.L., Ryan, P.A., Biberhofer, J., Ciborowski, J.J.H.: Fuzzy cognitive mapping as a tool to define management objectives for complex ecosystems. Ecol. Appl. **12**(5), 1548 (2002)
16. Tikhonov, A., Arsenin, V.: Solutions of Ill-Posed Problems. Wiley, New York (1977)
17. Bevilacqua, M., Ciarapica, F.E., Mazzuto, G.: Fuzzy cognitive maps for adverse drug event risk management. Saf. Sci. **102**, 194–210 (2018)
18. Boguñá, M., Castellano, C., Pastor-Satorras, R.: Nature of the epidemic threshold for the susceptible-infected-susceptible dynamics in networks. Phys. Rev. Lett. **111**(6), 068701 (2013)

Basic Approaches to Creating Automated Design and Control Systems in a Machine-Building Industry

Georgy Burdo(✉)

Tver State Technical University, A. Nikitin Emb. 22, Tver, Russia
gbtms@yandex.ru

Abstract. The paper focuses on a comprehensive solution of the problem of improving the efficiency of industrial engineering systems. At the same time, automated design systems, process control and product quality systems ensure time reduction and quality improvement of fitting-out in multiproduct machine building industry. The paper proposes the concept of organizational and technological design in production process development. The concept takes into account the agility of production systems and performance goals. The paper shows the order of implementing the proposed approaches and their successful execution conditions.

Keywords: Preproduction engineering · Multiproduct mechanical engineering production · Objective function · Production processes · Management · Product quality · Product life cycle

1 Introduction

Structural changes in machine building industry are related to the demand of consumers to have goods with the properties they need at the right time. Therefore, this predetermines the fact that production has become multiproduct, and its services were not quite ready.

For this reason, reducing the duration of technical and organizational preparation of production is an important scientific and technical problem. The industry requires to solve it. Further, the paper gives ways to solve a number of problems related to this problem.

2 Improving Decision-Making Procedures in Automated Systems

One of the most important parameters that determine the efficiency of production systems, are the time expenditures on the implementation of product life cycle stages implemented in the enterprise framework. We consider measures to reduce the production time of a new product sample.

This work was supported by RFBR, project 17-01-00566.

S. Kovalev et al. (Eds.): IITI 2019, AISC 1156, pp. 281–288, 2020.
https://doi.org/10.1007/978-3-030-50097-9_29

1. Automation of process management procedures in terms of their hierarchy and a time factor. It should be implemented as a system with a service-oriented architecture, with specific access rights for each user.
2. Automation of engineering works when implementing processes related to designing preproduction, with the preproduction engineering, with the organizational preparation of production, product manufacturing and testing. A special aspect for the conditions of the considered production systems is mobility and the possibility to correct decisions fast using them. For this purpose, software (computer-aided design systems—CAD, automated process control systems—APCS and automated enterprise control systems—AECS) should be informationally integrated with each other at the user level (designing preproduction CAD with preproduction engineering CAD, APCS with preproduction engineering CAD, APCS with production departments), and at the top level they should be integrated into AECS.
3. Parallel execution of works from different product life cycle stages (or from different steps of one stage). It is clear that the processes of design preparation start a project of creating a new product. However, already at the stage corresponding to the draft design, it is possible to carry out preproduction engineering: workability analysis, technology preplanning, etc.
4. Development of intelligent decision support systems for process implementation. This is primarily about decisions made and implemented by representatives of various functional units.
5. Creation a background for effective decision making at subsequent product life cycle stages.
   For example, commonality of parts and their elements simplifies both product design and the preproduction engineering due to using universal technological solutions. If processing operations are reserved for the least loaded equipment (design based on the current production situation), it reduces the occurrence probability of bottlenecks and simplifies the technological process management.
6. The construction of all enterprise processes (not only production) according to the lean manufacturing idea, which means doing what is necessary right now and reducing the waste of labor and funds. This requirement is the most important for enterprise processes implemented by top management [1].

3 The Concept of Organizational and Technological Design of Production Processes in Preproduction Engineering CAD

The studies of professors V.S. Korsakov, A.A. Matalin et al. [2–5] show that the production process design should be based on technical and economic principles.

Such approach to preproduction engineering is typical for enterprises that produce goods in large batches with fixed nomenclature. Unfortunately, it is often entirely transferred to of multiproduct machine building enterprises. However, both in the technical and scientific literature they neglect the fact that the current state and changes in production systems must change technology design and control algorithms.

The abovementioned needs of modern multiproduct machine-building industry require another principle – organizational. It determines the conformity of the designed process to operating conditions of a production system. This new principle forms the essence of the concept of organizational and technological design of production processes.

The majority of authors [5–8 and others] reduces production system state accounting to updating databases (little dynamic attribute), and does not take into account organizational aspects (high dynamic attribute). The proposed concept of organizational and technological approach to production process design can be determined by the following provisions [9, 10]:

1. Taking into account the current equipment loading situation in production units.
2. The subordination of the objective function of the designed production processes to the goals achieved by the company when performing a specific order.
3. Fast response of the processes of preproduction engineering.
4. Creating a background for designing effective production processes.
5. Creating a background for effective management of technological processes at the design stage of production processes.

The meaning of the first principle is to assign technological operations to work stations that are less busy at the moment.

The meaning of the second principle is to establish the objective function designed for this order production processes.

The third principle is fast response of the processes of preproduction engineering. It assumes the availability of operational information communication: a) between units engaged in preproduction engineering and the process control system (data on equipment loading); b) between units engaged in the preproduction engineering and the organization management system (the type of an order objective function).

It is reasonable to create a background for designing effective production processes when managing the processes of designing preproduction. For this purpose, there should be measures for product design unification, which leads to unification of process engineering and determines the possibility of implementing advanced technologies on modern equipment. This ensures a labor saving of design procedures, an improvement in the quality of design solutions by optimizing technology elements provided for standard parts and their surfaces [11].

Creating a background for effective management of production processes at the design stage means that design goals are subordinated to production system goals in the context of contract execution timelines, and "fixing" bottlenecks in technology divisions (see also principles 2 and 3).

It is clear that the concept implementation is impossible without using production process CAD and APCS, and in the information integration production process CAD and APCS based on the proposed principles.

4 Principles for Creating Integrated Production Process CAD – APCS

In order to take into account specific features of multiproduct machine building industry, we have defined the principles that determine creating integrated production process CAD - APCS (information system). They are the following:

1. Information integration should ensure the minimal response time of a production system to production situation changes, which follows from the Just in Time principle [12].
2. An integrated system should provide the ability to manage a product life cycle.
3. An integrated system should operate within the framework of a product quality management system.
4. An integrated system should be managed by an organization management system based on direct and reverse information links. It also should have external information links with the designing preproduction system, the controlled objects are production system workplaces.
5. The interaction between the production process CAD and APCS should be as follows. Information from the production process CAD (machining content, routes for parts passing through workplaces) is input for APCS. At the same time, the synthesis of this information takes into account information about the real situation in a production system from APCS (feedback). It creates the background for effective management at the stage of production processes design.

The modern trends in the development of machine-building industry [13, 14] include the following components: total quality control or total quality management; total productive maintenance; Just in Time production system. Just in Time system operates depending on the work of the automated process control system, which should make decisions aimed at reducing the general run-out production cycle. Therefore, it makes sense to consider the principles of creating such automated systems separately. These principles should ensure the semantic unity of planning and management processes of the procedures implemented in automated process control systems.

The 1st principle of creating an automated process control system is the principle of system unity and interconnection.

This implies considering all agents of an automated process control system separately, and makes an automated process control system an independent entity that preserves the properties of integrity.

The 2nd principle of creating an automated process control system is the correspondence of a plan hierarchy and control accuracy to the accuracy of enterprise plans. In the context of the system approach, the accuracy principle means that the accuracy of plans should increase as they become more detailed.

The third principle of creating an automated process control system is the continuity of control due to the feedback from an organization management system and technology divisions, and efficiency. Management decisions should have the minimum allowed response time. The frequency of tracking the unbalanced plans (feedback) and the speed of calculations are determined by the permissible response time of the automated process control system.

The 4th principle of creating an automated process control system is ensuring plan reserves. Implementation roadmaps developed by an automated process control system must have reserves, and the calculation algorithms must include situation assessment in technology divisions. This implies that the automated process control system has a volume planning agent, who builds a reserve power of technology units into schedules. It is important to understand that they cannot be too large, otherwise plans will be inaccurate; and too small reserves will cause frequent changes in the plans of an enterprise in general, which blurs the goals of company activities.

The 5th principle implies that an automated process control system should include an interactive hybrid mode that means the involvement of a person into a management process and obtaining additional management resources. Planning and control in the automated process control system has several stages: volume planning, scheduling, operational planning and dispatching (management). The first three stages are performed iteratively. The system suggests possible solutions at each iteration stage and at the end of a stage. An expert group analyze and assess them. At the dispatching stage, experts analyze and assess possible management decisions after completing each operational planning interval. One most rationale decision is made using the apparatus of fuzzy sets [15, 16].

The 6th principle of creating an automated process control system is complexity, i.e. taking into account the most important factors that influence the control problem solution in their interconnection, and the assessment of mutual influence of taken decisions. This implies to consider the production process management problem in broader terms, starting from the creation of a background for efficient operation of an automated process control system by observing a number of principles when creating technology divisions, temporal coordination of start-up preparation processes, and the implementation of controls using production process CAD.

5 Automated Product Quality Management Systems

Product quality is one of the fundamental issues of the machine building industry. Product quality is ensured at all stages of creating products. Therefore, decision making in designing preproduction CAD, preproduction engineering CAD, APCS and other automated product quality management systems must be performed according to both local criteria of each system and quality assurance [17]. The principles of creating an automated product quality management system considered in [19, 20] make this system to provide: a) products of good quality; b) implementation of decision-making procedures corresponding to the processes existing in a real production system, c) information integration with a higher level organization management system, d) integration of an automated product quality management system into automated systems for product life cycle support [18], e) compliance of decision making procedures with human logic and algorithms of specialists' actions, e) reducing the total time (cycle) for producing a high technology product sample. The principles are formulated as follows.

1. The principle of iterative decision making.
2. The system hierarchy.
3. The accuracy of system functioning.
4. Artificial intelligence elements [16, 21].
5. Combination of the work stages in time.
6. The derivativeness of product quality assessment criteria from a technical design specification, from the expected product performance indicators during its industrial production, and from the world experience in designing such products.
7. The multivariance principle.
8. Principles of forming decision assessment criteria:
   - a. The principle of the boundary type of criteria;
   - b. The principle of the relative type of criteria;
   - c. he principle of criteria complexity.
   - d. The criteria clarification principle.

6 Hierarchical Organization of Automated Systems

The hierarchical organization of automated systems related to the production at a machine-building enterprise can be represented as follows. The first level is an automated enterprise management system, an automated enterprise management system that organizes management of all processes of enterprise activities. The second level is an automated product quality management system that integrates production processes directly. The third level includes series-parallel automated systems for research work, designing preproduction, preproduction engineering, and automated process control systems. Such system hierarchy should ensure unconditional product quality assurance (automated product quality management systems) while ensuring good technical-and-economic indexes due to optimizing according to the criteria in designing preproduction CAD, preproduction engineering CAD and automated process control systems.

7 Conclusion

The measures considered below determine the stages of implementation. They cover not only the considered area of automated systems, but also some problems that provide a comprehensive solution of the problem of improving production system efficiency.

1. The order of works. First of all, it is reasonable to carry out works related to the design and technological unification. It is possible to carry out measures for balancing production capacity in parallel [22]. Next is the development of mechanisms for managing preproduction engineering in order to shorten a production system cycle. Then there goes the introduction of the concept of organizational and technological design of production processes.
2. Training of shop floor management personnel, especially line masters and managers of dispatch services. The main task of personnel training is to teach people the manage technological cycles and production facilities.

3. Training of design, technological and management service workers; the task is to teach them to work within the framework of ongoing processes and in accordance with certain criteria of decision assessment.
4. The phased introduction of information-integrated automated quality management systems, design preproduction systems, production process design systems, process control systems, which significantly increases the efficiency of decisions.
5. Develop an understanding a corporate responsibility sense for the result of all project participants.

The paper intentionally did not consider the issues of defining an enterprise mission, optimizing a functional structure of enterprise services and departments, coordinating work with contracting organizations, analyzing the readiness of a corporate management team for serious restructuring. These issues are the subject of a separate case.

References

1. Burdo, G.B., Stoyanova, O.V., Isaev, A.A., Chaplygin, S.A.: Hybrid integrated automated systems for design and process control. In: Congress on Intelligent Systems and Information Technologies “IS & IT 2017”, pp. 94–101 (2017). (in Russian)
2. Korsakov, V.S.: Mechanical Engineering Technology. Mashinostroenie, Moscow (1977)
3. Suslov, A.G., Dalsky, A.M.: Scientific Foundations of Mechanical Engineering Technology. Mashinostroenie, Moscow (2002)
4. Matalin, A.A.: Mechanical Engineering Technology. Mashinostroenie, Leningrad (2006)
5. Optimization of Design and Technological Preproduction at a Machine-Building Enterprise: Solution Catalog of the ASCON Company. ASCON (2004)
6. Kapustin, N.M.: An Automated System for Designing Technological Processes. Mashinostroenie, Moscow (1979)
7. Tsvetkov, V.D.: System-Structural Modeling and Design Automation of Technological Processes. Nauka I Tekhnika, Minsk (1979)
8. Kondakov, A.I.: CAD of Technological Processes. Akademiya, Moscow (2007)
9. Burdo, G.B., Palukh, B.V.: Theoretical foundations of an integrated automated system for designing and managing technological processes in multiproduct manufacture. Vestnik Samara State Tech. Univ. (Tech. Sci. Ser.) **4**(127), 44–54 (2010). (in Russian)
10. Burdo, G.B., Isaev, A.A.: Organizational and technological design in production process CAD of multiproduct manufacture. In: International Science and Technology Conference on “High Technologies in Mechanical Engineering”, Kurgan, Russia, pp. 137–140 (2012). (in Russian)
11. Bazrov, B.M.: Modular Technology in Mechanical Engineering. Mashinostroenie, Moscow (2001)
12. Womack, J.P., Jones, D.T.: Lean Thinking: Banish Waste and Create Wealth in Your Corporation. Productivity Press, New York (1996)
13. Womack, J.P., Jones, D.T.: Lean Thinking: Banish Waste and Create Wealth in Your Corporation, 2nd edn. Free Press, New York (2003)
14. Shingo, Sh.: Kaizen and the Art of Creative Thinking. Enna Product Corp., PCS Inc, Bellingham (2007)
15. Burdo, G.B., Palyukh, B.V., Semenov, N.A.: 15th National Conference on Artificial Intelligence with International Participation (KII 2016), Smolensk, Russia, vol. 1, pp. 239–247 (2016). (in Russian)

16. Zadeh, L.: Fuzzy sets. Inf. Control **8**, 338–353 (1965)
17. Shubenkova, E.V.: Total Quality Management, 2nd edn. Plekhanov Russian University of Economics, Moscow (2005)
18. Kolchin, A.F., Ovsyannikov, M.V., Strekalov, A.F., Sumarkov, S.V.: Product Lifecycle Management. Anarchist, Moscow (2002)
19. Burdo, G.B.: Intelligent planning and management procedures in production systems of geophysical industry. Softw. Syst. **3**(95), 107–110 (2011)
20. Burdo, G.B., Sorokin, A.Yu.: A model of an automated quality management system in multiproduct machine-building manufacture. Softw. Syst. **4**(104), 248–253 (2013)
21. Eremeev, A.P., Kurilenko, I.E.: Temporal output tools for real-time intelligent systems. In: Intelligent Systems. Collective Monograph, vol. 4, pp. 222–252. Fizmatlit, Moscow (2010)
22. Burdo, G.B.: Improving technological preparation of machine-building production. Mach. Tools (STIN) **7**, 2–8 (2016)

Principles of Organization of the Strategies of Content-Based Analysis of Aerospace Images

Denis R. Kasimov(✉), Valeriy N. Kuchuganov, and Aleksandr V. Kuchuganov

Kalashnikov Izhevsk State Technical University, Izhevsk, Russia
kasden@mail.ru

Abstract. The infinite variety of image subjects, the dependence of analysis algorithms and decision rules on the shooting conditions and image quality lead to the need to configure and retrain the computer vision system for almost every next series of images. The paper proposes the principles of organization and high-level language for description of strategies of content-based analysis of aerospace images. The decision maker specifies the strategy for processing and analyzing the image as a sequence of points of selection of actions or subtasks. In general, each action can be performed by different software modules, which require their own data structures, restrictions and rules. Accordingly, the results of the action will vary. The solver, which is controlled by the given strategy, selects variants of actions, data and constraints for each subtask, builds a decision tree, and monitors the progress of the decision. Examples of object detection strategies and results of their work on urban area aerial images characterized by a very high spatial resolution are given. Applied semantic models of actions and resources make the process of structuring and describing more visual and, at the same time, machine-readable. The process of describing the image analysis strategy is transferred from the level of specifying instructions/commands to the level of planning works and resources.

Keywords: Aerial image · Object detection · Semantic image analysis · Context-sensitive strategy · Decision tree · Strategy Description Language · Solver

1 Introduction

The infinite variety of image subjects, the dependence of analysis algorithms and decision rules on the shooting conditions and image quality lead to the need to configure and retrain the semantic analysis system for almost every next series of images. At the same time, the subject of images and tasks to be solved often change – temporary changes, mapping, detection of objects or control of their condition. Methods to reduce the semantic gap between low-level functions, such as color and simplest

This work is supported by the Russian Science Foundation under grant No. 18-71-00109.

S. Kovalev et al. (Eds.): IITI 2019, AISC 1156, pp. 289–299, 2020.
https://doi.org/10.1007/978-3-030-50097-9_30

shapes, and high-level human semantics include the use of object ontology to define concepts, their properties and ranges, machine learning, fuzzy logic, inference rules that describe the logic of a user's visual interpretation.

There are approaches aimed at automating the development of computer vision systems. For example, an approach based on descriptive image algebras is widely used for searching and describing objects [1, 2]. Within it, the image is described in terms of descriptors – the results of the calculation of the corresponding functions, or objects are recognized as integral patterns. However, the structural analysis of image objects is difficult, since the descriptors mainly calculate the integral characteristics or local features of the image.

The most popular strategies for the semantic analysis of aerospace images are ontological reasoning strategies based on knowledge-driven semantic rules.

The paper [3] describes a universal software environment for the classification of satellite images. The classification of satellite images is considered as the process of grouping pixels/regions into meaningful categories. Ontology is used to represent expert domain knowledge in a machine-readable format. Images are segmented and areas are grouped into meaningful categories using domain knowledge and classification rules. The classification of regions is performed iteratively by comparing the extracted values with the permissible ranges of spatial and temporal characteristics of the concepts of a higher level of the ontology.

In [4] a three-stage workflow is introduced: ontological modeling, initial classification based on the data-driven machine learning method and semantic classification of image objects based on knowledge-driven semantic rules. The results show that the method with ontology compared to the classification based on the decision tree without the use of ontology gave some statistical improvements in terms of accuracy for the specific images used in the work.

A promising direction for the further improvement of semantic image analysis systems is the creation of tools for describing flexible analysis strategies and mechanisms for automatic solution searching controlled by a given strategy.

Formalization and representation of a problem solving strategy in traditional programming languages (C, Pascal, etc.) is a very laborious process due to the insufficient declarativeness of these languages and the lack of built-in mechanisms for finding solutions among a variety of alternatives. In this respect, the Prolog language, which is based on the principles of automatic theorem proving, occupies a slightly more advantageous position. For example, in [5] a Prolog-program is created that effectively solves the problem of searching for objects of a given structure in the image. However, in our opinion, description of problem conditions and the solving strategy in a purely logical style is not a trivial task. The decision maker, as a rule, does not have the experience of logic programming and the corresponding thinking style.

The programming language GOLOG [6] for creating intelligent agents expands the traditional imperative programming by logical reasoning about the actions that an agent can perform on the basis of situational calculus. In [7] it is noted that this language causes difficulties for inexperienced users who do not have knowledge of mathematical logic or Prolog.

The planning languages STRIPS, ADL and PDDL [8] are focused on the synthesis of sequences of actions leading to the goal. These languages express planning concepts well (objects, schemes of actions, states, goals, etc.), but do not provide means for a more natural binding to other elements of the semantics of the problem being solved. In addition, in our opinion, the means to control the solution search process are not sufficiently transparent for the developer.

The goal of this work is to increase the reliability of decisions made by the system of semantic image analysis by creating tools for describing and mechanisms for implementing flexible analysis strategies based on the principles of context consideration and adaptation to image content.

We propose the principles of organization and high-level language for description of strategies of content-based analysis of aerospace images. The decision maker specifies the strategy for processing and analyzing the image as a sequence of points of selection of actions or subtasks. In general, each action can be performed by different software modules, which require their own data structures, restrictions and rules. Accordingly, the results of the action will vary. The solver, which is controlled by the given strategy, selects variants of actions, data and constraints for each subtask, builds a decision tree, and monitors the progress of the decision.

2 Principles of Organization of Image Analysis Strategies

In the course of experimental studies, three areas of adaptation of aerospace image analysis methods were identified:

- adaptation to the specifics of the problem – selection of tools, constraints and rules based on the domain ontology and tasks solved on the images;
- adaptation to image quality and shooting conditions (for example, weather);
- contextual adaptation – according to the situation in the analyzed area of the image.

Accordingly, the following principles of organizing image analysis strategies can be formulated:

The Principle of Taking into Account the Dependence of Semantics on Context. Consider this principle on an example. Suppose there is a color region that has a small size, close to a rectangular shape and a fairly uniform texture. These characteristics allow to refer the region only to the generalized category of objects of artificial origin. To further clarify the class of this object, it is necessary to analyze its relationship with other objects in the image. Depending on the situation, the region may be:

- a free-standing private house if it casts a shadow, which corresponds to the adjacent area having a low lightness, a rectilinear shape, a common border of a sufficient length and lying in the direction of sunlight incidence;
- a fragment of a large building if it borders other fragments that are similar in color and there is a shadow near the group of these fragments;
- a freshly repaired road section if it does not cast a shadow, lies within or is adjacent to an area reliably classified as a road;
- a lawn if it has no shade, is adjacent to a road and/or a sidewalk, has a greenish or yellowish tint, etc.

The Principle of Operational Change (Adaptation) of the Degree of Feature Approximation. Depending on the tested hypothesis about the class of the object, a different level of approximation of its features may be required. For example, when detecting roads, it is advisable to carry out a stronger approximation of the edges of candidate regions, since in this case, the straightness characteristic is of great importance and it must be assessed in conditions of great length.

The Principle of Structuring and Restructuring of Image Objects in the Process of Analysis. To simplify the process of contextual analysis of situations in which composite objects appear, consisting of several color areas (for example, a roof with two or more ramps), it is advisable to consider simpler representations of composite objects. One way to assembling is to build a non-convex hull around parts of a composite object in order to determine the outer boundary of the object.

The Principle of Iterative Refinement of (Shape) Features of Objects. Automatic color segmentation can result in regions that contain fragments of two or more objects. The facts that a region has characteristic appendages, narrow "isthmuses", straight sections along with winding ones clearly indicate that it covers objects of different classes and should be divided into parts. By dissecting the area into parts in appropriate places, it is possible to eliminate unnecessary parts that are not related to the target object. As a result, the objects are separated from each other, their boundaries are narrowed and features are refined. This allows to obtain color regions with more adequate quantitative and qualitative characteristics.

The Principle of Direct and Reverse Proof. In some cases, the method of proof by contradiction is effective, for example, eliminating false buildings by checking the presence of cars inside the area of the potential building, the location of the building directly on the road, etc.

The Principle of Clustering Objects and Structures. The remaining unclassified areas should be clustered to identify groups of areas with similar characteristics and to form reasonable hypotheses about the classes of objects.

We will interpret the term "***strategy***" as a sequence of points of selection of actions, resources and constraints to solve a problem and optimize the results.

Actions, as in other areas of planning, are divided into three categories: auxiliary, main, organizational. In the field of automation of interpreting satellite imagery, the auxiliary actions are the procedures of filtering, smoothing, contour extraction, segmentation, skeleton determination, approximation, etc. The main actions include: calculating feature values, fuzzification, defuzzification, clustering, grouping, classification, and other actions that are intended for information analysis and visualization of the results. The organizational actions are conditions, constraints, cycles, transitions, recursions and iterations. The conditions and constraints determine the areas of application of the main and auxiliary actions, ranges of features and dependencies.

3 Language for Description of Image Analysis Strategies

The created by us high-level language SDL (Strategy Description Language) is intended to describe strategies of processing and analyzing images and is focused on use by a decision maker with only the most basic programming skills. Below we consider the main constructs of the language SDL.

Main plan contains a sequence of main stages of the strategy. In its body, each stage of the strategy is represented as a call of a named sub-plan, just as traditional programs decompose into subprograms (procedures). Some sub-plans can solve similar tasks and their names can match. Such related subplans can be called in a batch manner using the special iteration syntax (the header of the sub-plan can contain the number of the iteration on which it should be called). In the body of the sub-plan, the logic of the corresponding strategy stage is described with the help of the constructs considered below.

The *Select* operator is similar to the logical quantifier of existence (∃). This operator consists of two parts (head and body):

```
Select(What = … From = … Which = … How = … Result = …)
{ .. }
```

The semantics of selection is determined by the set of attribute values. In accordance with the specified method of selection, one of the set of valid variants of values is formed and considered. Using this operator, it is possible to make a selection of: (a) a region from a set of regions, (b) a value from a list (or a range) of values, (c) a subset of regions from a set of regions, (d) a group of regions from a set of groups.

If several selection variants are expected and it is necessary to find the optimal one among them, then the actions of processing and evaluating each possible variant should be specified in the body of this operator. If the choice does not imply alternative variants, then the body of the operator should be left empty, and the necessary actions with the result of the selection should be performed outside of this operator.

When the select operator is executed, a branch point is created in the state tree. The solver generates and processes alternative variants until one of them is marked as successful. This variant is considered as a promising solution way that needs to be explored further. All unsuccessful variants are dead-end solutions. In the process of analyzing variants, it is possible to cut off unpromising solutions according to the branch and bound method. To do this, one should specify an action of numerical assessing the current variant and set a threshold for the assessment. If the variant's score does not exceed the threshold, then the variant is immediately discarded.

By using the action *Classify*, regions can be attributed to target (or auxiliary) object classes. The set of areas to be classified is specified in the *What* attribute. In the parameter *How*, a set of decision rules of the form IF-THEN is specified, for example:

```
IF Straightness >= Large, AvgWidth <= Small,
   HasNeighbor[Color = Concrete, Squareness >= Large]
THEN Sidewalk(Reliability = 52%)
```

In the course of the classification, each of the input areas is checked for compliance with the conditions of the specified rules. When a rule is triggered, the area receives a classification specified in the conclusion of the rule, with a given degree of reliability. Conditions of classification rules can be multi-level: first, general requirements, then specific options and special cases. They are described and interpreted in a similar way as description logic formulas [9].

The action *Sort* is used to arrange elements of a set in some order. This action should be applied before a selection operator or a cycle to ensure that the most promising elements are considered first, thereby reducing the search space.

With the help of the action *FindAdjacentGroups*, it is possible to identify families of adjacent areas. It has the following parameters: *Among* which set of regions should adjacent families be identified; *GroupSize* – a limit on the number of members in a group; *Result* – a designation of the set of resulting groups.

The action *Assemble* with the attributes *What*, *From*, *How* and *Result* allows to create a new object (region) from other objects. One of the ways to implement it is to form a non-convex hull around a group of regions.

The action *Split* has several possible interpretations: (a) split a set into several subsets; (b) split a color region into sub-regions; (c) split the edge of a color region into several sub-chains. The attributes of this action are as follows: *What* – source object to be split; *HowMany* – number of resulting parts; *How* – method and criteria of splitting; *Results* – designations of the resulting parts.

In the case of splitting a color region into sub-regions, the splitting criteria can be, for example, the following:

- the resulting sub-regions should be close in size – the ratio of their areas tends to 1;
- the place of splitting should be narrow enough – the ratio of the cross-section length to the perimeter of the sub-region tends to 0;
- the sub-regions must be regular figures – the squareness (triangleness/trapezoidalness) of the sub-region is high.

The action *Approximate* is used to reduce redundant information in order to simplify and speed up subsequent analysis of the shape of color regions. The attribute *What* specifies the source region or the set of regions whose edges are to be processed. In the parameter *How* the method and intensity (weak, moderate, strengthened) of the approximation are set.

By applying the action *Clusterize*, one can determine subgroups of objects that have similarities in a certain set of parameters. The attribute *What* specifies the source set of regions to be clustered. The parameter *HowMany* specifies the estimated number of resulting subgroups. In the attribute *ByFeatures* the names of features are given, by which the similarity of regions will be evaluated. The attribute *How* specifies the clustering method along with the necessary operational parameters.

The action *ColorSegmentation* extracts color regions from a given area of the image. In the attribute *Where* one can specify either the exact coordinates of an image area of interest or a previously extracted color region requiring refinement. The attribute *HowMuch* specifies the maximum resulting number of distinct colors. In the attribute *How* segmentation parameters are set (the degree of discretization of the HSV color space). In the parameter *Result* one should specify the designation of the resulting set of color regions.

With the help of the action *Fuzzify*, one can define qualitative values of features (e.g. small, medium, large, very large) along with the rules for converting quantitative values into qualitative ones. Using this tool, one can operate with different qualitative values of the same feature at different stages of the strategy. This makes it possible to adapt to different situations. In the action's attributes, the user specifies the set of regions of interest, the name of the feature in question, the names of the qualitative values of the feature, as well as the rules according to which the quantitative value of the feature will be converted into one or more qualitative ones. To facilitate the process of developing conversion rules, the system has a tool for constructing histograms of the distribution of numerical values of features. In the process of interpretation, the quantitative values are fuzzified using simple built-in trapezoidal functions, which are automatically scaled to the width of the user-defined ranges.

4 Experiment

The basic context-sensitive strategy for the analysis of aerospace images consists of two stages:

1. Preprocessing (considered in detail in [10, 11]), calculation of the features of the color regions, and preclassification.
2. Local analysis of regions (objects) classified with insufficient reliability.

As a result of image segmentation, a house with a multi-slope roof is usually represented not by a single color segment, but by a variety of color regions, which greatly complicates the detection process. The strategy for identifying such houses is presented below. It is described in the developed language SDL. The strategy involves assembling an object from several parts and checking for the presence of a shadow.

```
Select(What = Subset,
       From = Regions,
       Which = (Class = Undefined OR Reliability < 70%,
                ContourPointDensity < Large,
                Squareness >= Large OR
                Triangleness >= Large OR
                Trapezoidalness >= Large),
       Result = $Regs)
PerformAction FindAdjacentGroups(
   Among = $Regs,
   GroupSize >= 2,
   Result = $Groups)
PerformForEach(What = $G, From = $Groups) {
    Select(What = Subgroup,
           From = $G,
           Which = (Area = Small OR Medium,
                    SimilarHues = Yes,
                    Squareness >= Large),
           How = (AllPossibleSubsets,
                  InDescendingOrderOfNumberOfElements),
           Result = $S) {

      PerformAction Assemble(
         What = Region,
         From = $S,
         How = NonConvexHull,
         Result = $O)
      PerformAction Classify(
         What = $O,
         How = (IF HasNeighbor[Class = Shadow,
                AngleToNeighbor = SunlightIncidenceAngle,
                CommonEdgeLength > VerySmall]
                 THEN (Building(Reliability = 70%),
                       ChoiceIsSuccessful)))
    }
}
```

The presented approach, based on flexible analysis strategies, was experimentally tested on images of urban areas taken from the Inria Aerial Image Labeling benchmark dataset [12]. The selected images provide a good field for working out context-sensitive analysis strategies because they have a very high spatial resolution (0.3 m). At this resolution, there is a large variety of shapes, textures and combinations of objects [13], which adversely affects the quality of the work of standard detection methods that ignore the context.

Figure 1 shows an example of the result of detection of buildings on the aerial image. As can be seen, the system successfully detects buildings of different types and configurations, distinguishing them from concrete and paved areas and road parts, which are similar to buildings not only in spectral characteristics, but also in geometric shape. This is achieved by taking into account the context (environment) of objects.

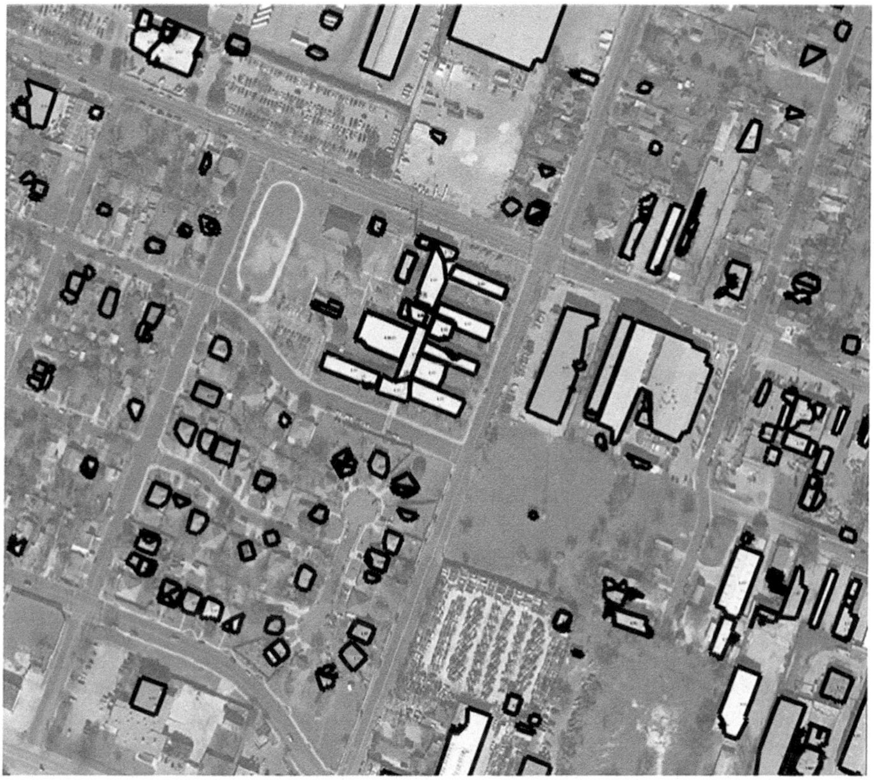

Fig. 1. The result of building detection on the very high-resolution image of an urban area

For the numerical evaluation of the system's effectiveness, widely-known object-oriented metrics were used: *Recall* – the proportion of targets that were successfully detected; *Precision* – the proportion of true objects in the detection result; *F1-measure* – the integrated characteristic of recall and precision.

In the test images, small buildings prevail in the total number of objects of interest. Many of them are partially covered by trees, and because of this the shape features lose their informativeness. With objects of this kind the system does not cope very well yet. For successful detection it is necessary to develop special analysis methods and strategies. Taking this moment into account, as well as the fact that the evaluation metrics consider all objects (large and small) with the same weight, at the current study stage we excluded partially obscured buildings from the assessment process.

Taking into account the conditions described above, the following rates of the building detection quality were obtained: *Recall* = 0.788, *Precision* = 0.853, *F1-measure* = 0.819. The results suggest that the proposed approach to the automatic analysis of aerospace images is quite workable and very promising. The quality of the results was partially influenced by the presence in the images of trucks, building materials and other similar objects, which by many parameters (shape, shadow) are similar to small buildings and therefore were mistakenly classified as buildings.

5 Conclusion

We propose the principles of organizing strategies for the structural analysis of the aerial image content as well as the high-level language SDL for their description.

Image analysis strategies are constructed from such actions as color image segmentation, approximation of the edges of color segments, refinement of object boundaries by the methods of optimal partitioning and assembling, calculation of object features, generation of variants of groups/values, formation of a hypothesis about the category of the object, classification of objects into target classes using expert rules, etc.

The advantage of the applied semantic models of actions and resources is that on their basis the visibility of the process of structuring and describing strategies increases with the preservation of the possibility of machine interpretation. The process of describing the image analysis strategy is transferred from the level of specifying instructions/commands to the level of planning works and resources.

In general, the SDL language allows the user to focus on strategy development and not to be distracted by routine, because it consists of a unified set of functions and quantifying operators over them and is reduced to specifying the points of choice of methods and their parameters. Interpretation of a given strategy is carried out by the solver module, which selects variants of actions, data and constraints for each subtask, builds a decision tree, and monitors the progress of the decision.

Due to the context-sensitive strategy, the system is automatically, depending on the content of a particular area, adapted to shooting conditions, the texture of artificial and natural objects, the nature of the boundaries between them and situations. Thereby, the relevance of the results of automatic analysis and description of image objects increases, and the decisions made by the system become more reasonable.

References

1. Gurevich, I.B., Trusova, Y.O., Yashina, V.V.: The algebraic and descriptive approaches and techniques in image analysis. In: Proceedings of the 4th International Workshop on Image Mining. Theory and Applications (IMTA-4-2013), pp. 82–93 (2013)
2. Gurevich, I.B., Yashina, V.V.: Descriptive image analysis: genesis and current trends. Pattern Recogn. Image Anal. **27**(4), 653–674 (2017)
3. Abburu, S., Golla, S.B.: A generic framework for multiple and multilevel classification and semantic interpretation of satellite images. World Eng. Appl. Sci. J. **7**(2), 107–113 (2016)

4. Gu, H., Li, H., Yan, L., Liu, Z., Blaschke, T., Soergel, U.: An object-based semantic classification method for high resolution remote sensing imagery using ontology. Remote Sens. **9**(4), 329 (2017)
5. Bychkov, I.V., Ruzhnikov, G.M., Fedorov, R.K., Avramenko, Y.V.: Interpretator yazyka SOQL dlya obrabotki rastrovykh izobrazheniy [The interpreter of the SOQL language for processing raster images]. Vychislitel'nyye tekhnologii Comput. Technol. **21**(1), 49–59 (2016). (in Russian)
6. Levesque, H.J., Reiter, R., Lespérance, Y., Lin, F., Scherl, R.B.: GOLOG: a logic programming language for dynamic domains. J. Logic Program. **31**(1–3), 59–83 (1997)
7. Ferrein, A., Steinbauer, G., Vassos, S.: Action-based imperative programming with YAGI. In: Proceedings of the 8th International Cognitive Robotics Workshop at AAAI 2012, pp. 24–31 (2012)
8. Russell, S.J., Norvig, P.: Artificial Intelligence: A Modern Approach, 3rd edn. Prentice-Hall, Upper Saddle River (2009)
9. Borgwardt, S., Peñaloza, R.: Fuzzy description logics – a survey. In: Moral, S., Pivert, O., Sánchez, D., Marín, N. (eds.) Scalable Uncertainty Management, SUM 2017. LNCS, vol. 10564, pp. 31–45. Springer, Cham (2017)
10. Kasimov, D.R.: Techniques for improving color segmentation in the task of identifying objects on aerial images. In: 2019 24th Conference of Open Innovations Association (FRUCT), Moscow, Russia, pp. 148–155 (2019)
11. Kasimov, D.R., Kuchuganov, A.V., Kuchuganov, V.N., Oskolkov, P.P.: Approximation of color images based on the clusterization of the color palette and smoothing boundaries by splines and arcs. Program. Comput. Softw. **44**(5), 295–302 (2018)
12. Maggiori, E., Tarabalka, Y., Charpiat, G., Alliez, P.: Can semantic labeling methods generalize to any city? The inria aerial image labeling benchmark. In: IEEE International Geoscience and Remote Sensing Symposium, IGARSS 2017 (2017)
13. Qin, R., Fang, W.: A hierarchical building detection method for very high resolution remotely sensed images combined with DSM using graph cut optimization. Photogram. Eng. Remote Sens. **80**(9), 873–883 (2014)

Parametrization of Functions in Multiattribute Utility Model According to Decision Maker' Preferences

Stanislav V. Mikoni[1(✉)] and Dmitry P. Burakov[2]

[1] St. Petersburg Institute for Informatics and Automation of the Russian Academy of Sciences, Saint-Petersburg, Russia
smikoni@mail.ru

[2] Emperor Alexander I St. Petersburg State Transport University, Saint-Petersburg, Russia

Abstract. In the article, the method of staged clarification of a choice model depending on decision maker' (DM) preferences is proposed. The method consists of several stages. At the first stage, the DM specifies an alternative that is best by its opinion. This information is used to detect a set of criteria defining the Pareto set that includes the specified alternative. At the next stage, the DM clarifies one-dimensional utility functions that have automatically been created for criteria, specifying the risk propensity or aversion. Basing on this information, the generalizing function and weight vector are selecting in order to the specified by DM alternative will be first in the resulting rating. At the last stage, if there is information about preferences on a set of alternatives, the one-dimensional utility functions are parametrized in the way that object ranks be in concordance with the specified preferences.

Keywords: Decision making · Multiattribute utility theory · Utility function · Preferences · Choice model

1 Introduction

One way to solve the task of multidimensional evaluation of alternatives (objects) is to evaluate them by aggregative utility [1]. As the object' utility, in these tasks mean degree of compliance of object to a target specified by the decision maker (DM). In scalar multidimensional optimization methods, the object' utility is calculated by aggregation the values of partial one-dimensional utility functions (UF) defined for each attribute characterizing quality of evaluated objects.

When DM defines the decision making task, he/she has a finite set of alternatives $X = \{x_1, x_2, \ldots, x_N\}$ and an objective as an image of an "ideal" object. The vector estimation value $\mathbf{y}_i = (y_1(x_i), y_2(x_i), \ldots, y_n(x_i))$ is assigned to each object x_i. To find best object x^*, it is necessary specify a linear order on the set X, which represents DM's

The studies were carried out with the financial support of the RFBR grant No. 17-01-00139, No. 19-08-00989 within the framework of the budget theme No. 0073-2019-0004.

S. Kovalev et al. (Eds.): IITI 2019, AISC 1156, pp. 300–308, 2020.
https://doi.org/10.1007/978-3-030-50097-9_31

preferences on a scales of each attribute. There is assumed that DM' preference relation is rational, i.e. satisfies the Edgeworth—Pareto axiom: if the object x_i is not worse that the object x_k by all of the attributes and at least by one of them it is better, the DM cannot prefer x_k, and object ranks in the rating are related the following: $\rho(x_i) < \rho(x_k)$. When the objects evaluated by aggregative utility value $u_O(x)$, the object x_i is preferable the object x_k, if the following conditions are true:

$$x_i \succ x_k \Leftrightarrow u_O(x_i) > u_O(x_k) \Leftrightarrow \rho(x_i) < \rho(x_k).$$

The results of ordering objects by aggregative utility $u_O(x)$ are depend on one-dimensional utility functions $u_j(y_j(x))$ forms, type of aggregative function (AF) that is used and priorities of evaluated attributes. In the case when DM has a point of view on the expected rating of objects, there is solved an inverse problem implied creation of a choice model corresponds DM' preferences. In the Multiattribute Utility Theory (MAUT) [2], it is divided to three different problems: attribute prioritization, choice of aggregative function, and creation of one-dimensional utility functions. A way to solve the first problem is in detail considered in [3]. The recommendations about solving the second problem are proposed in the paper [4]. The solution of the third problem is the most high-tech and time consuming. The given article has a purpose to simplify the solving of this problem. Before its solution, we consider the well-known approaches to the creation of one-dimensional utility functions.

2 Methods for Creating One-Dimensional Utility Functions

To one of first methods to create one-dimensional utility functions is belonging creating it by points [2], when the utility values are defined on some set of values of attribute scale as results of DM' polling. After that, creation of the function graphic is basing on the given utility values. In common case, the UF is non-linear; moreover, it can have negative values that represent risks of the losses. A measure of DM' risk propensity or aversion is represented by the form of UF: it is convex at the intervals where the DM has a risk propensity, linear at intervals where DM is indifferent to the risk, and concave at the intervals where the DM has a risk aversion (i.e. He\She is careful).

The continuous utility (value) functions were in particular studied in [5]. There was considered several variants of a concave utility function in point of view the ratio of risk measurements and an objective. A condition, when the risk of the exceeding of the target specified at an attribute' scale can be acceptable, has been proposed.

In source [6], relations between different types of value functions of several attributes in conditions of certainty and risk are studied. The work [7] is devoted to the early stages of creating functions of many attributes in the absence of a priori assumptions about the structure of DM preferences. It proposes a practical method for estimating interactions of monotonic functions. The proposed approach is tested by an experiment in which the subjects were asked to evaluate mobile phones in three attributes.

In a paper [8] is considered an alternative for non-parametric segmental concave forms that satisfy to theoretical UF properties related to monotone and concaving.

It proposes to use the approximation to arbitrary UF form based on a smooth mixing of Cobb—Douglas functions. In order to estimate how the created UF corresponds to DM preferences, the Bayesian approach is used. For the experimental test using the Monte Carlo method, where the McFadden Symmetric Generalization is used as the true function form. The proposed method has been used on a big set of banking data.

The Bayesian approach has been also used in the paper [9]; there was proposed a way to estimate how DM preferences correspond to UF created by points basing on DM answers about his preferences. A probabilistic view of DM preferences including the risk attitude is realized using Gaussian random processes. On two real data sets there was obtained the results with lower RMS errors than in source data.

Analysis of the cited publications leads to the conclusion: creation of the utility function is characterized by a big volume of expert work and checking of its correspondence to DM' preferences require a complex mathematical apparatus. These problems are especially aggravated for a large number of attributes inherent in complicated objects.

Because of procedures of one-dimension utility functions creation are difficult to implement in practice, to solve many application problems of evaluation of complicated objects with many attributes use simplified methods for constructing one-dimensional UF. For example, it is possible to interpret utility function as a function evaluating the partial achievement of a certain target. In this case DM specifies a certain target value on an attribute' scale (the target value can be a point or an interval); in a correspondence with the target value it is evaluated the DM' risk propensity or aversion. Such approach allows replacing the labor-intensive way to create the utility function by a set of points with parametrization of a reasonable according to risk propensity or aversion typical utility function [3].

3 Parametrization of Utility Functions

In this article, the following monotonic UF that defined in [3] as typical, are used:

$$u_P^{\max}(y;\alpha) = \left(\frac{y - y_{\min}}{y_{\max} - y_{\min}}\right)^{\alpha}, \tag{1}$$

$$u_P^{\min}(y;\alpha) = \left(\frac{y_{\max} - y}{y_{\max} - y_{\min}}\right)^{\alpha}, \tag{2}$$

$$u_L^{\max}(y;c,\beta) = (1 + \exp(-\beta \cdot (y - c)))^{-1}, \tag{3}$$

$$u_L^{\min}(y;c,\beta) = (1 + \exp(\beta \cdot (y - c)))^{-1}. \tag{4}$$

Here (1) and (2) are power monotonic functions, where power value $\alpha > 0$ specifies nonlinearity of the function: when $\alpha > 1$, functions are convex, and when $\alpha \in (0; 1)$ they are concave. The interval of attribute scale $[y_{\min}; y_{\max}]$ is domain of these functions and the interval of utility scale [0; 1] is the functions range. Functions (3) and (4), also known as *logistic functions*, also are monotonic and have the inflection point when

$y = c$, $c \in [y_{min}; y_{max}]$, $u_L(c; c, \beta) = 0.5$. The parameter β specifies the degree of nonlinearity of the function and defined as the following:

$\beta = m^{-1} \cdot \ln(d^{-1} - 1)$, where $d \in (0; 0.5)$ specifies maximum deviation from the limit values 0 and 1 at the ends of the interval $[c - m; c + m]$, that is symmetric relatively the point c, $[c - m; c + m] \subseteq [y_{min}; y_{max}]$.

Graphics of the utility functions (1)–(4) that defined at the interval [1, 9] are shown at Fig. 1.

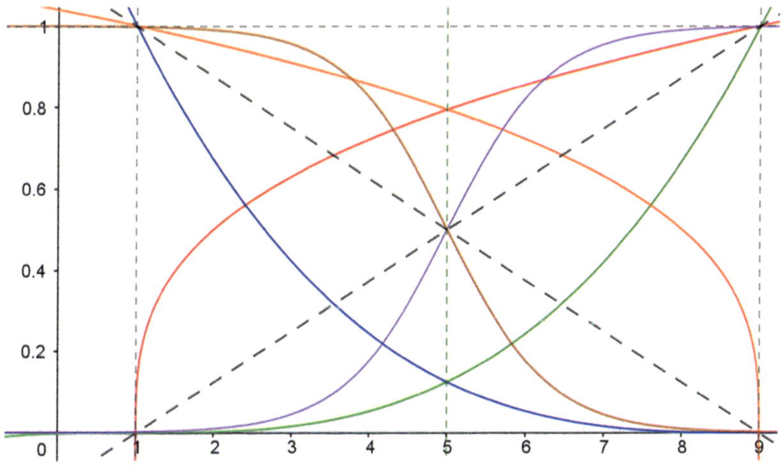

Fig. 1. Monotonic utility functions

When selection of the UF is based on the DM' risk propensity or aversion, the question of setting the parameters characterizing the nonlinearity of the UF remains open. Consider a way to make UF parameters more precision based on known by DM rating of the objects. The required order of objects in the rating in this case can be obtained by variation of UF nonlinearity parameters.

As the source data to solve this problem, we use:

- One-dimensional utility functions selected from the list of typical UFs basing on the known target values and DM' risk propensity or aversion at attribute' scale intervals that are neighbor for the target value,
- Expected rating on a control set of objects with greater confidence about ranks of best and worst objects.

It is required to come near the expected rating from the rating obtained on the set of objects within initial evaluation conditions by variation values of parameters of typical UFs (1)–(4) specified by DM and using the additive aggregative function (AAF) [3, 4] for obtaining the generalized utility values.

Coming the rating to the expected can be made by sequential exchange of ranks between objects with nearest values of generalized utility. For that, it is necessary to define the conditions of exchanging the ranks in a rating.

Because the aggregative functions are monotonic relative the Pareto dominance relation [4], exchanging of ranks between objects x_i and x_k is possible only if they are mutually non-dominated, i.e. are belonging to one dominance level. Therefore, at the first stage we need to know that an analyzed pair of objects belongs to one Pareto dominance level. If so, there is a way to exchange them ranks by variation of ratio of their contributions at each attribute into value of AAF, by changing of weights (priorities) and/or utility function values.

Let at the initial conditions specified by DM, object utilities are related as follow: $u_O^{(0)}(x_i) > u_O^{(0)}(x_k)$, i.e. $x_i \succ x_k$. Consider the conditions when the relation is changed to inverse $u_O^{(1)}(x_i) < u_O^{(1)}(x_k)$ due to variation of utility values $u_l(y_l(x_i))$ and $u_l(y_l(x_k))$ of these objects at l-th attribute. To do that, separate the contribution of l-th attribute utility in the generalized utility value calculated as multi-dimensional additive estimation:

$$u_O(x) = \sum_{j=1, l \neq j}^{n} w_j \cdot u_j(y_j(x)) + w_l \cdot u_l(y_l(x)).$$

If we need to obtain the inverse preference $x_i \prec x_k$, it is necessary to change the contributes ratio in favor to object x_k, by changing the utility values of x_i and x_k at the l-th attribute: $\Delta u_{li} = u_l^{(1)}(y_l(x_i)) - u_l^{(0)}(y_l(x_i))$ and $\Delta u_{lk} = u_l^{(1)}(y_l(x_k)) - u_l^{(0)}(y_l(x_k))$. These differences can be either positive or negative. As the weight coefficient w_l is constant here, *prerequisite* of the contribution ratio changing is ***positive difference*** of utility values changes of objects x_k and x_i at l-th attribute:

$$\Delta u_{lk} - \Delta u_{li} > 0. \tag{5}$$

Using the condition (5) for difference of initial generalized utility values $u_O^{(0)}(x_i) - u_O^{(0)}(x_k)$, obtain the prerequisite of exchanging of ranks between objects x_k and x_i due to variation of ratio of them utilities at l-th attribute:

$$w_l \cdot (\Delta u_{lk} - \Delta u_{li}) > u_O^{(0)}(x_i) - u_O^{(0)}(x_k). \tag{6}$$

But the condition (6) may be not enough to exchange ranks between objects x_i and x_k, because variation of contribute into multidimensional utility of objects x_i and x_k at l-th attribute entails variation of relative contributes $V_j(x)$ for all other attributes; it is a consequent of the expression:

$$V_j(x) = \frac{w_j \cdot u_j(y_j(x))}{u_O(x) + w_l \cdot \Delta u_{lx}}. \tag{7}$$

Addition of the weighted utility by l-th attribute in the denominator of formula (7) decreases contributes of other attributes into scalar utility value of an object. The marked variation of ratio of relative contributes $V_j(x)$ by other attributes takes into consideration a difference of new generalized utility values $u_O^{(1)}(x_k) - u_O^{(1)}(x_i)$. This difference should be not less than difference of initial utility values of these objects:

$$u_O^{(1)}(x_k) - u_O^{(1)}(x_i) > u_O^{(0)}(x_i) - u_O^{(0)}(x_k). \tag{8}$$

Therefore, the condition (8) is more powerful then condition (6). The condition (8) is necessary and sufficient for exchanging of ranks between objects x_i and x_k.

Because the values $y_l(x_k)$ and $y_l(x_i)$ at scale of l-th attribute are unchanged, then if the utility function u_l type is constant, the difference in the increments of the utility Δu_{lk} and Δu_{li} can be achieved only by changing the parameters that affect its nonlinearity.

As an adjustable parameter M, we choose a parameter affects to nonlinearity of utility function u_l. So, for a power function as an M parameter the power value is used ($M = \alpha$). For a convex function the initial parameter value is $M^{(0)} = 2$; for a concave function it is $M^{(0)} = 0.5$. Nonlinearity of a convex function is increases together with increasing M, and for a concave function, it is increases together with decreasing $M > 0$. For a logistic function that depends on a three different parameters, we will make adjustment of the parameter m, which affects to velocity of its variation near the target value c. Average degree of logistic function nonlinearity corresponds to the value

$$M^{(0)} = \frac{|m - c|}{y_{\max} - y_{\min}} = 1/3.$$

According the right part of the condition (6), a biggest possibility to exchange their ranks have objects x_i and x_k with the minimum initial difference between multidimensional utility values:

$$(x_i, x_k) = \underset{\substack{i \in \overline{1, N-1};\\ k \in \overline{i+1, N}}}{\arg\min} \left| u_O^{(0)}(x_i) - u_O^{(0)}(x_k) \right|. \tag{9}$$

Such objects are objects with neighbor ranks in a rating obtained basing on multidimensional utility values using initial nonlinearity parameter values $M_j^{(0)}$, $j = \overline{1,n}$. In order to detect a pair (x_i, x_k) that meets the condition (9), it is necessary to make $N \cdot (N - 1)/2$ operations of multidimensional utility values $u_O(x_i)$ comparison, $i = \overline{1,N}$.

At that, the condition (6) is necessary but not sufficient to detect the possibility to exchange ranks between objects x_i and x_k, because it overlooks the new multidimensional utility values $u_O^{(1)}(x_i)$ and $u_O^{(1)}(x_k)$, and consequently their differences $\Delta u_O(x_i) = u_O^{(1)}(x_i) - u_O^{(0)}(x_i)$ and $\Delta u_O(x_k) = u_O^{(1)}(x_k) - u_O^{(0)}(x_k)$. These differences can be either positive or negative depending on variation of UF u_l nonlinearity. They depend on the initial nonlinearity of the used UF, which affected by the initial value of parameter $M_l^{(0)}$, and on direction of the nonlinearity variation.

It is obvious that variation of the nonlinearity parameter M_l for utility function $u_l(y_l(x))$ until the limit value $M_{l,\max}$ (or $M_{l,\min}$) may not lead to the fulfillment of condition (8) for a pair of objects (x_i, x_k) that meets the condition (6). In this case, it is necessary to go to analysis the utility function u_j for the other j-th attribute, $j = \overline{1,n}$, $j \neq l$. The maximum number of tries to satisfy the condition (5) with bidirectional variation of the parameter M_j using its limit values is equal to $2 \cdot n$. Of course, this number is increasing when also used the intermediate values of the parameter.

Thus, to satisfy the condition (5) for a pair of objects (x_i, x_k) found by the condition (9), it is necessary to variate nonlinearity of n utility functions of type (1)–(4).

Consider the ways to reduce the busting volume of the parameter values. According to the left part of the condition (6), a maximum possibility to exchange ranks between objects x_i and x_k has an attribute, which provides the maximum difference between the utility value differences when the nonlinearity parameter M is under variation:

$$l^* = \arg\max_{l \in \overline{1,n}} (\Delta u_{lk} - \Delta u_{li}). \tag{10}$$

To find the l^*-th attribute that meets the condition (10), it is necessary to calculate $2 \cdot n$ differences of utility variations at l-th attribute, intending that the parameter M_l value can be changed to both sides relatively its initial value $M_l^{(0)}$.

The amount of computations can be reduced, if there is previously make analysis of variants of utility function value changes Δu_{lk} and Δu_{li} for objects x_k and x_i when the parameter M_l value is changed from $M_l^{(0)}$ to $M_l^{(1)}$. There are four options of such changes combination:

1. $\Delta u_{lk} > 0$, $\Delta u_{li} > 0$.
2. $\Delta u_{lk} > 0$, $\Delta u_{li} < 0$.
3. $\Delta u_{lk} < 0$, $\Delta u_{li} > 0$.
4. $\Delta u_{lk} < 0$, $\Delta u_{li} < 0$.

When a difference Δu_l is positive, the utility value at the l-th attribute is increased; when the difference is negative, the corresponding utility value is decreased.

The most interesting options are two and three with oppositely directed changes in utility values. The second option is most prospectively for ranks exchanging because it increases difference between the changes of utility values. As a result, the difference $u_l(y_l(x_k)) - u_l(y_l(x_i))$ in utility of objects x_i and x_k at l-th attribute increases in favor the object x_k.

The oppositely directed change of utility values is observed only for the logistic UF (3) or (4), if attribute values $y_l(x_i)$ and $y_l(x_k)$ are placed at the *different sides* relatively its inflection point c (see Fig. 1). If these values are placed at one side relatively this point, the analysis is reduced to 1st and 4th options that are also typical for functions (1) and (2) without inflection points. These options represent the unidirectional change of utility values. In these cases, the result depends on value of the difference (5) regardless of the simultaneous increase or decrease in the utility values. Thus, from four different options of changes in utility values for objects x_k and x_i at l-th attribute, it is considered only three options, and, at first, the second option as most prospectively.

Basing on the above, formulate the algorithm for analysis the possibility to change objects ranks in a rating. Source data for the algorithm are values $y_j(x)$ of objects $x \in X$ by n attributes, and values of utility function $u_j(y_j(x))$ of these attributes with initial values $M_j^{(0)}$ of nonlinearity parameters M_j, $j = \overline{1,n}$. The algorithm consists of the following steps:

1. Specifying the Pareto dominance relation $R_P \subset X \times X$ on a set of evaluated objects X; then the graph of this binary relation is ranked to $p \geq 1$ levels.

2. Determining the subsets X_r of objects belonging to r-th ranking level, and thereby mutually non-dominated in relation to each other, $X = \bigcup_{r=1}^{p} X_r$. For any pair of objects belonging to one of these sets, they ranks can be exchanged.
3. Calculating multidimensional utility values $u_O^{(0)}(x_i)$, $i = \overline{1, N}$.
4. Ranking the objects by calculated multidimensional utility values.
5. Finding a pair (x_i, x_k) of objects that meets the condition (9). These objects are claiming to exchange their ranks, $i, k \in \{1, \ldots, N\}$.
6. Selecting the attribute y_l, $l \in \{1, \ldots, n\}$, with different values for specified objects.
7. Performing the analysis of possibility to exchange ranks between objects from the pair (x_i, x_k) due to variation (increasing or decreasing) value of the nonlinearity parameter M_l for the utility function $u_l(y_l(x))$.
8. If the possibility exists, calculating new multidimensional utility values $u_O^{(1)}(x_i)$ and $u_O^{(1)}(x_k)$ with a new value of the nonlinearity parameter $M_l^{(1)}$.
9. If the condition (9) is satisfied, the pair (x_i, x_k) is considered to be exchangeable. If the condition (9) is not satisfied for the l-th attribute, going to selecting the next, $l + 1$-th attribute.

Because the above algorithm performs an analysis of the possibility of changing ranks for all attributes, and changes of utility functions are applied to all objects, it allows to detect all possible changes of the initial rating of objects within limits of its stability that defined by the dominance relation R_P.

For the experiments, the program System of Choice and Ranking ('SVIR' in Russian abbreviation) [10, 11] was used. The described algorithm has been successfully tested using the example of evaluating seven electronic books by five attributes. The exchange of object' ranks has been achieved by finding those attributes whose changing nonlinearity of the utility function influenced the objects rating.

4 Conclusion

Nonlinearity of a monotonic continuous function can be interpreted as a continuous change of the preference value in all dots of the attribute' scale. This fact allows using nonlinearity parameters of the function for changing the ratio of preference values of comparable attributes. In the paper, conditions for exchanging ranks between objects in the rating by changing the parameters of the function nonlinearity are formulated and justified.

The time-consuming procedure for constructing a one-dimensional utility function of the attribute is simplified by setting its target value and then selecting one of standard utility functions that satisfy the DM' risk propensity or aversion depending on his desire to distance or bring together estimations of objects in different ranges of the attribute' scale. The nonlinearity parameters of typical utility functions are specified at the final stage of the procedure, when the rating of multidimensional utility values is coming on to the expected rating. This process may interpret as education of the model.

The proposed approach makes it possible to significantly reduce the complexity of the initial stage of creating a multidimensional utility model by transfer of refinement of intermediate utility points to the model debugging stage.

References

1. Velasquez, M., Hester, P.T.: An analysis of multi-criteria decision making methods. Int. J. Oper. Res. **10**(2), 56–66 (2013)
2. Neumann, J., Morgenstern, O.: Theory of Games and Economic Behavior, Science edn. Wiley, Hoboken (1944)
3. Mikoni, S.V.: Management Decision Theory. SPb, St. Petersburg (2015)
4. Burakov, D.P., Garina, M.I.: Analysis of decision maker's preferences structure with standard aggregative functions. Tomsk State Univ. J. Control Comput. Sci. **38**, 11–16 (2017). https://doi.org/10.17223/19988605/38/2
5. Ehrgott, M.: Multicriteria Optimization, 2nd edn. Springer, Heidelberg (2005). 323 p. ISBN 3-540-21398-8
6. Beccacece, F., Borgonovo, E., Buzzard, G., Cillo, A., Zionts, S.: Elicitation of multiattribute value functions through high dimensional model representations: Monotonicity and interactions. Eur. J. Oper. Res. **246**(2), 517–527 (2015)
7. Karakaya, G., Koksalan, M., Ahipasaoglu, S.D.: Interactive algorithms for a broad underlying family of preference functions. Eur. J. Oper. Res. **265**(1), 248–262 (2018)
8. Tsionas, M.G., Izzeldin, M.: Smooth approximations to monotone concave functions in production analysis: an alternative to nonparametric concave least squares. Eur. J. Oper. Res. **271**(3), 797–807 (2018)
9. Gu, M., Bhattacharjya, D., Subramanian, D.: Nonparametric estimation of utility functions. https://arxiv.org/abs/1807.10840v1. Accessed 05 Sep 2018
10. The Choice and Ranking System ('SVIR' in Russian abbreviation). http://www.mcd-svir.ru/refer07.html. Accessed 05 May 2019
11. Mikoni, S.V.: Application of the universal decision support system SVIR to solving urban problems. In: Chugunov, A.V., et al. (eds.) DTGS 2016. CCIS, vol. 674, pp. 1–14. Springer, Cham (2016)

Translation of Cryptographic Protocols Description from Alice-Bob Format to CAS+ Specification Language

Liudmila Babenko and Ilya Pisarev(✉)

Information Security Department, Southern Federal University, Taganrog, Russian Federation
lkbabenko@sfedu.ru, ilua.pisar@gmail.com

Abstract. Cryptographic protocols are an important part of any protected system. Verifying their security is an important task to ensure the security of such systems. To enable security analysis, various verifiers are used. To do this, it is necessary to describe the protocol in specific specification languages, specify the verification objectives, and then directly conduct a security analysis. The paper presents a general approach for translating the presentation of the protocol from the Alice-Bob form to any specification language. The main elements allocated for the protocol when describing it in the specification language are given. Roles, data transmission channels, initial knowledge of roles, model of an attacker, initial knowledge of an attacker, message transfer are highlighted. A description of the CAS+ specification language for use in the Avispa verifier is provided. The algorithm for translating the Alice-Bob view to the CAS+ specification is given. An algorithm is presented for both automated verification of the protocol originally described in the Alice-Bob format and partially automated using the Avispa graphical interface. An example of the operation of the translation algorithm and automated verification is shown.

Keywords: Translator · CAS+ · Avispa · C# · Cryptographic protocols · Verification · Analysis

1 Introduction

After creating a cryptographic protocol, it is necessary to analyze its security. This is achieved through the use of formal verification tools. For this opportunity, it is necessary to describe the protocol in a special specification language and then verify it. However, the specification languages are rather complicated and take time to describe the selected protocol. The simplest form of a protocol description is the Alice-Bob representation. To describe the protocol in such a presentation does not require any special skills. Thus, the development of a translator from the simplest representation of Alice-Bob to the chosen specification language is relevant. In our case, the CAS+ [1] specification language was chosen for the automated Avispa [2] protocols security verifier. It is possible to use it for verification cryptographic protocols implementations. So cryptographic protocols structure extracted from source code, for example by parsers for languages C [3–5], Java [6–8], F# [9–11], C# [12] and after this with help of our translator it is possible to verify protocol by Avispa.

S. Kovalev et al. (Eds.): IITI 2019, AISC 1156, pp. 309–318, 2020.
https://doi.org/10.1007/978-3-030-50097-9_32

2 Importance of Work

The purpose of this article is to translate from the simplest description of the protocols in the Alice-Bob language to the CAS+ specification language for the Avispa verifier. This translator will be an integral part of automated cryptographic protocols implementation verifier developed by the authors. First you need to get the source code of the system. The verification scheme is shown in Fig. 1. From the source code it is necessary to extract the structure of the cryptographic protocol. This is done using the cryptographic protocols extractor developed by the authors [12]. Currently, it is possible to extract protocol structures from the source codes of the C# programming language. The result is a protocol description in the form of Alice-Bob. Next, the protocol is translated into specification languages for automated protocol security verifiers, such as Avispa [2], Scyther [13], ProVerif [14]. As a result of formal verification, a list of attacks on this protocol will be obtained.

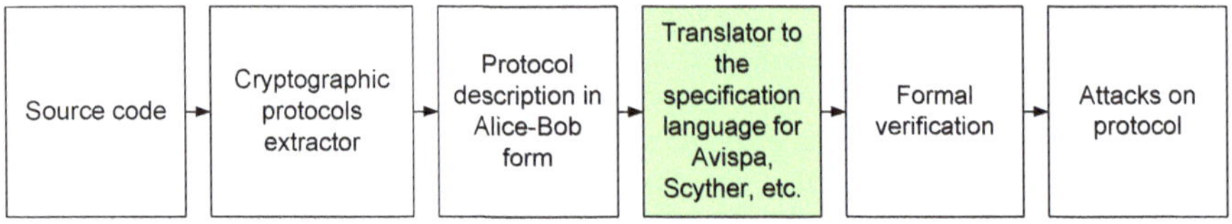

Fig. 1. Scheme of automated cryptographic protocols implementation verifier

The current paper deals with the Avispa verifier. It works with the HLPSL language. The format of the HLPSL language is extremely complicated, so Avispa has a translator from the CAS+ language to HLPSL. Accordingly, for security verification in this paper, translation into the CAS+ specification language is considered. For the CAS + language, the Alice-Bob format from which you can get a formal description is sufficient. For more complex formal languages, such as SPS [15], more information about the protocol is needed for formal description. In the future, for such languages, extraction of the extended structure of the cryptographic protocol will be used, containing additional information from the source codes to allow a formal description of the protocol.

3 General Approach for Translating Alice-Bob Views to Specification Languages

In formal verifiers, there are similar concepts that operate when verifying the security of cryptographic protocols. These concepts are: Roles, data transfer channels, initial knowledge of roles, the attacker's model, initial intruders' knowledge, messages. The task is an automated way to translate the description of the Alice-Bob form protocol into the specification language of the used verifier. To perform this task, a parser for describing the presentation of the Alice-Bob form protocol is required, which will allow to obtain separate parts of the protocol from this description. As an example, consider the approach proposed by the authors to translate into an abstract specification language. For clarity, the well-known Needham-Schroeder protocol is described, supplemented by message 4.

Alice-Bob presentation:

```
1. A -> B: AE_Pka(Na, A)
2. B -> A: AE_Pkb(Na, Nb, B)
3. A -> B: AE_Pka(Nb, K)
4. B -> A: SE_K(M1)
```

First you need to define a list of roles. Roles will be all parties involved in receiving or transmitting at least one message. To find the list of roles, it is necessary to search for the elements of strings on both sides of the "->" message transmission symbol. In this case, these are two sides *A* and *B*. Next, you need to typify the elements of the message. The main one is the discovery of encryption keys and their type. For the sake of simplicity, it is assumed that in the abstract form, *AE* and *SE*, respectively, will initially be set in differently asymmetric and symmetric encryption. Thus, the keys on which encryption is performed using *AE* will be considered public, and with symmetric encryption, *SE* will be considered a symmetric key. Thus, in our case, it is necessary to search for the designation "AE_" and "SE_", on the right side of which there will be keys. The keys *Pka, Pkb* are public keys, the key *k* is a symmetric key. Next, you need to determine the contents of the messages themselves. This can be done by searching all items that are not message indices, using the arrow "->" and the sides along its edges, as well as cryptographic functions with keys, in this case, "AE_Pka", "AE_Pkb", "SE_K". All other elements are directly elements making up the content of messages. In this case, it is: [*Na, Nb, A, B, K, M1*]. All elements that are designated as well as the name of the role are its identifier. In our case, message elements 1 and 2, *A* and *B*, are role identifiers. Next, a match is sought with a previously compiled list of encryption keys. In our case, in message 3, the symmetric key *K* is transmitted. Often used in the context of cryptographic protocols, the method of authenticating parties using a request-response scheme and using random numbers will allow the detection of random numbers. In the event that side 1 sends one element of the message to the other side 2, then in subsequent messages this side 2 sent the same element to either side 1 or a function from this element known to both sides (hash, addition with some value), it is considered that the scheme was found request-response. In this case, the elements involved in this are random numbers. In our case of such constructions 2: the first in message 1 side *A* sends a random number *Na* to side *B*, after which in message 2 side *B* returns to side *A* this random number *Na*, the second construct in message 2 sends from side *B* to *A* a random number *Nb*, after in message 3, side *A* returns this random number *Nb* to side *B*. Thus, in our case, the elements *Na, Nb* are random numbers. All other elements of the message that were not typed in the previous cases are merely semantic data. In our case, this is the element *M1*. It can be just a single message to the other side, some file, etc. As a result, at the moment there is:

1. Roles: *A, B*
2. Public keys: *Pka, Pkb*
3. Symmetric keys: *K*
4. Identifiers: *A, B*
5. Random numbers: *Na, Nb*
6. Semantic data: *M1*

To determine the knowledge of the parties requires a number of assumptions. In particular, to assume that the entire list of party identifiers and public keys for asymmetric encryption are available to all legal parties. In this case, these lists of identifiers and keys automatically fall into the initial knowledge of the parties. In addition, as the initial knowledge can be a symmetric key. To detect this, it is necessary to search for the first messages that were encrypted on symmetric keys. If the encryption key was not previously transmitted in previous messages, this indicates that both parties know this key initially, and it is thus recorded in the initial knowledge of these parties. As a result, the knowledge of the parties in our case:

```
A: A, B, Pka, Pkb
B: A, B, Pka, Pkb
```

Most of the verifiers use the Dolev-Yao intruder model. In it, the attacker is able to perform all sorts of operations with the channel and messages, intercept them, duplicate, block the channel, substitute, be an authorized user. However, the intruder can not access data, for example, encrypted on unknown keys or guess the values of random numbers. Thus, the knowledge of the attacker will include all open data that is accessible to all parties, including authorized ones. In our case, this data: *A, B, Pka, Pkb.*

As verification goals, verifiers most often use verification of the parties' authentication and the secrecy of the transmitted data. To verify the secrecy, it is necessary that all transmitted message elements be encrypted with keys unknown to the intruder, therefore, as a check target, each message element must be checked for secrecy. Authentication is most often associated with random numbers, such as in the CAS+ [1] language. In this case, the checks are set as one side authenticates the other by a random number. Therefore, it is enough to write a check for all random numbers of the current protocol. Thus for verification purposes it is indicated:

1. A authenticates B by Na
2. B authenticates A by Nb
3. Secrecy A, B, Na, Nb

Having determined all the above data and designations, it is possible to describe the protocol in any specification language for the use of formal verification, for example, for Avispa [2], Scyther [13], ProVerif [14].

4 CAS+ Language

The CAS+ language is a specification language with which the Avispa automated cryptographic protocols security verifier can work. More precisely Avispa works with HLPSL [16] language, but it has special translator from CAS+ to HLPSL. An example of the protocol description shown in the previous paragraph:

```
1    protocol TEST;
2    identifiers
3    A,B : user;
4    Na,Nb,M1  : number;
5    Pka,Pkb : public_key;
6    K : symmetric_key;
7
8    messages
9    1. A -> B: {Na, A}Pka
10   2. B -> A: {Na, Nb, B}Pkb
11   3. A -> B: {Nb, K}Pka
12   4. B -> A: {M1}K
13
14   knowledge
15   A : A,B,Pka,Pkb;
16   B : A,B,Pka,Pkb;
17
18   session_instances
19    [A:arole,B:brole,Pka:pkakey,Pkb:pkbkey]
20    [A:i,B:brole,Pka:ki,Pkb:pkbkey]
21    [A:arole,B:i,Pka:pkakey,Pkb:ki];
22
23   intruder_knowledge
24    arole,brole,pkakey,pkbkey,ki;
25
26   goal
27    A authenticates B on Na;
28    B authenticates A on Nb;
29    secrecy_of A [A,B];
30    secrecy_of K [A,B];
31    secrecy_of Na [A,B];
32    secrecy_of B [B,A];
33    secrecy_of M1 [B,A];
34    secrecy_of Nb [B,A];
```

Line 1 specifies the name of the protocol, followed by the identifiers section. In line 3, roles are assigned, in line 4 - random numbers and semantic data, in line 5 - public keys, in line 6 - symmetric key. In the area of lines 8–12 messages are set. In the field of lines 14–16, knowledge of roles is set. In the area of lines 18–21, sessions are specified in which a possible session is indicated, in which the attacker is represented by the other party. In lines 23–24, the attacker's knowledge is set. In the area of lines 26–28, verification objectives are set.

5 CAS Translation Algorithm from Alice-Bob View to CAS+ Specification Language

5.1 Set Definitions

At the beginning of processing, ordered sets are created, which will contain: names of roles, public keys, symmetric keys, random numbers, message elements, semantic data, secrecy goals, authentication goals, and the attacker's knowledge. Each line, which is a message, is read and divided into parts by delimiters. Here is an example of a split list of data for the first message and the list of elements is given in Table 1.

```
1. A -> B: AE_Pka(Na, A)
```

Table 1. Spitted list

Index	Value
[0]	"1."
[1]	"A"
[2]	"->"
[3]	"B"
[4]	"AE_Pka"
[5]	"Na"
[6]	"A"

5.2 Message Elements and Secrecy Goals

The contents of the message are all items with an index of 4 or more inclusive. From indexes 1 and 3 from all messages, the names of roles are taken and only unique ones are added to the set of role names. All data after 4 or more indexes that do not contain the names of cryptographic functions with keys are added to the set of message elements, in our case these are "AE_" and "SE_". However, if the key is found in the content of the message, then it will be a message element. Role IDs are not message items and are removed from the result set. The multitude of the secrecy goal is filled with all elements of messages with postscript between which parties the data should be secret.

5.3 Keys

Next comes the filling of sets of public and symmetric keys. This is done by searching for the names of cryptographic functions. For our protocol, only "AE_" and "SE_". As a result, for the first message this part of the message with the index 4. Now it is necessary to extract the key by dividing element 4 by the separator into two parts. In the first part there will be an encryption type - asymmetric, in the second part there will be a public key "Pka".

5.4 Authentication Goals

Next, you need to identify the purpose of authentication. Most often, for authentication, the principle of request-response is used, in which the party sends a random number to the other party, after which it receives in response the same number or some function known to both parties from this number. Thus, there are moments in which role 1 sent an element of a message to another role 2, after which, in some subsequent message, role 2 sent this element or a function of role 1 to it in response. Such moments would indicate that the parties are authenticated on the principle of request-response. Moreover, in addition to the random number itself, the number of authentication purposes is recorded and who sent it to whom for the first time. If the initiator of the interaction is role 1, then information will be written to the set that role 1 authenticates role 2 by the specified random number. In addition, the message elements participating in this exchange will be considered random numbers. All other non-random numbers, as well as keys transmitted in the body, which will be used to encrypt the message, will be considered semantic data. In our case, the element "M1" will be considered semantic data.

5.5 Filling CAS+ Specification

Next begins the construction of the CAS+ specification itself. Roles, numbers, public and symmetric key are filled from the corresponding sets (lines 1–6). Then messages from the Alice-Bob view with modified brackets and the position of the cryptographic functions are inserted into the message transfer section (lines 8–12). For example, in message 1, a translation is made from the form "1. A -> B: AE_Pka (Na, A)" to the form "1. A -> B: {Na, A} Pka".

5.6 Roles Knowledge

Further the section of knowledge of the parties is filled (lines 14–16). As knowledge of the parties, by default, all identifiers and public keys are entered, as well as symmetric keys that were not elements of the message when they were transmitted.

5.7 Sessions

Next, the sessions section is filled (lines 18–21). In the first session, parameters are set containing the knowledge necessary for the interaction of the parties. In our case, these are roles and public keys. Moreover, for each parameter, the received value is indicated through ":". The translator generates a name based on the name of the parameter and applying a conversion of all characters to lower case (see line 19). To be able to detect attacks, you must specify the sessions in which an attacker can replace the legal side. For this, it is necessary to single out the cases when an attacker replaces one or another side (lines 20–21).

5.8 Intruder Knowledge

Next, you need to highlight the intruder's knowledge section (lines 23–24). Here are well-known values, such as public keys and identifiers of the parties, as well as the public key of the attacker to simulate the situation when he acts as a legal party.

5.9 Verification Goals

The last section are verification targets (lines 26–34). They are filled from previously formed sets of authentication and secrecy purposes, changing the format to the required CAS+ language. The result is a complete description of the protocol.

The translator was implemented in the C # programming language as a console application, which has two parameters: the source file with the description of Alice-Bob and the file for saving the description in the language of the CAS+ specifications.

6 Automated Verification with Avispa Tool

Avispa Verifier works with the HLPSL specification language. This language is more complex than CAS+, but Avispa allows you to translate code from CAS+ to HLPSL. Thus, usually the protocol is first described in the CAS+ language. Figure 2 shows a verification of the test protocol, which was described on Alice-Bob specification language and with help of our translator was analyzed by Avispa.

The security verification procedure for the protocol originally described in the Alice-Bob form:

- Protocol description in Alice-Bob format
- Translation of Alice-Bob into the CAS+ specification language using the algorithm presented in this paper, implemented in the C# programming language.
- Translation of CAS+ to HLPSL using the built in Avispa translator.
- Protocol security verification.

You can see that an authentication attack was found. To be able to automate the process, you can use a script to call the programs sequentially, which will look like this:

1. Calling a translator from Alice-Bob to CAS+
2. Calling a translator from CAS+ to HLPSL
3. Calling a translator from HLPSL to IF
4. Call the backend of the OFMC [17] module for IF [18] (intermediate format) to verify security.

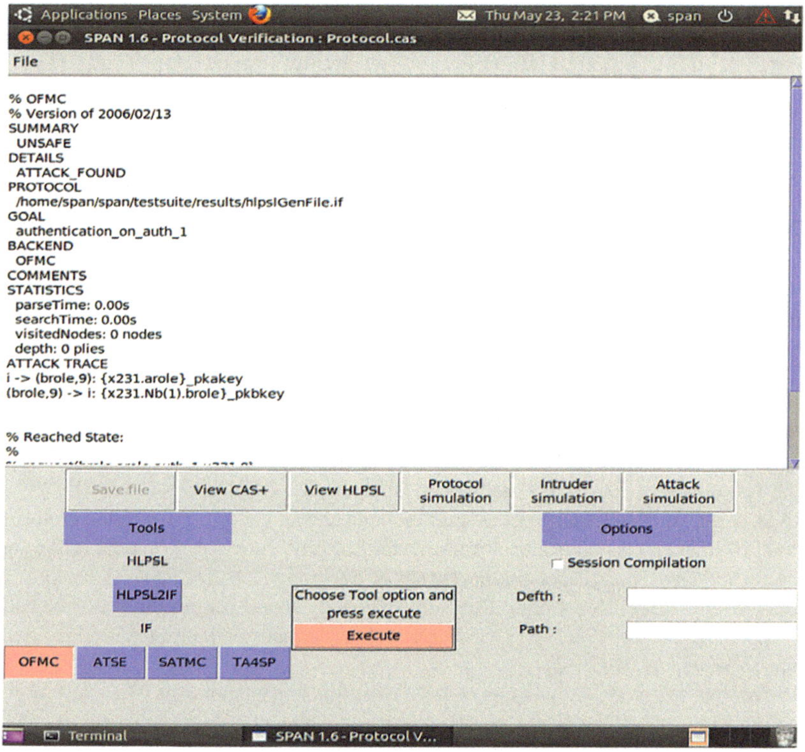

Fig. 2. Results of test protocol verification

7 Conclusion

In this paper, a general approach was described for translating the presentation of a protocol from the Alice-Bob form to any specification language. A description of the CAS+ specification language for use in the Avispa verifier is provided. The algorithm for translating the Alice-Bob view to the CAS+ specification is given. An algorithm for automated verification of the protocol described in the Alice-Bob format is given. An example of the operation of the translation algorithm implemented in the C # programming language is shown. The implemented translator correctly translates the protocol description from the Alice-Bob format to the CAS+ format. Demonstrated a sequential verification procedure using the Avispa test protocol using the translator developed by the authors. The presented translator can be used to facilitate the verification of protocol security using Avispa.

The work was supported by the Ministry of Education and Science of the Russian Federation grant № 2.6264.2017/8.9.

References

1. Saillard, R., Genet, T.: CAS+, March 21 2011–2018
2. Viganò, L.: Automated security protocol analysis with the AVISPA tool. Electron. Notes Theor. Comput. Sci. **155**, 61–86 (2006)
3. Chaki, S., Datta, A.: ASPIER: an automated framework for verifying security protocol implementations. In: 22nd IEEE Computer Security Foundations Symposium 2009, CSF 2009, pp. 172–185. IEEE (2009)
4. Goubault-Larrecq, J., Parrennes, F.: Cryptographic protocol analysis on real C code. In: International Workshop on Verification, Model Checking, and Abstract Interpretation, pp. 363–379. Springer, Heidelberg (2005)
5. Goubault-Larrecq, J., Parrennes, F.: Cryptographic protocol analysis on real C code. Technical report, Laboratoire Spécification et Vérification, Report LSV-09-18 (2009)
6. Jürjens, J.: Using interface specifications for verifying crypto-protocol implementations. In: Workshop on Foundations of Interface Technologies (FIT) (2008)
7. Jürjens, J.: Automated security verification for crypto protocol implementations: verifying the Jessie project. Electron. Notes Theor. Comput. Sci. **250**(1), 123–136 (2009)
8. O'Shea, N.: Using Elyjah to analyse Java implementations of cryptographic protocols. In: Joint Workshop on Foundations of Computer Security, Automated Reasoning for Security Protocol Analysis and Issues in the Theory of Security (FCS-ARSPA-WITS 2008) (2008)
9. Backes, M., Maffei, M., Unruh, D.: Computationally sound verification of source code. In: Proceedings of the 17th ACM Conference on Computer and Communications Security, pp. 387–398. ACM (2010)
10. Bhargavan, K., et al.: Cryptographically verified implementations for TLS. In: Proceedings of the 15th ACM Conference on Computer and Communications Security, pp. 459–468. ACM (2008)
11. Bhargavan, K., Fournet, C., Gordon, A.D.: Verified reference implementations of WS-security protocols. In: International Workshop on Web Services and Formal Methods, pp. 88–106. Springer, Heidelberg (2006)
12. Babenko, L.K., Pisarev, I.A.: Algorithm for analysis of initial code C# for extracting the structure of cryptographic protocols. Cybersecurity Issues (4), 28 (2018)
13. Cremers, C.J.F.: The scyther tool: verification, falsification, and analysis of security protocols. In: International Conference on Computer Aided Verification, pp. 414–418. Springer, Heidelberg (2008)
14. Küsters, R., Truderung, T.: Using ProVerif to analyze protocols with Diffie-Hellman exponentiation. In: 22nd IEEE Computer Security Foundations Symposium 2009, CSF 2009, pp. 157–171. IEEE (2009)
15. Almousa, O., Mödersheim, S., Viganò, L.: Alice and Bob: reconciling formal models and implementation. In: Bodei, C., Ferrari, G., Priami, C. (eds.) Programming Languages with Applications to Biology and Security, pp. 66–85. Springer, Cham (2015)
16. The AVISPA team, The High Level Protocol Specification Language (2006). http://www.avispa-project.org/
17. Basin, D., Mödersheim, S., Viganò, L.: OFMC: a symbolic model-checker for security protocols. Int. J. Inf. Secur. **4**, 181–208 (2005)
18. Armando, A., Basin, D., Boichut, Y., Chevalier, Y., Compagna, L., Cuéllar, J., Hankes Drielsma, P., Heám, P.C., Kouchnarenko, O., Mantovani, J., Mödersheim, S.: The AVISPA tool for the automated validation of internet security protocols and applications. In: International Conference on Computer Aided Verification, pp. 281–285. Springer, Heidelberg, July 2005

Approach to Conceptual Modeling National Scientific and Technological Potential

Alexey B. Petrovsky[1,2,3,4(✉)] and Gennadiy I. Shepelev[1]

[1] Federal Research Center "Informatics and Control", Russian Academy of Sciences, Prospect 60 Letiya Octyabrya, 9, Moscow 117312, Russia
pab@isa.ru
[2] Belgorod State National Research University, Belgorod, Russia
[3] V.G. Shukhov Belgorod State Technological University, Belgorod, Russia
[4] Volgograd State Technical University, Volgograd, Russia

Abstract. The paper presents a conceptual multilevel model of national scientific and technological potential in order to form and select options for the strategy of innovative development of the country. The model is based on the methodology of group verbal decision analysis and multidimensional assessment of innovations. The elements of the information-logical model and the intensity of connections between elements of different levels are evaluated by experts on qualitative criteria with verbal scales.

Keywords: Scientific and technological potential · Information-logical model · Group verbal decision analysis · Qualitative criteria

1 Introduction

The elaboration and justification of recommendations on the creation of promising high technologies that ensure innovative development of the country's economy are closely related to the forecast of the development of scientific and technological potential, assessment of the basic and applied significance of research results. Building methodological tools for multidimensional analysis of the state and trends of development of the national innovation system and the scientific and technological potential remains an important and still unsolved scientific problem [7, 8].

The paper describes a conceptual multilevel information-logical model of the national scientific and technological potential, which is based on the methodology of verbal decision analysis and a multidimensional assessment of innovation. The model allows creating and analyzing options of strategy for innovative development of the country. Qualitative criteria with verbal scales for expert evaluation of the model elements and connections between its elements are proposed.

S. Kovalev et al. (Eds.): IITI 2019, AISC 1156, pp. 319–329, 2020.
https://doi.org/10.1007/978-3-030-50097-9_33

2 Conceptual Model of Scientific and Technological Potential

The important role of scientific advances and technologies in the modern and future world dictates the need for a reasonable choice of priorities. Priorities should ensure the maximum contribution of science and technology to the achievement of national goals. This implies a balanced strategy, optimal allocation of resources, and concentration of main efforts in actual research areas [1, 5, 12]. Priorities should be harmonized with the competitive advantages of the country and global trends in the socio-economic development. The complexity of solving this problem is exacerbated by the constant increasing the number of promising and breakthrough scientific fields and the potential points of growth generated by them, which leads to an expansion of the list of possible options for scientific and technological development.

Among existing approaches to assessing the level of innovation development, the most developed are the evaluation of technological competitiveness, proposed in [13], and the integral estimation of the national innovation system used by Eurostat [15], in which the innovation system is considered as a black box. The system output is the competitiveness of products and services, characterized by indicators of the technological state of production and the country export capacity for high-tech products and services. At the level of macroeconomic analysis, the system inputs are specially designed synthetic characteristics and the corresponding indicators. Input indicators are based on quantitative factors and expert estimates, which are converted into averaged scores, that is, the so-called quantitative approach is used. At the same time, quantitative models are too simplified for the analysis of the state and trends in the development of the country's scientific and technological potential [4, 16].

Multidimensional assessment of the means of achieving the goals of national innovation policy, directions and results of scientific research, high-tech technologies and areas of their practical application are related to the so-called poorly formalized and ill-structured decision problems, where there are both quantitative and, mainly, qualitative indicators with verbal grading scales. Methods of verbal decision analysis are most suitable for solving such problems [6, 10].

The proposed approach to structuring the subject area, developing a conceptual model and criteria for evaluating the national scientific and technological potential is based on a system analysis of the modern innovation system of Russia. The method of Multilevel Information-Logical Structures (MILS) is focused on expert evaluation and analysis of options for strategic decisions [6, 8]. The main ideas of the MILS method are as following.

1. Construct a conceptual model of the subject area as a multilevel information-logical structure.
2. Form lists of elements of each level of the hierarchy.
3. Develop indicators and criteria for evaluating elements at all levels of the hierarchy and the intensity of connections between elements of different levels.
4. Collect factual data and expert assessments of each element at all levels of the hierarchy.
5. Collect expert assessments of elements' connections at different levels of the hierarchy.

6. Build decision rules of selection at each level of the hierarchy.
7. Select the best or acceptable solutions, taking into account the decision rules and requirements for the intensity of connections between elements of different levels.

The conceptual multilevel information-logical model of the scientific and technological potential, designed to form and select alternative options for strategic decisions on the innovative development of the country, includes the following blocks (Fig. 1):

- goals of scientific and technological development;
- innovations in sectors of the economy;
- critical technologies;
- scientific and technical directions;
- resources ensuring the achievement of innovation development goals;
- mechanisms contributing to the achievement of innovation policy goals.

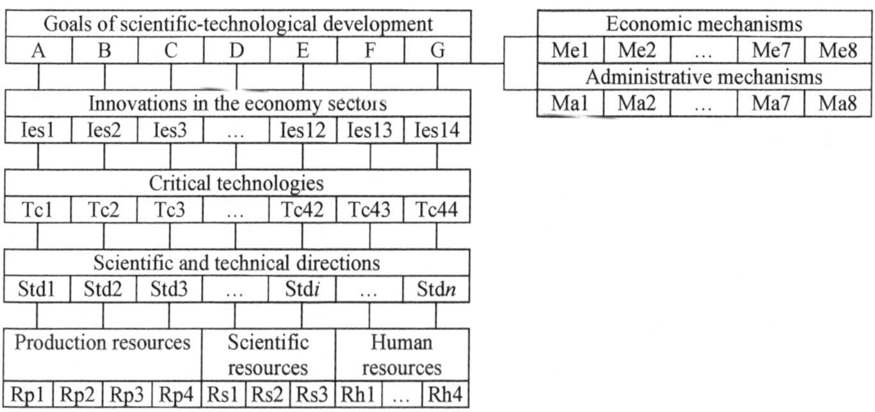

Fig. 1. Information-logical model of scientific and technological potential.

The purpose of the scientific and technological development of the Russian Federation is to ensure the independence and competitiveness of the country through the creation of an effective system for building up and the most complete utilization of the national intellectual potential [9]. The goals of the national innovation policy are the approved priority directions of scientific and technological development:

A. Transition to advanced digital, intellectual production technologies, robotic systems, new materials and methods of design, creation of systems for processing large volumes of data, machine learning, and artificial intelligence.
B. Transition to environmentally friendly and resource-saving energy, increasing the efficiency of extraction and deep processing of hydrocarbon raw materials, the formation of new sources, methods of energy transportation and storage.
C. Transition to personalized medicine, high-tech health care and health-saving technologies, including through the rational use of drugs (primarily, antibacterial).

D. Transition to a highly productive and environmentally friendly agricultural and aquatic economy, development and implementation of systems for the rational use of chemical and biological protection of agricultural plants and animals, storage and efficient processing of agricultural products, the creation of safe and high-quality, including functional, food products.
E. Counteraction to technological, biogenic, socio-cultural threats, terrorism and ideological extremism, as well as cyber threats and other sources of danger to society, economy and the state.
F. Connectivity of the territory of the Russian Federation through the creation of intelligent transport and telecommunication systems, as well as the occupation and retention of leadership positions in the creation of international transport and logistics systems, the development and use of space and airspace, the oceans, the Arctic and Antarctic.
G. Possibility of an effective response of Russian society to major challenges, taking into account the interaction of man and nature, man and technology, social institutions at the present stage of global development, including applying methods of the humanitarian and social sciences.

Innovations represent possible ways to achieve the goals of innovation development and are distributed across sectors of the economy. The list of innovations is formed by experts.

An innovation is defined by the Organization for Economic Cooperation and Development (OECD) as the application of new significantly improved products (goods and services), processes, new market methods or new organizational methods in business practice, in organizing workplaces or in establishing external relations [2]. According to Russian GOST, an innovation is "the final result of innovation activity that has been realized in the form of a new or improved product sold on the market, or a new or improved technological process used in practical activities" [3]. It is customary to distinguish the following types of innovation:

by a focus of action - basis innovations that implement major discoveries and inventions; improving innovations that implement small and medium-sized inventions; rationalizing innovations aimed at partial improvement of outdated generations of equipment and technology;
by a type of parameters - product innovations; process (technological) innovations; organizational and managerial (non-technological) innovations;
by a scale of distribution - the whole world; a country; an industry; a company.

The socio-economic and production-technological platform, where high-tech innovations are practically used, is the interbranch and industry-specific complexes: 1. Mining industry; 2. Energy; 3. Metallurgy; 4. Mechanical engineering and instrument making; 5. Defense industry; 6. Chemistry, forestry and biotechnology; 7.Agro-industrial complex; 8. Light industry; 9. Construction, transport, communications, information and communication technologies; 10. Environmental protection; 11. Health and welfare; 12. Education, science, culture, sports; 13. Trade and services; 14. Housing and household.

In order to become a successful innovation, a good idea must go through several stages of the "life cycle": the idea emergence - the possibility of using a scientific achievement for commercial purposes; the idea evolution - the development of a technology for the production of a new product that can be commercially implemented; the sample demonstration – the creation and presentation of a prototype to potential investors and customers; the product promotion - the creation of a demand in the market for new products; the consolidation in the market - the acquisition of confidence that a new product or technology will have a long and successful future in the existing market.

Critical technologies are the technologies that are important for the socio-economic sphere, national defense and state security. The list of critical technologies is approved by decree of the President of the Russian Federation and is periodically reviewed. Currently, the list includes 44 critical technologies.

Scientific and technical directions create the foundation for the development of critical technologies and include research in the field of understanding the processes occurring in society and nature, the development of nature-like technologies, human-machine systems, climate and ecosystem control; research related to the ethical aspects of technological development, changes in social, political and economic relations; basic research caused by the internal logic of the development of science, ensuring the country's readiness for great challenges that have not yet manifested and not received wide public recognition, the possibility of timely assessment of risks arising from scientific and technological development. The list of directions is formed by experts.

Resources ensuring the achievement of innovation development goals are divided into production, scientific and human resources.

Production resources include:

Rp1. Production facilities for the production of high technology products.
Rp2. Production capacity for the production of components and component base.
Rp3. Modern technological equipment, accessories, devices, tools.
Rp4. Functioning market of services for technological support of manufacturers.

Scientific resources include:

Rs1. Results of revolutionary scientific research that can dramatically affect the development of science and technology.
Rs2. Results of promising basic and applied research that can be quickly used in high-tech areas.
Rs3. Scientific and technical results of the possible borrowing of new knowledge and the reproduction of advanced promising technologies.

Human resources have components:

Rh1. Scientists and highly qualified specialists.
Rh2. Engineering and technical workers.
Rh3. Workers, employees and support workers.
Rh4. Administrative and management personnel.

Mechanisms contributing to the achievement of innovation policy goals are located at the level of objectives and are divided into economic and administrative.

Economic mechanisms are aimed at creating and mastering innovations, stimulating the production of high-tech products. These mechanisms include:

Me1. Demand for high-tech and high-tech products.
Me2. Demand for promising scientific and technical results.
Me3. Innovative activity of enterprises in the real sector of economy.
Me4. Innovative activity of small enterprises.
Me5. Function of the capital market.
Me6. Domestic investment in high-tech manufacturing.
Me7. External investments in high-tech production.
Me8. Transfer of knowledge and high technology to the domestic and global markets.

Administrative mechanisms are aimed at creating conditions that ensure the implementation of innovation, the economy's susceptibility to innovation. These mechanisms include:

Ma1. National strategy of innovation and scientific and technological development.
Ma2. Legislative and regulatory framework for the regulation of innovation.
Ma3. Public-private partnership in the implementation of innovations.
Ma4. Direct government support for small innovative enterprises.
Ma5. Support for basic and applied research, experimental development by large public and private corporations.
Ma6. Sectoral and regional venture funds, innovation financing agencies with the state participation.
Ma7. Support science cities, technopolises, science and technology parks.
Ma8. Information support of innovation activity.

3 Assessment and Analysis of Innovation Development Strategies

Elements of the information-logical model of the scientific and technological potential and connections between the elements are evaluated by several independent experts on many criteria and indicators, which have scales with detailed verbal formulations of quality gradations.

The innovation is characterized by the following indicators:

I1. Focus of innovation (basis; improving; rationalizing).
I2. Type of innovation (product; process; organizational and managerial).
I3. Scale of innovation (global; national; sectoral; intra-company).
I4. Significance of innovation for the development of the Russian economy (high; medium; low; difficult to estimate).
I5. Competitiveness of innovation (high; medium; low; difficult to estimate).
I6. Stage of development of innovation (idea emergence; idea evolution; sample demonstration; product promotion; consolidation in the market).
I7. Feasibility of innovation (less than 3 years; 3–7 years; more than 7 years; difficult to estimate).

The technology assessment criteria are:

T1. Focus of technology (basis innovation; improving innovation; rationalizing innovation).
T2. Importance of technology for the innovation creation (high; medium; low; difficult to estimate).
T3. Stage of technology development (fully developed; prototype developed; technical documentation developed; initial stage of development).
T4. Feasibility of the technology (less than 3 years; 3–7 years; more than 7 years; difficult to estimate).

The scientific and technical direction is estimated by the following criteria:

D1. Impact of the results obtained in the direction on the creation of critical technology (strong; moderate; weak; difficult to estimate).
D2. Change of the direction impact on the creation of critical technology in the future (will increase; will not change; will decrease; difficult to estimate).

The production, scientific or human resource is estimated by the following criteria:

R1. Resource accordance with the needs of the innovative development of the economy (fully; partially; not relevant)
R2. Resource availability (fully available; partially available; initial stage of formation; absent).
R3. Resource change in perspective (will increase; will not change; will decrease; difficult to estimate).
R4. Resource impact on the innovative development of the economy (strong; moderate; weak; difficult to estimate).
R5. Change of the resource impact on the innovative development of the economy in the future (will increase; will not change; will decrease; difficult to estimate).

The economic or administrative mechanism is evaluated by the following criteria:

M1. Mechanism accordance with the goals f the innovation policy (fully; partially; not relevant).
M2. Mechanism availability (fully available; partially available; initial stage of formation; absent).
M3. Mechanism change in perspective (will increase; will not change; will decrease; difficult to estimate).
M4. Mechanism impact on the achievement of innovation policy goals (strong; moderate; weak; difficult to estimate).
M5. Change of the mechanism impact on the achievement of innovation policy goals in the future (will increase; will not change; will decrease; difficult to estimate).

Mutual connections between elements of the information-logical model at different levels of the hierarchy are evaluated by the following criteria:

C1. Intensity of the elements' connection (high; moderate; low; absent).
C2. Change in the intensity of the elements' connection during the short term less than 3 years (will increase; will not change; will decrease; difficult to estimate).
C3. Change in the intensity of the elements' connection during the medium term from 3 to 7 years (will increase; will not change; will decrease; difficult to estimate).
C4. Change in the intensity of the elements' connection during the long term over 7 years (will increase; will not change; will decrease; difficult to estimate).

The peculiarity of group expertise procedures is the presence of many judgments that do not coincide with each other. The inconsistency of individual opinions is due to the ambiguity of the understanding of the problem being solved by different people, the difference in assessments of the same objects made by different persons, the specificity of the knowledge of the experts themselves, and many other circumstances. The combination of such assessments may have a complex structure in the attribute space, which is rather difficult to analyze in this space. It is not easy to introduce a metric for the objects' comparison. These difficulties can be overcome if we use another way of representing multi-attribute objects based on the formalism of the multiset theory [10, 11]. Multisets allow us to take into account simultaneously various combinations of the values of qualitative attributes, as well as their polysemy.

A multidimensional analysis of the research results' impact on the creation of promising high technologies, the choice of the best or acceptable strategy for innovative development of the Russian economy for a given time horizon, taking into account changing resource constraints, suggests that there are several acceptable alternatives and decision rules, which allow to compare the quality of alternatives. Variants of innovative development of the economy are constructed as a combination of multi-criteria expert assessments of the model elements. Different decision rules linking the elements of the model are formed by the decision maker (DM) or the head of the planning body. The decision rule is an algorithm of moving from the directive goals of the planning body to the sets of tools and resources necessary to achieve the goals. The decision rule is constructed by sequentially selecting, at each level of the structure, of the subsets of the model elements ensuring the implementation of the elements of the upper level.

The selection of elements and their inclusion in the "supporting subset" is based on the preferences of the planning body or decision maker. Depending on the specifics of the problem being solved, various methods of forming a "supporting subset" can be used, for example, by setting certain estimates by criteria for innovations. At the same time, each policy option will have its own set of tools and resources necessary to implement the policy, and its own set of mechanisms that contribute to the achievement of goals. As a result, several qualitatively different development scenarios can be obtained. For the final comparison of options, it is necessary to use other methods, in particular, to analyze the coincidences and differences of "providing subsets" for selected innovations, to assess the degree of different mechanisms' impact on the goal achievement. According to the analysis and comparison of scenarios, lists of problems that need to be solved are formed for promising directions of research, educational programs, legislative activities, etc., ensuring the innovative development of the economy.

The choice of alternative options for innovative development can be considered as a two-way process in which the transition takes place either from the current state to a possible future one ("direct process"), or from the required future state to the current one ("reverse process"). The "direct process" - the so-called bottom-up planning – starts from the capabilities currently available and goes into the state determined by the "natural" course of events in the process of implementation with traditional resources. The "reverse process" - the so-called top-down planning – starts from the needs (that is, the desired future state) and goes through element-by-element decomposition into a list of measures and resources necessary today to achieve the desired state, if it is possible.

Sometimes, direct and reverse planning processes are also referred as research and normative approaches, respectively. The normative approach is used when it is necessary to structure the problem as a whole, take into account the requirements of the external environment of the system, the goals of the planning body. That is why various methods of program planning, focused on solving "breakthrough" revolutionizing problems, it is based on this approach. Techniques based on the research approach do not require usually an exhaustive structuring of the problem. In this case, the choice of a development strategy can be reduced to the selection of the most significant research results within the framework of any policy.

The multilevel information-logical model of the scientific and technological potential allows us to "pass" through the hierarchical structure in different directions: top-down (from the given goals to the most appropriate set of means to achieve them); bottom-up (from disposable resources to possible goals); from the middle (from any level up to the goals and down to more detailed means). The model provides opportunities to identify: basis innovations that influence at the formation of new economy sectors and possible markets; improving innovations affecting the development of many economy sectors and existing markets; unique breakthrough and promising technologies with the potential for rapid distribution and application; replacing technologies and removable high technologies. Using such tools allows us to explore innovative processes within a single system of interrelated goals and means to achieve them in the production, implementation and dissemination of scientific knowledge, clarify the role of elements of the national innovation system in the transfer of knowledge and technology, evaluate the effectiveness of the impact of scientific and innovation activities on the economic development of the country.

4 Conclusions

We proposed a scientific and methodological toolkit for building multilevel information-logical structures that is based on the methodology of verbal decision analysis and aimed at the analysis of various strategies to achieve the goals of national innovation policy. The best or acceptable goals of innovation development and the most appropriate means to achieve them, provided with appropriate resources, are selected on the basis of multi-criteria expert assessments of the model elements and intensity of connections between elements at different levels. The numerical coefficients of the criteria importance and the numerical factors of the options value for strategic decisions are not calculated, as well as the qualitative estimates are not converted into any numerical indicators. The final results are described by verbal attributes, convenient to understand in natural language familiar to humans.

Multilevel information-logic modeling can be used in the development of forecasts, programs, plans for solving three types of problems: the determination of a collection of the means necessary to achieve the given goals (a choice of the subset from the existing set of means ensuring the achievement of the specified goals); the analysis of resource allocation options (a definition of a set of goals that can be achieved with available resources); the analysis of the possibility to achieve the given goals under the specified resource constraints. The proposed approach allows us to find the best collection of goals and means, that is, in a multi-criteria sense, the best option for a development strategy with available resources.

Acknowledgments. This work was supported by the Russian Foundation for Basic Research (projects 16-29-12864, 17-07-00512, 17-29-07021, 18-07-00132, 18-07-00280, 19-29-01047).

References

1. Boychenko, V.S., Petrovsky, A.B., Sternin, M.Yu., Shepelev, G.I.: The choice of priorities for scientific and technological development: the experience of the Soviet Union. Proc. Inst. Syst. Anal. RAS (Trudy Instituta sistemnogo analiza RAN) **65**(3), 3–12 (2015). (in Russian)
2. Dynamising National Innovation Systems. OECD Publishing, Paris (2002)
3. GOST R 56261-2014. Innovative management. Innovation. The main provisions. Introduction date 01.01.2016
4. Grants in science: accumulated potential and development prospects. In: Petrovsky, A.B. (ed.) Poly Print Service, Moscow (2014). (in Russian)
5. Koshkareva, O.A., Mindeli, L.E., Ostapyuk, S.F.: System aspects of the procedure for selecting and updating the priorities of the development of science. Innovations (Innovatsii) **6**, 20–31 (2015). (in Russian)
6. Larichev, O.I.: Theory and Methods of Decision Making. Logos, Moscow (2002). (in Russian)
7. Larichev, O.I., Minin, V.A., Petrovsky, A.B., Shepelev, G.I.: Russian fundamental science in the third millennium. Bull. Russ. Acad. Sci. (Vestnik Rossiyskoy Akademii Nauk) **71**(1), 13–18 (2001). (in Russian)
8. Minin, V.A.: On modeling logical connections between elements of the astronomical research process. Astronomical council of the USSR Academy of Sciences. Sci. Inf. (Astronomicheskiy sovet Akademii Nauk SSSR. Nauchnyye informatsii) **36**, 27–41 (1975). (in Russian)
9. The official website of the Council under the President of the Russian Federation on science and education. http://www.snto.ru/
10. Petrovsky, A.B.: Decision Making Theory. Publishing Center "Academiya", Moscow (2009). (in Russian)
11. Petrovsky, A.B.: Indicators of similarities and differences of multi-attribute objects in metric spaces of sets and multisets. Sci. Tech. Inf. Process. **45**(5), 331–345 (2018)

12. Petrovsky, A.B., Boychenko, V.S., Sternin, M.Yu., Shepelev, G.I.: The choice of priorities of scientific and technological development: the experience of foreign countries. Proc. Inst. Syst. Anal. RAS (Trudy Instituta sistemnogo analiza RAN) **65**(3), 13–26 (2015). (in Russian)
13. Porter, M., Snowdon, B., Stonehouse, G.: Competitiveness in a globalized world: Michael Porter on the microeconomic foundations of the competitiveness of nations, regions, and firms. J. Int. Bus. Stud. **37**, 163–175 (2006)
14. Record, R., Clarke, G., Lowden, R.: Investment Climate Assessment 2014. World Bank Publications, Washington (2015)
15. Sustainable Development in the European Union. Eurostat Publications, Luxembourg (2016)
16. Wilson, K.: An investigation of dependence in expert judgement studies with multiple experts. Int. J. Forecast. **33**, 325–336 (2016)

The Intelligent Technology of Integrated Expert Systems Construction: Specifics of the Ontological Approach Usage

Galina V. Rybina(✉), Elena S. Fontalina, and Ilya A. Sorokin

National Research Nuclear University MEPhI, Moscow, Russia
galina@ailab.mephi.ru

Abstract. This paper studies the results of developing the basic components of the AT-TECHNOLOGY workplace intelligent software environment, predetermined for automation and rendering of intelligent processes for building integrated expert systems (IES), based on a problem-oriented methodology. The article itself presents a review of the scientific, methodological and technological experience in the implementation and use of IES training. Also, the creation of a single ontological space of knowledge and skills for the automated creation of competence-based models of experts in the field of methods and technologies of artificial intelligence in the field of training in the object of study "Software Engineering", obtained at the Department of Cybernetics NRNU MEPhI.

Keywords: Intelligent software environment · AT-TECHNOLOGY workbench · Tutoring integrated expert systems · Problem-oriented methodology · Intellectual training

1 Introduction

The application of the new results conceived in the field of science, and more specifically in the field of artificial intelligence (AI), acts as an ethic of preparing study time on AI technology at NRNU MEPhI. Therefore, the theory of constructing IES of various architectural types is used. It is based on a problem-oriented methodology [1] and the intellectual software environment of the AT-TECHNOLOGY tool [2]. It provides absolute automated support for processes, development and maintenance of a spacious class of static and dynamic IES, which includes training, furthermore web-oriented (web-IES). Problems associated with the use of the AT-TECHNOLOGY problem-oriented methodology and tools for *intellectual learning* [2] by stablishing intelligent learning systems (ITS) based on the IES learning architecture and IES web, which were widely broadcasted in various passed years, for example [3–5], and in other cases. Nowadays, the effective practical application of this approach to the construction of ITS is due to two factors.

1. Effective production of the conceptual foundations of a problem-oriented methodology, which allows you to implement a fairly powerful functional required by modern ITS (building developed student models, adaptive learning models, problem area models, explanation models, teacher models, applied ontology models for

S. Kovalev et al. (Eds.): IITI 2019, AISC 1156, pp. 330–340, 2020.
https://doi.org/10.1007/978-3-030-50097-9_34

courses/disciplines/specialties, etc.) based on the ascendable architecture of the IES [1].

2. The use of intelligent software technologies (based on AT-TECHNOLOGY tools) for automated support of IES-based processes at all stages of the life cycle, equipping archiving of invaluable expert and methodical experience of subject teachers, lowering the intellectual load on knowledge engineers, reducing the time for developing IES and web- tutoring IES [3, 4]. Since 2008, tutoring IES and web-IES, developed in the laboratory "Intelligent Systems and Technologies" of the Department of Cybernetics of NRNU MEPhI, are actively used for automated support of core courses/disciplines in the area of applied mathematics and Information and software engineering. For all courses and disciplines adopting the basic tools of the AT-TECHNOLOGY workspace, applied ontologies appear and dynamically progress, which it together creates a generalized ontology "Intelligent Systems and Technologies". It allows users to create a single ontological space of knowledge and skills. Through integration with ontologies of basic courses/disciplines on programming technology and for the implementation of a whole set of functional tasks unique to intelligent technology [2, 3].

These training opportunities of IES and web IES absolutely correspond to the functional and technological views of modern foreign ITS, in particular [6, 7], and adaptive training systems [8], and create the prerequisites for further research on the implementation of promising ways of approaches in the form of intellectual observation and intellectual collective learning, as well as for the semantic integration of individual tutoring IES with their parallel use in the learning process. In modern conditions, the practical implementation and operation of ITS of any architectural complexity is impossible without instrumental software support for the design and maintenance of ITS at all stages of the life cycle. However, at present, there are no standard and generally accepted technologies for the development of intelligent systems, including ITS. For these purposes, general-purpose tools and platforms are used or specialized tools are created.

A representation of this is, the concept and general architecture of the IACPaaS Internet complex [9] which is aimed at supporting universal technological principles for the development and use of applied and instrumental intelligent systems and its management. The IACPaaS cloud platform [10] has been designed to support the creation, management and remote use of applied and instrumental multi-agent cloud services and their components for a variety of subject areas, in particular, on its basis, an intelligent learning environment for the diagnosis of acute, chronic diseases was implemented [11].

On the other hand, several foreign studies demonstrate an inclination to create problem-oriented tools and technologies for the development of intelligent systems of different classes, for example [12, 13]. Domestic works [14, 15], and others present frequent exciting projects and approaches.

The theoretical support ground of the new approach, which is used to provide instrumental support for the problem-oriented methodology of building the IES, is the concept of a "model of the intellectual environment" [1, 2]. As of now, it highlights the introduction, experimental research and active use, for example, to support the educational process at the Department of Cybernetics of the National Research Nuclear University MEPhI and other universities, the intellectual development of software for the automated construction of IES, the combination of knowledge engineering approaches, ontological engineering, intellectual planning and traditional programming [2, 4] and others.

2 Features of Prototyping Processes of Applied IES

One important feature of the problem-oriented methodology and intelligent software environment at the AT-TECHNOLOGY workplace is the provision of intelligent, complex and labor-intensive prototyping processes in the applied IES at all degrees of the life cycle, from requirements analysis to the creation of the IES prototypes series. In order to reduce the intellectual load on knowledge engineers, to lower possible erroneous actions and transient risks when creating an IES prototype, according to [2] it is aimed to use a *technological knowledge base* (KB) containing a significant number *of standard design procedures* (SDP) and *reusable components* (RUCs) that reflect the knowledge engineers' experience in developing IES applications.

Consequently, a secondary presentation of the problem of intelligent planning of IES prototyping processes is taken in account in the context of the IES prototyping process model in the following form:

$$M_{proto} = <T, S, Pr, Val, A_{IES}, PlanTask_{IES}> ,$$

where T is the set of problem domains for which applied IES are created; S is the set of prototyping strategies; Pr is the set of created prototypes of IES based on a problem-oriented methodology; Val is the function of expert validation of the prototype of the IES, determining the need and/or the possibility of creating subsequent IES prototypes for a particular problem domain; A_{IES} is the set of all possible actions of knowledge engineers in the prototyping process; $PlanTask_{IES}$ is the function of planning knowledge engineer's actions to achieve the current prototype of IES for a particular problem domain.

In order to provide effective implementation of the *PlanTaskIES* component of the *Mproto* model, modern methods of intelligent planning of the state space are used to study the mechanisms of some actions of knowledge engineers in building architectural models (*MIES*) of various IES [1, 2] at the commencing stages of the life cycle (requirements analysis, general and detailed design) were analyzed. The experiments displayed that the best results are achieved with the formation search space by modeling the knowledge engineer's actions when constructing fragments of the MIES model using the appropriate SDPs (graph theory is used to formally describe this process by reducing the problem). On the problem of covering the MIES model, presented in the form of a labeled graph with SDP fragments).

Therefore, the concept of the *MIES* model plays a fundamental role in constructing plans for prototyping IES. In [1, 2], the methods for creating *MIES*, which are based on the mapping of interactions of real systems using structural analysis, are extensive due to the inclusion of a certain element "informal operation" (operation NF), which draws attention to the need to attract experts and/or other sources of knowledge. Hence, the *MIES* model of the IES prototype is built in the form of a hierarchy of advanced data flow diagrams (EDFD) [2], which provides a detailed description of all *Mproto* components with an indication of some of the main components of the intelligent software environment model.

Features of the implementation of important components of the intellectual software environment of the AT-TECHNOLOGY workstation and IES development technologies based on it [1] are considered in detail in number works. From here it follows a brief description of the software tools of the intelligent software environment: kernels, including the smart scheduler, the user interface subsystem, and the extension library for interacting with operational RUCs. The kernel implements all the basic functionalities of the automatic support for IES prototype development, project file management, extension management, and other functions.

Technological KB consists of an extension library that stores operational skills in the form of plug-ins that implements the corresponding operational RUC and declarative part. The user interface subsystem has a convenient graphical interface on the basis of which the RUC interacts with the knowledge engineer using on-screen forms. The intelligent scheduler completes functions that are related to the planning of IES prototyping processes: using the EDFD hierarchy preprocessor, the EDFD hierarchy is processed in advance by converting it into one established diagram with maximum detail; The task of covering detailed EDFD with affordable SDP is done using a global plan generator. Based on the design bureau and the constructed generalized EDFD, coverage construction is performed; a detailed plan generator based on a given EDFD and KB coverage achieves detailing of each coating element (thus forming a preliminary integrated plan); Based on the analysis of the existing RUC, a detailed plan is developed in the plan interpretation component, where absolutely every task is associated with a specific RUC, and then with the guidance of the plan building component the required representation of the plan is generated.

To date, the technology for implementing IES prototypes has been experimentally studied using the latest version of the smart scheduler, three SDPs, as well as a set of operational and informational RUCs and other tools of the intelligent software environment at the AT-TECHNOLOGY workstation.

This work is devoted to some issues related to the SDP “Building IES for Learning and Web IES”, the risky elements of which are building an applied ontology of courses/disciplines and competency-oriented student models.

3 Basic Model of Applied Ontology of Course/Discipline

An important feature of IES training, which was developed on the basis of a problem-oriented methodology and AT-TECHNOLOGY workbench, is the ability to flexibly form an applied ontology of each course/discipline [2, 3] based on the use of an already built hierarchical system. The structure of the relevant courses/disciplines that reflect the level of knowledge of the teacher [1]. Conclusion of the course/discipline ontology model is a semantic network, where each element of the course/discipline is a node, and the relations between the elements are arcs [2]:

$$M_e = <V_e, U_e, C, K_e, RK, I_e, S_e> ,$$

where V_e is a set of course/discipline elements (sections, topics, sub-topics, etc.) presented as $V_e = v_1...v_n$, n is a number of course/discipline elements, and each element v_i is a set of three $v_i = <T_i, W_i, Q_i>$, $i = 1...n$, where T_i is an element name of the structure of the course/discipline, $W_i = [0...10]$ is a node weight v_{ei}, Q_i is a set of questions presented in the form $Q = <F_{ij}, S_{ij}, I_{ij}>$, $j = 1...q$, where F_{ij} is an enunciation of the question, $S_{ij} = s_{ij1},...,s_{ije}$ is a set of answers, I_{ij} is an identifier of the correct answer. A set of links between the elements of the course/discipline is defined as $U_e = u_j = <V_{kj}, V_{lj}, R_j>$, $j = 1...m$, where V_{kj} is a patent node, V_{lj} is a child node, R_j is a type of a link, at that $R = R_z$, where $z = 1...Z$, R_1 is a part-to-whole relationship (aggregation) meaning that the child node is a part of the parent vertex; R_2 is a link of the type of "association", meaning that to know the concept of the parent node, one needs to comprehend the notion of a child node; R_3 is a "weak" link, meaning that for possession of the concept of the parent node, possession of the concept of a child node is desirable, but not necessary. The set $C = C_i$, $i = 1...a$ contains hierarchical links between the elements of the course/discipline, where $C_i = <V_k, V_l>$, where V_k is a parent element, V_l is a child element in the hierarchical structure of course/discipline (for example, V_k is a section, and V_l is a topic of that section); In target competence model $K_e = <K, CK>$ component $K = \{K_i\}$, $i = 1...b$ is a set of target competence, at that $K_i = <NC_i, S_i>$, where NC_i is a name, but S_i is a code of the attribution of K_i, a $CK = \{CK_i\}$, $i = 1...c$ is a set of hierarchical links between competence, at that $CK_i = <K_{ki}, K_{li}>$, where K_{ki} is a parent-attribution from aggregate K, but K_{li} is a child-attribution from aggregate K, $k = 1...d$, $l = 1...e$. A set $RK = \{RK_i\}$, $RK_m = <V_{ki}, K_{li}, W_{cj}>$, $m = 1...f$ represents elements of the course/discipline links and competence, where V_{ki} is an element of V_e set, K_{lj} an element of K_e set, W_{cij} is a weight coefficient of attribution K_{lj} corresponding to the course/discipline element V_{ki}.

Element $I_e = <TR, CH>$ is a set of models of training impact [5, 6]; $TR = <T_e, RT>$ is an educational-training tasks (ETT) model following [5], where $T_r = \{T_{ei}\}$, $i = 1,...,c$ is a set of ETT, at that $T_{ei} = <D_a, C, V, V_u, O_v, P_a>$, where D_a is an initial data, C is limitations that must be taken into account when executing ETT, V is correct answers, $V_u = \{V_1,...,V_n\}$ is a description of the method of input of the result, where V_1 is a numerical value or an interval, V_2 is a set of alternative options, V_3 is a set of options, V_4 is filling in blanks in the text, V_5 is a selection of solution components from the list, V_6 is text labeling, V_7 is construction links between elements of the graphical representation; O_v is a function of result evaluation $O_v (V_s, V) => R$, where R is a set of estimates, V_s is an Input result; $RT = \{RT_i\}$, $i = 1...y$ is a set of links between the ontology of the course/discipline and the subset of the ETT; $CH = <C_h, RC>$ is a model of HT-textbook, where $C_h = \{C_{hi}\}$, $i = 1...d$ is a set of chapters of the hypertext textbook [1], at that $C_{hi} = \{M_1, M_2\}$, where M_1 is an HTML-model of HT, M_2 is an XML-model of HT, and $RC = \{RC_i\}$, $i = 1...g$ is a set of links between the element of the course/discipline and the subset of the head of the textbook.

Component $S_e = <PA, FA, SA>$ is an aggregate of models for elicitation student skills/abilities, where PA is a model of the process of elicitation students' abilities to simulate strategies of direct/reverse reasoning, FA is a model of the process of elicitation of the students abilities to simulate the simplest situations of the problem domain with frames, *SA* is a model of the process of elicitation the students' abilities to simulate the situations of the problem domain with semantic networks. In its turn, $PA = \langle PS, PR \rangle$, where *PS* is a production system in accordance with [2], $PR = \{PR_i\}$, $i = 1...m$ is an aggregate of links between the ontology elements of the course/discipline and the components of *PS*; $FA = \langle F, FR \rangle$, where *F* is an aggregate of procedures and reference prototype frames in notation FRL [2], $FR = \{FR_j\}$, $j = 1...n$ is a set of links between the ontology elements of the course/discipline and the components *F:* $SA = \langle S, SR \rangle$, where *S* is an aggregate of procedures and reference fragments of semantic networks, $SR = \{SR_k\}$, $k = 1...r$ is a set of links between the ontology elements of the course/discipline and the components *S*.

Mainly, if a question arises about a specialty/specialization or a field of training, an ontology specializes as a generalized ontology $O = \{Oi\}$, $i = 1...n$, combining a set of separate ontologies of *Oi* courses/disciplines, and determining the capacity of specific sections, subsections, topics of concepts used in the ontology of each course/discipline. Correspondingly, in the context of the development of IES training, the model of the generalized ontology Mo is provided in the form: $Mo = \{Moi\}$, where $Moi = Mei$, $i = 1...n$, n is a series of courses/disciplines. Specializations for which the *Oi* ontology is built. According to [2], each ontology of the course/discipline *Oi* is represented as: $Oi = \langle Me, Fe \rangle$, where Me is the model ontology of the course/discipline described above; $Fe = \{Fs, Fq, Fam, Fk, Fke\}$ a set of operations (procedures) for constructing the ontology (*Oi*) of the course/discipline, where *Fs* - procedures for structuring the course/discipline; *Fq* - procedures for formulating questions to selected elements of the course/discipline with one level of hierarchy; *Fam* - procedures for implementing the adaptive method of repertory grids (AMRG) [1] to classify the relationships between the elements of the course/discipline; *Fk* - procedures for constructing a model of target competence; *Fke* is a procedure for determining the relationship between attribution and course/discipline elements. The main and basic tools of the AT-TECHNOLOGY workspace allows its users to implement and visualize applied ontologies of the course/discipline and generalized ontologies.

A common basis for the main problems of intellectual learning [2–4] is the use of IES of different architectural typologies, processes of dissection of knowledge (declarative knowledge for a particular course/discipline) and skills (procedural knowledge that allow demonstrating how declarative knowledge trainees are used in practice). At the moment when these processes are carried out in the training IES, the current competency-based student model is dynamically created [1], based on the analysis of answers to questions from special web tests. Generation of test case variants is carried out before the start of web testing by applying the genetic algorithm to a particular ontology of the course/discipline or its fragment [2, 3], following the curriculum for conducting control measures. Then, the current student model is compared with the ontology of the course/discipline, which, as a result, the so-called "problem areas" are identified in the students' knowledge of the individual sections and topics of the course/discipline and the corresponding current competency. Consequently, ontologies of courses/disciplines are vital for identifying students' knowledge and building competency-based student models.

It is needed to propose and consider the place and role of ontologies in the processes of computer identification of student skills to solve learning problems. For training IES and web-IES, working on the basis of generalized ontologies, "Intelligent Systems and Technologies" occupy a certain place in the methods of detecting skills for solving learning problems that are related to modeling the thinking of a person (student) and others. Approaches related to including methods and means of traditional ES and IES [2, 16]. For example, one day, the study of specialized courses/disciplines in the field of tutoring "Software Engineering" is impossible without the availability of skills and abilities of students to solve the following tasks [2, 3, 5]: skills to build "self-study". Expert models of the simplest situations of a problem area based on frames and semantic networks, modeling direct/reverse thinking strategies in an expert system, building components of a linguistic model for the sublanguage of business prose, and others.

4 Ontological Approach to the Dynamic Formation of Competence-Oriented Models of Specialists

The generation of the modern scientific and educational appearance of the department in the field of AI is associated with the historical, conceptual and technological integration of all three of the above specialties within the main specialty "Applied Mathematics", and the creation and active education refer to the periods 1971–1990 and 1991–2014. Since 2015, this development phase has been continuing within the framework of the "Software Engineering" training direction (undergraduate and graduate programs).

It should be noted that the field of "Software Development" is a natural development of programming technology, and a modern understanding of this area appeared much later than the term "programming technology", appeared in our country (in 2004 in the core of knowledge SWEBOK, and in 2014 was published SWEBOK v. 3.0, which received international recognition as Technical Report ISO IEC 197596: 2015).

From the point of view of the methodology, it is important to note that by creating a generalized ontology "Intelligent Systems and Technologies" (which currently includes about 900 peaks from 8 courses/disciplines), it was possible to create a single space of ontological knowledge and skills that allows you to:

- To complete the training cycle, following a specific curriculum and the attached training materials (lectures, practical exercises, etc.);
- To guarantee and ensure the full-fledged construction of competence-oriented models of students for the entire period of study (bachelor, master) and, as the final result, the formation of models of future specialists, including individual psychological portraits);
- Make a comparison of the current competencies of the students with the target ones, identify the "problem areas" in knowledge and think out ways of learning in the form of solving specific practical problems for each student in order to achieve a higher level of competencies, etc.

- Providing the best opportunities for intellectual learning, namely: individual planning of the methodology for studying individual training courses; intellectual analysis of educational tasks; intelligent decision support, etc.

As a necessary informational and methodological resource for creating models of *professional competencies*, in particular, for such professions as "software engineer", "systems analyst", "IT systems specialist", "software architect", etc., Professional standards for information Industry technologies [18] were used quite efficiently. The main attention is paid to ontologies of such disciplines as "Technology for programming cybernetic systems", "Design and architecture of software systems", "Designing systems based on knowledge", "Dynamic intelligent systems", etc., in the framework of which students receive basic theoretical knowledge and practical techniques specific to the development of traditional software systems and intelligent systems of various architectural typologies, including the life cycle and methodology of design and development of software mnyh systems; various software system architectures; modeling in languages such as UML, testing, verification and certification of software; CASE-tools, workbench systems and other types of tools to automate the process of developing software systems, etc.

At present, there is no unique classification of competencies. But it should be noted that the generally accepted point of view is the distribution of professional and universal competencies. Further classification depends on the specifics of the profession, the traditions of the university that trains specialists in this field, and other features.

Thus, paying attention to the long-standing traditions of the Department of Cybernetics of NRNU MEPhI, we are talking about the integration of systems engineering and software engineering with methods and technologies of artificial intelligence, therefore, according to the Federal State Educational Standard 3+, the following two competencies are used as the basis for the training of knowledge engineers: *the formalization capacity in his or her subject area, with the view to the limitations on study methods in use; the ability to use methods and instrumental means of study of professional business items.* The achievement of these target competencies is facilitated by the common ontological space of knowledge and skills, which is formed by the applied ontologies of courses/disciplines of several tutoring IES and web-IES.

As for the information necessary for the formation of *social and personal competencies* (from the group of universal competencies), taking into account the personal characteristics of the trainees, it is possible to partially use the information presented in professional standards in the description of "self-development" for each specialty. In addition, there are many psychological tests, surveys that specialize in websites and other sources available to determine personal characteristics. The main problem is the search and careful selection of expert information, signaling the degree of manifestation of a specific competency for each of the personal characteristics.

As the experience of the development and use of tutoring IES and web-IES in the educational process revealed, the main problems in the formation of professional and universal competencies are:

- The choice for each at each stage of training of knowledge, skills and abilities that students should receive (using applied ontologies of courses/disciplines, generalized ontologies of individual areas of learning);

- Improving the methods of control and testing carried out both for the formation of modern competence-oriented models of students and at the end of training (using web-testing of students with the generation of options based on the genetic algorithm);
- Actively taking into account the personal characteristics of trainees when selecting and shaping learning strategies and influences, including the development of particular corrective learning influences aimed at developing personal characteristics in an individual trainee;
- The use of additional training based on identifying gaps in knowledge and skills, etc. (sets of educational interactions are used for different groups of students).

Based on the experience of building competency-based models of graduates (bachelors and masters) in the field of "Software Engineering", it is necessary to ensure the solution of the following urgent problems in the organization of the modern educational process:

- Conceptual understanding of the programming process and instilling computational and logical thinking skills for the implementation of specific tasks in the form of a program and/or a software system;
- The selection of various programming paradigms, including OO-oriented, functional, logical, environment-oriented, and several types of programming languages, as well as the development of competence in the use of a particular language;
- The use of new approaches in the field of AI by presenting general knowledge in the field of programming in the form of decision plans to help students experiencing difficulties in moving from a method or algorithm for solving a problem to its software implementation.

For example, the implementation of automated construction processes for competency-based models of specialists in the field of knowledge engineering based on the ontology "Intelligent Systems and Technologies" is presented in [5].

5 Conclusion

Therefore, to conclude, the ontological approach, which is based on the integration of various knowledge, data, documents, etc., allows to effectively organize the educational process in the direction of "Software Development" using training IES and web-IES for competency-based -oriented models of specialists. In the field of AI methods and technologies. This approach lays the foundation in relations with employers and potential clients, but also allows you to plan targeted training of specialists in various fields, starting with junior courses. This article provides the latest results related to the development of the ontology model proposed earlier in [3, 4], as well as to the improvement of software tools to support the new model. Based on experiments with a new ontology model and software, a new applied ontology was implemented for training at the Institute of Electric Welding at the course "Introduction to Intelligent Systems".

Acknowledgement. The work was performed with the support of the Russian Foundation for Basic Research (Project No. 18-01-00457).

References

1. Rybina, G.V.: The theory and technology of integrated expert systems construction. Nauchtekhlitizdat (2008). (Teoriya i tekhnologiya postroeniya integrirovannyh ekspertnyh sistem. Monografiya). (in Russian)
2. Rybina, G.V.: Intellectual systems: from A to Z: a series of monographs in three books. b.1: knowledge-based systems. Integrated expert systems. Nauchtekhlitizdat (2014). (Intellektual'nye sistemy: ot A do YA. Seriya monografij v tryoh knigah. Kniga 1. Sistemy, osnovannye na znaniyah. Integrirovannye ekspertnye sistemy). (in Russian)
3. Rybina, G.V.: Intellectual technology of construction of training integrated expert systems: new opportunities. Open Educ. **21**(4), 43–57 (2017). (Metody i programmnye sredstva intellektual'nogo planirovaniya dlya postroeniya integrirovannyh ekspertnyh sistem. Iskusstvennyj intellekt i prinyatie reshenij). (in Russian)
4. Rybina, G.V., Rybin, V.M., Blokhin, Yu.M., Sergienko, E.S.: Intelligent support of educational process basing on ontological approach with use of tutoring integrated expert systems. In: Proceedings of the Second International Scientific Conference «Intelligent Information Technologies for Industry» (IITI 2017). Advances in Intelligent Systems and Computing, vol. 680, pp. 11–20. Springer, Berlin (2018)
5. Rybina, G.V., Fontalina, E.S.: Automated construction of young specialists models with the use of tutoring integrated expert systems. In: Proceedings of IV International Conference on Information Technologies in Engineering Education, pp. 41–44 (2018)
6. Nye, B.D.: Intelligent tutoring systems by and for the developing world: a review of trends and approaches for educational technology in a global context. Int. J. Artif. Intell. Educ. **25**, 177–203 (2015)
7. Rahman, A.A., Abdullah, M., Alias, S.H.: The architecture of agent-based intelligent tutoring system for the learning of software engineering function point metrics. In: 2nd International Symposium on Agent, Multi-Agent Systems and Robotics, ISAMSR 2016, pp. 139–144 (2016)
8. Sosnovsky, S., Mitrovic, A., Lee, D., Brusilovsky, P., Yudelson, M.: Ontology-based integration of adaptive educational systems. In: 16th International Conference on Computers in Education (ICCE 2008), pp. 11–18 (2008)
9. Gribova, V.V., Kleshchev, A.S., Krylov, D.A., Moskalenko, F.M., et al.: The IACPaaS Project. Complex for intelligent systems based on cloud computing. Artif. Intell. Decis.-Making **1**, 27–35 (2011)
10. Gribova, V.V., Kleshchev, A.S., Krylov, D.A., Moskalenko, F.M., et al.: Basic technology for the development of intelligent services on the cloud platform IACPaaS. Part 1. Development of knowledge base and problem solver. Softw. Eng. **12**, 3–11 (2015)
11. Gribova, V.V., Ostrovsky, G.E.: Intellectual learning environment for the diagnosis of acute chronic diseases. In: Proceedings of the Fifteenth National Conference on Artificial Intelligence with International Participation of CII-2016, T3, pp. 171–179. Universum, Smolensk (2016)
12. Burita, L.: Intelligent software ATOM for knowledge systems development. In: Proceedings of the IADIS International Conference Intelligent Systems and Agents 2013, ISA 2013, Proceedings of the IADIS European Conference on Data Mining 2013, ECDM 2013 (2013)

13. Gharaibeh, N., Soud, S.A.: Software development methodology for building intelligent decision support systems. In: Doctoral Consortium on Software and Data Technologies – Proceedings of the Doctoral Consortium on Software and Data Technologies, DCSOFT 2008, In Conjunction with ICSOFT 2008, pp. 3–14 (2008)
14. Telnov, Yu.F., Kazakov, V.A.: Ontological modeling of network interactions of organizations in the information and educational space. In: Proceedings of the Fifteenth National Conference on Artificial Intelligence with International Participation of CII-2016, T1, pp. 106–114. Universum, Smolensk (2016)
15. Trembach, V.M.: Systems of management of databases of the evolving knowledge for the solution of problems of continuous education. MESI (2013)
16. Gavrilova, T.A.: Knowledge Engineering. Models and Methods. Textbook for Universities. SPb:Lan (2016)
17. Lavrishcheva, E.M.: Software Engineering. Paradigms, Technologies and CASE-Tools. Textbook for Universities. Yurayt Publishing House, Moscow (2016)
18. Development of professional standards for the information technology industry. The committee on education in the IT field. The website APIT. http://www.apkit.ru/default.asp?artID=5573

Experiment Planning Method for Selecting Machine Learning Algorithm Parameters

Maria Tutova, Marina Fomina(✉), Oleg Morosin, and Vadim Vagin

Department of Application Mathematics, National Research University "MPEI", Moscow, Russia
m_fomina2000@mail.ru, vagin@appmat.ru

Abstract. The paper deals with the problem of selecting the parameters of machine learning algorithms. The selection of the most important factors affecting the success of the machine learning algorithm can be done conducting a full factorial experiment. A comparison of the methods of organizing a full-factor experiment is given and the choice of the planning method for the Box-Wilson experiment is substantiated, which can significantly reduce the number of experiments to be carried out when building a model that relates the factors to the target parameter. The Box-Wilson experiment planning method is implemented in software. The results of software modeling obtained for real problems are given. These results confirm the reduction in time spent on optimizing model hyperparameters.

Keywords: Machine learning · Selection of parameters · Generalization · Planning of experiments

1 Introduction

Machine learning is an important field in data mining and aims to automatically formulate a way to solve a certain problem, for example, extracting rules representing generalized descriptions of situation classes and states of a complex object. Many methods of machine learning were developed as an alternative to classical statistical approaches and are closely related to the field of knowledge mining (Data Mining). Machine learning is located at the junction of various disciplines, but also has its own specifics related to the problems of computational efficiency and retraining.

Machine learning is not only a mathematical, but also a practical, engineering discipline. The theory, as a rule, does not immediately lead to methods and algorithms that are applicable in practice, which is why it is necessary to invent additional heuristics to compensate for the discrepancy between the assumptions made in the theory and the conditions of real problems. Virtually no research in machine learning can do without experiment on model or real data confirming the practical performance of the method [1].

Currently, machine learning algorithms are used in many areas that require processing and building forecasts based on a large amount of data. However, despite the variety of machine learning algorithms, the effectiveness of many of them depends on how well the most important parameters of the algorithm, or hyper parameters, are chosen.

S. Kovalev et al. (Eds.): IITI 2019, AISC 1156, pp. 341–351, 2020.
https://doi.org/10.1007/978-3-030-50097-9_35

The hyperparameters of the machine learning algorithm are the input parameters of the algorithm that influence its behavior. For example, such parameters often include sensitivity to errors or the number of iterations. The hyperparameter can also be a qualitative sign, for example, the kernel function for the support vector machine (SVM) algorithm.

The nontrivial problem of selecting parameters completely falls on the researcher. Well-known heuristics related to an algorithm or data, or methods aimed at selecting such parameters by testing different values help him in this. The combination of these two strategies in the theory allows to achieve high accuracy of the resulting model, but in real conditions the researcher always has to find a compromise between the quality of the model and the speed of its receipt. In this paper, a method is proposed that significantly reduces the search space to find suboptimal values of hyperparameters. Particularly an acute question of the selection of parameters is in the automatic learning systems in the compositions of real-time systems [2]. Thus, the relevance of the study is due to the growing demand for tools and methodologies for the effective optimization of hyperparameters of machine learning algorithms.

1.1 Overview of the Methods of Selecting Parameters of Machine Learning Algorithms

A number of methods are known that allow selecting the best of the set of essential parameters: the grid search method, its modification—RandomSearch method [3], Bayesian optimization method [4], gradient optimization method [5], evolutionary optimization method [4] spectral approach [6] and the radial basis function method [7]. The last two are highly efficient, but narrowly focused methods, poorly expanding on a variety of machine learning algorithms.

The GridSearch and RandomSearch methods are very effective, but require a lot of enumeration. Bayesian optimization methods show better results in fewer iterations compared with the GridSearch and RandomSearch iteration methods [7–9]. However, these methods are difficult to build and apply from a software point of view when it comes to non-classical algorithms and custom modifications.

These methods are collectively called methods of hyperparameter optimization. Each of them has its advantages and disadvantages. It is also important to understand that the concept of "optimization" in its classical sense is not applicable to machine learning algorithms. In this case, optimization is more likely the search for a more suitable option, and it is not necessary to speak about the optimality of the solution obtained due to the complexity of the problem.

If the hyperparameters of machine learning algorithms are perceived as stochastic elements, then it would be useful to describe the effect a particular parameter has on the system. Then you can get at your disposal some heuristic methods to determine how to configure the algorithm as a system. For this purpose, a method is needed that allows determining the influence of each of the essential parameters on the system for a small number of experiments. This method can serve as a method for planning a Box-Wilson experiment [10].

The idea of the Box-Wilson method is as follows: the experimenter is asked to carry out successively small series of experiments, in each of which all parameters simultaneously vary according to certain rules. The series are organized in such a way that after the mathematical processing of the previous series of results, it was possible to choose the conditions for the next series. So consistently, step by step, the region of local optimum can be achieved [11].

Comparing the Box-Wilson method with the existing methods of hyperparametric optimization, we see that it is a combination of simplified Bayesian optimization (due to statistical evaluation) and gradient (due to the use of steep ascent) one.

2 Problem Statement

The main goal of the machine experiment is to establish the relationship between the values of the parameters of the machine learning algorithm and the success of this algorithm. The success of the machine learning algorithm depends on a number of parameters (factors), with each factor having a certain impact on the results achieved. To evaluate this effect, you need to build a mathematical model that reflects the relationship of various factors with the target parameter. This target parameter (also called the optimization parameter) reflects the most important property for the researcher of the classification model constructed by the algorithm.

Experiment planning is a procedure for choosing the conditions and the number of experiments necessary and sufficient to solve the problem with the required accuracy. Let's consider the idea of a full factorial experiment.

Let $x_1, x_2, \ldots, x_n$ be a factors of an experiment and y be a target parameter. A mathematical model is an equation of the form $y = (x_1, x_2, \ldots, x_n)$, that links the optimization parameter y to the factors x_i. The function $y = (x_1, x_2, \ldots, x_n)$ is also called the response function. To examine the influence of factors on the value of y, experiments are carried out according to a definite plan that allows to realize all possible combinations of factors. Suppose that in the planned experiments each specific factor may take values on a certain interval (interval of variation). All factors will be considered at two fixed levels - upper and lower, in accordance with the boundaries of the interval. Since factors can be of different nature, they can be preliminarily brought to the same scale (+1 and −1). In the following, we will use the following notation: the "+" sign indicates that during experience the value of the factor is set up at the top level (maximum), and the sign "−" indicates that the value of the factor is set up at the bottom level (minimum).

During the experiment, one need to get the value of y for each possible combination of all factors. Having obtained the results for all combinations of factor values, we can proceed to the construction of the function $y = (x_1, x_2, \ldots, x_n)$. This function is represented in the multidimensional factor space (variables) and may be non-linear. When conducting a full factorial experiment in the factor space, a fixed (starting) point is selected with the specific values of all factors, relative to which the grid of possible experimental points is constructed. The construction of the experiment plan is reduced to the choice of experimental points that are symmetrical about the starting point (ground level). The levels of the factors are depicted by two points on the coordinate line, symmetrical with respect to the main level (the main level is marked as 0).

A full factorial experiment (FFE) is an experiment in that all possible combinations of levels of factors are realized. It is convenient to present the experimental conditions in the form of a table—a planning matrix, where the rows correspond to different experiments, and the columns represent the values of the factors. When performing the FFE, each factor varies exactly at two levels: top level (maximum) and bottom level (minimum).

To represent the response function $y = (x_1, x_2, \ldots, x_n)$, a first-degree polynomial will be used. Indeed, for a nonlinear function, at almost any of its points, one can select a sufficiently small neighborhood, within which this function is close to linear and, therefore, can be approximated by a plane tangent to this point. Then the response function will take the form:

$$\boldsymbol{y} = b_0 + b_1 x_1 + b_2 x_2 + \ldots + b_n x_n \tag{1}$$

and the construction of the model will consist in finding the coefficients $b_1, b_2, \ldots, b_n$. On the one hand, such a polynomial contains information about the direction of the gradient, on the other, it has the minimum possible number of coefficients for a given set of factors [11].

The coefficients $b_1, b_2, \ldots, b_n$ calculated from the results of the experiment indicate the strength of the influence of factors on target y. The coefficient value corresponds to the contribution of this factor to the value of the optimization parameter y when varying the values of the factor between the upper and lower levels relative to the starting point. If the coefficient b_i is positive, then with an increase in the value of the factor i, the optimization parameter y increases; otherwise, y decreases.

Let $\boldsymbol{y}$ depend on three factors x_1, x_2, x_3. In this case, the matrix of the three-factor experiment will be supplemented with three columns labeled x_1x_2, x_1x_3, x_2x_3. Then the model will look like this:

$$\boldsymbol{y} = b_0 + b_1 x_1 + b_2 x_2 + b_3 x_3 + b_{12} x_1 x_2 + b_{13} x_1 x_3 + b_{23} x_2 x_3 \tag{2}$$

The method of least squares is used to find the values of the coefficients of a polynomial. The coefficients calculated from the results of the experiment indicate the strength of the factor influence. If the value of the found coefficient b_{ik} is significant, it is possible to speak about the presence of the interaction effect of two factors.

After the model has been built (the coefficients of the model have been calculated), it is necessary to check its suitability, such a check in the theory of experimental design is called a model adequacy check [10]. A model is called adequate if, in a certain subdomain, including the coordinates of the experiments performed, with the help of this model, it is possible to predict the response value with a given accuracy.

To estimate the significance of individual factors from the results of experiments, we calculate the residual variance that characterizes the dispersion of experimental data relative to the regression equation. We use for this purpose the residual sum of squares. The residual variance, or the adequacy variance of s^2_{ad}, is the residual sum of squares divided by the number of freedom degrees:

$$s_{ad}^2 = \frac{\sum_{i=1}^{N} \Delta y_i^2}{f} \tag{3}$$

Here N is the experiment number, Δy_i^2 is defined by the difference between the result of the experiment and the theoretical result obtained using the assessment model, f is the number of freedom degrees. In statistics, the number of freedom degrees is the difference between the experiment number and the coefficient number of the regression model that are calculated from the results of these experiments independently of each other.

For checking hypotheses about the model adequacy, there is a criterion developed in statistics. It is called Fisher F-criterion and is defined by the following formula:

$$F = \frac{s_{ad}^2}{s_{\{y\}}^2} \tag{4}$$

$s_{\{y\}}^2$ is the reproducibility dispersion for the model $\boldsymbol{y}$. In relation to (4), the adequacy variance characterizes the discrepancy between the experiment results and the values of the output value y, calculated using a regression equation.

The reproducibility variance $s_{\{y\}}^2$ characterizes the average scatter of the results of repeated measurements in all experiments with respect to their mathematical expectations. If in each experiment from N experiments, the number of repeated measurements is the same and it is equal to n, the reproducibility dispersion is calculated by the formula:

$$s_{\{y\}}^2 = \frac{\sum_{j=1}^{N} \sum_{i=1}^{n} (y_{ij} - \bar{y}_j)^2}{N(n-1)} = \frac{\sum_{j=1}^{N} \Delta \hat{y}_i}{N} \tag{5}$$

In formula (5), $(y_{ij} - \bar{y}_j)^2$ determines the difference between the result of a particular measurement and the average result from the j-th experiment. Further we will denote $(y_{ij} - \bar{y}_j)^2/(\mathrm{n}-1)$ as $\Delta \hat{y}_i$.

The calculated value of F_{exp} is compared with the table value F_{table}, that is selected at a given level of significance α, and the set values of the freedom degrees f for the numerator and denominator. If relation (6) is satisfied, then the model is considered adequate and can be used to describe the object. Otherwise, the model is not adequate.

$$\mathrm{F_{exp}} < \mathrm{F_{table}}(\alpha,\ \mathrm{f},\ \mathrm{N}) \tag{6}$$

The significance verification of each coefficient is carried out independently using confidence intervals. When using a full factorial experiment or regular fractional replicas (in this case only some part of the FFE table is used), the confidence intervals for all coefficients (including interaction effects) are equal to each other [10].

The calculation of a confidence interval for the coefficients is fairly easy. First of all, it is necessary to find the variance $s^2_{\{bi\}}$ of the regression coefficient b_i. It is determined by the formula: $s^2_{\{bi\}} = \frac{s^2_{\{y\}}}{N}$.

It can be seen from the formula that the dispersions of all coefficients are equal to each other, since they depend only on the experimental error $s^2_{\{y\}}$ and the experiment number N.

Now it is easy to build a confidence interval: $\Delta b_j = \pm t * s^2_{\{bj\}}$.

Here, t is the tabular value of the Student's criterion (Student's criterion is used to test the hypothesis about the equality of the coefficient b to a specific number a) with the number of degrees of freedom with which $s^2_{\{y\}}$ was determined and the chosen level of significance (the level of significance is understood as the probability to take the correct hypothesis as an incorrect; usually the value of this value is chosen to be 0.05); $s^2_{\{bj\}}$ is the square error of the regression coefficient.

A coefficient is significant if its absolute value is greater than the confidence interval. Insignificant coefficients suggest that this factor does not significantly affect the optimization parameter.

3 Selecting Parameters for Machine Learning Algorithm

To apply the Box-Wilson experiment planning method, we will stick to the following algorithm:

1. To determine the hyperparameters of the model.
2. Highlight the essential parameters.
3. For each significant parameter, determine the top and bottom levels.
4. Create an experiment plan (planning matrix).
5. Conduct an experiment.
6. Assess the quality of the resulting model.
7. If the model is adequate: choose the most significant factor as a base factor, define the step of variation and perform movement along the response surface in the direction of the linear approximation gradient. At the point which, in a given direction, corresponds to the largest value of y, a new series of experiments is put and a new direction of movement is chosen. Such a process (steep ascent) continues until a local extremum is reached.

Essential parameters will be considered not so much those that influence on the model (since all parameters affect the model to one degree or another), but those whose influence on the model must be investigated. For example, the main idea of the AdaBoost algorithm is to build a strong (basic) classifier (that is, a recognition algorithm with a high classification accuracy) based on the composition of several classifiers that individually have a low classification accuracy (weak classifiers) [12].

It is important to note that after the experiment, nothing prevents the researcher from changing the set of essential parameters and conducting a new experiment. Comparison of the results of such experiments may also be of research interest.

As an example, let us consider the solution of the classification problem on one of the data sets that are offered on the Kaggle website [13] to researchers who want to test the success of their machine learning algorithms in comparison with the results achieved by other developers. The data set for this competition contains 369 anonymized attributes. Anonymization of features means that the researcher does not know the nature of the feature: the feature has an abstract name (for example, "var173") and contains numerical data. All that is known is that this competition is conducted by the Santander Group bank and the objective function is to determine the value of the transaction of potential bank customers. Information about what numeric signs mean is not disclosed in a dataset. For solving the problem, we'll use the AdaBoost algorithm (the SAMME.R realization for real numbers [14]).

For it, we select the following three parameters:

- **max_depth** - the maximum depth of the decision tree for the weak classifier, vary from 1 to 10.
- **n_estimators** - the number of weak classifiers participating in Voting (from 100 to 900);
- **learning_rate** - "learning speed", antigradient multiplier in calculating the voice weight of a weak classifier (vary from 0.01 to 0.9);

The first parameter is selected as the key hyperparameter for a weak classifier in this algorithm. The greater depth of the tree can give higher accuracy, but it is at greater risk of retraining.

The number of weak classifiers and the learning rate are two key parameters of the AdaBoost algorithm, between which we have to look for a compromise. As a rule, in the presence of a large number of classifiers, even with a low learning rate, the best classification accuracy can be obtained; and vice versa: a small number of classifiers can be compensated for by a high learning rate. For each specific task, the combination of these parameters must be selected anew.

We have produced a full-factor experiment, varying the factors according to Table 1. We use cross-validation in 10 blocks to avoid retraining. To assess the quality of the model, the training is carried out every time on one of 10 subsamples, then the accuracy of the classification of test examples is estimated by the classifier based on such a subsample. At this stage, one of the remaining 9 sub-samples is chosen as the source of test cases. At the end, the results of all experiments are averaged. The standard deviation of the experience will be the mean of the over all repeated experiments. As a metric, that is used to assess the quality of the model, it is proposed to use the average value of the area metric under the error curve (AUC ROC), built for repeated experiments [15]. The planning matrix and the results of the experiment are shown in Table 1.

Table 1. Planning for the AdaBoost algorithm to solve the problem of identifying dissatisfied customers of Santander Bank

№	Planning matrix								Metrics	Mean square deviation
	x_0	x_1	x_2	x_3	x_1x_2	x_1x_3	x_2x_3	$x_1x_2x_3$		
1	+	-	-	-	+	+	+	-	71.87%	0.056
2	+	-	-	+	+	-	-	+	82.76%	0.0101
3	+	-	+	-	-	+	-	+	76.17%	0.0183
4	+	-	+	+	-	-	+	-	82.13%	0.0129
5	+	+	-	-	-	-	+	+	76.74%	0.0117
6	+	+	-	+	-	+	-	-	72.17%	0.0206
7	+	+	+	-	+	-	-	-	76.27%	0.0107
8	+	+	+	+	+	+	+	+	72.70%	0.0156

The average value of the optimization parameter y at the starting point of the multidimensional factor space is equal to the contribution of the free term, that is, the coefficient $b0$: 0.76355. In Table 2, in addition to b0, all coefficients of the regression model are presented, reflecting the influence of both individual factors and combinations of the interaction of factors on the resulting parameter. To find the coefficients of the model, the least squares method was used. When these coefficients are found, a regression model is built [10].

Table 2. Regression model coefficients

$b0$	$b1$	$b2$	$b3$	$b1b2$	$b1b3$	$b2b3$	$b1b2b3$
0.76355	−0.01880	0.00466	0.01089	−0.00451	−0.03124	−0.00490	0.00741

The next step is to check the adequacy of the regression model, that is, you must make sure that the model is close to a real multidimensional function and can be successfully used to predict y values. The verification is carried out on the basis of using the F - Fisher criterion (formulas (3), (4), (5)), by calculating the variance of the adequacy and variance of reproducibility [10]. To do this, we calculate, in accordance with the results of the experiments presented in Table 2, the variance of the reproducibility $s^2_{\{y\}}$ and the variance of the adequacy of s^2_{ad}:

$$s^2_{\{y\}} = \frac{\sum_{i=1}^{N} \Delta\hat{y}_i}{N} = 0.06, \; s^2_{ad} = \frac{\sum_{i=1}^{N} \Delta y_i^2}{f} = 0.112 \tag{7}$$

where $f = 4$ – number of freedom degrees.

Let's check the adequacy of the regression model using the Fisher's F-test by the relation (6). The value of F_{exp} (4) is compared with the tabular value $F_{table}(\alpha, f, N)$, where $\alpha = 0.05$ is the significance level, N is the number of freedom degrees of reproducibility: $F_{exp} = \frac{s^2_{ad}}{s^2_{\{y\}}} = 1.86 < F_{table} = 6.4$.

The implementation of the inequality makes it possible to conclude that the model constructed is adequate, that means it can be used to analyze the process.

The next step is to find a confidence interval. The confidence interval is determined by the formula 7 with the known $s^2_{\{y\}}$: $\Delta b = t * s^2_{\{bi\}} = t * \frac{s^2_{\{y\}}}{N} = 0.0006227$, where $t = 2.778$ is the tabular value of the Student criterion with the number of degrees of freedom $f = N - k - 1 = 4$, where N = 8 is the number of experiments and k = 3 is the number of factors. The level of significance $\alpha = 0.05$. Student's criterion is used to test the hypothesis about the equality of the coefficient b to a specific number α.

The coefficient b of the regression model is significant if its absolute value is greater than the confidence interval. Comparing the coefficients of the regression model presented in Table 2, with a confidence interval, we can see that even the coefficients of the interaction effects are significant. This is an expected result: as mentioned when choosing factors, the learning rate and the number of weak classifiers are parameters whose mutual influence is significant.

Analyzing further the values of the coefficients, we see that the effects of $b1$, $b3$ and $b1b3$. are most significant. The effects of $b1$, and $b1b3$ have negative coefficients, which means that the value of the max_depth factor should be left at the lower level. The effect of $b1b3$ is negative and more significant than the positive effect of $b3$, which may indicate that a smaller value of the factor learning_rate will rather lead to a better result than a large one.

To test this hypothesis, we will make a steep ascent using two parameters: learning_rate and n_estimators. Select the steps for learning_rate = 0.01 and for n_estimators = 50.

Figure 1 shows the scatterplot for the results obtained during a steep ascent. A total of 320 experiments were conducted. The star on the scatter diagram shows values greater than 84%, values greater than 83% - rhombus, 82% - square, 80% - a triangle, and all that is less - circles.

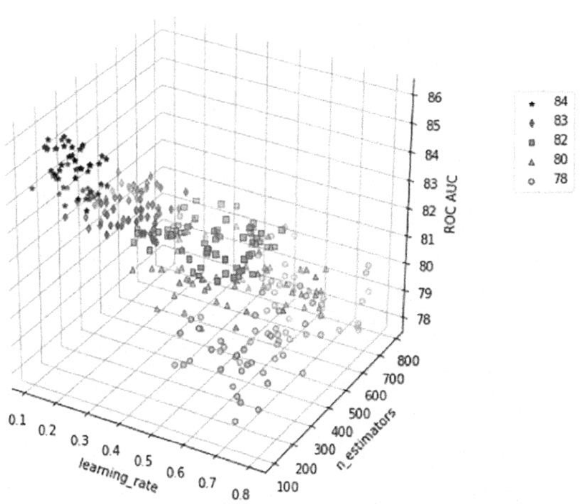

Fig. 1. A steep climb in the parameters of **learning_rate** and **n_estimators**.

In the scatter Fig. 1 we will see that the best values of the metric are achieved with smaller values of learning_rate and n_estimators. This confirms our hypothesis, derived from the values of the effect coefficients.

The best value of the experiments conducted during a steep ascent is 83.10% with a deviation of 0.012% with the parameters n_estimators = 250 and learning_rate = 0.1.

As a test, we will optimize using GridSearch, using the same steps to vary the parameters. For a complete search, 1350 experiments will be required (taking into account repeated experiments as part of cross-validation). We obtain the best accuracy of 83.094% with a deviation of 0.014 with the parameters n_estimators = 250 and learning_rate = 0.1 that coincides with the result obtained due to the full factor experiment and steep ascent.

The performance of the full factor experiment took 10,268.93 s along with the ascent, while the same optimization using the GridSearch method took 58,320.56. The result was the same, which suggests that the use of the Box-Wilson experiment planning method can significantly reduce the time for optimizing model hyperparameters.

4 Conclusion

Based on the obtained results, it can be concluded that the experiment planning method is suitable for the initial fluent analysis of the algorithm optimum region with a sufficiently high accuracy. The accuracy of the experiment planning method, however, is lower than the accuracy of the GridSearch method, however, the analysis speed allows localizing the optimum region sufficiently for the subsequent application of the GridSearch method on a narrower interval even at the stage of algorithm selection or data preprocessing. Such an advantage will be useful not only to researchers, but also in the composition of automatic learning systems as a naive decision-making algorithm regarding model optimization.

Acknowledgments. This work was supported by grants from the Russian Foundation for Basic Research № 18-01-00201, 17-07-00442, 18-29-03088.

References

1. Witten, I.H., et al.: Data Mining: Practical Machine Learning Tools and Techniques. Morgan Kaufmann, Burlington (2016)
2. Olson, R.S., et al.: Automating biomedical data science through tree-based pipeline optimization. In: European Conference on the Applications of Evolutionary Computation, pp. 123–137. Springer (2016)
3. Bergstra, J., Bengio, Y.: Random search for hyper-parameter optimization. J. Mach. Learn. Res. **13**(Feb), 281–305 (2012)
4. Bergstra, J.S.: Algorithms for hyper-parameter optimization. In: Advances in Neural Information Processing Systems, pp. 2546–2554 (2011)
5. Foo, C., Do, C.B., Ng, A.Y.: Efficient multiple hyperparameter learning for log-linear models. In: Advances in Neural Information Processing Systems, pp. 377–384 (2008)

6. Hazan, E., Klivans, A., Yuan, Y.: Hyperparameter optimization: a spectral approach. arXiv preprint arXiv:1706.00764 (2017)
7. Diaz, G.I.: An effective algorithm for hyperparameter optimization of neural networks. IBM J. Res. Dev. **61**(4), 9:1–9:11 (2017)
8. Snoek, J., Larochelle, H., Adams, R.P.: Practical Bayesian optimization of machine learning algorithms. In: Advances in Neural Information Processing Systems, pp. 2951–2959 (2012)
9. Thornton, C.: Auto-WEKA: combined selection and hyperparameter optimization of classification algorithms. In: Proceedings of the 19th ACM SIGKDD International Conference on Knowledge Discovery and Data Mining, pp. 847–855. ACM (2013)
10. Box, G.E., Wilson, K.B.: On the experimental attainment of optimum conditions. In: Breakthroughs in Statistics, pp. 270–310. Springer (1992)
11. Adler, Y.P., Markova, E.V., Granovskii, Y.V.: Experiment Planning in the Search for Optimum Conditions. Science, Moscow (1976). (in Russian)
12. Freund, Y., Schapire, R.E.: A decision-theoretic generalization of on-line learning and an application to boosting. J. Comput. Syst. Sci. **55**, 119–139 (1997)
13. https://www.kaggle.com/c/santander-value-prediction-challenge
14. Zhu, J., Zou, H., Rosset, S., Hastie, T.: Multi-class AdaBoost (2009)
15. https://en.wikipedia.org/wiki/Receiver_operating_characteristic

Logical Approaches to Anomaly Detection in Industrial Dynamic Processes

Vera V. Ilicheva(✉), Alexandr N. Guda, and Petr S. Shevchuk

Rostov State Transport University (RSTU), Rostov-on-Don, Russia
vilicheva@yandex.ru, guda@rgups.ru,
petr.shevchuk@mail.ru

Abstract. This paper proposes novel methods for anomaly detection in industrial dynamic processes based on logical and algebraical approaches. The paper contains an overview of some intensively developed approaches to dynamic processes modeling in terms of their suitability for the detection of anomalies. Main models of industrial dynamic processes using classical theory of stability, models with Pareto distribution, wavelet analysis, the problem of change-point, models with data mining methods are considered. Two logical approaches with discrete time are proposed: first one is using classical predicate logic and second one is using dynamic description logic.

Keywords: Anomaly detection · Stability theory · Wavelet analysis · Pareto distribution · Change-point problem · Data mining · Predicate logic · Dynamic description logic

1 Introduction

The relevance of the anomaly detection problem in industrial dynamic processes caused by the need to ensure the safety and stability of vital processes in the context of the continuous growth of the information flows volume and complexity.

In this paper we discuss some intensively developed approaches to the dynamic processes modeling in terms of their suitability for detecting anomalies. Anomalies are understood as the deviations of process states from the "normal" behavior: emergency, crisis situations, a sharp growth or fall in some indicators, changes in values that led to a loss of system stability. Next, we consider models of classical stability theory, models with Pareto distribution, wavelet analysis, the problem of change-point and data mining models. Deficiencies of "non normality" detection methods based on changes in statistical indicators dynamics lead to the idea hybrid models creating that use formalized expert knowledge to specify the detected emergencies. Fuzzy logic is often used as such a formal tool.

The rest of this paper is organized as follows. Section 2 considers related work in anomaly detection area that concerns to industrial dynamic processes research area. In Sect. 3 we propose using the methods of classical and extended description logic to diagnose anomalies in dynamic processes with a discrete time. The proposed specification tools have computable and effectively realizable semantics. The tools of classical predicate logic provide the diagnosis of contradictions and incompleteness (uncertainty)

S. Kovalev et al. (Eds.): IITI 2019, AISC 1156, pp. 352–361, 2020.
https://doi.org/10.1007/978-3-030-50097-9_36

of descriptions too. The approach using the classical logic of predicates is based on the description of the process in the form of logical formulas (axioms) of a certain type, allowing for an effective interpretation. However, formulations such as "find the maximum possible number", "there is no more than one follower", "event2 immediately following an event1" that have a simple expression in a natural language require some effort to express them in the language of executable logic theories. In such cases, the use of dynamic descriptive logics often gives a more adequate description of the situations.

2 Related Work

2.1 Stability Models

The theory of stability provides the mathematical foundations for the analysis of equilibria and the conditions for exiting them. The standard model of a dynamic process is described by a differential equation of the form: $\dot{x} = f(x, t), x(0) = c$, where x is generally a vector of variables. One of the main problems of modeling is the identification of parameters and their values at which the system keeps steady state. Usually we are talking about Lyapunov stability [1]: if $f(0, t) = 0$ is an equilibrium point, the system is stable if with small perturbations the process trajectories stay close to this point for all time points after t. In practice, a stronger assertion on Lyapunov's asymptotic stability is often used, requiring that at $t \to \infty$ the system returns to the equilibrium point $x(t) \to 0$, which is a feature of the type of center or stable node. The problem may be the identification of all equilibrium points for f, which may be strange attractors, limit cycles, etc. [2]. Scenarios of equilibrium changes are determined in the theory of bifurcations [3] and its varieties – the theory of catastrophes [4]. Here we consider not only points, but also more complex equilibria, for example, periodic closed trajectories or Lorentz attractors.

Knowing the general behavior of stable solutions, it is possible to predict the development of information objects even when there is no exact idea of the specific mechanisms that determine their dynamics, and such predictions may be more correct than those obtained by traditional expert methods [5]. Bifurcations – variants of development very often arise when a system moves from apparent stability to chaos. During these periods, social and information systems are most sensitive to impacts that can play a "fatal" role in choosing a new attractor. Therefore, an important task of modeling is finding the bifurcation points of the process and the formation of fluctuations that determine the choice of the desired trajectory (attractor). "A small fluctuation can serve as the beginning of evolution in a completely new direction" [6]. Stability is analyzed for technical devices, biological systems, neural networks, to diagnose and predict the stability reserves of large technical systems [7].

2.2 Models with Pareto Distribution

Statistical distributions are often used when analyzing data. According to the classical Gaussian distribution, which is the basis of engineering calculations and technical norms, large disasters are so rare that they can be neglected. The most commonly used is a continuous one-dimensional two-parameter $(x_m, k > 0)$ distribution of a random variable X, known as the Pareto distribution:

$$F_X(x) = F(X < x) = 1 - \left(\frac{x_m}{x}\right)^k, \ \forall x \geq x_m; \text{ with density } f_X(x) = \begin{cases} \frac{kx_m^k}{x^{k+1}}, & x \geq x_m, \\ 0, & x < x_m. \end{cases}$$

For small x and any k, the function F increases unlimitedly, therefore the values of x, smaller than the threshold x_m, are not taken into account. The main problem of the Pareto distributions is the divergence of high-order moments: the mathematical expectation exists for $k > 1$, the dispersion is for $k > 2$, for the other values they are infinite. For the correctness of the models, the expression for the density is limited in the range of large values x ($x < x_{max}$), then x_{max} is an additional parameter.

Gaussian distribution curve decreases very quickly, without taking account of rare but large events; the Pareto curve decreases much more slowly, given such events, forming a "heavy tail". Distributions with a heavy tail play a significant role in assessing the probability of catastrophes: earthquakes, hurricanes, floods, market crashes, damage from information leaks, etc. For large x, the formula gives results that are many orders of magnitude different from the similar estimate for the Gaussian distribution, even if the tail is short (25–30 observations above the threshold x_m). Pareto dependencies are manifesting in the dynamics of physical, biological, socio-economic systems, they are used to model and manage technological risks [8, 9]. But the question of how and when the Pareto distribution arises remains open [10].

2.3 Wavelet Models

Wavelet analysis is used in problems related to the analysis of non-stationary time series, spatial fields with a complex multiscale structure, time signals with varying spectral composition [11]. The basis of the analysis is a linear transformation with local parameters in time and frequency. They are generated by one basic function – the mother wavelet through its shifts and stretches along the time axis. Continuous wavelet transform for the function (signal) $f(t)$: $W(a,b) = (f(t), \psi(t)) = \frac{1}{\sqrt{a}} \int_{-\infty}^{\infty} f(t)\psi^*\left(\frac{t-b}{a}\right)dt$, where $\psi(t)$ is the parent wavelet, ψ^* are basis functions, a and b are scale (inverse to frequency) and shift parameters. Discrete wavelet transforms are also used. There are many types of wavelets, for example, Haar, Gauss, Shannon, Morlet, Paul, Daubechies wavelets. Analyzing wavelet should have zero averages, all functions should be obtained from it by means of scale transformation and shift, have inverse transformation.

Methods based on wavelet analysis have the unique ability of detailed frequency analysis in time – to select "frequency-time windows". Compared with the known window Fourier transform, the wavelet transform allows to simultaneously analyze the high-frequency and low-frequency components of the process. The width of the window varies with the magnitude of the scale. Increasing the width of the window leads to the selection of lower frequencies (global information, large scale), with a decrease – high (detailed information, small scale). Coefficients of wavelet transform are calculated by the scalar product of the studied data with different wavelet shifts at different scales and characterize the degree of proximity of the two functions – the analyzed process and the analyzing wavelet. Such technology allows to restore information about the dynamics of the original process, to detect various kinds of anomalies, periods of stability violations and to predict them.

Wavelet analysis is widely used in image processing, pattern recognition, neural network training as a hierarchical basis, well suited for describing the dynamics of complex non-linear processes with interaction of disturbances in wide ranges of spatial and temporal frequencies, for solving detection problems network traffic anomalies – virus, hacker attacks, software failures, signal areas with local jumps, breaks [12, 13]. Traditional correlation and spectral analysis cannot identify trends and cycles in non-stationary time series, the dependence of the frequency of cycles on time. Wavelet coefficients may to be used as a measure of proximity between the data in data mining, the search for knowledge, clustering [14].

Performing a wavelet transform highlights the most powerful information and ignores less useful noise, which is important for training neural networks and analyzing big data. Due to the different frequency "vision", it is possible to detect the instability of states, invisible to other models. The quality of the analysis depends on the successful choice of the type of wavelet, the a priori non-obviousness of this choice belongs to the disadvantages of the approach. The second problem is that the selected wavelet can have a large computational complexity.

2.4 Analysis in the "Change Point" Problem

This class includes probability-statistical models (continuous and discrete) in the tasks of quickest detection of anomalies – "change points" of casual processes. The object of study is stochastic systems and processes, multidimensional time series. A disorder is characterized by a change in the statistical properties of a series – breaks, fractures, growth in scatter, cycles and other deviations from the normal course. The methods of analysis in this field are supported by the mathematical results of many researchers. The tasks of quickest detection are formulated as optimal stopping problems, the stopping time corresponds to the moment of raising the "alarm" about the appearance of "change point". There were proposed the theory of optimal stopping rules and the basis for solving problems of quickest detection of spontaneous changes in probability characteristics (change point moment), based on these rules [15]. The most commonly sequential methods of detecting abrupt changes are considered, in which at each step of observation all the information obtained in the previous steps is used [16]. Possible detection errors are false alarms, when a disorder decision is made prematurely, and a delay – a late detection. Main efforts in research are aimed at minimizing the number of false operating and the moments of missing the start of changes. The quality of resolving procedures directly depends on the selected statistical characteristics of the observed process. For sequential approaches, the average time between neighboring moments of false alarms and the average delay time in change point detecting are considered. At the time of the disorder, the average, spectral and dispersion properties of the process may change. The problem of detecting changes in the distribution in a sequence of random variables has been studied well. To solve it, the following methods are used: moving average, exponential smoothing, cumulative amounts. Especially autoregressive models and stochastic difference equations [17] are used. Another task related to the detection of the change point moment is the estimation of unknown parameters of the processes. For autoregression-type models, methods of least squares, maximum likelihood method, and stochastic approximation are well known. For processes with continuous time, the asymptotic properties of estimates, the limit distribution, the rate of estimates convergence to the true values of the parameters are studied also.

The approach can be used both to detect anomalies in highly dynamic processes: video streams, network traffic, surveillance systems, various signals, and to diagnose diseases, the emergence of epidemics. Interest in the problem of quickest change point detecting increases, models become more complicated, for example, the dependence of observations is allowed. The advantages of the approach include the possibility of mathematical justification of the solutions. However, the resolving procedures construct and their analytical study can be very difficult. The accuracy of change point detecting depends on the selected statistical characteristics, the complexity and the degree of noise distortion of the investigated process.

2.5 Models with Data Mining

In most variants of this approach, the initial information is presented in the form of time series. Anomalous states and their possible indicators are presented in the form of segments of series – patterns that need to be found in the information flow. Detection and prediction of the appearance of anomalous patterns is complicated by non-stationarity, stochastic processes, poorly structured input data containing noise, interference, inaccurate measurements. For anomalous pattern detection, artificial intelligence methods are usually used such as classification and clustering. Classification determines the identity of the presented pattern to one of the constructed classes based on any signs. For the classification of time series, the methods of *k*-nearest neighbors, support vectors, decision trees, hidden Markov models, autoregression models, neural networks, Bayes classifier are used. In the clustering method, objects differ based on their similarity with each other. A measure of similarity may be, for example, a correlation. Patterns, their characteristics, often – sample distributions that correspond to the explicitly chosen series states can be clustering [18].

Clustering-based anomaly detection may increase the frequency of false functioning, for example, in network traffic due to the use of an inappropriate proximity measure. Different statistical approaches for comparing normal values and anomalies can produce different results. In the review of methods for detecting network attacks [19], a number of difficulties were noted in the application of data mining. This is the impossibility of detecting previously unseen attacks (they were not in the training set), which leads to a large number of malfunctions and missed attacks. In addition, the problem is the definition of traffic normality: an anomaly can be diagnosed even in the case of its legitimate change. The change in the traffics character may not show significant deviations of statistical parameters but be a deliberated attack. A common problem with this approach is the choice of features that characterize normal behavior and anomalies. In [19], the expediency of using the rules that make it possible to justify why an attack signal was fixed at a particular time was noted. Hybrid models are created, integrating expert knowledge and data mining techniques. Knowledge is often described in fuzzy logic language. In [20] such a hybrid with a neural network is presented, in [21] a method of hybridization of a stochastic Markov model with fuzzy logic productions is proposed.

3 Proposed Logical Approaches to Anomaly Detection

3.1 Models Based on Predicate Logic

Logical modeling, as an artificial intelligence tool, is more often associated with predicate logic. We suggest using the approach proposed in [22, 23] to diagnose anomalies. The model description is set by the executable specification in the first-order predicate calculus, time is represented by a finite set of points. The specification consists of a multi-sorts signature containing sets of names of predicates, functions (parameters, system attributes), constants, and the theory T - a set of axioms (logical statements about objects). Constants encode domain objects and some fixed values. The theory of $T = T_{fact} \cup T_{def} \cup T_{rest}$ is composed of three groups of axioms: a set of facts T_{fact}, initiating a logic inference; a set of definitions T_{def} of relations and functions (dependencies between objects and their attributes), perceived by the interpreter as rules for making direct logic derivation and building a model; a set of restriction T_{rest} on the properties and behavior of objects, tested on the constructed model. Axioms may include negations and equality of terms. When interpreting the $T_{fact} \cup T_{def}$, undefined, redefined function values and logical contradictions are diagnosed, the analysis of the T_{rest} allows detecting emergency situations, anomalies of structure and behavior. Time is considered as a finite discrete set of points. For example, the interpretation of the axiom from the T_{def}: $\forall t,\ t_0 \in \tau(signal(t_0) \wedge t \geq t_0 \rightarrow state(t))$ will set the state *mode* for the time points after the time t_0 occurrence of the *signal* event. The falsity of the T_{rest} axiom: $\forall x,\ t(reception(\text{'Z'}, x, t) \rightarrow \neg reception(\text{'T'}, x, t))$ identify an emergency of simultaneous reception of 'Z' and 'T' signals. Possible anomalies are diagnosed by the interpreter during the truth check of T_{rest}, the undefined values of functions and contradictions are detected during logical inference from $T_{fact} \cup T_{def}$. The approach was used for programs semantic analysis, development of space docking simulators, modeling of technological processes at railway stations. The descriptions are simple, easily modified, tested and integrated. The interpreter is universal, its speed depends only on the type of formulas. An important advantage of the approach is to obtain a complex, previously unknown "global picture" of relationships and behavior from the observed "local" dependencies and situations described by axioms. Natural non-deterministic descriptions are well suited for describing parallel processes. However, such tasks as describing a *possible* sequence of events, identifying the *immediate* predecessor, the moment of obtaining the *optimal accumulated* value require some effort to express them in the proposed language.

3.2 Approach Based on Dynamic Description Logic

For a subtler description of the processes, we propose an approach to the anomalies detection based on the use of description logic (*DL*), extended by actions. Such type of logic we define as dynamic description logic (*DDL*). DL originates from frames, semantic networks, *KL-ONE* system representing knowledge as structured networks with inheritance. The knowledge bases in the *DL* architecture consist of *TBox* (terminology, schemes) and *ABox* (facts, data), and the knowledge domain itself is described in terms of *C* concepts, roles of *R*, and individuals *v*. *DL* have solvable formal semantics and complete, correct inference procedures with an optimized implementation. In relation to classical logic, *DL* is described as "solvable fragments of predicate logic".

Traditional *DL's* are well applicable to the description of static domains and ontologies, but they are of little use for describing dynamic processes. The development of *DL* leads to a description of the actions. So, in [24] determined actions were introduced in the *DL* as follows. For the knowledge base described in *DL* and designated as $KB = (T, A)$, where T is *TBox*, A is *ABox*, the action is defined as $ACt \equiv (P_{ACt}, E_{ACt})$, where P_{ACt} is the set of formulas that define the preconditions for actions, E_{ACt} is the set of formulas that determine the effects of actions. In terms of the dynamic variant of the *DL* in *ALC*, actions can be combined according to the following rules: $\alpha_i, \alpha_j \rightarrow ACt(v_1, \ldots, v_n)|\alpha_i; \alpha_j|\alpha_i \cup \alpha_j|\alpha^*|\varphi?$, where α_i, α_j is an atomic action; ";", "?", "*" – the process of action execution, sequential, branching, and iterative, respectively; "φ?" is an action to check the truth of the *DL* formula. The formulas φ, φ' of the *ALC* are written according to the following rules: $\varphi, \varphi' \rightarrow C(v)|R(v_i, v_j)|\varphi \vee \varphi'|\bar{\varphi}|\langle\alpha\rangle\varphi$, where v_i, v_j are individuals; C – concept, R – role; $\langle\alpha\rangle\varphi$ – means that the action α is performed under the condition that the formula φ is true.

Let us describe an approach to modeling one type of dynamic system based on *DDL*. The dynamic system S is described as a determinate discrete-event system $S = \langle X, \Sigma, \delta, X_0 \rangle$, where X is a finite set of system states; Σ is a finite set of the event alphabet; $\delta : X \times \Sigma \rightarrow X$ is a function of state transitions, $X_0 \in X$ is the initial state. Denote by N_C, N_R, N_V finite sets of names of concepts, roles, individuals (variables), respectively. Each state $x \in X$ of system S has an interpretation of $I(x) = \left(\Delta^I, C_0^{I(x)}, \ldots, R_0^{I(x)}, \ldots, v_0^{I(x)}, \ldots\right)$, where $C_i^{I(x)} \subseteq \Delta^I$, $C_i \in N_C$; $R_i^{I(x)} \subseteq \Delta^I \times \Delta^I$, $R_i \in N_R$; $v_i^{I(x)} \in \Delta^I$, $v_i \in N_V$. Each action α in the system S is interpreted as a binary relation $\alpha^I \subseteq X \times X$. The change of the system states due to actions with a concrete set of individuals (variables $v_1, \ldots, v_n$) takes into account the previously considered preconditions and effects from actions for each input set of formulas $\varepsilon = (v_1, \ldots, v_n) = \{\varphi_1(v_1), \ldots, \varphi_n(v_n)\}$. A sequence of events $e = \langle e_1, \ldots, e_j, \ldots, e_k \rangle, e \in \Sigma, k = 1, 2\ldots$ is formed, and the function $\delta_\alpha(x_i, e_j, \langle\alpha\rangle\varphi)$ of the transition to the state i of the system is fulfilled under the condition that the formula φ is true.

The problem of detecting anomalies in a system modeled in *DDL* is formulated as follows. Let $M = \langle W, A, \Delta, W_o, \Gamma \rangle$ be a dynamic discrete-event system with a finite set of states W, a finite set of actions A, a function of state transitions Δ, an initial state W_0 and $\Gamma \subseteq X$ – a finite set of marked states of the system. We mark the system states, dividing them into normal, marked with a "+" and anomalous, marked with a "−": $\Gamma = \Gamma^+ \cup \Gamma^-$, $\Gamma^+ \cap \Gamma^- = \varnothing$, by selecting a subset of actions $A_{norm} \subseteq A$, $A_{norm}(v_1, \ldots, v_n) = (P_{norm}, E_{norm})$, for which the preconditions are truth in W, that is, $W \models P_{norm}$ and the subsets of actions A_{anorm}, for which the preconditions are false in W, that is, $W \models \overline{P_{norm}}$. The solution of the problem can be based on the methods presented in [25]. The proposed approach is tested to diagnose the effects of the implementation of high-speed rail processes.

4 Conclusion

All the variety of approaches to modeling dynamic systems that take into account the anomalous effects demonstrates essentially different possibilities of analysis.

Continuous models in the form of differential equations are easier to study in mathematical terms. With the dimension of the variables vector $n \leq 2$, these models are already well studied; for large n, the model can show the chaotic behavior of the object. An important question is the stability of solutions with respect to the numerical values of the paramcters. The problem may also be the lack of proof of the existence and uniqueness of the solution for certain types of equations.

Wavelet analysis is constructed to best approximate the process and is therefore suitable for predicting behavior. By studying the time-frequency decomposition of the process dynamics, it is possible to accurately determine the stable states and the appearance of anomalous development. Wavelet analysis methods are suitable for non-stationary dynamic processes and are increasingly using in data mining systems to determine the measure of the clusters proximity, getting rid of noise. The success of predictions depends on the choice of wavelet and parameter values; the difficulties of applying this approach are also associated with the computational complexity of some wavelets.

Methods for solving a change point problem are aimed at identifying the exact moment of the occurrence of a failure, their quality closely depends on the choice of statistical characteristics of the process and criteria that reduce the number of false alarms and the delay time of the reaction to the anomaly.

Data mining approaches are often criticized for focus on the samples previously encountered in the training set. Here, as in the models of statistical analysis, the correct choice of monitored parameters is important: too little – the model is incomplete – skipping anomalies, too many – false operating.

The logical approach is focused on the detection and analysis of various deviations. Logical modeling in the classical predicates logic assumes an explicit formulation of the anomalies checked by the interpreter, it also allows you to get an accurate diagnosis of the uncertainty of important parameters and contradictions in the implemented knowledge system. However, the method does not allow to adequately specify a sequential chains, accumulate optimal values. The use of dynamic description logic can become more suitable for solving such problems. This approach requires more testing.

To summarize, we can say that almost all of the models considered can predict the development of the process and therefore claim to be predictive. The problem is – is the prediction true? It is known that different models of the same process can give different, even contradictory forecasts. The accuracy of diagnosis depends on the experience and qualification of the researcher, the choice of characteristics that are essential for identifying anomalies, the speed and testability of decision-making procedures. The "normality" of behavior evolves, new data lead to retraining, and the question "is this an anomaly?" may become insoluble. Adapting the model to possible modifications of the "good" and "bad" anomalies can be very difficult. Due to the lack of universal detection procedures, the detection process is poorly automated.

Acknowledgement. The reported study was funded by the Russian Foundation for Basic Research, projects № 19-07-00329-a, 18-08-00549-a.

References

1. Barbashin, E.A.: Lyapunov Functions. Nauka, Moscow (1970)
2. Casti, J.: Connectivity, complexity, and catastrophe in large-scale systems. In: International Institute for Applied Systems Analysis, Brisbane, Toronto. Wiley, Chichester, New York (1979)
3. Marsden, J., McCracken, M.: The Hopf Bifurcation and its Applications. Springer, New York (1976)
4. Arnold, V.I.: Catastrophe Theory, 3rd edn. Springer, Berlin (1992)
5. Dodonov, A.G., Lande, D.V.: Vitality of Information Systems. "Naukova dumka" Publisher, Kiev (2011)
6. Prigogine, I., Stengers, I.: Order Out of Chaos. Man's New Dialogue with Nature. Heinemann, London (1984)
7. Pogosov, A.U.: Diagnostics of the Hidden Dynamics of Processes in NPP Reactor Installations. "Nauka i Tehnika", Odessa (2013)
8. Ostreykovsky, V.A., Pavlov, A.S., Shevchenko, E.N.: Technology-related risk modeling in complex dynamic systems using the Pareto distribution. Proc. Cybern. **1**, 39–44 (2016)
9. Malinetskii, G.G.: Risk control and rare disastrous events. Matem. Mod. **14**(8), 107–112 (2002)
10. Chernavskii, D.S., Nikitin, A.P., Chernavskaya, O.D.: Origins of Pareto distribution in nonlinear dynamic systems. Biophysics **53**(2), 158–163 (2008)
11. Astaf'eva, N.M.: Wavelet analysis: basic theory and some applications. Phys. Usp. **39**(11), 1085–1108 (1996)
12. Shelukhin, O.I., Filinova, A.S.: Comparative analysis of algorithms for detecting traffic anomalies using discrete wavelet analysis. T-Comm: Telecommun. Transp. **8**(9), 89–97 (2014)
13. Barford, P., Kline, J., Plonka, D., Ron, A.: A signal analysis of network traffic anomalies. In: Proceedings of the 2nd ACM SIGCOMM Workshop on Internet Measurement, pp. 71–82 (2002)
14. Burnayev, E.V., Olenyev, N.N.: A measure of proximity for time series based on wavelet coefficients. In: Proceedings of the XLVIII Scientific Conference of MIPT, Dolgoprudny, pp. 108–110 (2005)
15. Shiryaev, A.N.: On conditional-extremal problems of the quickest detection of nonpredictable times of the observable Brownian motion. Theory Probab. Appl. **53**(4), 663–678 (2009)

16. Kozinov, I.A., Maltsev, G.N.: Modified algorithm for detecting change-point of a random process and its application in the processing of multispectral data. Inf. Control Syst. **3**(58), 9–17 (2012)
17. Vorobeichikov, S.E., Konev, V.V.: On the detection of change points in dynamical systems. Autom. Remote Control **51**(3), 324–334 (1990)
18. Lee, K., Kim, J., Kwon, K.H., Han, Y., Kim, S.: DDoS attack detection method using cluster analysis. Expert Syst. Appl. **34**(3), 1659–1665 (2008)
19. Branitskiy, A.A., Kotenko, I.V.: Analysis and classification of methods for network attack detection. SPIIRAS Proc. **45**(2), 207–244 (2016)
20. Toosi, A.N., Kahani, M.: A new approach to intrusion detection based on an evolutionary soft computing model using neuro-fuzzy classifiers. Comput. Commun. **30**(10), 2201–2212 (2007)
21. Kovalev, S.M., Sukhanov, A.V.: Special temporal pattern recognition technique based on hybrid stochastic model. Izvestiya SFedU. Eng. Sci. **4**(153), 142–149 (2014)
22. Ilyicheva, O.A.: Means of effective preliminary treatment of errors for systems of logic prototyping. In: Automation and Remote Control, № 9, pp. 185–196 (1997)
23. Guda, A.N., Ilicheva, V.V., Chislov, O.N.: Executable logic prototypes of systems engineering complexes and processes on railway transport. In: Proceedings of the Second International Scientific Conference "Intelligent Information Technologies for Industry" (IITI 2017), vol. 2, pp. 161–170 (2017)
24. Huang, H., Shi, Z., Wang, J., Huang, R.: DDL: Embracing actions into Semantic Web. IFIP Int. Fed. Inf. Process. **228**, 81–90 (2006)
25. Chang, L., Lin, F., Shi, Z.: A dynamic description logic for representation and reasoning about actions. In: Knowledge Science, Engineering and Management. LNCS, vol. 4798, pp. 115–127. Springer, Heidelberg (2007)

The Problem of Updating Adjustable Parameters of a Grain Harvester in Intelligent Information Systems

Valery Dimitrov(✉), Lyudmila Borisova, and Inna Nurutdinova

Don State Technical University, Rostov-on-Don, Russian Federation
kaf-qm@donstu.ru, borisovalv09@mail.ru, nurut.inna@yandex.ru

Abstract. The problems of updating adjustable parameters of a grain harvester tools operating in changing environmental conditions is considered. For technological adjustment of such hierarchical multilevel systems, intelligent information systems are applied. While performing technological adjustment of the harvester in the process of harvesting, the incoming quantitative, qualitative, and estimating data are analyzed. When considering semantic spaces of environmental factors and the harvester adjustable parameters, different kinds of uncertainty stipulate application of logical-linguistic approach and mathematical apparatus of fuzzy logics. The paper considers the questions of creating a knowledge base for updating adjustable parameters in cases when a value deviation of the quality harvesting indices from the standard one is observed. Since there are a lot of reasons for a fault, and it is unknown beforehand which of them has resulted in deviation, there are quite a number of ways of responding them. Interrelations between indices of functional efficiency and adjustable parameters are established with the help of empirical rules obtained on the basis of expert data. In order to optimize the fuzzy inference mechanism operation of the intelligent information system there appears the necessity for establishing relevance of the applied rules of knowledge base. To solve this problem a game-theory approach has been used, the concepts of efficiency indices matrix and risk matrix of ineffective decision-making have been introduced. An example of choosing a strategy of searching for adequate response to the occurrence of the harvesting indices fault has been given. Laplace criterion, expected-value criterion, and Savage test used for decision-making in "games with nature" have been applied. The analysis of the obtained results has been carried out, conditions and the sphere of application of the suggested approach have been discussed.

Keywords: Intelligent information system · Technological adjustment · Grain harvester · Expert knowledge · Membership function · Linguistic variable · Decision-making

S. Kovalev et al. (Eds.): IITI 2019, AISC 1156, pp. 362–371, 2020.
https://doi.org/10.1007/978-3-030-50097-9_37

1 Introduction

Search The efficiency of harvesting is defined to a large extent by the preset adjustable parameters of a grain harvester. In a number of papers [1–4] the problems of preliminary technological setting of the harvester adjustable parameters have been considered in more detail, so, in [2, 5] the algorithm of the preliminary technological adjustment unit operation of an intelligent information system (IIS) of the harvester control has been described. Prompt updating of technological settings in case of finding out some faults of harvesting quality is no less important problem. Degradation of harvesting quality indices being revealed in sufficient losses, shattering of grain etc. can be caused not only by insufficiently accurate preliminary technological setting of the harvester working tools, but it can be the result of breakdown of its units or a change in environmental conditions of harvesting. The promptly determined cause of performance fault and the proper updating of the harvester operating parameters in the field will make it possible to eliminate sufficient losses in financial and labour resources [6, 7]. These factors stipulate actuality of the problem of creating an updating unit in IIS, which is designed for determining disturbances of harvesting quality and prompt updating of technological adjustments of the harvester tools.

The paper is devoted to the problems of forming expert information for the updating unit of IIS intended for a decision support on the parameters of the harvester technological adjustment in the field.

2 Problem Solution

2.1 Problem Statement

A harvester is referred to multilevel hierarchical systems operating in changing environmental conditions. Expert data on environmental conditions and also on interrelations of these conditions and adjustable harvester parameters are fuzzy by character. It is obvious that to describe such systems, the use of conventional mathematical approaches such as regression models [8–10], experimental and statistical methods [7, 10], is not very effective by virtue of the fact that it is difficult or even impossible to optimize the consequent bulky mathematical constructions in them. In addition, we should note two more significant restrictions for using similar approaches. First, regression and statistical models are only applicable in the range of values under consideration; second, decisions on the updating parameters should be made promptly in the field. To describe decision-making processes and to control technological processes in similar complex systems a mathematical apparatus of the theory of fuzzy sets is used [11]. It allows to operate on fuzzy constraints and aims, set them with the help of linguistic variables. The efficiency of decision-making on the basis of fuzzy models significantly depends on how much the expert data is adequate to a real situation. With regard to the considered problem of updating adjustable parameters of the harvester, expert data adequacy requirements include a large variety of aspects. In the first place, estimation of causes for quality fault in technological process of harvesting. In the second place, the recognition of possible variants of response, i.e. entering the rules of fuzzy produce. In the third place, efficiency estimation of each of the response strategies in the corresponding conditions.

In the hierarchy of solving the problem of technological adjustments prompt updating, the first level deals with the revealing the causes of performance quality faults. The system of interrelations among external attributes of the harvesting process fault, reasons leading to these attributes, and the ways of faults elimination has complex and not always unambiguous character. As a result of investigations, an identification of interrelations is presented in [12], 40 external attributes of the technological process fault are indicated. It has been established that some deviations in operation quality indices depend on 7 and more parameters, so, for example, one external attribute of the thresher performance fault is influenced, on average, by 6 parameters. But not only values of adjustable parameters that influence performance indices, the parameters of technical condition are also essential, for example, for the reaping part it is necessary to take into account 13 parameters of technical condition. The existence of the second group of factors complicates the problem of technological updating.

The next hierarchical level is the choice of the response strategy to the given fault in operation. The difficulty of decision-making is stipulated by a number of circumstances, the most essential of them are possible availability of several faults in performance quality, possible existence of several reasons for the same fault, availability of several variants of the fault elimination, uncertainty of the exact cause of the fault.

2.2 Methods

The decision-making process formalization on the updating parameters of the harvester technological adjustments has a hybrid character: it is intended to use an expert approach, based on fuzzy expert knowledge, for building a hierarchical decision tree, application of probability-theoretical models, and also use of the "games with nature" criteria for estimation of efficiency of the chosen decision-making strategies.

The problem of decision-making concerning the technological parameters updating refers to the class of non-formalized problems of decision-making in fuzzy conditions for the multilevel hierarchical systems, combines are referred to. The problem of updating parameters is complicated by the presence of cross-dependencies among adjustable parameters and different indices of the performance quality. Figure 1 presents a diagram, depicting complexity of the problem of technological adjustments updating. Let us consider "separator fan rotational speed – SFRS" as an adjustable parameter. The vertical scale presents the recommended bounds range of SFRS values change for the certain crop. Let us assume, that in accordance with the certain aims of harvesting and external environment a SFRS value corresponds to the point A. The horizontal scales (for the purpose of simplified representation of the problem statement) present only two quality indices – "hopper grain dockage" and "excessive losses of loose grain with chaff". Each of these scales presents bounds of quality index variability, and the bounds of the accepted values of the given index in accordance with the agricultural requirements (tolerance) are conditionally represented.

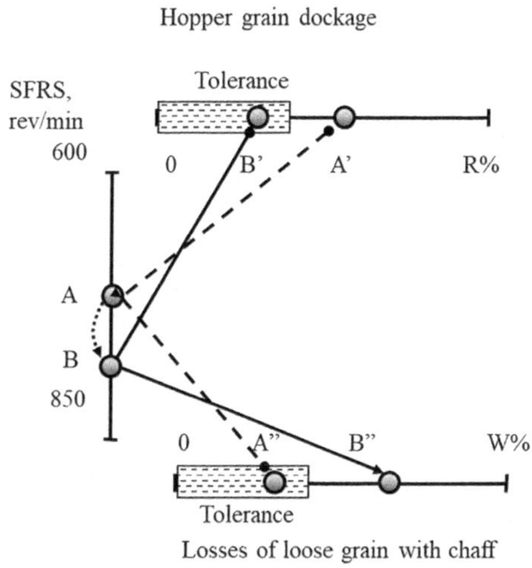

Fig. 1. The diagram of the interrelation: adjustable parameter – quality index

In order to build a mathematical model of a real system and processes of technological adjustment occurring in it, it is necessary to establish a reasonable degree of abstraction. Therefore, assume that the system possesses the following properties:

1. Appearance of more than one external attribute of the technological process fault, i.e. out-of-bounds values of an output parameter, during sufficiently small interval of time is impossible;
2. Meaningfulness (validity) distribution of impact of the system different technical parameters into the possibility of appearing an external attribute of the technological process quality fault is specified a priori.

Property 1 provides the possibility of using expert data on the forms of response to the fault attribute in some system of preferences. The existence of more than one fault attribute requires creation of considerably more complex system of expert data which takes into account possible correlation. At the initial stage there is no necessity in it because of the small interval of time under consideration. Property 2 supposes the availability of knowledge base on the dependence of the technological process quality fault attributes appearance upon technical parameters of the setting. Such a hierarchically-structured data is formed on the basis of expert knowledge and theoretical considerations.

When choosing solutions in fuzzy conditions, expert data is presented in the form of the system of conditional fuzzy statements correlating between the values of input and output parameters of the process of decision-making. In the problem under consideration the system of linguistic fuzzy statements represents empirical rules which establish interrelation between one of the quality indices of the technological process of harvesting and the collection of adjustable parameters of the harvester working tools.

Let us consider the problem where, depending upon possible values of output situation (B_j), an expert states a supposition on possible input situation (A_i), i.e. possible values of input parameters. Let X, Y, Z,... denote a set of values of input parameters, adjustable parameters of the harvester working tools, which substantially influence the magnitude of the output parameter designated as V. Let us introduce linguistic variables (LV's) for input and output parameters. According to logic-linguistic approach [12, 13] we have developed the models of input and output characteristics X, V in the form of semantic spaces and corresponding them membership functions (MF's):

$$\{X_i, T(X_i), U, G, M\}, \mu_R(x_1, x_2, \ldots, x_{i,}) \in (0; 1),$$
$$\{Y_j, T(Y_j), U, G, M\}, \mu_R\left(y_1, y_2, \ldots, y_{j,}\right) \in (0; 1),$$
$$\ldots\ldots\ldots\ldots\ldots\ldots\ldots\ldots\ldots\ldots\ldots\ldots\ldots\ldots\ldots\ldots$$
$$\{V_k, T(V_k), U, G, M\}, \mu_R(v_1, v_2, \ldots, v_{j,}) \in (0; 1),$$

where X, Y, V are names of the LV, T - set of its values, or terms, which are the names of the LV's defined over the set U, G - syntactic procedure describing the process of deriving the new values of LV's from the set T, M - semantic procedure, which allows to map a new value generated by the procedure G into fuzzy variable, μ - MF's.

Taking into account the example chosen for consideration, we cite a linguistic description of the external attribute of the harvester performance quality indices fault "Shattering of grain" and one of the adjustable harvester parameters which is connected with the given attribute "Clearance uniformity between a threshing drum and a concave".

A tuple of the LV "Shattering of grain" has the form:

$$< \text{Shattering of grain}, \ \% \{\text{Negligible, Average, Increased}\}, \ [0 - 3] > .$$

A tuple of the LV "Clearance uniformity between a threshing drum and a concave along the width of the threshing and separating unit (CUTDC)" has the form:

$$< \text{CUTDC}, \ \% \{\text{Non-uniform, Uniform}\}, \ [0 - 50] > .$$

The graphs of MF's of linguistic variables "Shattering of grain" and "Clearance uniformity between a threshing drum and a concave" are presented in Fig. 2.

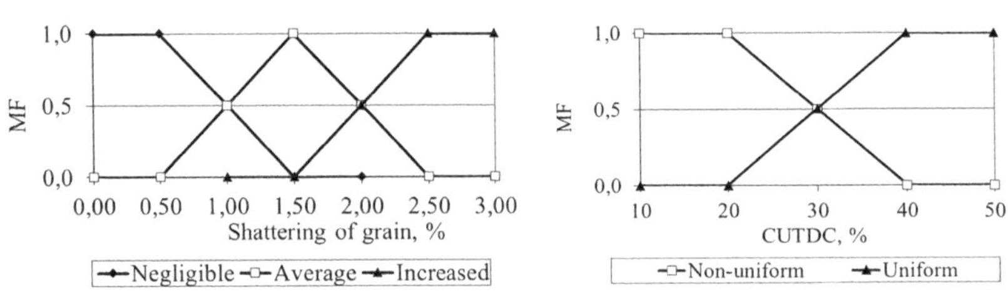

Fig. 2. Membership functions of LV's: a) «Shattering of grain»; b) «Clearance uniformity between a threshing drum and a concave».

The result of the analysis is a generalized model of the domain «technological adjustment» [12–14] in the form of composition of fuzzy relations of the semantic spaces under consideration:

$$R_1 \circ R_2 \, for \, \forall x \in X; \; \forall y \in Y; \; \forall v \in V;$$
$$\mu_{R_1 \circ R_2}(x, v) \vee \left(\mu_{R_1}(x, y) \wedge \mu_{R_2}(y, v)\right)$$

where $\mathrm{R_1}$ is fuzzy relation "harvesting factors – adjustment parameters," $\mathrm{R_1\{X_i, T(X_i), U, G, M\} \times \{Y_j, T(Y_j), U, G, M\}; \forall(x, y) \in X \times Y;}$ $\mathrm{R_2}$ is fuzzy relation among adjustment parameters and the harvester performance indices, $\mathrm{R_2\{Y_j, T(Y_j), U, G, M\} \times \{V_k, T(V_k), U, G, M\}; \forall(y, v) \in Y \times V}$.

Let us consider the system of logic statements, depicting an expert's experience on updating the harvester adjustable parameters at the appearance of the external attribute of the technological process fault. Let us write the statement $\mathrm{E_{ji}}$: <β_W is α_{Eii}> where β_W - is a generalized LV, determined on the set $\mathrm{W = X \times Y \times Z}$ and taking basic values α_{Eji} with MF: $\mu_{Eji}(w) = \min\{\mu_{Xji}(x), \mu_{Yji}(y), \mu_{Zji}(z)\ldots\}$.

Let us designate the statements <β_W is α_{Wj}> and <β_V is α_{Vj}> through $\tilde{A}_j$ and $\tilde{B}_j$. Then the system fuzzy statements will be written in the form of:

$$\tilde{L}^{(1)} = \begin{cases} \tilde{L}_1^{(1)}: & <\text{if} \quad \tilde{A}_1, \quad \text{then} \quad \tilde{B}_1>, \\ \tilde{L}_2^{(1)}: & <\text{if} \quad \tilde{A}_2, \quad \text{then} \quad \tilde{B}_2>, \\ \ldots & \\ \tilde{L}_m^{(1)}: & <\text{if} \quad \tilde{A}_m, \quad \text{then} \quad \tilde{B}_m> \end{cases}$$

Fuzzy statements correspond to the general form:

$$\tilde{A}_j :< \beta_W \quad \text{is} \quad \alpha_{W_j} > \quad \text{и} \quad \tilde{B}_j :< \beta_V \quad \text{is} \quad \alpha_{V_j} >,$$

where $\alpha_{V_1}, \alpha_{V_2}, \alpha_{V_3}$ are corresponding values of the terms of the output LV [14, 15].

Let us consider in detail the mechanism of forming an expert knowledge base which is structured in order of importance and sequence of application.

Suppose, in the process of the grain harvester operation, a deviation of performance indices is found. This deviation might be stipulated both by changes of external conditions in which the harvester operates, and by changes of its technical condition. A lot of such prerequisites which result in the given deviation of quality indices will be designated as $S = \{B_j\}|_{j=1}^{n}$. An expert analysis of empirical data and theoretical considerations make it possible to estimate actuality of each of the reasons for deviation and determine possible ways of response so as to eliminate faults in operation. Each of the ways of response will be called as an admissible strategy, and the whole set of admissible strategies will be designated as $G = \{A_i\}|_{i=1}^{m}$. The impact efficiency of each of the strategies depends both on the reason of the fault, i.e. B_j, and on many other ones, among them random factors, collective action of which can be presented as an efficiency of application of the mentioned strategy. Let us consider *m* by *n* matrix C, each element c_{ij} of which determines the result of application of strategy A_i to eliminate faults caused by reason B_j. Element magnitudes of matrix C change in the range from 0 to 1, and the more efficient application of the strategy A_i to eliminate the fault caused by the reason B_j, the closer the value of the element c_{ij} to unit. We will call matrix C as a matrix of efficiency indices and consider its elements as initial data of the decision-making procedure for choosing optimal strategy of technological adjustment. This approach allows to use criteria applied in single-stage procedures of decision-making under uncertainty [16].

Since it is unknown beforehand which of the reasons has caused deviation of the harvester performance indices, then there is no necessary information for conclusion that the probabilities of these reasons are different. On the basis of the principle of insufficient reason [16] we can assume that the probabilities of reasons are the same. Then the problem of choice of technological adjustment optimal strategy may be regarded as the problem of decision-making under risk when solution $A^* \in G = \{A_i\}_{i=1}^{m}$ is chosen which provides the greatest anticipated value of efficiency. In this case it is reasonable to apply Laplace criterion [16]:

$$\max_i L(i) = \max_i \frac{1}{m} \sum_{j=1}^{m} c_{ij}. \tag{1}$$

Upon availability of a priory values of possibilities p_j of reasons presence $\{B_j\}$, obtained, for example, by the method of expert estimation or as a result of the analysis of statistical information, this criterion acquires the form:

$$\max_i M(i) = \max_i \sum_{j=1}^{m} c_{ij} p_j. \tag{2}$$

Let us introduce the concept of risk matrix R of ineffective decision, the elements r_{ij} of which are obtained in the following manner. It is obvious, that with the known reason B_j of the performance quality fault, the most efficient strategy A_i would be used wherein efficiency index takes the greatest value, it is the maximal element in column j, we designate it β_j. In order to obtain the elements of risk matrix r_{ij}, it is necessary to deduct a true value of efficiency index from β_j:

$$r_{ij} = \beta_j - c_{ij}. \tag{3}$$

We use Savage test [16] of minimal risk where as an optimal strategy we choose the one wherein risk magnitude in the worst conditions is minimal:

$$\min_i S(i) = \min_i \max_j r_{ij}. \tag{4}$$

The considered approach allows us to eliminate great risk in decision-making. The application of optimality criteria makes it easy to choose decisions under the conditions of uncertainty. Which of the criteria should be used in a certain situation depends on presence or absence of expert estimations of rejection reason probabilities, risk significance in decision-making and also some other factors.

2.3 Example

To illustrate this, we consider the criteria application features when choosing the way of eliminating faults during the harvester operation in the field. Let us consider, for example, an external attribute of the harvester operation fault «Shattering of grain». It has been determined [12] that the reasons for this fault may be inadequately set parameters or technical condition of the harvester working tools: B_1 – rotational speed of the threshing drum; B_2 - rasps and concave condition; B_3 - clearance uniformity between a threshing drum and a concave along the width of the threshing and separating unit; B_4 - drum – concave clearance. There are possible strategies of action to eliminate the given fault: A_1 - moderate decrease of the threshing drum rotational speed; A_2 - considerable decrease of the threshing drum rotational speed; A_3 - moderate increase of the drum – concave clearance; A_4 - considerable increase of the drum – concave clearance; A_5 - providing uniformity of the threshing drum – concave clearance by the way of repairing or replacement; A_6 - repairing of rasps. Matrix C of the calculated values of the efficiency index is presented in Table 1.

Table 1. Matrix C of the efficiency index values, Laplace criterion $L(i)$ and risks matrix.

Strategy number	Matrix C				$L(i)$	Matrix R			
1	0,8	0,1	0,4	0,6	0,475	0,05	0,7	0,35	0,2
2	0,85	0,1	0,2	0,5	0,4125	0	0,7	0,55	0,3
3	0,55	0,1	0,6	0,8	0,5125	0,3	0,7	0,15	0
4	0,6	0,05	0,5	0,7	0,4625	0,25	0,75	0,25	0,1
5	0,2	0,6	0,75	0,4	0,4875	0,65	0,2	0	0,4
6	0,3	0,8	0,3	0,2	0,4	0,55	0	0,45	0,6
$\max_i c_{ij}(r_{ij})$	0,85	0,8	0,75	0,8		0,65	0,75	0,55	0,6

Without any a priori values of reason probabilities which have caused a fault of the technological process of harvesting, we apply Laplace criterion (1). The criterion values for each of the strategies are indicated in the sixth column of Table 1. It is obvious that the optimal strategy for this criterion is strategy A_3 - moderate increase of the drum – concave clearance.

To determine an optimal strategy according to Savage test we produce the risks R matrix of ineffective decision-making. We find maximal elements in each column (they are indicated in the end line of Table 1) and calculate the elements of the risks matrix from the formula (3). As a result we obtain the risks matrix (Table 1).

In the end line of Table 1 maximal elements are indicated column wise. When choosing the minimal element we obtain the best strategy according to Savage test (4); it is A_2 – considerable decrease of the threshing drum rotational speed.

Upon availability of a priori values of reason probabilities, which have caused the given fault of the harvester performance indices, we apply the criterion of maximal mathematical expectation (2). Let us use for the probabilities B_j the following values: p_1 = 0,28; p_2 = 0,25; p_3 = 0,18; p_4 = 0,2. Note, that the sum of values of probabilities p_j is not equal to 1, since the estimates of probabilities are obtained on the basis of expert estimates and the analysis of statistical data. Besides, there may be some other factors which have caused appearance of the external attribute of the harvester operation fault «Shattering of grain».

The values of the criterion of mathematical expectation calculated by formula (2): $M(1) = 0{,}441$; $M(2) = 0{,}399$; $M(3) = 0{,}447$; $M(4) = 0{,}371$; $M(5) = 0{,}421$; $M(6) = 0{,}378$. Therefore, the optimal strategy according to the criterion of maximal mathematical expectation is the strategy A_3 - moderate increase of the drum – concave clearance.

The obtained sequences of criteria values are designed to use in updating algorithm in IIC to determine the sequence of operations so as to eliminate the given fault. Sorting of rules in the knowledge base in accordance with their efficiency is achieved. The problem of sorting rules consists in choosing the strategy wherein the most probably correct conclusion on the condition that the base has the rules with efficiency less than 100%.

3 Conclusion

The suggested approach based on the models of expert knowledge makes it possible to unite the available heterogeneous data (determinate, statistical, linguistic) characterizing external environment, parameters of the harvester technical condition, and its performance indices, and use them when forming the decision support systems in the sphere of harvester operation. Implementation of the considered in the paper algorithm for searching the most effective strategy of response to the technological process faults is based on the application of optimality criteria, it allows to structure the expert knowledge base in accordance with efficiency of production rules, and herewith does not require considerable "computational resources". Introduction, as a subsystem, of the expert system inference mechanism for harvester technological adjustment of the considered rational decisions search procedure will make it possible to obtain variants of decisions which are adequate for certain harvesting conditions, harvester trademarks, technical condition of the certain harvester. A significant advantage is cutting time for decision-making due to object-orientated use of the knowledge base rules.

The suggested approach can be used in different IIS with feedback where there appears the necessity to put in order the knowledge base in accordance with efficiency of empirical rules. In addition, it will be helpful while solving such problems as search for fault reasons in different technical systems and machines.

References

1. Borisova, L., Dimitrov, V., Nurutdinova, I.: Intelligent system for technological adjustment of the harvesting machines parameters. In: Advances in Intelligent Systems and Computing, vol. 680, pp. 95–105 (2018)
2. Dimitrov, V., Borisova, L., Nurutdinova, I.: Intelligent support of grain harvester technological adjustment in the field. In: Advances in Intelligent Systems and Computing, vol. 875, pp. 236–245 (2019)
3. Omid, M.: Design of an expert system for sorting pistachio nuts through decision tree and fuzzy logic classifier. Expert Syst. Appl. **38**, 4339–4347 (2011)
4. Craessaerts, G., de Baerdemaeker, J., Missotten, B., Saeys, O.: Fuzzy control of the cleaning process on a combine harvester. Biosyst. Eng. **106**, 103–111 (2010)
5. Borisova, L., Dimitrov, V., Nurutdinova, I.: Algorithm for assessing quality of fuzzy expert information. In: Proceedings of IEEE East-West Design & Test Symposium (EWDTS 2017), Serbia, pp. 319–322 (2017)
6. Rybalko, A.G.: Osobennosti uborki visokourojainih zernovih kultur (Some Features of Harvesting High-Yield Crops). Agropromizdat, Moscow (1988). (in Russian)
7. Yerokhin, S.N., Reshetov, A.S.: Vliyanie tekhnologicheskih regulirovok na poteri zerna za molotilkoj kombajna Don-1500 (Influence of technological adjustments on grain loss behind the thresher of the combine Don-1500). Mech. Electrif. Agric. **6**, 18–19 (2003). (in Russian)
8. Vetrov, E.F., Genkin, M.D., Litvin, L.M., et al.: Optimizaciya tekhnologicheskogo processa po statisticheskim dannym (Optimization of technological process according to statistic data). Mashinovedenie **5**, 48–55 (1986). (in Russian)
9. Litvin, L.M., Zhalkin, E.V., Vetrov, E.F.: Obobshchennaya ocenka zonal'nyh pokazatelej raboty zernouborochnyh kombajnov (Generalized estimation of zone operational performance of combine harvesters). Mach. Agric. **5**, 41–45 (1989). (in Russian)
10. Tsarev, Y.A., Kharkovsky, A.V.: Perspektivy ispol'zovaniya elektronnoj sistemy upravleniya v kombajnah «Don» i «Niva» (The prospects of using electronic control system in combine harvesters "Don" and "Niva"). Tract. Agric. Mach. **1**, 37–38 (2005). (in Russian)
11. Zadeh, L.: Knowledge representation in fuzzy logic. In: Yager, R.R., Zadeh, L.A. (eds.) An Introduction to Fuzzy Logic Applications in Intelligent Systems. The Springer International Series in Engineering and Computer Science, vol. 165, pp. 1–27. Springer, New York (1992)
12. Borisova, L.V., Dimitrov, V.P.: A linguistic approach to the solution of the problem of technological adjustment of combines. Mordovia Univ. Bull. **27**(2), 181–193 (2017)
13. Borisova, L.V., Nurutdinova, I.N., Dimitrov, V.P.: Approach to the problem of choice of values of the adjustable parameters harvester based on fuzzy modeling. Don State Tech. Univ. Bull. **5-2**(81), 100–107 (2015)
14. Borisova, L.V., Nurutdinova, I.N., Dimitrov, V.P.: Fuzzy logic inference of technological parameters of the combine-harvester. WSEAS Trans. Syst. **14**, 278–285 (2015)
15. Dimitrov, V., Borisova, L., Nurutdinova, I.: The problem of choice of optimal technological decisions on harvester control. In: MATEC Web of Conference XIV International Scientific-Technical Conference "Dynamic of Technical Systems" (DTS-2018), vol. 226, p. 04023 (2018)
16. Taha, H.A.: Vvedenie v issledovanie operacij (Operations Research: Introduction). Vil'yams, Moscow (2005). (in Russian)

The Development the Knowledge Base of the Question-Answer System Using the Syntagmatic Patterns Method

Nadezhda Yarushkina, Aleksey Filippov(✉), and Vadim Moshkin

Ulyanovsk State Technical University, Ulyanovsk, Russia
{jng, al.filippov, v.moshkin}@ulstu.ru

Abstract. This paper presents an ontological model of a text document of a large electronic archive. In addition, the article contains original algorithms for extracting syntagmatic patterns from text documents. The paper also describes the developed search algorithms in the knowledge base of the electronic archive using the mechanism of syntagmatic patterns. In conclusion, the results of the experiments on the developed question-answer (QA) system in comparison with existing information systems are presented.

Keywords: QA-systems · Ontology · Syntagmatic patterns · Knowledge base

1 Introduction

Getting data from large repositories of unstructured information is a complex task and requires the use of intelligent algorithms.

Currently, intelligent question-answer systems [1–6] based on the UIMA architecture [4–7] are being developed to extract the necessary knowledge from large repositories.

The information system should perform the following steps to get an answer to a user's question:

- preprocess the question in a natural language;
- "understand" the question, ie conduct a series of analyzes (syntactic, statistical, semantic, cognitive);
- to formulate a question to the knowledge base (electronic repository). The repository of knowledge can be represented in various forms: a database of text documents, a wiki resource, a database, etc.;
- conduct selection and evaluation of the responses received;
- convert the data to natural language;
- display a response to the user.

S. Kovalev et al. (Eds.): IITI 2019, AISC 1156, pp. 372–383, 2020.
https://doi.org/10.1007/978-3-030-50097-9_38

Thus, the work of question-answer systems includes the resolution of important problems [7–9]:

- the need to structure information storage to increase the speed and efficiency of information retrieval;
- the need to attract specialists in a specific subject area for learning the question-answer system.

In this paper, we proposed an original ontological model of a text document of a large electronic archive. In addition, we proposed search algorithms in the knowledge base of a question-answer system using the mechanism of syntagmatic patterns [10].

A syntagmatic pattern is a pattern that is used to search semantically independent word combinations in text data. These combinations are called syntagmatic units.

The basic principles that form the basis of the proposed approach are:

1. Ontological structuring and classification of text resources allows you to create the necessary knowledge base for a question-answer system.
2. The knowledge base consists of syntagmatic units. Each syntagmatic unit can be retrieved using a pattern search (syntagmatic patterns).

2 Model of Applied Ontology of Knowledge Base Text Documents

Building an ontology in the classification of documents is necessary to take into account the characteristics of the subject area of the organization and increase the speed of searching for the necessary data. The ontology defines a semantic scale for defining a set of documents from one class.

Formally, the applied ontology of the knowledge base is:

$$O_{ARC} = \langle T, T_{ORG}, Rel, F \rangle,$$

where T is a set of terms of knowledge base documents;

T_{ORG} is a set of terms in a problem area of an organization;

Rel is a set of ontology relationships. A set of relationships includes the following components:

$$Rel = \{R_H, R_{PartOF}, R_{ASS}\},$$

where R_H is a hierarchy relation;

R_{PartOF} is a part-whole relationship;

R_{ASS} is an relation of association.

Formally, a set of terms for documentation of a knowledge base of an information system can be represented as follows:

$$T = \left(T^{D_1} \cup T^{D_2} \cup \ldots \cup T^{D_k}\right) \cup T^{ARC},$$

where $T^{D_i}, i = \overline{1,m}$ - is the set of terms of the i-th problem area;

T^{ARC} is a set of terms of a problem area obtained from documents of an electronic archive of an organization.

Formally, the functions of interpretation of the subject ontology are presented as follows:

$$F = \{F_{T_{ORG}T}, F_{T^{ARC}T^D}\},$$

where $F_{T_{ORG}T}: \{T_{ORG}\} \rightarrow \{T\}$ is an interpretation function that defines the correspondence between the terms of the problem area of the organization and the terms of the electronic archive documentation;

$F_{T^{ARC}T^D}: \left\{T^{ARC}\right\} \rightarrow \left\{T^D\right\}$ is an interpretation function that defines the correspondence between the terms of the problem domain obtained from documents of the electronic archive of the organization and the terms of the problem domain [11].

An example of the knowledge base structure of a QA-system using the example of an electronic documentation archive is shown in Fig. 1.

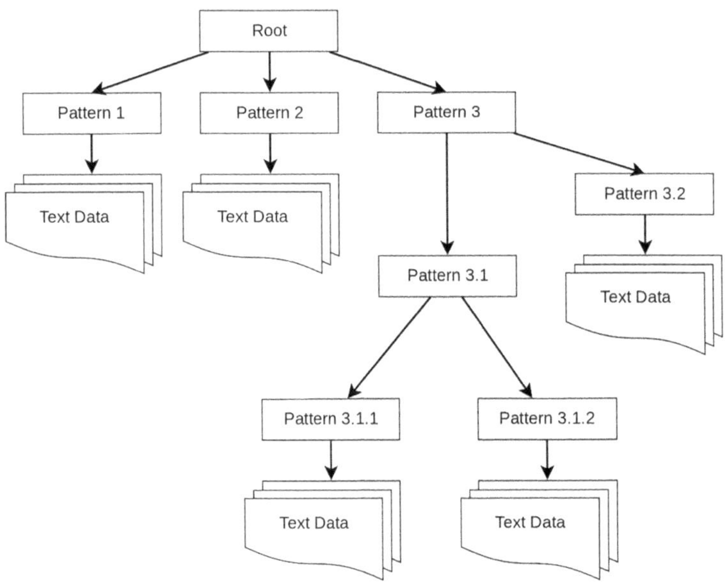

Fig. 1. An example of the knowledge base structure of a QA-system.

3 The Algorithm for Extracting Syntagmatic Patterns

Currently, natural language processing methods [1, 11–15] are used to develop complex question-answer systems. First of all, the following types of text data analysis are used: graphematic, morphological, syntactic and semantic.

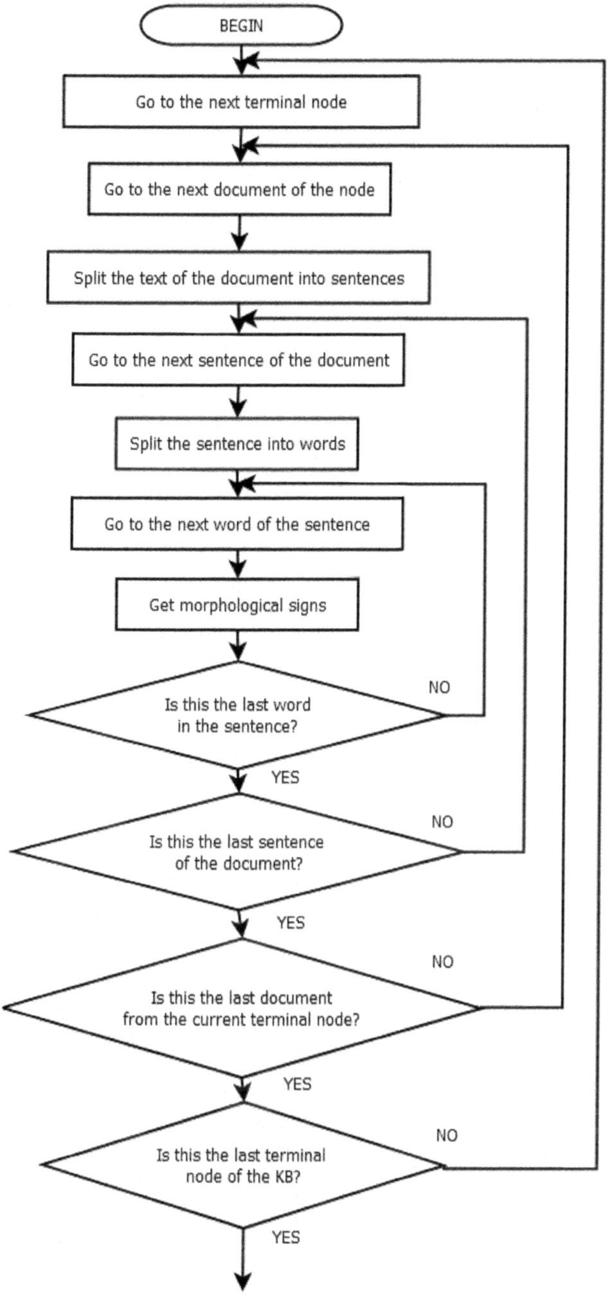

Fig. 2. The algorithm for extracting syntagmatic patterns (Part 1).

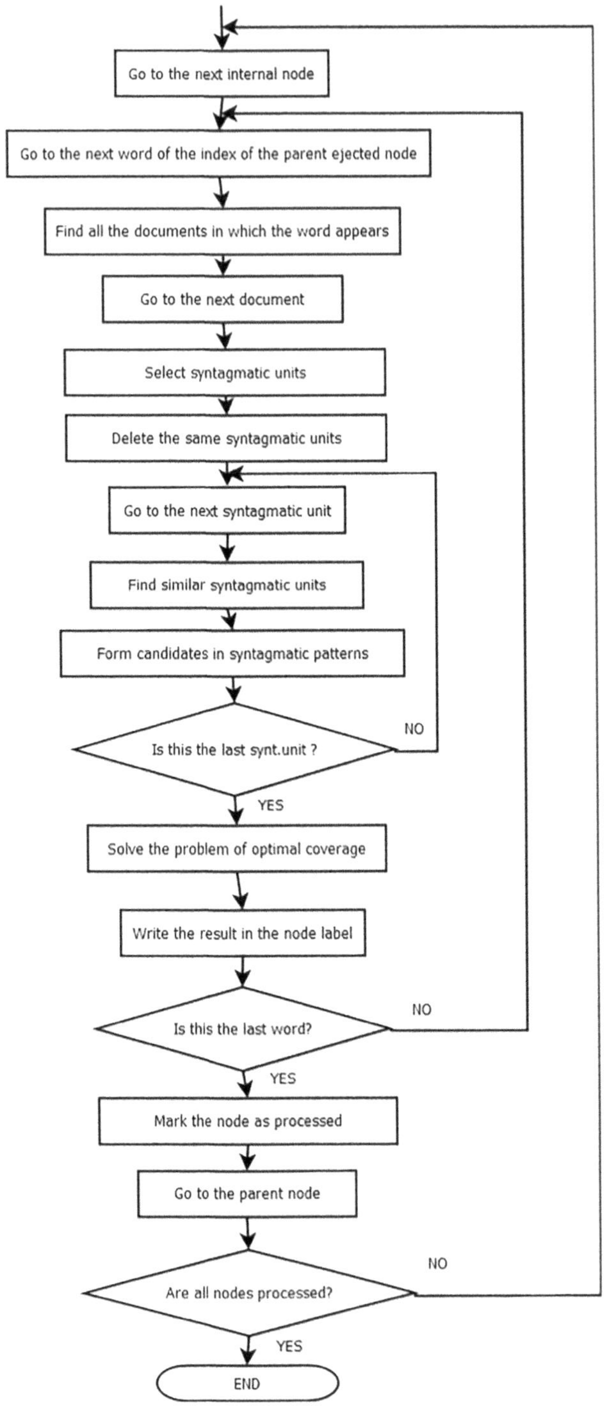

Fig. 3. The algorithm for extracting syntagmatic patterns (Part 2).

The approach proposed in this paper is to obtain syntagmatic patterns, with the help of which the question-answer system finds syntagmatic units in the knowledge base.

The following types of analysis must be performed to extract syntagmatic patterns from text fragments:

- graphematic
- morphological.

The algorithm for extracting syntagmatic patterns is shown in Fig. 2 and Fig. 3.

Thus, internal nodes of the knowledge base tree are marked with syntagmatic patterns as a result of this algorithm.

4 The Search for the Answer to the Question in the Knowledge Base

Preprocessing text resources to find answers to questions involves extracting syntagmatic patterns. In this way, text documents are structured and ready to search for syntagmatic units.

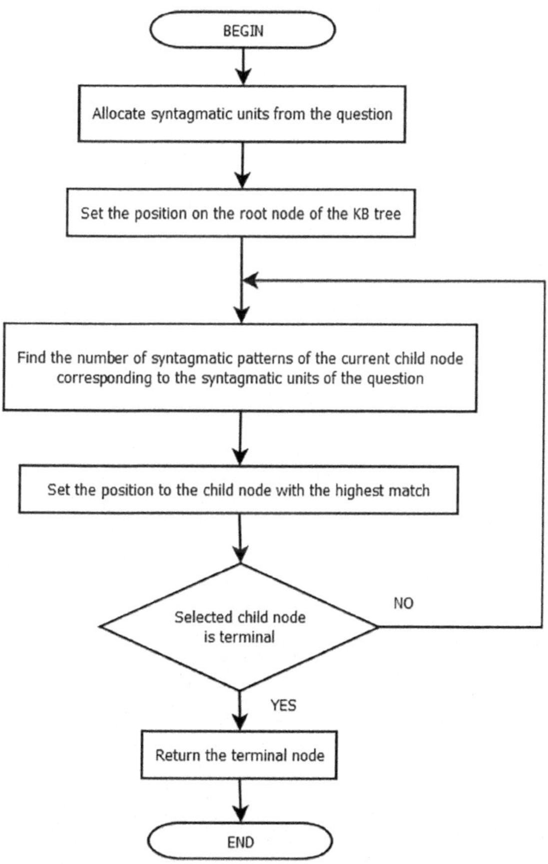

Fig. 4. The search algorithm for the most relevant terminal node of the KB tree

The algorithm for finding the answer to the question using syntagmatic patterns is presented in Fig. 4.

It is necessary to find in the text documents the most relevant sentence after finding the terminal node. The answer to the question is the most relevant sentence [14].

The search algorithm for the most relevant sentence from the text documents of the terminal node found is represented in Fig. 5:

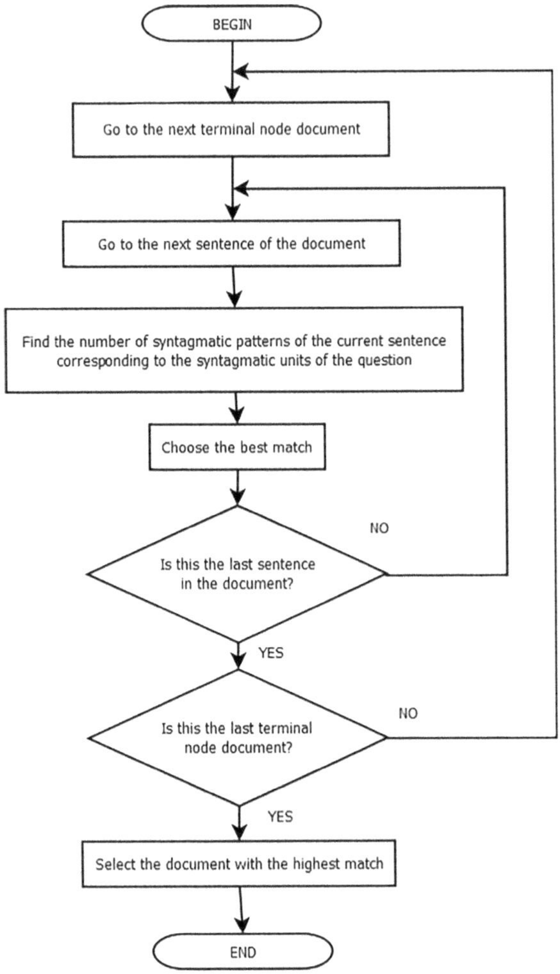

Fig. 5. The search algorithm for the most relevant sentence from the text documents

Thus the two algorithms presented above make it possible to organize the search for the most relevant answer to an incoming question [14].

5 Experiments

Some existing question-answer systems that are modules of home automation systems were considered. Also, an analysis of information retrieval approaches in the KB and the generation of responses was carried out. The main result of the question-answer system is the answer to the question asked. To evaluate the response, you must use the following criteria:

1. Relevance: the answer must answer the question.
2. Correctness: the answer must actually be correct.
3. Expressiveness: the answer should not contain extraneous and irrelevant information.
4. Completeness: the answer must be complete (and not contain only part of the answer to the question).
5. Validity: the answer must be accompanied by an appropriate context, allowing the user to understand why this particular answer was chosen.

Answers received by question-answer systems are conventionally divided into "long" (250 characters) and "short" (50 characters). The main metrics used in evaluating the effectiveness of question-answer systems are accuracy, completeness, F-measure, and also average mutual rating. The listed metrics have the following models:

$$Accuracy = \frac{Number\ of\ correct\ answers}{Number\ of\ answered\ questions}$$

$$Completness = \frac{Number\ of\ correct\ answers}{Total\ number\ of\ questions}$$

$$F\text{-}measure = \frac{2 * (Accuracy * Completness)}{Accuracy + Completness}$$

The average mutual rating is a statistical measure used to evaluate systems that provide a list of possible answers to a question:

$$MRR = \sum_{i=1}^{n} \frac{1}{r_i}$$

where n is the number of questions, and r_i is the number of the first correct answer to the question.

The experimental environment consisted of 200 technical issues, all of which were covered by a hierarchical classification of questions. All data for the experiments were taken from the KB of the TREC conference.

QA-System LASSO. This question-answer system uses a combination of syntactic and semantic analysis of the question. The heuristic rules for finding the keywords used to identify an answer are determined during the analysis process. In addition, filtering and indexing of paragraphs is applied. Table 1 shows the results of testing the system:

Table 1. The results of experiments with the QA-system LASSO

	MRR (soft score)	MRR (strict score)
Short answer	68.1%	55.5%
Long answer	77.7%	64.5%

QA-System FALCON. This QA system is an advanced version of LASSO using WordNet for semantic processing of questions. A module for checking and screening responses has been added so that the system at the output provides only one answer to overcome the main drawback of LASSO. Table 2 shows the results of testing the system:

Table 2. The results of experiments with QA-system FALCON

	MRR (soft score)	MRR (strict score)
Short answer	59.9%	58.0%
Long answer	77.8%	76.0%

QA-System QA-LaSIE. Information search functionality interacts with a natural language processing system that performs linguistic analysis in this software system. The information retrieval system processes the question as a search query, and returns a set of ordered documents or passages. The natural language processing system additionally analyzes the question and the documents and adds their semantic representation to them. These views were used to extract the answer.

Testing was conducted using two different search engines, Table 3 shows the best of the results.

Table 3. The results of experiments with a QA-system QA-LaSIE

	MRR (soft score)	MRR (strict score)
Short answer	26.67%	16.67%
Long answer	53.33%	33.33%

The proposed approach was not very successful, since only two thirds of the test questions were processed. Also, the system was limited to a small number of domain ontologies.

ASQA QA System. The system uses phrases that are automatically extracted from the question, as part of search queries. In the process of extracting phrases, techniques such as the recognition of named entities, lists of noise words and recognition of parts of speech are used.

The system was tested on two data sets. One was the filing of technical documentation, the other was open wiki sources from the Internet. Testing was conducted in two modes: fully automatic extraction of phrases and extraction of phrases, followed by manual adjustment. Test results showed that the overall accuracy of working with the Internet-based documents was lower than with the documentation. Manual adjustment of requests increased the quality of the system, but not by much. Table 4 shows the best results.

Table 4. The results of experiments with the QA-system ASQA

	MRR (soft score)	MRR (strict score)
Short answer	53.2%	54.0%
Long answer	71.8%	67.0%

Our Project QA-System. The question-answer system is based on hybrid algorithms for semantic-cognitive analysis of various types of unstructured and semi-structured information resources using natural language processing and knowledge engineering approaches within a single software environment.

The knowledge base of the QA-system implements the methods of adaptation to various problem areas through automated training (wiki-resources, technical and methodological documentation, PCU) using the principles of knowledge engineering. Among the question-answer systems considered, this system showed the best result, especially in the "soft evaluation" section of the answers (Table 5).

Table 5. The results of experiments with the QA-system

	MRR (soft score)	MRR (strict score)
Short answer	65.9%	64.0%
Long answers	80.8%	79.0%

Of course, many question-answer systems are closed and satisfy the needs of one particular company for a specific subject area, so conducting experiments on all systems is not possible. Despite this, the results of the experiments can be considered objective and prove the effectiveness of the developed and software-implemented semantic analysis algorithms for semi-structured resources within the home automation system.

6 Conclusion

Thus, the work of question-answer systems involves the resolution of important problems:

- the need to structure information storage to increase the speed and efficiency of information retrieval;
- the need to attract specialists in a specific subject area for training the question-answer system.

In this work, an ontological model of a text document of a large electronic archive was proposed for solving the tasks. This electronic archive is a knowledge base for question and answer systems.

An algorithm was developed to extract syntagmatic patterns from text fragments and search for syntagmatic units using them. As the results of experiments showed, this approach is effective in comparison with the question-answer systems FALCON, LASSOQ, A-LaSIE, ASQA when searching for answers to open questions.

In the future, it is planned to develop algorithms for searching syntagmatic patterns in various languages.

Acknowledgments. This paper has been approved within the framework of the federal target project "R&D for Priority Areas of the Russian Science-and-Technology Complex Development for 2014-2020", government contract No 05.604.21.0252 on the subject "The development and research of models, methods and algorithms for classifying large semistructured data based on hybridization of semantic-ontological analysis and machine learning".

References

1. Ahmad, S.: Smart metering and home automation solutions for the next decade. In: International Conference on Emerging Trends in Networks and Computer Communications (ETNCC) (2011)
2. Balsamo, D., Gallo, G., Brunelli, D.: Non-intrusive Zigbee power meter for load monitoring in smart buildings. In: Sensors Applications Symposium (SAS). IEEE (2015)
3. Farhangi, H.: The path of the smart grid. IEEE Power Energy Mag. **8**(1), 18–28 (2010)
4. Jurafsky, D., Martin, J.H.: Speech and language processing [Electronic resource], 14 April 2020. https://web.stanford.edu/~jurafsky/slp3/28.pdf
5. Berant, J., Chou, A., Frostig, R., Liang, P.: Semantic parsing on freebase from question-answer pairs. In: Proceedings of the 2013 Conference on Empirical Methods in Natural Language Processing (EMNLP), pp. 1533–1544 (2013)
6. Bordes, A., Chopra, S., Weston, J.: Question answering with subgraph embeddings [Electronic resource], 14 April 2020. https://arxiv.org/pdf/1406.3676.pdf
7. Epstein, E.A., Schor, M.I., Iyer, B., Lally, A., Brown, E.W., Cwiklik, J.: Making Watson fast. IBM J. Res. Dev. **56**(3.4), 15–19 (2012)
8. Ferrucci, D., Brown, E., Chu-Carroll, J., Fan, J., Gondek, D., Kalyanpur, A.A., Lally, A., Murdock, J.W., Nyberg, E., Prager, J., et al.: Building Watson: an overview of the DeepQA project. AI Mag. **31**(3), 59–79 (2010)
9. Gallagher, S., Zadrozhny, W., Shalaby, W., Avadhani, A.: Watsonsim: overview of a question answering engine [Electronic resource], 14 April 2020. https://arxiv.org/pdf/1412.0879.pdf
10. Ferrucci, D., Lally, A.: UIMA: an architectural approach to unstructured information processing in the corporate research environment. Nat. Lang. Eng. **10**(3–4), 327–348 (2004)
11. Bollacker, K., Evans, C., Paritosh, P., Sturge, T., Taylor, J.: Freebase: a collaboratively created graph database for structuring human knowledge. In: Proceedings of the 2008 ACM SIGMOD International Conference on Management of Data, pp. 1247–1250 (2008)
12. Chen, D., Manning, C.D.: A fast and accurate dependency parser using neural networks. In: Proceedings of the 2014 Conference on Empirical Methods in Natural Language Processing (EMNLP), pp. 740–750 (2014)

13. Zarubin, A., Koval, A., Filippov, A., Moshkin, V.: Application of syntagmatic patterns to evaluate answers to open-ended questions. In: Proceedings of the 2017 Communications in Computer and Information Science (CITDS), pp. 150–162 (2017)
14. Zarubin, A.A., Koval, A.R., Filippov, A.A., Moshkin, V.S.: The applying of syntagmatic patterns for the development of question-answer systems. In: The Collection: Information Technologies and Nanotechnologies ITNT-2018 Proceedings, pp. 2422–2428. Academician S.P. Korolev Samara National Research University (2018)
15. Boyarskiy, K.K., Kanevskiy, Ye.A.: Semantic and syntactic parser SemSin. Nauchno-tekhnicheskiy Vestn. informatsionnykh tekhnol. mekhanikiioptiki **5**, 869–876 (2015)
16. Artemov, M.A., Vladimirov, A.N., Seleznev, K.E.: Review of Russian NLP systems [Electronic resource], 14 April 2020. http://www.vestnik.vsu.ru/pdf/analiz/2013/02/2013-02-31.pdf
17. Automatic text processing [Electronic resource], 14 April 2020. http://aot.ru
18. Lally, A., Prager, J.M., McCord, M.C., Boguraev, B., Patwardhan, S., Fan, J., Fodor, P., Chu-Carroll, J.: Question analysis: how Watson reads a clue. IBM J. Res. Dev. **56**(3.4), 2–14 (2012)
19. Mikhaylov, D.V., Kozlov, A.P., Emelyanov, G.M.: An approach based on analysis of n-grams on links of words to extract the knowledge and relevant linguistic means on subject-oriented text sets. Comput. Opt. **41**(3), 461–471 (2017). (in Russian). https://doi.org/10.18287/2412-6179-2017-41-3-461-471
20. Mikhaylov, D.V., Kozlov, A.P., Emelyanov, G.M.: Extraction the knowledge and relevant linguistic means with efficiency estimation for formation of subject-oriented text sets. Comput. Opt. **40**(4), 572–582 (2016). (in Russian). https://doi.org/10.18287/2412-6179-2016-40-4-572-582

Clustering on the Basis of a Divisive Approach by the Method of Alternative Adaptation

Gennady E. Veselov, Boris K. Lebedev, and Oleg B. Lebedev(✉)

Southern Federal University, Rostov-on-Don, Russia
gev@sfedu.ru, lebedev.b.k@gmail.com,
lebedev.ob@mail.ru

Abstract. The paper presents a combined approach to clustering based on the integration of the divisive method with the methods of collective alternative adaptation. This group of methods is characterized by the sequential separation of the original cluster consisting of all objects, and the corresponding increase in the number of clusters. The objective is to enhance the convergence of the algorithm and the ability to exit from local optima, which allows you to work with large-scale problems and get high-quality results in a reasonable time. In this paper, the process of finding a solution is represented as an adaptive system. Under the influence of a series of adaptive actions, all objects (collective) are successively redistributed between the clusters. The goal of a specific object m_i is to reach a state in which the total vector of forces acting on it from all objects placed in the same cluster with mi had the maximum value. The goal of the collective of objects is to achieve such a separation of objects into clusters, at which the minimum distance between a pair of objects belonging to different clusters has a maximum value. To implement the adaptation mechanism, each object m_i is assigned an adaptation machine *AAi*. Studies have shown that the time complexity of the algorithm at one iteration has an estimate of $O\ (n^2)$, where n is the number of objects.

Keywords: Pattern recognition · Clustering · Collective alternative adaptation · Automatic adaptation · Hybrid algorithm

1 Introduction

The development and dissemination of computer processing of information led to the emergence in the mid-twentieth century of the need for technology that allows machines to recognize in the information they process. Development of methods for machine recognition allows you to expand the range of tasks performed by computers and make machine processing of information more intelligent. Examples of recognition applications include text recognition, machine vision, speech recognition, fingerprints, and more. Despite the fact that some of these problems are solved by a person at a subconscious level with great speed, no computer programs have yet been created that

This work was supported by the grant from the Russian Foundation for Basic Research the project № 19-07-00645 a.

S. Kovalev et al. (Eds.): IITI 2019, AISC 1156, pp. 384–392, 2020.
https://doi.org/10.1007/978-3-030-50097-9_39

solve them in an equally general form. Existing systems are designed to work only in special cases with a strictly limited scope [1–5].

The main, traditional tasks of the theory of pattern recognition are the choice of informative features, the choice of decision functions, the preliminary classification of objects (taxonomy). An image can be defined using a set of different implementations, which is called a training set. The algorithm that solves the recognition problem in accordance with the principle laid down in it, classifies the object specified by some code. The real object code is the finite discrete sequence of the encoder reference. This approach allows us to consider the code as a point in the feature space. The number of axes in space will be called dimension. We assume that in the feature space some points are given. These points are codes of some real objects. In the future, speaking of the images of objects, we will imply their code in the feature space [6, 7].

Cluster analysis (Data clustering) is the task of partitioning a given sample of objects (situations) into disjoint subsets, called clusters, so that each cluster consists of similar objects, and the objects of different clusters are significantly different [1–8].

Clustering goals:

- Understanding the data by identifying the cluster structure. Dividing the sample into groups of similar objects allows you to simplify further data processing and decision-making by applying to each cluster its own analysis method (divide and conquer strategy).
- Data compression. If the initial sample is excessively large, then you can reduce it, leaving one of the most typical representative from each cluster.
- Detection of novelty (novelty detection). There are atypical objects that cannot be attached to any of the clusters.

In all these cases, hierarchical clustering can be used, when large clusters are broken up into smaller ones, which in turn are split up even smaller, etc. Such tasks are called taxonomy tasks. The taxonomy results in a tree-like hierarchical structure. In addition, each object is characterized by listing all the clusters to which it belongs, usually from large to small. Visually, the taxonomy is represented as a graph, called a dendrogram.

One of the widespread approaches to the representation of objects is their representation in the form of points of Euclidean space, which is constructed as follows [3, 4].

Given a set of objects $M = \{m_i \mid i = 1, 2, ..., n_i\}$. Each object is described by a combination of some features (properties, characteristics, parameters): $X = \{x_j \mid j = 1, 2, ..., n_j\}$, whose sets are the same for all objects. The set X_i of the characteristic values of the object m_i determines in some way its description $X_i = \{x_j (m_i \mid j = 1, 2, ..., n_j\}$ [3].

Signs can be expressed in terms of yes/no, yes/no/unknown, numerical values, values from a set of possible options, etc. If the signs are represented by real numbers, then we can consider the vectors of images as points of n-dimensional Euclidean space.

The paper deals with the deterministic formulation of the recognition problem, which implies the problem of finding a partition in the feature space X of a set of objects M into L mutually disjoint subsets M_l, $\cup\ M_l = M$, each of which corresponds to a certain class K_i [8, 9].

The following main types of clustering rules can be distinguished [1–3]:

- decisive (discriminant) functions;
- distance functions;
- likelihood functions.

When constructing clustering rules based on distance functions, the premise is that the natural indicator of the similarity of objects is the proximity of the points describing these objects in the Euclidean space. For each pair of objects, the "distance" between them is measured – the degree of similarity. There are many metrics, here are just the main ones: The most common function of distance is the geometric distance in multidimensional space:

$$\rho(x, x') = \sqrt{\sum_{i}^{n} (x_i - x_i')^2} \tag{1}$$

To give more weight to more distant objects, the square of the Euclidean distance is used. This distance is calculated as follows:

$$\rho(x, x') = \sum_{i}^{n} (x_i - x_i')^2 \tag{2}$$

It is important to note that when classifying on the basis of the distance function, as in other methods and on the basis of other approaches, the Euclidean distance between a pair of objects reflects the degree of similarity and is constant and therefore cannot change when grouped.

A number of groups of approaches can be identified [11–13]. There are probabilistic approaches (K-means, K-medians, EM-algorithm, FOREL family algorithms, discriminant analysis), approaches based on artificial intelligence systems (C-means fuzzy clustering method (C-means), Kohonen neural network, genetic algorithm), logical approach (the construction of a dendrogram is carried out using a decision tree, graph-theoretic approach (graph clustering algorithms), hierarchical approach (divided into agglomerative (unifying) and divisive (dividing). There are other methods not included in previous groups.

There are a large number of clustering algorithms, however, if the initialization failed, the convergence of the algorithm may be slow. In addition, the algorithm can stop at a local minimum and give a quasi-optimal solution.

The paper presents a combined approach to clustering based on the integration of the divisive method with the methods of collective alternative adaptation. This group of methods is characterized by the sequential separation of the original cluster consisting of all objects, and the corresponding increase in the number of clusters. At the beginning of the algorithm, all objects belong to one cluster, which is divided into smaller clusters in subsequent steps, resulting in a sequence of splitting groups. The goal is to enhance the convergence of the algorithm and the ability to exit from local optima, which allows you to work with large-scale problems and get high-quality results in a reasonable time.

2 Formal Statement of the Clustering Problem

Let M be the set of objects, K the set of numbers (names, labels) of clusters [10]. Given the distance function between each pair of objects $\rho(m_i, m_j)$. There is a finite sample of objects $M = \{m_1, ..., m_n\}$. It is required to split the sample M into L disjoint $K = \{K_l| l = 1, 2, ..., L\}$, called clusters, so that each cluster $K_l \in K$ consists of objects close in the metric ρ, and the objects of different clusters are significantly different. Moreover, each object m_j is assigned a cluster number. The clustering algorithm is a function a: $M \rightarrow K$, which assigns a cluster number to any $m_j \in M$ object. The number of clusters – $|K|$. In some cases, it is known in advance, but more often the task is to determine the optimal number of clusters from the point of view of one or another quality criterion of clustering [2–7]. Constructing the required partition, i.e. training is based on the training sequence [13, 14]. The task of clustering can be considered as the construction of an optimal partitioning of objects into groups. At the same time, optimality can be defined as the requirement to minimize the root-mean-square split error.

The first step is to separate the objects of the first class K_1. The initially randomly selected object $m_j \in M$ is placed in cluster K_1, and the remaining objects are placed in *Kend(1)*. Next, the formation of K_1 continues using the collective alternative adaptation algorithm. After the formation of K_1 is completed, the splitting is repeated on the set *Kend(1)* of the remaining images in order to separate the second class K_2 from *Kend(1)* etc. The partitioning process is completed after the separation of the next subset K_j from *Kend(j)* becomes impossible.

In this paper, the process of finding a solution is presented in the form of an adaptive system operating under partial or complete uncertainty and changing external conditions and the information obtained in the process of working on these conditions is used to improve work efficiency. This situation is characterized by two factors: the state of the environment in which the object is located, and the object of adaptation itself. The process of search adaptation has a consistent multi-stage nature, at each stage of which is determined by the adaptive effect on the object, increasing its efficiency and optimizing the quality criteria.

3 Adaptive Clustering Algorithms

Imagine the process of formation of the next cluster in the form of an adaptive system that works on the basis of modeling the collective behavior of adaptation machines. The presentation of the initial formulation of the problem in the form of an adaptive system based on ideas of collective behavior implies the solution of the following problems [16–18]:

a) the formation of models of the environment and objects of adaptation;
b) the formation of local goals of the adaptation objects and the global goal of the team;
c) the development of alternative states of the object of adaptation, the structure of the learning adaptation machine and the mechanisms of *AA* transitions;

d) development of methods for generating control signals for reward or punishment in the process of the adaptive algorithm;
e) development of the overall structure of the adaptive search process.

At each iteration, taking into account the adaptive effect, a group redistribution of elements between the clusters is performed, i.e. transition to a new solution. Elements are objects of adaptation. We will consider the elements as material points, which are acted upon by the forces of attraction and repulsion. The state of the object in the environment corresponds to the total force acting on the object from other objects.

The state of the environment is characterized by the composition of the elements in the clusters and, as a result, the values of the forces of attraction and repulsion acting on each element. The nature of these forces varies depending on the chosen optimization criteria. Between objects m_i and m_j placed in the same cluster are the force of attraction.

The work of an object under the action of an adaptive influence consists in moving from one cluster in which it is placed to the neighboring cluster. The nature and magnitude of the adaptive effect on each object are individual. Under the influence of a series of adapting actions, the nature and magnitude of which changes at each iteration, all objects (collective) are successively redistributed between the clusters.

The goal of a specific object m_i is to reach a state in which the total vector of forces acting on it from all objects placed in the same cluster with m_i had the maximum value.

The goal of the collective of objects is to achieve such a separation of objects into clusters, at which the minimum distance between a pair of objects belonging to different clusters has a maximum value.

At each stage t, the next cluster K_t is formed. There is a selection of *Mend(t)* objects not included in the clusters formed in the previous steps. The randomly selected sample object *Mend(t)* is included in the K_t cluster. For each element $m_i \in$ *Mend(t)*, there are only two alternatives A_{1i} and A_{2i} of separation across the clusters K_t, *Kend(t)*. A particular solution of the problem is a set of alternatives for $m_i \in$ *Mend(t)* objects, according to which objects are divided into classes. The set of all possible partial solutions make up the solution space.

The process of finding the optimal solution in the solution space is presented in the form of an adaptive system operating in conditions of uncertainty.

At each iteration, under the action of the adaptive action, alternatives for objects are re-selected. The state of the environment is determined by the set of selected alternatives for objects and, as a consequence, by splitting the sample of objects *Mend(t)* into two classes K_t, *Mend(t)*. The objects of adaptation are the objects of the sample *Mend(t)*.

Let S_i^1 be the total cost of the distances between $m_i \in$ *Mend(t)* and the *Mend(t)* sample objects included in K_t.

Let S_i^2 be the total cost of the distances between and *Mend(t)* objects that are not included in K_t.

The local goal of each adaptation object m_i is to achieve a state in which the preferred alternative coincides with the realized one. That corresponds to the achievement of a state in which S_i^2 is significantly greater than S_i^1.

The global goal of the collective of automata adaptation is the formation of a cluster of K_t, including objects as close as possible to each other.

To implement the adaptation mechanism, each object m_i is assigned an AA_i automatic adaptation with two groups of states $\{C_i^1$ and $C_i^2\}$ corresponding to two alternatives A_i^1 and A_i^2. The number of states in a group is given by the parameter Q_i, called the depth of the memory. The signal "encouragement" or "punishment" is sent to the input of the AA_i adaptation machine depending on the state of the adaptation object m_i in the environment. In Fig. 1 shows the graph diagram of transitions automatic adaptation. The "+" symbol indicates transitions under the "encouragement" signal, and the "–" symbol indicates transitions under the "punishment" signal. If the AA_i adaptation machine is in one of the states of the C_i^1 group, alternative A_i^1 is implemented, according to which object m_i is turned on in the cluster K_1. If the adaptation automaton AA_i is in one of the states of the C_i^2 group, then the alternative A_i^2 is realized, in accordance with which the object m_i is included in the K_2 cluster.

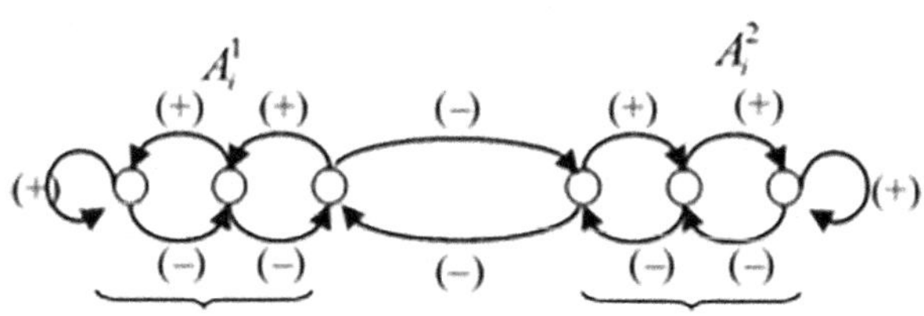

Fig. 1. Graph-scheme transitions automatic adaptation

The method of generating control signals for AA_i is as follows.

For each object m_i, the total cost of the distances S_i^1 between the object m_i and objects m_j that are part of K_t, $m_j \in K_t$

$$S_{1i} = \rho(m_i, m_j), \{j \mid m_j \in K_1\}.$$
$$S_{2i} = \rho(m_i, m_j), \{j \mid m_j \in K_2\}.$$

Next, the coefficient h_i is calculated as the ratio of the values of S_i^1 and S_i^2: $h_i = S_i^2 / S_i^1$. The number of clusters in some cases is known in advance, but more often the task is to determine the optimal number of clusters from the point of view of one or another subjective criterion of the quality of clustering. There is no uniquely best quality criterion for clustering. There are a number of heuristic criteria [10–14], as well as a number of algorithms that do not have a clearly defined criterion but implement a fairly reasonable clustering "by construction". All of them can give different results [10–14].

The clustering problem can be posed as a discrete optimization problem: it is necessary to assign cluster numbers y_i to objects m_i in such a way that the value of the chosen quality functional takes the best value. There are many types of clustering functionals, but there is no "most correct" functional [15]. In fact, each clustering method can be considered as an exact or approximate algorithm for searching for the optimum of a certain functional.

The work of the presented adaptive algorithm has two goals.

The sum of the average intracluster distances should be as small as possible.

The sum of intercluster distances should be as large as possible.

The task of clustering can be considered as the construction of an optimal partitioning of objects into groups. At the same time, optimality can be defined as the requirement to minimize the root-mean-square split error: where c_j is the "center of mass" of the cluster j (a point with average values of characteristics for this cluster) [8].

At each step of the adaptive system, the process of collective adaptation is carried out for four cycles [16–18]. Pre-selected values of the two threshold parameters T_1 and T_2 such that $T_1 < T_2$.

1. At the first step, according to the calculated environment parameters, the states of the objects m_i in the environment are estimated. For this, the parameters S_i^1, S_i^2, $h_i = S_i^2/S_i^1$ are calculated.
2. In the second cycle for each adaptation machine located in one of states of C_i^1:
   if $h_i \geq T_2$, a "reward" (+) signal is generated;
   if $h_i < T_1 < T_2$, then a "punishment" (−) signal is produced;
   if $T_1 < h_i < T_2$, then with probability $S_i^2/(S_i^2 + S_i^1)$ a signal is produced.
   "Rewards" (+), and with probability $S_i^1/(S_i^2 + S_i^1)$ a "punishment" signal is produced.
   For each adaptation machine that is in one of the states of the C_i^2 group:
   if $h_i \geq T_2$, then a "punishment" (−) signal is produced;
   if $h_i < T_1 < T_2$, then a "reward" (+) signal is generated;
   if $T_1 < h_i < T_2$, then with probability $S_i^2/(S_i^2 + S_i^1)$ a signal is produced.
   "Punishment", and with a probability $S_i^1/(S_i^2 + S_i^1)$ a "reward" (+) signal is generated.
   Parameters T_1 and T_2 (thresholds of significance) are controllers and are set a priori, with $T_1 \leq T_2$.
3. In the third cycle, under the action of "encouragement" and "punishment" signals, the automata *AA* adaptations are moving to new states.
4. On the fourth cycle for each object m_i, alternatives are implemented, in according to *AA* conditions.

By changing the thresholds of the distance T_1 and T_2, it is possible to control the degree of difference in the resulting clusters. As a criterion for stopping the operation of the algorithm, we consider the minimum change in the root-mean-square error. It is also possible to stop the operation of the algorithm if at some iteration there are no objects moving from cluster to cluster.

4 Experiments

An adaptive clustering algorithm was implemented in *C++*. During the experimental studies, two objectives were pursued: finding the best combination of values of control parameters such as *q*, *AA* memory depth, and *N*, the number of iterations; the study of the efficiency of the algorithm. Studies have shown that the time complexity of the algorithm at one iteration has an estimate of $O(n)$, where n is the number of objects, and the maximum efficiency of the adaptive search is provided for the values of the control parameters: $q = 2$, $N = 80$, where q is the memory of AA and N is the number of iterations. The probability of obtaining the optimal solution after a single run with the joint operation of the algorithms was 0,9. The use of probabilistic values of control signals and probabilistic methods of implementing alternatives based on the annealing simulation method increases the possibilities of the algorithm to exit the "local wells". To analyze the accuracy of the solutions obtained, a number of examples were synthesized with a priori known optimal value of the objective function. Investigations were exposed to examples in which the training sample contained up to 1000 examples. Comparison with known algorithms [19–22] showed that with less.

5 Conclusion

Despite the rather large number of developed models and cluster analysis algorithms, when solving applied problems, researchers often encounter a number of problems, including the difficulty in justifying the quality of analysis results that take into account the specifics of a particular task, as well as the problem of finding a global extremum in the classes.

From this we can conclude that it is expedient to further develop the methods of cluster analysis, which allow solving these problems. Promising trends include the development of hybrid clustering methods. The paper considers new principles for solving the clustering problem based on the integration of models of adaptive behavior of biological systems. To implement the adaptation mechanisms, an adaptive system architecture has been developed, as well as the structure and behavior mechanisms of the adaptation machine. The key problem that was solved in this paper is the method of generating control signals for AA_i "encouragement" or "punishment" depending on the state of the object of adaptation m_i in the environment. The source of improvement can be a more detailed study of the issues of reflexive behavior, based on the pairwise interaction of the elements being placed, which will speed up the process of the elements reaching the target state.

Acknowledgements. This research is supported by grants of the Russian Foundation for Basic Research of the Russian Federation, the project № 19-07-00645 A.

References

1. Han, J., Kamber, M.: Data Mining: Concepts and Techniques, pp. 148–169. Morgan Kaufmann Publishers, Burlington (2001)
2. Witten, I.H., Frank, E., Hall, M.A.: Data Mining: Practical Machine Learning Tools and Techniques, 3rd edn., pp. 457–482. Morgan Kaufmann Publishers, Burlington (2011)
3. Lbov, G.S., Berikov, V.B.: Stability of Decisive Functions in Pattern Recognition and Analysis of Different Types of Information. Publishing House of the Institute of Mathematics, Novosibirsk (2005). 218 p.

4. Zhuravlev, Yu.I., Ryazanov, V.V., Senko, O.V.: Recognition. Mathematical Methods. Software System Practical Applications. Fazis, Moscow (2006). 346 p.
5. Schlesinger, M., Glavach, B.: Ten lectures on statistical and structural recognition. Naukova Dumka, Kiev (2004). 187 p.
6. Hastie, T., Tibshirani, R., Friedman, J.: The Elements of Statistical Learning. COMPACT - Comparative Package for Clustering Assessment. A Free Matlab Package, pp. 327–344. Springer (2006)
7. Berkhin, P.: Survey of Clustering Data Mining Techniques. Accrue Software (2002). 233 p.
8. Vorontsov, K.V.: Algorithms for clustering and multidimensional scaling. Lecture Course. Moscow State University (2007). 156 p.
9. Lebedev, B.K., Lebedev, V.B.: Evolutionary learning procedure in pattern recognition. News SFU. Publishing house SFU, no. 8, pp. 83–88 (2004)
10. Kotov, A., Krasilnikov, N.: Data Clustering (2006). 246 p.
11. Berikov, V.S., Lbov, G.S.: Current trends in cluster analysis. All-Russian competitive selection of review and analytical articles in the priority area "Information and Telecommunication Systems" (2008). 126 p.
12. Fern, X.Z., Brodley, C.E.: Clustering ensembles for high dimensional data clustering. In: Proceedings of the International Conference on Machine Learning, pp. 186–193 (2003)
13. Fred, A., Jain, A.K.: Combining multiple clusterings using evidence accumulation. IEEE Trans. Pattern Anal. Mach. Intell. **27**, 835–850 (2005)
14. Jain, A., Murty, M., Flynn, P.: Data clustering: a review. ACM Comput. Surv. **31**(3), 264–323 (1999)
15. Kureichik, V.M., Lebedev, B.K., Lebedev, O.B.: Search adaptation: theory and practice. Fizmatlit, Moscow (2006). 289 p.
16. Tsetlin, M.L.: Research on the theory of automata and modeling of biological systems. Science, Moscow (1969). 245 p.
17. Kureichik, V.M., Lebedev, B.K., Lebedev, O.B.: Splitting based on modeling adaptive behavior of biological systems. Neurocomput.: Dev. Appl. **2**, 28–34 (2010)
18. Arthur, D., Vassilvitskii, S.: How slow is the k-means method?. In: Proceedings of the 2006 Symposium on Computational Geometry (SoCG) (2006)
19. Gorban, A.N., Zinovyev, A.Y.: Principal graphs and manifolds (chap. 2). In: Soria Olivas, E., et al. (eds.) Handbook of Research and Development, Methods, and Techniques, pp. 28–59. IGI Global, Hershey (2009)
20. Mirkes, E.M.: K-means and K-medoids applet. University of Leicester (2011). 262 p.
21. Coates, A., Ng, A.Y.: Learning feature representations with k-means. Stanford University (2012). 188 p.
22. Vyatchenin, D.A.: Fuzzy methods of automatic classification. Technoprint, Minsk (2004). 190 p.

Matrix-Like Representation of Production Rules in AI Planning Problems

Alexander Zuenko(✉), Yurii Oleynik, Sergey Yakovlev, and Aleksey Shemyakin

Kola Science Centre of the Russian Academy of Sciences, Institute for Informatics and Mathematical Modelling, Apatity, Russia
zuenko@iimm.ru

Abstract. The paper presents the AI planning technology developed to study poorly formalized subjects domains, the knowledge of which is of a quantitative and qualitative character. The technology implements constraint programming paradigm and supports the subject domain model being open for operative modifications, which allows inclusion or exclusion of constraints, quality criteria, as well as setting the initial and goal states specified by subdefinite parameters. The originality of this work lies in the fact that a new type of constraints, namely smart table constraint of D-type, is proposed for representation and efficient processing of the production rules, with their processing being carried out by the authors' methods of non-numerical constraints satisfaction, which gives the substantial gain in the performance against the conventional algorithms of table constraints propagation.

Keywords: Constraint programming technology · Constraint satisfaction problem · Matrix-like structures · Qualitative constraints · AI planning

1 Introduction

AI planning is a field of artificial intelligence, which is of great interest nowadays [1]. There are distinguished two fields of research [2]: AI planning and scheduling. The paper follows the viewpoint accepted in the AI community that the scheduling problem is one of the varieties of AI planning problem.

AI planning includes selecting the actions to be used in the problem solution and arranging the actions into a sequence that helps reach a solution.

There are two trends among the main methods of AI planning: 1) search methods based on the use of the planning graph [3], and 2) planning methods based on logical reasoning [4].

The reported study was funded by RFBR, project number 18-07-00615-a.

S. Kovalev et al. (Eds.): IITI 2019, AISC 1156, pp. 393–402, 2020.
https://doi.org/10.1007/978-3-030-50097-9_40

Classical methods are inapplicable to many real-world planning problems, because such methods assume the availability of complete and correct information, as well as a deterministic, fully observable environment. As a rule, in practical planning problems, the environment is not completely observable (incomplete description of the state vector is possible), and the actions of the performer are non-deterministic. In the modeling of such environments, the relationships between directly observable and deducible variables can have the character of not only quantitative but also logical dependencies. The relationships between actions can be complex and modeled, for example, by production rules or logical formulas. As a result, it is required to co-process heterogeneous quantitative and qualitative information. Giving the classical AI planning methods new opportunities is associated with increasing their computational complexity.

Recently, an approach to solve the AI planning problem using the constraint programming technology is being developed [5–7].

The Constraint Satisfaction Problem (CSP) is the tuple (V, D, C), where: $V = \{v_1, ..., v_n\}$ is a set of variables; $D = \{D_1, ..., D_n\}$ is a set of domains of the corresponding variables; $C = \{C_1, ..., C_m\}$ is a set of constraints. Constraint C_i is a relation determined on a subset of the values of all the variables, i.e., $C_i \subseteq D_1 \times ... \times D_n$. The solution of the CSP is such a substituting for all the variables under which all the constraints are satisfied.

The constraint programming technology used in the solving of combinatorial search and combinatorial optimization problems, is widely applied in scheduling problems [8–10]. However it is still not widely applied in the solving of more general problems of AI planning.

Constraint programming methods are extension of the methods based on logical reasoning. In the early 90s, progress has been made in the field of creation of effective reasoning algorithms for propositional logic. Developed SAT-solvers use the representation of the problem in the form of SAT problem (Boolean satisfiability problem) [4]. The disadvantages of the approach are: 1) SAT-planners require huge amounts of memory (gigabytes) to represent tasks of relatively small dimension, since all possible actions and statements are explicitly presented for each discrete time point; 2) the representation of actions that have different duration or are connected by the time constraints is not supported; 3) while representing problem in a form of a SAT problem, a part of information about the problem's structure, which can be used to accelerate the infection process, is lost.

In framework of the constraint programming technology, these disadvantages can be successfully overcome. The constraint programming technology allows organizing inference with undetermined parameters, which makes it possible to model aspects of uncertainty and the non-determinism of the actions of the performer.

The constraint programming technology is very suitable for modeling of poorly formalized subject domains. The model is constructed step by step and may include the types of regularities and optimization criteria, which could not be taken into account a priori. On the other hand, if the solution of problem does not exist, a possibility should be provided to make the problem "easy to solve" by eliminating some regularities and (or) optimization criteria from consideration. Under these conditions, the application of conventional methods of the operations research theory is problematic.

The development of suitable data structures to represent the CSP problem is one of the important issues in the development of AI planning methods, since the speed of solving problems strongly depends on those data structures.

We suggest to represent the qualitative constraints, namely production rules, as specialized matrix structures (smart tables of D-type), with their processing being carried out by the authors' methods of matrix-like constraints satisfaction.

2 The Formal Apparatus for Matrix-Like Representation of Production Rules

A very important type of so-called global constraints in constraint programming technologies is *table constraints* [11]. A table form is very suitable for different non-numerical (qualitative) relations of the subject domain to be described. There are many studies devoted to the development of constraint propagation algorithms for such constraints [12].

However, the complexity of tables processing increases exponentially with the size of tables. So, of special actuality for table constraints are the development of the ways to compactly represent these tables in the computer memory, and the development of effective methods of their processing.

There is a wide spectrum of studies dealing with compact representation of table constraints in the world literature [11,13]. One of the forms of compact representation of table constraints is *compressed tables* [13]. The components of compressed tables are not individual values but subsets of values of the domains of the corresponding variable. Each row of a compressed table is interpreted as a Cartesian product of subsets, and an entire compressed table – as a union of Cartesian products corresponding to the table rows.

The article [11] proposes to generalize the notion of *compressed tables*, allowing simple arithmetic expressions to be represented as components of tuples. The generalized table constraints were proposed to be called *smart table constraints*. For instance, let there be the following set of tuples $\{(a, b, a), (a, c, a), (b, b, b), (b, c, b), (c, b, c), (c, c, c)\}$ on variables $\{X_1, X_2, X_3\}$ with domains $\{a, b, c\}$ This set of tuples can be represented as a smart table containing only one row:

X_1	X_2	X_3
$=X_3$	$\neq$ a	$*$

Symbol "$*$" means that the corresponding component of the table contains all possible values of its domain.

Note that even before the introduction of the concept of "smart table constraint", such structures were introduced in the work [14].

However, in the term of the described structures it is not convenient to describe some kinds of knowledge, for example, production rules. Using compressed and smart tables of the types mentioned above, it is convenient to model logical formulas that can be easily represented as a disjunction of some conjunctions.

Unlike the articles mentioned above [11–13], in the earlier papers of one of the authors of this article [15] two types of matrix-like structures (the C-systems and the D-systems) are proposed to represent the compressed tables. We have not met any analogues of D-systems in the constraint programming.

The studies in [15,16] propose the constraints propagation algorithms, which are based on the equivalent transformations of the C- and D-systems. Evaluation of the effectiveness of the original inference methods is presented in [17].

However, production rules modeling based on the compressed table constraints of D-type is problematic in the following cases: 1) the domains of variables are not finite sets; 2) not only one-place predicates are used as elementary statements in expressions.

The originality of this work lies in the fact that a new type of constraints, namely smart table constraint of D-type, is proposed for representation and efficient processing of the production rules.

Let there be three real variables X, Y, Z and the domains of the variables be equal to the interval "[2,4]". Let the following set of production rules be also given (the semantics of the rules are not significant here):

$$(X \in [2,3]) \wedge (Y > X) \to (Z = 4);$$
$$(X \in [2,3]) \wedge (Y < X) \to (Z = 2);$$
$$(X \in [2,3]) \wedge (Y = X) \to (Z = 3).$$

Replacing the implication in the formulas, we have:

$$(X \notin [2,3]) \vee (Y \leq X) \vee (Z = 4);$$
$$(X \notin [2,3]) \vee (Y \geq X) \vee (Z = 2);$$
$$(X \notin [2,3]) \vee (Y \neq X) \vee (Z = 3).$$

This set of rules can be represented as the following smart table of D-type:

$$\begin{matrix} & X & Y & Z & YX & \\ \Bigg] & \notin [2,3] & \varnothing & = 4 & \leq & \Bigg[\\ & \notin [2,3] & \varnothing & = 2 & \geq & \\ & \notin [2,3] & \varnothing & = 3 & \neq & \end{matrix}.$$

The matrices of D-type will be written using inverted square brackets. Designation ("$\varnothing$") describes the component which does not contain any values.

A distinctive feature of the proposed smart table of D-type from compressed table of D-type is the presence of complex attributes in the header of the matrix. In the matrix shown as an example, there is a complex attribute YX. The domain of this attribute contains all kinds of comparison of pair of attributes Y and X, namely $\{>, <, =, \neq, \leq, \geq\}$.

Inference (propagation) on these structures, as well as in the case of compressed tables of D-type (D-systems), is carried out using Affirmation 1–5, but has its own distinguishing features due to the need to process complex attributes.

The affirmations used to propagate the constraints represented in the form of smart tables of D-type are below:

Affirmation 1: If, at least, one tuple of the smart table of D-type is empty (all the components are empty), the smart table is empty (the corresponding system of constraints is inconsistent).

Affirmation 2: If all the components of an attribute are empty, the attribute can be eliminated from the smart table of D-type (all the components in the corresponding column are removed) and the pair "the eliminated attribute – its domain" is included into the partial solution.

Affirmation 3: If in the smart table of D-type there is a tuple (row) containing only one nonempty component, all the values not included into this component, are deleted from the corresponding domain. After that the tuple is eliminated from the smart table of D-type.

Affirmation 4: If a tuple of the smart table of D-type contains, at least, one full component, this tuple is removed.

Affirmation 5: If the component of an attribute of the smart table of D-type contains the value not belonging to the corresponding domain, this value is deleted from the component.

Let, in addition to the three rules above, the following unary constraints be specified for the variables X and Y: $X \in (2,3)$, $Y \in (3,4)$. When the constraints propagating, these sets become domains of the variables X and Y. According to Affirmation 5, all values are removed from the components of the first column, since they are not contained in the new domain. Thus, all components of the first column become empty. Further calculations illustrate the features of processing of the complex attribute YX. From the set of relations $\{>, <, =, \neq, \leq, \geq\}$, for $X \in (2,3)$, $Y \in (3,4)$, only the relation "$>$" is consistent. So, the domain of the attribute YX is reduced and becomes equal to $\{>\}$. After "adjustment" of the smart table to the new domain of the complex attribute YX (using Affirmation 5), we have:

$$\begin{array}{cccc} X & Y & Z & YX \end{array}$$
$$\left]\begin{array}{cccc} \varnothing & \varnothing & =4 & \varnothing \\ \varnothing & \varnothing & =2 & > \\ \varnothing & \varnothing & =3 & > \end{array}\right[.$$

Given that the components YX of the second and third rows are equal to the domain of the attribute YX, these rows can be eliminated from further consideration (according to Affirmation 4). The first row contains only one non-empty component (in the attribute Z). According to Affirmation 3, the new domain of the attribute Z will be equal to the singleton $\{4\}$ (the variable Z is equal to 4). Then, according to Affirmation 4, the first row is deleted from the matrix. All rows of the matrix were eliminated without the generation of empty rows – this is a marker of the successful completion of the constraint propagation. Finally, we have: $X \in (2, 3)$, $Y \in (3, 4)$, $Z = 4$. After propagation the value of the variable Z becomes defined.

3 Example of AI Problem

As an example of an applied problem of AI planning, we consider the problem of planning of localizing the consequences of on-land oil products spills (Fig. 1). Working-out of such plans is carried out under many constraints. The peculiarities of the object and its environment (organizational structure, technological processes, topography and infrastructure), regulatory requirements, generic accident scenarios, equipment and force dislocation are taken into account. Different optimization criteria and forms of the planning results representation are possible.

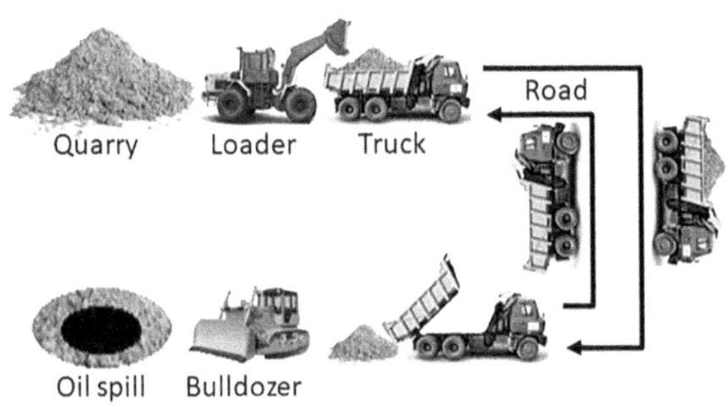

Fig. 1. Localizing the consequences of on-land oil products spills

Let's consider that the localizing technology is based on the use of a dam. The material used as clean sand is taken from a sand quarry. There are three types of participants considered within the problem:

1. Dump trucks to carry sand to build a dam.
2. Loaders to load sand into dump trucks.
3. Bulldozers to construct the sand dam.

There are both actions common to all the participants ("moving", "waiting") and actions specific to the participant type ("loading", if it is a loader, "dam constructing", if it is a bulldozer).

The constraints of the CSP considered here, among which there are both quantitative and qualitative regularities, can be divided into several types.

1. Constraints defining the optimization criteria. For instance, the costs on localizing must not exceed some value, or localizing must be completed in 6 hours from the moment of getting the information about the spill. The similar constraints are proposed to represent in a form of global constraints [1], which substantially accelerates the process of the CSP solution. A global constraint is a set of simpler constraints whose joint consideration allows working-out a more effective satisfaction procedure than the general-purpose algorithms.

2. Constraints dictated by the peculiarities of the accident localization technology applied. For instance, a loader cannot load two dump trucks simultaneously, or the dam can be constructed simultaneously with sand transportation, etc. The constraints defining the transition between i and $i+1$ stages of the plan also related to this group of constraints. These constraints specify the admissible sequence of actions depending on the type of the participants and the type of the point of the territory and can be represented in the form of production rules.

3. Constraints defining the particular participants or the subtypes of the participants taking part in localizing. For instance, the dump trucks of a certain type may be loaded only by the loaders of defined types, or it can move between the points specified at a speed which is lower than the standard one, etc. These constraints also are of a qualitative character.

4 The Problem Representation Within the Constraint Programming Paradigm

All the participants of localizing are the subclass of an abstract class "*Participant of localizing*" including the interfaces for a dump truck, a loader and a bulldozer.

To define the characteristics of the territory, two program classes were developed: class "*Waypoint*" defining each point significant in making a plan, and class "*Road*" defining the connection between the two neighboring waypoints.

The actions may be either common for all the participants, or depend on the type of the participant and the type of the waypoint where these actions are to be made. Thus, the actions can be divided into a number of subclasses.

Consider the simplest situation when there are only two waypoints, such as: $point_1$ is the point being a quarry and a parking, simultaneously, and $point_2$ is the point of the spill.

There is also a road path connecting these. There is one dump truck tr and one loader ld in the territory. To plan possible actions, the following subclasses are used:

start is the subclass defining start of working, based on which the particular actions (instances st_1^{ld} and st_1^{tr}) are generated; the instances st_1^{ld} and st_1^{tr} are the actions to start working in the point $point_1$ for ld and tr, respectively;

stop is the subclass defining the end of working, based on which the particular actions (instances sp_1^{ld} and sp_1^{tr}) are generated; the instances sp_1^{ld} and st_1^{tr} are the actions to finish working in the point $point_1$;

move is the subclass defining moving of the participants from one point to another, based on which particular actions (instances m_{12}^{ld}, m_{21}^{ld}, m_{12}^{tr} and m_{21}^{tr})

are generated; the instances m_{12}^{ld}, m_{21}^{ld}, m_{12}^{tr} and m_{21}^{tr} are the actions of the participants ld and tr which move from the point $point_1$ to the point $point_2$, in both sides;

wait is the subclass defining waiting of the participant in the point, based on which particular actions (instances w_1^{ld}, w_2^{ld}, w_1^{tr} and w_2^{tr}) are generated; the instances w_1^{ld}, w_2^{ld}, w_1^{tr} and w_2^{tr} are the actions of waiting in the points $point_1$ and $point_2$, respectively;

Sand loading by the loader ld into the dump truck tr in the $point_1$ is described by two actions: l_{tr} is the action of the loader, f_{ld} is the action of the dump truck.

Further, consider in detail the peculiarities of the constraint representation based on the approach proposed, i.e., with the help of matrix-like structures.

Introduce the Variables:

X_i^{ld} is the action carried out by the loader ld at the i-th stage of the plan (corresponds to the program structure $ld.status[i]$);

X_{i+1}^{ld} is the action carried out by the loader ld at the $(i+1)$-th stage of the plan;

The designations D_i^{ld} and T_i^{ld} used below correspond to the duration and start time of the action X_i^{ld}, respectively.

X_i^{tr} is the action carried out by the dump truck tr at the i-th stage of the plan (corresponds to the program structure $tr.status[i]$);

The domain of the variables X_i^{ld}, X_{i+1}^{ld} is a set of actions of the loader ld: $\{st_1^{ld}, sp_1^{ld}, m_{12}^{ld}, m_{21}^{ld}, w_1^{ld}, w_2^{ld}, l_{tr}\}$. The domain of the variable X_i^{tr} is a set $\{st_1^{tr}, sp_1^{tr}, m_{12}^{tr}, m_{21}^{tr}, w_1^{tr}, w_2^{tr}, f_{ld}\}$. The domains of the rest of variables consist of integer values denoting the moments of the time are the stages of modelling. Assume that a stage of the plan lasts *Steptime* minutes.

As an example, consider two constraints imposing the conditions for the beginning of a new action in the first point:

$$\begin{aligned}((X_i^{ld} = w_1^{ld}) \vee (X_i^{ld} = st_1^{ld}) \vee (X_i^{ld} = m_{21}^{ld}) \vee (X_i^{ld} = l_{tr})) \wedge \\ (T_i^{ld} \leq a_i - D_i^{ld}) \leftrightarrow ((X_{i+1}^{ld} = m_{12}^{ld}) \vee (X_{i+1}^{ld} = w_1^{ld}) \vee \\ (X_{i+1}^{ld} = sp_1^{ld}) \vee (X_{i+1}^{ld} = l_{tr})) \wedge (T_{i+1}^{ld} = a_i)\end{aligned} \tag{1}$$

$$\begin{aligned}(X_{i+1}^{ld} = l_{tr}) \rightarrow \\ ((X_i^{tr} = st_1^{tr}) \vee (X_i^{tr} = m_{21}^{tr}) \vee (X_i^{tr} = w_1^{tr})) \wedge (T_i^{tr} \leq a_i - D_i^{tr})\end{aligned} \tag{2}$$

Here a_i is the constant value that equals to $Steptime * i$.

Listed in constraint (1) are the actions $\{w_1^{ld}, st_1^{ld}, m_{21}^{ld}, l_{tr}\}$ which could be finished at the $point_1$ at the i-th stage of the plan, and it is pointed out that if the actions finish, carried out at the $(i+1)$-th stage of the plan is any of the actions $\{m_{12}^{ld}, w_1^{ld}, sp_1^{ld}, l_{tr}\}$. The constraint (2) specifies the conditions for the action l_{tr}. to start.

Being represented in such a way, the constraints are not effectively processed by most current constraints programming systems. Let's illustrate the way these constraints are represented with the help of the matrix apparatus proposed.

The constraint (1) can be represented by the smart table constraint of D-type $K_1[X_i^{ld}, T_i^{ld}D_i^{ld}, X_{i+1}^{ld}, T_{i+1}^{ld}]$:

$$\left]\begin{array}{cccc} X_i^{ld} & T_i^{ld}D_i^{ld} & X_{i+1}^{ld} & T_{i+1}^{ld} \\ \{w_2^{ld}, m_{12}^{ld}, sp_1^{ld}\} & T_i^{ld} > a_i - D_i^{ld} & \{w_1^{ld}, sp_1^{ld}, m_{12}^{ld}, l_{tr}\} & \varnothing \\ \{w_2^{ld}, m_{12}^{ld}, sp_1^{ld}\} & T_i^{ld} > a_i - D_i^{ld} & \varnothing & = a_i \\ \varnothing & T_i^{ld} \le a_i - D_i^{ld} & \{w_2^{ld}, m_{21}^{ld}, st_1^{ld}\} & \neq a_i \\ \{w_1^{ld}, st_1^{ld}, m_{21}^{ld}, l_{tr}\} & \varnothing & \{w_2^{ld}, m_{21}^{ld}, st_1^{ld}\} & \neq a_i \end{array}\right[.$$

The constraint (2) can be represented as $K_2[X_{i+1}^{ld}, X_i^{tr}, T_i^{tr}D_i^{tr}]$:

$$\left]\begin{array}{ccc} X_{i+1}^{ld} & X_i^{tr} & T_i^{tr}D_i^{tr} \\ \{st_1^{ld}, sp_1^{ld}, m_{12}^{ld}, m_{21}^{ld}, w_1^{ld}, w_2^{ld}, l_{tr}\} & \{st_1^{tr}, m_{21}^{tr}, w_1^{tr}\} & \varnothing \\ \{st_1^{ld}, sp_1^{ld}, m_{12}^{ld}, m_{21}^{ld}, w_1^{ld}, w_2^{ld}, l_{tr}\} & \varnothing & T_i^{tr} \le a_i - D_i^{tr} \end{array}\right[.$$

Explain the way the smart tables of D-type defined above are formed, with the constraint (2) taken as an example. Constraint (2) can be rewritten as:

$$[\neg(X_{i+1}^{ld} = l_{tr}) \vee ((X_i^{tr} = st_1^{tr}) \ \vee (X_i^{tr} = m_{21}^{tr}) \vee (X_i^{tr} = w_1^{tr}))] \wedge$$
$$[\neg(X_{i+1}^{ld} = l_{tr}) \vee (T_i^{tr} \le a_i - D_i^{tr})].$$

Each of the "multiplied" square brackets is correlated with the corresponding row of the smart table $K_2[X_{i+1}^{ld}, X_i^{tr}, T_i^{tr}D_i^{tr}]$.

This representation of the initially set constraints (1) and (2) allows each of the newly produced constraints to be processed by specialized highly effective algorithms-propagators.

5 Conclusion

Application of the constraint programming technology for solving AI planning problems allows us to overcome many of the shortcomings of schedulers based on classical methods and to accelerate computations.

The authors' studies showed that processing of qualitative constraints represented in the form of logical expressions and production rules, is not sufficiently effective in the systems like those mentioned above, and cannot be implemented for comprehensible time even at a rather small dimension of problem. A new type of constraints, namely smart table constraint of D-type, is proposed for representation and efficient processing of the production rules. Constraint processing is carried out by the authors' methods of non-numerical constraints satisfaction, which gives the substantial gain in the performance against the conventional algorithms of table constraints propagation. Testing of the constraint propagation algorithms was performed using the Choco library. The application of the author's approach to the representation and processing of production rules made it possible to obtain a performance gain of 33% (in the worst case) compared to the standard representation methods and algorithms of the Choco library.

References

1. Russel, S., Norvig, P.: Artificial Intelligence: A Modern Approach, 3rd edn. Prentice Hall, Upper Saddle River (2010)
2. Smith, D.E., Frank, J., Jónsson, A.K.: Bridging the gap between planning and scheduling. Knowl. Eng. Rev. **15**(1), 47–83 (2000)
3. Blum, A., Furst, M.: Fast planning through planning graph analysis. Artif. Intell. **90**, 281–300 (1997)
4. Rintanen, J.: Planning and SAT. In: Biere, A., van Maaren, H., Heule, M., Walsh, T. (eds.) Handbook of Satisfiability, pp. 483–504. IOS Press, Amsterdam (2009)
5. Steger-Jensen, K., Hvolby, H.-H., Nielsen, P., Nielsen, I.: Advanced planning and scheduling technology. Prod. Planning Control **22**(8), 800–808 (2011)
6. Bit-Monnot, A.: A constraint-based encoding for domain-independent temporal planning. In: Proceedings of 24th International Conference "Principles and Practice of Constraint Programming", pp. 30–46, Lille. (2018). https://doi.org/10.1007/978-3-319-98334-9
7. Rossi, F., van Beek, P., Walsh, T.: Handbook of Constraint Programming. Elsevier, Amsterdam (2006)
8. Margaux, N., Artigues, C., Lopez, P.: Cumulative scheduling with variable task profiles and concave piecewise linear processing rate functions. Constraints **22**(4), 530–547 (2017)
9. Kreter, S., Rieck, J., Zimmermann, J.: Models and solution procedures for the resource-constrained project scheduling problem with general temporal constraints and calendars. Eur. J. Oper. Res. **251**(2), 387–403 (2016)
10. Letort, A., Carlsson, M., Beldiceanu, N.: Synchronized sweep algorithms for scalable scheduling constraints. Constraints **20**(2), 183–234 (2015). https://doi.org/10.1007/s10601-014-9172-8
11. Mairy, J., Deville, Y.,Lecoutre, C.: The smart table constraint. In: Michel, L. (eds.) Integration of AI and OR Techniques in Constraint Programming, CPAIOR 2015, Lecture Notes in Computer Science, vol. 9075. Springer, Cham (2015)
12. Xia, W., Yap, R.H.C.: Optimizing STR algorithms with tuple compression. In: Schulte, C. (ed.) CP 2013. LNCS, vol. 8124, pp. 724–732. Springer, Heidelberg (2013)
13. Perez, G., Regin, J.-C.: Improving GAC-4 for table and MDD constraints. In: O'Sullivan, B. (ed.) CP 2014. LNCS, vol. 8656, pp. 606–621. Springer, Heidelberg (2014)
14. Zuenko, A., Fridman, A.: Development of n-tuple algebra for logical analysis of databases with the use of two-place predicates. J. Comput. Syst. Sci. Int. **48**(2), 254–261 (2009). https://doi.org/10.1134/S1064230709020099
15. Zuenko, A.: Matrix-like structures for representation and processing of constraints over finite domains. Advances in Intelligent Systems and Computing, vol. 875, pp. 428–438. Springer Nature Switzerland AG (2019). https://doi.org/10.1007/978-3-030-01821-4_45
16. Zuenko, A., Lomov, P., Oleynik, A.: Applying constraint propagation methods to speed up the processing of ontology requests. SPIIRAS Proc. **1**(50), 112–136 (2017). https://doi.org/10.15622/sp.50.5. (In Russian)
17. Zuenko, A., Oleynik, Y.: Programming of algorithms of matrix-represented constraints satisfaction by means of choco library. In: Advances in Intelligent Systems and Computing, vol. 875, pp. 1–10. Springer Nature Switzerland AG. (2019). https://doi.org/10.1007/978-3-030-01821-4_46

LP Structures Theory Application to Building Intelligent Refactoring Systems

Sergey Makhortov(✉) and Aleksandr Nogikh

Voronezh State University, Universitetskaya pl. 1, Voronezh 394018, Russia
msd_exp@outlook.com

Abstract. An approach of automatized object-oriented code refactoring is described that applies lattice-based algebraic structures for type hierarchy representation and optimization. A distinctive feature of these algebraic structures is their ability to model aggregation not as a relation between two independent sets of types and attributes, but as a relation between two specific types. The property makes it possible to perform a more careful optimization of type hierarchy. The described approach focuses on redundant attributes removal and on the relocation of identical attributes into their common superclasses ("Pull Up Field" technique). In this paper it is demonstrated how the adopted algebraic structures can be extended to model a wide range of type hierarchies. Also, they are shown to be able to perform transformations allowing for external constraints. Such constraints may represent some additional knowledge of the type hierarchy or of the refactoring process itself. The described approach employs only the fundamental ideas of object-oriented programming. Supplemented with language-specific features it may be used as a basis for building intellectual systems that facilitate object-oriented code refactoring.

Keywords: Object-oriented programming · Type hierarchy · Refactoring · LP structure · Decision support system · Software development tool

1 Introduction

Software maintenance is estimated to be the most costly phase during the software development life cycle [1]. During the maintenance phase, a software system is modified to resolve its flaws and to meet emerging requirements. It forces a software system to leave its original design, and that unavoidably contributes to code quality degradation and makes further modifications more and more problematic.

Refactoring helps to cope with such negative effects. It is the process of changing the software system in such a way that it does not alter the external behavior of the code yet improves its internal structure [2]. Refactoring methods overview can be found in [2].

Obviously, refactoring can be problematic for large software systems since they can be too complex to be adequately understood and analyzed by a human. It justifies the need for tools that can aid the developer in the refactoring process. Such tools adopt various mathematical methods for code analysis and transformation. One of the methodologies widely used for software refactoring is Formal Concept Analysis (FCA). The approach models two-dimensional structures with the "object-feature" semantics.

S. Kovalev et al. (Eds.): IITI 2019, AISC 1156, pp. 403–411, 2020.
https://doi.org/10.1007/978-3-030-50097-9_41

More about FCA can be found in [3]. FCA has been applied to many areas, including study and optimization of CRUD (Create, Read, Update, Delete) matrices [4], object-oriented systems refactoring [5, 6] and problems of knowledge engineering [7].

In [8] it was demonstrated that models for dealing with refactoring problems can also be obtained using LP (Lattice-Production) structures theory. The paper presented an approach for the automatized optimization of type hierarchy, including relocation of identical attributes into their common superclasses (described as "Pull Up Field" in [2]) and redundant attributes removal. This model was further developed in [9], where it served as the basis for discovering a new method of refactoring, the "merging of attributes".

A distinctive feature of the method described in [8] is how aggregation is reflected in the model. In the case of FCA, types and attributes are treated as distinct sets [6], while in the case of LP structures, attributes are closely related to type hierarchy. This property makes it possible to put more information into the model and perform a more thorough optimization of type hierarchy.

The model suggested in [8] has a simplified semantics of "obtaining access to attributes and services". This limitation prevents adequate modeling of hierarchies where a type has multiple attributes of the same type. Also, there were no means of external control over the attribute uplifting. This paper presents an extended version of the model that overcomes those limitations. As a result, it opens new possibilities for building intelligent tools for automatized refactoring.

The text below only presents the model and a generalized approach of refactoring automatization based on the model. The application of the approach to specific programming languages and practical implementation issues are out of the scope of the paper.

2 Method Description

The [8] paper introduced LP structures on type lattices. Below are the basic definitions and important results from that paper which are relevant for Sects. 3 and 4.

Definition 1. LP (Lattice-Production) structure is an algebraic structure consisting of a lattice $\mathbb{F}(\leq, \wedge, \vee)$ and a binary relation over the lattice (denoted by $\leftarrow$). The relation must have the following properties:

- reflexivity: $a \leftarrow a$;
- transitivity: if $a \leftarrow b$ and $b \leftarrow c$, then $a \leftarrow c$ $(a, b, c \in \mathbb{F})$;
- distributivity (for the discussed topic – $\vee$-distributivity): if $b_1 \leftarrow a$ and $b_2 \leftarrow a$, then $b_1 \vee b_2 \leftarrow a$;
- $\leq \subseteq \leftarrow$.

In object-oriented programming, there are at least two kinds of relationships between types (classes): inheritance and aggregation. Types and inheritance links are represented by the lattice $\mathbb{F}$: if a type represented by a (for simplicity, "represented by" in the case of $\mathbb{F}$ is omitted later in the text) is a parent of type b, then $b \leq a$. To make the lattice bounded, $\mathbb{F}$ is supplemented with two special elements: a fictitious common ancestor I (in case it is absent) and a fictitious successor of all types O.

On the lattice $\mathbb{F}$, we build a relation R, corresponding to aggregation. If a type $a \in \mathbb{F}$ has an attribute of type $b \in \mathbb{F}$, then $(b, a) \in R$. Note that ($\mathbb{F}$,R) is not yet an LP structure since R is not guaranteed to have the properties stated in Definition 1.

Both relations ($\leq$ and R) have common semantics – in the case of $b \leq a$ and $(b, a) \in R$ type b obtains access to attributes and services provided by type a. It is clear from the semantics that such relation of "obtaining access to attributes and services" is reflexive and transitive. A closure of R with respect to there requirements of Definition 1, will be denoted by $\overset{R}{\longleftarrow}$.

$\vee$-distributivity of $\overset{R}{\longleftarrow}$ (if $b_1 \overset{R}{\longleftarrow} a$ and $b_2 \overset{R}{\longleftarrow} a$, then $b_1 \vee b_2 \overset{R}{\longleftarrow} a$) can be interpreted in the following way. If type b_1 has access to attributes and services provided by a and the same is true for b_2, then the least common ancestor of b_1 and b_2 also has access to attributes and services provided by a. If an attribute of type a is lifted to $b_1 \vee b_2$, then b_1 and b_2 will possess such an attribute by means of inheritance, which corresponds to the "Pull Up Field" refactoring.

Refactoring process aims to improve the internal structure of a software system, and code quality degradation as a result of refactoring cannot be tolerated. In order to avoid such situations, $\vee$-distributivity of $\overset{R}{\longleftarrow}$ must not be unconditional. In [8] it is restricted so that the resulting type hierarchy satisfies the following rules.

- If an attribute is inherited from only one ancestor type, it must not be inherited from multiple.
- A type should not be aware of its child types, i.e. if $(a, b) \notin \overset{R}{\longleftarrow}$ and $b \leq a$, then type a must not have an attribute of type b.

Below are the definitions that were used in [8] in order to build a model that accommodates the rules specified above.

A pair of elements $a, c \in \mathbb{F}$ is said to be transitive in R if $(a, c) \in R_1^*$, where R_1^* is the transitive closure [10] of relation $R_1 = R \backslash \{(a, c)\}$.

Definition 2. Let R be a relation over $\mathbb{F}$. Two pairs of the kind $(b_1, a), (b_2, a) \in R$ are called $\vee$-compatible in R if there exist $c_1, c_2 \in \mathbb{F}$ such that $b_1 \vee b_2 \leq c_1 \vee c_2$, $(c_1 \vee c_2) \wedge a = O$, and $(c_1, a), (c_2, a) \in R$ are not transitive in $R \cup \leq$. (c_1, c_2, a) will be referred to as a $\vee$-distributive triple.

Definition 3. Consider some $\vee$-distributive triples: $T = (c_1, c_2, a)$ and $T' = (c_1', c_2', a')$. T is called neutralizing for T' if one of the following conditions is satisfied:

1. $a = a'$, $c_1 \vee c_2 \neq c_1' \vee c_2'$ and at least one of the inequalities $c_i' < c_1 \vee c_2$ holds;
2. $a < a'$, $c_1 \vee c_2 \neq c_1' \vee c_2'$ and at least one of the inequalities $c_i' \leq c_1 \vee c_2$ holds;
3. $a < a'$, $c_1 \vee c_2 = c_1' \vee c_2'$.

Definition 4. A $\vee$-distributive triple $T' = (c_1', c_2', a')$ is called conflictless if there are no neutralizing $\vee$-distributive triples for it.

Definition 5. Two $\vee$-compatible pairs are called conflictless $\vee$-compatible if the corresponding $\vee$-distributive triple is conflictless.

Definitions 6, 7, and 8 complete the construction of LP structure on type lattice.

Definition 6. A relation R is called logical if it contains $\leq$, is transitive and for all $\vee$-compatible pairs $(b_1, a), (b_2, a) \in R$ it is true that $(b_1 \vee b_2, a) \in R$.

For a partially ordered set, we distinguish between concepts of the minimal element (no lesser element exists) and the least element (the least element among all).

Definition 7. Logical closure of R is the *least* logical relation containing R and the set of its conflictless $\vee$-compatible pairs.

Definition 8. Two relations R_1 and R_2 defined over the same lattice are called equivalent ($R_1 \sim R_2$) if their logical closures coincide. Logical reduction of R is a *minimal* R_0 such that $R_0 \sim R$.

Let T_{pairs} be a set of conflictless $\vee$-distributive triples of R. Consider a relation $\tilde{R} = R \cup \leq \cup \{(c_1 \vee c_2, a) | (c_1, c_2, a) \in T_{pairs}\}$.

The following theorems from [8] are relevant for the following discussion.

Theorem 1. Logical closure of R exists and is equal to the transitive closure $\tilde{R}^*$ of $\tilde{R}$.

Logical closure existence enables automatized equivalent transformation of type hierarchies.

Theorem 2. Let $\tilde{R}$ already be constructed for R. Then the logical reduction of R equals $R^0 \setminus \leq$, where R^0 is a transitive reduction [10] of $\tilde{R}$.

Logical reduction computation corresponds to the process of refactoring the modeled type hierarchy.

In [8] it was demonstrated that the described transformations can be performed in polynomial time.

3 Extending the Model

This section describes the proposed solutions for the issues defined in the introduction section. First, a new model parameter is introduced that enables control over attribute uplifting. Then, a generalized type lattice is suggested that removes any limitations of aggregation modeling capabilities.

There is a point that needs to be clarified before proceeding to the model extension. Logical reduction of a relation does not have to be unique and not all logical reductions can be considered as valid results of type hierarchy refactoring. In order to be more precise, later in the text logical reduction will be assumed to be computed as stated in Theorem 2.

Remark 1. From Theorem 1 and Theorem 2 follows: if $(t, a) \in R$ then $(t, a) \in (R_0 \cup \leq)^*$, where $*$ is a transitive closure operation and R_0 is a logical reduction of R.

It is suggested that the model be extended with a new parameter – a predicate of the form $P : \mathbb{F} \times \mathbb{F} \to \{0, 1\}$. The predicate is meant to control the refactoring process in the following way: if $P(t, a) = 0$, then t must not aggregate a as a result of refactoring. In order to ensure that this predicate does not conflict with the original type hierarchy, the set of possible predicates needs to be restricted. As a part of its definition, P is required to have the following property: $P(t, a) = 1$ if t aggregated a before the refactoring process.

The new parameter can be integrated into the model as an additional condition for pairs to be $\vee$-compatible. Two pairs $(b_1, a) \in R$ and $(b_2, a) \in R$ will not be considered $\vee$-compatible and the corresponding triples (c_1, c_2, a) will not be considered $\vee$-distributive if $P(c_1 \vee c_2, a) = 0$.

Theorem 3. Let $\tilde{R}$ be a logical relation with predicate P, and let R_0 be a logical reduction of it. Then R_0 does not contain pairs (t, a), such that $P(t, a) = 0$.

Proof. Let $(x, a) \in \tilde{R}$ and $P(x, a) = 0$.

If (x, a) is transitive in $\tilde{R}$, then $(x, a) \notin R_0$ (otherwise R_0 would not be minimal). Therefore we later only consider cases when (x, a) is not transitive in $\tilde{R}$. A situation when $(x, a) \in R$ and $P(x, a) = 0$ is impossible due to restrictions on P.

The only case left is when $(x, a) \in \tilde{R}$ is true because there exists a conflictless $\vee$-compatible pair $(b_1, a), (b_2, a) \in \tilde{R}$, where $b_1 \vee b_2 = x$. Then (by Definition 2), $\exists c_1, c_2 : x \leq c_1 \vee c_2$, where $(c_1, a), (c_2, a) \in \tilde{R}$. If $x < c_1 \vee c_2$, then (c_1, c_2, a) is a neutralizing triple for all (b_1, b_2, a) where $b_1 \vee b_2 < c_1 \vee c_2$, and $(b_1, a), (b_2, a)$ is not conflictless. $x = c_1 \vee c_2$ is impossible due to $P(x, a) = 0$. Therefore $(b_1, a), (b_2, a) \in \tilde{R}$ cannot be conflictless $\vee$-compatible pairs. Then either (x, a) is transitive (which contradicts the currently considered case), or $(x, a) \notin \tilde{R}$.

If $(x, a) \notin \tilde{R}$, then $(x, a) \notin R_0$. Otherwise $R \sim R_0$ will not hold. □

Note that the predicate $P(t, a)$ is only guaranteed to affect cases when an attribute of type a belongs to the type t. In order to ensure that t will not have an attribute of a itself and will not inherit it, $P(t', a)$ must be 0 for all $t' : t \leq t'$.

The rest of the chapter is dedicated to the discussion of modeling aggregation.

In the original paper [8], aggregation is modeled by means of a binary relation over the type lattice. It significantly restricts the set of type hierarchies that can be refactored by following this approach.

As a solution to the problem, it is suggested that attributes be added as subtypes into the type hierarchy. A type lattice that contains both elements representing types and elements representing attributes, later in the text is called extended type lattice.

Let A be the set of attributes of the original (before refactoring) type hierarchy. Below are the steps required for building the extended type lattice.

- Type lattice $\mathbb{F}$ is built as described in [8].
- For each $t \in \mathbb{F}$ let $A_t \subseteq A$ be the set of attributes of the type t. A_t is expected to be divided into disjoint sets $\{A_t^{(i)}\}_{i=1}^N$ according to the following rule. If $a, b \in A_t^{(i)}$, then it is acceptable to replace a and b by a single attribute in their common parent. The division is supposed to be done before the model construction and application.
- For each $A_t^{(i)}$ from $\{A_t^{(i)}\}_{i=1}^N$ an element (denoted by $t_{agg}^{(i)}$) is added into the type lattice as a child type of t (i.e. $t_{agg}^{(i)} \leq t$). Let $\tilde{T}$ be the set of all such added elements.

Binary relation R on the extended type lattice is suggested to be constructed in the following way. Let $a \in A$ be an attribute of the original type hierarchy. Also let $A_t^{(i)}$ be an attribute subset such that $a \in A_t^{(i)}$, let $t_{agg}^{(i)}$ be an element representing $A_t^{(i)}$ in the type lattice, and let a belong to type t'. Then attribute a is represented in R by a pair $\left(t', t_{agg}^{(i)}\right) \in R$.

Below is an example of extended type hierarchy for a simple messaging system. Its UML class diagram is depicted on the left of Fig. 1. There are messages sent from one user to another (represented by the PrivateMessage class) and messages from one user to all (the PublicMessage class). User class is not present in the diagram since its content is not relevant for the example.

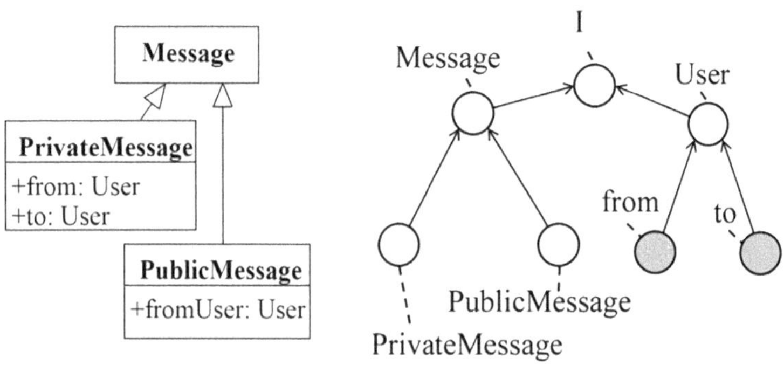

Fig. 1. UML diagram and the corresponding extended type lattice.

On the right in Fig. 1 there is an extended type hierarchy for the system. The fictitious successor of all types O is not included in order to simplify the illustration. Links are directed from descendant elements to ancestor elements. Elements of the set $\tilde{T} = \{\text{from, to}\}$ are highlighted in gray.

Binary relation R has tuples (PrivateMessage, from), (PrivateMessage, to), (PublicMessage, from) that correspond to the field "from" of PrivateMessage, field "to" of PrivateMessage and field "from User" of PublicMessage respectively.

The described approach can be used to model aggregation without any restrictions on the acceptable type hierarchies. The quality of transformations suggested by the model depends heavily on the accuracy of decisions made during extended type lattice construction.

4 Model Application

This section describes a generalized process of object-oriented software refactoring that is based on the application of LP structures on type lattices.

Step 1. Constructing and Configuring the Model

Firstly, this step involves the construction of a type lattice $\mathbb{F}$ and a binary relation over it.

Depending on the specific goal, the model can be constructed using the whole type hierarchy or only a part of it. As an example, for refactoring C++ code it may be reasonable to exclude links of virtual inheritance since lifting an attribute to a virtually inherited class may change the software system behavior.

Model configuration may involve the following.

- Controlling whether it is permissible to combine specific attributes in their common parent, when an extended type lattice is used. Let a_1 and a_2 be some attributes of a currently refactored software system. They can only be combined if they are put into the same subset of attributes $A_t^{(i)}$.
- Specifying whether an attribute can be uplifted to specific parent types. This can be done with a predicate $P : \mathbb{F} \times \mathbb{F} \to \{0, 1\}$ described in Sect. 3.

Step 2. Obtaining a New Set of Attributes

There exist at least two methods of obtaining a new set of attributes.

- Performing "Pull Up Field" refactoring only. This can be done primarily by building a set of conflictless $\vee$-distributive triples. Then, for each such triple (c_1, c_2, a) an attribute of type a must be assigned to the type $c_1 \vee c_2$ and all attributes a must be removed from types t such that $t < c_1 \vee c_2$. In the case of extended type lattice, a would be an element representing some attribute subset (by the construction of R).
- Performing a complete type hierarchy optimization. In this case, a logical reduction of R must be found as stated in Theorem 2.

Step 3. Matching Sets of Attributes Before and After the Refactoring Process

In addition to constructing the model and obtaining a new set of attributes, two more problems need to be solved to complete the refactoring process.

1. Updating definitions. In order to complete the task it is needed to extract some more information from the model. Each attribute a of the new type hierarchy must be matched with those original attributes that were combined into a. The mapping would make it possible to consider properties of existing attributes while generating new code.
2. Updating references. All references to attributes that were replaced or removed must be replaced with references to the attributes of the new type hierarchy. It also requires construction of the corresponding mapping between old and new attribute sets. It is also necessary to take into account that in the case of complete type hierarchy optimization a reference to one attribute of previous type hierarchy might be replaced by a chain of references to new attributes.

Let us formulate the problems in terms of the model. Note that during the transformations described in step 2, it is only a binary relation that is modified, while $\mathbb{F}$ remains unchanged. Let R be a binary relation over $\mathbb{F}$ before any transformations, and let R' be the binary relation after the transformations. Solutions to the tasks will be considered given the following assumptions.

- The model is constructed in such a way that before the transformations it is possible to unambiguously match each $(t, a) \in R$ with an attribute of the original type hierarchy.
- Each $(t', a') \in R'$ corresponds to an attribute of the new type hierarchy.
- R' is a logical reduction of R.

The solution to the second problem can be represented as a function $f_{new} : R \to SEQ_{R'}$, where $SEQ_{R'}$ is a set of sequences $\{(a_i, b_i)\}_{i=1}^{N}$. Such sequences correspond to valid "chains" of references to attributes. They consist of pairs $(a_i, b_i) \in R'$ such that $b_i \leq a_{i+1}$ (for $i < N$). Let $x = (t, a) \in R$ and $f_{new}(x) = \{(a_i^x, b_i^x)\}_{i=1}^{N}$. For valid functions f_{new}, the following boundary conditions must hold: $t \leq a_1^x$ and $a = b_n^x$.

A valid f_{new} is constructed below. Let $(x, y) \in R$ be an argument for which it is needed to calculate the value of f_{new}. It follows from Remark 1 (Sect. 3) that there must be a sequence $P = \{(x_i, y_i)\}_{i=1}^{M}$, where $(x_i, y_i) \in R' \cup \leq$, $x_i = y_{i-1}$ for $i > 1$, $x_1 = x$, $y_m = y$. Due to the transitivity of $\leq$, a sequence $P' = P \setminus \leq$ has all the properties required in the previous paragraph. The choice of $(x, y) \in R$ was arbitrary, and therefore the constructed f_{new} is valid.

The solution for the second problem can be used to solve the first one. Formally, the task is to find a function $f_{prev} : R' \to P(R)$, where $P(R)$ is a power set of pairs of R. Such pairs, according to the assumptions currently used, can be unambiguously matched with attributes of the original type hierarchy.

Let last_pair: $SEQ_{R'} \to R'$ be an auxiliary relation that for a sequence $\{(a_i, b_i)\}_{i=1}^{N} \in SEQ_{R'}$ returns its last element (a_N, b_N). Then the solution to the second problem can be described as follows $f_{prev}(x) = \{y \in R \mid \text{last\_pair}(f_{new}(y)) = x\}$.

Thus, both problems can be successfully resolved when the described model is used for type hierarchy transformations.

5 Conclusion

The issues of LP structures theory application to the refactoring automatization are considered in this paper.

Extensions to the model are suggested that facilitate the use of the model for the analysis and refactoring of a wide range of object-oriented software systems. The process of employing the model for the refactoring purposes is considered, including model construction and configuration, transformations, the process of obtaining and applying the results. At the higher level these stages can be interpreted not as a complete sequence of actions, but as a part of a complex refactoring process that gradually accomplishes its various objectives.

The model suggested in the paper has extensive capabilities of formal description and transformation of type hierarchies. That makes it possible to consider the model as a basis for building intellectual refactoring systems.

The reported study was supported by RFBR project 19-07-00037.

References

1. Nexhati, A.: Justification of software maintenance costs. Int. J. Adv. Res. Comput. Sci. Softw. Eng. **7**(3), 15–23 (2017)
2. Fowler, M.: Refactoring: Improving the Design of Existing Code. Addison-Wesley, Boston (1999). ISBN: 0-201-48567-2
3. Ganter, B., Wille, R.: Formal Concept Analysis: Mathematical Foundations, p. 284. Springer, Heidelberg (1999)
4. Torim, A.: A visual model of the CRUD matrix. In: Henno, J., et al. (eds.) Information Modelling and Knowledge Bases XXIII, pp. 313–320. IOS Press, Amsterdam (2012)
5. Huchard, M.: Analyzing inheritance hierarchies through Formal Concept Analysis. A 22-years walk in a landscape of conceptual structures. MASPEGHI: MechAnisms on SPEcialization, Generalization and inHerItance, July 2015, Prague, Czech Republic, pp. 8–13 (2015). https://doi.org/10.1145/2786555.2786557
6. Godin, R., Valtchev, P.: Formal concept analysis-based class hierarchy design in object-oriented software development. Lecture Notes in Computer Science, vol. 3626, pp. 304–323. Springer, Heidelberg (2005)
7. Rouane-Hacene, M., Huchard, M., Napoli, A., Valtchev, P.: Using formal concept analysis for discovering knowledge patterns. In: Proceedings of the 7th International Conference on Concept Lattices and Their Applications, Sevilla, Spain, 19–21 October 2010, pp. 223–234 (2010)
8. Makhortov, S.D.: LP structures on type lattices and some refactoring problems. Program. Comput. Softw. **35**(4), 183–189 (2009). https://doi.org/10.1134/S0361768809040021
9. Makhortov, S.D., Shurlin, M.D.: LP-structures analysis: substantiation of refactoring in object-oriented programming. Autom. Remote Control **74**(7), 1211–1217 (2013). https://doi.org/10.1134/S0005117913070126
10. Aho, A.V., Garey, M.R., Ulman, J.D.: The transitive reduction of a directed graph. SIAM J. Comput. **1**(2), 131–137 (1972)

Hybrid Approach for Bots Detection in Social Networks Based on Topological, Textual and Statistical Features

Lidia Vitkova(✉), Igor Kotenko, Maxim Kolomeets, Olga Tushkanova, and Andrey Chechulin

St. Petersburg Institute for Informatics and Automation of the Russian Academy of Sciences, 14-th Liniya, 39, St. Petersburg 199178, Russia
{vitkova,ivkote,kolomeec,tushkanova, chechulin}@comsec.spb.ru

Abstract. The paper presents a hybrid approach to social network analysis for obtaining information on suspicious user profiles. The offered approach is based on integration of statistical techniques, data mining and visual analysis. The advantage of the proposed approach is that it needs limited kinds of social network data ("likes" in groups and links between users) which is often in open access. The results of experiments confirming the applicability of the proposed approach are outlined.

Keywords: Social network analysis · Visual analysis · Data mining · Statistics · Bots detection

1 Introduction

Currently, social networks have a huge impact on their wide audience. This is primarily due to the aggregating capabilities of social networks, unified interfaces, easy access to information, and the ability of each member to express publicly their point of view. Social networks have also proven to be a news and advertising platform for private companies and government institutions, as they provide wide range of tools for managing content posted, followed by an analysis of its relevance to the Internet audience and an assessment of its response.

At the same time, the tools for detecting and controlling potentially harmful information, targeted by violators in the continuously growing information space of social networks, are not sufficiently developed from a technical and legal point of view. These factors give an attacker the opportunity to use social networks to implement information and psychological attacks on untrained users. As Guardian columnist Natalie Nougayrède has observed: "The use of propaganda is ancient, but never before has there been the technology to so effectively disseminate it" [11]. Also a new phenomenon of the 21st century is the emergence of "fake adherents". Such a phenomenon can be attributed to "information disorder" [20]. Thus, the analysis of social networks in order to find anomalies in information interaction is an important and topical task.

S. Kovalev et al. (Eds.): IITI 2019, AISC 1156, pp. 412–421, 2020.
https://doi.org/10.1007/978-3-030-50097-9_42

In the paper, the authors investigate and propose methods for analyzing social networks designed to detect suspicious activity, which operate even with the minimum amount of available basic data. The scientific novelty of the paper is in the proposed hybrid method for analyzing social networks, which through the use of statistical methods, data mining, and visual analysis allows one to obtain results even under limitations. The paper shows the progress of the authors' research aimed at development of models and algorithms for decision-making support in countering the spread of unwanted information in social networks [7, 10, 12].

The paper is structured as follows. The second section provides an overview of the related works. The third section describes the proposed approach. The fourth section presents the experiments and compares the results. The fifth section analyzes the results and discusses possible areas of future research.

2 Related Works

Finding the best solutions to detect bots in social networks is a crucial task today. However, the concept of the "bot" in 2003–2005 was more often associated with the Botnet [13]. Over time, the concept of the "bot" flowed into network science and social network analysis.

In [17] the method for detecting game bots, i.e., autoplaying game clients, is proposed. The bot detection method is based on behavior analysis. Today, platforms for online games, e.g. Steam, are also social networks. So perhaps the method proposed in [17] was the first attempt to recognize bots. A large number of papers were published by researchers on this topic in 2011–2012. And further the number of works on this topic increases exponentially. In [1], the authors raise the question of how popular a bot can become without a completed profile and attempts to copy human behavior and how it can affect the choice of new social connections among other users of social platforms. In [14], political campaigns on microblogging platforms are explored when multiple profiles are created to imitate wide support for the candidate. However, in that period the works on the identification of bots are not yet so common.

After the scandals associated with election campaigns in 2016, the number of papers devoted to the detection of the bots in social networks has increased significantly. In [5], the authors state that society is puzzled by the development of advanced methods for the automatic detection of the social bots, but the efforts of the academic community in this direction just started. In [5], available and well-known methods and tools for identifying the bots are described. In [16], the issue of detection of fake news is raised. The authors are based on the fact that fake news is written with the purpose of misleading the reader. And, since there is such a goal for the one who creates the news, it is therefore necessary to introduce an auxiliary function – to analyze the social connections of users with fake news and with fake profiles. Also in the proposed method, the authors [16] come from the fact that fake users produce large, incomplete, unstructured, and noisy data.

The service BotOrNot, which still functions today, is presented in [3]. The purpose of the service is to provide an opportunity to check the degree of similarity of the bots with the Twitter users, including when calculating an advertising company. In [3], bot profiles are called sybil accounts. In [2], the issue of detecting Sybil accounts' BIRDS is raised due to their cohesion and connectedness. Such an approach has become widespread in further research, when the topology and relatedness of fake profiles are considered. The paper [18] proposes a framework to detect such entities on Twitter. The authors are testing a classification structure derived from publicly accessible Twitter dataset. They use labeled and enriched sample to train their own model. According to the estimates obtained in [18], from 9% to 15% of active Twitter accounts are bots. Next, the authors analyze the connections between the bots and show that simple bots, as a rule, interact with the bots that exhibit a more human-like behavior. Also, according to the results of the analysis of content flows, the authors demonstrate retweet strategies and mentions taken by bot-farms as a template for interacting with various target groups.

In [19], the authors proceed to the discovery of information companies using the modified framework previously described in [18]. They set themselves the task of determining whether the news, the message, the meme is artificially advancing at the very moment when it becomes popular. Classifying such information is quite a difficult task, but such announcements cause attention spikes that can easily be confused with organic trends. The authors [19] developed a machine learning structure for classifying memes that were marked as trending on Twitter. Twitter hashes are used to detect memes. However, the authors have still not proposed a generalized method. Although combining the meme detection approach and evaluating the sources of information, identifying the bots among active sources, seems to be a promising task. The paper [4] uses a large dataset of tweets collected during the UK-EU referendum campaign. The dataset was divided into 2 populations – users and bots. As part of the work, it was demonstrated that the statistics of correlations and the probability of a joint burst differ in the selected populations.

And yet, the most approaches to detect the bots in social networks use large amounts of data. Building a graph of relationships between millions of users in a social network requires a long time or very large computing power. For the decision support system, the speed, simplicity, and validity of the results obtained in the analysis process are important. But it is not always possible to use high technologies. The question arises of how to detect a bot account in a social network with incomplete or insufficient data availability.

3 Proposed Approach

In the proposed approach we proceed from the fact that the bots can be created by an attacker for the purpose of cheating the rating, and for the purpose of making a profit. It is assumed that there are two types of bots on a social network: (1) fake "likers" profiles [15]; (2) communicative capitalist profiles [6]. Both are shown through likes to posts, comments, and photos. At the same time, to detect two types of bots, it is enough open data and incomplete information about who put like without indicating the like time and the content analysis. Collecting indicators information for the user allows one to select a general population characteristic of some object of observation, for example, a group on a social network. Analysis and evaluation of the general population are carried out after its division into stratified sampling.

In order to formalize the process of segmentation of the general population into representative samples in the proposed method, we introduce the following strata: (1) E-fluentials; (2) Sub-E fluentials; (3) Activist; (4) Sub-Activist; (5) Observer.

The separation of strata is made according to the following algorithm. In the first step, users are sorted in descending order of the number of likes from them. The average number of likes per user is calculated. All users with activity below the average are classified as the group (5) users. At the second step, the average number of likes per user is calculated for the other most active users. The users that have less likes then the average in the active group are cut off at the group (4). In the third step, the characteristic of the average value of the sum (1), (2), (3) is revealed. The group of activists who are not leaders in the number of likes is identified. In the fourth step, the user group is segmented into (1) and (2), as in the previous steps. In the fifth step, the subscriber/follower connection between the object of observation (a group on social network) and the user is checked. At the sixth step, the following parameters are calculated for each segment using the source data: Count (total number of users in the segment); Median (median of user likes by segment); Var (dispersion of user likes by segment), Q1–Q3 (respectively, the upper and lower quantiles of the distribution); StdDev (standard deviation characteristic of the segment); Range (range of likes of a segment); Qutliers and PctQutliers (number and percentage of surges, respectively). At the seventh step, the surges in each group are selected to analyze user profiles. At the eighth step, the E-fluentials sample is analyzed completely, due to the highest number of outgoing likes from them in the general population, which may indicate a sign of behavior characteristic of communicative capitalists. At the ninth step in the Observer sample, user pages are selected that are not related to the object of observation, but are included in the general population, and a separate analysis is performed to identify clusters of fake likers. At the tenth step for each segment, the topology of the friends' relationship between participants in a representative sample is investigated. Also the relatedness of the user's profile with a cluster of bots from Not Member Observer is analyzed.

The proposed algorithm differs from the existing ones because it does not require a complicated data collection. Baseline data are user likes and their relationships with each other. Then, for the selected segments, algorithms are launched that allow revealing the signs (any, except for the actual likes), characteristic for each segment. The last step of the method is the use of visual models to display the extracted information and visual analysis of data by the system operator.

4 Experiments

As a research object for the experiment, Vkontakte social network group "Life in Kudrovo" was chosen. There are 52721 subscribers of the group. A total of 46093 users have liked something within the group during its existence, 22332 of them are not subscribed to the group and are not members of it.

4.1 Statistics Approach to Group Analysis

The highest percentage of users who are not in the group belong to the Observer segment (Table 1). By stratified sampling, a table was built with statistical indicators for 4 segments for subsequent demonstration on plot-boxes as part of decision support (Table 2).

Table 1. Summary table for the group "Life in Kudrovo"

Stratified sampling	Count	Member	Not member	Member	Not member
Observer	40388	18539	21849	45,90%	54,10%
SubActive	4483	4104	379	91,55%	8,45%
Active	900	824	76	91,56%	8,44%
SubE-fluentials	231	210	21	90,91%	9,09%
E-fluentials	90	83	7	92,22%	7,78%

Table 2. Statistical indicators

Statistical indicator name	E-fluentials	Sub-E fluentials	Activist	Observer*
Count	90	231	900	18539
Median	1536	665	225,5	3
Var	2,6168e+06	24392,1	7461,01	22,601
Q1	1279,25	565,5	174	2
Q3	2313	798	302	7
StdDev	1617,65	156,18	86,3771	4,75405
Range	10895	625	328	19
Qutliers	8	0	0	1358
PctQutliers	0,0888889	0	0	0,073521

*The Observer group includes only the users that are subscribed to the studied social network group, due to the high level of noise from non-group profiles.

Thus, during comparing statistical parameters of stratified sampling, it is clear that there are anomalous surges in E-fluentials and Observer groups. Users demonstrating abnormal activity can be further analyzed during the process of bots detecting. However, with the support of decision-making, the fact of detecting signs of abnormal surges is important for the choice of further activities.

Also within the framework of the experiment, at the initial stage, prior to the visualization, Not Member users were grouped in the Observer segment, which are conditionally assigned to fake likers prior to the topology analysis. Users included in E-fluentials are assigned to the communicative capitalists.

The Sub-E fluentials and Activist segments are the most resistant; they have no surges of anomalous activity, which may be a sign of pages being "humane".

4.2 Machine Learning Approach to Group Analysis

At the next step of the developed methodology, it is supposed to use automatic classification of user accounts with further automatic identification of bots among them. The users of the Vkontakte social network, previously assigned to the E-fluentials, Sub-E fluentials, and Active user groups, were used to build the classifier, a total of 1221 user accounts. Each user account in the sample is described by the following features: (1) features extracted directly from the user profile (sex, country, religion, whether user interests are listed in the profile or not, whether user favorite movies are listed in the profile or not, whether user favorite music is listed in the profile or not, whether user favorite books are listed in the profile or not); (2) aggregated features (the number of posts on the "wall" of the user profile, the number of photos in user profile, the number of videos in user profile, the number of user friends, the number of user groups, the number of user comments in the studied group in the social network).

After 2018, access to the API of many social networks was limited, in addition, users are now able to completely hide information about themselves in the social network. In this regard, the profile data that was collected during the study is very limited, which corresponds to the real situation in the field of identifying anomalous activity. There is significant amount of missing values in the attributes describing the user profile in the collected dataset. The missing values could appear in the dataset during data collection for two reasons: either the user did not provide relevant information about himself in the profile, or he/she chose to hide it from a wide range of users on the social network. In both cases, this information is considered to be missing in the profile, therefore during the analysis such missing values were replaced with the value "Not specified". To build a classifier, the dataset was divided into two parts in the ratio of 70% to 30%. In the first part of the sample, a cross-validation with four blocks was used to select the optimal parameters for each classifier. The second part of the sample was used to evaluate the classifier with optimal parameters using accuracy and F-measure metrics.

During a series of experiments, the following classifiers, implemented in the Scikit-learn Python library [9], were investigated: logistic regression implemented in the LogisticRegression class; support vector method implemented in the SVC class; naive Bayes classifier implemented in the GaussianNB class; decision tree implemented in the DecisionTreeClassifier class; random forest implemented in the RandomForestClassifier class; classifier based on the boosting, implemented in the AdaBoostClassifier class; voting of several classifiers, implemented in the VotingClassifier class.

As expected, none of the classifiers could significantly improve the accuracy compared to the baseline (Table 3).

Obviously, this is due to the limited data obtained through the open API. However, the best in the F-measure VotingClassifier (which implements the "hard" voting strategy of the LogisticRegression, RandomForestClassifier and AdaBoostClassifier classifiers) can be used as a means of accounts filtering before further in-depth statistical analysis.

At the next stages, to solve the problem of user accounts classifying, it is planned to use a larger training sample, as well as to consider other classes of user accounts. It is assumed that the number of features used and the percentage of values missing do not

Table 3. Classifiers estimation

Classifier	Accuracy	F-measure
Baseline	0.589	0.604
SVC	0.755	0.669
LogisticRegression	0.784	0.719
GaussianNB	0.669	0.651
RandomForestClassifier	0.755	0.713
DecisionTreeClassifier	0.763	0.666
AdaBoostClassifier	0.763	0.737
VotingClassifier	0.771	0.724

change much, so it is necessary to look for ways to build more accurate classifiers using minimum amount of source data received from the social network API.

4.3 Visual Analytics Approach to Group Analysis

One of the methods used for analyzing groups is visual analytics, which allows the operator to assess the types of structures that certain users constitute. Consider an example based on the visualization of graphs of friends of various classes of users. The visualization module was implemented using the D3.js Force Layout Graph library. Force layout is useful when analyzing small samples, since unlinked vertices of the graph repel from the center and fly apart, while the connected component remains in the center of the screen.

The basis of visual analytics is that different classes of users can constitute different structures. For the experiment, the most active user classes were selected from Table 1 (more active than the Active class). The resulting images and their correspondence to the class are defined in Table 4 and are available at [8].

Table 4. Classes of sets and their figures

Stratified sampling	Member type	Count	Figure/Class
E-fluentials	NotMember	7	A
SubE-fluentials	NotMember	21	B
Active	NotMember	76	C
E-fluentials	Member	83	D
SubE-fluentials	Member	210	E
Active	Member	824	F
E-fluentials and SubE-fluentials and Active	NotMember	104	G
E-fluentials and SubE-fluentials and Active	Member	1117	H
E-fluentials	NotMember and Member	90	I
SubE-fluentials	NotMember and Member	231	K
Active	NotMember and Member	900	L
E-fluentials and SubE-fluentials and Active	NotMember and Member	1221	M

With visualization, the operator can assess the type of structure and its size. On the resulting images, we can distinguish three types of graph structures: clouds of unrelated users, k-plex (almost fully connected graphs) and structures similar to trees.

Consider the resulting graphs [8] in more detail. In classes A–C, users are not connected, which is typical for non-group users. The exception is a five-point star in class A (Fig. 1). In the center of the star there is a local popular activist, and its presence can be explained by the appearance of an opinion leader in the sample.

In class D, there is obviously a strongly connected community among users with a small sample size (Fig. 2). At the same time, in the class E (Fig. 3) there is no a large connected component, despite the fact that in E the number of users is greater than D. This suggests an anomaly that can be investigated. This community may turn out to be a group of activists, bots or a closely connected community. In class F, there is a tree-like connected component (Fig. 4), which naturally occurs on relatively large samples among loosely coupled users.

Fig. 1. Star structure [8]

Fig. 2. Strongly connected component [8]

Fig. 3. Small linear graphs can appear on small sets [8]

Fig. 4. Tree-like graphs can appear on bigger sets [8]

Classes G and H are the union of classes A–C (notMember) and D–F (Member), respectively, and therefore provide similar images. Classes I–L give similar images to classes D–F. Class M is a generalization of all classes, but it already has a tree structure and strongly connected component.

Based on this analysis, we can draw the following conclusions: (1) for classes A–C, a typical structure is a disjointed graph. Even small trees and stars as in Fig. 3 can be considered as anomalies; (2) for classes D–E, the disjointed graph is also a typical structure. Anomalies can be considered large trees and strongly connected components as in Fig. 4; (3) for class F, the appearance of a weakly connected component like a tree is typical; (4) classes G and H are generalizations of classes A–C and D–F, respectively, and can be used for preliminary analysis; (5) classes I–L are identical to classes D–F; (6) class M is a generalization of all figures and can be used for preliminary assessment of structures.

The obvious fact is that as Member type sample grows, the likelihood of the tree structure grows, and the appearance of a strongly connected component is an anomaly. For a sample of the notMember type, practically any deviation from a disconnected graph already indicates an anomaly. Thus, visual analytics can be applied to a preliminary analysis of a social network, creating and confirming hypotheses regarding the degree of connectedness of users and the presence of anomalies in their structures. For different classes of samples, it is possible to determine certain patterns of structures, the deviation from which may be a signal for a deeper study of the anomalies that have been discovered. The data on the structures themselves and the typical patterns for the classes can also be used as input parameters for machine learning methods and as parameters in the statistical analysis of groups.

5 Conclusion

The paper presents an approach to hybrid analysis of social networks. This approach provides means to collect more heterogeneous information regarding users profiles for future analysis of possible inappropriate information sources, repeaters and aggregators. The difficulty in identifying the disseminators of such information and in countering them on social networks lies in the requirements that today apply to monitoring systems. First of all, such systems work with an extremely large amount of data, with a huge number of messages created by users. Traditional methods of analysis of potential sources, repeaters and aggregators and counteraction to them do not cope with their tasks. The proposed hybrid approach adds new features, on the basis of which measures can be taken to protect the audience of social networks from inappropriate content.

In further work it is planned to conduct additional experiments aimed at analyzing information flows. Currently, the authors explore the possibilities of combining the approach with previous approaches that were focused on information flow analysis.

Acknowledgements. This research was supported by the Russian Science Foundation under grant number 18-71-10094 in SPIIRAS.

References

1. Aiello, L.M., et al.: People are strange when you're a stranger: impact and influence of bots on social networks. In: International AAAI Conference on Weblogs and Social Media (2012)

2. Cook, D.M.: Birds of a feather deceive together: the chicanery of multiplied metadata. J. Inf. Warfare **13**(4), 85–96 (2014)
3. Davis, C.A., et al.: Botornot: a system to evaluate social bots. In: Proceedings of the 25th International Conference Companion on World Wide Web, International World Wide Web Conferences Steering Committee, pp. 273–274 (2016)
4. Duh, A., Rupnik, S.M., Korošak, D.: Collective behavior of social bots is encoded in their temporal twitter activity. Big Data **6**(2), 113–123 (2018)
5. Ferrara, E., et al.: The rise of social bots. Commun. ACM **59**(7), 96–104 (2016)
6. Gavra, D.P., Dekalov, V.V.: Communicative capitaland communicative exploitation in digital society. In: 2018 IEEE Communication Strategies in Digital Society Workshop (ComSDS), pp. 22–26. IEEE (2018)
7. Gorodetsky, V., Tushkanova, O.: Data-driven semantic concept analysis for user profile learning in 3G recommender systems. In: 2015 IEEE/WIC/ACM International Conference on Web Intelligence and Intelligent Agent Technology (WI-IAT), Singapore, pp. 92–97 (2015). https://doi.org/10.1109/wi-iat.2015.80
8. http://comsec.spb.ru/files/iiti2019/vis2D.html
9. https://scikit-learn.org/
10. Kotenko, I., Chechulin, A., Komashinsky, D.: Categorisation of web pages for protection against inappropriate content in the internet. Int. J. Internet Protoc. Technol. (IJIPT) **10**(1), 61–71 (2017)
11. Nougayrede, N.: In this age of propaganda, we must defend ourselves. Here's how, The Guardian (31/01/18) (2018). Accessed 28 Mar 2018.https://www.theguardian.com/commentisfree/2018/jan/31/propaganda-defend-russia-technology
12. Pronoza, A., Vitkova, L., Chechulin, A., Kotenko, I.: Visual analysis of information dissemination channels in social network for protection against inappropriate content. In: 3rd International Scientific Conference on Intelligent Information Technologies for Industry, IITI 2018. Sochi, Russian Federation, 17–21 September 2018, Advances in Intelligent Systems and Computing, vol. 875, pp. 95–105 (2019)
13. Puri, R.: Bots & botnet: an overview. SANS Inst. **3**, 58 (2003)
14. Ratkiewicz, J., et al.: Detecting and tracking political abuse in social media. In: Fifth International AAAI Conference on Weblogs and Social Media (2011)
15. Satya, B.P.R., et al.: Uncovering fake likers in online social networks. In: Proceedings of the 25th ACM International on Conference on Information and Knowledge Management, pp. 2365–2370. ACM (2016)
16. Shu, K., et al.: Fake news detection on social media: a data mining perspective. ACM SIGKDD Explor. Newsl. **19**(1), 22–36 (2017)
17. Thawonmas, R., Kashifuji, Y., Chen, K.-T.: Detection of MMORPG bots based on behavior analysis. In: Proceedings of the 2008 International Conference on Advances in Computer Entertainment Technology (ACE 2008), pp. 91–94. ACM, New York (2008). https://doi.org/10.1145/1501750.1501770
18. Varol, O., et al.: Online human-bot interactions: detection, estimation, and characterization. In: Eleventh International AAAI Conference on Web and Social Media (2017)
19. Varol, O., et al.: Early detection of promoted campaigns on social media. EPJ Data Sci. **6**(1), 13 (2017)
20. Wardle, C., Derakhshan, H.: Information disorder: towards an interdisciplinary framework for research and policy-making. Council of Europe (2017). https://firstdraftnews.com/resource/coe-report/

Absolute Secrecy Asymptotic for Generalized Splitting Method

V. L. Stefanuk[1,2(✉)] and A. H. Alhussain[1]

[1] Peoples' Friendship University of Russia,
Miklucho-Maklaya str. 6, 117198 Moscow, Russia
amanie.alhussain@mail.ru

[2] Institute for Information Transmission Problems,
Bolshoi Karetny per. 19, 127051 Moscow, Russia
stefanuk@iitp.ru

Abstract. Generalized integer splitting differs from the plain one in use of a new random number at each step of the process. It assumed that the receiver is informed on these numbers and the splitting level k, and hence he is able to restore an original text using some known procedures from the number theory.

The present paper contains a probabilistic analysis of the information secrecy of the generalized integer splitting method with respect to an unauthorized access to transmission channel. A lemma has been proved in the paper stating that the probability of the successful information restoration is exponentially reduced with increase of the splitting level.

Keywords: Integer splitting · Generalized integer splitting · Unauthorized text restoration · Integer division · Modular arithmetic · Asymptotic secrecy

1 Introduction

Method of integer splitting we proposed in our publications [1–4] and patented as an invention in [5] as one of the new ways for application of modular arithmetic to provide information secrecy. It consists in the replacement of each text symbol with a chain of integers constructed in accordance with some rules. This method was described in [3], with main definitions and concepts allowing replacement of any integer on the base of some other integer with a sequence of k integers. This procedure was referred to as k-level integer splitting method. In present research the procedure was analyzed with respect a possibility of unauthorized access to the information transmitted over communication channels.

Our paper is organized in the following way. The Sect. 2 contains definitions and theoretical grounds for so-called generalized integer splitting. In Sect. 3 a lemma was formulated and proved that connected with security level provided. Conclusions are provided at the end of the paper right, before the reference list.

The research work was partially supported by Russian Fond for Basic Research, grants № 18-07-00736A and № 17-29-07053.

S. Kovalev et al. (Eds.): IITI 2019, AISC 1156, pp. 422–431, 2020.
https://doi.org/10.1007/978-3-030-50097-9_43

2 Main Definitions and Concepts

The integer splitting method, described in papers [1–4], allows several definitions. The special feature of the generalized splitting methods is the use of some base of vector of integers $\vec{r} = (r_1, r_2, \ldots, r_n)$, satisfying some inequality requirements.

Principles of transmission was described in our previous publications. These principles assumes the symmetrical way of safe data transmission and storage [1–4].

Definition 1. The generalized integer splitting of integer a over the vector base $\vec{r} = (r_1, r_2, \ldots, r_l)$, $l = k - 1$, is the following representation of the number a as a sequence of integers $a_1, a_2, a_3, \ldots, a_{k-1}, a_k$ in the following manner:

$$
\begin{aligned}
a_1 &= \delta^{(2)}, \text{where } \delta^{(2)} = r_1 \bmod a, r_1 > a, \\
a_2 &= \delta^{(3)}, \text{where } \delta^{(3)} = r_2 \bmod q^{(2)}, q^{(2)} = \left\lfloor \frac{r_1}{a} \right\rfloor, r_2 > q^{(2)}, \\
a_3 &= \delta^{(4)}, \text{where } \delta^{(4)} = r_3 \bmod q^{(3)}, q^{(3)} = \left\lfloor \frac{r_2}{q^{(2)}} \right\rfloor, r_3 > q^{(3)}, \\
&\ldots\ldots \\
a_{k-1} &= \delta^{(k)}, \text{where } \delta^{(k)} = r_{k-1} \bmod q^{(k-1)}, q^{(k-1)} = \left\lfloor \frac{r_{(k-2)}}{q^{(k-2)}} \right\rfloor, \\
r_{k-1} &> q^{(k-1)}, \\
a_k &= q^{(k)}, \text{where } q^{(k)} = \left\lfloor \frac{r_{k-1}}{q^{(k-1)}} \right\rfloor.
\end{aligned}
\tag{1}
$$

Here the symbols δ denote the residue obtained during integer division r_i/a, the symbol $\lfloor\ \rfloor$ means the resulting integer part in this division, and the term k is referred to as the splitting level.

Restoration rule for the original symbol a after application of the generalized splitting method and with known values for the level k and vector $\vec{r}$, consisting in sequential restoration of the values from (1) in reverse order in comparison to their arriving to the receiving communication unit, is the following one:

$$
\begin{cases}
\frac{\left(r_i - \delta^{(j)}\right)}{q^{(j)}}, \textit{where} \quad j = k, k-1, \ldots, 3, 2 & \textit{when } k > 1 \\
\qquad\qquad\qquad\quad i = k-1, \ldots, 2, 1 & \textit{when } k > 1
\end{cases}
\tag{2}
$$

3 The Main Lemma and Its Proof

In accordance with (1) after receiving $(a_1, \ldots, a_k)$ at the input one has to compute $(q^{(2)}, \ldots, q^{(k)})$ based only on the known values $(r_1, \ldots, r_{k-1})$ and k. However, the last two values are naturally unknown for the unauthorized person who is able only to obtain the sequence $(a_1, \ldots, a_k)$.

Let $\Pr(A \mid C, k)$ be the probability of successful restoration of original text A using only the secured text C.

Thus, the analysis of probability of unauthorized text restoration will be done under suggestion that the hacker knows the formal procedures of splitting (1) and symbol restoration rule (2) yet the secret keys used for information safety is not known.

The main our result was formulated as the following lemma:

Lemma 1. The probability of unauthorized text restoration of the original text A on the base of the result of generalized integer splitting C is exponentially reduced with increase of k in accordance with the following expression:

$$\Pr(A \mid C, k) = \left(\sum_{i=2}^{k} L^{(i-1) \times \left\lfloor \frac{N}{i} \right\rfloor} \right)^{-1}. \tag{3}$$

In (3) the value N is the size of the secured text C, the sequence C is the result of generalized splitting of original text A, and L is the number of possible events during full search in the space of gammas from l values of the numbers $\{\tilde{r}_1, \tilde{r}_2, \ldots, \tilde{r}_L\}$, where $l = k - 1$ in accordance with definition (1).

The Proof of Lemma 1

Potential infringer is having in his disposal the sequence C of N integers $\{c_1, c_2, \ldots, c_N\}$. He knows the splitted text in communication channel as well as the rule of symbol restoration. Yet, the level of generalized splitting level and the gammas, used in the process of splitting, remains to him unknown.

As it is supposed that the infringer knows the method of security, described in the definition (1), and he knows model of symbol restoration (2), he begin with construction of a suggested sequence from L independent quasi random numbers.

$$R = \{\tilde{r}_1, \tilde{r}_2, \ldots, \tilde{r}_L\}. \tag{4}$$

From the definition (1) it is obvious that unlike of the case of simple splitting [1–4] the number of gammas needed for generalized splitting may vary and depends on the level of splitting k. However instead of random search it is more natural for the infringer to use a deterministic full search procedure over the set of all the parameters unknown to him.

The proof will be based on the full mathematical induction procedure.

a) *The proof for the case* $k = 2$

It will be shown that in case $k = 2$ the following is true:

$$\Pr(A \mid C, 2) = \left(L^{\left\lfloor \frac{N}{2} \right\rfloor} \right)^{-1} \tag{5}$$

In an attempt to restore the transmitter symbol our infringer has to take several steps.
First Step: The infringer has to subdivide the text C into collection of integer pairs. Resulting collection of sequentially observed pairs to be considered by the infringer[1] is equal to

$$X_2 = \left\lfloor \frac{N}{2} \right\rfloor \tag{6}$$

Second Step: The infringer begins the full search of values from the set R (see (4)) (the full search is made with return). From the definition (1) we conclude, that in case of generalized splitting for $k = 2$ each pair from two elements is computed with use only one value from $r_i = 1, 2, \ldots, L, \quad r_i \in \overrightarrow{r}$.

The infringer at this step will perform full search of values from $\left\lfloor \frac{N}{2} \right\rfloor$ elements with the return from the values of the set R of L size. The number of all possible outcomes follows to the next expression:

$$x_2 = L^{\left\lfloor \frac{N}{2} \right\rfloor}. \tag{7}$$

The third step is to attempt extracting the original text with the rule (2) using pairs of integers, described in the first step above, and the sets of numbers (gammas) built in the process of full search with elements of R at the second step, when various values were considered.

Let's consider an event s_2. It is the case of successful illegal extraction of original A following the result of generalized splitting, i.e. the line C.

The probability $\Pr(s_2 \mid C, 2)$ of such an event is defined with the formula:

$$\Pr(s_2 \mid C, 2) = \frac{e_2}{w_2}, \tag{8}$$

where e_2 is the number of events of successful restoration by the intruder of meaningful[2] source text A provided that the splitting level is two $k = 2$; w_2 – is the complete number of all possible attempts to restore a symbol from source text A under the condition $k = 2$.

Let us start with finding the value w_2, which is the total of all attempts to obtain the current symbol of the source text A for $k = 2$. From expressions (6) and (7) the number w_2 is defined with the following expression:

$$w_2 = L^{\left\lfloor \frac{N}{2} \right\rfloor}. \tag{9}$$

[1] For simplicity it is assumed that the first pair of values in the space C corresponds to the start of the symbol transmission after its splitting.

[2] A meaningful text helps to decide whether the current symbol is correctly recognized. For instance, it might be the symbol «e» at the very end of the text A: «Mom soap fram_».

From the literature on ciphering [6–8] it is known, that $e_2 = 1$, as the only one of enciphered expressions might be true, i.e. from all the attempts of restoration of source text only one case would produce the meaningful text corresponding to the original that was ciphered. This will be the case, when the gamma chosen at the transmission time corresponds to the gamma used at the time of deciphering. Hence,

$$e_2 = 1. \tag{10}$$

Applying the obtained values w_2 and e_2 in the formula (8) one obtains:

$$\Pr(A \mid C, 2) \leq \frac{1}{L^{\lfloor \frac{N}{2} \rfloor}} = \left(L^{\lfloor \frac{N}{2} \rfloor}\right)^{-1} \tag{11}$$

Note 1. The above inequality symbol changes to equal sign, if the text A is meaningful, i.e. it allows solving the problem discussed in the footnote 2. From the other side, if the transmitted text is meaningless for the intruder, then the full search approach would not be appropriate at all at this splitting level.

Presently we will consider the text A to be always meaningful. And in what follows, the inequality $\leq$ will be always replaced with equality sign.

Continuing the proof of the Lemma 1, note that intruder in his logic not only performs full search through the values of the set R in an attempt to open the text A. Also in case of missed success for $k = 2$ he will start a new similar search for the case $k = 3$ and again in case of missed success he will try to consider higher values of splitting k.

b) *Proof for* $k = 3$.

The proof for this case is a certain development of the full search for the previous case $k = 2$ in the sense that previous amount of search will be added to the current amount of search. Similar consideration will be used in what follows for the proof of the Lemma 1 in general case.

For instance, we will show that $\Pr(A \mid C, 3)$, i.e. the probability of unsanctioned extraction of the symbol a in case $k = 3$, will be defined with next expression:

$$\Pr(A \mid C, 3) = \left(\sum_{i=2}^{3} L^{(i-1) \times \lfloor \frac{N}{i} \rfloor}\right)^{-1}. \tag{12}$$

This will be explained again as a number of steps that the intruder should perform, which are analogous to the steps considered in case $k = 2$ previously.

The first step is the division of the text C i.e. the result of representation it as sequential groups containing tree integers. The number of such groups that will be studied by the infringer is equal to

$$X_3 = \left\lfloor \frac{N}{3} \right\rfloor. \tag{13}$$

The second step is when the intruder starts to perform full search of the values in the gammas $\{\tilde{r}_1, \tilde{r}_2, \ldots, \tilde{r}_L\}$ (the search is performed with the return).

In accordance with definition (1) in case of generalize splitting under $k = 3$ each triples will be computed using two values $\tilde{r}_1$ and $\tilde{r}_2$.

At this step the intruder performs the search through $2 \times \lfloor \frac{N}{3} \rfloor$ values of elements (with return) from the set R of the size L. The number of possible outcomes is provided with following expression

$$x_3 = L^{2\times \lfloor \frac{N}{3} \rfloor}. \tag{14}$$

The third step is aimed to an attempt to extract the original text using rules (2) with the help of combinations of integers, described in the first step above, and the combination of integers (gammas), obtained during the full search through the values of with the help of values obtained in the search of elements of set R with the combination of numbers of the first step and with the collection of numbers (gammas) obtained in the search process performed in the second step.

Now let us consider an event s_3 that corresponds to successful extraction of unauthorized source text A using the result of the generalized splitting C for $k = 3$.

In this event probability $\Pr(s_3 \mid C, 3)$ will be defined with the following expression:

$$\Pr(s_3 \mid C, 3) = \frac{e_3}{w_3}, \tag{15}$$

where e_3 is the number of events of restoration by intruder of a meaningful text A for $k = 3$; w_3 is the total of possible attempts to restore the text A for $k = 3$.

Let us find first the value of w_3 that is the total of possible attempts to restore the text A for $k = 3$. As obviously the intruder was not able to extract “a meaningful” symbol at the previous level of splitting, i.e. in case $k = 2$, then the number w_3 represents summation of the following values

$$w_3 = w_2 + x_3, \tag{16}$$

Where the number w_2 is the total of all the possible attempts to obtain the source text A assuming that $k = 2$, and x_3 is the total of all the events connected with the possible attempts assuming $k = 3$.

Substituting w_2 и x_3 from (9) and (14) into (16) we will obtain next expression:

$$w_3 = L^{\lfloor \frac{N}{2} \rfloor} + L^{2\times \lfloor \frac{N}{3} \rfloor} = \sum_{i=2}^{3} L^{(i-1)\times \lfloor \frac{N}{i} \rfloor}. \tag{17}$$

From the publication on ciphering [5–7] it is well known that $e_3 = 1$.

$$e_3 = 1. \tag{18}$$

Substituting w_3 и e_3 from Eqs. (17) и (18) into formula (15), we obtain final result as follows:

$$\Pr(A \mid C, 3) = \frac{1}{w_3} = \frac{1}{\sum_{i=2}^{3} L^{(i-1)\times \lfloor \frac{N}{i} \rfloor}} = \left(\sum_{i=2}^{3} L^{(i-1)\times \lfloor \frac{N}{i} \rfloor} \right)^{-1} \tag{19}$$

c) *The proof of the Lemma 1 for arbitrary k.*

Let assume due to the principle of mathematical induction that Lemma 1 is true for $(k - 1)$, i.e. the following expression is true

$$\Pr(A \mid C, k-1) = \left(\sum_{i=2}^{k-1} L^{(i-1)\times \lfloor \frac{N}{i} \rfloor} \right)^{-1} \tag{20}$$

In analogy with (17) и (19), we conclude that

$$w_{k-1} = \sum_{i=2}^{k-1} L^{(i-1)\times \lfloor \frac{N}{i} \rfloor}. \tag{21}$$

Similar to relation we obtain (16)

$$w_k = w_{(k-1)} + x_k, \tag{22}$$

Where $x_{\bar{k}}$ is computed by intruder in two steps:

The first step is the representation of the split text C as sequential combinations from k integers. Then the number of combinations considered by the intruder will be equal to

$$X_k = \left\lfloor \frac{N}{k} \right\rfloor. \tag{23}$$

The second step is the case when the intruder starts the full search using the set R (with the return).

From the definition (1) we conclude that in case of *generalized splitting* each combination from k integers is computed with the use of $(k-1)$ values $\tilde{r}_1, \ldots, \tilde{r}_{k-1}$.

In this step the intruder will search through the values from $(k-1) \times \left\lfloor \frac{N}{k} \right\rfloor$ elements (with the return) from the set R of the size L. The number of possible outcomes in this situation is shown by the following expression:

$$x_k = L^{(k-1) \times \left\lfloor \frac{N}{k} \right\rfloor}. \tag{24}$$

Substituting x_k, w_{k-1} from Eqs. (24) and (21) into expression (22) we will obtain next formula:

$$w_k = \sum_{i=2}^{k-1} L^{(i-1) \times \left\lfloor \frac{N}{i} \right\rfloor} + L^{(k-1) \times \left\lfloor \frac{N}{k} \right\rfloor}.$$

Or, otherwise, we have

$$w_k = \sum_{i=2}^{k} L^{(i-1) \times \left\lfloor \frac{N}{i} \right\rfloor}. \tag{25}$$

Consider the event s_k that corresponds to successful unauthorized restoration of the source text A from the result of the generalized splitting C with the splitting level k. Hence, the probability $\Pr(s_k \mid C, k)$ of this event is defined with the following:

$$\Pr(s_k \mid C, k) = \frac{e_k}{w_k}, \tag{26}$$

where e_k is the number of events of successful restoration by the intruder of the meaningful text A in case of the slitting depth is equal to k; the w_k represents the total of all the events of possible attempts of restoration of source text A.

As it was mention in our proof before, the references [5–7] and other publications inform that only one single solution of extracted solutions in ciphering may be true i.e. $e_k = 1$.

Substituting this e_k and w_k from Eq. (25) into (26), we obtain (3). Thus, the collection of points a), b) and c) and the induction principle are proving the Lemma 1.

The Behavior Property of $\Pr(A \mid C, k)$

The curve in next Fig. 1 illustrates the behavior of the expression (3) for the probability of unauthorized restoration of source text for various k.

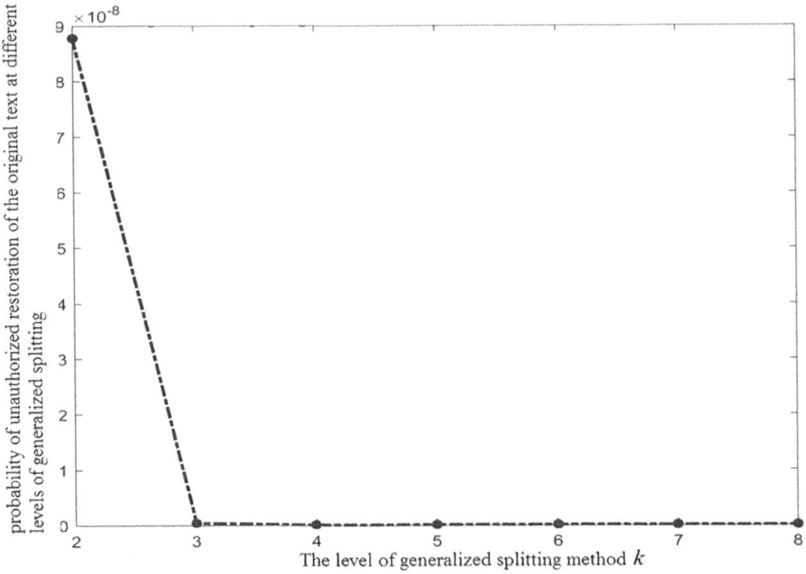

Fig. 1. The probability of possible unauthorized restoration of source text for various levels *k* of the general splitting

4 Conclusion

From the Lemma 1 one may conclude that under $k \to \infty$ the probability of restoration of the secured original text tends to zero if the security key is not known. From Lemma 1, one may conclude that

$$\Pr(A \mid C, k) \to 0 \,\text{When}\, k \to \infty. \tag{29}$$

It means that the method of generalized splitting has the property of asymptotical absolute secrecy. This result is rather new, though it may be considered as an extension of the result by C.E. Shannon [9] concerning the use of gamming for binaries.

The Fig. 1, which has been built upon results of the current paper, demonstrates an exponential speed of approaching to the absolute secrecy border. It means that generalized splitting method may be effectively used with relatively small values of the splitting level *k*. This property is very important for practical applications of new approach to information safety as the increase of splitting level *k* is obviously leads to a delay of information transmission.

References

1. Stefanuk, V.L., Alhussain, A.Kh.: Symmetry ciphering based on splitting method (Symmetrichnoe shifrovanie na osnove metoda rassheplenija). Estestvennye i technicheskie nauki **93**(3), 130–133 (2016)
2. Stefanyuk, V.L., Alhussain, A.Kh.: Symmetric encryption on the base of splitting method. Bull. PFUR Ser. Math. Inf. Sci. Phys. **24**(2), 53–61 (2016)
3. Stefanuk, V.L., Alhussain, A.Kh.: Control of degree of information safety level with integer splitting method. (Kontrol stepeni zashity informatsii metodom tselochislennogo raschep-lenija). Iskusstvenny intellect i prinyatie reshenii (4), 86–91 (2016)
4. Alhussain, A.Kh., Stefanuk, V.L.: Probability properties of the splitting procedure. (Veroyatnostnye svoystva protsedury rassheplenia). Iskusstvenny intellect i prinyatie reshenii (3), 49–57 (2017)
5. Stefanyuk, V.L., Alhussain, A.Kh.: A ciphering tool with splitting method, invention patent № 2643502, priority from 8 December 2015. (Sposob shifrovania metodom rassheplenia. Federalnii institut promyshlennoy sobstvennosty, Gosudarstvenny Reestr Izobretenii Rossiskoy Federatsii, patent № 2643502 na izobretenie, prioritet from 8 December 2015)
6. Schneier, B.: Applied Cryptography: Protocols, Algorithms, and Source Code in C, 2nd edn. p. 758. Wiley, Hoboken (1996)
7. http://www.cse.iitm.ac.in/~chester/courses/16e_cns/slides/02_Classical.pdf
8. http://cse.iitkgp.ac.in/~debdeep/courses_iitkgp/Crypto/slides/shannonI+II.pdf
9. Shannon, C.E.: Communication theory of secrecy systems. Bell Syst. Tech. J. **28**(4), 656–715 (1949)

Simulation of a Daytime-Based Q-Learning Control Strategy for Environmental Harvesting WSN Nodes

Jaromir Konecny[1(✉)], Michal Prauzek[1], Jakub Hlavica[1], Jakub Novak[1], and Petr Musilek[2]

[1] Department of Cybernetics and Biomedical Engineering, VSB-Technical University of Ostrava, Ostrava-Poruba 70800, Czech Republic
jaromir.konecny@vsb.cz

[2] Department of Electrical and Computer Engineering, University of Alberta, Edmonton, AB, Canada

Abstract. Environmental wireless sensor networks (EWSN) are designed to collect environmental data in remote locations where maintenance options are limited. It is therefore essential for the system to make a good use of the available energy so as to operate efficiently. This paper describes a simulation framework of an EWSN node, which allows to simulate various configurations and parameters before implementing the control system in a physical hardware model, which was developed in our previous study. System operation, namely environmental data acquisition and subsequent data transmission to a network, is governed by a model-free Q-learning algorithm, which do not have any prior knowledge of its environment. Real-life historical meteorological data acquired in the years 2008–2012 in Canada was used to test the capabilities of the control algorithm. The results show that setting of the learning rate is crucial to EWSN node's performance. When set improperly, the system tends to fail to operate by depleting its energy storage. One of the aspects to consider when improving the algorithm is to reduce the amount of wasted harvested energy. This could be done through tuning of the Q-learning reward signal.

Keywords: Q-learning · Environmental harvesting · Wireless sensor network · Simulation

1 Introduction

Wireless sensor networks (WSN) find their use across many application areas. This paper focuses on Environmental WSN (EWSN) that are specifically designed to collect ambient environmental data, such as temperature measurements, light monitoring, soil monitoring or gas analysis [7]. EWSN are often deployed in remote locations without access to mains electricity or possibility

S. Kovalev et al. (Eds.): IITI 2019, AISC 1156, pp. 432–441, 2020.
https://doi.org/10.1007/978-3-030-50097-9_44

for regular maintenance. Therefore, there is significant need for advanced control strategies to achieve a reliable and location-adaptive functionality of EWSN nodes.

Various EWSN intelligent control strategies were introduced in previous researches. One of the fundamental principles of reliable operation is called energy neutrality [2]. Generally, this design principle says that the node cannot, on average, consume more energy than its energy harvesting module can deliver. The energy neutrality condition can be fulfilled by three basic node topologies [8]: 1. an autonomous harvesting system without energy storage, 2. an autonomous hybrid harvesting system with energy storage, and 3. a battery-supplemented harvesting system.

Such hardware topologies present implementation challenges of intelligent control strategies where the core of the problem lies in balancing data measurement and transmission intervals [4]. There are several research challenges including state of charge estimation of energy storage devices [1] or implementation of reconfigurable system [6]. Also, soft-computing methods are applicable such as fuzzy logic-based control strategies [13] or optimization approaches, e.g. differential evolution [5].

In this contribution, a simulation-based research of EWSN node with daytime-driven Q-learning algorithm is presented. For evaluation purposes a simulation framework to test proposed approach on historical weather data is used. The proposed solution is compared with timer-based controller published in our previous research article [3].

This paper is organized into five sections. Section 2 describes an EWSN model of EWSN node which is used for setting simulation parameters. The proposed Q-learning controller strategy is described in Sect. 3, results are then analyzed in Sect. 4. The final section brings major conclusions and outlines possible directions for future work.

2 Hardware Model of EWSN

Framework introduced in this study simulates behaviour of a hardware model of an environmental wireless sensor node. The latest hardware model results from incremental improvements on the first prototype described in our previous work [9]. It is enclosed in IP67 casing and features 32-bit RISC microcontroller ARM Cortex-M0+, a photovoltaic cell for harvesting solar energy which is then stored in a supercapacitor, auxiliary batteries, an SD card for acquired data and a wireless transmitter (IEEE 802.15.4).

The hardware model comprises 3 PCBs for the microcontroller, photovoltaic cell and signal pre-processing, respectively. Other features are further detailed in [3].

EWSN's controller switches between four modes of operation - Sleep, Standby, Sensing and Transmission modes, wherein the sensing and transmission rates are adapted by Q-learning agent described in the next section.

An estimate of current energy availability (state of energy storage, or SoES) is made in the Standby mode. In the Sensing mode, environmental data is acquired through sensors and stored in a small non-volatile internal memory. The Transmission mode comprises data acquisition, data transfer from the non-volatile memory to the SD card and data transmission using the wireless transmitter. In case that none of the above modes is activated by the controller, the EWSN device switches back to the Sleep mode. All scenarios are further detailed in our previous study [3].

The hardware model has the solar panel area of 648 mm^2 with energy efficiency of 21%. The power supply voltage is 3.3 V and current draw in the Sleep mode is 190 µA. The total energy buffer capacity is 750 J. The system actions require the following amounts of energy – data storing: 8 mJ, data transmission: 30 mJ, data sensing: 2 J. These values were obtained through standard operation measurements using an experimental setup and were subsequently used as input parameters for the simulation framework described in our previous paper [3].

3 Q-Learning EWSN Controller Strategy

Q-learning algorithm was used to design optimal control strategy implemented in the EWSN node described in previous sections. First introduced in [11], Q-learning is a model-free reinforcement learning algorithm belonging to the domain of machine learning methods.

One of the main characteristics of reinforcement learning algorithms is learning from system's interaction with its environment. The learning agent needs to learn how to map situations (so-called states) to actions. This mapping is called a policy. When the agent acts (i.e. selects action) in certain state, the environment responds back with reward signal, telling the agent how good or bad the selected action was. The goal of learning agent is to maximize cumulative reward in a long run. Negative reward signal is a basis for altering the policy. Mathematical formalization of the decision making problem consisting of states S, actions A and rewards R is known as finite Markov decision process.

Unlike in supervised learning, there is no need for the agent to have a training dataset provided by an expert. By learning from interactions and discovering features of the environment possibly unknown to an expert, the reinforcement learning agent can outperform supervised learning methods in many applications [10]. One of the major advantages of Q-learning algorithm is that there is no need for a model of the environment. This is particularly useful in case the increased environment complexity.

The Q-learning algorithm uses so-called Q-table as a memory, in which the quantitative values of actions are stored. The size of the table is defined by the number of states and actions [12]. Essentially, the idea of Q-learning is to make estimation of future reward (represented by $Q(S_t, A_t)$) for taking action A at state S while following the optimal policy. The basic principle of the Q-learning algorithm can be described by the following equation:

$$Q(S_t, A_t) \leftarrow Q(S_t, A_t) + \alpha \left[R + \gamma \max_a Q(S_{t+1}, a) - Q(S_t, A_t) \right], \qquad (1)$$

where $Q(S_t, A_t)$ represents an estimated value of reward in Q-table for current action A_t and state S_t, α is a step-size parameter (known as learning rate), which controls convergence speed of learning. $\alpha = 0$ means that the algorithm uses only previously obtained knowledge (i.e. previous estimates of the reward signal), $\alpha = 1$ the algorithm uses only new knowledge. R represents a received reward signal. γ is a discount rate determining whether the agent seeks to maximize the immediate reward (in case $\gamma = 0$) or to maximize the future cumulative reward (in case $\gamma = 1$), thus becoming more farsighted when estimating future values of the reward signal R.

As for the agent's policy (commonly denoted as Π), one of possible implementations is, for instance, so-called ϵ-greedy policy in which the agent tries to maximize the cumulative reward by acting greedily (i.e. exploits previously obtained knowledge by selecting the action with maximum estimated reward). However, with certain small probability ϵ the agent selects previously untested action to explore whether it could receive possibly higher reward. If so, the corresponding value estimation in the Q-table is updated. The following algorithm represents the Q-learning procedure.

Algorithm 1. Q-learning procedure

Initialize $Q(s, a)$, for all $s \in S$, $a \in A(s)$
Initialize expert constants α and γ
Initialize S
while EndOfData == FALSE **do**
  Choose A from $S_{(t-1)}$ using policy derived from Q
    (e.g. ϵ-greedy)
  Take action $A_{(t-1)}$ and observe R, S
  $Q\left(S_{(t-1)}, A_{(t-1)}\right) \leftarrow Q\left(S_{(t-1)}, A_{(t-1)}\right)$
    $+\alpha\left[R + \gamma \max_a Q(S, a) - Q\left(S_{(t-1)}, A_{(t-1)}\right)\right]$
  $S_{(t-1)} \leftarrow S$; $A_{(t-1)} \leftarrow A$
end while

In the beginning the Q-table is initialized (e.g. in this experiment all elements were set to zeros), as well as the learning rate α and the discount rate γ. In the third step the simulation environment is initialized so that the learning agent starts to receive current representation of the environment's state S. The agent then iteratively selects action A using policy derived from the Q-table. After taking action the agent receives reward R and observes new state of the environment. Using these inputs the corresponding reward value estimation $Q(S_t, A_t)$ in the Q-table is updated by the formula (1) described previously. The current action A and state S are reassigned as previous action $A_{(t-1)}$ and state $S_{(t-1)}$ variables and the entire process is repeated.

The implementation described in this paper differs from the standard Q-learning algorithm in that the environment is not reset in case of a system failure (i.e. depleted energy storage) or after achieving the target.

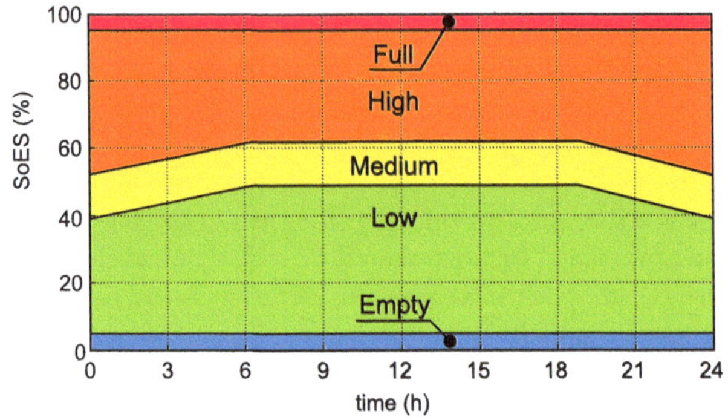

Fig. 1. Energy storage states definition according to the time of day

In the simulation environment described in the previous sections, the daytime-based states are used, taking into account diurnal cycles (i.e. day and night periods). Five states of energy storage are defined, along with percentage ranges of energy storage corresponding to each energy state (see Fig. 1). The *Empty* state is defined in range from 0% to 5%. The *Low* state is defined in range from 5% to (39%–49%). The *Medium* state is between Low and High states intervals (39%–49%) and (52%–62%). The *High* state is defined in range from (52%–62%) to 95% and *Full* state is defined in range from 95% to 100%. The location-specific average sunrise and sunset times were calculated (6:27 am and 6:46 pm). Naturally, the state, at which the system is, largely depends on daytime.

The rewards are defined with respect to states of energy storage (SoES). Preferably, the EWSN node should operate in *Medium* energy storage state. The *Full* and *Empty* states are not optimal for the EWSN node operation. When the energy storage is full, it is not possible to store more energy, which leads to waste of harvested energy. On the other hand, when the EWSN node lacks energy, there is significant risk of a system failure due to depleted energy storage. When in operation, the EWSN node will be switching the energy states among *High*, *Medium* and *Low*. If EWSN node is in *Low* state than it should take actions (i.e. reduce measurement rate and transmission cycles) to save energy until there is enough newly harvested energy in which case the system will switch to *Medium* energy state. Similarly, when in the *High* energy state the system should utilize excessive energy (by increasing sampling rate as well as transmission rate) until the energy state becomes *Medium* again.

The penalties and rewards need to be designed according to the targeted functionality. In this EWSN simulation framework, the (positive) reward signal dynamically changes whenever the energy states are heading towards the *Medium* state, which is considered to be optimal. Otherwise, the penalty (i.e. negative reward) strategy is implemented to dynamically penalize changes of energy state in the opposite direction (e.g. from *Medium* to *Empty*).

The definitions of rewards and penalties strategy are as follows:

$$R(S,A) = \begin{cases} -1 & S \text{ is } Empty \wedge \text{SoES} \downarrow \\ -0.78 & S \text{ is } Full \wedge \text{SoES} \uparrow \\ -0.75 & S \text{ is } Low \wedge \text{SoES} \downarrow \\ -0.75 & S \text{ is } High \wedge \text{SoES} \uparrow \\ -0.25 & Medium \rightarrow High \vee Medium \rightarrow Low \\ 0.2 & Medium \rightarrow Medium \wedge \text{SoES} \uparrow \\ 0.5 & Medium \rightarrow Medium \wedge \text{SoES} \downarrow \\ 0.67 & S \text{ is } High \wedge \text{SoES} \downarrow \\ 0.75 & S \text{ is } Low \wedge \text{SoES} \uparrow \\ 1 & S \text{ is } Empty \wedge \text{SoES} \uparrow \\ 1.19 & S \text{ is } Full \wedge \text{SoES} \downarrow \end{cases} \quad (2)$$

4 Results

The goal of the conducted experiment was to evaluate the soft computing method-based control algorithms implemented in the EWSN node. A timer-based controller's performance was compared to Q-learning controller having different operational settings, namely the learning rate (α) described in the previous section. The discount rate (γ) was experimentally set to 0.97 and fixed during the entire experiment, thus stimulating the system to prefer long-term cumulative rewards rather than immediate rewards. The simulation time throughout the experiment was 8 s comprising in total 2 538 721 steps.

Three operational variables, enabling the assessment of a proper functionality of the EWSN node, were measured and analyzed during the testing procedure. The most important variable was a fail count, representing the amount of cases in which the EWSN node did not have enough energy for its operations (environmental data sensing and transmission) due to depleted energy storage. The other monitored variables were an average sensing and a transmission rate per day. In addition, another factor of interest was the amount of unused energy, whose reduction might be considered as a room for improvement in the future research. As mentioned in Sect. 3, the dataset comprises 5 years of real-life meteorological data acquired in Canada between 2008 and 2012.

The core of the timer-based control lies in defining a time interval in which certain actions are executed. The length of the time interval of a particular action depends on state of the energy storage and on predefined linear function. The time from the last taken action is measured and compared with the time interval. If the interval is exceeded, the parameters of this action are set.

Comparison of results is shown in Table 1. The timer-based controller provided by far the highest sensing rate (an average of 510 samples per day) with rather low average transmission rate (about 10 transmissions per day). Also, it had the least unused energy (about 50% compared to other algorithms). On the other hand, it failed to operate in 1 669 cases (over the period of 5 years).

As for the performance of the Q-learning algorithm implementation, it is evident from the results that the learning rate α has a substantial impact on

Table 1. Simulation results comparison

Parameter	Q-learning				Timer-based
	$\alpha = 0.30$	$\alpha = 0.50$	$\alpha = 0.68$	$\alpha = 0.90$	
Simulation step (sec)	60	60	60	60	60
Total days simulated	1 763	1 763	1 763	1 763	1 763
Total steps count	2 538 721	2 538 721	2 538 721	2 538 721	2 538 721
OK state count	1 471 485	1 970 903	2 538 721	2 538 721	2 537 052
FAIL state count	1 067 236	567 818	0	0	1 669
Supercap. full-capacity count	186 328	233 924	249 327	264 019	180 590
Supercap. total unused energy (J)	571 321	748 368	799 847	859 840	424 687
Sensing cycles count	757 455	728 781	735 674	712 126	899 391
Transmission cycles count	757 455	728 781	735 674	712 126	17 143
Avg. sensing rate per day	428.5	413.4	417.3	403.9	510.1
Avg. transmission rate per day	428.5	413.4	417.3	403.9	9.7

the fail count. The experiment showed that the optimal value of α was 0.68 with respect to number of fail counts and sampling rate. With this setting the Q-learning EWSN controller never failed to operate during the simulation (time frame of 5 years) and it provided satisfactory results.

It is also evident from the results that the Q-learning controller with $\alpha = 0.30$ had the highest transmission and sensing rates compared to the other Q-learning settings, as well as the lowest amount of unused energy. However, it manifested the highest failure rate (approx. 1 million failures over the course of 5 years). It failed on average in every third iteration, which is inadmissible in real-life environment. The Q-learning algorithm having $\alpha = 0.50$ had significantly lower amount of failures, nevertheless it failed in every fifth step.

The optimal Q-learning setting of $\alpha = 0.68$ reduced the number of failures to zero. However, the amount of unused energy slightly grew up compared to other Q-learning settings (see Supercapacitor full-capacity count in Table 1). The value of α above 0.68 did not perform better – while the number of failures remained zero, the amount of unused energy grew up. As a consequence the average sensing and transmission rate went slightly down.

Figure 2 shows the simulation output for the first two years, starting on Jan 1, 2008. The top chart shows the maximum (green), minimum (red) and average/medium (yellow) values of stored energy for each day. In the summer period, SoES varied between 45% and 100%. When the SoES reached the full capacity (100%), the harvested energy was not fully utilized, thus making the system's operation inefficient, particularly in the evening hours when the energy storage was fully charged. During summer night time, the minimal charge varied between 9% and 50%.

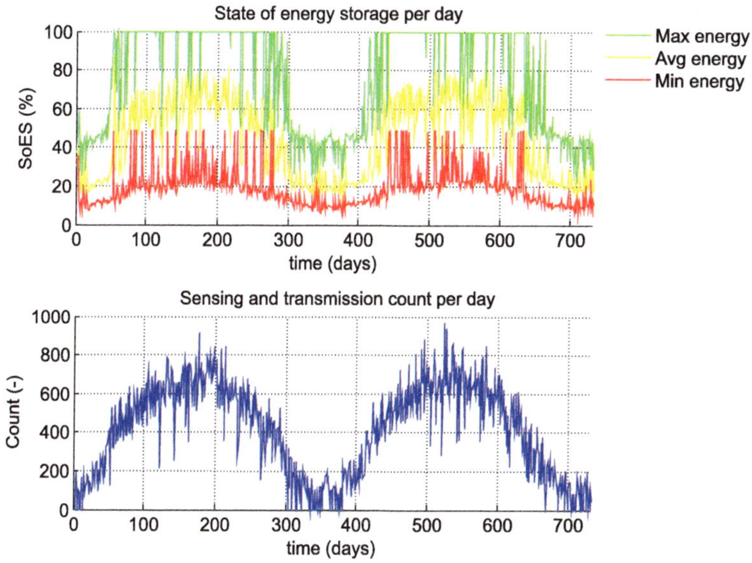

Fig. 2. Two years simulation output for $\alpha = 0.68$: Top graph – SoES maximum, minimum and average level in each day; Bottom graph – total transmission and sensing cycles per a day.

Compared to the summer season having the minimum SoES of 45%, the maximum SoES in the winter was approx. 46%. Therefore, the amount of available energy during the winter was significantly decreased which resulted in daily sensing cycles reduction by 75% (approx. 200 per day compared to 800 in the summer). EWSN device's performance on two selected days (day no. 191 in the summer 2008 and day no. 340 in the winter 2008) is depicted in Fig. 3. As expected, the level of SoES in the summer was on average substantially higher compared to the SoES in the winter (see top charts). 24-h sensing and transmission rates on the same days in the bottom charts show long periods of approx. 60 sensing and transmission operations per hour during daytime in the summer compared to short-term maximum of 30 sensing and transmission operations per hour during daytime in the winter. The distribution of sensing and transmission cycles is not uniform. Particularly, there were no measurements and transmissions during night time in winter and all measurement and transmission cycles were concentrated only to afternoon hours. This implies that the future work should focus on better tuning of the system's operation during night time, e.g. through adjustment of reward and penalty strategies.

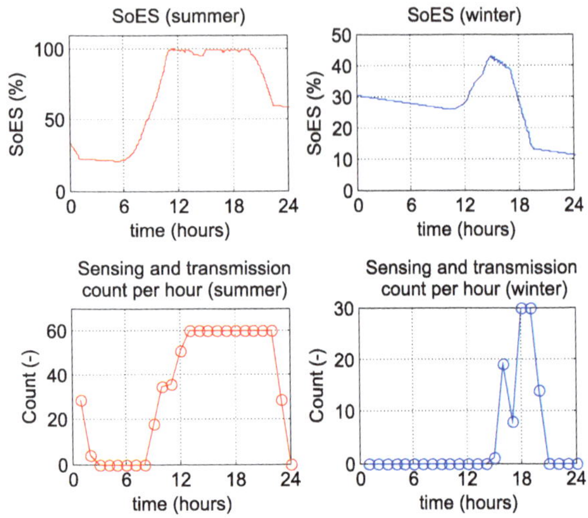

Fig. 3. Simulation output for two selected days for $\alpha = 0.68$ (winter – no. 191 and summer – no. 340)

5 Conclusion and Future Work

This paper describes an environmental wireless sensor network (EWSN) node simulation framework, in which the system is controlled by daytime-based Q-learning algorithm. The simulation uses real-life historical meteorological data to emulate functionality of EWSN node deployed, for instance, in remote areas without possibility of frequent maintenance.

The conducted experiments show that the Q-learning algorithm can control the EWSN node behavior, particularly setting node's data acquisition and data transmission rates depending on available harvested energy or energy in the energy storage. One of the main advantages of the Q-learning algorithm is that the EWSN node does not require any knowledge (e.g. model) of its environment.

It was found that learning rate (α) parameter has a significant impact on the node's performance. When set improperly, the system tends to fail to operate (i.e. the energy storage is depleted), which leads to environmental data losses. The algorithm generally performed well during daytime operations (irrespective of the season). On the other hand, there is room for improvement in regard to night time operation, especially in the winter season. The system's tuning could be done particularly through reward signal optimization.

Our future work will now focus on adding new inputs to the controller, for instance weather forecast data to allow for prediction of the harvested energy in the near future. Also, increasing knowledge of season time is in consideration. We assume that this will help the controller to understand the environment's interdependencies. One of the focuses will also be tuning sensing and transmission distribution, particularly during night time.

Acknowledgement. This work was supported by the European Regional Development Fund in the Research Centre of Advanced Mechatronic Systems project, project number CZ.02.1.01/0.0/0.0/16_019/0000867 within the Operational Programme Research, Development and Education s the project SP2019/107, "Development of algorithms and systems for control, measurement and safety applications V" of Student Grant System, VSB-TU Ostrava.

References

1. Castagnetti, A., Pegatoquet, A., Belleudy, C., Auguin, M.: An efficient state of charge prediction model for solar harvesting WSN platforms. In: 2012 19th International Conference on Systems, Signals and Image Processing, IWSSIP 2012, pp. 122–125 (2012)
2. Kansal, A., Hsu, J., Zahedi, S., Srivastava, M.B.: Power management in energy harvesting sensor networks. ACM Trans. Embed. Comput. Syst. **6**(4), 38 (2007)
3. Konecny, J., Prauzek, M., Borova, M., Janosova, K., Musilek, P.: A simulation framework for energy harvesting in wireless sensor networks: single node architecture perspective. In: 2019 23rd International Conference Electronics, ELECTRONICS 2019 (2019)
4. Krömer, P., Prauzek, M., Musilek, P.: Harvesting-aware control of wireless sensor nodes using fuzzy logic and differential evolution. In: Workshop on Energy Harvesting Communications, IEEE International Conference on Sensing, Communication, and Networking (SECON 2014), June 2014
5. Kromer, P., Prauzek, M., Musilek, P.: Harvesting-aware control of wireless sensor nodes using fuzzy logic and differential evolution. In: 2014 11th Annual IEEE International Conference on Sensing, Communication, and Networking Workshops, SECON Workshops 2014, pp. 51–56 (2014)
6. Li, Y., Jia, Z., Xie, S.: Energy-prediction scheduler for reconfigurable systems in energy-harvesting environment. IET Wirel. Sens. Syst. **4**(2), 80–85 (2014)
7. Musilek, P., Prauzek, M., Krömer, P., Rodway, J., Barton, T.: Intelligent energy management for environmental monitoring systems. Smart Sens. Netw.: Commun. Technol. Intell. Appl. 67–94 (2017). https://doi.org/10.1016/B978-0-12-809859-2.00005-X
8. Prauzek, M., Konecny, J., Borova, M., Janosova, K., Hlavica, J., Musilek, P.: Energy harvesting sources, storage devices and system topologies for environmental wireless sensor networks: a review. Sensors (Switzerland) **18**(8), 2446 (2018)
9. Prauzek, M., Musilek, P., Watts, A.G.: Fuzzy algorithm for intelligent wireless sensors with solar harvesting. In: Proceedings of the IEEE SSCI 2014 - 2014 IEEE Symposium Series on Computational Intelligence - IES 2014: 2014 IEEE Symposium on Intelligent Embedded Systems, pp. 1–7 (2014)
10. Sutton, R.S., Barto, A.G.: Reinforcement Learning, 1st edn. The MIT Press, Cambridge (2018)
11. Watkins, C.J.C.H.: Learning from delayed rewards. Ph.D. thesis, Cambridge University (1989)
12. Watkins, C.J.C.H., Dayan, P.: Technical note: Q-learning. Mach. Learn. **8**(3), 279–292 (1992). www.scopus.com. Cited By: 4441
13. Watts, A.G., Prauzek, M., Musilek, P., Pelikan, E., Sanchez-Azofeifa, A.: Fuzzy power management for environmental monitoring systems in tropical regions. In: 2014 International Joint Conference on Neural Networks (IJCNN), pp. 1719–1726, July 2014

Evolutional Modeling

Adaptive Time Series Prediction Model Based on a Smoothing P-spline

Elena Kochegurova(✉), Ivan Khozhaev, and Elizaveta Repina

National Research Tomsk Polytechnic University, Lenina Avenue, 30, 634050 Tomsk, Russia
kocheg@mail.ru

Abstract. One major task of modern short-term forecasting is to increase its speed without deteriorating the quality. This is especially relevant when developing real-time forecasting models. The hybrid forecasting model proposed in this paper is based on a recurrent *P*-spline and enables adaptation of parameters by evolutionary optimization algorithms. An important characteristic of the proposed model is the use of a shallow prehistory. Besides, the recurrent *P*-spline has a cost-effective computational scheme; therefore, the forecast speed of the model is high. Simultaneous adaptation of several parameters of the *P*-spline allows forecast accuracy control. This leads to the creation of various versions of forecasting methods and synthesizing hybrid mathematical models with different structures.

Keywords: Time series prediction · Hybrid model · Evolutionary algorithms

1 Introduction

Forecasting time series is an urgent task with a variety of applications in a different areas [1–3].

However, recent forecasting strongly associates with large data sets due to the growing use of GPS, including information from positioning systems of vehicles and other objects, mobile phones, fitness bracelets, the Internet of Things (IoT) sensors, etc. Data collected by various systems are time series. Time series data are always considered as a whole and not as separate numerical fields [4]. These data are dynamic and their analysis makes it possible to assess trends, which warrants the forecasting properties of time series [5, 6].

Depending on the number of predicted values (forecast horizon), forecasting can be classified into short-term, medium-term and long-term. The short-term forecast horizon includes 1–3 values ahead [7] and is used in a variety of applications. Short-term forecasting is relevant to predict man-made and natural emergency situations, various natural phenomena; to develop renewable energy sources and predict loads in traditional power systems; to forecast demand and prices, as well as financial markets; to identify outbreaks of infections.

S. Kovalev et al. (Eds.): IITI 2019, AISC 1156, pp. 445–455, 2020.
https://doi.org/10.1007/978-3-030-50097-9_45

2 Background and Related Work

There are several classifications of short-term forecasting methods. According to one of them, all forecasting methods can be divided into intuitive and formalized models.

Formalized models, in turn, include statistical and structural forecasting methods [8] (or methods based on artificial intelligence as referred to in [9]).

According to [10], forecasting methods can be grouped according to the parametric and non-parametric approaches. The parametric approach combines methods that use prior knowledge of data distribution: exponential smoothing based on autoregression and moving averages. Non-parametric approach is often further classified into global or local.

However, recently, many hybrid methods appeared that combine different approaches to time series analysis. Such methods are commonly used in real-life applications. According to [10], the frequency of using hybrid approaches exceeded 21% against 54% of the use of non-parametric methods and 25% of the use of parametric methods. Therefore, it is advisable to group the methods of short-term forecasting into three categories: statistical methods, Structural forecasting models, Hybrid methods.

Statistical methods use the information about the distribution of data to obtain forecast models. This makes the methods dependent on the parameters that after optimization significantly affect the forecast results. Statistical methods are divided into the following groups based on their mathematical complexity [7]:

- exponential smoothing models (simple exponential smoothing, Holt exponential smoothing and Holt-Winters exponential smoothing).
- regression models, which are based on the influence of regressors on the output variable. Regression models are mainly implemented by the least squares method (LSM).
- autoregressive models are based on the fact that the current value of a predicted variable is determined by the function of the past values of the same variable. ARIMA model is the most commonly used and includes the stages of autoregression, integration and estimation of the moving average. The ARIMA model has several modifications: ARIMAX includes some exogenous factors, SARIMA takes into account data seasonality, VARIMA is based on multi-vector time series.

Structural forecasting models are based on machine learning and have no restrictions on the nature of input data. These methods are reliable when applied to both complex and highly nonlinear data.

There are many options for using artificial neural networks (ANN) in short-term forecasting. Neural networks may differ in architecture, the number of layers and neurons, the use of a specific activation function, or the learning method. The following types of ANNs are used in forecasting:

- feed-forward neural networks;
- recurrent neural networks (RNN, LSTM, GRU);
- convolutional neural networks [11].

Support vector machines (SVM) are widely used in regression analysis and classification problems [12, 13]. When applied to forecasting complex and non-linear data, SVM requires that not simple separation planes, but complex core functions are used.

This group of hybrid forecasting methods is currently the most promising when creating forecasting models to be applied in complex processes.

Hybrid methods combine methods from different groups, typically one statistical and one structural. For example, the combination of fuzzy logic and wavelets has become very attractive [14]. Fuzzy logic reduces data complexity and uncertainty, while a wavelet transform is used to analyze non-stationary signals to highlight their local fragments.

Similarity-based methods are some most effective hybrid methods. The k-nearest-neighbors algorithm (kNN) has proven well in the problems of classification and clustering [15] and is also effective as part of hybrid forecasting methods. kNN in combination with approximation functions (for example, weighted or local average) are effectively used by a number of authors for nonlinear and complex time series [16, 17].

Another combination of forecasting methods is based on metaheuristic optimization and an approximation approach to modeling and forecasting time series. Moreover, in this case, simple approximation functions replace complex regression relations. Piecewise polynomial spline functions are the most promising from the position of describing local behavior of complex time series. The choice of spline nodes to obtain individual spline fragments is actually a way to create a pattern to find the similarity of a part of the time series in the future. In addition, some types of spline functions allow real-time implementation, and therefore enable forecasting.

3 Contribution

The choice of the type of spline function as a time series model is largely determined by the applied task of time series analysis. These can be regression splines, B-splines, P-splines, smoothing splines, etc.

Smoothing splines are an effective tool to approximate noisy data. The most well-known smoothing spline is a cubic smoothing spline [18], which is especially effective in a posteriori smoothing.

Regression splines use a specific set of basic functions with a reduced number of nodes and are based on the least squares method. There are no penalties introduced for a non-smooth spline.

P-splines [19] and B-splines [20] have high performance only at the optimal choice of nodes. The choice of nodes is a way to adapt the spline to time series [21]. Adaptation can significantly increase the effectiveness of splines.

Another type of adaptation is associated with the estimation of the penalty parameter based on regression and probabilistic approaches [22].

P-splines have the properties of both regression and smoothing splines: a reduced number of nodes and control over the smoothness of the spline.

The proposed modification of *P*-spline implements a variational approach to individual time series fragments (data groups) [23, 24]. Creating a group of input data, on one hand, solves the problem of selecting nodes. On the other hand, it identifies a time series fragment (shapelet) and thereby implements a hybrid approach to forecasting based on similarity. A compact form of estimating the parameters of the recurrent spline function makes it possible to implement the spline in real time.

To enable the real-time use, the extremal functional (B. W. Silverman, 1994) was modified [23]. The functional is defined separately for a group of h input samples for each i-th spline segment. A discretization step Δt is introduced to control the dimension of the functional.

$$J(S) = (1-\rho)(h\Delta t)^2 \int_{t_0^i}^{t_h^i} [S''(t)]^2 dt + \rho \sum_{j=0}^{h} [S(t_j^i) - y(t_j^i)]^2 \tag{1}$$

In (1), a smoothing factor $\rho \in [0,1]$ normalized by analogy with [18] was introduced, but only for each i-th spline segment. This narrows the range of choice and gives the smoothing parameter a physical meaning. *P*-spline for the i-th segment in real time

$$S_i(\tau) = a_0^i + a_1^i \cdot \tau + a_2^i \cdot \tau^2 + a_3^i \cdot \tau^3, \; -q \leq \tau \leq h - q. \tag{2}$$

The ratio of the moments of computation (τ) and conjugation (q) for the i-th spline segment generates several computational schemes of the recurrent spline, which were studied in detail in [24]. The current *P*-spline is most adapted for real time forecasting, Fig. 1.

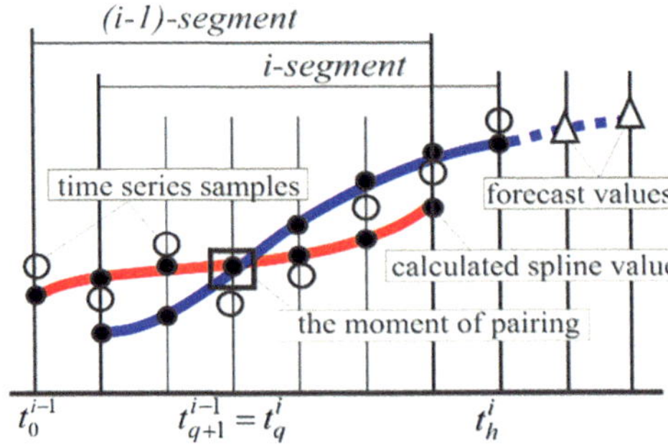

Fig. 1. Forecasting based on P-spline

Coefficients of a cubic spline in a recurrent form were obtained [23].

$$\begin{aligned} & a_0^i = a_0^{i-1} + a_1^{i-1} + a_2^{i-1} + a_3^{i-1}; && A = 6(1-\rho)h^4 + \rho H_5; \\ & a_1^i = a_1^{i-1} + 2a_2^{i-1} + 3a_3^{i-1}; && B = 4(1-\rho)h^3 + \rho H_4; \\ & a_2^i = \frac{\rho(F_1^i C - F_2^i A)}{BC - A^2}; && C = 12(1-\rho)h^5 + \rho H_6; \\ & a_3^i = \frac{\rho(F_2^i B - F_1^i A)}{BC - A^2}; && F_1^i = \sum_{k=0}^{h} y(t_k^i)k^2 - a_0^i H_2 - a_1^i H_3; \\ & H_n = \sum_{k=0}^{h} k^n. && F_2^i = \sum_{k=0}^{h} y(t_k^i)k^3 - a_0^i H_3 - a_1^i H_4. \end{aligned} \quad (3)$$

In (3), q is the reference number inside the i-th segment ($j = \overline{0, h}$; $q = \overline{0, h-1}$), where continuous derivatives of the spline $S^{(k)}(t_{q+1}^{i-1})_+ = S^{(k)}(t_q^i)_-$ conjugate, $k = 0, 1$ for defect 2 and $k = 0$ for defect 1. Breaking coefficients a_2^i, a_3^i are found from the condition of minimizing the functional $\frac{\partial J(S)}{\partial a_2^i} = 0, \frac{\partial J(S)}{\partial a_3^i} = 0$.

Formulas (3) are local inside the spline segment and recurrent with respect to its coefficients. In fact, the group of samples of the i-th spline segment h defines the time series fragment with common dynamic properties. Therefore, the choice of the size of the samples group h is the task of adaptation and significantly improves the quality of forecasting.

The model of a time series fragment in the form of a P-spline makes it possible to obtain forecast values for $\tau > h$ in real time. Figure 1 shows the forecast horizon for two values $\tau = h + 1$ and $\tau = h + 2$.

4 Adaptation of *P*-spline Parameters

Different metaheuristic algorithms have been repeatedly used to train many structural forecasting models, including ANN [25]. The group of evolutionary algorithms has been most successfully used in many scientific and technical problems among the meta-heuristic algorithms [26]. Structural schemes of genetic (GA) and artificial immune algorithms (AIS) are very similar and are schematically presented in the combined Fig. 2.

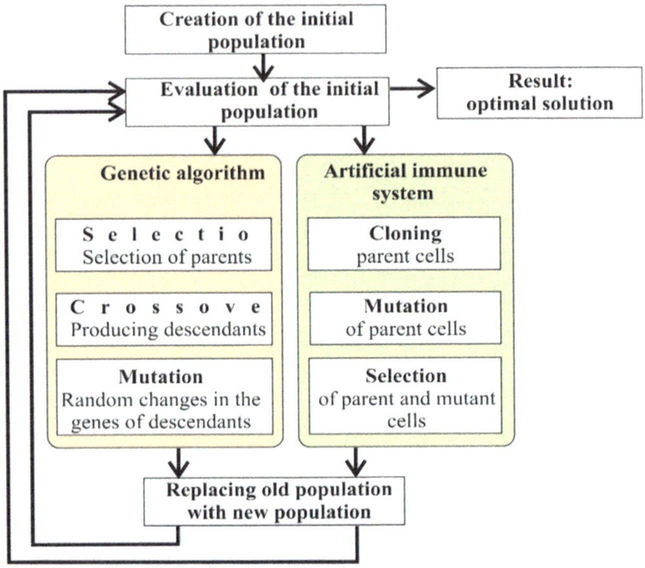

Fig. 2. Structural scheme of genetic and AIS algorithms

The most popular application of evolutionary algorithms is vector optimization; besides, they are able to bypass local optima and therefore are easy to implement. In this study, GA and AIS evolutionary algorithms were used to optimize the parameters of the P-spline by analogy with ANN. Both algorithms have shown good performance.

AIS uses the ability of a natural immune system to produce new types of antibodies and select the most suitable of them to interact with antigens in the body [26].

GA ensures the survival of the strongest genes from the many generated based on the evolutionary principles of heredity, variability and natural selection [3]. The use of GA in optimization problems requires less computational resources. However, the advantage of AIS is the possibility to implement parallel distributed search.

When adapting parameters of the *P*-spline, the main advantage of evolutionary algorithms is used, i.e. simultaneous optimization of several parameters. In this case, two basic parameters of the P-spline were studied simultaneously: the smoothing factor $\rho \in [0, 1]$ and the segment length $h \in [3, 20]$ ($h = 3$ corresponds to the use of 4 samples to construct a third-order polynomial; the spline is difficult to use in real-time systems at $h > 20$).

5 Computational Results

The quality of the forecast is traditionally evaluated by a set of indicators [1, 10] that reflect the time and accuracy characteristics of extrapolation. The following requirements were set for the main indicator of forecast accuracy:

- independence of the time series scale;
- resistance to overshoots;
- symmetric assessment.

Root Mean Squared Percentage Error (RMSPE) was used as the main accuracy parameter

$$RMSPE = \sqrt{\frac{1}{n-1}\sum_{i=1}^{n}\left(\hat{y}_i - y_i\right)^2 \cdot \frac{100}{y_{\max} - y_{\min}}}, [\%]. \qquad (4)$$

In (4) $y_{\max} - y_{\min}$ is an error between the observed y_i and the forecasted value $\hat{y}_i$.

Forecasting capabilities of the proposed hybrid algorithm were evaluated on two test (synthetic) models and two real time series. RMSPE was estimated for random and optimal parameters of the P-spline obtained by evolutionary algorithms.

The Table 1 shows the optimal parameters of the hybrid P-spline-based forecasting algorithm that were obtained on the basis of genetic and AIS algorithms. Significantly, for different values of the optimal parameters, RMSPE is approximately the same at different noise levels of the input data σ_ξ. A random signal $\xi(t) : M\{\xi(t)\} = 0; M\{\xi^2(t)\} = \sigma_\xi^2$ was selected as noise.

Table 1. Optimal parameter values and RMSPE for test time series

σ_ξ, %	0%		5%		10%		15%		20%	
Test-1 $y_1(t) = 10 \cdot \sin\left(\frac{2\pi \cdot t}{100}\right)$										
	GA	AIS	GA	AIS	GA	AIS	GA	AIS	GA	AIS
h_{opt}	3	3	8	5	3	4	3	4	3	4
ρ_{opt}	0.99	0.99	0.61	0.72	0.49	0.68	0.42	0.37	0.35	0.3
$RMSPE_{min}$	0.006	0.006	0.69	0.65	1.17	1.18	1.66	1.58	2.01	2.06
Test-2 $y_2(t) = \sin\left(\frac{\pi \cdot t}{20}\right) \cdot e^{0.02 \cdot t} + 3$										
	GA	AIS	GA	AIS	GA	AIS	GA	AIS	GA	AIS
h_{opt}	3	4	12	11	3	3	3	3	3	3
ρ_{opt}	0.99	0.99	0.97	0.83	0.62	0.58	0.54	0.48	0.45	0.43
$RMSPE_{min}$	0.025	0.015	1.4	1.32	2.38	2.47	3.39	3.05	4	3.84

Figure 3 shows the minimum RMSPE depending on the noise level of the input data σ_ξ.

The main purpose of the test data forecast is to assess the trend component of time series in terms of noise. The analysis of the forecast error for samples 3a), 3b) shows a significant reduction in the input data noise at the forecast horizon of one step ahead. The error of input data decreases by a mean of 5–10 times depending on the type of input function. In terms of computational complexity, the analyzed optimization algorithms have no advantage over each other.

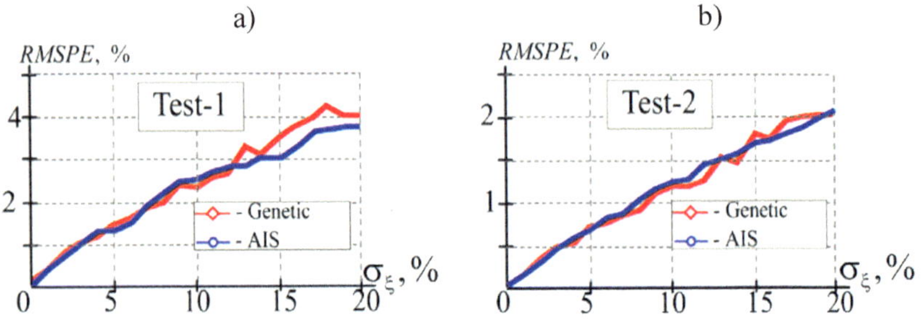

Fig. 3. RMSPE forecasting error at changing input data noise level σ_ξ

Comparing the performance of genetic and AIS algorithms, it should be noted that the running time of the immune algorithm is much higher than that of the genetic one and excesses it by a mean of 1.5–1.75 times. Therefore, a genetic algorithm was used in further experiments to adjust the *P*-spline parameters to be used for real data.

The following two data sets were used as real data: the closing prices of the shares of Megafon and Sberbank in 2017–2018. Box-and-whiskers plots were used to display the results that are presented in Figs. 4a) and b) for Megafon and Sberbank.

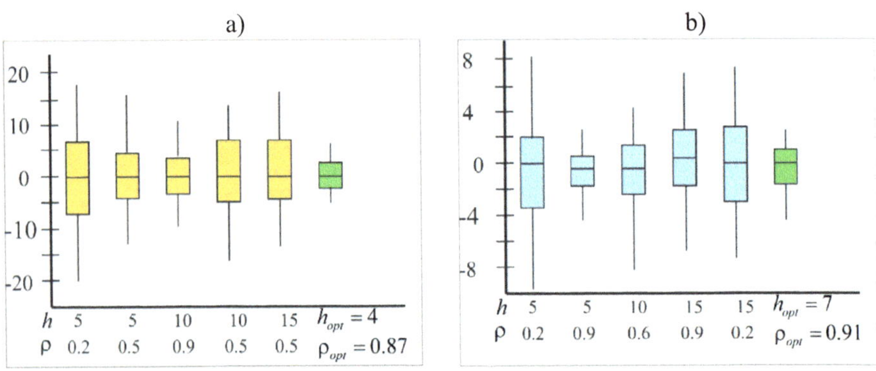

Fig. 4. Spread of RMSPE for random and optimal values of *P*-spline parameters

A graphic display of real time series and corresponding forecast models (Fig. 5) revealed a number of additional properties of the *P*-spline in a hybrid forecasting model.

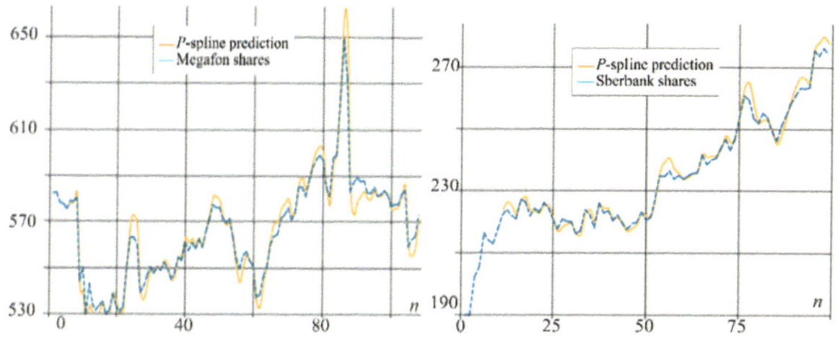

Fig. 5. Real time series and forecasted values

According to the MSPE-assessment, the quality of forecasting based on P-spline is in the range of 2–4%. This is a good indicator in comparison with other forecasting methods [10]. However, some hybrid methods are more accurate. For example, a combination of neural networks and regression methods (ARIMA) gives a forecast error of about 1% [27]. A hybrid method based on kNN has a forecast error of less than 1% when used for multidimensional time series [1]. However, the prehistory depth in this case is 49900 data points and the one-step-ahead forecast is carried out based on the last 100 data points in the time series. Therefore, this type of forecasting is impossible in real time. The hybrid method based on P-splines provides an acceptable ratio of forecast quality and speed; the type and nature of input data does not matter and does not require preliminary processing.

6 Conclusion

This study considered a hybrid approach to time series forecasting, which reproduces the process dynamics in the form of a recurrent P-spline model.

The proposed model was tested on time series of closing prices of shares from two well-known companies. Comparison with the results of other hybrid methods (based on ANNs and kNNs) showed that the forecast accuracy of the proposed method is lower (by an average of 2 times); however, it is acceptable for engineering applications. The obtained results proved that model parameters should be adapted in order to obtain effective forecasts. When using genetic and immune optimization algorithms, the preference is given to the genetic one: it enables the forecast speed that is on average 1.5 times higher at a comparable accuracy.

An important characteristic of the proposed forecasting model is a shallow pre-history of about 10 time series data points. This makes the main contribution to the forecast speed and makes this model suitable for real time use.

In future studies, it is proposed to investigate the properties of this hybrid model using a vector forecast efficiency criterion. Multi-criteria effectiveness should include accuracy measures (RMSPE, MAPE), calculation speed and the change in forecast trends.

Acknowledgment. The reported study was funded by RFBR according to the research project № 18-07-01007.

References

1. Yin, Y., Shang, P.: Forecasting traffic time series with multivariate predicting method. Appl. Math. Comput. **291**(1), 266–278 (2016)
2. Zhang, K.Q., Qu, Z.X., Dong, Y.X., Lu, H.Y., Leng, W.N., Wang, J.Z., Zhang, W.Y.: Research on a combined model based on linear and nonlinear features—a case study of wind speed forecasting. Renewable Energy **130**, 814–830 (2019)
3. Sbrana, G., Silvestrini, A., Venditti, F.: Short-term inflation forecasting: the M.E.T.A. approach. Int. J. Forecast. **33**, 1065–1081 (2017)
4. Fu, T-c: A review on time series data mining. Eng. Appl. Artif. Intell. **24**(1), 164–181 (2011)
5. Wang, H., Zhangc, Q., Wud, J., Panf, S., Chene, Y.: Time series feature learning with labeled and unlabeled data. Pattern Recogn. **89**, 55–66 (2019)
6. Box, G.E.P., Jenkins, G.M., Reinsel, G.C., Ljung, G.M.: Time Series Analysis: Forecasting and Control, 5th edn. Wiley, Hoboken (2015)
7. Montgomery, D.C., Jennings, C.L., Kulahci, M.: Introduction to Time Series Analysis and Forecasting. Wiley, Hoboken (2015)
8. Parmezan, A., Lee, H., Wu, F.: Metalearning for choosing feature selection algorithms in data mining: proposal of a new framework. Expert Syst. Appl. **75**, 1–24 (2017)
9. Yang, J.M.: Power system short-term load forecasting. Ph.D. thesis. Elektrotechnik und Informationstechnik der Technischen Universität, Germany, Darmstadt (2006)
10. Parmezan, A., Souza, V., Batistaa, G.: Evaluation of statistical and machine learning models for time series prediction: identifying the state-of-the-art and the best conditions for the use of each model. Inf. Sci. **484**, 302–337 (2019)
11. Ta, X., Wei, Y.: Research on a dissolved oxygen prediction method for recirculating aquaculture systems based on a convolution neural network. Comput. Electron. Agric. **145**, 302–310 (2018)
12. Utkin, L.V.: An imprecise extension of SVM-based machine learning models. Neurocomputing **331**, 18–32 (2019)
13. Vapnik, V.N.: The Nature of Statistical Learning Theory. Information Science and Statistics, 2nd edn. Springer, New York (1999)
14. Lu, C.: Wavelet fuzzy neural networks for identification and predictive control of dynamic systems. IEEE Trans. Industr. Electron. **58**(7), 3046–3058 (2011)
15. Zhang, M.L., Zhou, Z.H.: A k-nearest neighbor based algorithm for multi-label classification. In: 1st IEEE International Conference on Granular Computing, pp. 718–721. IEEE, New York (2005)
16. Chernoff, K., Nielsen, M.: Weighting of the k-Nearest-Neighbors. In: 20th IEEE International Conference on Pattern Recognition (ICPR), Istanbul, Turkey, pp. 666–669. IEEE (2010)
17. Liu, H., Zhang, S.: Noisy data elimination using mutual k-nearest neighbor for classification mining. J. Syst. Softw. **85**(5), 1067–1074 (2012)
18. de Boor, C.A.: Practical Guide to Splines. Springer, New York (2001)
19. Tharmaratnam, K., Claeskens, G., Croux, C., Salibián-Barrera, M.: S-estimation for penalized regression splines. J. Comput. Graph. Stat. **9**(3), 609–625 (2010)
20. Budakçı, G., Dişibüyük, Ç., Goldman, R., Oruç, H.: Extending fundamental formulas from classical B-splines to quantum B-splines. J. Comput. Appl. Math. **282**, 17–33 (2015)

21. Eilers, P.H.C., Marx, B.D.: Splines, knots, and penalties. Comput. Statistics **2**(6), 637–653 (2010)
22. Aydin, D., Memmedli, M.: Optimum smoothing parameter selection for penalized least squares in form of linear mixed effect models. Optimization **61**(4), 459–476 (2012)
23. Kochegurova, E.A., Gorokhova, E.S.: Current estimation of the derivative of a non-stationary process based on a recurrent smoothing spline. Optoelectron. Instrum. Data Process. **52**(3), 280–285 (2016)
24. Kochegurova, E.A., Kochegurov, A.I., Rozhkova, N.E.: Frequency analysis of recurrence variational P-splines. Optoelectron. Instrum. Data Process. **53**(6), 591–598 (2017)
25. Martín, A., Lara-Cabrera, R., Fuentes-Hurtado, F., Naranjo, V., Camacho, D.: EvoDeep: a new evolutionary approach for automatic Deep Neural Networks parametrization. J. Parallel Distrib. Comput. **117**, 180–191 (2018)
26. Yang, X.S.: Nature-Inspired Optimization Algorithms, 1st edn. Elsevier, Amsterdam (2014)
27. Khashei, M., Bijari, M.: A novel hybridization of artificial neural networks and ARIMA models for time series forecasting. Appl. Soft Comput. **11**(2), 2664–2675 (2011)

Improving the Safety System of Tube Furnaces Using Genetic Algorithms

M. G. Bashirov(✉), N. N. Luneva, A. M. Khafizov, D. G. Churagulov, and K. A. Kryshko

Salavat Branch, FSBEI of Higher Education "Ufa State Petroleum Technological University", Ufa, Republic of Bashkortostan, Russian Federation
eapp@yandex.ru

Abstract. Modern tube heating furnaces of the oil and gas industry are hazardous production facilities that have high accident risks. Most accidents occur as a result of depressurization of the coil and intrusion of the end product into the combustion space of a tube furnace. The process of coking in tube furnaces in this article is considered by the example of a tube furnace of box type (wide chamber) used in an atmospheric oil separation unit. The software product UniSim Design of Honeywell Corporation was used as a medium for creating a computer predictive model of the furnace process. The value of the estimated criterion as the rate of coke production (more than 0.8 kg/day per unit area of the inner surface of 1 m^2) corresponding to the maximum permissible level of coking the tube furnace coil is determined. The article proposes to monitor one of the parameters of coke production as the rate of coke production using the capabilities of APCS systems. The rate of coke production is calculated by means of virtual quality analyzers using the parameters of the coke production (fractional composition of the product, temperature, flowrate, pressure, product density). There has been developed a program to improve the APS and ESD systems of the furnace in which the genetic algorithm searches for solutions of Power-Law equations with several unknowns.

Keywords: Tube furnace · Emergency · Genetic algorithm · Computer simulator · Tubular coil · The rate of coke production · Virtual analyzer

1 Statistics and Causes of Accidents in Tube Furnaces

Modern tube heating furnaces of the oil and gas industry are hazardous production facilities that have high accident risks. Analysis of official statistics on accidents at production facilities of the oil and gas industry from 2007 to 2017 showed that tube heating furnaces accounted for 11.6% of all accidents where 85% were fires and explosions (see Fig. 1; Table 1) [1, 2].

The main causes of accidents in tube furnaces can be classified as follows:

– as technical ones: unsatisfactory condition of technical devices, structures, buildings; imperfection of the technological process or shortcomings in the design of the tube furnace;

S. Kovalev et al. (Eds.): IITI 2019, AISC 1156, pp. 456–464, 2020.
https://doi.org/10.1007/978-3-030-50097-9_46

- as organizational ones: procedural violation; unsatisfactory organization of repair work; improper production control; deliberate disconnection of emergency automatic protection; low level of knowledge of safety and industrial safety requirements at the enterprise; violation of industrial discipline;
- as others: intentional damage or failure of technical devices; alcoholic (drug) intoxication of workers; spontaneous natural phenomenon.

Often fires and explosions in furnaces lead to the destruction of equipment, which is located in the immediate vicinity. Most accidents occur as a result of depressurization of the coil and intrusion of the end product into the combustion space of the tube furnace. Depressurization of the coil often occurs due to destruction in areas of concentration of mechanical stress or burnout in the internal sediments (coking). Timely cleaning the coil and predicting the onset of the ultimate stress-strain state prevent the occurrence of explosions or fires [3].

To ensure the safe operation of tube furnaces, the most advanced information and control systems are used to date consisting of an automated process control system that implements automated process control and warning protection system (WPS) to prevent and avoid emergency situations when the maximum permissible values of parameters are reached displaying the state of the technological process, Emergency Shutdown System (ESS) ensuring safe stop setting or transferring the process to a safe state according to a given "hard" logic. The WPS and ESS systems use special certified sensors, actuators, measuring instruments and programmable logic controllers (PLC). PLCs have a redundant architecture which increases their reliability and safety of the process. The controllers of the ESS form a control action on the actuator to resolve the emergency situations. Automation technology must comply with the safety integrity level (SIL). The safety integrity level (SIL) is a discrete level which specifies requirements for functional safety (see Table 2).

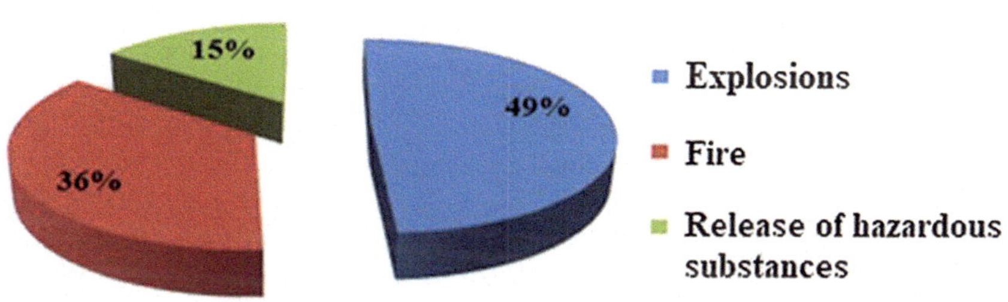

Fig. 1. Statistics of accidents in tube furnaces

The safety integrity level (SIL) for technological blocks of oil and gas production of I and II categories should correspond to SIL 3 (SIL 1 is the smallest, easily achievable; SIL 4 is the largest, difficult to achieve) [3, 4].

Table 1. Statistical data on accidents in the oil and gas industry

Equipment	Number of accidents, %
1 Process pipelines	31,2
2 Pumping stations	18,7
3 Bulk capacity vessels	15,0
4 Tube furnaces	11,6
5 Distillation, vacuum and other columns	11,2
6 Industrial sewage	8,5
7 Tank farms	3,8

Table 2. The probability of failure/fault-free performance of all the automation equipment on the SIL

Safety integrity level (SIL)	Mode of operation with a low intensity of requests (the average probability of failure of a function upon request)	Mode of operation with a high demand rate or continuous request mode (probability of dangerous failures per hour)
1	from 10^{-2} to 10^{-1}	from 10^{-6} to 10^{-5}
2	from 10^{-3} to 10^{-2}	from 10^{-7} to 10^{-6}
3	from 10^{-4} to 10^{-3}	from 10^{-8} to 10^{-7}
4	from 10^{-5} to 10^{-4}	from 10^{-9} to 10^{-8}

The analysis of the technological regulations of tube furnaces of existing industry and special scientific literature showed that the WPS systems used react to dangerous coking according to the following criteria: there is a difference in the raw material temperature readings between parallel streams at the output of the radiant furnace coils (this is also possible if the instrumentation fails and with unsteady feed of raw materials); there is a significant increase in pressure at the entrance of the raw material to the furnace at its constant consumption. The temperature of the flue gases on the bridge wall increases with the same amount of fuel burned.

2 Practical Implementation of Proposals to Ensure the Safety of Tube Furnaces

Currently, methods are used to determine the burnout of the tubular coil, which make it impossible to assess the damage to the metal of the coil and control its wall thickness during operation of the furnace. Existing ESS systems react and stop furnaces in the event of a coil burnout in the following situations: the temperature at the bridge wall or that of the flue gases at the exit of the furnace rises significantly (also accompanied by a decrease in the oxygen concentration in the flue gases at the furnace output); product pressure is reduced at the exit of the furnace. Under plant conditions, the first method of implementation of the ESS is mainly used.

The process of coking in tube furnaces is considered by the example of a tube furnace of box type (wide-chamber) used in an atmospheric oil separation unit. The formation of coke in the furnace coils is a complex process; its speed depends on many factors. Ideally, the equation for the rate of coke production should be based on the physical and chemical theory of coking, taking into accounts the specifics of the specific furnace (based on the analysis of the furnace for a long time, on expert data). The article uses empirical formulas for calculating the rate of coke production, based on the use of virtual quality analyzers and genetic algorithms to demonstrate the basic idea of improving automatic control systems, WPS and ESS.

Let us consider in more detail the use of genetic algorithms (GA) to improve the WPS and ESS systems. The genetic algorithm (GA) is an optimization method based on the concepts of natural selection and genetics. In this approach, the variables characterizing the solution are represented as genes in the chromosome. GA operates with a finite set of solutions (population) and generates new solutions as different combinations of parts of the population solutions using operators such as selection, recombination (crossover) and mutation. The new solutions are positioned in the population according to their position on the surface of the studied function [5]. The first step is to set a certain number of initial populations whose number of genes is equal to the number of sensors (controlled technological parameters). After that, the fitness of the population is determined (how far they are from the solution). If at least one of the obtained populations satisfies the specified accuracy, the solution is found and the general cycle of the genetic algorithm is completed. If the solution is not found, the population is sorted by fitness, the most unsuitable is mutated, the rest are subjected to crossing operations. After these operations, the obtained new chromosomes are also checked for entering the specified accuracy. If the condition is met, the algorithm terminates and the program starts to calculate the level of coke deposition, if not, the cycle is repeated again. The cycle will be repeated until the condition of the specified accuracy of the solution is met, or the limit on the number of completed cycles is reached [4].

The software product UniSim Design of Honeywell Corporation (see Fig. 2) was used as a medium for creating a computer predictive model of the furnace process. The model of the furnace coil with the length of 60 m of low-alloy heat-resistant steel 15X5M has been developed. With the help of the built-in utilities UniSim Design, the calculation of the process of formation of coke deposition inside the coil of the tube furnace was made depending on the operating modes, the values of technological parameters and the duration of operation. The value of the estimation criterion as the rate of coke production (more than 0.8 kg/day per area of the inner surface of 1 m^2) corresponding to the maximum permissible level of coking tube furnace coil is determined. The model in UniSim Design allows to analyze the operation of the furnace in real time, to verify the algorithms and proposed solutions to improve the level of fire and industrial safety [6].

When improving the information-management systems of furnaces the experience of implementing "Advanced Process System and Safety (APCS)" systems was taken into consideration. The basis of APCS is a multi-dimensional predictive control system with optimization that solves certain tasks: control of the process unit in a stationary mode (APCS compensates for noise faster and better than operator); optimization of the unit in dynamic mode (see Fig. 3).

In fact, the APCS-system is a multi-dimensional predictive controller and controls a set of process parameters (manipulated variables), setting the PID-regulator setpoints or, less often, affecting the control valves directly. At the same time, the system monitors that other parameters of technological process (controlled variables) correspond to the required values (setpoints), unilateral or bilateral restrictions. If there are disturbing parameters that are measured, but for some reason are not included in the operation of the APCS-system, they are compensated in the way that would have been done in the automated process control system for controlling the disturbance (see Fig. 4).

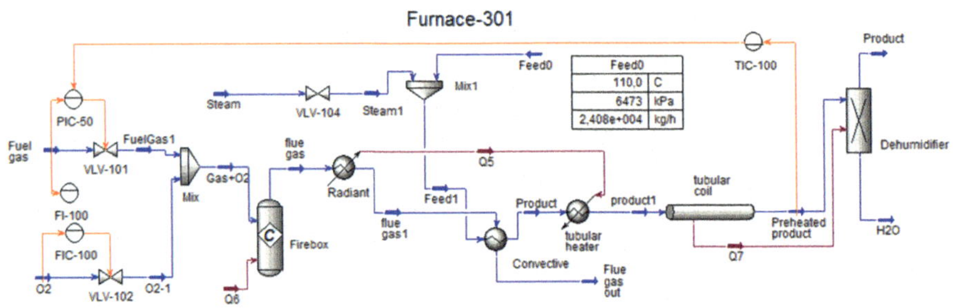

Fig. 2. Tube furnace model in the software product UniSim Design

In technological processes of oil and gas production, manipulated variables (MV) are those that, as a rule (but not necessarily), are controlled by operators, for example, the mass flow of raw materials, the temperature of the product, the pressure of the fuel gas in the furnace, the partial pressure of the reacting hydrocarbons, the rotational rate of the smoke exhauster (fan). Controlled variables (CV) are dependent parameters of technological process, i.e. the variables depend on MV. These include: firstly, the variables included in the optimization task, and, secondly, the specifications of the products. As examples of the first variables, the temperature of the furnace bridgewall, the position of the control valves, the pressure drop of the furnace feed can be specified. Examples of the second variables are the flash point of the fuel gas, the flame temperature during combustion in oxygen, the inhibitory ability of the reagents used, the rate of coke production, the residence time of the raw material in the tubular coil. Examples of disturbing variables (DV) can serve as the flow rate, density, viscosity, fractional composition, the temperature of the raw material at the entrance to the furnace.

It should be noted that many controlled variables in APCS systems (the product specifications mentioned above) may not be measured in a real tube furnace, but obtained by analytical modeling using “virtual analyzers” (VA).

An example of such a problem is the minimization of the quadratic objective function to optimize the temperature equalization of raw material flows in parallel furnace coils:

$$J = \sum_{i=1}^{3} A_i^2 \cdot (T_{1\_i} - T_{1\_it})^2 + \sum_{q=1}^{3} B_q^2 \cdot (T_{2\_q} - T_{2\_qt})^2 + \sum_{k=1}^{3} C_k^2 \cdot (T_{3\_k} - T_{3\_kt})^2 \quad (1)$$

Where, A_i, B_q and C_k are the coefficients of the objective function (initial values at the stage of starting the controller are equal to 1);

T_{1_it}, T_{2_qt}, T_{3_kt} are target values of the temperature deviations of the raw material flows from their average values (initial values at the stage of starting the controller are equal to 0).

Similar empirical dependences are compiled for each virtual analyzer. The article proposes to monitor one of the parameters of coke production as the rate of coke production using the capabilities of APCS systems. The rate of coking is calculated by means of virtual quality analyzers using the parameters of the coking process (fractional composition of the product, temperature, flow rate, pressure, product density). The obtained value of the rate of coking will be an average value due to the fact that it is not constant along the length of the coil. The sharp increase and achievement of the critical value of the rate of coke production indicates a suboptimal operation of the tube furnace. Considering the avalanche-like nature of coking change over time, it is possible to predict the time of maximum coking of the furnace coils and their burnout. The value of the rate of coke production is used by the system "advanced process control

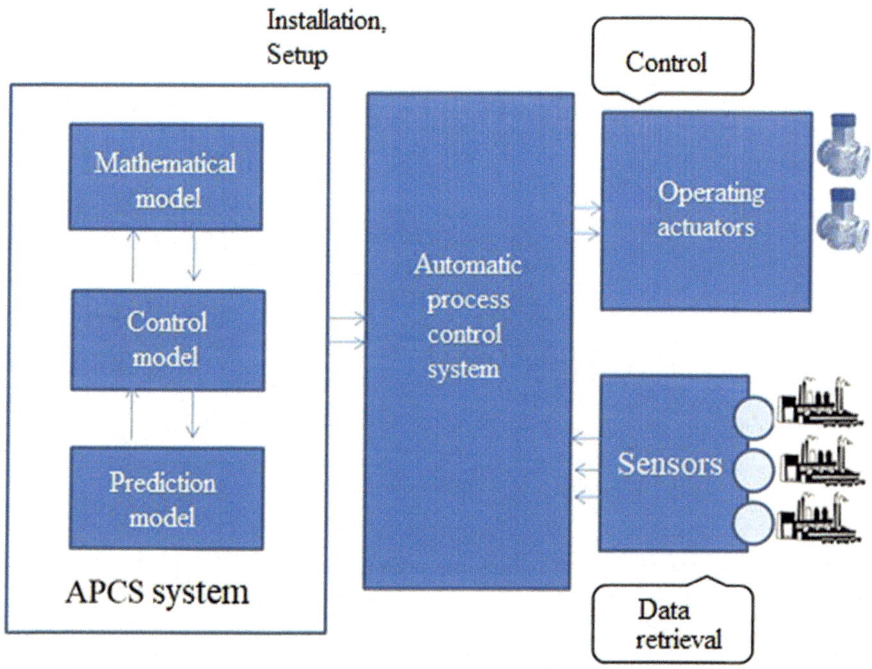

Fig. 3. The principles of the information management system

and safety (APCS)" of the furnace to form control actions on the technological process aimed at reducing this parameter. A sharp increase in the rate of coking will be a condition for the operation of the WPS system signaling a potentially dangerous mode of operation of the tube furnace. The achievement of a certain critical value of the rate of coke production in combination with the achieved limit level of coking of the coil is perceived by the ESS system as a potentially dangerous situation and will lead to the implementation of the function of stopping the furnace, preventing the occurrence of fire or explosion.

The APCS system is implemented in the form of specialized software which is installed on the DCS of the process unit configured in a special way and customized to the requirements of the process.

To improve the WPS and ESS systems of the furnace, a software is developed where the genetic algorithm searches for the solution of power equations with several unknowns. To calculate the level of coil coking the level change pattern of coke deposits was empirically determined [7]:

$$U = Y^{\left(-\frac{T_{now}}{F-1}\right)} - 1 \tag{2}$$

Where, U is the sediment level in % of the maximum value;

Y is the base of the exponential function, which is calculated based on the standard time of use of the tubular coil;

F is the fitness function;

T_{now} is the current time of the coil use.

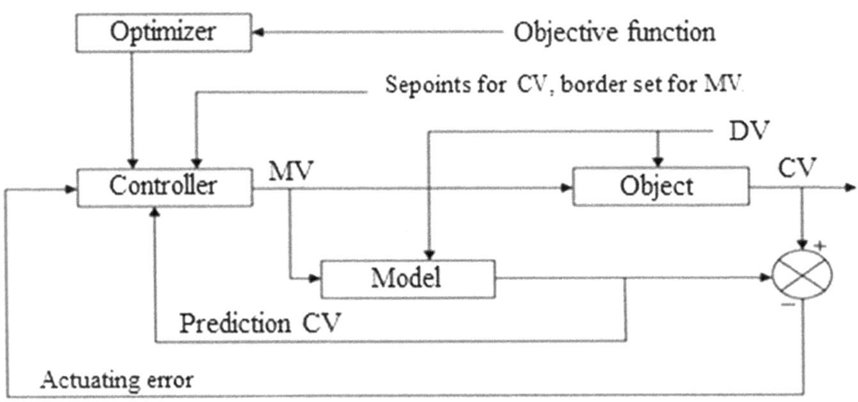

Fig. 4. Diagram of the model predictive control system

The basis of the exponential function is calculated by the formula:

$$Y = 101^{\left(-\frac{1}{T_{norm}-1}\right)} \tag{3}$$

Where, T_{norm} – standard usage time of the coil (the time from burn to burn according to the regulations).

The fitness function is calculated by the formula:

$$F = K_{in}^2 \cdot T_{in} + K_{out}^3 \cdot T_{out} + K_{pass} \cdot T_{pass}^2 + K_{prod}^2 \cdot F_{prod} + K_{prod} \cdot \upsilon_{prod}^2 \\ + K_p^2 \cdot \Delta P + K_{fr}^2 \cdot M_{fr} + K_{den}^2 \cdot \rho \tag{4}$$

Where, K is the coefficients of the fitness function, selected by the genetic algorithm;

T_{in}, T_{out}, T_{pass} are sensor readings of temperature at the inlet, the outlet, pass of the tube furnace, respectively;

F_{prod}, υ_{prod} are readings of the flow sensors and virtual analyzer of flow rate, respectively;

ΔP is the difference of the readings of the pressure sensors at the inlet and outlet of the furnace coil;

M_{fr} is fractional composition of naphthenic and alkaline groups influencing coking;

ρ is the sensor density.

Constraints:

- $2 > F > 1{,}2$;
- if there are no readings from any sensor, then the coefficient $K = 0$;
- the number of days of operation of the tube heating furnace T_{norm} must be at least 2 days.

Criterion:

- it is necessary to control the coking rate (it depends on the coke production rate) and to prevent the achievement of the critical value of the coke production rate.

The termination criteria for the GA operation is: finding a solution of the required quality (with an accuracy of 0.00001); reaching a certain number of generations (1000); the expiration of the time allowed for evolution (10 min); getting solutions that hit the global maximum.

Interaction between the user and the program is handled by GUI (buttons, icons, etc.). After the end of the genetic algorithm, the user is presented with a graph on the screen with the level of coil coking as a percentage (see Fig. 5).

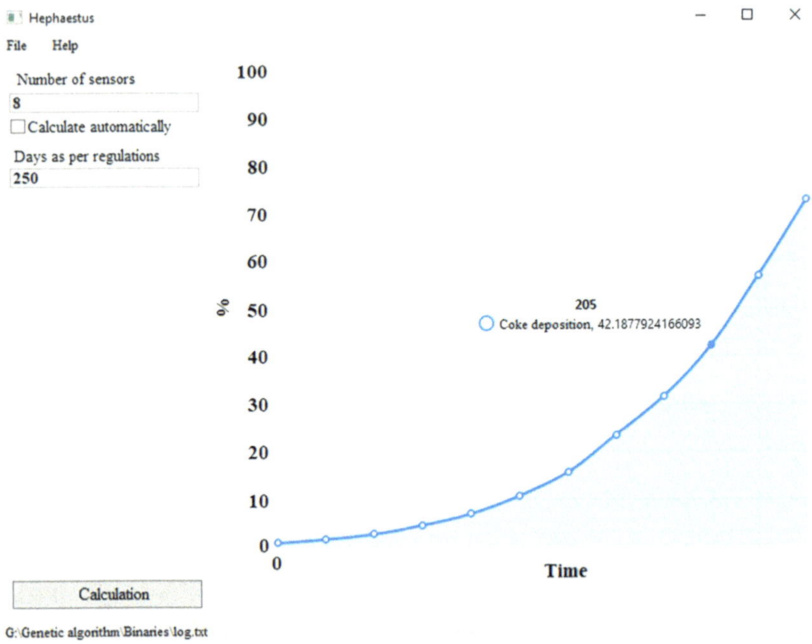

Fig. 5. The result of the program showing the level of coking in the coil at any given time

References

1. Bashirov, M.G., Pavlova, Z.H., Zakirnichnaya, M.M., Khafizov, A.M.: Improvement of automatic control systems and the emergency shutdown system of tube furnaces on the basis of monitoring parameters of the coke production process. In: Network Edition "Oil and Gas Business", no. 1, pp. 120–144 (2018)
2. Khafizov, A.M., Bashirov, M.G., Churagulov, D.G., Aslaev, R.R.: System development of "advanced control" to assess the tube furnace service life and increasing the effectiveness of the automatic emergency shutdown system. In: Fundamental Research, no. 12–3, pp. 536–539 (2015)
3. Bashirov, M.G., Khafizov, A.M., Kryshko, K.A.: Improvement of the tube furnace safety information-management system. In: Integration of Science and Education in Universities of Oil and Gas – 2018: Proceedings of the International Scientific and Methodological Conference, pp. 197–199. Publishing House of USPTU, Ufa (2018)
4. Genetic algorithm. http://www.codenet.ru/progr/alg/Smart/Genetic-Algorithms.php. Accessed 7 Jan 2019
5. Connolly, B.: Survival of the fittest: natural selection of algorithms. https://msdn.microsoft.com/ru-ru/library/dd335954.aspx. Accessed 7 Jan 2019
6. Khafizov, A.M., Fomichev, S.S., Aslaev, R.R., Bashirov, M.G.: Development of automated systems for monitoring technological processes and electrical equipment of oil and gas enterprises. In: Book: K. Tinchurin Readings, pp. 24–25 (2015)
7. Bashirov, M.G., Khafizov, A.M., Kryshko, K.A.: Certificate of state registration of computer programs No. 2019610698. Program for the implementation of the algorithm to assess the coil coking level of tube furnaces in real time using the genetic algorithm. Registered 16 January 2019

Integrated Approach to the Solution of Computer-Aided Design Problems

L. A. Gladkov(✉), N. V. Gladkova, D. Y. Gusev, and N. S. Semushina

Southern Federal University, Taganrog, Russia
leo_gladkov@mail.ru, nadyusha.gladkova77@mail.ru, {dmgusev, semushina}@sfedu.ru

Abstract. This article describes an integrated approach to the placement and routing of elements of digital computing devices of large dimensions. The approach is based on the joint decision of the problem of designing such devices using fuzzy control methods, multiagent systems and parallel computing. The authors describe the problem under consideration and show a brief analysis of existing approaches to its solution. The article describes the following main points: the structure of the proposed algorithm and its main stages; developed genetic operators crossover; proposed model of the formation of the population of solutions; developed heuristics, operators and search strategies for optimal solutions. The structure of the parallel search algorithm is developed. A scheme for parallelizing the computation process and on the basis of the island model is proposed. The article shows the results of computational experiments. These experiments confirm the effectiveness of the proposed method. The authors comment on the brief analysis of the results.

Keywords: Computer-aided design · Optimization tasks · Bioinspired algorithms · Hybrid methods · Multiagent systems · Parallel computing

1 Introduction

Methods of computer-aided design are very important in creating a new electronics. The tendency of growth of the integration degree of integrated circuits leads to a substantial increase in the complexity of their design, which is caused by increasing the dimension of the design problems solved.

One of the important stages of the design process is the stage of design engineering. Moreover, for obtaining the exact solution is required to perform an exhaustive search, it is not possible. Therefore, in practice, various meta-heuristic algorithms are developed to solve such problems, which make it possible to find solutions close to optimal (quasi-optimal) [1–3].

Problems of placement and tracing hold a special role at the stage of construction engineering. Traditionally they are resolved at different stages of different methods,

S. Kovalev et al. (Eds.): IITI 2019, AISC 1156, pp. 465–475, 2020.
https://doi.org/10.1007/978-3-030-50097-9_47

which lead to an increase in the cost of time and computing resources. Therefore, it is advisable to develop integrated methods for solving problems of placement and routing. The development of complex methods is of special concern. The integration of these methods opens the opportunity to solve these problems in one cycle, taking into account mutual limitations and current results [4, 5].

One of the approaches that make it possible to successfully solve the problem of increasing the efficiency and quality of the solutions obtained is the integration of various scientific methods of computational intelligence [6–10]. Bioinspired algorithms are one of the ways of increasing the efficiency of the decision-aided design tasks of complex technical systems, including large-scale integrated circuits. They are successfully used to solve the problems of structural and parametric optimization, resulting in the design of complex technical systems.

A new stage of development of the theory of genetic algorithms have become a hybrid system based on a combination of various scientific fields. There are various methods of hybridization. One of them is fuzzy genetic algorithms where fuzzy math methods used to adjust the parameters of the genetic algorithm.

In this paper, we propose a new approach to solving design engineering problems based on the integration of various approaches, such as bioinspired algorithms, fuzzy control, multiagent organization and parallel computations [11–14].

2 Problem Definition

Consider the problem of the initial placement using the principle of maximum selection. Optimization is performed in two stages: the construction of the initial placement and its improvement. In the process of improving the initial placement, we will use the quality assessment of obtained tracing of compounds. This assessment includes the following two criteria as the number of tracks and the total failure to reach contacts. On the basis of the evaluation of trace quality designed program can improve the result by placement stage, or displays the result on the screen.

Any optimization problem can be described in a tuple as follows: <X, D, Q> . We can interpret this tuple for solving problem of placement and tracing as follows:

X is the set of chromosomes of all populations, D – restrictions imposed on the set X to obtain feasible solutions. Suppose P_t is some populations in step t, $P_t = \{h_1, h_2 ... h_s\}$, where h_s is the chromosome of this population, t = [1, N]; s = [1, M] (N – number of population, M – number of chromosomes in the population). Then the set of all solutions is given by:

$$X = \{P_t; t = 1, 2, \ldots, NumIt\} = \{h_{st}; t = 1, 2, \ldots, N; s = 1, 2, \ldots, M\}.$$

The objective function Q is the normalized additive criterion that includes an assessment of the number of connections and non-tracing total chain length:

$$Q = k_1 Q_1 + k_1 Q_2,$$

where k_1–k_2 are weights of local criteria by which we take into account the importance of a criterion of the quality of solutions; Q_1 is the quantity criterion of non-traced connections, Q_2 is the criterion for evaluating the total length of the chains.

Optimization problem is reduced to minimize the criterion values Q, i.e. $Q(X) \to \min Q(h_{\text{опт}}) = \min Q(h_{ij})$, where $h_{ij} \subset S$.

The most complex and responsible tasks of the design phase of designing electronic equipment are placement and routing tasks. The most complex and responsible in terms of the quality of future products are the placement and tracing tasks. These tasks are closely interrelated with each other, since the result of solving the task of placing the elements is the initial information for the routing task, and the quality of the solution of the allocation problem directly affects the complexity and quality of the routing task.

3 Description of the Algorithm

The authors propose two modified research models and the formation of a new population in order to improve search efficiency [15, 16].

The model of formation of population based on the minimum gap between the generations. For the crossover operator, a pair of parents selected based on one solution as "elite", and one as random. Initial solutions obtained after the crossover forms subpopulations, from which the best individuals are selected. "Roulette wheel" method can be used in the process of random selection. In this case, the mutation operator is not applicable.

The model of population formation is based on a generalization of generations. The initial set of decisions is selected from the starting population. Genetic operators are applied to the subset thus formed. A model based on generalization of generations preserves the solutions obtained at the previous iteration.

Algorithm 1 Creation of population.	
1:	selection of two parents
2:	application of the crossover operator to create offspring, o_{new}.
3:	evaluation of o_{new}.
4:	call p_{LS} for o_{new}.(getting o_{new} with help of p_{LS})
5:	if $u(0; 1) < p_{LS}$ then
6.	finding the best chromosome in the population, c_{best}.
7.	execution of crossover operator "hill-climbing" (o_{new}; c_{best}; n_{off} ; n_{it}).
8.	c^1_{xhc} and c^2_{xhc} Returns chromosomes with the best objective function value (c^1_{xhc} is the best)
9.	replacement $c_{best} \to c^1_{xhc}$.
10.	pasting the best chromosome in the population c^2_{xhc}.
11.	pasting c_{new} in the population.

where P_{LS} is one of the presented strategies: minimum gap between the generations and generations of generalization, $c^{1}{}_{xhc}$ and $c^{2}{}_{xhc}$ are chromosomes selected from the population.

Modified genetic operators play an important role in the implementation of fuzzy genetic algorithm. The main mechanism for obtaining new solutions is the crossover operator. The algorithm uses a modified operator «hill-climbing» based on the gradient of local search methods [15, 16].

The authors propose a modified crossover operator, i.e. extremum search operator «hill-climbing» .

Suppose $X = (x_1, \ldots, x_n)$ and $Y = (y_1, \ldots, y_n)$ are chromosomes with real coding, selected on the basis of certain strategies for the implementation of the operator crossover. The result is offspring:

$$Z_1 = \left(z_1^1 \ldots z_n^1\right) \text{ and } Z_2 = \left(z_1^2 \ldots z_n^2\right),$$

where z_i^1 – randomly chosen number in the interval $[l_i^1,\ u_i^1]$, and besides:

$$l_i^1 = max\{a_i, x_i - I * \alpha\} \text{ and } u_i^1 = min\{b_i, x_i + I * \alpha\},$$

a z_i^2 is choosing on the interval $[l_i^2,\ u_i^2]$ and

$$l_i^2 = max\{a_i, y_i - I * \alpha\} \text{ and } u_i^2 = min\{b_i, y_i + I * \alpha\},$$

where $I = \lfloor x_i - y_i \rfloor$.

The main advantages of the operator is given the diversity of the population and making the degree of "closeness" to the newly obtained solutions parent.

Two modified crossover operator can also be used in the implementation of the algorithm: arithmetic and linear [15, 16].

Suppose $C1 = \left(c_1^1, c_2^1, \ldots, c_n^1\right)$ and $C2 = \left(c_1^2, c_2^2, \ldots, c_n^2\right)$ are two chromosomes chosen for execution of genetic operator. At the same time, assume that the following conditions are truth $c_k^1 \leq c_k^2$ and $f(C_1) \geq f(C_2)$.

As a result of the arithmetic crossover operator creates three offspring:

$H_1 = \left(h_1^1, \ldots, h_n^1\right)$, $H_2 = \left(h_1^2, \ldots, h_n^2\right)$, $H_3 = \left(h_1^3, \ldots, h_n^3\right)$, where $h_k^1 = wc_k^1 + (1 - w)c_k^1$, $h_k^2 = wc_k^2 + (1 - w)c_k^2$, $h_k^3 = wc_k^3 + (1 - w)c_k^3, k = 1, \ldots, n, w$ is a constant, defined on the interval [0; 1].

Here is an example of changing the location of elements as a result of the statement. The resulting solution will lie on a straight line drawn through given points. We get three different descendants, looking for the nearest seat; the best solution for the crossover will be the first child.

Three descendant also created after the linear crossover $H_q = \left(h_1^q, \ldots, h_k^q, \ldots, h_n^q\right), q = 1, 2, 3$, where $h_k^1 = 0.5c_k^1 + 0.5c_k^2, h_k^2 = 1.5c_k^1 - 0.5c_k^2, h_k^3 = -0.5c_k^1 + 1.5c_k^2$.

Applying the operator in the development of crossover «hill-climbing» combination of gradient methods, we shall compare the results of such statements as follows: arithmetic, linear and «hill-climbing» .

Note that the best results are obtained by using the crossover operator «hill-climbing» . Besides the advantage of using the gradient method is ease of implementation and the possibility to start the process of search and optimization with any valid and not necessarily with reference solutions.

Modified non-uniform mutation operator is also provided. Non-uniform mutation operator is the operator using which the decisions in the algorithm are modified in such a way that at the initial stage of the search algorithm provides a uniform, and then tries to improve results locally optimal solutions.

Incest strategy is used as a mechanism of self-adaptive mutation operator. It lies in the fact that the density of mutations (each mutation probability gene) is determined for each child based on its close genetic parents. For example, it may be the ratio of the number of matching gene parent chromosomes to the total number of genes. As a result of incest, in the initial stages of the algorithm is very small with a high diversity of the gene pool of a population probability of mutation, i.e., almost the crossover will occur. With a decrease in diversity arising in the event of falling into local optimum algorithm, the probability will increase the mutation. Obviously, the full descent of the population will be a stochastic algorithm, thus the probability of exit from the local population will increase the optimum.

Fuzzy logic controller is used to speed up the search process. Using a set of chromosomes and their objective functions, we can set a certain range of values of the objective function, which we need. The changes in control algorithm parameters are specified by using the values of linguistic variable «Magnitude» . In this case, a fuzzy logic controller provides the possibility of a reverse control action on the parameters of the genetic algorithm in order to correct them promptly.

To assess the current state of the population, the fuzzy logic controller uses the following values: the best (f_{best}), worst (f_{worst}) and average (f_{ave}) values of the FF at the current iteration (t), as well as the comparison of these values with similar values at previous iterations (t−1) [4–10, 17]:

$$e_1(t) = \frac{f_{ave}(t) - f_{best}(t)}{f_{ave}(t)}; e_2(t) = \frac{f_{ave}(t) - f_{best}(t)}{f_{worst}(t) - f_{best}(t)};$$

$$e_3(t) = \frac{f_{best}(t) - f_{best}(t-1)}{f_{best}(t)}; e_4(t) = \frac{f_{ave}(t) - f_{ave}(t-1)}{f_{ave}(t)}.$$

The output parameters are the probabilities of crossover, mutation and migration, respectively - $Pc(t), \Delta Pm(t), \Delta Pmg(t)$ [17]:

$$Pc(t) \in [0;\ 1]; Pm(t) \in [0;\ 1]; Pmg(t) \in [0;\ 1].$$

We use the parameters e_i as the input variables $\overline{x}_i$. The obtained values of $\overline{y}$ will be equivalent to the parameters $\Delta Pc, \Delta Pm, Pmg(t)$ [17].

The development of hybrid approaches and systems based on the integration of various scientific directions gives reason to believe that using the parallel structure of calculations and constructing a scheme of the evolution process, it is very promising to use the principles of constructing multi-agent systems [14]. Evolutionary modeling and multi-agent methods are closely interrelated. On the one hand, the application of the principles of evolutionary development allows solving the problems of adaptation of multi-agent systems to changes in the external environment. On the other hand, evolution can be used as a driving force, as a mechanism and as a motivation for achieving the set goals.

Evolutionary theory and modeling together with fuzzy logic allow us to create an algorithm for determination of the agents interaction. The agents are characterized by the parameters defined in the interval [0, 1]; thus, using the fuzzy logic, we can modify the genetic operators and mutation operator in the algorithm. Resulting from the algorithm of crossing-over the parent-agents, we obtain the child-agents, which compose a family (an agency) in conjunction with the parent-agents. We can identify some parallels (correlations) between the basic concepts of evolutionary modeling and the theory of multi-agent systems (Table 1) [17].

Table 1. The relationship between the concepts of the theory of multi-agent systems and the theory of evolution

Evolutionary modeling	Theory of multi-agent systems
Gene	Agent property
Chromosome	Set of properties
Individual (solution)	Agent
Family (2 parents and 1 offspring)	Society of agents
Population	Evolving multi-agent system

A parallel multipopulation genetic algorithm (Fig. 1) is used to jointly solve the placement and routing problems [11–13]. It assumes the parallel fulfillment of evolutionary processes on several populations. For the exchange of individuals, the island model of the parallel genetic algorithm are used.

In the island model, asynchronous processes are synchronized at migration points. The migration operator is used to exchange individuals between populations. The selection of individuals for migration is performed from a certain number of chromosomes of the population that have the best FF value. The selection is based on an estimate of the number of non-routed connections.

An important point in the development of a parallel algorithm is the selection of the frequency of migration. An increase in the frequency of migrations leads to a degeneration of populations, and its decrease, on the contrary, to a decrease in convergence. The hybrid migration operator allows migration only when it is necessary, for example, when there is a threat of premature convergence.

For each placement variant described by the chromosome, a routing is performed. Then, from one population to another, a certain number of chromosomes with the best value of this index are copied. In this case, the same number of chromosomes with the worst value of the indicator is removed from the populations. Figure 1 shows a diagram of a model of a parallel genetic algorithm performed on two populations. In practice, the number of populations can be much larger.

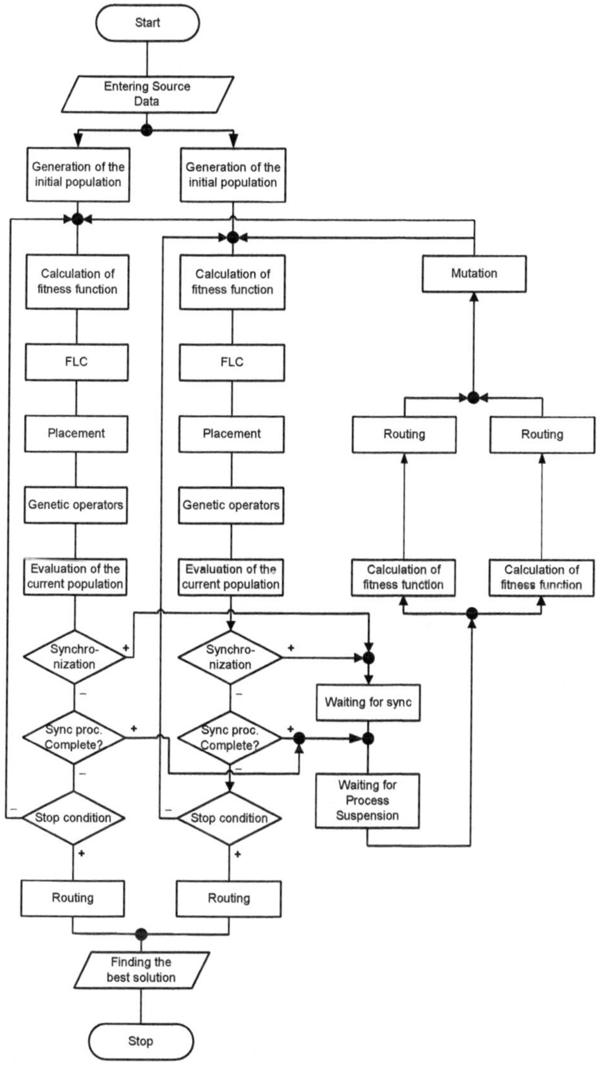

Fig. 1. The scheme of the integrated parallel algorithm

4 Results of Experiments

Algorithms for placement and routing are implemented in the form of generalized algorithms that can handle various input data [17]. The input of the algorithm is given data on the topology of the printed circuit board.

The developed hybrid parallel algorithm is investigated on the basis of two hardware configurations: Intel® Core (TM) i7-3630QM CPU @ 2.40 GHz, RAM – 8 Gb (configuration 1); Intel® Core (TM) 2 Quad CPU Q8200 @ 2.40 GHz, RAM – 4 Gb (configuration 2). The researches include several experiments with the number of elements from 100 to 3000 and the interval 100. The number of connections, iterations and the population size were constant. The results of the experiments are represented as the diagrams of the average execution time and the common number of elements in Fig. 2.

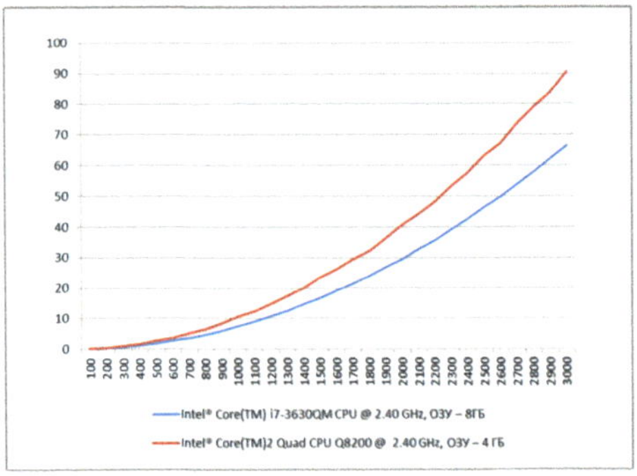

Fig. 2. The dependence of the execution time of the algorithm on the number of elements

On the basis of the proposed algorithms, methods and procedures of the authors we have developed a software system for integrated solutions of problems of placing and tracing. Search for optimal solutions is carried out using two variants of the search strategy. In addition, possibility of parallelization of computational experiments series performed, and for each option, you can choose own parameters [13, 14]. The results of experimental series are displayed on the screen (Fig. 3).

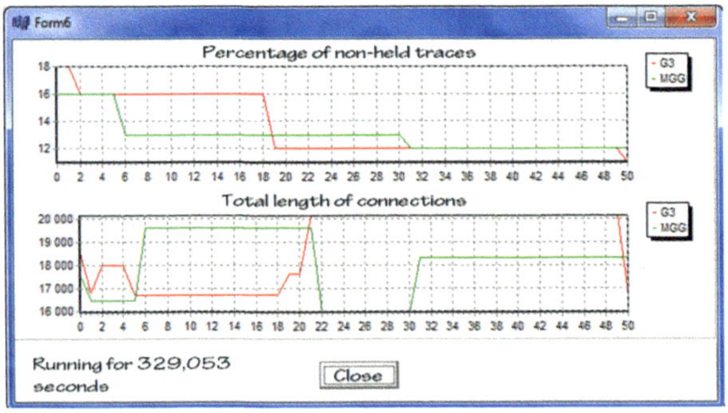

Fig. 3. Graphs of output parameters change of fuzzy controller

Research has been done on the dependence of the running time of the algorithm using various models of population formation (model based on the minimal generation gap (MMG) and model based on the generalized generation gap (G3)), various crossover operators (linear and arithmetic operators, and operator based on "hill-climbing" procedure). The results of these studies are reflected in Tables 2, 3 and 4. Analysis of the obtained experimental data will allow choosing the most effective combinations of operators for solving problems of different dimensions.

The developed «hill- climbing» operator on average 2.5% faster than classical operators, such as arithmetic and linear. If we compare the results of different models, we can conclude, that the model is based on a generalization of generations operator «hill- climbing» allows to obtain an average result of 1.5% faster. Comparing results of the models G3 and MMG in the number of non-held connections: for schemes with a large number of elements is more effective to use the model of the minimum gap between the generations.

Table 2. The effectiveness of population formation strategies

Number of elements	The percentage of failed connections								
	300	450	600	750	900	1050	1200	1350	1500
MMg	0	0	11	21	16	19	18	15	12
G3	0	0	10	20	30	18	19	12	14

Table 3. The number of failed connections for a different number of elements

Number of elements	The percentage of failed connections			
	300	600	900	1200
"Hill-climbing"	0	5	18	18
Arithmetic crossover	6	11	21	27
Linear crossover	5	18	16	24

Table 4. The number of failed connections for different numbers of iterations

Number of iterations	The percentage of failed connections			
	150	200	300	400
"Hill-climbing"	18	17	16	12
Arithmetic crossover	22	23	21	16
Linear crossover	20	20	18	13

To improve the effectiveness and control the parameters of the genetic algorithms, we used the fuzzy logical controller (FLC). The execution time while using the FLC exceeds the execution time without using the FLC not more than 0,5% in terms of the same number of the elements. However, the FLC allows us to improve the quality of the obtained decisions by 25% on the average in comparison with the sequential genetic algorithm with the same number of iterations (Fig. 3, 4). The effectiveness of the FLC is improved after introduction of the training block on the basis of the artificial neural network model.

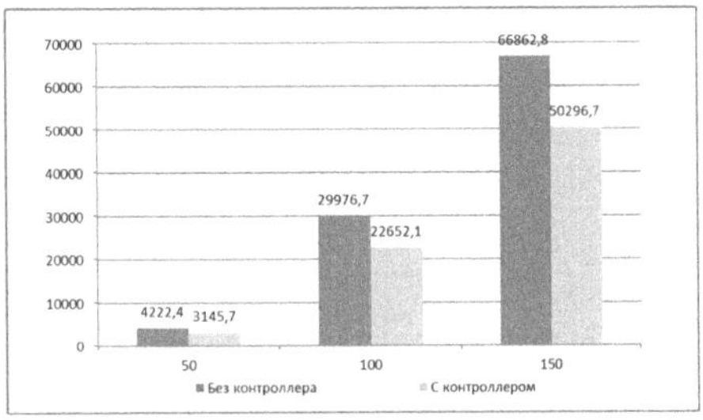

Fig. 4. The quality of the solutions obtained with and without the FLCr

5 Conclusions

Analysis of the developed hybrid models and systems based on the integration of various scientific fields allows us to conclude that they are promising. Good results can be obtained by combining parallel computing and the principles of building multi-agent systems.

Comparison of the results with the existing test tasks (benchmarks) are also carried out in the analysis of the results. These results demonstrate that the proposed method is not inferior in performance to existing models, but at the same time allows for the joint solution problems of placement and tracing. At the same time, it is necessary to continue research in this area [15].

Acknowledgment. The reported study was funded by the Russian Foundation for Basic Research, project number 18-07-01054, 20-01-00160.

References

1. Cohoon, J.P., Karro, J., Lienig, J.: Evolutionary algorithms for the physical design of VLSI circuits. In: Ghosh, A., Tsutsui, S. (eds.) Advances in Evolutionary Computing: Theory and Applications, pp. 683–712. Springer Verlag, London (2003)
2. Alpert, C.J., Mehta, D.P., Sapatnekar, S.S.: Handbook of Algorithms for Physical Design Automation. CRC Press, New York (2009)
3. Shervani, N.: Algorithms for VLSI Physical Design Automation. Kluwer Academy Publisher, USA, p. 538 (1995)
4. Glagkov, L.A., Glagkova, N.V., Leiba S.N.: Electronic computing equipment schemes elements placement based on hybrid intelligence approach. In: Proceedings of the 4th Computer Science On-line Conference 2015 (CSOC 2015). Software Engineering in Intelligent Systems, vol. 2, no. 348, pp. 35–45 (2015)
5. Gladkov, L.A., Gladkova, N.V., Leiba, S.N., Strakhov, N.E.: Development and research of the hybrid approach to the solution of optimization design problems. In: Abraham, A., Kovalev, S., Tarassov, V., Snasel, V., Sukhanov, A. (eds.) Proceedings of the Third International Scientific Conference "Intelligent Information Technologies for Industry" (IITI 2018). Advances in Intelligent Systems and Computing, vol. 875, pp. 246–257. Springer, Cham (2018)

6. Michael, A., Takagi, H.: Dynamic control of genetic algorithms using fuzzy logic techniques. In: Proceedings of the 5th International Conference on Genetic Algorithms, pp. 76–83. Morgan Kaufmann (1993)
7. Im, S.-M., Lee, J.-J.: Adaptive crossover, mutation and selection using fuzzy system for genetic algorithms. Artif. Life Robot. **13**(1), 129–133 (2008)
8. Herrera, F., Lozano, M.: Fuzzy adaptive genetic algorithms: design, taxonomy, and future directions. Soft. Comput. **7**, 545–562 (2003)
9. Herrera, F., Lozano, M.: Adaptation of genetic algorithm parameters based on fuzzy logic controllers. In: Herrera, F., Verdegay, J.L. (eds.) Genetic Algorithms and Soft Computing, pp. 95–124. Physica-Verlag, Heidelberg (1996)
10. King, R.T.F.A., Radha, B., Rughooputh, H.C.S.: A fuzzy logic controlled genetic algorithm for optimal electrical distribution network reconfiguration. In: Proceedings of 2004 IEEE International Conference on Networking, Sensing and Control, Taipei, Taiwan, pp. 577–582 (2004)
11. Rodriguez, M.A., Escalante, D.M., Peregrin, A.: Efficient distributed genetic algorithm for rule extraction. Appl. Soft Comput. **11**, 733–743 (2011)
12. Alba, E., Tomassini, M.: Parallelism and evolutionary algorithms. IEEE T. Evolut. Comput. **6**, 443–461 (2002)
13. Zhongyang, X., Zhang, Y., Zhang, L., Niu, S.: A parallel classification algorithm based on hybrid genetic algorithm. In: Proceedings of the 6th World Congress on Intelligent Control and Automation, Dalian, China, pp. 3237–3240 (2006)
14. Tarasov, V.B.: Ot mnogoagentnykh sistem k intellektual'nym organizatsiyam. Editorial URSS, M (2002)
15. Deb, K., Joshi, D., Anand, A.: Real-Coded Evolutionary Algorithms with Parent-Centric Recombination. Kanpur Genetic Algorithms Laboratory (KanGAL), Kanpur, PIN 208 016, India
16. Lozano, M., Herrera, F., Krasnogor, N., Molina, D.: Real-coded memetic algorithms with crossover hill-climbing. Evol. Comput. **12**(3), 273–302 (2004)
17. Gladkov, L.A., Gladkova, N.V., Leiba, S.N., Strakhov, N.E.: Development and research of the hybrid approach to the solution of optimization design problems. In: Advances in Intelligent Systems and Computing, vol. 875. International Conference on Intelligent Information Technologies for Industry IITI 2018, vol. 2, pp. 246–257. Springer Nature Switzerland AG (2019)

Mind Evolutionary Computation Co-algorithm for Optimizing Optical Systems

Maxim Sakharov[1(✉)], Thomas Houllier[2,3], and Thierry Lépine[2,4]

[1] Bauman MSTU, 5/1 2-ya Baumanskaya, 105005 Moscow, Russia
max.sfn90@gmail.com
[2] Univ-Lyon, Laboratoire Hubert Curien, UMR CNRS 5516, 18 rue Benoît Lauras, 42000 Saint-Etienne, France
[3] Sophia Engineering, 5 Rue Soutrane, 06560 Sophia Antipolis, France
[4] Institut d'Optique Graduate School, 18 rue Benoît Lauras, 42000 Saint-Etienne, France

Abstract. This paper presents the modified global optimization co-algorithm based on the Mind Evolutionary Computation (*MEC*) algorithm for optimizing optical system designs. This kind of systems requires high precision. As a result optimization algorithms tend to converge slowly to a local optimum while trying to guarantee high quality of solutions. The concept of a co-algorithm helps to overcome this issue by identifying the most promising search areas and allocating there more computational resources. The outline of the proposed co-algorithm as well as its software implementation are described in the paper. The algorithm was utilized to optimize the structure of an optical Cooke triplet lens and helped identify various designs with best optical properties. All results are presented in the paper.

Keywords: Global optimization · Optical system design · Mind Evolutionary Computation · Co-evolution algorithms

1 Introduction

In recent years, so called population based algorithms have become a powerful optimization methods for solving global optimization problems. However, the performance of those algorithms may heavily depend on the values of their free parameters. In the majority of cases there are no recommendations on selecting numeric values of those free parameters. Usually they are selected based on the characteristics of a problem at hand [1, 2]. Experimental studies [3] also suggest, a population based algorithm performs better when it contains more information on a problem at hand but it is not feasible to tune an algorithm to every task.

In recent years, many researchers in the field of population based optimization methods have proposed different ways to increase algorithms' performance by means of either meta-optimization or hybridization. Meta-optimization implies the tuning of free parameters' values that would improve the efficiency of an algorithm being used [1, 4]. A development of hybrid algorithms, on the other hand, implies a combination of

S. Kovalev et al. (Eds.): IITI 2019, AISC 1156, pp. 476–486, 2020.
https://doi.org/10.1007/978-3-030-50097-9_48

various or same methods with different numeric values of the free parameters in such a way that the advantages of one method would overcome the disadvantages of another one.

Optimization methods are a crucial part of optical systems design. Several optimization algorithms are typically included in optical design software. As a general rule, optical designers define a scalar objective that comprises all the optical quality and geometrical constraints specifications and optimize the system parameters. Optical system parameters are generally radii of curvature of lenses/mirrors, relative geometrical position of optical elements in the system, choice of glass for the lenses, lists of surface parameters for more complex optical surfaces, etc.

These optical optimization problems often show non-linear behavior, ill conditioning and many local minima. With the advent of freeform optics [5], in which surfaces become non-rotationally symmetric and are described with polynomial coefficients such as Chebyshev, Legendre, Zernike, Forbes (among others), these problems present a growing number of parameters, easily in the hundreds. This kind of application requires high precision and in turn slows down the convergence of population based algorithms if they don't adapt to a problem.

In this work the Mind Evolutionary Computation co-algorithm (*CoMEC*) was studied [6]. It was originally presented by the authors in [7] and proved more efficient than the traditional MEC algorithm while maintaining the same level of computational expenses. The *CoMEC* algorithm was modified in this paper to provide the required high precision when optimizing optical systems.

The proposed modification was presented in this work along with its modular software implementation. They were later utilized to optimize the Cooke triplet, an optical system often used to showcase optimization methods in optical design [8–10]. Despite its relatively low dimensionality, the Cooke triplet example is already non-trivial, as can be seen from the variety of results for the same problem depending on the algorithm being used [11]. The *CoMEC* algorithm allowed the identification of better minima than those that were found with any of the methods presented in [11].

2 Problem Statement and Simple *MEC*

We consider a deterministic global constrained minimization problem

$$\min_{X \in D} \Phi(X) = \Phi(X^*) = \Phi^*. \tag{1}$$

Here $\Phi(X)$ is the scalar objective function, $\Phi(X^*) = \Phi^*$ is its required minimal value, $X = \left(x_1, x_2, \ldots, x_{|X|}\right)$ – $|X|$-dimensional vector of variables.

Feasible domain D is determined with inequality constraints

$$D = \left\{X | x_i^{min} \le x_i \le x_i^{max}, i \in [1 : |X|]\right\} \subset R^{|X|}. \tag{2}$$

A concept of the *MEC* algorithm was first proposed in 1998 [12, 13]; it imitates several aspects of human behavior in society; every individual represents an intelligent agent which operates within a group of similar individuals. In order to succeed within its group, an individual has to learn from the most successful individuals in this group. And groups themselves should operate according to the same principle to stay alive in the intergroup competition.

The *MEC* algorithm can be easily considered as a population-based algorithm with a single multi-population made of leading groups $S^b = \left(S_1^b, S_2^b, \ldots, S_{|S^b|}^b\right)$ and lagging groups $S^w = \left(S_1^w, S_2^w, \ldots, S_{|S^w|}^w\right)$, which include $|S^b|$ and $|S^w|$ subpopulations respectively. In the traditional *MEC* algorithm the number of individuals within each subpopulation is set to be the same and is equal to $|S|$; however, this was modified in other works [7].

Each of subpopulations S_i^b, S_j^w has its own communication environment named a local blackboard and denoted as C_i^b, C_j^w correspondingly. Besides, a multi-population as a whole $S = \{S^b, S^w\}$ has a common global blackboard C^g.

This algorithm is composed of three main stages: initialization of groups, similar-taxis and dissimilation and was studied in detail by the authors in [7]. Operations of similar-taxis and dissimilation are repeated iteratively while the best obtained value of an objective function $\Phi(X)$ is changing. When the best obtained value stops changing, the winner of the best group from a set of leading ones is selected as a solution to the optimization problem.

3 Co-algorithm Based on *MEC*

In this work we propose a modification of the co-evolution *MEC* algorithm proposed in [7]. An idea of co-evolution is based on the simultaneous evolution of several populations within one search domain, which solve one problem with different values of free parameters and compete for certain common resource. The *CoMEC* algorithm uses a total number of individuals as resource for competition. As evolution proceeds more successful subpopulations get more agents for search exploration, taking them from less successful subpopulations. In such a manner, the total number of agents remains the same. The modification proposed in this work was designed to meet the requirements of high precision: the accuracy of the algorithm is increased while the number of iterations grows.

The dissimilation stage of the *MEC* algorithm was modified in accordance with a concept of co-evolution while the similar-taxis stage was modified to provide the required precision. A general scheme of the modified algorithm, named *CoMEC*, can be described as follows.

1. Initialization of groups within the search domain D.
   (a) Generate a given number γ of groups $S_i, i \in [1:\gamma]$, where γ – free parameter of the algorithm.
   (b) Generate a random vector $X_{i,1}$, whose components are distributed uniformly within the corresponding search subdomain. Identify this vector with the individual $s_{i,1}$ of the group S_i.

(c) Determine the initial coordinates of the rest of the individuals in the group $S_{i,j}, j \in [2:|S|]$ following the formula

$$X_{i,j} = X_{i,1} + N_{|X|}(0, \sigma), \tag{3}$$

in other words, they are placed randomly around the main individual $s_{i,1}$ in accordance with $|X|$-dimensional normal distribution law $N_{|X|}(0, \sigma)$, with zero mathematical expectation along all $|X|$ coordinates and standard deviation σ (another free parameter of the algorithm).

(d) Calculate the scores of all individuals in every population S_i and put them on the corresponding local blackboards.

(e) Create leading $\boldsymbol{S}^b$ and lagging $\boldsymbol{S}^w$ groups on the basis of obtained information.

2. Similar-taxis operation is performed in every group.
   (a) Take information on the current best individual $s_{i,j}, j \in [1:|S_i|]$ of the group S_i from the blackboard C_i.
   (b) The value of parameter σ used for generating new agents decreases depending on the number of iterations:

$$\sigma = \begin{cases} \sigma_0, \; if\, \lambda < \hat{\lambda}, \\ \frac{1}{\left(\lambda - \hat{\lambda}\right)^{\theta}} + \varepsilon, if \lambda \geq \hat{\lambda}. \end{cases} \tag{4}$$

Here $\hat{\lambda}$ is the threshold number of iterations; when $\lambda \geq \hat{\lambda}$, the standard deviation σ start decreasing; σ_0 is the initial value of the standard deviation; θ is the speed parameter (the recommended value $\theta = 0.2$) ε is the tolerance used to identify the stagnation.

   (c) Create leading groups $\boldsymbol{S}^b = \left(S_1^b, S_2^b, \ldots, S_{|S^b|}^b\right)$ and lagging groups $\boldsymbol{S}^w = \left(S_1^w, S_2^w, \ldots, S_{|S^w|}^w\right)$ on the basis of obtained information.
   (d) Put information on the new winners in every group S_i of the multi-population on the corresponding local and global blackboards.

3. Dissimilation operation.
   (a) Read the scores of all groups $\Phi_i^b, \Phi_j^w, i \in \left[1:\left|S^b\right|\right], j \in [1:|S^w|]$ from the global blackboard C^g (scores of the best individuals in the groups).
   (b) Compare those scores. If the score of any leading group S_i^b appeared to be less than the score of any lagging group S_j^w, then the latter becomes a leading group and the former becomes a lagging one. If the score of a lagging group S_k^w is lower than the scores of all leading groups for ω consecutive iterations, then it's removed from the population.
   (c) A number of agents $\left|S_j^w\right|$ in every lagging group S_j^w is reduced proportionally to their scores at the global blackboard. A number of individuals in the best group from S_j^w is reduced by one, and a number of individuals in the worst group is reduced by a value of p_A, which can be calculated as follows

$$p_{|S^w|}^A = \left\lfloor \frac{(1 - |S_{min}|) \cdot |S|}{\omega} \right\rfloor. \quad (5)$$

Here ω – remove frequency of lagging groups, in other words if within ω iterations a group remains a lagging one, it is removed from a population; $\lfloor \cdot \rfloor$ – a symbol of nearest least whole number; $|S_{min}|$ is the minimum allowed size of a group (the recommended value – 20% of the initial size).

(d) A number of individuals, being removed from intermediate groups is calculated using a linear approximation

$$p_k^A = \left\lfloor \frac{(k-1) \cdot \left(p_{|S^w|}^A - 1\right)}{|S^w| - 1} + 1 \right\rfloor, \; k \in [2 : |S^w| - 1]. \quad (6)$$

(e) When a number of agents in a particular group is below a minimum allowed value $|S_{min}|$ this number is set to be $|S_{min}|$.

(f) A number of individuals in all leading groups except for the best one doesn't change.

(g) A number of agents in the best group is increased by the number of individuals removed from other groups $\sum_{k=1}^{|S^w|} p_k^A$.

(h) Using the initialization operation and formula (3) each removed group is replaced with a new one.

Evaluate the termination criteria. If either a number of stagnation iterations λ_{stop} or maximum allowed number of iterations λ_{max} exceed their limits then the computational process should be stopped, and the best current individual is set to be a solution X^* to the optimization problem. Otherwise it continues and goes to point 2. In such a manner, the total number of agents within a whole multi-population remains unchanged in the progress of the computational process.

4 Optimizing Optical Systems

The Cooke triplet is a three lenses rotationally symmetric objective lens which is very popular in photography. It works for an object at infinity and finite image position. The optical surfaces it comprises are spherical. Taken as a mathematical optimization problem, it has input parameters and an output scalar evaluation score that must be defined. We defined the exact same problem as in [11].

The input parameters in this system can be the radii of curvature of the spherical optical surfaces, the refraction index of the glasses at the chosen wavelength and the distances between the apex of successive surfaces. For the present study, we took fixed refraction indexes and all the radii and distances as problem inputs.

The output scalar score that indicates the optical performance of the optical system must be built using either ray tracing or analytical aberration theory. For the sake of simplicity, we chose the more academic analytical Seidel third order optical aberration theory. We also added a constraint on the system total track. Note that our analysis is simply monochromatic, i.e. we analyzed the optical aberrations only of a single wavelength passing through the optical system. A diagram of the optical system is shown in Fig. 1 while its numerical parameters are presented in Table 1. We use the optical design software Zemax to draw the system layout.

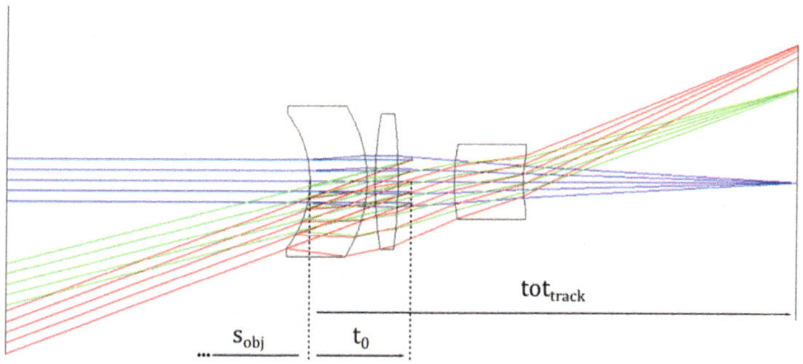

Fig. 1. Cooke triplet diagram and paraxial quantities definitions. The layout was drawn using the optical design software Zemax. The fields represented here are 0; +14; +20° in blue; green; red respectively.

The refraction index of the air is 1.0, that of all the glasses is 1.62, independent of wavelength. The object is numerically at infinity. The paraxial entrance pupil is at distance t_0 from the apex of the first surface and its diameter is 1 mm to simplify the computations, in accordance with [14]. The first five thicknesses are system inputs, the last one is chosen to complete the sum of distances to $Tot_{track,target}$. The choice of variable bounds for distances makes it impossible for the sum to exceed $Tot_{track,target}$ but we included this total track component for the sake of completeness. All the curvatures for the six optical surfaces are problem inputs. We have in total 11 inputs for this optimization problem with associated search space bounds ($|X| = 11$).

Table 1. Cooke triplet case study parameters

Parameter name	Values
Refraction indices	1.0; 1.62; 1.0; 1.62; 1.0; 1.62; 1.0
s_{obj}	+1e6 mm (+ or – infinity are equivalent here)
t_0	2.5 mm
$Tot_{track,target}$	12 mm
Entrance pupil diameter	1 mm
Distance between surface vertices	d_0, d_1, d_2, d_3, d_4 in range [0.2; 2.0] mm
Radii of curvature	$r_0, r_1, r_2, r_3, r_4, r_5$
Curvature bounds	[−0.5; 0.5] mm^{-1}

Note that optimizing on curvatures rather than radii of curvatures makes the search space easier to express. Optical surfaces with radii around 0 are infeasible as opposed to curvatures around 0 which are feasible. These quantities are inverse of each other.

Our objective function, or merit function (MF), is computed using the paraxial aberration computation highlighted in Born and Wolf [14] (chapter 5.5 equation 24). We used the height of a paraxial skew ray through the system, this is slightly different than the Zemax method of using a "real" skew ray. The paraxial method is less accurate but simpler in implementation. The exact merit function we used is then simply:

$$MF = B^2 + C^2 + D^2 + E^2 + F^2 + \left(Tot_{track} - Tot_{track,target}\right)^2. \tag{7}$$

With B, C, D, E, F the well-known Seidel third order aberration coefficients of the system corresponding to spherical, astigmatism, field curvature, distortion and coma respectively. The total track component is redundant for the specific search space we chose here, but this component can be used for other problem setups. This MF score is to be minimized. Of course, all orders should be taken into account ideally, as optical design software do. Here, we limit ourselves to the third order, which is easier to implement.

4.1 Numerical Experiments

The proposed *CoMEC* algorithm was implemented by the authors in *Wolfram Mathematica* 11.3. The software implementation has a modular structure, which helps to modify algorithms easily [15]. In addition, the merit function to be minimized was also implemented in the same software.

Taking into account that performance of the algorithms significantly depends on the random initial position of individuals, the multi-start method was utilized for this study with 100 launches [16]. The best obtained value of objective function Φ^* was used as the main performance index along with the average value of objective function $\bar{\Phi}$ based on the results of all launches and the number of iterations of the algorithm. All computational experiments were carried out using a personal computer with CPU *Intel Core 2 Duo* 2,53 GHz and 4 GB of RAM. For this set of experiments the following values of the algorithm's free parameters were used: initial number of individuals in a group $|S| = 20$; number of groups $\gamma = 50$; the tolerance $\varepsilon = 10^{-12}$; $\hat{\lambda} = 500$; $\lambda_{stop} = 2000$; the remove frequency $\omega = 20$ iterations; the initial standard deviation $\sigma_0 = 0.1$; the numbers of leading and lagging groups are equal $|S^b| = |S^w|$.

Results of the numerical experiments are presented in Table 2. To put them in perspective, let us compare them with the results presented in [11], where the same problem was studied with five different algorithms [1, 2]: Particle Swarm Optimization, Nelder-Mead Method, Gravity Search, Cuckoo Search and Covariance Matrix Adaptation Strategy. Out of 100 runs with 5000 evaluations each, for each algorithm, totaling 500 runs and 2.25E + 06 evaluations, the very best MF that were found were greater than 10E − 6. With the *CoMEC* algorithm, we find on average, at the end of each run, a better system than the best system in [11]. Moreover, we have found with the *CoMEC* a global minimum of the problem at 4.95E − 18, which was never encountered in [11].

4.2 Analyzing Results

Additional analysis of the obtained results helped to identify several completely different designs of the Cooke triplet lens which are worth further investigation from the practical perspective. They were visualized and analyzed using Zemax OpticStudio software.

Table 2. Obtained results of the numerical experiments

Name	Values
Best obtained value $\Phi(X^*)$	4.95E − 18
Average value $\bar{\Phi}(X)$ based on all 100 launches	4.27E − 07
Worst obtained value $\Phi(X)$	3.75E − 05
Average number of *CoMEC* iterations $\bar{\lambda}$	9.78E + 03

The design which corresponds to the best obtained value $\Phi(X^*) = 4.95\text{E} - 18$ is presented in Fig. 2. On the left side, the Seidel diagram illustrates 3rd order geometric aberrations. Concerning the columns, 2 corresponds to the first diopter (input face of the first lens), 3 to the 2nd diopter, etc. Although some diopters can create a significant level of aberrations, the contributions of the different diopters compensate each other and the overall objective has a very low level of 3rd order aberrations.

On the right side, we traced the lenses and 3 beams for 3 different fields. The plots take into account all orders of geometric aberrations, which is why, especially in the fields, the beams do not converge towards a quasi-point zone. So, even if this result is not perfect, it could be an excellent starting point for a new optimization in a commercial software like Code V or Zemax OpticStudio. This is very important because, in general, it is very difficult to obtain "good" starting points, i.e. starting points that lead to high performance systems after optimization.

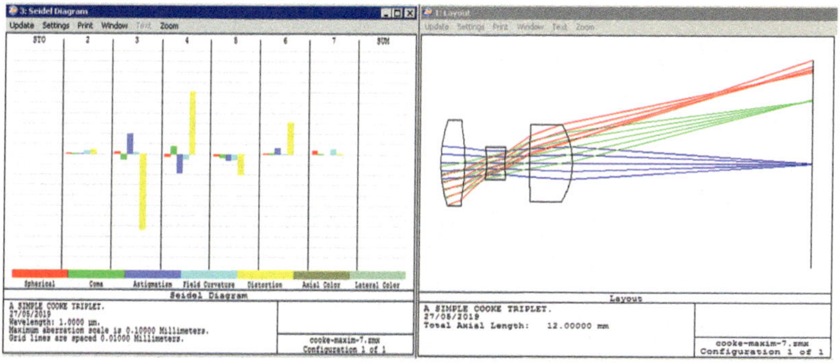

Fig. 2. The Cooke triplet design with the lowest found MF value: $\Phi(X^*) = 4.95\text{E} - 18$

Figures 3, 4 and 5 demonstrate different designs of the Cooke triplet lens with small values of the objective function that also can be used as starting points for further optimization from the perspective of optical design. In [11] it was already showed that using search algorithms other than those included in the commercial tools could improve significantly the optimization performance, at least for the specific problem under study of the Seidel third order aberrations in the Cooke triplet lens.

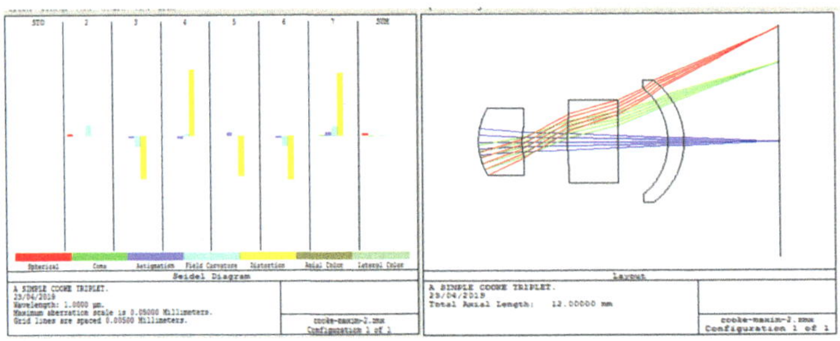

Fig. 3. The Cooke triplet design where $\Phi(X^*) = 4.16\text{E} - 09$

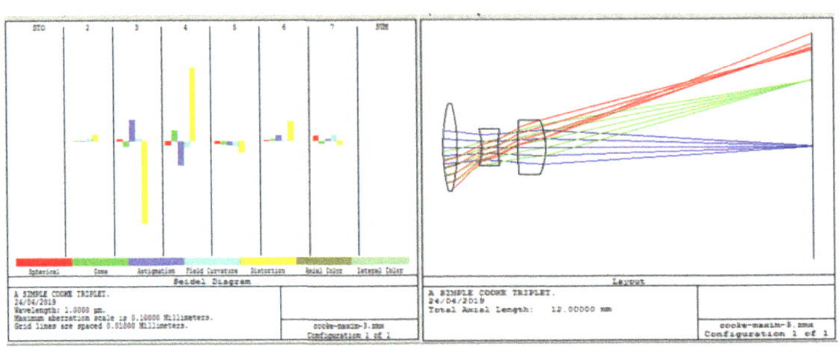

Fig. 4. The Cooke triplet design where $\Phi(X^* =)\ 2.25\text{E} - 11$

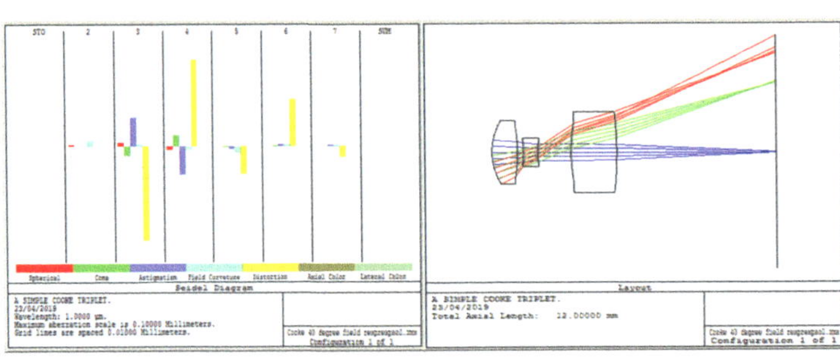

Fig. 5. The Cooke triplet design where $\Phi(X^*) = 2.58\text{E} - 06$

5 Conclusion

This paper demonstrates that applying advanced optimization techniques, such as the *CoMEC* algorithm, can improve results even further. On the other hand, this approach leads to increased computational expanses (larger number of evaluations).

The fact that optical designers can benefit from using advanced search algorithms like the *CoMEC* algorithm, even on relatively simple optical systems like the Cooke triplet lens, leads us to believe that this benefit could exist a fortiori for complex freeform optical systems with MF based on ray tracing where optimization performance is currently insufficient in our opinion. This will be studied in further works.

References

1. Karpenko, A.P.: Modern algorithms of search engine optimization. In: Nature-Inspired Optimization Algorithms, p. 446. Bauman MSTU Publication, Moscow (2014)
2. Weise, T.: Global Optimization Algorithms - Theory and Application. University of Kassel, p. 758 (2008)
3. Sakharov, M., Karpenko, A.: Multi-memetic mind evolutionary computation algorithm based on the landscape analysis. In: 7th International Conference Theory and Practice of Natural Computing, TPNC 2018, Dublin, Ireland, 12–14 December 2018, pp. 238–249, Proceedings. Springer (2018). https://doi.org/10.1007/978-3-030-04070-3
4. Agasiev, T., Karpenko, A.: The program system for automated parameter tuning of optimization algorithms. Procedia Comput. Sci. **103**, 347–354 (2017). https://doi.org/10.1016/j.procs.2017.01.120
5. Fuerschbach, K.: Freeform, φ-polynomial optical surfaces: optical design, fabrication and assembly, PhD Thesis. http://hdl.handle.net/1802/28531
6. Chengyi, S., Yan, S., Wanzhen, W.: A survey of MEC: 1998–2001. In: 2002 IEEE International Conference on Systems, Man and Cybernetics IEEE SMC2002, Hammamet, Tunisia, 6–9 October 2002, vol. 6, pp. 445–453. Institute of Electrical and Electronics Engineers Inc. (2002)
7. Sakharov, M., Karpenko, A.: Performance investigation of mind evolutionary computation algorithm and some of its modifications. In: Proceedings of the First International Scientific Conference "Intelligent Information Technologies for Industry" (IITI 2016), pp. 475–486. Springer (2016). https://doi.org/10.1007/978-3-319-33609-1_43
8. Lakshminarayan, H., Banerjee, S.: Genetic algorithm in the structural design of Cooke triplet lenses. In: Design and Engineering of Optical Systems II, vol. 3737. International Society for Optics and Photonics (1999). https://doi.org/10.1117/12.360005
9. Vasiljevic, D.M.: Optimization of the Cooke triplet with various evolution strategies and damped least squares. In: Optical Design and Analysis Software, vol. 3780. International Society for Optics and Photonics (1999). https://doi.org/10.1117/12.363779
10. Bociort, F., Van Driel, E., Serebriakov, A.: Networks of local minima in optical system optimization. Opt. Lett. **29**(2), 189–191 (2004). https://doi.org/10.1364/OL.29.000189
11. Houllier, T.: Search algorithms and optical systems design. In: Zemax ENVISION 2019, Paris, France, 26–28 March 2019
12. Jie, J., Zeng, J.: Improved mind evolutionary computation for optimizations. In: Proceedings of 5th World Congress on Intelligent Control and Automation, Hang Zhou, China, pp. 2200–2204 (2004)

13. Jie, J., Han, C., Zeng, J.: An extended mind evolutionary computation model for optimizations. Appl. Math. Comput. **185**, 1038–1049 (2007)
14. Born, M., Emil, W., Bhatia, A.B.: Principles of Optics, seventh (expanded) edn. Cambridge University Press, Cambridge (1999)
15. Sakharov, M., Karpenko, A.: Parallel multi-memetic global optimization algorithm for optimal control of polyarylenephthalide's thermally-stimulated luminescence. In: Optimization of Complex Systems: Theory, Models, Algorithms and Applications. WCGO 2019. Advances in Intelligent Systems and Computing, vol. 991, pp. 191–201. Springer, Cham (2020). https://doi.org/10.1007/978-3-030-21803-4_20
16. Sakharov, M.K., Karpenko, A.P.: Adaptive load balancing in the modified mind evolutionary computation algorithm. Supercomput. Front. Innovations **5**(4), 5–14 (2018). https://doi.org/10.14529/jsfi180401

Coverage with Sets Based on the Integration of Swarm Intelligence and Genetic Evolution

Boris K. Lebedev, Oleg B. Lebedev(✉), and Artemiy A. Zhiglaty

Southern Federal University, Rostov-on-Don, Russia
lebedev.b.k@gmail.com, lebedev.ob@mail.ru,
artemiy.zhiglaty@gmail.com

Abstract. The composite architecture of a multi-agent bionic search system based on swarm intelligence and genetic evolution is proposed for solving the problem of covering with sets. Two approaches to hybridization of the search by particle swarm and genetic search are considered: sequential and combinatorial. The link of this approach is a single data structure describing the solution of the problem in the form of a chromosome. New ways of coding solutions and chromosome structures have been developed to represent solutions. The key problem that was solved in this paper is related to the development of the structure of the affine space of positions (solutions), which allows displaying and searching for solution interpretations with integer parameter values. In contrast to the canonical particle swarm method, to reduce the weight of affinity bonds, by moving the p_i particle to a new position of the affine solution space, a directed mutation operator was developed, the essence of which is to change the integer values of genes in the chromosome. The overall estimate of time complexity lies within $O(n^2)$–$O(n^3)$.

Keywords: Set coverage · Particle swarm · Genetic evolution · Affine space · Integer parameters · Integration · Directional mutation operator

1 Introduction

In this paper, the objects of study are bionic search systems based on swarm intelligence and genetic evolution for solving the problem of covering by sets. The coverage problem is as follows [1]. The source data of the set coverage problem is a finite set X, as well as the family of its subsets $F = \{X_i \backslash i = 1, 2, \ldots, n\}$ such that $X_i \subset X$ and $\cup X_i = X$. The problem of minimal set coverage is to find the set $P \subseteq F$ with the minimum number of $X_i \in P$ and $X_i \in F$ subsets such that $\cup X_i = X$. [1, 2]. The result of the ongoing search for the most effective methods was the use of bionic methods and algorithms for intellectual optimization, based on the modeling of collective intelligence. In this regard, currently one of the main ways to improve the efficiency of methods for solving global search problems is the development of hybrid algorithms [3]. In hybrid algorithms, the advantages of one algorithm can compensate for the

This work was supported by the grant from the Russian Foundation for Basic Research the project № 18-07-00737 a.

S. Kovalev et al. (Eds.): IITI 2019, AISC 1156, pp. 487–496, 2020.
https://doi.org/10.1007/978-3-030-50097-9_49

shortcomings of another [4]. A composite architecture of a multi-agent bionic search system based on swarm intelligence and genetic evolution is proposed [5].

In genetic algorithms, the genotype of a solution is represented as a chromosome. When solving combinatorial problems, genes in chromosomes usually have integer values [3]. The position of a particle in the search space (position) is equivalent to a genotype in evolutionary algorithms.

The canonical paradigm of particle swarm involves the use of real values of parameters in multidimensional, real, metric spaces [6–8]. However, in most genetic algorithms, genes in chromosomes have discrete integer values. In turn, chromosomes are some interpretations of solutions that transform into solutions by decoding chromosomes. This does not allow direct use of the canonical paradigm of a swarm of particles. In this connection, the development of a modernized structure of the search space, a data structure for representing solutions and positions, and modernized mechanisms for moving particles in a search space is relevant.

2 Affine Search Space

Let there be a linear vector space (*LVP*) whose elements are n-dimensional points (positions). To each of any two points p and q of this space, we uniquely associate a unique ordered pair of these points, which we will call further a geometric vector (vector). $p, q \in V(p, q)$ is a geometric vector (an ordered pair). The set of all points of an *LVP*, supplemented by geometric vectors, is called a point vector or affine space. Affine space is n-dimensional if the corresponding *LVP* is also n-dimensional.

The affine-relaxation model (*ARM*) of a particle swarm is the vertex graph of which correspond to the positions of the particle swarm, and the arcs correspond to affine connections between the positions (points) in the affine space. Affinity is a measure of the proximity of two agents (particles). At each iteration, each agent p_i moves in the affine space to a new state (position), at which the weight of the affinity connection between the agent p_i and the base (best) agent p^* decreases.

The positions of the best particles from the point of view of the objective function are declared “center of attraction”. The displacement vectors of all particles in affine space rush to these centers. The transition is possible taking into account the degree of proximity to one base position or to a group of neighboring positions taking into account the probability of transition to a new state.

3 Search by Particle Swarm

In heuristic algorithms of swarm intelligence, the multidimensional search space is populated by a swarm of particles [9]. In the process of searching by the particle swarm method, each particle moves to a new position. The new position in the canonical paradigm of particle swarming is defined as:

$$x_i(t+1) = x_i(t) + v_i(t+1), \tag{1}$$

where $v_i(t + 1)$ is the velocity of a particle moving from position $x_i(t)$ to position $x_i(t + 1)$. The initial state is defined as $x_i(0)$, $v_i(0)$ [10]. The above formula is presented in vector form. For a separate dimension j of the search space, the formula takes the form:

$$x_{ij}(t+1) = x_{ij}(t) + v_{ij}(t+1), \tag{2}$$

where $x_{ij}(t)$ is the position of the particle p_i in dimension j, $v_{ij}(t + 1)$ – is the velocity of the particle p_i dimension j.

We introduce the notation:

- $x_i(t)$ is the current position of the particle, $f_i(t)$ is the value of the objective function of the particle p_i at position $x_i(t)$;
- $x_i^*(t)$ is the best position of the particle p_i, that it has visited since the first iteration, and $f_i^*(t)$ is the value of the target function of the particle p_i in this position (the best value since the start of the life cycle of the particle p_i.

$x^*(t)$ is the position of the swarm particle with the best value of the objective function $f^*(t)$ among the swarm particles at time t.

Then the velocity of the particle p_i in step $(t + 1)$ in dimension j is calculated as:

$$v_{ij}(t+1) = w \cdot v_{ij}(t) + k_1 \cdot rnd(0,1) \cdot (x_{ij}^*(t) - x_{ij}(t)) + k_2 \cdot rnd(0,1) \cdot (x_j^*(t) - x_{ij}(t)), \tag{3}$$

where $rnd(0, 1)$ is a random number on the interval (0, 1), (w, k_1, k_2) are some coefficients. The speed $v_i(t + 1)$ is considered as a means of changing the solution.

An analogue of the $v_i(t + 1)$ rate is the directed mutation operator (*DMO*), the essence of which is to change the integer values of the genes in the chromosome $H_i(t)$. Moving the p_i particle to a new position means the transition from the chromosome $H_i(t)$ to the new one – $H_i(t + 1)$ with the new integer values of the genes obtained after applying the *DMO*.

As mentioned above, the positions are set by chromosomes. The positions $x^*(t)$, $x_i(t)$, $x_i^*(t)$, $x_i^c(t)$ correspond to the chromosome $H^*(t) = \{g_l^*(t)||l = 1, 2, ..., n_l\}$, $H_i(t) = \{g_{il}(t)|l = 1, 2, ..., n_l\}$, $H_i^*(t) = \{g_{il}^*(t)|l = 1, 2, ..., n_l\}$ $H_i^c(t) = \{g_{il}^c(t)|l = 1, 2, ..., n_l\}$. The number of axes in the solution space is equal to the number n of genes in the chromosomes $H_i(t)$, $H_i^*(t)$, $H^*(t)$. The starting points on each axis l are the integer values of the genes. As an estimate of the degree of closeness between the two positions $x_i(t)$ and $x_z(t)$, we will use the distance S_{iz} (the weight of the affinity link) between the chromosomes $H_i(t)$ and $H_z(t)$. The purpose of moving the chromosome $H_i(t)$ in the direction of the chromosome $H_z(t)$ is to reduce the distance between them.

To account for the simultaneous particle p_i to the best to the position $x^*(t)$, among the particles of the swarm at the moment of time t, and to the best position $x_i^*(t)$ of the particle p_i, which she visited since the beginning of the first iteration, a virtual center is formed (position) attraction $x_i^c(t)$ of the particle p_i. The formation of the virtual position $x_i^c(t)$ is carried out by applying the procedure of virtual movement from the position $x_i^*(t)$ to the virtual position $x_i^c(t)$ towards the position $x^*(t)$. After determining the center of attraction $x^{*}(t)$, the particle p_i moves in the direction of the virtual position $x_i^c(t)$ from the position $x_i(t)$ to the position $x_i(t+1)$. After moving the particle p_i to the new position $x_i(t+1)$, the virtual position $x_i^c(t)$ is eliminated.

The local goal of moving the particle p_i is to reach the position with the best value of the objective function. The global goal of a particle swarm is the formation of an optimal solution to the problem.

4 Setting the Coverage Problem

The algorithms known in the literature [1, 2] of the coating optimize the following indicators: the total cost of the cells covering the scheme; total number of cells required for the implementation of the scheme; the number of types of cells used; the number of intercell connections, the total number of elements included in the covering cell set.

Let the set $E = \{e_i | i = 1, 2, ..., n\}$ be given the types of elements included in the covered functional scheme. The quantitative composition of the scheme according to the types of elements is described by the vector $V = \{v_i | i = 1, 2, ..., n\}$, where v_i is the number of elements of type e_i that make up the scheme. Let a set of cells $T = \{t_j | j = 1, 2, ..., m\}$ be given. The quantitative composition of the cells is described using the matrix $A = \|a_{ij}\|_{n \times m}$, where a_{ij} is the number of elements of type e_i in a cell of type t_j. With the help of the vector $C = \{c_j | j = 1, 2, ..., m\}$, its cost c_j is set for each cell t_j. If all the elements of the circuit will be implemented by the elements contained in the set T, then the circuit is considered covered.

We introduce the integer variable x_j, which determines the number of cells of type t_j that are included in the covering set. The coverage task is formulated as follows:

$$\begin{aligned} &\text{minimize} && F = \sum_{j=1}^{m} x_j \cdot c_j \\ &\text{with restrictions} && \sum_{j=1}^{m} a_{ij} \cdot x_j \geq v_i, i = 1, 2, \ldots, n;\ x_j \geq 0, j = 1, 2, \ldots, m; \\ &&& a_{ij} \geq 0. \end{aligned} \tag{4}$$

Thus, the solution of the problem is a set of values of the parameters $C = \{c_j | j = 1, 2, ..., m\}$, at which the function F (the total cost of the cells of the covering set) has the minimum value. If as an indicator c_j take the total number of elements included in the cell t_j, i.e. $c_j = \sum_{i=1}^{n} a_{ij}$ the function F determines the total number of elements included in the covering set of cells.

5 Formation of Decision Space

For convenience, we will carry out the process of forming the solution space using the example of [6]. Let the covered scheme be composed of 3 types of elements: $E = \{e_1, e_2, e_3\}$. The quantitative composition of the scheme is described by the vector $V = \{30,$

10, 21}. The set of covering cells includes 5 types: $T = \{t_1, t_2, t_3, t_4, t_5\}$. The matrix A, describing the quantitative composition of the cells, has the form shown in Fig. 1.

We introduce the matrix B, which defines the boundary requirements for the quantitative composition of the elements covered by cells of each type. $B = ||b_{ij}||_{n\times m}$, where b_{ij} is the minimum number of elements of type e_i, which must necessarily be covered by cells of type t_j, $b_{ij} \geq 0$, b_{ij} is an integer. At the same time for the implementation of the full coverage of all elements, in accordance with the requirements of the matrix B, the following restrictions must be met:

$$\sum_{j=1}^{m} b_{ij} = v_i, i = 1, 2, \ldots, n. \tag{5}$$

For the above example, one of the possible variants of the matrix B has the form shown in Fig. 1.

Matrix B uniquely corresponds to the covering cell set, which is defined as follows. First, the matrix $D = \|d_{ij}\|_{n\times m}$; the element of the matrix $d_{ij} = \lceil b_{ij}/a_{ij} \rceil$ is the ceiling of the number b_{ij}/a_{ij} – the smallest integer not less than b_{ij}/a_{ij} and actually equal to the minimal number of cells of type t_j, neccssary to cover b_{ij} elements of type e_i.

Then, within each j-th column of the matrix D, the maximum number $d_{j\max}$, is found, which is the minimum number $x_j = d_{jmax}$ of the cells of type t_j, providing coverage v_1 of elements of type e_1; v_1 elements of type e_2, …; δ_n of elements of type e_n in accordance with the requirements of matrix B. Moreover, except for elements of type e_i, for which $d_{ij} = d_{jmax}$, the other types of elements will be covered with excess. For our case, the matrix D and the covering set X has the form shown in Fig. 2.

	Item types	Types of covering cell				
		t_1	t_2	t_3	t_4	t_5
$A =$	e_1	2	1	2	3	2
	e_2	3	2	2	1	2
	e_3	1	2	3	1	2

	Item types	Types of covering cell					$v_i = \sum b_{ij}$
		t_1	t_2	t_3	t_4	t_5	
$B =$	e_1	8	4	4	9	5	$v_1 = 30$
	e_2	3	2	3	1	1	$v_2 = 10$
	e_3	5	2	7	6	1	$v_3 = 21$

Fig. 1. Matrix A and B

	Item types	Types of covering cell				
		t_1	t_2	t_3	t_4	t_5
	e_1	$\lceil 8/2 \rceil=4$	$\lceil 4/1 \rceil=4$	$\lceil 4/2 \rceil=2$	$\lceil 9/3 \rceil=3$	$\lceil 5/2 \rceil=3$
$D =$	e_2	$\lceil 3/3 \rceil=1$	$\lceil 2/2 \rceil=1$	$\lceil 3/2 \rceil=2$	$\lceil 1/1 \rceil=1$	$\lceil 1/2 \rceil=1$
	e_3	$\lceil 5/1 \rceil=5$	$\lceil 2/2 \rceil=1$	$\lceil 7/3 \rceil=3$	$\lceil 6/1 \rceil=6$	$\lceil 1/2 \rceil=1$
	$d_{j\max}$	5	4	3	6	3

	x_1	5
	x_2	4
$X =$	x_3	3
	x_4	6
	x_5	3
	$\sum x_j$	21

Fig. 2. Matrix D and cover set X

In this paper, the solution space is represented by the set R of matrices B. The search for a solution is reduced to the search for such a matrix $B^* \in R$, i.e. to search for the totality of such values of the elements b^*_{ij} of the matrix B^*, which optimize the quality indicator (criterion).

6 Chromosome Structure and Decoding Principles

As mentioned above, the matrix B is considered as a solution. The chromosome structure is designed so that the genes in the same chromosome loci are homologous. We represent the matrix $B = \|b_{ij}\|_{n \times m}$ as a set of B_i rows $B = \{B_i | i = 1, 2, 3, ..., n\}$, where $B_i = \{b_{ij} | j = 1, 2, ..., m\}$. Each row B_i of the matrix B corresponds to the chromosome H_i and vice versa. Each H_i is a set of $(m-1)$ genes, the values of which can vary within the range defined by the parameter v_i:

$$0 < g_{il} < v_i, \qquad g_{il} \in H_i \tag{6}$$

In turn, matrix B is represented by a set of n chromosomes H_i. Consider the mechanisms of encoding and decoding on the example of a single line. Let there is a list $B_i = \langle b_{ij} | j = 1, 2, ..., m \rangle$ with a fixed sum of the values of the elements $v_i = \sum_j (b_{i\,j})$. Such lists underlie the interpretation of solutions in the problems of coverage [11], resource allocation [8, 9], and others. The chromosome $H_i = \{g_{il} | l = 1, 2, ..., (n-1)\}$ is a set of $(n-1)$ genes g_{il}, the values of which can vary within the range defined by the parameter v_i: $g_{il} \in H_i$, $0 \leq g_{il} \leq v_i$.

The process of transition from chromosome H_i to the list is as follows. First, genes are ordered by increasing their values, that is, if $g_{il} \in H_i$ and $g_{il+1} \in H_i$, then $g_{il+1} \geq g_{il}$. Values of g_{il} ranging from 0 to v_i are considered as coordinates of reference points on a segment of length v_i (from 0 to v_i), dividing the segment into intervals. The length of the interval between two adjacent reference points is the value of the corresponding element b_{ij} of the list B_i, and the sum of the values of the elements of the list B_i is v_i. For example. Suppose there is a chromosome $H_i = \{8, 12, 16, 25\}$, $v_i = 30$, $(m-1) = 4$. On a segment of length $v_i = 30$, anchor points with coordinates <0, 8, 12, 16, 25, 30> are plotted, dividing the segment into intervals of length <8, 4, 4, 9, 5> . These values are the values of the B_i list. B_i = <8, 4, 4, 9, 5> , $v_i = 30$.

7 Moving Particles in the Search Space

Let there be a swarm of particles $P = \{p_k | k = 1, 2, ..., K\}$. Each particle p_k in step t is located at position $X_k(t)$. Since matrix B_k contains n rows, position $X_k(t)$, corresponding a particle p_k, is determined by a set of n chromosomes corresponding to n rows of the matrix

B_k, $X_k(t) = \{H_{ki}(t)|i = 1, 2, ..., n\}$. The chromosome $H_{ki}(t) = \{g_{kil}|l = 1, 2, ..., (n-1)\}$ is a collection of $(n\text{-}1)\}$ genes g_{kil}, whose values can vary within the range defined by the parameter v_i: $g_{kil} \in H_{ki}$, $0 \le g_{kil} \le v_i$.

The search space includes the number of axes equal to the number $(n-1)$ of genes in the chromosome $H_{ki}(t)$. Each i-th gene in the chromosome $H_{ki}(t)$ corresponds to an i-th axis (axis number). The starting points on each X_i axis are integers ranging from 0 to v_i. The movement of the particle p_k from the position $X_k(t)$ to the position $X_k(t + 1)$ under the influence of attraction to the position $X_z(t)$ is carried out by applying the directional mutation operator (DMO) to the chromosomes of the position $X_k(t)$ as follows.

As an estimate of the degree of proximity between the two chromosomes $H_{ki}(t)$ $X_k(t)$ and $H_{zi}(t) \in X_z(t)$, we will use the S_{kzi} value of the distance between the chromosomes:

$$S_{kzi} = \sum_{l=1}^{m-1} |g_{kil} - g_{zil}|. \tag{7}$$

As an estimate of the degree of closeness between two positions $X_k(t)$ and $X_z(t)$ the distance between $X_k(t)$ and $X_z(t)$ is used.

$$Q_{kz} = \sum_{i=1}^{n} S_{kzi}. \tag{8}$$

The essence of the transfer procedure implemented by *DMO* is the change in the difference between the values of each pair of genes (g_{kil}, g_{zil}) of two positions $X_k(t)$ and $X_z(t)$: $i = 1, 2, ..., n$; $l = 1, 2, ..., (m-1)$. The chromosomes $H_{ki}(t)$ and $H_{zi}(t)$, chromosomes are sequentially viewed (starting from the first), and the corresponding genes are compared. If in the course of sequential viewing of loci in the current locus l, a "mutation" event occurs with probability P, the $g_{kil} \in H_{ki}(t)$ gene mutates. Let $R_{kzi}(t)$ be the number of loci in the $H_{ki}(t)$ and $H_{zi}(t)$ chromosomes, in which the values of the $g_{kil}(t) \in H_{ki}(t)$ and $g_{zil}(t) \in H_{zi}(t)$ genes do not coincide. The probability of mutation π depends on the number of $R_{kzi}(t)$ mismatches between the positions, and is determined as follows:

$$\pi = \alpha \cdot R_{kzi}(t)/(m-1), \tag{9}$$

where α is the coefficient, $(m-1)$ is the length of the chromosome. Thus, the greater the number of $R_{kzi}(t)$ mismatched genes between chromosomes $H_{ki}(t)$ and $H_{zi}(t)$, the greater the likelihood that the value of $g_{kil}(t) \in H_{ki}(t)$ will be changed.

A simple lottery $L(1,\pi,0)$ is a probabilistic event that has two possible outcomes 1 and 0, the probabilities of which we denote by π and $(1-\pi)$, respectively. In other words, with the probability π lottery $L(1,\pi,0) = 1$, and with probability $(1-\pi)$ the lottery $L(1,\pi,0) = 0$. The values of the genes of the new position of the particle p_i as a result of the mutation are defined as:

$$g_{kil}(t+1) = g_{kil}(t), \text{ if } (g_{kil}(t) = g_{zil}(t));$$
$$g_{kil}(t+1) = g_{kil}(t) + L(1,\pi,0), \text{ if } (g_{kil} < g_{zil}(t)); \quad (10)$$
$$g_{kil}(t+1) = g_{kil}(t) - L(1,\pi,0), \text{ if } (g_{kil} > g_{zil}(t)).$$

Example. Let $b_{ki} = 23$, $m = 6$. $H_{ki}(t) = <1,2,4,8,8>$, $B_{ki}(t) = <1,1,2,4,0,15>$, $H_{zi}(t) = <3,3,4,6,7>$, $B_{zi}(t) = <3,0,1,2,1,16>$, $S_{kzi}(t) = 6$.

Let with some probability in $H_{ki}(t)$ mutated genes in *1* and *4* loci. As a result of directed mutation. $H_{ki}(t+1) = <2, 2, 4, 7, 8>$, $B_{ki}(t+1) = <2, 0, 2, 3, 1, 15>$, $S_{kzi}(t+1) = 4$.

The movement of particles in the search space can be described as a sequence steps.

Step 1. Formulate the initial formulation of the coverage problem.
Step 2. Based on the initial formulation of the coating problem, form a swarm of particles $P = \{p_k\ k = 1, 2, ..., K\}$.
Each particle p_k corresponds to the matrix $B_k = \|b_{kij}\|_{n \times m}$
Step 3. Set the number of iterations T. Set: $t = 1$.
Step 4. For each particle p_k, form the position $X_k(t)$ in the $(m-1)$ – dimensional search space.
Step 5. For each position $X_k(t)$, calculate the value of the objective function $f_k(t)$.
Step 6. Find at step t the position of the $X^*(t)$ swarm with the best value of the objective function $f_k(t)$.
Step 7. To identify for each particle pk, the best position $X^*k(t)$, which she visited from the beginning first iteration.
Step 8. For each particle p_k, establish an estimate of the degree of proximity between the positions $X_k(t)$ and $X^*(t)$.
Step 9. Moving each particle p_k from the occupied position $X_k(t)$ to the position $X_k(t+1)$ under the influence of attraction to a better position $X^*(t)$ of the swarm (social center of attraction) by applying the directed mutation operator (*DMO*) to the chromosomes of the $X_k(t)$ position.
Step 10. If $t < T$, then $t = t + 1$ and go to step 5, otherwise go to step 11.
Step 11. The end of the algorithm.

8 Genetic Search Mechanisms

At the first stage of the hybrid algorithm, the genetic evolution of the population of positions $X = \{X_k | k = 1, 2, ..., K\}$ is performed, in which a swarm of particles $P = \{p_k | k = 1, 2, ..., K\}$ is placed. The position $X_k(t)$, corresponding to the particle p_k, is determined by a set of n chromosomes corresponding to the n rows of the matrix B, $X_k(t) = \{H_{ki}(t) | i = 1, 2, ..., n\}$.

At each generation, crossing-over operators and mutations are first implemented, and then the expanded population is reduced by selective selection, i.e. decrease to initial volume [4]. After selecting a pair of parent positions $X_1(t)$ and $X_2(t)$, pairs of corresponding single rows of the chromosome matrix $(H_{1i}(t), H_{2i}(t))$, $H_{1i}(t) \in X_1(t)$, $H_{2i}(t) \in X_2(t)$, $i = 1, 2, \ldots, n$. Crossover is performed for each pair of chromosomes $(H_{1i}(t), H_{2i}(t))$. Since chromosomes are homologous, crossing over is performed by the mutual exchange of genes between a pair of chromosomes in randomly selected loci. Each chromosome position undergoes a mutation. The mutation operator is performed by changing the relative position of two (randomly selected) genes in the chromosome.

9 Experimental Studies

Based on the considered paradigm, the ***SWARM COVERING*** program has been developed. To carry out studies of the program, a procedure was used to synthesize test cases with a known optimum F_{opt} by analogy with the well-known method [8]. Quality assessment is the value of F_{opt}/F – "degree of quality", where F is the evaluation of the solution obtained. Based on the results of experimental studies, an average dependence of the degree of quality on the number of iterations was constructed (Fig. 3).

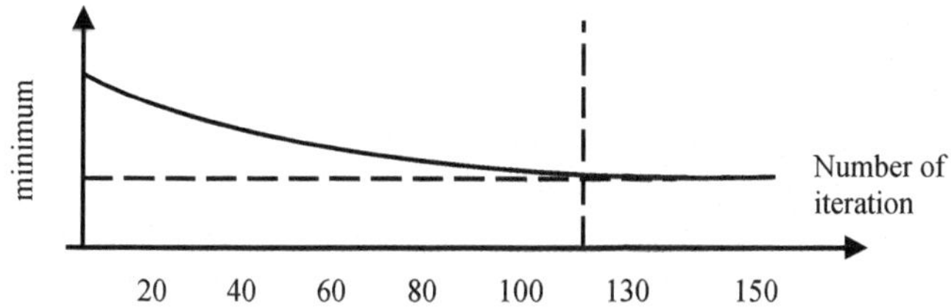

Fig. 3. - Dependence of the degree of quality of the ***SWARM COVERING*** algorithm

The algorithm converges at an average of 125 iterations. As a result of the research conducted, it was established that the quality of the decisions of the hybrid algorithm is 10–15% better than the quality of the solutions of the genetic and swarm algorithms separately. Comparison of the values of the criterion obtained by the hybrid on test examples with a known optimum showed that the probability of obtaining a global optimum was 0.96. On average, the result obtained by the proposed algorithm at the 130th iteration differs from the exact one by 0.15%. On average, the launch of the program provides a solution that differs from the optimal one by less than 2%. A comparison with known algorithms [7–11] showed that, with shorter operating time for the solutions obtained using the *SWARM COVERING* algorithm, the deviation of the objective function from the optimal value is less on average by 6%. The tasks for which the developed algorithm was tested are available in the OR object library [7–11]. The time complexity of the algorithm with fixed values of M and T lies within $O(n)$. The overall estimate of the time complexity lies within $O(n^2)$–$O(n^3)$.

10 Conclusion

The paper uses a symbolic representation of the solution to the problem of coverage in the form of a compact matrix of boundary requirements. A composite architecture of a multi-agent bionic search system based on the integration of swarm intelligence and genetic evolution is proposed.

The key problem that was solved in this paper is related to the development of the structure of the affine space of positions with integer parameter values. In contrast to the canonical particle swarm method, a directed mutation operator was developed to reduce the weight of affinity bonds by moving the particle to a new position of the affine solution space. New chromosome structures have been developed to represent solutions. The probability of obtaining a global optimum was 0.96. The estimate of complexity lies within $O(n^2)–O(n^3)$.

Acknowledgements. This research is supported by grants of the Russian Foundation for Basic Research of the Russian Federation, the project № 18-07-00737 a.

References

1. Zabinyako, G.I.: Implementation of algorithms for solving the problem of covering sets and analyzing their efficiency. Comput. Technol. **12**(6), 50–58 (2007)
2. Karpenko, A.P.: Modern search engine optimization algorithms. Algorithms inspired by nature: a tutorial, p. 446. Publishing House MSTU, M (2014)
3. Wang, X.: Hybrid nature-inspired computation method. Doctoral Dissertation, Helsinki University of Technology, TKK Dissertations, Espoo, p. 161 (2009)
4. Clerc, M.: Particle Swarm Optimization, p. 246. ISTE, London (2006)
5. Lebedev, B.K., Lebedev, O.B.: Hybrid bioinspired algorithm for solving a symbolic regression problem. In: News SFU. Technical science, no. 6(167), pp. 28–41. SFU publishing house, Rostov-on-Don (2015)
6. Lebedev, B.K., Lebedev, V.B.: Coating by the particle swarm method. In: Fizmatlit, M. (ed.) Proceedings of the VI International Scientific and Practical Conference "Integrated Models and Soft Calculations in Artificial Intelligence", pp. 611–619 (2011)
7. Lebedev, B.K., Lebedev, O.B., Lebedeva, E.M.: Resource allocation based on hybrid models of swarm intelligence. Sci. Tech. J. Inf. Technol. Mech. Opt. **17**(6), 1063–1073 (2017)
8. Cong, J., Romesis, M., Xie, M.: Optimality, scalability and stability study of partitioning and placement algorithms. In: Proceedings of the International Symposium on Physical Design, Monterey, CA, pp. 88–94 (2003)
9. Hang, N.M.: Application of the genetic algorithm for the problem of finding coverage of a set. In: Works of the Institute of System Analysis of the Russian Academy of Sciences, vol. 33, pp. 206–221 (2008)
10. Konovalov, I.S., Fathi. V.A., Kobak. V.G.: Application of the genetic algorithm for solving the problem of covering sets. Vestnik of the Don State Technical University, no. 3(86), pp. 125–132 (2016)
11. Esipov, B.A., Muravev, V.V.: Investigation of algorithms for solving the generalized minimum coverage problem. In: Proceedings of the Samara Scientific Center of the Russian Academy of Sciences, vol. 16, no. 4(2). pp. 35–48 (2014)

Fuzzy Models and Systems

Development of a Diagnostic Data Fusion Model of the Electrical Equipment at Industrial Enterprises

Anna E. Kolodenkova(✉), Elena A. Khalikova, and Svetlana S. Vereshchagina

Samara State Technical University, Samara, Russia
anna82_42@mail.ru

Abstract. Continuous monitoring and diagnostics of the equipment technical condition are needed to improve reliability, to prevent possible failures, to ensure the service life extension of electrical equipment (EE) at industrial enterprises. In the present work, the authors suggest to use a diagnostic data fusion model developed for the EE technical condition diagnosis. To test the model, a scenario for searching the EE failure state was made and implemented. A diagnostic data fusion model is necessary to process the increasing amount of information produced by various EEs for subsequent analysis. The proposed data fusion model uses the levels of the Joint Directors of Laboratories (JDL) model, Data Mining technology, the ontology storage and EE diagnostics and prediction models and methods based on probabilistic statistical methods and soft computing methods. A detailed description of a fault detection model for EE at an oil company is considered. The developed diagnostic data fusion model will make it possible to identify EE faulty states and failures, as well as to increase the efficiency of making diagnostic decisions under the conditions of heterogeneous data obtained from a lot of EEs.

Keywords: Electrical equipment · Heterogeneous data · Data fusion · Soft computing methods

1 Introduction

EE failures at industrial enterprises that occur at the operation stage may become sources of severe EE damages, power supply failure for consumers, as well as emergency situations followed by significant economic damage [1–3]. Such failures are due to the processes of gradual deterioration of their properties leading to the expired service life and loss of efficiency. One of the ways to address this problem is the automation of the monitoring and diagnosing process of which involves detection of defects, failures and malfunctions, analysis of the EE technical condition, forecasting the values of EE parameters in various EE types and others.

The increase in the number of different EE types and the heterogeneous diagnostic data has resulted in the need to use data fusion technology which is becoming increasingly popular. The development of data fusion technology for EE diagnosis

S. Kovalev et al. (Eds.): IITI 2019, AISC 1156, pp. 499–506, 2020.
https://doi.org/10.1007/978-3-030-50097-9_50

gives significant advantages in comparison to separate processing due to the expansion of the volume of heterogeneous data and synergistic effect.

However, despite the numerous applications of heterogeneous data fusion technology, there is a number of key problems such as data heterogeneity, equipment accuracy and the choice of data fusion method [4, 5].

Diagnosis of EE, in turn, provides solving a wide range of various tasks: to identify the current condition of the equipment; detect existing malfunctions; determine the causes of malfunctions; predict the technical condition; make a scientifically based decision on equipment malfunction and provide recommendations on their elimination. However, such a diagnosis is significantly complicated due to the following problems: the complexity of the monitored equipment; inadequate introduction of modern methods of "soft computing"; limited information available about EE; heterogeneous information [6–8].

For these reasons, the authors suggest the development of a diagnostic data fusion model of electrical equipment at industrial enterprises which will make it possible to detect EE faulty states and reduce the time for making decisions when diagnosing EE.

2 Analysis of the Problem and Research Objectives

Currently, a fairly large number of structural models of data fusion have been proposed, they are divided into three types (data-based model, action-based model and role-based model) and can be widely used in EE diagnosing in any industry (Fig. 1) [4, 9].

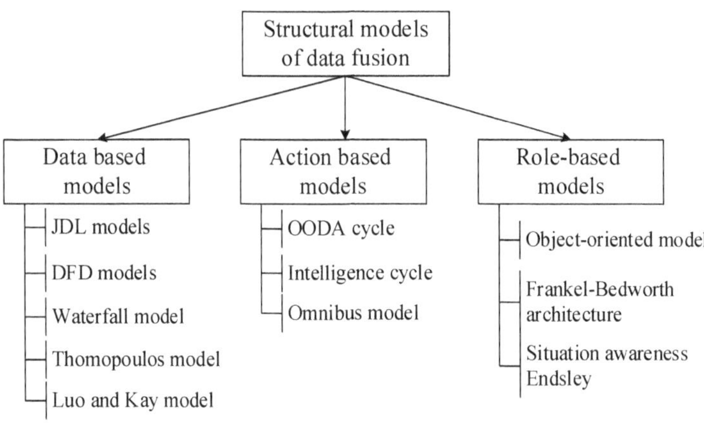

Fig. 1. Classification of data fusion models.

However, it is rather difficult to define a universal data fusion model for a specific use of EE, since collecting data from several different types of equipment is a complex challenge. For this reason, one can use several fusion models or their combinations for efficient data collection and processing. A typical example is the Complex model which is a combination of three data fusion models: the OODA (O – observe, O – orient, D – decide, A – act) loop, the Waterfall model and the DFD (Data-Feature-Decision).

Also, some aspects should be taken into account, for example, what kind of data will be collected and at what stage; what is the purpose of the use of a particular model in terms of requirements and future use; at what stage the model will make decisions. For example, the JDL model and the Waterfall model can be applied for work at a high level, but their use is also possible at a low level; OODA loop should be used for the activities of individuals, individuals and organizations in a competitive environment; Endsley's situation awareness model is used when a specialist providing effective safety management of an organization has a deep knowledge of the current situation (situation awareness).

The EE diagnosis is studied in the papers of many Russian and foreign authors [10–13]). However, despite the significant number of works, this problem is still open. First, the existing approaches are based only on the use of statistical information (limited application of artificial intelligence technologies), thereby neglecting experience and knowledge which leads to a lack of accurate and complete information about all EE parameters; secondly, not all models take into account environmental factors (for example, climatic conditions) affecting EE operation; thirdly, in the majority of models the actual values of the diagnosed parameters (not their dynamics) are considered; fourth, most of the developed methods are designed for a specific EE.

The systemic nature of this problem determines the necessity to develop a diagnostic data fusion model for EE at industrial enterprises which could provide more complete and accurate data on the operating system and support making effective diagnostic solutions at the EE operation phase.

3 Diagnostic Data Fusion Model of the Electrical Equipment

In the developed diagnostic data fusion model of EE, the levels of the JDL model are used (a detailed description of which can be found in [4, 14]), as well as Data Mining technologies which provide detecting hidden patterns and unknown knowledge in raw data for interpretation and useful at making managerial decisions (decision trees; nearest and k-nearest neighbor methods; support vector machine; correlation and regression analysis) and methods and models for EE diagnosing and predicting based on probabilistic-statistical methods (Bayesian networks, Dempster-Shafer's theory) and soft computing methods (fuzzy set theory, neural networks). Next, we determine the levels of the proposed model for the heterogeneous data merge (that are simple and understandable) as applied to the diagnosis of the EE. In order to understand further the material in Fig. 2 presents a model for merging the heterogeneous data of EE which was considered in detail in the authors earlier work [15].

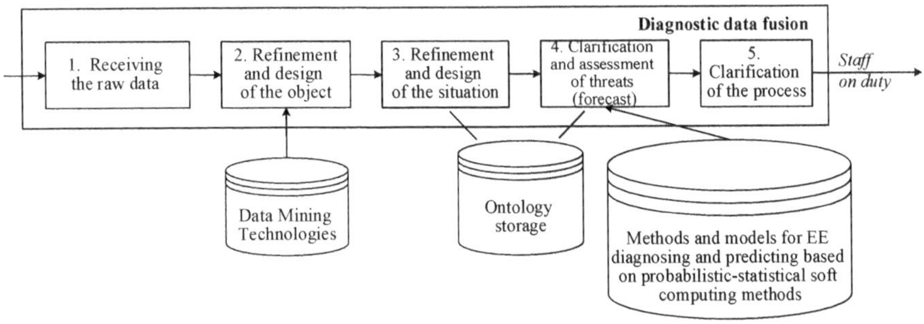

Fig. 2. A diagnostic data fusion model of EE.

The first level includes the obtaining the raw data from the EE, regulatory documentations, State Standards, technical data sheets, etc.

At the second level, the necessary data for EE diagnostics (parameter names, parameter values, etc.) is extracted from the data obtained at the first level. Next, this data is analyzed to reduce the data size without reducing the informational value, if there is measurement repeatability. This level also includes the determination of the data collection method.

Data analysis includes the following:

1. filtering: elimination of redundant data obtained from different types of EE and elimination of gross errors;
2. normalization: debugging, bringing numerical values to the single unit of measurement and the order of magnitude;
3. data classification using Data Mining technology.

Note that each specific data format should have its own specific, appropriate methods. The situation is understood as a set of linked data that is related to the specific equipment.

At the third level, a contextual description of the relationship between the EE parameters and the observed situation is carried out using the ontology storage. At this level, a priori information, knowledge and information about the environment are used. Note that the ontology storage which includes fuzzy and fuzzy-functional models is not considered in this article.

At the fourth level, an assessment of the current situation (identification of possible threats, generation of new knowledge) is carried out. At this stage, probabilistic statistical methods, soft computing methods and ontology storage are used. This task is difficult because it is connected to complex calculations.

The fifth level shows the results of the two previous levels in the form of graphs, tables and diagrams to staff on duty. Note that the proposed model considers the diagnostic data fusion from the point of view of a systematic approach, and it is not attached to any industry. To test the proposed diagnostic data fusion model, a scenario of EE faulty state detecting was made and implemented, relevant to the levels of the JDL model using knowledge of the staff on duty, methods and models of EE faulty state detecting based on probabilistic methods and soft computing methods (Fig. 3).

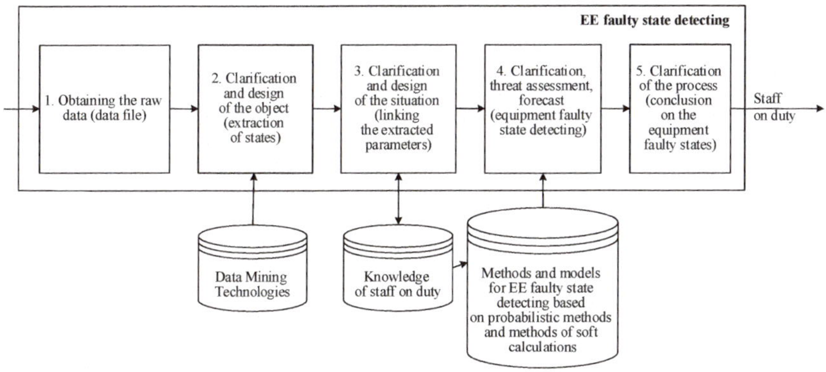

Fig. 3. Fault detecting model of EE.

At the first stage, a file in the.xls format (Fig. 4) is obtained which is a table that contains the state of a specific piece of EE (test time, names of equipment parameters and their values).

	A	B	C	D	E	F	G	H
1	Время	Маркир.	$U_{A(1)}$, B	$U_{B(1)}$, B	$U_{C(1)}$, B	$I_{A(1)}$, A	$I_{B(1)}$, A	$I_{C(1)}$, A
2	12.10.2018 10:30:00	да	226,39	206,77	194,73	47,79	45,04	78,43
3	12.10.2018 10:30:03	да	226,69	206,14	192,98	47,07	44,55	80,17
4	12.10.2018 10:30:06	да	226,59	206,24	193,19	47,05	44,48	80,69
5	12.10.2018 10:30:09	да	226,11	205,65	192,6	49,35	46,05	82,44
6	12.10.2018 10:30:12	да	225,94	205,56	192,14	48,05	44,76	81,15
7	...							
2401	12.10.2018 12:29:51	нет	219,74	211,34	210,04	49,95	47,76	65,87
2402	12.10.2018 12:29:54	нет	219,72	211,63	209,9	50,12	48,07	65,97
2403	12.10.2018 12:29:57	нет	219,95	211,92	209,95	50,1	48,24	66,26
2404	12.10.2018 12:30:00	нет	219,91	211,99	209,51	49,8	47,14	65,9

Fig. 4. Fragment of the table with the raw data.

The raw data was taken at the oil company and the values of actual measurements of voltage and current (the tests were carried out at different times of the day) were transmitted to the substation on the low-voltage side.

At the second stage, the relevant equipment parameters (names of equipment parameters and their values) necessary for further EE diagnosis based on the experience and knowledge of the staff on duty are extracted from the file (Fig. 5).

	A	C	F
1	Время	$U_{A(1)}$, B	$I_{A(1)}$, A
2	12.10.2018 10:30:00	226,39	47,79
3	12.10.2018 10:30:03	226,69	47,07
4	12.10.2018 10:30:06	226,59	47,05
5	12.10.2018 10:30:09	226,11	49,35
6	12.10.2018 10:30:12	225,94	48,05
7	...		
2401	12.10.2018 12:29:51	219,74	49,95
2402	12.10.2018 12:29:54	219,72	50,12
2403	12.10.2018 12:29:57	219,95	50,1
2404	12.10.2018 12:30:00	219,91	49,8

Fig. 5. Fragment of extracted parameters and their values.

At the third stage, the extracted data is supplemented and supported by knowledge from the duty on staff.

At the fourth stage, the detection of the EE faulty state is carried out using the method based on the Dempster-Shafer theory described in the paper [16].

The concept of the method is to calculate the degree of confidence $\beta_j(x_i)$ for the j-EE that allows one to evaluate the state of the equipment $x_i(i = 1, \ldots, N, j = 1, \ldots, M)$ and the probability mass $m_j(x_i)$ of troubleshooting diagnosis for the equipment using the normalized weighting factor which reflects the sensitivity of the equipment $w_j^{nor} = [w_{j1}, w_{j2}, \ldots, w_{ji}]$ (w_{ji} is a measure of similarity between the i-type of malfunction and the j-sensor) formed on the basis of some preliminary knowledge of each piece of EE and is presented in the form of intervals, fuzzy triangular and trapezoidal numbers in order to combine the mass probability values $m_1(x_i) \oplus m_2(x_i) \oplus \ldots \oplus m_j(x_i)$.

Table 1 shows a fragment of the results of probability masses and their combination calculations for each state based on weight factors.

Table 1. Calculated probability mass values.

Equipment state	$\beta_1(x_i)$	$m_1(x_i)$	$\beta_2(x_i)$	$m_2(x_i)$	$\beta_3(x_i)$	$m_3(x_i)$	$m_1(x_i) \oplus m_2(x_i) \oplus m_3(x_i)$
x_1	0.279	0.051	0.401	0.023	0.385	0.026	0.11
x_2	0.758	0.189	0.954	0.041	0.533	0.073	0.24
…	…	…	…	…	…	…	…
x_{23}	0.513	0.011	0.673	0.113	0.289	0.271	**0.42**
x_{24}	0.632	0.045	0.974	0.078	0.594	0.347	0.36

At the fifth stage, information about the EE faulty state is produced.

In the Table 1 you can see that the EE faulty state in the oil industry cannot be detected when considering three different pieces of equipments separately. After the data fusion, it is revealed that the faulty state is x_{23} with the mass of probability 0.42. Note that the state with the highest expected value is selected as the EA faulty state.

4 Conclusion

This paper presents a diagnostic data fusion model based on the levels of the JDL model, and it is used as a general framework for defining the levels of data processing and their interconnection. To test the proposed diagnostic data fusion model, a scenario for finding the EE faulty state of the oil industry was made and implemented.

When implementing the scenario using the knowledge obtained from the staff on duty and the developed method based on the Dempster-Shafer theory, the faulty state of the equipment was detected. Further studies will be focused on the development of scenarios for diagnosing and predicting EE using an ontology storage which includes fuzzy and fuzzy-functional models. The developed diagnostic data fusion model will provide obtaining more comprehensive and accurate data on EE and making scientifically based management decisions.

Acknowledgement. The work was supported by RFBR (Grants No. 19-07-00195, No. 19-08-00152).

References

1. Abramov, O.V.: Monitoring and forecasting of the technical condition of systems of responsible appointment. Inform. Control Syst. **2**(28), 4–15 (2011)
2. Voloshin, A.A., Voloshin, E.A.: Forecasting the technical condition of the equipment and managing the stability of the energy system through technology of the internet of things for monitoring in electric networks of low. Int. J. Humanit. Nat. Sci. **12**, 128–134 (2017)
3. Horoshev, N.I., Eltishev, D.K.: Integrated assessment and forecasting of technical condition of the equipment of electrotechnical complexes. Inform. Control Syst. **4**(50), 58–68 (2016)
4. Kovalev, S.M., Kolodenkova, A.E., Snasel, V.: Intellectual technologies of data fusion for diagnostics technical objects. Ontol. Designing **9**(1), 152–168 (2019)
5. Alofi, A., Alghamdi, A., Alahmadi, R., Aljuaid, N., Hemalatha, M.: A review of data fusion techniques. Int. J. Comput. Appl. **167**(7), 37–41 (2017)
6. Khramshin, V.R., Nikolayev, A.A., Evdokimov, S.A., Kondrashova, Y.N., Larina, T.P.: Validation of diagnostic monitoring technical state of iron and steel works transformers. In: IEEE NW Russia Young Researchers in Electrical and Electronic Engineering Conference (EIConRusNW), pp. 596–600 (2016)
7. Saushev, A.V., Sherstnev, D.A., Shirokov, N.V.: Analysis of methods for diagnosing high-voltage apparatus. Bull. Admiral Makarov State Univ. Maritime Inland Shippin **9**(5), 1073–1085 (2017)
8. Bulac, C., Tristiu, I., Mandis, A., Toma, L.: On-line power systems voltage stability monitoring using artificial neural networks. In: International Symposium on Advanced Topics in Electrical Engineering, pp. 622–625 (2015)
9. Nakamura, E.F., Loureiro, A.A., Frery, A.C.: Information fusion for wireless sensor networks: methods, models, and classifications. ACM Comput. Surv. **39**(3), 55 (2007)
10. Pareek, S., Sharma, R., Maheshwari, R.: Application of artificial neural networks to monitor thermal condition of electrical equipment. In: 3rd International Conference on Condition Assessment Techniques in Electrical Systems (CATCON), pp. 183–187 (2017)
11. Eltyshev, D.K.: On the development of intelligent expert diagnostic system for assessing the conditions of electrical equipment. Syst. Methods Technol. **3**(35), 57–63 (2017)

12. Modern methods of diagnostics and assessment of the technical condition of electric power equipment. https://niitn.transneft.ru/u/section_file/246601/22.pdf. Accessed 10 May 2019
13. Vdoviko, V.P.: Methodology of the high-voltage electrical equipment diagnostics system. Electricity **2**, 14–20 (2010)
14. Garcia, J., Rein, K., Biermannn, J., Krenc, K., Snidaro, L.: Considerations for enhancing situation assessment through multi-level fusion of hard and soft data. In: 19th International Conference on Information Fusion (FUSION), Heidelberg, pp. 2133–2138 (2016)
15. Kolodenkova, A., Khalikova, E., Vereshchagina, S.: Data fusion and industrial equipment diagnostics based on information technology. In: International Multi-Conference on Industrial Engineering and Modern Technologies (FarEastCon), Vladivostok, Russia, pp. 1–5 (2019)
16. Kolodenkova, A.E., Dolgiy, A.I.: Diagnosing of devices of railway automatic equipment on the basis of methods of diverse data fusion. Adv. Intell. Syst. Comput. **875**, 277–283 (2019)

Temporal Reachability in Fuzzy Graphs for Geographic Information Systems

Alexander Bozhenyuk[1(✉)], Stanislav Belyakov[1], Margarita Knyazeva[1], and Vitalii Bozheniuk[2]

[1] Southern Federal University, Nekrasovsky 44, Taganrog 347922, Russia
avb002@yandex.ru, beliacov@yandex.ru, margarita.knyazeva@gmail.com

[2] RWTH Aachen University, Templergraben 55, 52056 Aachen, Germany
square.nabla@gmail.com

Abstract. In this paper the idea of temporal reachability in fuzzy directed graph is introduced. Temporality in fuzzy graphs is a generalization of a fuzzy graph on temporal relations, where the incidence property of graph vertices is changed in the discrete time. Reachability in such temporal graph structures refers to the ability to get from one vertex to another within different time moments and the reachability relation on fuzzy directed graph is the transitive closure on the edge set. Efficient processing queries and graph structured information under uncertainty is an important research topic especially in XML databases, geographical information systems (GIS), web mining and etc. Fuzzy temporal graph models are usually applied in geographical information systems to formalize relations between objects and to check connections between them. A method for searching temporal reachability in fuzzy graph structure is proposed, an example of finding a fuzzy reachability set in temporal graph is considered as well.

Keywords: Fuzzy temporal graph · Geographic information systems · Reachability degree · Fuzzy path · Fuzzy transitive closure

1 Introduction

Geographical information technologies (or GIS) are aimed at creating information models of objects, events and features of the real world by visualization techniques and they give an opportunity to manipulate graphical data. Large directed graphs stands as data source, environmental model and graphical user interface for specialists. Representation of objects and further manipulation with them is impossible without temporal variables.

The analytical methods used for geographical information systems (GIS) include discrete and continuous methods of representing and manipulating objects [1, 2]. Graphs and graph theory methodology are used for visualization, analysis and synthesis of GIS models [3, 4] as a universal way of relation-oriented modeling.

Geographical information models are usually temporally dynamic. Temporal scenarios presuppose changes in spatial coordinates, time labels, modelling durations of actions and provide dynamics characteristics for GIS models. Uncertainty and

S. Kovalev et al. (Eds.): IITI 2019, AISC 1156, pp. 507–516, 2020.
https://doi.org/10.1007/978-3-030-50097-9_51

incompleteness of the data, the process map construction that involves several sources of cartographic information and geodatabases usage, data acquisition from unreliable sources outside GIS and other factors can introduce categories of uncertainty and fuzziness into GIS model [5, 6].

When using graphs, it is traditionally assumed that the reachability of vertices is uniquely determined by a given graph invariant. In some cases this is not so [7]. The weights of the individual edges of the graph can change in time, in the process of modeling, which forces us to take a different look at the procedures for analyzing and synthesizing graphs, algorithms for solving optimization problems on graphs.

2 Fuzzy Temporal Graph and Geographic Information Systems

Let us consider an example illustrating the listed features of statement of tasks. In Fig. 1 shows a map, on the basis of which the scheme is constructed.

Fig. 1. Fragment of the network and road map

According to the scheme, a graph of the road network can be constructed, which will solve the problems of finding the shortest path, evaluate flows, find service centers, determine connectivity, and solve other problems known from graph theory [3, 4]. At the same time, for a given train schedules, it is impossible to solve the task of transporting an additional (out-of-schedule) composition between a given pair of stations in the minimum time in the traditional way. Individual sections of the path at some time intervals in the network graph are not included. Here the following tasks can be posed:

- what is the shortest path from the vertex *A* to the vertex *B* if the time interval of the beginning of the motion from the vertex *A* is given;
- what is the shortest path from the vertex *A* to the vertex *B* if the time interval of arrival at the vertex *B* is given.

The above statements are in practice related to the uncertainty of the time and metric characteristics of the transport network.

The complexity of graph models grows in the case of solving logistics problems. In Fig. 2 shows an example of a logistics network describing possible directions for the movement of material flows. The weight of each arc is determined in a complex manner, based on the properties of the object being moved, the spatial environment of the transport network, and the dynamics of processes affecting transportation.

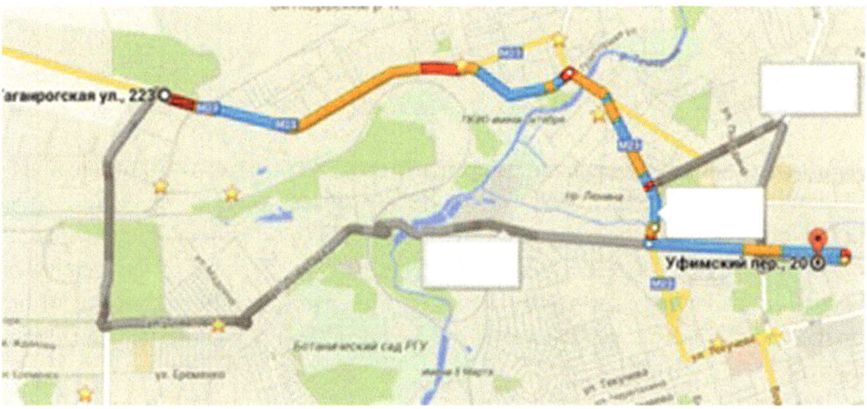

Fig. 2. Example of logistics network for road transport

Thus, the solution of a number of problems using geoinformation models requires a specific - temporal description of the graph [8, 9]. The behavior is described by a function of time. We can specify the following models for describing behavior:

- analytical model. A well-known drawback of this approach is the need to apply simplifications that make it possible to apply mathematical formalisms. Simplifications result in model inadequacy;
- statistical model. This model is applicable in cases where the researcher has the necessary means of observation under unchanging conditions. In practice, this is difficult to achieve. But improving the technical means of Earth observation makes it possible to talk about the prospects of such an approach;
- expert logical model. This model is built on the basis of individual experience of experts. Its drawback is subjectivism. Nevertheless, expert knowledge in certain cases gives a more adequate result than a complete formalization [10];
- cart metric model. A measurement on a map is a generalized operation that involves estimating the relationships between objects on the map. An electronic map is considered as a source of metric and non-metric data. The modern structural

organization of GIS makes it possible to associate with any cartographic objects any complex behavior, maintains communication with any external data sources of local and global networks. There is an idea of models of a combined nature, which includes the collection of real information, the accumulation of experience and the use of the logic of experts.

The latest version of the model is the most promising. Its implementation requires the study of the apparatus of analysis and synthesis of the theory of fuzzy temporal graphs, which, on the one hand, would have the property of fuzziness, and on the other hand, would change in time [11, 12]. It should be noted that the concept of temporal graph itself is interpreted in the literature in a rather wide range - from time plots to oriented acyclic graphs and Petri nets [13–18].

In this paper, a fuzzy temporal graph is considered, in which fuzzy connections between the vertices of the graph [12] vary in discrete time [19].

3 Basic Definitions

Definition 1. [20, 21] A temporal fuzzy graph is a triple $\tilde{G} = (X, \tilde{U}_t, T)$, where X is the set of vertices of the graph c with the number of vertices $|X| = n; T = \{1, 2, \ldots, N\}$ is the set of natural numbers defining (discrete) time; $\tilde{U}_t = \{<\mu_t(x_i, x_j)/(x_i, x_j)>\}$ - fuzzy set of edges, where x_i, $x_j \in X$; $\mu_i(x_i, x) \in [0,1]$ is the value of the membership function μ_t for the edge (x_i, x_j) at time $t \in T$. Moreover, for different instants of time for the same edge (x_i, x_i), the values of the membership function are, in general, different.

Example 1. Consider a fuzzy temporal graph $\tilde{G} = (X, \tilde{U}_t, T)$ for which the set of vertices $X = \{x_1, x_2, x_3, x_4, x_5, x_6\}$, time $T = \{1, 2, 3\}$, $n = 6$, $N = 3$, and fuzzy set of edges is given in the form:

$$\tilde{U}_t = \{<0,8_1/(x_1, x_2>, <1_2/(x_1, x_2)>, <0,5_1/(x_2, x_1)>, <1_2/(x_2, x_1)>, <0,7_2/(x_2, x_3)>,$$
$$<0,8_3/(x_2, x_3)>, <1_1/(x_3, x_2)>, <1_3/(x_3, x_2)>, <0,6_1/(x_3, x_5)>, <1_3/(x_3, x_5)>, <0,8_1/(x_5, x_3)>,$$
$$<0,9_3/(x_5, x_3)>, <0,9_2/(x_5, x_6)>, <1_3/(x_6, x_5)>, <0,8_1/(x_1, x_4)>, <1_3/(x_1, x_4)>, <0,8_1/(x_4, x_1)>,$$
$$<1_2/(x_4, x_1)>, <0,9_2/(x_4, x_5)>, <0,6_1/(x_5, x_4)>, <1_3/(x_5, x_4)>\}.$$

The graphically fuzzy temporal graph can be defined as a fuzzy oriented graph, on the edges of which the values of the membership function μ_t at the instants $t \in T$ are indicated.

The graph considered in Example 1 has the form shown in Fig. 3.

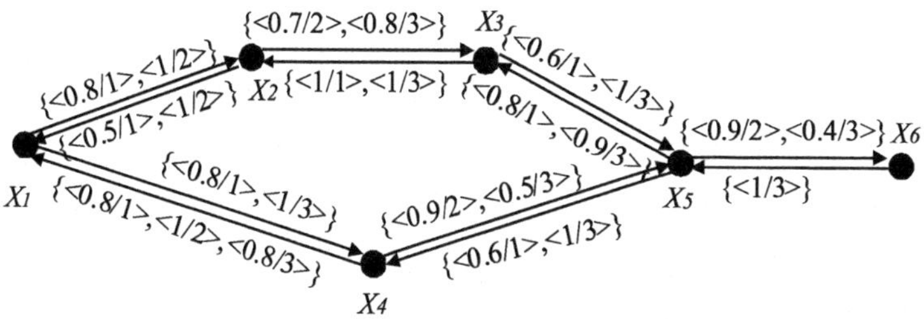

Fig. 3. Example of fuzzy temporal graph for $T = \{1,2,3\}$.

Definition 2 [22]. The vertex x_j is a fuzzy adjacent vertex x_i with respect to the time instant $t \in T$ if the following condition is satisfied: $\mu_t(x_i, x_j) > 0$.

The directed fuzzy path $\tilde{L}(x_i, x_k)$ of the fuzzy temporal graph is called the directed sequence of fuzzy edges leading from the vertex x_i to the vertex x_k, in which the finite vertex of any edge different from the last one is the initial vertex of thc next edge:

$$\tilde{L}(x_i, x_k) = \, <\mu_{t_1}(x_i, x_1)/(x_i, x_1)>, <\mu_{t_2}(x_1, x_2)/(x_1, x_2)>, <\mu_{t_k}(x_{k-1}, x_k)/(x_{k-1}, x_k)>, \quad (1)$$

for which the following conditions are satisfied:

$$\mu_{t_1}(x_i, x_1) > 0, \mu_{t_2}(x_2, x_3) > 0, \ldots, \mu_{t_k}(x_{k-1}, x_k) > 0, \quad (2)$$

and for the instants of time $t_1, t_2, \ldots, t_k \in T$ the following inequality holds:

$$t_1 \leq t_2 \leq \ldots \leq t_k. \quad (3)$$

In other words, if in a fuzzy path (1) each successive vertex is an adjacent previous vertex by a moment of time no less than the moments at which all the previous vertices in this sequence are fuzzy adjacent.

The conjunctive strength of the path $\tilde{L}(x_i, x_k)$ is determined by the expression:

$$\mu_{\&}(\tilde{L}(x_i, x_k)) = \underset{t_1, t_2, \ldots, t_k}{\&} \mu_{t_j} < x_i, x_j > .$$

Definition 3. A fuzzy path $\tilde{L}(x_i, x_k)$ will be called a simple path between the vertices x_i and x_k if its part is not any other fuzzy path between the same vertices x_i and x_k.

It is obvious that this definition coincides in form with the definition for a crisp graph.

The vertex x_k is fuzzy achievable from the vertex x_i in the fuzzy temporal graph if there is an oriented fuzzy path $\tilde{L}(x_i, x_k)$ from the vertex x_i to the vertex x_k.

Definition 4. The value t_k is called the reachability time of the vertex x_k from the vertex x_i, and the value $\mu_{\&}(\tilde{L}(x_i, x_k))$ is called the reachability degree by the path $\tilde{L}(x_i, x_k)$.

Let there exist several sequences $\tilde{L}$ of the form (1) from the vertex x_1 to the vertex x_k, then the values $i_{k-1}^{(j)}$ for each sequence can be different. The smallest of these values is called a minimal time reachability $t_{\min}(x_1, x_k)$ vertex x_k from the vertex x_1, that is:

$$t_{\min}(x_1, x_k) = \min_{j=\overline{1,L}}\{i_{k-1}^{(j)}\}, \tag{4}$$

and the corresponding value $\mu(t_{\min})$ will be called the degree of attainability at the minimum time.

Example 2. In the fuzzy temporal graph shown in Fig. 3, the vertex x_3 is reachable from the vertex x_1 by the sequence $seq_1 = (x_1, x_2, x_3)$ with a reachability degree 0.7 with reachability time $t = 2$, and with a reachability degree 0.8 with reachability time $t = 3$; is achievable with the sequence $seq_2 = (x_1, x_4, x_5, x_3)$ with a reachability degree 0.8 for $t = 1$ and with a reachability degree 0.9 for $t = 3$. Therefore, the value $t_{\min}(x_1, x_3) = 1$, and value $\mu(t_{\min}) = 0.8$.

Let $\{\tilde{L}(x_i, x_j)\}$ be a family of fuzzy paths with the help of which the vertex x_j is accessible from the vertex x_i. We denote by $\alpha_t^{i,j}$ the greatest degree of attainability of some path from vertex x_i to vertex x_j with time $t \in \overline{1,T}$.

Definition 5. A fuzzy set $\tilde{A}(x_i, x_j) = \{<\alpha_t^{i,j}/t> | t \in \overline{1,n}\}$ is called a fuzzy temporal set of reachability of the vertex x_j from the vertex x_i.

We will assume that each vertex is attainable from itself with degree 1 at any instant of time $t \in \overline{1,T}$. That is, the following is true:

$$(\forall i \in \overline{1,n})[\tilde{A}(x_i, x_i) = \{<1/t> | t \in \overline{1,T}\}].$$

Definition 6. A family of fuzzy sets $\tilde{\Gamma}(x_i) = \{\tilde{A}(x_i, x_j) | j = \overline{1,n}\}$ is called the temporal transitive closure of the vertex x_i.

The set $\tilde{\Gamma}(x_i)$ determines the temporal degree of reachability from the vertex x_i to each vertex $x_j \in X$.

We introduce the operation of temporal intersection of fuzzy sets to calculate a fuzzy temporal set of reachability.

Let $\tilde{C}_1 = \{<\mu_1(t)/t>\}$ and $\tilde{C}_2 = \{<\mu_2(t)/t>\}$ be fuzzy sets, $t \in \overline{1,T}$.

Definition 7. The temporal intersection of the sets $\tilde{C}_1$ and $\tilde{C}_2$ is the set $\tilde{C}_1 \cap_t \tilde{C}_2 = \{<\mu_{\cap t}(t)/t>\}$ in which the membership function $\mu_{\cap t}$ is defined as: $\mu_{\cap t}(t) = \min\{\max_{\tau=\overline{1,t}}\{\mu_1(\tau)\}, \mu_2(t)\}$.

Property. $\tilde{C}_1 \cap_t \tilde{C}_2 \neq \tilde{C}_2 \cap_t \tilde{C}_1$.

Example 3. Let $\tilde{C}_1 = \{<0.8/1>, <0.6/2>, <0.4/3>\}$ and $\tilde{C}_2 = \{<0.3/1>, <0.7/2>, <0.9/3>\}$. Then $\tilde{C}_1 \cap_t \tilde{C}_2 = \{<0.3/1>, <0.7/2>, <0.8/3>\}$ and $\tilde{C}_2 \cap_t \tilde{C}_1 = \{<0.3/1>, <0.6/2>, <0.4/3>\}$.

4 Method for Determining the Temporal Reachability in Fuzzy Graph

Consider a method for finding a transitive closure of fuzzy temporal graph. Let us consider this method using the example of the graph presented in Fig. 3. The incidence matrix of this graph is shown in Fig. 4:

	x_1	x_2	x_3	x_4	x_5	x_6
x_1	{1/1,1/2,1/3}	{0.8/1,1/2}	∅	{0.8/1,1/3}	∅	∅
x_2	{0.5/1,1/2}	{1/1,1/2,1/3}	{0.7/1,0.8/2}	∅	∅	∅
$R = x_3$	∅	{1/1,1/3}	{1/1,1/2,1/3}	∅	{0.6/1,1/3}	∅
x_4	{0.8/1,1/2,0.8/3}	∅	∅	{1/1,1/2,1/3}	{0.9/2,0.5/3}	∅
x_5	∅	∅	{0.8/1,0.9/3}	{0.6/1,1/3}	{1/1,1/2,1/3}	{0.9/2,0.4/3}
x_6	∅	∅	∅	∅	{1/3}	{1/1,1/2,1/3}

Fig. 4. Incidence matrix of fuzzy temporal graph $\tilde{G}$

We introduce three columns. The first column A corresponds to the elements of $\tilde{A}(x_1, x_i)$ sets. Elements of the second column Pr indicate that the line has already been viewed (1), or not yet (0). The elements of the third column Y contain the number of the previous vertex. The initial values of these columns are shown in Fig. 5.

A	Pr	*Y*
∅	0	0
∅	0	0
∅	0	0
∅	0	0
∅	0	0
∅	0	0

Fig. 5. Initial values of columns

- According to the algorithm, we select an arbitrary vertex, for example x_1. Assign value $A(1) := \{<1/1>, <1/2>, <1/3>\}$, $Y(1) := 1$.
- We select the first row of the matrix R for which Pr (i) = 0 (the row has not yet been viewed) and $A(i) \neq \emptyset$ (there is some path from the vertex x_1). At this step, this is the

first row (i = 1). In the first row of the matrix R we find the elements (r_{12}, r_{14}), which are not equal $\emptyset$. The number of the previous vertex is written in the corresponding elements of the column Y: Y(2): = 1; Y(4): = 1. Then, in the corresponding elements of column A, we write down of reachability degree of the vertex from the vertex x_1:

$$A(2) := A(2) \cup A(Y(2)) \cap_t r_{12} = \{<0.8/1>, <1/2>\},$$

$$A(4) := A(4) \cup A(Y(4)) \cap_t r_{14} = \{<0.8/1>, <1/3>\}.$$

- Assign the value of Pr(1): = 1 (row 1 of the matrix has been viewed).
- Then we choose the first row of the matrix R for which Pr (i) = 0 (the row has not yet been viewed) and $A(i) \neq \emptyset$ (there is some path from the vertex x_1). At this step, this is the second row (i = 2). In the second row of the matrix R we find the elements, which are not equal $\emptyset$. This is the element r_{23}. The number of the previous vertex is written in the corresponding element of the column Y: Y(3): = 2. Then, in the corresponding element of column A, we write down of reachability degree of the vertex from the vertex x_1:

$$A(3) := A(3) \cup A(Y(3)) \cap_t r_{23} = \{<0.7/2>, <0.8/3>\}.$$

- Assign the value of Pr(2): = 1 (row 2 of the matrix has been viewed).
- The process continues until all the elements of the vector Pr become equal to 1 (all rows were viewed).
- Make the same steps and for all columns of the matrix R.

As a result, we get the column $A = \tilde{\Gamma}(x_1)$, which is shown in Fig. 6:

$A = \tilde{\Gamma}(x_1)$	Pr	Y
$\tilde{A}(x_1, x_1) = \{<1/1>, <1/2>, <1/3>\}$	1	0
$\tilde{A}(x_1, x_2) = \{<0.8/1>, <1/2>, <0.8/3>\}$	1	1
$\tilde{A}(x_1, x_3) = \{<0.7/2>, <0.8/3>\}$	1	2
$\tilde{A}(x_1, x_4) = \{<0.8/1>, <1/3>\}$	1	1
$\tilde{A}(x_1, x_5) = \{<0.8/2>, <0.8/3>\}$	1	2
$\tilde{A}(x_1, x_6) = \{<0.8/2>, <0.8/3>\}$	1	3

Fig. 6. Values of columns after the operation of algorithm

Here, the column A determines the temporal transitive closure $\widetilde{\Gamma}(x_1)$ of vertex x_1. The transitive closure of the vertex x_1 in particular shows that any vertex of the graph is reachable from the vertex x_1 with degree at least 0.8, but at different times.

5 Conclusions

The introduced concepts of fuzzy temporal set of reachability can serve as a basis for modeling complex processes in GIS, in which the elements have fuzzy relationships that vary in discrete time. It should be noted that the definition of a fuzzy temporal set of reachability refers to the problem of analysis. Another, more complex task is the problem of synthesis, which in general form can be formulated as follows: the degree of which edges of the graph, and at what time it is necessary to increase, so that the reachability of the fuzzy graph under consideration reaches a certain, predetermined value. The considered method of finding the transitive closure can serve as a basis for analyzing the connectivity of fuzzy temporal graphs.

Acknowledgments. This work has been supported by the Russian Foundation for Basic Research, Projects №. 18-01-00023 and №. 19-07-00074.

References

1. Malczewski, J.: GIS and Multicriteria Decision Analysis. Wiley, New York (1999)
2. Longley, P., Goodchild, M., Maguire, D., Rhind, D.: Geographic Information Systems and Science. Wiley, New York (2001)
3. Kaufmann, A.: Introduction a la theorie des sous-ensemles flous. Masson, Paris (1977)
4. Christofides, N.: Graph Theory: An Algorithmic Approach. Academic press, London (1976)
5. Goodchild, M.: Modelling error in objects and fields. In: Goodchild, M.F., Gopal, S. (eds.) Accuracy of Spatial Databases, pp. 107–113. Taylor & Francis, Basingstoke (1989)
6. Zhang, J., Goodchild, M.: Uncertainty in Geographical Information. Taylor & Francis, New York (2002)
7. Erusalimskiy, I.M.: Graph with attenuation on arcs and amplification in vertices and routing in information networks. Eng. J. Don, 1 (2015). ivdon.ru/ru/magazine/archive/n1y2015/2782
8. Belyakov, S., Bozhenyuk, A., Kacprzyk, J., Knyazeva, M.: Fuzzy modeling in the task of control cartographic visualization. In: Rutkowski, L., et al. (eds.) ICAISC 2019, LNAI, vol. 11508, pp. 261–272. Springer, Heidelberg (2019)
9. Bozhenyuk, A., Belyakov, S., Kacprzyk, J.: Optimization of jobs in GIS by coloring of fuzzy temporal graph. In: Advances in Intelligent Systems and Computing, vol. 896, pp. 25–32. Springer-Verlag, Heidelberg (2019)
10. Pospelov, D.: Situational Management: Theory and Practice. Nauka, Moscow (1986)
11. Monderson, J., Nair, P.: Fuzzy Graphs and Fuzzy Hypergraphs. Springer-Verlag, Heidelberg (2000)
12. Bozhenyuk, A., Gerasimenko, E., Kacprzyk, J., Rozenberg, I.: Flows in Networks Under Fuzzy Conditions. Springer, Heidelberg (2017)
13. Kostakos, V.: Temporal graphs. Proc. Physica A Stat. Mech. Appl. **6**(388), 1007–1023 (2008)

14. Barzilay, R., Elhadad, N., McKeown, K.: Inferring strategies for sentence ordering in multidocument news summarization. J. Artif. Intell. Res. **17**, 35–55 (2002)
15. Bramsen, P.: Doing Time: Inducing Temporal Graphs. Tech. rep., Massachusetts Institute of Technology (2006)
16. Baldan, P., Corradini, A., Konig, B.: Verifying finite-state graph grammars: an unfolding-based approach. Lecture Notes in Computer Science, vol. 3170, pp. 83–98. Springer (2004)
17. Baldan, P., Corradini, A., Konig, B.: Verifying a behavioural logic for graph transformation systems. In: Proceedings of COMETA 2003, vol. 104 of ENTCS, pp. 5–24. Elsevier (2004)
18. Dittmann, F., Bobda, C.: Temporal graph placement on mesh-based coarse grain reconfigurable systems using the spectral method. In: From Specification to Embedded Systems Application, vol. 184, pp. 301–310. Springer (2005)
19. Bershtein, L., Bozhenyuk A.: The using of temporal graphs as the models of complicity systems. Izvestiya UFY. Technicheskie nayuki. TTI UFY, Taganrog **4**(105), 198–203 (2010)
20. Bershtein, L., Belyakov, S., Bozhenyuk A.: The using of fuzzy temporal graphs for modeling in GIS. Izvestiya UFY. Technicheskie nayuki. TTI UFY, Taganrog **1**(126), 121–127 (2012)
21. Bozhenyuk, A., Gerasimenko, E., Rozenberg, I.: Method of maximum two-commodity flow search in a fuzzy temporal graph. In: Advances in Intelligent Systems and Computing, vol. 641, pp. 249–260. Springer, Heidelberg (2018)
22. Bozhenyuk, A., Belyakov, S., Knyazeva, M., Rozenberg, I.: Searching method of fuzzy internal stable set as fuzzy temporal graph invariant. In: Communications in Computer and Information Science, vol. 583, pp. 501–510. Springer-Verlag, Heidelberg (2018)

Improving Quality of Seaport's Work Schedule: Using Aggregated Indices Randomization Method

Vasileva Olga[1], Kiyaev Vladimir[2], and Azarov Artur[3(✉)]

[1] Saint-Petersburg State University, Saint-Petersburg, Russia
vasiljevaa@mail.ru
[2] Saint-Petersburg State University of Economics, Saint-Petersburg, Russia
[3] St. Petersburg Institute for Informatics and Automation of the Russian Academy of Sciences, 39 14-Th Line V.O, Saint-Petersburg, Russia
artur-azarov@yandex.ru

Abstract. Genetic algorithms can be used in combination with multi-agent monitoring and data collection technologies for compiling and adjusting the dynamically changing work schedule of an economic entity with complex internal infrastructure. In this case, the important issue is the quality of the drawn-up schedule and its monitoring. The article reviews the possibility of applying the method of randomized summary indicators for assessing the quality of the compiled and dynamically adjusted service schedules for ships in the seaport.

Keywords: Scheduling · Genetic algorithms · Multi-agent technologies · The method of summary indicators

1 Introduction

The effective work of any modern infrastructure system of an economic entity requires the construction of adequate mathematical models of control flows, work flows and information flows, allowing for the interaction and synchronization of actions of all services of this subject. Such complex infrastructural formations can include the life support systems of megacities, airports, sea and river ports, large road and rail hubs, logistics companies (intercity transportation, warehousing, etc.). The scheduling of such objects allows to accurately synchronize the work of all services. It is a well-known task solved by various mathematical and managerial methods. At the same time, developing mathematical methods, as well as increasing the informatization of the society as a whole, allow us to develop new approaches to the construction of complex dynamic schedules.

Works [7, 8] were devoted to the methods for constructing the schedule of the sea cargo sweat work on the basis of the genetic algorithm work. Among other things, these works considered weights systems reflecting the value of certain indicators of workflows in the servicing of sea vessels used to form the schedule. Obviously, not all weights can be established immediately and unambiguously, due to a fairly high degree

S. Kovalev et al. (Eds.): IITI 2019, AISC 1156, pp. 517–522, 2020.
https://doi.org/10.1007/978-3-030-50097-9_52

of uncertainty of the time periods of arrival and handling of vessels. The construction of unambiguous and unchangeable assessments can lead to incorrect accounting of real situations, which in some cases, as the rich statistics of incidents in sea and river ports show, can lead to an abnormal or emergency event with undesirable consequences.

In this regard, for a more accurate assessment of the weights and the application of these estimates in the genetic algorithm in the formation of the work schedule, it seems appropriate to involve the mathematical apparatus of the method of summary indicators (SMEs), namely, an effective type of aggregated indices randomization method (AIRM) [15].

2 Model's Description

In our works [7, 8], an approach for the formation of the seaport operations schedule based on processing data from various functional agents (in our case, intelligent software agents), based on a genetic algorithm, was proposed. For checking the effectiveness of the genetic algorithm, different weights were used, which are responsible for the formation of the final grade, reflecting the quality of the developed schedule. An important part of the algorithm is devoted to the formation of an estimate of the time spent on servicing a vessel in port. This assessment is based on data received from agents. Then, an estimate of the vessel's maintenance location freedom in the schedule In this case, two sets of ships are considered: $S_1 = \{s_i | s_i \in G\}$ — set of ships already on the port schedule, and $S_2 = \{s_i | s_i \in G\}$ — set of vessels, information about the time of arrival of which to the port has already received, but not yet included in the current schedule. We consider the schedule to be formed for 14 days. It is formed on the basis of a set of S_2 when making an assessment of the vessel's service location freedom in the schedule. This assessment is presented in the form

$$Y_i = \frac{T_i}{\max\limits_{S_2}(T_i)},$$

where Y_i — freedom evaluation service locations i vessel in schedule, T_i – parking time estimate i vessel in the seaport.

After assessing the i vessel's service location freedom in the schedule, the list of vessels from the set S_2 by ascending their assessments of location freedom:$Y_i \leq Y_{i+1}, i = 1..|S_2|$. Then, the resulting set of assessments is divided into subsets according to the cargo priority indicator P.

After sorting, the vessels with the lowest estimate of freedom and the highest cargo priority are added to the schedule, the following factors are taken into account: the terminal has a free berth and cranes of the required carrying capacity;

- the terminal has the necessary infrastructure for servicing the vessel;
- there is no "overlap" of the terminal operation time.

If the prerequisites are met, the quality of the ship servicing location in the schedule is assessed according to the following criteria: the appearance of unloaded time (windows) in the work of the terminals;

- redundancy of the terminal's loading and unloading capacity relative to the ship's needs;
- carrying out loading and unloading operations at night;
- the possibility of combining loading and unloading and servicing the vessel;
- speed of loading and unloading operations;
- disappearance of windows in the terminal operation;
- the absence of idle terminals in the presence of vessels requiring maintenance.

Note that the list of factors presented is not exhaustive, because depending on the complexity of a particular infrastructure, specific types of work, the presence or absence of necessary transport communications and storage facilities, the geographical location of the port, etc., the number and composition of these factors may vary significantly.

3 Aggregated Indices Randomization Method

Consider first the classical scheme of AIRM [10–15]. It can be presented as a sequence of the following steps.

1. Formed vector $x = (x_1, \ldots, x_n)$ of initial characteristics is formed, each of which is necessary, and all of them together are sufficient for a full, comprehensive assessment of the studied quality of the object. In the considered paradigm, the initial characteristics are the formal parameters of the vessel, according to which an estimate of the freedom of the location of the time (period) of serving the i vessel in the schedule is formed.

2. Formed vector $q = (q_1, \ldots, q_m)$ of individual indicators representing the $q_i(x)$, $i = 1, \ldots, m$ functions, initial characteristics vector $x = (x_1, \ldots, x_n)$ and evaluating various aspects of the object under study using different criteria. In our case, each individual indicator q_i is a function of one source characteristic x_i: $q_i = q_i(x_i)$, $i = 1, \ldots, m = n$. Also, the considered indicators are polarized (an increase in the value of the indicator q_i for fixed values of all other indicators q_j, $i \neq j$, increases the value of the composite indicator Q) and normalized ($q_i \in [0, 1]$). If the normalization condition is satisfied, the function $q_i = q_i(x_i)$ is naturally called a normalizing function. Under the function $q_i(x)$ means the value that can take variable x.

3. A type of synthesizing function $Q(q)$, is chosen that compares the vector of individual indicators $q = (q_1, \ldots, q_m)$ with a summary assessment characterizing the object under study as a whole, in our case it is about assessing the quality of the schedule of operation of a sea cargo port. At the same time, the synthesizing function $Q(q)$ depends on the vector $w = (w_1, \ldots, w_m)$ of non-negative parameters $w_1, \ldots, w_m$, which determine the significance of individual indicators $q_1, \ldots, q_m$ espectively, for the summary assessment Q:$Q = Q(q) = Q(q; w)$. Factors w_i discussed above.

4. The value of the vector of parameters $w = (w_1, \ldots, w_m)$, $w_i \geq 0$, usually interpreted as weighting factors ("weights"), specifying the degree of influence of individual indicators $q_1, \ldots, q_m$ on the summary estimate Q. Often used additional condition of normalization $w_1 + \ldots + w_m = 1$ uggests the value of the parameter w_i as an estimate of the relative weight of an individual indicator q_i.

Now it is possible to proceed to the consideration of the AIRM [15]. As mentioned earlier, there is no way to uniquely determine the weight of each variable. Therefore, for further work, it is necessary to fix the form of each normalizing function $q_i = q_i(x_i)$ and the synthesizing function $Q(q; w) = Q_+(q; w)$. Further, it can be argued that ordinal information on the correspondence of weights can be obtained from experts in the subject area under consideration. Such information can be presented in the form of system $OI = \{w_i > w_j,\ w_r = w_s,\ \ldots\}$ of inequalities and equalities for weights [6].

It can also be assumed that experts can also provide interval (fuzzy) information on weighting factors. Such interval information can be represented as system $II = \{a_i \le w_i \le b_i,\ \ldots\}$ of inequalities, defining the limits of possible variation of weighting coefficients.

Further, there are two options, depending on the sufficiency of information. The first option is that the combined information $I = OI \cup II$ is sufficient to uniquely determine the vector $w = (w_1, \ldots, w_m)$. In the second case, such information is not enough, then it can be argued about the incompleteness of information I.

Thus, the problem in question may contain non-numeric, non-exact and non-complete expert knowledge (NNN-knowledge, NNN-information) I, due to the uncertainty of the situation being assessed. If the presence of NNN-information I about weight coefficients allows reducing the set $W = \{w^{\theta},\ \theta \in \Theta\}$ of all possible weight vectors to the set $W(I)$ of all admissible weight vectors, satisfying the relation of the strictly inclusion of sets $W(I) \subset W$, then this information is nontrivial.

Having formed the set of all admissible weight vectors $W(I)$, to simulate such uncertainty of setting the weight vector according to the NNN-information I we will use a random weight vector $\tilde{w}(I) = (\tilde{w}_1(I), \ldots, \tilde{w}_m(I))$, uniformly distributed on the set $W(I)$. Randomization $\tilde{w}(I) = (\tilde{w}_1(I), \ldots, \tilde{w}_m(I))$ of the vector of weight coefficients $w = (w_1, \ldots, w_m)$ entails randomization of $\tilde{Q}_+(q; I) = Q_+(q; \tilde{w}(I))$ of the corresponding composite index $Q_+(q; w)$.

For random weight coefficients $\tilde{w}_1(I), \ldots, \tilde{w}_m(I)$ as estimates it is natural to use mathematical expectations $\bar{w}_i(I) = E\,\tilde{w}_i(I)$, $i = 1, \ldots, m$. To measure the accuracy of the estimates of $\bar{w}_i(I)$ standard deviations $s_i(I) = \sqrt{D\,\tilde{w}_i(I)}$, can be used where $D\,\tilde{w}_i(I)$ is the variance of the random weighting factor $\tilde{w}_i(I)$. The vector of mathematical expectations $\bar{w}(I) = (\bar{w}_1(I), \ldots, \bar{w}_m(I))$ can be interpreted as a numerical image of NNN-information I.

The reliability of the ordering of randomized weights $\tilde{w}_r(I)$, $\tilde{w}_s(I)$ is determined by the probability $p\,(r, s; I)$ of the stochastic inequality $\tilde{w}_r(I) > \tilde{w}_s(t)$.

For a randomized composite indicator synthesizing using NNN-information I indicators $q_1, \ldots, q_m$, it is possible to calculate the average rating $\bar{Q}_+(q; I) = E\,Q_+(q; \tilde{w}(I))$.

To measure the accuracy of this estimate, it is natural to use the standard deviation $S(q; I) = \sqrt{D\,\tilde{Q}_+(q; I)}$, where $D\,\tilde{Q}_+(q; I)$ is the variance of the random index $\tilde{Q}_+(q; I) = Q_+(q; \tilde{w}(I))$. The reliability of the ordering of randomized estimates $Q_+(q^{(j)}; \tilde{w}(I))$, $Q_+(q^{(l)}; \tilde{w}(I))$ can be estimated by the probability $P(j, l; I)$ of the stochastic inequality $Q_+(q^{(j)}; \tilde{w}(I)) > Q_+(q^{(l)}; \tilde{w}(I))$. It is this average score that can reflect the quality of the developed schedule.

4 Conclusion

The article presents a mathematical apparatus that can be applied when working with NNN-information on assessing the quality of the genetic algorithm, which forms the schedule of operation of the sea cargo port. The use of this device seems appropriate due to the impossibility of unambiguous determination of the weights of factors that are decisive in the formation of the quality assessment of the genetic algorithm. In our opinion, this will improve the accuracy of scheduling and, therefore, better perform the work of the sea cargo port. The described method can be applied to the organization of effective work of any large business entity.

AIRM, that was described in this article is one of the options for using the methods described by N.V. Khovanov. This method is an almost universal assessment tool under the conditions of uncertainty of complex multiparameter objects of different nature. So, for example, using the software implementations of the ASP methodology, it is possible to evaluate the effectiveness of technological processes, the combat potential of complex military-technical systems, the severity of damage to biological objects, the level of pathogenicity of geographic zones, the quality of various types of synthetic rubber, etc.

References

1. Dombi, J.: Basic concepts for a theory of evaluation: the aggregative operator. Eur. J. Oper. Res. **10**, 282–293 (1982)
2. Hovanov, N., Kornikov, V., Seregin, I.: Qualitative information processing in DSS ASPID-3 W for complex objects estimation under uncertainty. In: Proceedings of the International Conference "Informatics and Control", pp. 808–816 (1997)
3. Hovanov, N., Kornikov, V., Seregin, I.: Randomized synthesis of fuzzy sets as a technique for multicriteria decision making under uncertainty. In: Proceedings of the International Conference "Fuzzy Logic and Applications", pp. 281–288 (1997)
4. Hovanov, N., Kornikov, V., Tokin, I.: A mathematical methods system of decision making for developmental strategy under uncertainty. Global Environmental Change. Perspective of Remote Sensing and Geographic Information Systems, pp. 93–96 (1995)
5. Hovanov, N., Kolari, J.: Estimating the overall financial performance of Mexican banks using a new method for quantifying subjective information. J. Financ. Eng. **7**(1), 59–77 (1998)
6. Hovanov, N., Fedotov, Yu., Zakharov, V.: The making of index numbers under uncertainty. Environmental Indices: Systems Analysis Approach, pp. 83–99 (1999)
7. Vasileva, O., Kiyaev, V.: Generation of efficient cargo operation schedule at seaport with the use of multiagent technologies and genetic algorithms. In: Proceedings of the Third International Scientific Conference "Intelligent Information Technologies for Industry", pp. 401–409 (2018)
8. Vasileva, O., Kiyaev, V.: Monitoring and controlling the execution of the sea cargo port operation's schedule based on multi-agent technologies. In: CEUR Workshop Proceedings, 2nd International Scientific and Practical Conference "Fuzzy Technologies in the Industry – FTI", vol. 2258, pp. 243–248 (2018)
9. Wittmuss, A.: Scalarizing multiobjective optimization problems. Math. Res. **27**, 255–258 (1985)

10. Granichin, O.: Information management systems with a variable structure of the state space. Adaptive control with predictive models with variable state space structures, pp. 29–64 (2018). (in Russian)
11. Granichin, O., Amelina, N., Proskurnikov, A.: Adaptive management with predictive models with a variable structure of the state space with application to the systems of network motion control and automation of medical equipment. Adaptive control with predictive models with variable state space structures, pp. 5–28 (2018). (in Russian)
12. Kalmuk, A., Granichin, O.: A randomized control strategy with predictive models under conditions of non-definiteness based on an algorithm for eliminating areas of knowledge-codiating correlations. Adaptive control with predictive models with variable state space structures, pp. 65–82 (2018). (in Russian)
13. Trukhaev, R.: Models of decision making under uncertainty. Science (1981). (in Russian)
14. Homenyuk, V.: Elements of the theory of multipurpose optimization. Science (1983). (in Russian)
15. Hovanov, N.: Evaluation of complex objects in the context of a lack of information. In: Proceedings of the 7th international scientific school "Modeling and analysis of safety and risk in complex systems", pp. 18–28 (2008). (in Russian)

Assessment of the Information System's Protection Degree from Social Engineering Attack Action of Malefactor While Changing the Characteristics of User's Profiles: Numerical Experiments

Artur Azarov[1(✉)], Alena Suvorova[2], Maria Koroleva[3], Olga Vasileva[4], and Tatiana Tulupyeva[1]

[1] St. Petersburg Institute for Informatics and Automation of the Russian Academy of Sciences, 39 14-th Line V.O., Saint-Petersburg, Russia
artur-azarov@yandex.ru

[2] National Research University Higher School of Economics, 16 Soyuza Pechatnikov, Saint-Petersburg, Russia

[3] BMSTU, ul. Baumanskaya 2-ya, 5/1, Moscow, Russia

[4] Saint Petersburg State University, Saint-Petersburg, Russia

Abstract. The article describes an approach to the analysis of changes in the user's protection level from the social engineering attack actions of malefactor in the case of applying two strategies to increase the level of protection. The first deals with changing information system's users (dismissal/advanced training), and second is changes in user access policies to critical information stored in such information systems. Numerical experiment is also presented.

Keywords: Social engineering attacks · User's vulnerabilities profile · Access policies · User's social graph

1 Introduction

The modern world does not stand still. Smart homes, smart clinics, assistants in almost all areas of human life. Such innovations require serious information support, new information systems, which, among other things, contain various personal data and essential confidential corporate information. Thus, the protection of such information also arises everywhere. Such challenges pose serious challenges to information security specialists. More and more advanced data protection methods are being developed [7–17]. But the users ensuring the operation of such information systems, and, consequently, being a potential source of information leaks, remain without proper attention [10]. The reasons for the leaks can be insider attacks as well as external influences on users, the purpose of which is to obtain confidential data. Impacts of this

The results were partially supported by RFBR, project No. 18-37-00340, and Governmental contract (SPIIRAS) No. 0073-2019-0003.

S. Kovalev et al. (Eds.): IITI 2019, AISC 1156, pp. 523–530, 2020.
https://doi.org/10.1007/978-3-030-50097-9_53

kind can be combined by the term social engineering attack of the malefactor, and separately taken impact - the social engineering attack impact of the malefactor.

The article is devoted to the development of a method for analyzing changes in the level of protection of an information system against social engineering attacks aimed at users of such systems in the event of a change in information system users (dismissal/advanced training), as well as changes in user access policies to critical information stored in such information systems. The approach is based on a change in the degree of user vulnerabilities and a corresponding recalculation of the probabilities of social engineering attack actions success and the overall level of protection of the information system. In addition, possible losses from the point of view of the company's office work are considered, in the event of a user's dismissal, as well as in the case of changes in the availability of certain critical documents by various users.

2 Algorithm Description

The article deals with the models previously proposed in [1–4], as well as the approaches presented in [1–4]. A common complex consists of models of various elements of an information system as a whole, including, but not limited to, a user model, a document model, an informational system model, and a malefactor's model. We will dwell in more detail on the user model, whose parameters are the user's access to certain critical documents, the level of access to these documents, as well as communications between users. On the basis of data on relations between users, a graph of social connections of users with loaded bidirectional arcs and loaded vertices can be constructed. The weights of arcs are the probabilities of transition from user to user, formed on the basis of the type of user's connections, and the weight of each vertex is the total probability of success of the social engineering attack impact on the user of the information system, built on the basis of the severity of the user's vulnerabilities [1–4]. In this model, we will also include a skills module, which can be represented as $S = (S_1, \ldots, S_n)$. Each S_i is the degree to which a person has a certain skill, which can be useful in his work. Also in this model there is user access to various critical documents. This type of access can be read - create/modify - delete - full access, it seems appropriate for each type of access level to construct such a probability.

Methods for constructing probabilities that allow assessing the security of users, both for the case of user training and for the case of changing user access policies to critical documents, were presented in [1–4]. This paper proposes a combination of these methods, so a cumulative effect can be achieved, significantly increasing the level of user's protection level.

We will consider several options for assessing changes in the overall level of information system security, depending on the actions taken to work with users. The first option is assess the changes in the level of user security in accordance with the possibility of their training firstly, then change the availability of certain critical documents. The second is that at first there is a change in user access policies for critical documents, then there is an additional analysis of the possibility of user training.

Let's consider the methods used to change the level of security that are used in this work.

Based on the user's vulnerabilities profile, the full probability of a social engineering attack impact on the user can be constructed. Among all users of the information system, users with minimal security ratings are selected. Then an assessment of the overall security level of the information system will be made.

In this paper, we estimate the change in the security level following one of the two strategies

- Dismissing an employee and replacing him with a new employee;
- Training an employee, with a corresponding increase in his level of security.

It is clear that replacing an employee with a new one with a comparable level of vulnerability leads to an increase in the overall level of protection (at least for the first time): a new communication employee is weaker, respectively, the attack spread through him is less likely. But such a decision may be unprofitable in accordance with other criteria of business. Even under the assumption that 1) the level of security of a new employee is at least not lower than that of the person he replaces (you can first estimate on the basis of assessments of psychological characteristics and then simulate user vulnerabilities); 2) a new employee has the necessary skills to work (assessment of employee competencies and skills can be carried out with the help of professional testing or other skills assessment of a new employee), it will be necessary to take into account that an employee will need time to get acquainted with the activities of the organization and department, which will reduce his effectiveness. Moreover, weak ties with other team members, according to numerous studies, affect both the work of the team and the effectiveness of this particular employee [6]. In particular, a meta-analysis conducted in [6] shows that the success of an employee is interrelated with his position in the social network of the organization and the number of connections.

Therefore, for a more complete assessment of the impact of a decision, it is necessary to take into account both the changes in the security assessment and the changes in the overall performance of the team.

To analyze the changes in the probabilities of changes in the security of critical information stored in the information system, we suggest, as an example, to consider the probability of access to critical information with the "read" access type.

To analyze the changes in the probabilities of changes in the critical information protection level, that is stored in the information system, we suggest, as an example, to consider the probability of access to critical information with the "read" access type. The probability can be expressed as $p^r_{\mathrm{doc-acc}} = 1 - (1 - p^r_{\mathrm{doc1}})\ldots(1 - p^r_{\mathrm{docN}})$ for each type of access, where $p^r_{\mathrm{doc1}} \ldots p^r_{\mathrm{docN}}$ is the probability of full access to critical documents $1..N$ for users with the type of access "read". In this way, estimates of the probabilities estimates of critical information protection level can be obtained, both for each document separately and for all information in aggregate, if the overall probability of access to all documents is considered.

3 Modeling

We will conduct a numerical experiment as follows. First, when modeling the replacement of an employee for a new one, it is necessary to make changes in the indicators of security, the weight of the arcs that link the replaced user to the rest of the social graph, while the number of arcs remains unchanged. To simplify, we choose weights randomly from a uniform distribution on the interval $[0, a_i]$, where a_i is the weight of the corresponding arc of the previous employee. In general, the weights selection algorithm can be based on the minimum, average, or maximum weight of the arcs already contained in the graph of user social connections.

Secondly, we calculate the generalized skill of the team based on the individual skills of its members. In a number of studies, simple metrics are used to aggregate team skills — average, median, minimum (that is, assessment for the weakest), maximum (in situations where at least one participant is needed who can complete the task), span, etc. [5, 6]. However, a more reasonable assessment is taking into account the strength of the ties between the participants [5]. We will calculate the general skill of the team as a weighted in-degree of individual skills, where weights in the social relations graph (weighted in-degree) will be used as weights.

Third, we assume that employee training leads to a decrease in the degree of his vulnerability by bi%, where bi is chosen randomly from a uniform distribution over the interval [0, 100].

Comparing the significance of changing the general level of information system security and changing the level of provision of a company with the performance capabilities, it can be decided either to replace employee with a new one, or to train an employee with minimal security assessments. After the formation of such estimates and the adoption of appropriate decisions, the analysis of the graph can be continued until the employee being analyzed is either an employee that has been already analazyded or a new employee who has already been reviewed. Then a new assessment of the security of the entire information system can be obtained with new parameters of employees.

To assess the security of critical information stored in the information system, by changing access policies, it is proposed to consider a variety of documents, where is a set of critical documents $D = \left\{ \{Doc_i\}_{i=1}^{n}, \left\{ Us_i, \{UsR_j\}_{j=1}^{4} \right\}_{i=1}^{m} \right\}$, where $\{Doc_i\}_{i=1}^{n}$ is documents to which constant access should be provided, Us_i is a set of m users who should have access UsR_j to documents. At the same time, users can have several types of access at once. It should be noted that the optimization of access immediately to this level will cause a significant increase in the time flow of business processes in the organization, because users will be forced to request additional access each time they need a document to which they do not have a predetermined mandatory access. Formation of the probability of protection of critical information stored in the information system, based on the set D, will allow to form an estimate of the lower limit of acceptable optimization. This limit, which may be denoted as $Q_{\min}$, is a set of limits calculated for each type of access $Q_{\min} = \left\{ Q_{\min}^{r}, Q_{\min}^{c}, Q_{\min}^{d}, Q_{\min}^{f} \right\}$. At the same time, the owners of an information system can assign an acceptable level of risk, that is, the minimum acceptable level of protection of an information system, to achieve which the optimization process should strive. We denote this value as $Q_{\max}$, for this limit, the statement that this set of limits is also true, i.e. $Q_{\max} = \{Q^r_{\max}, Q^c_{\max}, Q^d_{\max}, Q^f_{\max}\}$. For each state of the system, a limit's estimate is formed $Q = \{Q^r, Q^c, Q^d, Q^f\}$; a comparison is also made for the fulfillment of the constraints of the set D. Then the criterion of checking is $Q_{\min} < Q < Q_{\max}$.

To form a computational experiment, it is proposed to consider a two-level graph, reflecting, on the one hand, communication between users, and on the other, communication between users and critical documents contained in the information system.

4 Experiment Design and Results

To explore the influence of the order of techniques application we conducted the series of experiments.

First, we randomly assigned initial values including user network structure, strength of connections between users, vulnerability degree for each user and two matrices that describe minimal rights and real rights. We explored system with 20 users and 15 documents. The connection between users was assign randomly according to binomial distribution with the probability of success equal to 0.16. The resulting graph structure is presented on Fig. 1.

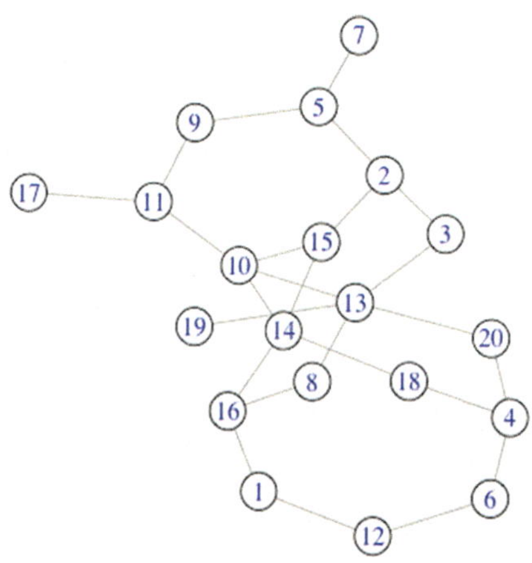

Fig. 1. User graph structure

The strength of each connection was assign according uniform distribution with maximum value equal to 0.2, the vulnerability for each user was also drawn from uniform distribution with maximum value equal to 0.07.

Finally, we assigned the access rights for the users, including minimal set of rights, randomly for each pair use-document assuring that each user had access to at least one document and each document could be accessed by at least one user. The structure is presented in Table 1, the green sells indicate minimal set of rights.

Table 1. Initial structure of access rights.

	u1	u2	u3	u4	u5	u6	u7	u8	u9	u1	u11	u12	u13	u14	u15	u16	u17	u18	u19	u2
doc1			1	1		1		1		1	1	1			1					
doc2	1				1		1	1	1		1			1		1		1	1	
doc3			1	1	1		1	1		1	1	1	1	1		1		1	1	1
doc4	1	1	1	1			1	1		1		1	1			1		1	1	
doc5	1	1	1		1	1	1		1		1	1	1	1	1	1	1	1		
doc6		1	1	1	1	1			1		1	1		1		1			1	1
doc7	1	1	1	1				1	1			1	1	1	1	1		1	1	
doc8	1	1	1	1			1		1		1			1		1				1
doc9	1				1	1				1	1		1	1	1	1	1			
doc10			1		1	1		1	1	1	1				1	1		1	1	
doc11		1	1	1	1	1		1	1	1	1		1	1		1			1	
doc12	1	1				1	1	1	1				1		1		1		1	1
doc13	1	1	1	1		1		1	1	1		1							1	
doc14		1		1	1	1						1		1			1	1	1	
doc15	1	1		1	1	1	1		1	1	1				1	1	1	1	1	

The total system vulnerability for real rights structure was equal to 0.99, the total system vulnerability for minimal rights structure was equal to 0.75.

After initializing step, we considered two strategies for reducing total vulnerability score for the system. The first one ("Rights - > Education") was to apply rights reduction first and then reduce the vulnerability by educating the most vulnerable users. The second strategy ("Education - > Rights") had reversed order: we educated users first and then optimize rights structure. As described above, the optimal rights structure was obtained using genetic algorithm with upper-limit system vulnerability score equal to 0.85, and we modelled "educating" by reducing user's vulnerability by bi% assigned uniformly from [25, 45].

We run the experiment 150 times for each strategy and then compared the average vulnerability results. Also we considered the situation when we choose only the half of users (most vulnerable ones) to educate and also repeated the experiment 150 times

The first strategy (Rights - > Education) showed better results compared to the second one (Fig. 2) and the difference was statistically significant (Student t-test, p-value < 0.01). Moreover, the same results were in the case then we considered educating not all users (Fig. 2).

However, the second strategy produced less strict rights structure, the number of user-documents pairs that indicate right to access the document was on average 10% higher for the second strategy compared to the first one based on initial rights structure.

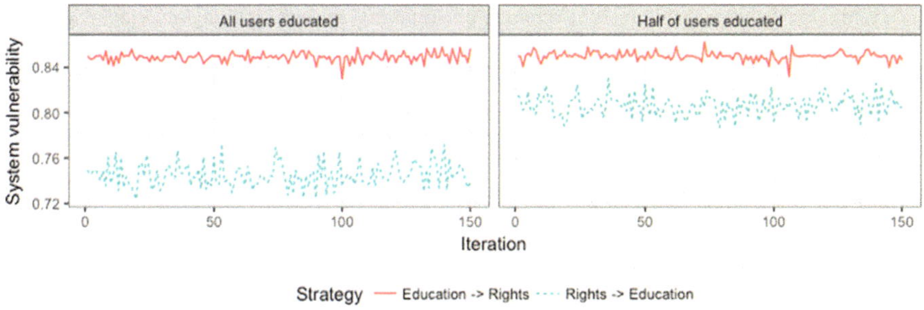

Fig. 2. The strategy comparison

5 Conclusion

The article describes an approach to the analysis of changes in the user's protection level from the social engineering attack actions of malefactor in the case of applying two strategies to increase the level of protection. It should be noted that when applying these strategies, the level of protection can be even higher if we use deeper user's training, i.e. change the conditions by which users are selected for who training should be conducted. Also a possible continuation of this study may be the use of various methods of user's training, depending on the identified psychological characteristics of users.

References

1. Azarov, A., Abramov, M., Tulupyev, A., Tulupyeva, T.: Models and algorithms for the information system's users' protection level probabilistic estimation. In: Proceedings of the First International Scientific Conference "Intelligent Information Technologies for Industry" (IITI 2016), vol. 2, pp. 39–46 (2016)
2. Azarov, A., Abramov, M., Tulupyeva, T., Tulupyev, A.: Users' of information system protection analysis from malefactor's social engineering attacks taking into account malefactor's competence profile. Biologically Inspired Cognitive Architectures (BICA) for Young Scientists, pp. 25–30 (2016)
3. Azarov, A., Suvorova, A., Tulupyeva, T.: Changing the information system's protection level from social engineering attacks, in case of reorganizing the information system's users' structure. In: II International Scientific-Practical Conference « Fuzzy Technologies in the Industry » , pp. 56–62 (2018)
4. Azarov, A., Suvorova, A.: Sustainability of the user's social network graph to the social engineering attack actions: an approach based on genetic algorithms. In: XXII International Conference on Soft Computing and Measurement (SCM 2018), pp. 126–129 (2018)
5. D'Innocenzo, L., Mathieu, J., Kukenberger, M.: A meta-analysis of different forms of shared leadership–team performance relations. J. Manag. **42**(7), 1964–1991 (2017)
6. Fang, R., Landis, B., Zhang, Z., Anderson, M., Shaw, J., Kilduff, M.: Integrating personality and social networks: A meta-analysis of personality, network position, and work outcomes in organizations. Organ. Sci. **26**(4), 1243–1260 (2015)

7. Gupta, B., Tewari, A., Jain, A., Agrawal, D.: Fighting against phishing attacks: state of the art and future challenges. Neural Comput. Appl. **28**, 3629–3654 (2017)
8. Huda, A.S.N., Živanović, R.: Accelerated distribution systems reliability evaluation by multilevel Monte Carlo simulation: implementation of two discretisation schemes. IET Gener. Transm. Distrib. **11**(13), 3397–3405 (2017)
9. Kharitonov, N., Maximov, A., Tulupyev, A.: Algebraic Bayesian Networks: The Use of Parallel Computing While Maintaining Various Degrees of Consistency. Stud. Syst. Decis. Control **199**, 696–704 (2019)
10. Kotenko, I., Chechulin, A., Branitskiy, A.: Generation of source data for experiments with network attack detection software. J. Phys: Conf. Ser. **820**, 12–33 (2017)
11. Liu, J., Lyu, Q., Wang, Q., Yu, X.: A digital memories based user authentication scheme with privacy preservation. PLoS ONE **12**(11), 0186925 (2017)
12. Schaik, P., Jeske, D., Onibokun, J., Coventry, L., Jansen, J., Kusev, P.: Risk perceptions of cyber-security and precautionary behavior. Comput. Hum. Behav. **62**, 5678–5693 (2017)
13. Shindarev, N., Bagretsov, G., Abramov, M., Tulupyeva, T., Suvorova, A.: Approach to identifying of employees profiles in websites of social networks aimed to analyze social engineering vulnerabilities. In: International Conference on Intelligent Information Technologies for Industry, pp. 441–447 (2017)
14. Struharik, R., Vukobratović, B.: A system for hardware aided decision tree ensemble evolution. J. Parallel Distrib. Comput. **112**, 67–83 (2018)
15. Suleimanov, A., Abramov, M., Tulupyev, A.: Modelling of the social engineering attacks based on social graph of employees communications analysis. In: 2018 IEEE Industrial Cyber-Physical Systems (ICPS), pp. 801–805 (2018)
16. Terlizzi, M., Meirelles, F., Viegas Cortez da Cunha, M.: Behavior of Brazilian banks employees on facebook and the cybersecurity governance. J. Appl. Secur. Res. **12**, 224–252 (2017)
17. Tulupyev, A., Kharitonov, N., Zolotin, A.: Algebraic Bayesian networks: consistent fusion of partially intersected knowledge systems. In: The Second International Scientific and Practical Conference "Fuzzy Technologies in the Industry – FTI 2018", pp. 109–115 (2018)

Method for Synthesis of Intelligent Controls Based on Fuzzy Logic and Analysis of Behavior of Dynamic Measures on Switching Hypersurface

Andrey A. Kostoglotov[1(✉)], Alexander A. Agapov[1(✉)], and Sergey V. Lazarenko[2(✉)]

[1] Rostov State Transport University, Rostov-on-Don, Russia
kostoglotov@icloud.com, aaalexander2794@gmail.com
[2] Don State Technical University, Rostov-on-Don, Russia
lazarenkosv@icloud.com

Abstract. The region of phase space with fuzzy boundaries is considered. The dynamics of the controlled system in this area is defined by a TS-model of the MISO type with a rule base up to the functions of each of the conclusions of the production rules. In contrast to the known solutions, this paper proposes to use controls synthesized based on the maximum condition of the generalized power function as the functions in each conclusion of the production rules. In this case to provide the various operation modes we need to build the corresponding set of switching hypersurfaces, which form is determined by the synthesizing function. To build the function we perform the analysis of dynamic measures on the switching hypersurface in the phase space. This function together with the use of fuzzy logic allows to develop a method for the synthesis of intelligent controls. The constructiveness of the developed method is confirmed by the results of the analysis of the solutions to the problem to control a nonlinear unstable object, obtained using the mathematical modeling.

Keywords: Synthesis · Intelligent control · Combined maximum principle · Fuzzy logic · Phase space

1 Introduction

The operating conditions of modern information control systems are characterized by uncertainties of various types. For example, it can be associated with the use of a relatively simple mathematical model of the object, which allows to apply the maximum principle of L.S. Pontryagin, leading to relay controls. Application of such control in practice often requires to smooth the regimes of frequented switching, as well as to adapt them to the inaccuracies of the switching points and the discrepancy between the chosen mathematical model and the actual dynamics of the processes. The solutions to the extremal synthesis problems constructed in this way are no longer optimal. This fact together with the nonlinearity and multimodality of the controlled objects motivates development of methods for the synthesis of quasi-optimal controls [1, 2].

S. Kovalev et al. (Eds.): IITI 2019, AISC 1156, pp. 531–540, 2020.
https://doi.org/10.1007/978-3-030-50097-9_54

A constructive approach to the synthesis of multimode control for nonlinear systems that does not require to solve a set of two-point boundary value problems is based on using the condition of maximum of the generalized power function [3–10], which allows to obtain a feedback structure up to the synthesizing function. To construct the structure we perform an analysis of dynamic measures on the switching hypersurface. As a result, both well-known optimal relay solutions for linear systems [3, 5] and new original quasi-optimal continuous controls for nonlinear systems [3, 4, 6–9] are obtained. To ensure the multi-mode control for such synthesized controls, the comprehensive information about the characteristics of the control object is required [10, 11].

In technical applications, the operation of the control systems involves the resolution of structural uncertainty, which is appeared as the result of intersections of the phase space regions with fuzzy boundaries. An effective mathematical apparatus for the synthesis of control systems with a limited incompleteness of information about the characteristics of the control object is the fuzzy logic. For example, using such logic leads to TS-models of MISO-type systems with bases of production rules containing the functions [11] in the right-hand side. In contrast to the known solutions, this paper proposes to use controls synthesized based on the maximum condition of the generalized power function in each conclusion of the production rules for different operating modes. This allows to build a new fuzzy inference algorithm to form a fuzzy control, which can be classified as intellectual [1, 11].

The paper aims to develop a method for the synthesis of intelligent controls based on the analysis of the behavior of dynamic measures on switching hypersurface in a phase space region with fuzzy boundaries.

2 Formulation of the Problem

Let us consider the Lagrangian controlled system characterized by a set of m modes of operation, and its phase space can be represented by a set of m regions. In each region the dynamics of the controlled system satisfies the Hamilton-Ostrogradskii principle. According to this principle, the extended functional has the form [12]:

$$S_i = \int_{t_{0i}}^{t_{ki}} (T + A_i)dt, i = \overline{1, m}, \tag{1}$$

here we assume that the kinetic energy is a quadratic form of the generalized velocities:

$$T = \frac{1}{2}a\dot{q}^2 \tag{2}$$

where a is the inertia coefficient; q is the generalized coordinate; the dot denotes the time derivative; $A_i = \int_{q(t_{0i})}^{q(t_{ki})} Q_i dq$ is the work of the generalized forces Q_i; $[t_{0i}, t_{ki}] \subset R$ is the specified time intervals.

In accordance with the Hamilton - Ostrogradskii principle, when the system moves from the initial state $q_{0i}(t_{0i}), \dot{q}_{0i}(t_{0i})$ to the final state $q_{ki}(t_{ki}), \dot{q}_{ki}(t_{ki})$ under the action of generalized forces on the true trajectory, the stationarity principle for the integral of the action (1) is true in each region [12]

$$\delta' \mathrm{S}_i = \int_{t_{0i}}^{t_{ki}} (\delta T + \delta' \mathrm{A}_i) dt = 0, \; i = \overline{1, m}, \tag{3}$$

where $\delta' A_i = Q_i \delta q$ - is the elementary work on virtual movements, and the sign δ' denotes an infinitesimal value.

From (3) for the multimode dynamic system [10, 12, 13] we can get the Lagrange equations of the second kind:

$$\begin{gathered} \frac{d}{dt}\left(\frac{\partial T}{\partial \dot{q}}\right) - \frac{\partial T}{\partial q} = Q_i \,, i = \overline{1, m}\,, \\ q(t_{0i}) = q_{0i}, \dot{q}(t_{0i}) = \dot{q}_{0i}. \end{gathered} \tag{4}$$

The structures of generalized forces Q_i as functions of the generalized coordinates q and the generalized velocities $\dot{q}$ with the definitional domains $D_i = \{(q, \dot{q}) | q \in [q_{0i}, q_{ki}], \dot{q} \in [\dot{q}_{0i}, \dot{q}_{ki}]\}$ can be found from the condition of the minimum of the objective functional

$$\mathrm{I}[q] = \int_{t_{0i}}^{t_{ki}} \mathrm{F}(q) dt \tag{5}$$

as solutions to inverse problems of dynamics with constraints

$$Q_i \in \bar{G}_i, \tag{6}$$

where $\mathrm{F}(q)$ is a continuous positive function with its partial derivatives; $\bar{G}_i$ are closed sets of admissible generalized forces.

We assume that the boundaries of the regions of the phase space are not well known and the domains of definition have the form $D_i' = \{(q, \dot{q}) | q \in M_{xi}, \dot{q} \in M_{yi}\}$, where M_{xi}, M_{yi} are fuzzy pairwise intersecting intervals: $\forall q_{0i}, q_{ki}, \forall q \in [q_{0i}, q_{ki}]$, $m_{xi}(q) \geq \min(m_{xi}(q_{0i}), m_{xi}(q_{ki}))$; $\forall \dot{q}_{0i}, \dot{q}_{ki}$, $\forall \dot{q} \in [\dot{q}_{0i}, \dot{q}_{ki}]$, $m_{yi}(\dot{q}) \geq \min(m_{yi}(\dot{q}_{0i}), m_{yi}(\dot{q}_{ki}))$; $m_{xi}(q)$, $m_{yi}(\dot{q})$ are the membership functions. Triplets (x, T_x, X) and (y, T_y, Y) specify the linguistic variables x and y; $T_x = \{T_x^1, \ldots, T_x^m\}$, $T_y = \{T_y^1, \ldots, T_y^m\}$ are the term sets of the linguistic values on the fuzzy sets $X \subseteq \cup_{i=1}^m M_{xi}$, $Y \subseteq \cup_{i=1}^m M_{yi}$.

In $m-1$ zones of uncertainty of the operational modes $D'_j \cap D'_{j+1}, j = \overline{1, m-1}$ the structure of the generalized forces is not completely known. Let the additive and multiplicative structural uncertainty take place, and the desired control force in fuzzy regions of the phase space $D'_j \cap D'_{j+1}$ has the following form

$$u_j = \eta_{j+1} Q_{j+1} - \eta_j Q_j \tag{7}$$

where η_j are unknown functions taking into account the structural uncertainties of the control object;

$$0 \leq \eta_j \leq 1. \tag{8}$$

Then the fuzzy inference is determined by the *TS*-model for a *MISO*-type system with a rule base as follows [11]:

$$\begin{aligned}
&R^1: \text{если } (q \text{ есть } M_{x1}) \text{ и } (\dot{q} \text{ есть } M_{y1}), \text{ то } (u = Q_1(q, \dot{q})), \text{ иначе}\\
&R^2: \text{если } (q \text{ есть } M_{x2}) \text{ и } (\dot{q} \text{ есть } M_{y2}), \text{ то } (u = Q_2(q, \dot{q})), \text{ иначе}\\
&\dots\dots\dots\dots\dots\dots\dots\dots\dots\dots\dots\dots\dots\dots\dots\dots\\
&R^m: \text{если } (q \text{ есть } M_{xm}) \text{ и } (\dot{q} \text{ есть } M_{ym}), \text{ то } (u = Q_m(q, \dot{q})).
\end{aligned} \tag{9}$$

Let the set of rules (9) is given up to the functions $Q_i(q, \dot{q}), i = \overline{1, m}$, of each conclusion.

The problem of synthesis of fuzzy control is as follows: for each of the m conclusions of the set of rules (9), it is required to find functions $Q_i(q, \dot{q}), i = \overline{1, m}$, as solutions to the extremal problems (4) (5) under constraints (3), (6).

3 Maximum Condition for the Function of Generalized Power

The solution to m extremal problems (4), (5) under constraints (3), (6) can be found based on the study of the extended objective functional in each region [4–8]

$$\mathrm{I}_1[q, Q_i] = \mathrm{I}[q] + \lambda \mathrm{S}_i[Q_i], \tag{10}$$

where λ is the Lagrange multiplier.

Suppose that there exists the generalized force $\tilde{Q}_i \in \bar{G}_i$ and the corresponding trajectory $\tilde{q}$ that delivers a minimum of (5). Variation of the generalized force leads to the generalized force $Q_i = \tilde{Q}_i + \delta Q_i$ to which the motion along a roundabout path corresponds, here δQ_i is a variation such as $Q_i \in \overline{G}_i$. Then the following inequality holds [4–8]:

$$\mathrm{I}_1(q, Q_i) - \mathrm{I}_1(\tilde{q}, \tilde{Q}) \geq 0, \quad \forall \quad Q_i \neq \tilde{Q}. \tag{11}$$

The study of Eq. (1) leads to inequality [4–8]:

$$[\lambda \tilde{Q}_i + \tilde{V}]\dot{\tilde{q}} \geq [\lambda Q_i + V]\dot{q}, \tag{12}$$

where $V = \frac{\partial F}{\partial q}$ is fictitious generalized force.

Then to achieve an extremum (5) it is necessary that the function of generalized power

$$\Phi_i = [\lambda Q_i + V]\dot{q} \tag{13}$$

reaches its maximum [4–8]

$$\max_{Q_i \in \bar{G}_i} \Phi_i = \max_{Q_i \in \bar{G}_i} [\lambda Q_i + V]\dot{q}. \tag{14}$$

Using condition (14) allows to determine the switching hypersurface in each of the m operational modes. The condition of the maximum of the generalized power function (14) implies that on a set of admissible generalized functions the switching curve is determined up to the synthesis functions $\mu_i(q, \dot{q})$ by the following equation [4–10]:

$$\mu_i(q, \dot{q})\dot{q} - V = 0, \tag{15}$$

that determine the switching hypersurfaces in the phase space

4 Construction of the Synthesis Function Based on the Analysis of the Dynamic Measures on the Switching Hypersurface

Suppose that in each of the regions of the phase space the synthesizing function is the same. To construct the function the work [3] proposes the method of phase trajectories. According to this method we represent the generalized force in the form $Q = Q' + U$, where Q' are the components of the generalized force independent of the control generalized forces and U are the controls (the control generalized forces). Then the Lagrange equations of the second kind (4) take the form

$$\frac{d}{dt}\frac{\partial T}{\partial \dot{q}} - \frac{\partial T}{\partial q} = Q' + U, \tag{16}$$

and the expression (15) can be reduced to the form

$$\lambda \frac{dp}{dt} - \frac{\partial}{\partial q}[\lambda(T + A) - F] = \mu \dot{q} \tag{17}$$

where p is generalized momentum.

For the representative point on the switching line, the following condition is met:

$$\lambda(T + A) - F = 0. \tag{18}$$

Taking this into account, we can write the expression (15) as

$$\mu\dot{q} = -\lambda\frac{dp}{dt}. \tag{19}$$

Following the developed method [3], we replace in (19) the time derivative by the derivative with respect to the generalized coordinate in accordance with the Whittaker theorem [13]. Thus we obtain:

$$\mu = -\lambda\frac{dp}{dq} \tag{20}$$

From Eq. (20) we obtain the following equation of the switching hypersurface [3, 10]:

$$-\lambda a\left|\frac{d\dot{q}}{dq}\right|\dot{q} - V = 0. \tag{21}$$

5 Example. The Control Synthesis in a Phase Space Region with Fuzzy Boundaries

Let us study the following Lagrange equations of the second kind

$$\begin{aligned} &\ddot{q} = a\,\sin(q) + U, \\ &q(0) = q_0, \dot{q}(0) = \dot{q}_0; \end{aligned} \tag{22}$$

and let $a = 3, q_0 = 20, \dot{q}_0 = 10$.

In the phase space with fuzzy boundaries, a *TS*-model of a *MISO* type system with a rule base:

$$\begin{aligned} &R^1 : \text{если } q \text{ есть } A_1 \text{ и } \dot{q} \text{ есть } B_1, \text{ то } u = Q_1(q, \dot{q}), \text{ иначе} \\ &R^2 : \text{если } q \text{ есть } A_2 \text{ и } \dot{q} \text{ есть } B_2, \text{ то } u = Q_2(q, \dot{q}), \end{aligned} \tag{23}$$

is given up to the functions $Q_i \in G_i, i = \overline{1, m}$, of each conclusion.

The uncertainty about the coordinates of the phase space points, where the operational modes of dynamic system (22) functions change, can be described by the input linguistic variables - "position on the phase plane" with the term set $T_x = \{\text{"близко"}, \text{"далеко"}\}$ and "speed on the phase plane" with term set $T_y = \{\text{"медленно"}, \text{"быстро"}\}$ with membership functions shown in Fig. 1 [11, 14].

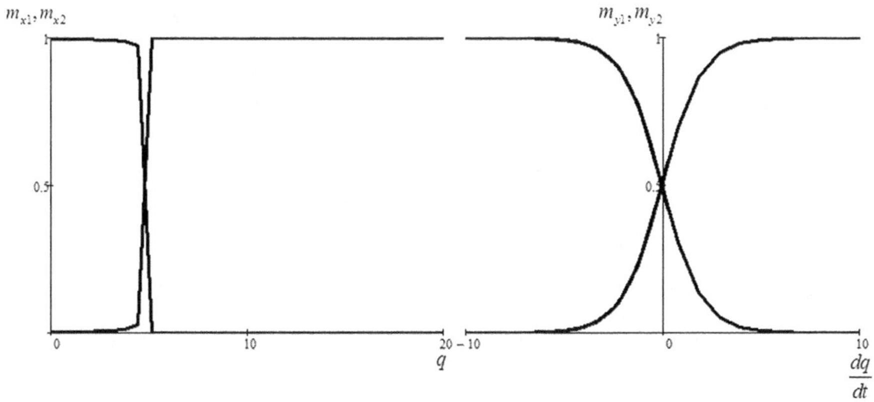

Fig. 1. The input membership functions.

For (23) the logical conclusion is determined by the expression [11]

$$u = \sum_{i=1}^{2} \eta_i Q_i \tag{24}$$

where

$$\eta_1 = \frac{\alpha_1}{\alpha_1 + \alpha_2}, \eta_2 = \frac{\alpha_2}{\alpha_1 + \alpha_2} \tag{25}$$

here $\alpha_1 = m_{x1}(q^*) \wedge m_{y1}(\dot{q}^*)$, $\alpha_2 = m_{x2}(q^*) \wedge m_{y2}(\dot{q}^*)$ are the cutoff levels for the premises of each of the rules, calculated for the crisp values of the generalized coordinate q^* and the generalized velocity $\dot{q}^*$.

The problem of synthesis of the fuzzy control is to find for each of the two conclusions of the rule (23) functions $Q_i(q, \dot{q}), i = \overline{1, 2}$, as solutions to the extremal problems (22) (5) with $F(q) = \frac{1}{2}q^2$, constraint (3), if Q_1 is a bounded piecewise constant function, and $Q_2(q, \dot{q})$ is a bounded piecewise continuous function.

From (14), (15), (21) it follows that for m = 1 [3]

$$Q_1 = |U|\text{sign}(-\mu\dot{q} - q), \tag{26}$$

where $|U|$ is a non-negative constant.

However, in practice chatter modes of operation are undesirable. To exclude them a saturation functions can be used when approximating the relay control laws [15]. For m = 2 from (26) we have

$$Q_2 = \text{sat}(-\mu\dot{q} - q) = \begin{cases} |U|\text{sign}(-\mu_2\dot{q} - q), |-\mu_2\dot{q} - q| > |U|, \\ \frac{-\mu\dot{q} - q}{|U|}, |-\mu\dot{q} - q| \leq |U| \end{cases} \tag{27}$$

During calculations the differential relation (21) can be replaced by the following expression [5]:

$$\mu = -\lambda \left| \frac{\dot{q}}{Lq + \varepsilon} \right|, \tag{28}$$

where L, ε are coefficients.

In accordance with (24)–(28) the value of the fuzzy output is determined by the formula:

$$u(q, \dot{q}) = \frac{\alpha_1 |U|}{\alpha_1 + \alpha_2} \operatorname{sign}\left(-\lambda \left| \frac{\dot{q}}{Lq + \varepsilon} \right| \dot{q} - q \right) + \frac{\alpha_2}{\alpha_1 + \alpha_2} \operatorname{sat}\left(-\lambda \left| \frac{\dot{q}}{Lq + \varepsilon} \right| \dot{q} - q \right) \tag{29}$$

The efficiency of the synthesized intelligent control is evaluated using mathematical modeling. We compare the piecewise constant control (26) and piecewise continuous control (27), which, unlike the control (29), do not allow to take into account division of the phase space into regions in accordance with the modes of operation, but they provide constraint on the magnitude of the feedback.

Figure 2 shows the laws of the generalized coordinate and velocity variations. The dots show the results of mathematical modeling (26): the dotted curve is Eq. (27), the solid curve is Eq. (29). The advantage of the intelligent speed control is obvious.

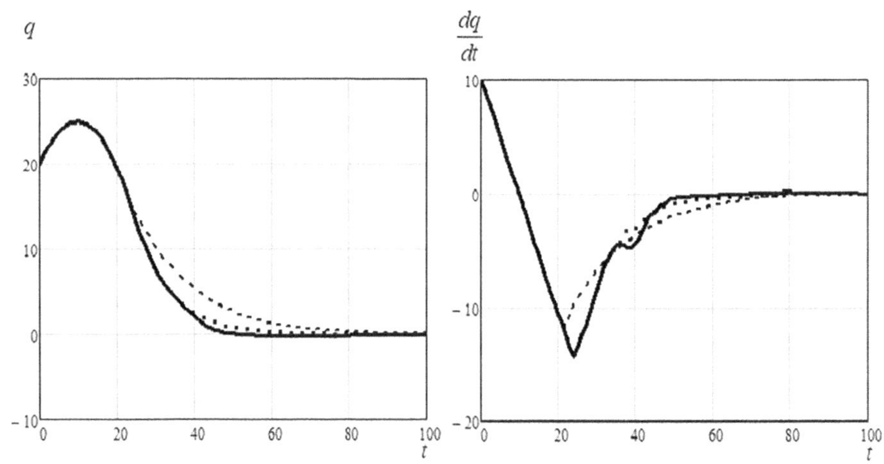

Fig. 2. The system dynamics in a phase space with fuzzy boundaries.

Figure 3 shows the timing diagrams of the synthesized controls. The notations are kept the same. The calculations show that in terms of quality

$$I = \int_{t_0}^{t_k} \frac{1}{2} q^2 dt \tag{30}$$

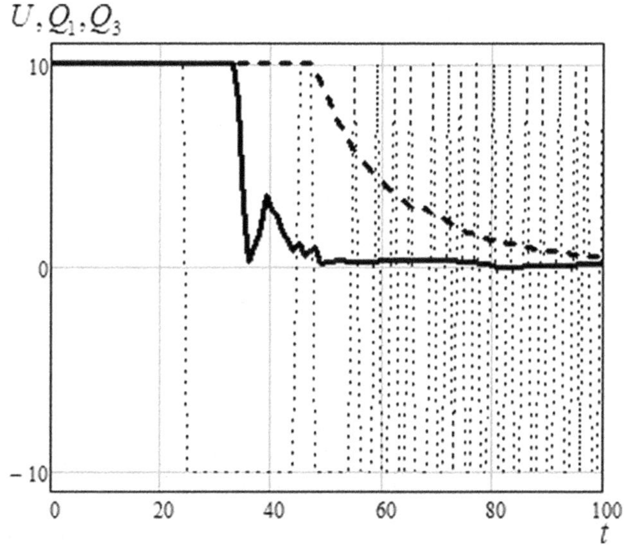

Fig. 3. The controls.

the intellectual control loses 0.2% to the piecewise constant control and wins 6.7% from the control (27). At the same time, Fig. 3 shows that in terms of the control cost the obtained solution to the synthesis problem is more efficient than (26) and (27).

6 Conclusion

The method for the synthesis of intelligent controls based on the analysis of the dynamic measures on the switching hypersurface in a phase space region with fuzzy boundaries is developed. The carried out mathematical modeling shows that the fuzzy control synthesized based on this solution in comparison with the piecewise constant and piecewise continuous controls built based on the combined maximum principle gives the gain in the transient process time and the control costs with similar values of the quality indicator (28).

The work is carried out under the RFBR grants No. 18-01-00385 A, No. 18-08-01494 A.

References

1. Hung, L.C., Chung, H.Y.: Decoupled sliding-mode with fuzzy-neural network controller for nonlinear systems. Int. J. Approximate Reasoning **46**, 74–96 (2007)
2. Vasiliev, S.N., Kudinov, Yu.I, Pashchenko, F.F., Durgaryan, I.S., Kelina, A.Yu., Kudinov, I. Yu., Pashchenko, A.F.: Intelligent control systems and fuzzy regulators. Part II. Learning fuzzy regulators, fuzzy PID regulators. Sens. Syst. **3**(211), 3–12 (2017)

3. Kostoglotov, A.A., Lazarenko, S.V., Lyaschenko, Z.V.: Intellectualization of measuring systems based on the method of structural adaptation in the construction of tracking filter. In: Proceedings of 2017 20th IEEE International Conference on Soft Computing and Measurements (SCM 2017), pp. 568–570 (2017)
4. Kostoglotov, A.A., Kostoglotov, A.I., Lazarenko, S.V.: Joint maximum principle in the problem of synthesizing an optimal control of nonlinear systems. Autom. Control Comput. Sci. **41**(5), 274–281 (2007)
5. Kostoglotov, A., Lazarenko, S., Agapov, A., Lyaschenko, Z., Pavlova, I.: Designing the knowledge base for the intelligent inertial regulator based on quasi-optimal synthesis of controls using the combined maximum principle. Adv. Intell. Syst. Comput. **874**, 190–200 (2019)
6. Kostoglotov A., Lazarenko S., Penkov A., Kirillov I., Manaenkova O.: Synthesis of adaptive algorithms for estimating the parameters of angular position based on the combined maximum principle. In: Proceedings of the Third International Scientific Conference "Intelligent Information Technologies for Industry" (IITI 2018). Advances in Intelligent Systems and Computing, vol 874. Springer, Cham (2019)
7. Deryabkin, I.V., Kostoglotov, A.A., Lazarenko, S.V., Lyaschenko, Z.V., Manaenkova, O. N.: The synthesis of adaptive multi-mode regulators based on combined control of the combined maximum principle. Bull. Rostov State Transp. Univ. **3**(63), 124–132 (2016)
8. Kostoglotov, A., Lazarenko, S., Pugachev, I., Yachmenov, A.: Synthesis of intelligent discrete algorithms for estimation with model adaptation based on the combined maximum principle. Adv. Intell. Syst. Comput. **874**, 116–124 (2019)
9. Agapov, A.A., Kostoglotov, A.A., Lazarenko, S.V., Lyaschenko, Z.V., Lyaschenko, A.M.: Analysis and synthesis of non-linear multi-mode control laws using the combined maximum principle. Bull. Rostov State Transp. Univ. **1**(73), 119–125 (2019)
10. Pegat, A.: Fuzzy modeling and control. Moscow: BINOM. Laboratory of Knowledge, pp. 798 (2013)
11. Lur'e, A.I.: Analytical mechanics Moscow: GIFML, pp. 824 (1961)
12. Markeev, A.P.: Theoretical Mechanics, p. 416. Nauka, Moscow (1990)
13. Muzichenko, N.Y.: Synthesis of optimal linear meter in observations through correlated noise based on fuzzy logic algorithms. Radio Eng. Electr. **6**, 1–4 (2014)
14. Anan'evskij, I.M., Reshmin, S.A.: Continuous control of mechanical system based on the method of decomposition. Izvestiya RAN. Theory Control Syst. **4**, 3–17 (2014)

Synthesis of Multi-model Algorithms for Intelligent Estimation of Motion Parameters Under Conditions of Uncertainty Using Condition of Generalized Power Function Maximum and Fuzzy Logic

Andrey A. Kostoglotov[1], Igor V. Pugachev[2], Anton S. Penkov[1], and Sergey V. Lazarenko[1,2(✉)]

[1] Rostov State Transport University, Rostov-on-Don, Russia
kostoglotov@icloud.com, pencha_@gmail.com
[2] Don State Technical University, Rostov-on-Don, Russia
lazarenkosv@icloud.com, bakut_8536@mail.ru

Abstract. The article considers the problems relating the estimation of the motion parameters under conditions of uncertainty, which are caused by the lack of a priori information about the nature of the movement of the controlled object. Using traditional kinematic models may lead to divergence of the estimation process and failure of the computational procedure. The article shows that the constructive results of the synthesis of 2D dynamic models in the polar coordinate system can be provided using the maximum condition for the function of generalized power. Adaptation of the obtained models to different modes of motion is carried out using fuzzy logic. The constructiveness of the approach is confirmed by a comparative analysis of the results of mathematical modeling.

Keywords: Adaptation · Polar coordinate system · Maximum condition for function of generalized power · Fuzzy logic

1 Introduction

The structure of the filters for estimating the parameters of motion is determined by the mathematical models of the maneuvering observation object. Usually the target acceleration is represented by a random process without taking into account the inertial properties of the observation object. As studies [1] show, even the problems of estimation the parameters of the vehicles movement demands the multiparameter adaptation of the obtained algorithms based on a polynomial approximation of the motion trajectory. Dynamic models allow to reduce significantly the parametric uncertainty. The simplest models cover motion in a plane and require to use the nonlinear filtering methods. In this regard, the most promising are multi-model algorithms that admit a change in structure in accordance with one of the modes [2, 3], for example, a motion according to a known law, maneuvering, etc. The basis for constructing such solutions are the theory of statistical synthesis and the fuzzy logic [4, 5].

S. Kovalev et al. (Eds.): IITI 2019, AISC 1156, pp. 541–547, 2020.
https://doi.org/10.1007/978-3-030-50097-9_55

Under conditions of a priori uncertainty about the maneuver features of the observation object, the constructive results of the synthesis of dynamic models can be provided using the maximum condition for the function of generalized power. The solution obtained in this way is determined up to a synthesizing function that defines the switching surface. To construct the structure we perform an analysis of dynamic measures on the switching hypersurface [6–15]. The results of mathematical modeling show that the algorithms for estimating motion parameters based on them are advantageously stable and need a small amount of computational costs.

The aim of the work is to increase the efficiency of the 2D motion models under conditions of uncertainty in the target maneuverability.

So the problem to solve is the synthesis of the multi-model algorithms for intelligent estimation of the motion parameters under conditions of uncertainty using condition of generalized power function maximum and fuzzy logic.

2 Formulation of the Problem

In the observation space we define the target functional [15]:

$$J = \frac{1}{2}\int_{0}^{t_1} [\mathbf{y}(t) - \mathbf{h}(\hat{\mathbf{q}}, t)]^T \mathbf{N}^{-1} [\mathbf{y}(t) - \mathbf{h}(\hat{\mathbf{q}}, t)] dt = \int_{0}^{t_1} F(\mathbf{y}, \hat{\mathbf{q}}, t) dt, \tag{1}$$

where $\mathbf{N}^{-1}$ is the weight matrix characterizing the intensity of interference in the observation channel

$$\mathbf{y}(t) = \mathbf{h}(\hat{\mathbf{q}}) + \boldsymbol{\xi}(t), \tag{2}$$

Here $\mathbf{q} \in R^n$ is the vector of the generalized coordinates, $\mathbf{h}(\hat{\mathbf{q}})$ is the known function of the generalized coordinates, $\boldsymbol{\xi}(t) \in R^n$ is the vector of random effects on the observation channel of known intensity, $t \in [0, t_1] \subset R$, n is the number of degrees of freedom of the dynamical system.

The change of state satisfies the Hamilton-Ostrogradskii principle [16], and the extended action functional accordingly can be written as:

$$S = \int_{0}^{t_1} [\lambda(T + A) + F] dt, \tag{3}$$

where T is the kinetic energy, A is the work of generalized forces Q_s at the true trajectory,

$$\mathbf{Q} = [Q_1, Q_2, \ldots, Q_n]^T \in \overline{G}, \tag{4}$$

here $\overline{G}$ is the set of the generalized forces, admitting the observed motion; λ is the Lagrange multiplier. Let the vector of generalized forces (4) is unknown.

The problem to build a motion model is to find the vector of the generalized forces (4) from the condition of the minimum of the objective functional (3), which takes into account the constraints following from the Hamilton-Ostrogradskii principle.

3 Synthesis of Dynamic Motion Models Based on Maximum Condition

The solution of the problem is carried out on the basis of the methodology of the combined maximum principle [6–15], according to which the minimum of the objective functional can be found from the condition of maximum of the generalized power function

$$\Phi(\mathbf{q},\dot{\mathbf{q}},\mathbf{Q},\lambda) = \max\sum_{s=1}^{n}[\lambda\, Q_s + V_s]\dot{q}_s,\ s=\overline{1,n}, \tag{5}$$

where $V_s = \frac{\partial F}{\partial q_s}$.

From (5) it follows that the structure of the right-hand side of the Lagrange equations of the second kind [16]

$$\frac{d}{dt}\frac{\partial T}{\partial \dot{q}_s} - \frac{\partial T}{\partial q_s} = Q_s,\ s=\overline{1,n}, \tag{6}$$

under conditions of uncertainty of the dynamics of the object motion model is determined by the expression [6–15]

$$Q_s = \lambda^{-1}[\mu_s(\mathbf{q},\dot{\mathbf{q}})\dot{q}_s - V_s], s=\overline{1,n} \tag{7}$$

where $\mu_s(\mathbf{q},\dot{\mathbf{q}})$ is synthesizing function. To construct this function the work [6] proposes the method of phase trajectories. According to this method we represent the generalized force in the form $Q = Q' + U$, where Q' are the components of the generalized force independent of the control generalized forces and U are the controls (the control generalized forces). Then the Lagrange equations of the second kind (6) take the form

$$\frac{d}{dt}\frac{\partial T}{\partial \dot{q}} - \frac{\partial T}{\partial q} = Q' + U, \tag{8}$$

and the expression (8) can be reduced to the form

$$\lambda\frac{dp}{dt} - \frac{\partial}{\partial q}[\lambda(T+A) - F] = \mu\dot{q} \tag{9}$$

where p is generalized momentum.

For the representative point on the switching line, the following condition is met:

$$\lambda(T+A) - F = 0. \tag{10}$$

Taking this into account, we can write the expression (15) as

$$\mu\dot{q} = -\lambda\frac{dp}{dt}. \tag{11}$$

Following the developed method [6], we replace in (19) the time derivative by the derivative with respect to the generalized coordinate in accordance with the Whittaker theorem [17]. Thus we obtain:

$$\mu = -\lambda\frac{dp}{dq} \tag{12}$$

The dynamic motion model takes the form

$$\frac{d}{dt}\frac{\partial T}{\partial \dot{q}_s} - \frac{\partial T}{\partial q_s} = \lambda^{-1}\left[-\lambda\left|\frac{dp}{dq}\right|\dot{q}_s - V_s\right], s = \overline{1,n}. \tag{13}$$

4 Synthesis of Multi-model Algorithms for Intelligent Estimation of Motion Parameters Under Conditions of Uncertainty

It is reasonable to obtain the equations of motion using curvilinear coordinates, which are determined by the vector of observations. In this case the additional nonlinear transformations of the vector of random actions on the observation channel in (2) are excluded.

Let us consider the case when observations are made in the polar coordinate system.

$$\mathbf{y}(t) = \begin{bmatrix} h_1 \\ h_2 \end{bmatrix} + \begin{bmatrix} \xi_1 \\ \xi_2 \end{bmatrix} \tag{14}$$

The corresponding model of the kinetic energy for the target of a unit mass has the form [15, 16]:

$$T = \frac{1}{2}\left(\dot{h}_1^2 + h_1^2\dot{h}_2^2\right), \tag{15}$$

then the Lagrange equations of the second kind in the polar coordinate system up to the structure of the generalized forces can be written as follows:

$$\ddot{h}_1 = h_1\dot{h}_2^2 + Q_1,$$
$$\ddot{h}_2 = -\frac{2\dot{h}_1\dot{h}_2}{h_1} + \frac{1}{h_1^2}Q_2. \tag{16}$$

Now we solve the problem to find the generalized forces $Q_1 \in G, Q_2 \in G$ from the minimum condition (3) and the observation Eq. (14). In accordance with (13), (15) the 2D equation of motion takes the form

$$\ddot{h}_1 = h_1\dot{h}_2^2 + \lambda^{-1}\left[-\left|\frac{d\dot{h}_1}{dh_1}\right|\dot{h}_1 - N_{11}^{-1}(y_1 - h_1)\right] = U_{11},$$
$$\ddot{h}_2 = -\frac{2\dot{h}_1\dot{h}_2}{h_1} + \frac{\lambda^{-1}}{h_1^2}\left[-\left|\frac{d\left(h_1^2\right)}{dh_2}\dot{h}_2 + \frac{d\dot{h}_2}{dh_2}h_1^2\right|\dot{h}_2 - N_{22}^{-1}(y_2 - h_2)\right] = U_{12}. \tag{17}$$

Under conditions of low maneuver intensity, characterized by a time constant $\alpha \leq \frac{1}{60}c^{-1}$ [4], it is reasonable to use a simpler model of the kinetic energy

$$T = \frac{1}{2}\left(\dot{h}_1^2 + \dot{h}_2^2\right), \tag{18}$$

which, in accordance with [12], leads to the following quasilinear equations of motion:

$$\ddot{h}_1 = -\sqrt{\lambda^{-1}}\dot{h}_1 - \lambda^{-1}N_{11}^{-1}(y_1 - h_1) = U_{21},$$
$$\ddot{h}_2 = -\sqrt{\lambda^{-1}}\dot{h}_2 - \lambda^{-1}N_{22}^{-1}(y_2 - h_2) = U_{22}. \tag{19}$$

Then we perform the comparative analysis of (18) and (19) using the mathematical modeling of target motion along trajectories with different maneuver intensities and maneuver time constants. It is assumed that the distance to the locator is 18.3 km, the target speed is 300 m/s, the maximum overload is 4 g [4]; rms noise observations 140 m and $4 \cdot 10^{-3}$ rad. The results are summarized in Table 1.

The uncertainty with respect to the coordinates of the points of the phase space, where the structures of the dynamic system change, can be described by the input linguistic variable "motion mode" with a term set $T = \left\{"\alpha > \frac{1}{60}", "\alpha \leq \frac{1}{60}"\right\}$ with the membership S-function [18]. The multi-model algorithm for intelligent estimation of motion parameters is defined by a TS-model of a MISO type system with a rule base

$$R^1: \text{if } \ddot{q} \text{ is } A_1, \text{ then } \left(\ddot{h}_1 = U_{11} \text{ и } \ddot{h}_2 = U_{12}\right), \text{ else}$$
$$R^2: \text{if } \ddot{q} \text{ is } A_2, \text{ then } \left(\ddot{h}_1 = U_{21} \text{ и } \ddot{h}_2 = U_{22}\right), \tag{20}$$

where A_1, A_2 are the fuzzy sets corresponding to the motion of the maneuvering and non-maneuvering target.

Table 1. Trajectory 1

Intensity of maneuver $\sigma^2_{\tilde{h}_1}$, m^2/s^4, and $\sigma^2_{\tilde{h}_2}$, rad^2/s^4; the maneuver constant α s^{-1}	Calculations using (17): $\overline{(y_1 - h_1)^2}$, м^2	Calculations using (17): $\overline{(y_1 - h_1)^2}$, рад^2	Calculations using (19): $\overline{(y_1 - h_1)^2}$, м^2	Calculations using (19): $\overline{(y_1 - h_1)^2}$, рад^2
$\sigma^2_{\tilde{h}_1} = 460$ m^2/s^4, $\sigma^2_{\tilde{h}_2} = 1{,}4 \cdot 10^{-6}$ rad^2/s^4, $\alpha = 0.1$ (target of "airplane" class, slow turn [4])	11,025.3	$0,24 \cdot 10^{-6}$	11,019.0	$0,23 \cdot 10^{-6}$
$\sigma^2_{\tilde{h}_1} = 460$ m^2/s^4, $\sigma^2_{\tilde{h}_2} = 1{,}4 \cdot 10^{-6}$ rad^2/s^4, $\alpha = 0.5$ (target of "airplane" class, counter shooting maneuver [4])	12,524.7	$0,26 \cdot 10^{-6}$	14,156.1	$0,28 \cdot 10^{-6}$
$\sigma^2_{\tilde{h}_1} = 460$ m^2/s^4, $\sigma^2_{\tilde{h}_2} = 1{,}4 \cdot 10^{-6}$ rad^2/s^4, $\alpha = 1$ (target of "airplane" class, maneuver because of atmospheric turbulence [4])	13,157.2	$0,32 \cdot 10^{-6}$	15,741.0	$0,35 \cdot 10^{-6}$

5 Conclusion

A multi-mode dynamic mathematical model of motion constructed using fuzzy logic is based on a 2D dynamic model synthesized using the maximum condition for the function of generalized power. The results of mathematical modeling show that application of the model provides a gain in terms of the accuracy of the estimation of motion parameters with intensive target maneuver, on average above 8%, depending on the type of maneuver. The developed TS-model of a MISO-type system with a rule base defines a multi-model algorithm for intelligent estimation of motion parameters.

The work is carried out under the RFBR grants No. 18-01-00385 A, No. 18-08-01494 A, No 18-38-00937 mol_a.

References

1. Lyakhov, S.V.: Determination of the parameters of the dynamics of the movement of the vehicle during test maneuvers based on the experimentally calculated path. Mech. Mach. Mech. Mater. **3**, 74–96 (2014)
2. Li, X.R., Vesselin, P.G.: Survey of maneuvering target tracking. part I: dynamic models. IEEE Trans. Aerosp. Electr. Syst. **4**, 1333–1364 (2003)
3. Bar-Shalom, Y., Rong, Li X., Kirubarajan, T.: Estimation with applications to tracking and navigation. Wiley, New York (2001)
4. Singer, R.A.: Estimating optimal tracking filter performance for manned maneuvering targets. IEEE Trans. Aerosp. Electr. Syst. **4**, 473–483 (1970)

5. Grinyak, V.M.: Fuzzy determination of the motion nature with multi-model tracking of the craft trajectory by surveillance radar. Neurocomput. Dev. Appl. **6**, 13–20 (2013)
6. Kostoglotov, A.A., Kostoglotov, A.I., Lazarenko, S.V.: Synthesis of systems with optimal speed based on the combined maximum principle. J. Inf. Measuring Control Syst. **15**, 34–40 (2007)
7. Kostoglotov, A.A., Kostoglotov, A.I., Lazarenko, S.V.: The combined-maximum principle in problems of estimating the motion parameters of a maneuvering aircraft. J. Commun. Technol. Electr. **4**, 431–438 (2009)
8. Kostoglotov, A.A., Lazarenko, S.V.: Nonsmooth analysis in measurement processing. Meas. Tech. **2**, 117–124 (2009)
9. Andrashitov, D.S., Kostoglotov, A.A., Kuznetsov, A.A., Lazarenko, S.V., Andrashitov, D. S., Kostoglotov, A.A., Kuznetsov, A.A., Lazarenko, S.V.: Structural synthesis of lagrangian systems of automatic control with the use of first integrals of motion. J. Inf. Measuring Control Syst. **12**, 12–18 (2015)
10. Kostoglotov, A.A., Kuznetcov, A.A., Lazarenko, S.V., Losev, V.A.: Synthesis of the tracking filter with structural adaptation based on the combined maximum principle. J. Inf. Control Syst. **4**, 2–9 (2015)
11. Kostoglotov, A.A., Kuznetsov, A.A., Lazarenko, S.V.: Synthesis of model of process with non-stationary perturbations based on the maximum of the generalized power function. Math. Model. **12**, 133–142 (2016)
12. Kostoglotov, A.A., Lazarenko, S.V., Deryabkin, I.V., Manaenkova, O.N., Losev, V.A.: The method of optimal filtering based on the analysis of the behavior of invariants on characteristic trajectories in the phase space. Eng. Bull. Don **4**, 60 (2016)
13. Kostoglotov, A.A., Lazarenko, S.V., Lyaschenko, Z.V.: Intellectualization of measuring systems based on the method of structural adaptation in the construction of tracking filter. In: Proceedings of 2017 20th IEEE International Conference on Soft Computing and Measurements, pp. 568–570 (2017)
14. Kostoglotov, A., Lazarenko, S., Agapov, A., Lyaschenko, Z., Pavlova, I.: Designing the knowledge base for the intelligent inertial regulator based on quasi-optimal synthesis of controls using the combined maximum principle. Adv. Intell. Syst. Comput. **874**, 190–200 (2019)
15. Kostoglotov, A., Lazarenko, S., Penkov, A., Kirillov, I., Manaenkova, O.: Synthesis of adaptive algorithms for estimating the parameters of angular position based on the combined maximum principle. In: Proceedings of the Third International Scientific Conference "Intelligent Information Technologies for Industry" (IITI 2018). Advances in Intelligent Systems and Computing. vol. 874, pp. 107–115. Springer, Cham (2019)
16. Lur'e, A.I.: Analytical Mechanics Moscow: GIFML (1961)
17. Markeev, A.P.: Theoretical Mechanics. Nauka, Moscow (1990)
18. Pegat, A.: Fuzzy Modeling and Control. BINOM. Laboratory of Knowledge, Moscow (2013)

Biotechnical System for the Study of Processes of Increasing Cognitive Activity Through Emotional Stimulation

Natalya Filatova(✉), Natalya Bodrina, Konstantin Sidorov, Pavel Shemaev, and Gennady Vinogradov

Tver State Technical University, Lenina Avenue 25, Tver, Russia
nfilatova99@mail.ru

Abstract. The article discusses the structure and software of the biotechnical system used to study the processes of human cognitive activity during the long-term execution of the same type of computational operations.

The methodical features of the experiments are considered. For the analysis of brain electrical activity, spectral analysis and nonlinear dynamics methods were used. Based on the evaluation of the power spectra, as well as the entropy of signals, the correlation dimension, estimates of maximum vectors and density of points in the center of attractors reconstructed from fragments of electroencephalograph (EEG) signals, signal descriptions for a sliding computation window were created. Given the number of random factors that can cause changes in the characteristics of EEG signals, further analysis was performed using fuzzy algorithms. Emotional stimulation causing weak positive or negative reactions was used to enhance cognitive activity. Control of emotional reactions was carried out using an additional channel for recording signals of electrical activity of facial muscles (EMG signals). The greatest effect was observed in the stimulation of negative emotions, the speed of performing computational operations after emotiogenic stimulation increased in all subjects, and the number of errors decreased. To interpret the assessments of the subject's cognitive activity, Sugeno algorithm was used. The duration of emotion stimulation was determined using the Mamdani fuzzy inference algorithm.

The article presents the results of experimental studies of monitoring algorithms for 5 characteristics of the EEG and EMG signals, as well as the control algorithm for a specific type of cognitive activity.

Keywords: Algorithm · EEG · EMG · Fuzzy set · Stimulated emotion · Cognitive activity

---

This work was supported by grants from the Russian Foundation for Basic Research (№ 17-01-00742, № 18-37-00225)

S. Kovalev et al. (Eds.): IITI 2019, AISC 1156, pp. 548–558, 2020.
https://doi.org/10.1007/978-3-030-50097-9_56

1 Introduction

Emotional state of a person and his mental activity are interrelated. Everybody agrees with this fact. Moreover, at this stage there is a number of directions that implement this relationship.

Depending on the methods of human exposure, there are approaches that use drugs or direct current action on certain parts of the brain (transcranial magnetic or electric stimulation). However, these methods are used in the diagnosis and treatment of certain pathologies. Studying human emotional responses usually includes indirect stimuli (information objects — video, audio, or scent samples) [1–4]. All these approaches are based on monitoring testee's reactions to a presented stimulus, as well as his facial expressions or speech samples. During experimental studies, nonstationarity of testee's characteristics is taken into account, which makes it necessary to carry out long-term tests with the subsequent averaging of marks for each testee.

Studying cognitive activity includes a similar approach. Monitoring is based on electroencephalogram signals (EEG signals) [5–9]. During their analysis, there are attempts to determine frequency intervals in the spectrum of EEG signals, which may be associated with certain types of cognitive activity [6, 10].

An interesting research field is related to training a testee to control certain types of EEG signal rhythms [10]. It can be considered as the first step in creating a method of controlling mental activity.

Based on the experimental results and the results known from publications of other specialists, we have created a specialized system "EmoCognitive" that allows implementing separate variations of an information impact on a testee in order to indirectly control his cognitive activity during certain types of mental operations.

2 Basic Functions of "EmoCognitive"

The "EmoCognitive" system is intended for joint research of cognitive and emotional reactions of a person to external informational stimuli.

The engineering features of this system are presented in [11]. They are associated with using two autonomous channels for recording emotional states and cognitive activity of a testee. This allows the following:

- detecting emotional responses during calculation tasks,
- extracting from EEG signals for a cognitive activity analysis only fragments corresponding to the testee's neutral state,
- changing the type of stimulus material when there is no emotional response during stimulation.

The functionality of the "EmoCognitive" system is quite good and allows not only reproducing various experimental schemes, but also constantly expanding methods and algorithms to process results.

During the experiment, the system allows recording testee's brain electrical activity, cardiogram, heart rate, electromyograph signals (monitoring of muscular responses).

Speech recording is simultaneous. There might also be a recording of an experiment video protocol.

Informational stimuli can be presented using a screen with a video projector or a laptop.

A wide range of functions for processing experimental results includes calculating spectral and correlation characteristics for certain fragments of recorded signals, as well as reconstructing attractors and assessing their characteristics by calculating the correlation dimension, entropy, and other characteristics used to analyze signals by nonlinear dynamics methods.

There is a module for working with fuzzy estimates of these characteristics, as well as an experimental module for managing an information flow, which makes it possible to monitor cognitive activity and stimulate testee's emotions.

3 Methodology for Conducting Experiments with the "EmoCognitive" System

The objectives defined when developing a program of experiments with the "EmoCognitive" system are the following:

- to study the relationship between testee's cognitive activity when performing monotonous calculating operations and the characteristics of bioelectric signals recorded by two independent channels;
- to prove experimentally the relationship between known spectral characteristics [3–6, 12] and special indicators determined by attractors reconstructed using the same signal patterns after analyzing the same EEG signal fragments [8, 13];
- to assess how much the algorithm that controls the stimulator of emotions affects cognitive activity indicators of a testee.

To achieve these objectives, there is a methodology for conducting experiments including a number of mandatory requirements.

1. The duration of the experiment (T) associated with monitoring testee's cognitive activity under monotonous calculating load should not be less than 60 min (usually $T \leq 180$ min).
2. There is one type of arithmetic operation to assess cognitive activity in the experiment.
3. To record bioelectric signals, special sensors are placed on a testee's head, which create inconvenience and propel fatigue. Considering this fact, there are three experiment phases: familiarizing, assembling and working.
4. The duration of a familiarizing (T_f) and working (T_w) phases is subject to variation depending on the goal of the experiment; however the constraints of the type $T_w > T_f$, and $80\%\ T \leq (T_f + T_w)$ should be fulfilled.
5. The assembling phase is represented by a time interval that includes putting sensors on a testee's head. Usually this interval does not exceed 10 min.

6. During the familiarizing phase, a testee must solve about 200 tasks without recording the brain electrical activity and facial expressions. Actually, a testee has time to develop his own method to do the tasks.
7. The working phase undergoes further fragmentation; there is a pre-stimulation (T_{w1}), stimulation (T_{w2}) and post-stimulation (T_{w3}) stage.

At the pre-stimulation stage, a testee continues performing calculation tasks, but there is parallel recording of his brain electrical activity and an electromyogram of facial muscles. The duration of this stage is based on the results of cognitive activity monitoring $T_{w1} = f(ka)$ (where: T_{w1} is the duration of the pre-stimulation stage, ka is cognitive activity). One of the main conditions for a transition to the stimulation stage is a decrease in the cognitive activity $ka(t_j)$ compared with some basic level ($ka(t_j) < ka_{base}$).

The duration of the stimulation stage is determined by a special control algorithm. But in the performed experiments it is usually $T_{w2} \leq 20$ min. At this stage, instead of calculation tasks, a testee watches a video that may consist of several short fragments without pauses or breaks. The stimulus is focused on obtaining a response of a single sign (positive or negative emotion).

At the post-stimulation stage, a testee resumes performing calculation tasks. The duration of this stage also depends on the control algorithm.

8. To create individual markers with basic values of bioelectric signals, after completing the calculation experiment, there goes synchronous recording of EEG and EMG signals from a testee, who is in a quiet state (with closed eyes and then with open eyes), for 4–5 min.
9. A mandatory requirement of the experiment is testee's own assessment of the level of their emotional reactions when perceiving a stimulus, as well as the level of fatigue throughout the experiment and the complexity of tasks.

4 Experimental Results with the "EmoCognitive" System

The experiments involved 10 men aged 20–30 years. During the experiment a testee performed cognitive tasks – he did sums on the multiplication of two-digit numbers by one-digit number in his mind.

The success indicators of this task performance were:

- the number of incorrect answers when solving each group of tasks ("errors");
- the number of incorrect answers to the same question ("repeated mistakes");
- time for doing each group of 10 tasks ("time").

Changes in cognitive activity, as well as emotional response when perceiving stimulus material were determined by comparing the variations of individual characteristics found from fragments of EEG signals (cognitive activity) or EMG signals (emotional response).

Spectral characteristics are typically used during the analysis of bioelectric signals. But many papers [7, 8, 12, 14–16] show that the methods of nonlinear dynamics and characteristics of special phase portraits reconstructed by bioelectric signals (attractors) are used for these purposes. Our previous experience in analyzing EEG signals by the methods of nonlinear dynamics has shown good results; we have proposed new characteristics that showed changes in an attractor structure [7, 11–15].

To illustrate the efficiency of using nonlinear dynamics and new characteristics, we compared calculation results for the same fragments of EEG signals of characteristics of power spectra, signal entropy, correlation dimension, estimates of maximum vectors and point density in the center of attractors [7] reconstructed by EEG signals. The listed characteristics are determined for 15 leads, in which electrical activity of testee's brain was recorded during the performance of one task. The fragment illustrates the solution with the maximum calculation speed for this testee.

Comparative analysis of the results has shown that trend directions of nonlinear dynamics features coincide with trend directions of spectral characteristics in most leads (87%) (Fig. 1).

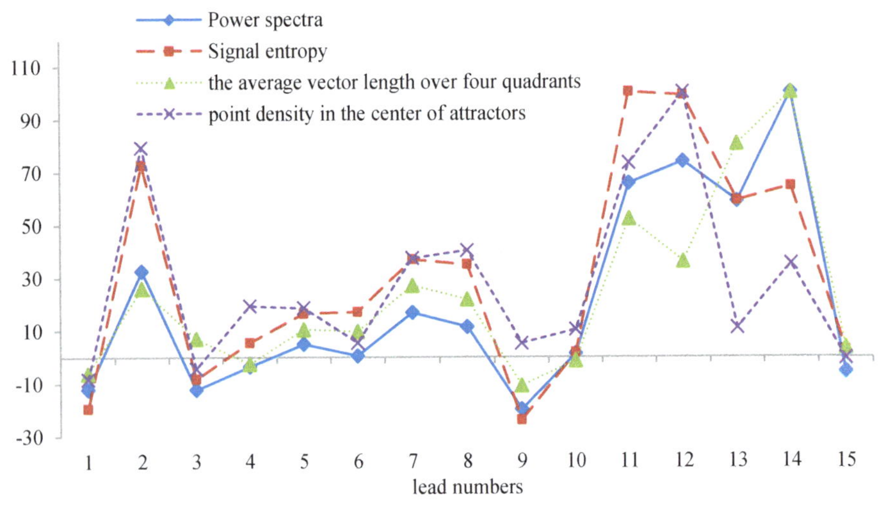

Fig. 1. Changes in spectral characteristics and nonlinear dynamics features by 15 leads of the electroencephalogram when solving calculation operations

The data on the effect of emotions on the indicated characteristics known from publications are confirmed. However, their sensitivity varies significantly (Table 1). The highest average estimates of changes (δ) in features by 15 leads are typical for nonlinear dynamics features (entropy and point density in the center of attractors). The feature determined by the power spectrum turns out to be the roughest compared with the characteristics of attractors reconstructed by signals at certain points of a testee's head.

After comparison of the task performance results at separate stages of the working phase we determined two patterns:

- the maximum effect was when negative emotion stimulation,
- all testees showed that the speed of performing calculating operations has increased after emotiogenic stimulation (Fig. 2).

Table 1. Calculated values

Characteristic	A lead with δ max	δ% (mean value considering a sign)	δ% (absolute value average)
Power spectrum area	F7-A1	20.6%	27.9%
Entropy	T4-A2	30.4%	37.3%
Maximum vector of an attractor	F7-A1	23.4%	26.3%
Density in the attractor center	T3-A1	27.9%	29.8%

After negative stimulation, the testees performed tasks by an average of 20% faster (Fig. 3), and the number of errors decreased by 16%. However, the number of repeated errors increased by 11%.

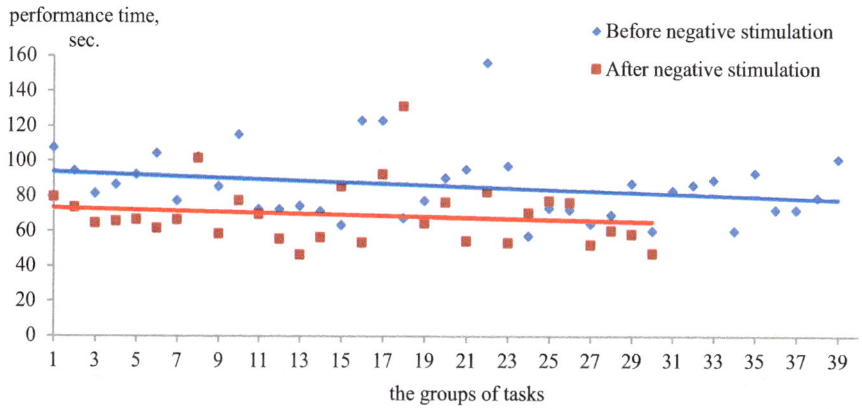

Fig. 2. The group task performance time at the stages before (T_{w1}) and after (T_{w3}) negative stimulation

The "EmoCognitive" system allows filtering bioelectric signals. Due to this system, the set of characteristics for monitoring EEG has extended. In particular, in addition to the average estimate of the power spectrum by all leads, a limited number of leads can be used to monitor cognitive activity (we distinguish leads P3-A1 and P4-A2).

Theta (4–8 Hz), Beta 1 (13–24 Hz) and Beta 2 (24–35 Hz) biorhythms seem to be the most informative frequency ranges, which show the variation of spectral characteristics of cognitive activity. According to the experiments, after stimulation procedures with a negative stimulus, the testees experienced an increase in the estimates of absolute power values – the area of the power spectrum graph for the specified frequency ranges.

The methods of nonlinear dynamics are also used to analyze biomedical signals recorded during experiments. The reconstruction of attractors for a sequence of fragments of electrical signals makes it possible to trace the dynamics of cognitive activity and emotional responses of a testee by changes in the characteristics of these structures. The monitoring algorithms use two attractor characteristics: the average vector length over four quadrants (R) and the total density at the center of the attractor (γ) [7, 15].

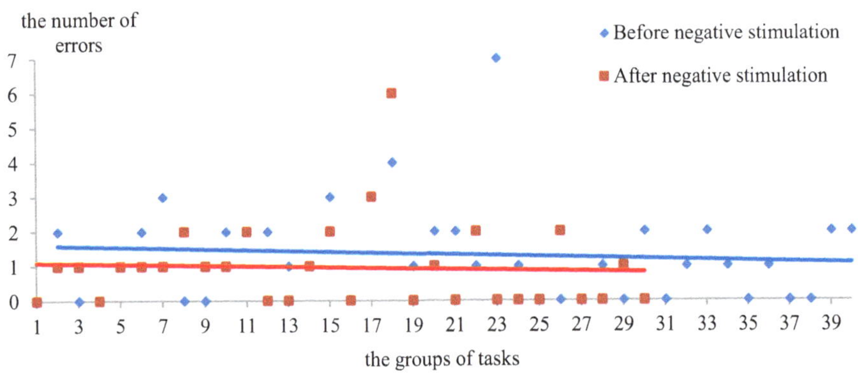

Fig. 3. The number of errors in groups of tasks before (T_{w1}) and after (T_{w3}) negative stimulation

5 The Cognitive Activity Control Algorithm

The algorithm for controlling testee's cognitive activity is based on the assumption that there is a certain average level of estimates of his characteristics (Yav), determined by EEG signals at a certain time interval (T). This interval corresponds to one session of testee's work when he performs one type of cognitive task.

Thus, setting up an algorithm for a new testee is reduced to the definition of the following:

- average estimates of the characteristics of EEG and EMG signals before the start of the working session,
- forming basic scales (LP) for a set of membership functions of the corresponding linguistic variables (input – characteristics of cognitive activity, KA; output – the duration of emotiogenic stimulation, Stimul-time),
- combining new LP scales with basic set of membership functions.

The session includes monitoring EMG and EEG signals and determining spectral characteristics for each calculation window, as well as defining entropy, density, and other characteristics of the attractors reconstructed by the corresponding signals. Based on these estimates, we form a series of $Y(t_j)$ vectors, where j is the number of the calculating window. The vector $Y(t_i)$ characterizes the average cognitive activity of the testee in the time interval (t_i). Then there goes the phasification of $Y(t_i)$ vector components and determination of testee's cognitive activity using the Sugeno algorithm. When cognitive activity decreases (KA < 0.58), the input information flow presented to a testee is replaced with video stimuli. At the initial stages of working with a testee, the stimulation duration (stimul-time) is determined by the Mamdani fuzzy inference algorithm (Fig. 4).

Considering the fact that unaccounted factors might affect the result of perceiving emotiogenic information, the algorithm provides two conditions.

1. If the analysis of EEG signals at t > stimul-time, cognitive activity increases, but the analysis of EMG signals reveals an emotional response, then there is switching to the target information channel and returning to cognitive operations.
2. If the first condition is not fulfilled, the stimulation time is doubled.

Let us consider a fragment of the algorithm while testee's cognitive activity reduces (Fig. 5a) against a gradual decrease in the average estimate of the EEG signal spectral power (Fig. 6).

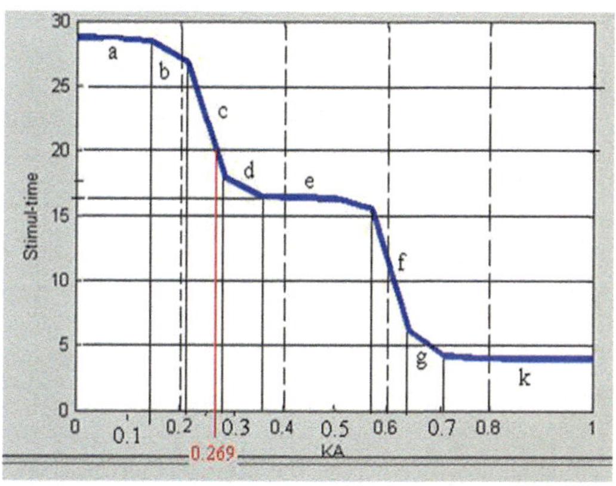

Fig. 4. Determination of stimulation time depending on the cognitive activity (KA) level

To assess cognitive activity, we have chosen the Sugeno algorithm since we do not have a physically based interpretation of the scale to measure this variable.

In this example, the stimulation time, which is determined using the Mamdani algorithm (Fig. 5b), should be at least 978 s (this is the minimum stimulation time).

After the stimulation is completed (Fig. 6), the average estimate of the spectral power increases, which transfers into a fuzzy set "Big" after the 16th fragment. Other characteristics have similar changes. As a result, the cognitive activity estimate changes: KA = Big.

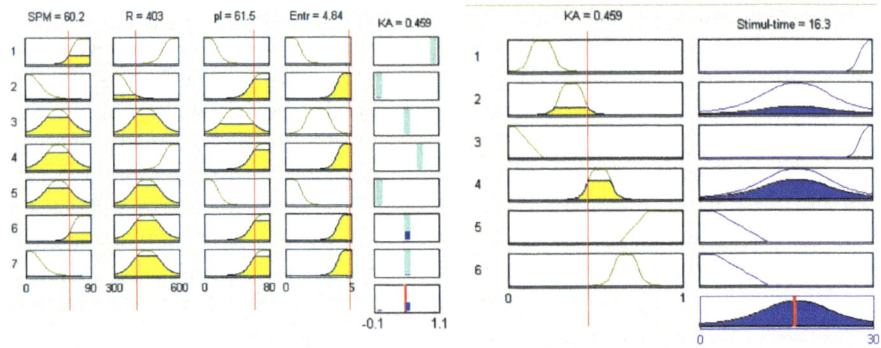

(a) (b)

Fig. 5. The diagrams of the algorithms for assessing KA (a) and the diagrams of the stimulation time (b)

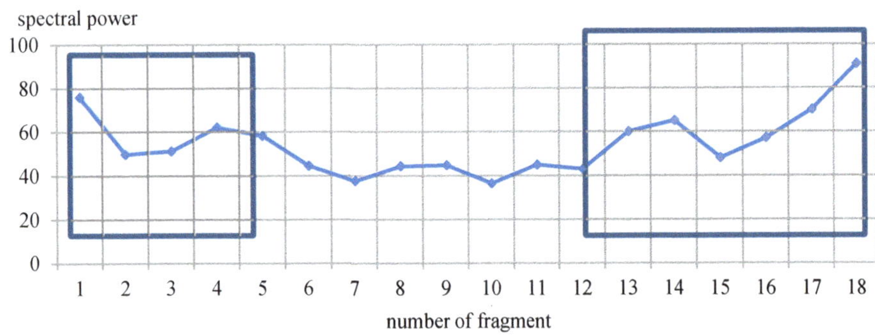

Fig. 6. Changes in the spectral feature during KA control

6 Conclusion

The experiments with the new algorithm have confirmed the possibility of not only monitoring cognitive activity, but also raising its level by non-drug and non-invasive actions. Stimulation of weak negative emotions arising from the perception of video information leads to an increase in the speed of performing computational operations and a reduction in certain types of errors in calculations. The results are preliminary and need further clarification. In particular, it is necessary to estimate the duration of the stimulation effect.

References

1. Lapshina, T.N.: EEG-indication of emotional states of a person. Bull. Moscow State Univ. Psychol. Series **2**, 101–102 (2004). (in Russian Vestnik MGU, seriya: Psihologia)
2. Davidson, R.J.: Affective style and affective disorders: perspectives from affective neuroscience. Cogn. Emot. **12**(3), 307–330 (1998). https://doi.org/10.1080/026999398379628
3. Pomer-Escher, A., Tello, R., Castillo, J., Bastos-Filho, T.: Analysis of mental fatigue in motor imagery and emotional stimulation based on EEG. In: Proceedings of the XXIV Brazilian Congress of Biomedical Engineering "CBEB 2014", pp. 1709–1712. Canal6, Brazil (2014)
4. Başar, E., Güntekin, B.: Review of delta, theta, alpha, beta, and gamma response oscillations in neuropsychiatric disorders. Suppl. Clin. Neurophysiol. **62**, 303–341 (2013). https://doi.org/10.1016/b978-0-7020-5307-8.00019-3
5. Soininen, H., Partanen, J., Paakkonen, A., Koivisto, E., Riekkinen, P.J.: Changes in absolute power values of EEG spectra in the follow-up of Alzheimer's disease. Acta Neurol. Scand. **83**(2), 133–136 (1991). https://doi.org/10.1111/j.1600-0404.1991.tb04662.x
6. Klimesch, W.: EEG alpha and theta oscillations reflect cognitive and memory performance: a review and analysis. Brain Res. Rev. **29**(2–3), 169–195 (1999). https://doi.org/10.1016/s0165-0173(98)00056-3
7. Filatova, N.N., Sidorov, K.V., Shemaev, P.D., Rebrun, I.A., Bodrina, N.I.: Analyzing video information by monitoring bioelectric signals. In: Abraham, A., et al. (Eds.): Proceedings of the Third International Scientific Conference "Intelligent Information Technologies for Industry" (IITI 2018), Advances in Intelligent Systems and Computing 875, vol. 2, pp. 329–339. Springer, Switzerland (2019). https://doi.org/10.1007/978-3-030-01821-4_35
8. Mirzakulova, S.A., Shuvalov, V.P., Mekler, A.A.: Studying network traffic using nonlinear dynamics methods. J. Theor. Appl. Inf. Technol. **95**(21), 5869–5880 (2017)
9. Grissmann, S., Faller, J., Scharinger, C., Spuler, M., Gerjets, P.: Electroencephalography based analysis of working memory load and affective valence in an n-back task with emotional stimuli. Front. Hum. Neurosci. **11**(616), 1–12 (2017). https://doi.org/10.3389/fnhum.2017.00616
10. Reiner, M., Rozengurt, R., Barnea, A.: Better than sleep: theta neurofeedback training accelerates memory consolidation. Biol. Psychol. **95**, 45–53 (2014). https://doi.org/10.1016/j.biopsycho.2013.10.010
11. Filatova, N.N., Bodrina, N.I., Sidorov, K.V., Shemaev, P.D.: Organization of information support for a bioengineering system of emotional response research. In: Proceedings of the XX International Conference "Data Analytics and Management in Data Intensive Domains" DAMDID/RCDL. CEUR Workshop Proceedings, pp. 90–97. CEUR. Moscow, Russia (2018)
12. Sidorov, K.V., Filatova, N.N., Shemaev, P.D.: An interpreter of a human emotional state based on a neural-like hierarchical structure. In: Abraham, A., et al. (Eds.): Proceedings of the Third International Scientific Conference "Intelligent Information Technologies for Industry" (IITI 2018), Advances in Intelligent Systems and Computing 874, vol. 1, pp. 483–492. Springer, Switzerland (2019). https://doi.org/10.1007/978-3-030-01818-4_48
13. Filatova, N.N., Sidorov, K.V.: Computer models of emotions: construction and methods of research. Tver State Technical University (2017). (in Russian Kompyuternye Modeli Emotsy: Postroenie i Metody Issledovaniya)

14. Filatova, N.N., Sidorov, K.V., Shemaev, P.D.: Prediction properties of attractors based on their fuzzy trend. In: Abraham, A., et al. (eds.) Proceedings of the Second International Scientific Conference "Intelligent Information Technologies for Industry" (IITI 2017), Advances in Intelligent Systems and Computing 679, vol. 1, pp. 244–253. Springer, Switzerland (2018). https://doi.org/10.1007/978-3-319-68321-8_25
15. Filatova, N.N., Sidorov, K.V., Shemaev, P.D., Iliasov, L.V.: Monitoring attractor characteristics as a method of objective estimation of testee's emotional state. J. Eng. Appl. Sci. **12**, 9164–9175 (2017). https://doi.org/10.3923/jeasci.2017.9164.9175
16. Rabinovich, M.I., Muezzinoglu, M.K.: Nonlinear dynamics of the brain: emotion and cognition. Adv. Phys. Sci. **180**(4), 371–387 (2010). https://doi.org/10.3367/ufnr.0180.201004b.0371. (in Russian Uspekhi Fizicheskikh Nauk)

The Methodology of Descriptive Analysis of Multidimensional Data Based on Combining of Intelligent Technologies

T. Afanasieva[1(✉)], A. Shutov[2], E. Efremova[2], and E. Bekhtina[2]

[1] Ulyanovsk State Technical University, Ulyanovsk, Russia
tv.afanasjeva@gmail.com
[2] Ulyanovsk State University, Ulyanovsk, Russia

Abstract. There are many intelligent technologies successfully used for descriptive analysis of multidimensional numerical data. The paper focuses on developing the methodology for complex descriptive analysis of such data by their multi-level granulation in groups meaningful for domain experts. For this goal the methodology to combine following intelligent technologies: clustering of numeric data, formal concept analysis, fuzzy scales and linguistic summarizing is proposed. The proposed methodology of analysis is useful for extraction of properties from multidimensional numerical data, starting with the formation of groups of objects similar in quantitative terms, and ending with their linguistic interpretation by propositions included qualitative properties. Basic definitions, problem statement, step by step representing of methodology for complex descriptive analysis of multidimensional numerical data and case study are provided.

Keywords: Descriptive analysis · Multidimensional numerical data · Intelligent technologies · Granulating · Formal concept analysis · Linguistic summarizing

1 Introduction

In the study of complex systems presented in the form of multidimensional models, methods and technologies of data mining are in demand. At the same time, at the first stage of their application, descriptive analytics, which allows to extract knowledge useful for making informed decisions in industry, production and management, becomes the most important. Often, the results of descriptive analytics are used in deeper analysis and predictive analytics methods. Specific tasks of descriptive analytics of multidimensional data include the task of extracting knowledge about groups of objects that have similarities in the values of indicators and about the properties of objects in the obtained groups. A separate task of descriptive analytics is the task of presenting the knowledge in the form of information granules and their linguistic interpretations. To solve these problems, a lot of intelligent technologies have been

The reported study was funded by RFBR, project 20-07-00672

S. Kovalev et al. (Eds.): IITI 2019, AISC 1156, pp. 559–569, 2020.
https://doi.org/10.1007/978-3-030-50097-9_57

developed and used, which, as a rule, are applied separately. Usually, the results of descriptive analytics of multidimensional data are presented in the form of a set of statistical characteristics of indicators and criteria that require further interpretation by experts in the applied domain.

To solve the problem of descriptive analytics of multidimensional data "from beginning to end" clustering, formal concept analysis, fuzzy models and their combinations are recently used, in which information granules are defined by object or/and attributes properties [1–11]. Merge of clustering and interactive visualization [1] was applied to represent meaningful information about patterns, relationships and outliers in clusters in large weather datasets. In the studies [2, 3] authors showed that combining clustering and classification of multidimensional data achieves better predictions than the best baseline. Clustering and formal concept analysis (FCA) [12] were successfully applied to analysis medical tactile images [8]. FCA is data analysis method based on applied lattice theory for extraction and representation of useful information, of some objects (attributes, and properties) and of data relations. In the study [8] FCA applied at data preprocessing step for data clustering, and the obtained results overperform k-means method ones. Although supervised and unsupervised models produce information granules, the latter should be interpreted by domain experts or by some technique of linguistic summary. One of the ways to deal with this task was proposed in [7] where generalizations of FCA to fuzzy context was used. Approach provided documents summarization using two-step clustering was proposed in the work [4]. The result of such summarization was presented as information granule in the form of the set of the best scored sentences. Since information granules summarize the knowledge by a set of judgments about data, they also found application in description of time series behavior [5, 11]. To represent the latter in the form of the propositions fuzzy sets were used, the studies of which were laid in the works [6, 9, 10]. From this point of view, combining clustering, fuzzy models and FCA seems appropriate in multidimensional data analysis to mine granules of different data properties and their levels.

However, the analysis of the relative works for descriptive analysis of multidimensional data allowed us to conclude that such combining of intelligent technologies with linguistic summary of its results to generate a set of judgments about the properties of formed groups is not sufficiently presented in the scientific research. This article focuses on bridging this gap in the descriptive analysis of multidimensional data using granulation methods such as clustering, fuzzy sets and linguistic summarization based on the works of Yager, Zadeh and Kacprzyk [9–11]. Each of applied technologies provides granulating of data of certain level, but together they could extract more useful and complete knowledge.

The rest of the paper is organized as follows. In the second Part the problem of complex descriptive analysis based on granulation of multidimensional data sets is formulated. In the third Part of the article, a technique based on combining of intelligent technologies for solving this problem is proposed, which enables to obtain a description of multidimensional data in the form of propositions. Case study and description of the proposed technique for granulation of multidimensional sets of numerical data is given in Part 4. As an experimental set of multidimensional data, the data of medical observation of male patients with cardiovascular disease was used. In conclusion, the obtained results are discussed, and future works are formulated.

2 Statement of the Problem of the Descriptive Analysis

Consider multidimensional numeric data in terms of many-valued and formal contexts according to FCA [12, 13]. Following to [13] the many-valued context is a set:

$$D = (G, M, W, J), \tag{1}$$

where G is a set of objects, M is a set attributes, W is a set of possible values, $w \in W, w \subseteq \mathbb{R}$ and J is a ternary relation $J \subseteq G \times M \times W$, such that for any $g \in G$, $m \in M$ there exists at least one value $w \in W$, satisfying $(g, m, w) \in J$. The expression $(g, m, w) \in J$ means "the object g has an attribute m with the value w". Suppose that for each attribute $m \in M$ on the set of its values W there is a set of its properties Y. Define the context D_m by the attribute m

$$D_m = \langle M, W, Y, Z \rangle, \tag{2}$$

where $Z \subseteq M \times W \times Y$ is a binary relation, such that for any value w of attribute $m \in M$ there exists at least one value y of its properties Y, satisfying $(m, w, y) \in Z$. Here$(m, w, y) \in Z$ determines granule: "a value w of an attribute m has a property y". Each property $y \in Y$ can be presented by a number, interval, linguistic term and is created on a set W of corresponding M. Suppose that in a set G there are at least two groups of objects having similarity in some subset on W of M. Then with each group and its objects we will associate a certain number $c \in C$ characterizing the type or class of objects. At the other hand, $c \in C$ expresses some abstract property of all objects belonging to some cluster. It allows to define the context D_g by object

$$D_g = \langle G, C, V \rangle, \tag{3}$$

where $V \subseteq G \times C$ is a binary relation such that for any $g \in G$ there exists at least one value $c \in C$, satisfying $(g, c) \in V$. Expression $(g, c) \in V$ determines granule: "the object g contains in the cluster c". The formal context for a many-valued context D is defined by an expression [12]

$$D_f = \langle G, Y, M, I \rangle, \tag{4}$$

where G is a set of objects, Y is a set of properties of attributes $m \in M$, $I \subseteq G \times Y \times M$ is a binary relation. The semantics of the expression $(g, m, y) \in I$ is described by the granule: "object g has property y by attribute m". Summarizing defines a certain general characteristic of a set of objects represented in the form of a many-valued context or in the form of a formal context with respect to an individual attribute or an individual property of this attribute. The result of the summary can be expressed in a quantitative and linguistic form. Summarizing in the form of a quantitative characteristic is the result of the aggregation function: for example, summing objects in each cluster, calculating the numerical values of the cluster centers or a percentage of data with a given property belonging to a certain cluster. Linguistic summarization allows you to move from quantitative characteristics to linguistic characteristics (linguistic assessments), understandable to human and expressing useful information. The result of linguistic summarization is described by a set of propositions. The kinds of abstract propositions which is proposed in the paper for descriptive analysis of multidimensional data are shown in Fig. 1.

P1: g 'contained in cluster' c
P2: 'in cluster' c Q_g g
P3 : g 'has' y 'by' Q_m m
P4 : g 'has' Q_y y
P5 : g 'has' Q_y y 'by' m
P6: Q_g g 'have' y
P7 : Q_g g 'have' y 'by' m
P8 : 'in cluster ' c Q_g g 'have' y
P9 : 'in cluster ' c Q_g g 'have' y 'by' m
P10: Q_g g 'has' y 'by' Q_m m
P11: 'in cluster' c Q_g g 'have' y 'by' Q_m m
P12: Q_g g 'have' Q_y y
P13 : Q_g g 'have' Q_y y 'by' m
P14 : 'in cluster ' c Q_g g 'have' Q_y y
P15: 'in cluster' c Q_g g 'have' Q_y y 'by' m
P16 : Q_g g 'have' Q_y y 'by' Q_m m
P17: 'in cluster' c Q_g g 'have' Q_y y 'by' Q_m m

Fig. 1. The kinds of abstract propositions

The abstract propositions used with taking into account the splitting of data into clusters and the decomposition of attributes by their properties. Along with the usual propositions considered in the form of granules, most of the used abstract propositions can be considered as macro-granules formed on the basis of lower-level granules (e.g., P8, P10-P17). Thus, the considered set of propositions describes multidimensional data at different levels of granulation. In Fig. 1, the following parameters of abstract propositions are provided: $Q = \{Q_g, Q_y, Q_m\}$ denotes a set of linguistic quantifiers represented by linguistic variables based on an estimate of the number of objects, attributes and properties studied; g is an object, m is its attribute, y denotes an attribute property; c determines the cluster number obtained after applying any clustering method. The value of linguistic quantifier is a fuzzy term (e.g., "most", "more than 50%", "very few", "all", "more than half", "less than average", etc.) of linguistic variable constructed on the interval of numerical values of change of the corresponding object of estimation. Therefore, each formed proposition has a degree of truth T, which is determined by the aggregation of degrees of membership of fuzzy terms of quantifiers included in the proposition. The degree of truth is necessary to select the most significant characteristics in the automatic summarization of the objects of study on the set of its attributes. One of the ways of fuzzy estimating of interval of possible values W with linguistic terms (ACL-scale) was described in the work [14].

Let us formulate the statement of the problem of a complex descriptive analysis of multidimensional numerical data. For a given multidimensional data set D, a set of properties Y of its attributes, some method of clustering F, a given threshold ε (ε ∈ [0.5, 1]) to form a collection of propositions P_T from abstract propositions P = {P1, …, P17} having a degree of truth $T \geq \varepsilon$.

Remark. For propositions in which there is no quantifier, the degree of truth is assumed to be 1.

The proposed set of abstract propositions (Fig. 1) in comparison with clusters obtained in the clustering of multidimensional data by traditional methods allows us to clarify the expert knowledge about the properties of a single object and cluster, and to gain an idea about new groups of objects.

3 Methodology of Descriptive Analysis of Multidimensional Data

In accordance with the statement of the problem of descriptive analysis of multidimensional objects, it is proposed to use the following intelligent technologies: teaching without a teacher to create clusters, the technology of linguistic scaling and evaluation of properties, the technology of forming a formal context, necessary to extract additional quantitative characteristics of propositions and technologies of linguistic summarizing to obtain linguistic quantifiers of propositions. Applying of each of the above mention technologies is considered as granulating multidimensional data at some level of abstraction and from different points of view, and enables to form groups of similar objects. Dividing a large amount of data with a large number of attributes into a smaller number of groups with linguistic interpretations makes it much easier to understand them by business and non-technical users. This type of granulation establishes a "generalization" relationship between the object and its granule. A variant of granulating (macro-granulating) of multidimensional data is granulating of previously created granules, as well as their combinations. The method of complex descriptive analysis of multidimensional data includes the following stages, each of which is considered as a granulating of data of a certain level. Inputs of proposed method include multidimensional data set D, a set of properties Y of its attributes, clustering technique F, the set of abstract propositions P and a threshold ε (ε ∈ [0.5, 1]). The output of the method is the set of propositions P_T with degree of truth $T \geq \varepsilon$. Below step by step methodology for solving the problem of complex descriptive analysis of multidimensional data is provided.

1. Creating linguistic scales Y with a set of properties $y \in Y$ for each attribute $m \in M$ on the set of intervals of its values W

$$f1 : W \rightarrow Y \quad (5)$$

2. The formation of the context D_m on attributes

$$f2 : M \rightarrow Y \tag{6}$$

3. The formation of the context on objects D_g. Application of the clustering technique F of objects G and labeling each object with a class C

$$f3 : G \rightarrow C \tag{7}$$

4. Creating a formal context D_f for objects and clusters

$$f4 : G \rightarrow Y \tag{8}$$

$$f5 : C \rightarrow Y \tag{9}$$

5. Generation the numerical estimates of the ratios of objects Kg by the properties Ky and by the attributes Km

$$f6 : D_f \rightarrow K \tag{10}$$

6. Construction of linguistic variables of quantifiers Qg, Qy, Qm on the set of numerical estimates $K = \{Kg,\ Ky,\ Km\}$, respectively.

$$f7 : K \rightarrow Q \tag{11}$$

7. Computing the degree of truth T of quantifiers Qg, Qy, Qm for abstract propositions P and output the propositions P_T having the degree of truth $T \geq \varepsilon$

$$f8 : D_f \rightarrow P_T,\ P_T \subseteq P. \tag{12}$$

4 Case Study

For the experimental study of the proposed methodology of descriptive analysis multidimensional numerical data D(100×29) of medical observation of males with chronic heart failure due to coronary artery disease were used. This study was performed in Department of Cardiology of Central Clinical Hospital, Ulyanovsk, Russian Federation. Follow up was 12 months. A sample of g patients ($g = 100$) was studied by M attributes ($M = 29$). Attributes were considered as indicators, characterizing patients from different point of view, they include

- Clinical and pathophysiology indicators (n = 19) (anemia, atrial fibrillation (AF), myocardial infarction, stroke, peptic ulcer, diabetes mellitus (DM), chronic kidney disease (CKD), arterial hypertension (AH), left ventricular ejection fraction (LVEF), triglyceride, cholesterol, creatinine, glomerular filtration rate (GFR), systolic blood pressure (SBP), diastolic blood pressure (DBP), age, body mass index (BMI), waist circumference, left ventricular mass index (LVMI), psychological and behavioral Indicators – 7 indicators (adherence, smoking, salt deprivation, physical activity, dementia, depression, anxiety) were assessed according to Russian [16], European [17] American [18] and International [19] Guidelines.
- Parameters of social status – 3 parameters (family status, level of education, employment (working/not working) were studied.
- Medical adherence, symptoms of depression and anxiety were investigated according to recently score [20].

In the studied sample, depending on age, patients were distributed: 42–55 years 22%, 56–66 years – 42%, 67–76 years – 26%, 77–85 years – 10%. At the first stage of applying the proposed methodology, linguistic scales were created with a set of properties Y for each indicator $m \in M$ on the set of intervals of acceptable values of W. The following liguistic terms were defined as linguistic labels for the properties: N - the liguistic term belongs to the range of values that are considered from the medical point of view "without pathology" ; A - the liguistic term belongs to the range of values that are considered from a medical point of view "pathology" ; H - the liguistic term belongs to the range of A values, which are considered from the medical point of view "pathology: slight deviation" ; F - the liguistic term belongs to the range of values A, which are considered from a medical point of view "pathology: deviation of the first degree" ; S - the liguistic term belongs to the range of values A, which are considered from a medical point of view "pathology: deviation of the second degree" ; T - the liguistic term belongs to the range of values A, which are considered from a medical point of view "pathology: deviation of the third degree".

For each indicator m ∈ M the range of possible values of W and corresponding linguistic terms Y were determined using ACL-scale [14] on the basis of expert knowledge and practice of medical observation. Using formed linguistic scales, a context D_m for the attributes was obtained, that is, each indicator was associated with its own set of linguistic labels. In the next step, the context D_g by objects was formed by clustering using the k-means technique. As the result of study of quality index 6 clusters was obtained. Study of the quality of clustering was performed on of McClain-RAO index [15], which is defined as the quotient between the mean within cluster and between-cluster distances. After that, a formal context D_f was formed for all objects in each cluster, which allowed to calculate quantitative estimates of Kg, Ky, Km of the presence of y ∈ Y properties for the entire sample of male patients and for the formed clusters. According to the proposed methodology, linguistic variables Qg, Qy, Qm were formed on the set of numerical estimates Kg, Ky, Km. The latter were presented in percentage. This made it possible to develop one linguistic scale with the same fuzzy terms for the quantifiers of propositions, presented in Table 1.

Table 1. Parameters of the triangular membership function of linguistic variable of quantifiers

Linguistic term for Qg, Qy, Qm	The parameters of the triangular membership function %	Linguistic term for Qg, Qy, Qm	The parameters of the triangular membership function %
all	95, 100,100	less than half	30, 40, 50
almost all	85, 91, 100	minority	10, 25, 35
most	65, 70, 90	very few	0, 10, 20
more than half	50, 60, 70	there are no	0, 0, 5
half	45, 50, 55		

As additional crisp linguistic terms for Qg, Qy, Qm characterizing the inter-cluster ratio *Kg*, *Ky*, *Km* the following set of terms was determined: maximum and minimum percentage. Table 2 shows some numerical estimates of the ratios of objects in six clusters for some properties of the attributes *M*, which made it possible to further construct informational granules in the form of the propositions *P*.

Table 2. Numeric estimates of *Kg*, *Ky*, *Km*

Num. cluster	Quantity of patients, *Kg*	Average age, *Km*	% smokers *Ky*	% patients with a maximum deviation of indicators, *Ky*
1	14	63	57%	21%
2	17	74	41%	47%
3	10	63	30%	30%
4	15	64	40%	60%
5	26	58	42%	4%
6	18	61	61%	33%

As a result of the proposed methodology of complex descriptive analysis, propositions P_T at the level of truth equal 0.8 were generated, some of them are given in Table 3. The resulting propositions express the knowledge hidden in multidimensional numerical data obtained through the integration of intelligent technologies, such as formal conceptual analysis, clustering, fuzzy scales and linguistic summarization. The most important limitation is the dependence of the proposed methodology on the quality of the performed clustering of multidimensional data and on relevance of linguistic scales for estimating properties of attributes.

Table 3. The result of complex descriptive analysis in the form of propositions

Proposition	Interpretation in the form of a sentence
$Q_g g$ *'have' y 'by' m*, Q_g = «Almost all» ; $y = A$ m = arterial hypertension	Almost all patients have arterial hypertension
$Q_g g$ *'have' y 'by' m*,Q_g = «Very few» ; $y = A$ m = adherence	Very few patients have adherence to follow medical recommendations
'In cluster' c $Q_g g$, $c = 5$; Q_g = «maximum percentage»	In cluster 5, the maximum percentage of patients
'In cluster' c $Q_g g$ *'have'* $Q_y y$ *'by'* Q_m m $c = 5$; Q_g = «smallest percentage» Q_y = «maximum percentage» , $y = A$; Q_m= «all»	In cluster 5, the smallest percentage of patients has the maximum percentage of deviations in all indicators
'In cluster' c $Q_g g$ *'have' y 'by' m* $c = 5$; Q_g = «all» , $y = N$; m = family	In cluster 5, all patients have a family
'In cluster' c $Q_g g$ *'have' y 'by'* $Q_m m$ $c = 2$; Q_g = «more than half» $y = A$; Q_m = «more than half»	In cluster 2, more than half of the patients have abnormalities in more than half of the indicators
'In cluster' c $Q_g g$ *'have' y 'by' m* $c = 2$; Q_g = «more than hal» $y = N$; m = smoking	In cluster 2, more than half of non-smoking patients
'In cluster' c $Q_g g$ *'have' y 'by'* $Q_m m$ $c = 4$; Q_g = «more than half», $y = A$; Q_m = «most»	In cluster 4, more than half of patients have abnormalities in most indicators
'In cluster' c $Q_g g$ *'have' y 'by' m* $c = 6$; Q_g = «maximum percentage», $y = A$; m = smoking	In cluster 6, the maximum percentage of smoking patients
'In cluster' c $Q_g g$ $c = 3$; Q_g = «smallest percentage»	In cluster 3 the smallest percentage of patients
'In cluster' c $Q_g g$ *'have' y 'by' m*, $c = 3$; Q_g = «smallest percentage», $y = A$; m = smoking	In cluster 3 the smallest percentage of smoking patients
'In cluster' c $Q_g g$ *'have' y 'by' m* $c = 6$; Q_g = «maximum percentage» $y = A$; m = smoking	In cluster 6, the maximum percentage of smoking patients
'In cluster' c Q g *'have' y 'by' m* $c = 6$; Q_g = «there are no» $y = N$; m = body weight	In cluster 6 there are no patients with recommended body weight
'In cluster' c Q_g g *'have' y 'by' m* $c = 6$; Q_g = «all», $y = S$; m = physical activity	In cluster 6, all patients have a minimum physical activity

5 Conclusions

The article proposes a methodology of complex descriptive analytics for multidimensional numerical data based on combining of intelligent technologies: clustering, formal concept analysis, fuzzy estimating and linguistic summarization. At the core of this is its focus on user-friendly and automated knowledge extraction in the form of multi-level granules, represented by propositions. In comparison to clustering of multidimensional numerical data the developed methodology in addition enables to make linguistic description of objects in obtained groups on their properties using formal concept analysis and fuzzy estimates. Moreover, it can search groups of objects, having some similarity in intensity of property without their clustering. In contrast to formal concept analysis, this method assumes that objects satisfy the similarity by properties in terms of both quantitative and qualitative attribute values. The latter are set by linguistic variables using the knowledge of domain experts. This enables to create the so-called 'immersion' of the 'applied' context in the formal and get macro-granules. The application of the descriptive analysis methodology is shown on the data of medical observations. Future work will be aimed at developing types of propositions that unite particular propositions, which will allow us to obtain their generalizations.

References

1. Kocherlakota, S.M., Healey Ch. G.: Interactive Visual Summarization of Multidimensional Data. In: Proceedings of the 2009 IEEE international conference on Systems, Man and Cybernetics (SMC 2009), pp. 362–369, San Antonio, TX, USA (2009)
2. Chakraborty, T.: Combining clustering and classification for ensemble learning. J. Latex Class Files **13**(9), 1–14 (2014)
3. Bini, B.S., Mathew, T.: Clustering and regression techniques for stock prediction. Procedia Technol. **24**, 1248–1255 (2016)
4. Deshpande, A.R., Lobo, L.M.R.J.: Text summarization using clustering technique. Int. J. Eng. Trends Technol. (IJETT) **4**(8), 3348–3351 (2013)
5. Kacprzyk, J., Wilbik, A., Zadrożny, S.: Linguistic summarization of time series under different granulation of describing features. In: Kryszkiewicz, M., Peters, J.F., Rybinski, H., Skowron, A. (eds) Rough Sets and Intelligent Systems Paradigms. RSEISP 2007, LNCS. vol. 4585, Springer, Heidelberg (2007)
6. Boran, E., Akay, D., Yager, R.R.: An overview of methods for linguistic summarization with fuzzy sets. Expert Syst. Appl. **61**(C), 129–144 (2016)
7. Ganter, B., Kuznetsov, S.O.: Pattern structures and their projections. In: Conceptual Structures: Broadening the Base, Lecture Notes in Computer Science, vol. 2120, pp. 129–142. Springer, Heidelberg (2001)
8. Nersisyan, S., Pankratieva, V., Staroverov, V., Podolskii, V.: A. greedy clustering algorithm based on interval pattern concepts and the problem of optimal box positioning. J. Appl. Math. **2017**, 1–9 (2017). Article ID 4323590
9. Yager, R.R., Ford, K.M., Cañas, A.J.: An approach to the linguistic summarization of data. information processing and management of uncertainty in knowledge based systems. In: An Approach to the Linguistic Summarization of Data. Springer-Verlag, Berlin (1991)

10. Zadeh, L.A.: A prototype-centered approach to adding deduction capabilities to search engines – the concept of a protoform. In: Proceedings of the Annual Meeting of the North American Fuzzy Information Processing Society (NAFIPS 2002), pp. 523–525 (2002)
11. Kacprzyk, J., Zadrozny, S.: Linguistic summaries of time series: a powerful tool for discovering knowledge on time varying processes and systems. Informatyka Stosowana **1**, 149–160 (2014)
12. Ganter, B., Wille, R.: Formal Concept Analysis: Mathematical Foundations. Springer Verlag, Berlin (1999)
13. Gugisch, R.: Many-valued context analysis using descriptions. In: 9th International Conference on Conceptual Structures, pp. 157–168, Stanford, CA, USA (2001)
14. Afanasieva, T., Yarushkina, N., Gyskov, G.: ACL-scale as a tool for preprocessing of many-valued. In: The Second International Workshop on Soft Computing Applications and Knowledge Discovery, pp. 2–11 (2016)
15. Charrad, M., Ghazzali, N., Boiteau, V., Niknafs, A.: NbClust: an R package for determining the relevant number of clusters in a data set. J. Stat. Softw. **61**(6), 1–36 (2014)
16. Kardiovaskulyarnaya profilaktika. Natsional'nyye rekomendatsii. Razrabotany komitetom ekspertov rossiyskogo obshchestva kardiologov. Kardiovaskulyarnaya terapiya i profilaktika, **10**(6) (2011)
17. Piepoli, M.F., Hoes, A.W., Agewall, S., Albus, C., Brotons, C., Catapano, A.L., Cooney, M.T., Corrà, U., Cosyns, B., Deaton, C., Graham, I., Hall, M.S., Hobbs, F.D.R., Løchen, M.L., Löllgen, H., Marques-Vidal, P., Perk, J., Prescott, E., Redon, J., Richter, D.J., Sattar, N., Smulders, Y., Tiberi, M., van der Worp, H.B., van Dis, I., Verschuren, W.M.M., Binno, S.: European guidelines on cardiovascular disease prevention in clinical practice. Eur. Heart J. **37**, 2315–2381 (2016)
18. Arnett, D.K., Blumenthal, R.S., Albert, M.A., Buroker, A.B., Goldberger, Z.D., Hahn, E.J., Himmelfarb, C.D., Khera, A., Lloyd-Jones, D., McEvoy, J.W., Michos, E.D., Miedema, M.D., Muñoz, D., Smith Jr., S.C., Virani, S.S., Williams Sr., K.A., Yeboah, J., Ziaeian, B.: ACC/AHA guideline on the primary prevention of cardiovascular disease. J. Am. Coll. Cardiol. **73**(12), 1494–1563 (2019)
19. Levin, A., Stevens, P.E., Bilous, R.W., Coresh, J., De Fran-cisco, A.L.M., De Jong, P.E., Griffith, K.E., Hemmelgarn, B.R., Iseki, K., Lamb, E.J., Levey, A.S., Riella, M.C., Shlipak, M.G., Wang, H., White, C.T., Winearls, C.G.: Kidney disease: improving global outcomes (KDIGO) CKD work group. KDIGO 2012 clinical practice guideline for the evaluation and management of chronic kidney disease. Kidney Int. Suppl. 3, 1–150 (2013)
20. Sheilini, M., Hande, H.M., Prabhu, M.M., Pai, M.S., George, A.: Impact of multimodal interventions on medication nonadherence among elderly hypertensives: a randomized controlled study. Patient Prefer Adherence **13**, 549–559 (2019)

Applied Systems

Method for Achieving the Super Resolution of Photosensitive Matrices

Aleksey Grishentsev[1], Artem Elsukov[1(✉)], Anatoliy Korobeynikov[1,2], and Sergey Arustamov[1]

[1] St. Petersburg National Research University of Information Technologies, Mechanics and Optics, St. Petersburg, Russia
agrishentsev@ya.ru, artemdeimon@gmail.com, sergey.arustamov@gmail.com
[2] Pushkov Institute of Terrestrial Magnetism, Ionosphere and Radio Wave Propagation of the Russian Academy of Sciences St.-Petersburg Filial, St. Petersburg, Russia
korobeynikov_a_g@mail.ru

Abstract. The authors present the method for increasing the spatial resolution of images obtained by photosensitive matrices during their registration. The method is based on the formation of a sequence of frames obtained by sequential displacement of the matrix. Due to the mathematical processing of the generated sequence, the resolution of the resulting image is increased. The authors propose a possible design of the device that implements the displacement of the matrix, due to the magneto-dynamic suspension. The main advantage of this method is an increase of the spatial resolution without reducing the linear dimensions of individual elements of the photosensitive matrix.

Keywords: Per resolution · Inverse convolution · Photosensitive matrix · Digital signal processing · Digital image processing

1 Introduction

One of the important characteristics of digital image recorders is their spatial resolution, the increase of which is an urgent task of improving the quality of electronic digital receiver devices for optical radiation. One possible solution is to reduce the linear dimensions of the individual elements of the photosensitive arrays. The disadvantage of this method is a reduction of signal-to-noise ratio and decrease the sensitivity of individual elements (pixels) of the optical radiation sensor [1]. Thus, there is a problem of increasing the spatial resolution of the device for registering digital images without changing the linear dimensions of the photosensitive elements.

With registering a photo image, a three-dimensional scene is displayed on two-dimensional space; in digital receiving devices of optical radiation, photosensitive elements are located in cells of regular two-dimensional grids [2].

S. Kovalev et al. (Eds.): IITI 2019, AISC 1156, pp. 573–580, 2020.
https://doi.org/10.1007/978-3-030-50097-9_58

In [2] it is stated that only three types of regular grids are possible on the plane: triangular, square and hexagonal. The paper presents a method for increasing the spatial resolution of digital optical radiation receiving devices, in which the photosensitive elements have a mutual arrangement in the form of a square regular grid, as the most common type of arrangement of elements in the sensors used for recording images. The set of all pixels located in the cells of a square grid forms a photosensitive matrix.

Examination of small details by a human eye is achieved if an eye is in continuous motion [3]. In this case, an eye movement can be arbitrary and involuntary. Arbitrary eye movements are carried out when moving the gaze from one point to another. Involuntary movements are continuous and are carried out automatically with a small amplitude. The normal acuity of human vision is reached if a person distinguishes the extreme points of an object at an angle of one minute, while an image of about 5 microns is formed on the retina, which corresponds to the linear dimensions of the three cones. Some people have visual acuity three or more times higher than normal. In [4–7], it was shown that the parameters of eye movement affect visual acuity. Therefore, it is possible to assume that the visual perception of images in living visual systems is influenced not only by direct information coming to the eye in the form of optical radiation, but also objects formed by the neural network of the brain due to eye movement and the observed spatial gradient of light radiation [8]. Such intelligent device sensory systems enable to ensure openness and interoperability [9].

The aim of the research is to develop a method for improving the spatial resolution of image recording devices, by forming a sequence of frames shifted relative to each other and subsequent processing of this sequence, aimed at restoring.

2 Implementation Method

Reception of the image by a photosensitive matrix is accompanied by a limitation of the spatial frequency (Fig. 1). The original image (Fig. 1(a)) falls on a photosensitive matrix (Fig. 1(b)), in which each element generates an output brightness level of a pixel, as the integral sum of the radiation incident on this element. In accordance with the theorem of V. A. Kotelnikov [10] an image is formed at the output of the photosensitive matrix (Fig. 1(c)), with a limited spatial frequency.

With successive displacements of the matrix along the horizontal and vertical axes by a step $\Delta x < l_x$, $\Delta y < l_y$, respectively, where l_x is the horizontal size of one element, l_y is the vertical size of one element, it is possible to get a sequence of different frames, each of which has a size $K_x \times K_y$ where K_x, K_y is the number of rows and columns of the photosensitive matrix. In the future, it will be assumed that the initial position of the matrix is its position with the minimum coordinate along the vertical and horizontal axis. If $\Delta x = \frac{l_x}{N}$ and $\Delta y = \frac{l_y}{M}$, where $N, M \in \mathbb{N}$ s the number of horizontal and vertical displacements,

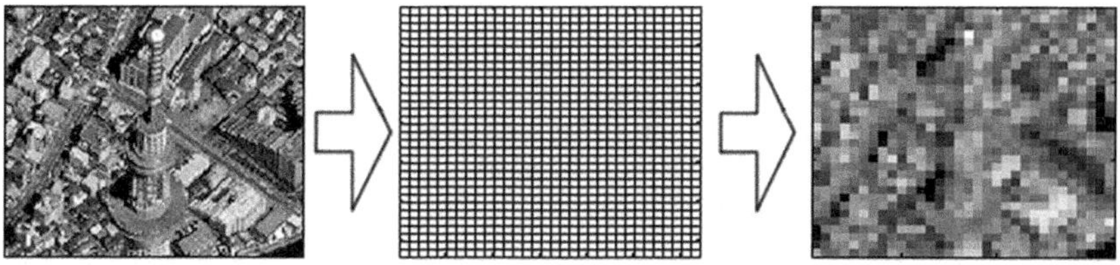

a b c

Fig. 1. Modeling the limitation of the spatial frequency when an image is received by a photosensitive matrix: a - is the original image, b - is the photosensitive matrix, c - is the image of the image

respectively, combining the resulting sequence into one frame g, according to the formula:

$$g[n,m] = q_{uv}\left[\left\lfloor \frac{n}{N} \right\rfloor, \left\lfloor \frac{m}{M} \right\rfloor\right],$$

enables to increase the number of samples of the resulting image, where $0 \leq u < N$ - depends on n and is equal to the remainder of d divided by N, $0 \leq v < N$ - depends on m and is equal to the remainder of dividing m by M, $\lfloor \frac{n}{N} \rfloor$ - rounding the number $\frac{n}{N}$ to a whole downwards. The $q_u v$ frame of size $K_x \times K_y$ is an image obtained when it is registered by a photosensitive matrix shifted by a step $u\Delta x$ and $v\Delta y$ n the horizontal and vertical direction relative to its original position. The image g of the size $NK_x \times MK_y$ may have a large number of samples of the original image falling on the photosensitive matrix in the form of optical radiation, which, according to V. Kotelnikov's theorem, allows g to keep higher spatial frequencies.

Figure 2 illustrates an example of the process of imaging g, with $M = N = 2$ and $K_x = K_y = 4$.

Figure 2(a) contains 4 images with the original image generated by optical radiation and with four different positions of the photosensitive matrix (black grid). After registering the frames in these four matrix positions, a 4×4 q_{uv} sequence of frames is obtained, that is converted into a $p_{uv}[m,n]$ image sequence (Fig. 2(c)) 8×8 in size, that is described as:

$$p_{uv}[n,m] = \begin{cases} q_{uv}\left[\frac{n}{N}, \frac{m}{M}\right], & \text{with } n == u (\text{mod } N) \\ & \text{and } m == v (\text{mod } M) \\ 0 & \text{all other cases} \end{cases}$$

The image g (Fig. 2(d)) is shaped as the sum of all the mappings in the p_{uv} sequence.

Although the g function contains more information, high frequencies are suppressed due to filtering performed by each element of the photosensitive matrix. Therefore, to obtain an image in higher resolution it is necessary to restore the high-frequency components.

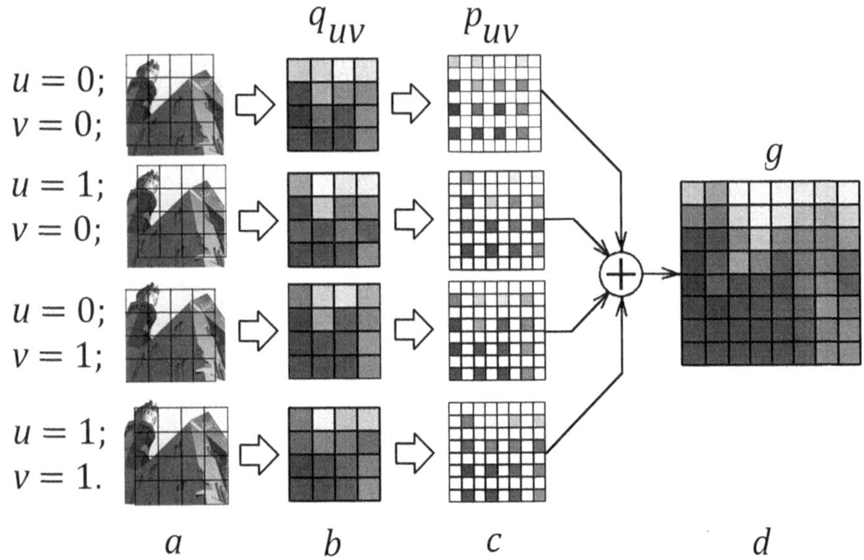

Fig. 2. Getting the sequence of shifted frames: a - the original image; b - matrix receiver of radiation sequentially registers the image at various displacements; c - the received displaced images; d - the sum of displaced images

Shaping the image $s(x, y)$ on the photosensitive matrix can be considered as the operation of convolving the original image $f(x, y)$ with the point spread function (PSF) $h(x, y)$ of each element of the matrix [2,11]:

$$s(x, y) = \int_{-\infty}^{+\infty}\int_{-\infty}^{+\infty} f(x - \alpha, y - \beta)h(\alpha, \beta)d\alpha d\beta = (f * h)(x, y)$$

in the continuous case and the expression:

$$s[n, m] = \sum_{i=0}^{K_x-1}\sum_{j=0}^{K_y-1} f[n - i, m - j]h[i, j] = (f * h)[n, m]$$

in the discrete case, where $(f * h)$ is a convolution operator of f with h.

The convolution of two images f and h in the spatial domain is equal to the product of the spectral images f and h the frequency domain:

$$\mathcal{F}\{f * h\} = \mathcal{F}\{f\}\mathcal{F}\{h\},$$

where $\mathcal{F}$ is the operator of the Fourier transform of two-dimensional functions.

Thus, the restoration of high frequencies in the image g can be carried out using a relation that implements the inverse convolution operation:

$$\mathcal{F}\{w\} = \frac{\mathcal{F}\{g\}}{\mathcal{F}\{h\}},$$

where w is the resulting high resolution image of $NK_x \times MK_y$ size.

When performing a reverse convolution, a significant problem may be the determination of edge information for the restoration of which it is possible to use the following solutions:

1. reducing the size of the edge elements of the photosensitive matrix to the size of the step offset;
2. use of darkening of the photosensitive matrix at the edges so that the extreme pixels are completely in the shadow and receive a signal of zero level or, more precisely, a signal equivalent to the noise level at zero level of the useful signal.

Point 1 contradicts the condition of preserving the size of the elements of the photosensitive matrix, therefore, we consider as an example point 2 - darkening of the edge pixels (Fig. 3). Darkening the edge pixels of the matrix is equivalent to darkening the edges of the original image so that the radiation does not fall on the edge pixels of the matrix. At the points where the values of $\mathcal{F}\{h\}$ are negligible, i.e. less than a certain threshold value, $\mathcal{F}\{h\}$ was assumed to be zero.

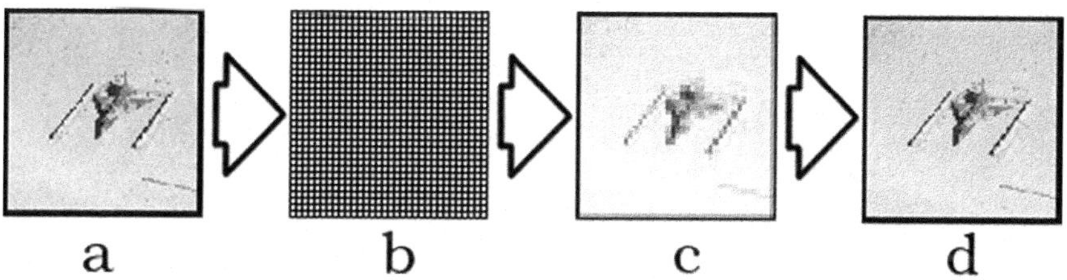

Fig. 3. The original signal-image (a) of the image edge is darkened, which is equivalent to darkening the edges of the matrix; a matrix radiation detector (b) sequentially registers an image at various displacements; three discrete displacements of the matrix along axes are used; i.e. total of nine possible positions; the total signal obtained at various displacements of the matrix (c); reconstructed image in the super resolution (d) using the operation of inverse convolution

3 Physical Basis for Implementation

The essential issue is the technology of displacement of the matrix. To obtain a high quality of the resulting image (in super-resolution), a high precision of the matrix displacement is required. For positioning the photosensitive matrix, magnetic suspensions can be used, similar to those used in optical disc drives (CD, DVD, Blu-Ray) for micromanipulations of the head lens. A diagram of the possible design of the matrix offset device is shown in Fig. 4.

In the passive state, i.e. in the state when the photosensitive matrix is not displaced (1), the platform (2), on which the photosensitive matrix is fixed, is attracted by permanent magnets (3). Their location is fixed by thin metal wires (4), that simultaneously serve as conductors for supplying inductors (5). When applying voltage to the coils of inductance (5), magnetic fields are induced, that

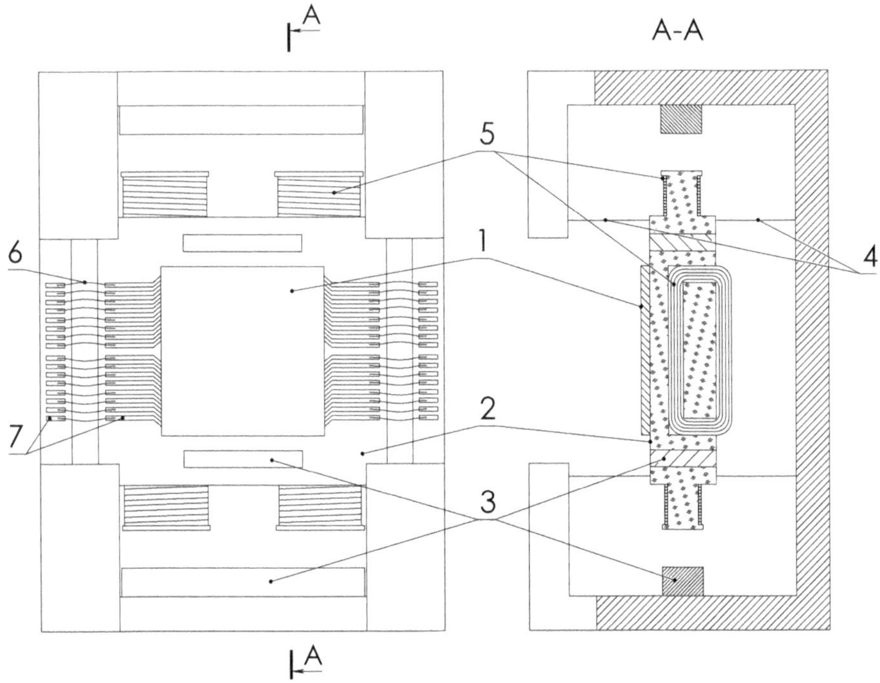

Fig. 4. General view of the magnetodynamic suspension of a photosensitive matrix

interact with the fields of permanent magnets, thereby changing the position of the platform with photosensitive matrix. By varying the voltage, and hence the currents flowing through the coils, we may control the strength of the magnetic interaction and, therefore, the magnitude of the displacement of the platform. Mutual orthogonal orientation of the axes of the inductor enables to change the position of the platform along two mutually orthogonal axes. The connection of the photosensitive matrix with the other devices is carried out using flexible conductors (6), which are rigidly fixed on the contact pads (7).

Sensor technology X3 (X3 Technology) produced by CMOS photosensitive arrays proposed by the Foveon company [12,13] deploys the property of electromagnetic waves, in our case in the optical range, to penetrate to different depths, in our case into the thickness of the semiconductor at different lengths of electromagnetic waves. Registration of various color components is carried out by a single pixel having a total area, and as a result, we do not need to use Bayer's RGB filter. Foveon research shows an increase in the spatial resolution of full-color images due to X3 technology as a result of the abandonment of the Bayer filter [14]. Matrices that do not use the Bayer filter are best suited for the implementation of super-resolution due to bias in the formation of full-color images; other matrices can be used to form monochrome images.

4 Conclusion

The authors present a method for increasing the spatial resolution of images obtained as a result of their registration by a photosensitive matrix. The method is based on sequential registration of shifted frames. Proposed physical and mathematical basis for the implementation of the method.

The main advantages of the considered method of increasing the spatial resolution are:

- the possibility of using photosensitive arrays with a large area of individual photosensitive elements, which ensures the preservation of the high sensitivity of the matrices and the possibility of registering weak signals;
- relative simplicity of hardware and software implementation based on existing technological solutions;
- the possibility of upgrading existing optical receivers without changing the optical channel signal conversion;
- the possibility of upgrading existing optical receivers without changing the optical channel signal conversion;
- the proposed technology is not limited to optical radiation sensors and may be implemented to obtain super resolution in the non-optical range, where matrix radiation receivers are used, for example, in antenna arrays [15].

References

1. Grishentcev, A., Elsukov, A.: Design and analysis of algorithm for smooth stiching of electronic navigation charts. In: 2018 IEEE International Conference on Electrical Engineering and Photonics (EExPolytech), IET – 2018, pp. 123–125 (2018)
2. Jahne, B.: Digital Image Processing, 6th edn. Springer, Berlin (2005)
3. Bates, W.H.: The Bates Method for Better Eyesight Without Glasses. Holt, Rinehart and Winston, New York (1981)
4. Yabrus, A.L.: Rol' dvizhenij glaz v processe zrenija (The role of eye movements in the visual process). Nauka, Moscow (1965). 166 p
5. Kumar, G., Chung, S.T.: Characteristics of fixational eye movements in people with macular disease. Invest. Ophthalmol. Vis. Sci. **55**(8), 5125–5133 (2014)
6. Siritkina, I.V., Fahretdinova, D.A., Koshelev, D.I.: Visual acuity and indicators fixing in violation of central vision various origins, vol. 12, no. 173, pp. 271–275 (2014)
7. Grishentsev, A.Y., Korobeynikov, A.G., Korikov, K.K., Velichko, E.N.: Method for compression of optical observation data based on analysis of differential structure. In: Optical Memory and Neural Networks (Information Optics), IET – 2016, vol. 25, no. 1, pp. 32–39 (2016)
8. Grishentcev, A.Y.: Efficient image compression on the basis of differentiating analysis. J. Radio Electron. (11) (2012). jre.cplire.ru/jre/nov12/index.html. (in Russian)
9. Velichko, E.N., Grishentsev, A., Korikov, C., Korobeynikov, A.G.: On interoperability in distributed geoinformational systems. Lecture Notes in Computer Science (including subseries Lecture Notes in Artificial Intelligence and Lecture Notes in Bioinformatics), IET – 2015, vol. 9247, pp. 496–5045 (2015)

10. Baskakov, S.I.: Circuit and Signals: Textbook, 4th edn. LENAND, Moscow (2016). 528 p
11. Smith, W.S.: Digital Signal Processing, 2nd edn, p. 688. California Technocal Publisher, San Diego (1999)
12. Foveon, Inc. http://www.foveon.com/. Accessed 21 Feb 2019
13. Lyon, R.F., Hubel, P.M.: Eyeing the Camera: Into the Next Century. Foveon, Inc., Santa Clara. www.foveon.com/files/CIC10_Lyon_Hubel_FINAL.pdf
14. Hubel, P.M., Liu, J., Guttosch, R.J.: Spatial Frequency Response of Color Image Sensors: Bayer Color Filters and Foveon X3. Foveon, Inc., Santa Clara. www.foveon.com/files/FrequencyResponse.pdf
15. Velichko, E.N., Grishentsev, A.Y., Korobeynikov, A.G.: Inverse problem of radiofrequency sounding of ionosphere. Int. J. Mod. Phys. A **31**(2–3), 1641033 (2016). IET – 2016

The Information Entropy of Large Technical Systems in Process Adoption of Management Decisions

Yury Polishchuk(✉)

Orenburg State University, Orenburg, Russia
Youra_Polishuk@bk.ru

Abstract. The task of monitoring the information entropy of large technical systems is considered, which consists in the automated control of the quantitative evaluation of the information entropy of the system and allows us to conclude that management solutions can be made based on stored the current data. This task is particularly relevant for large technical systems, as there are discrepancies between the stored factual information characterizing their state and the actual state of the system. The appearance of discrepancies is due to the scale of the controlled system and causes difficulties in the generation of management solutions developed by a group of decision makers. As a practical example, the paper considers the possibility of automated reduction of information entropy for the collection system of the Orenburg gas field consisting of five gas wells.

Keywords: Information entropy of the system · Large technical systems · Management of large technical systems

1 Relevance

The operation process of the large technical systems (LTS) creates the need to make management decisions, and is characterized by high complexity of management, the emergence of non-standard working situations, the need to adapt to environmental disturbances, discrepancies between the actual state of the controlled system and its display in the accompanying exploitation documentation (AED).

The LTS is managed by a group of decision makers (GDM) based on a review of the AED. The latter is due to the complexity of the LTS management process [1].

The process of filling the AED with factual data characterizing the state of the LTS occurs with a certain delay. For this reason, there is a discrepancy between the real state of the LTS and its mapping in the AED. In this case, the likelihood of the adoption of incorrect management decisions of GDM for LTS increases, which is caused by the growth of the information entropy of LTS [2].

S. Kovalev et al. (Eds.): IITI 2019, AISC 1156, pp. 581–588, 2020.
https://doi.org/10.1007/978-3-030-50097-9_59

2 Problem Statement

The filling of the AED with new factual data is realized by means of pulsed information flows (IF), in which factual data is transmitted in the content of electronic documents. Each IF is characterized by the pulse pitch, since the information on it enters the AED after a specified time interval. The duration of the pulse step of the IF is determined by the GDM and can be adjusted if necessary, for example, in the event of non-standard situations on the LTS such as an breakdown, repair, etc.

The factual data stored in the AED, identify the state of the LTS at a particular point in time. Among the AED data, there are two main types of parameters. The first type is static parameters that do not change for LTS depending on time. The second type is dynamic parameters. For this type of parameters, obsolescence is characteristic, i.e. discrepancy between the real value of the LTS parameter at the moment and stored in the AED, due to their dependence on time [3]. To assess the discrepancy between the values of dynamic parameters, GDM can use the obsolescence function. Thus, the novelty of the work is to take into account the value of information entropy when making management decisions for the LTS.

When implementing the management process, the LTS GDM forms an IF and determines for them the values of the impulse steps in such a way as to minimize the obsolescence of the factual data and thereby prevent the adoption of incorrect management decisions. However, the process of operating the LTS is accompanied by non-standard situations and various technological problems that lead to a divergence of the values of real data from the LTS and their display in the AED. In practice, this makes it difficult or even impossible for GDM to identify the state of the LTS at the time of the decision.

For example, consider the informational entropy diagram of the established process of operating a conditional LTS (Fig. 1).

Data on the conditional LTS comes in the form of IF, the number of which is equal, two of them are displayed in the upper diagrams of Fig. 1. According to the first and last IF, the AED receives information on the state of dynamic parameters having various obsolescence functions and pulse steps. The initial IFs make various contributions to the total entropy of the LTS, which is depicted in the lower diagram. For the rest of the IFs $2, \ldots, N_{IF} - 1$ the AED receives information on the state of static parameters that do not change over time, and therefore make a constant fixed contribution to the total entropy of the LTS. In the diagram of the total IF, the threshold level of admissible informational entropy, which determines the GDM, is highlighted. In cases when the graph of the total informational entropy is below the permissible level of the entropy of GDM, a sufficient amount of relevant information is stored in the AED to identify the state of the LTS and make correct management decisions. Otherwise, additional information is required on the state of the managed system.

Thus, the task of monitoring the information entropy of LTS is to automatically keep its values within acceptable limits (limits values).

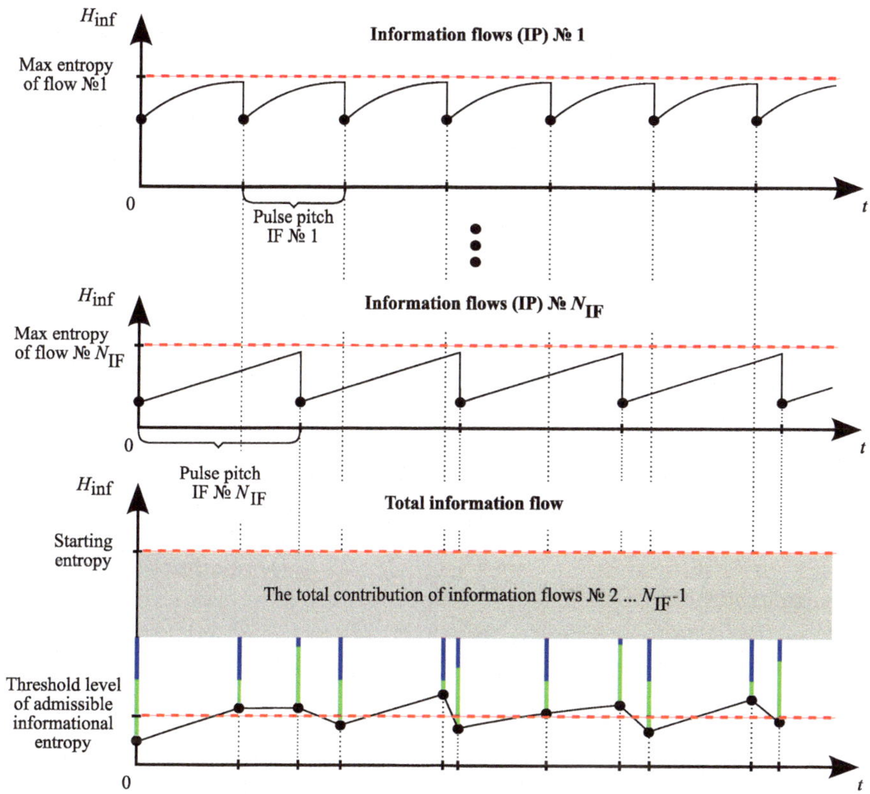

Fig. 1. Diagram of the information entropy of the established process of operation

To calculate the information entropy, apply the following expression [4]:

$$H_{inf} = -\log_2 \left(\frac{F_{st}(P^a_{st}, K_{st}) + F_{dyn}(P^a_{dyn}, F_{obs}, K_{dyn})}{F_{st}(GP^a_{st}(T_{ex}), K_{st}) + F_{dyn}(GP^a_{dyn}(T_{ex}), F_{obs}, K_{dyn})} \right), \quad (1)$$

where F_{st}, F_{dyn} – are the functions that calculate the sums of the products of the power of the sets of static and dynamic (taking into account the obsolescence function) parameters by weights; P^a_{st}, P^a_{dyn} – the set of known, respectively, static and dynamic (with the time of their existence) system parameters; F_{obs} – LTS obsolescence function; K_{st}, K_{dyn} – the set of weights corresponding to static and dynamic parameters; GP^a_{st}, GP^a_{dyn} – functions that generate sets of theoretically possible, respectively, static and dynamic parameters of the system for a specified period; T_{ex} – time of operation of the system.

3 Practical Realization

As an example, the process of controlling the information entropy of a collector-beam mining system (CBMS) for gas and condensate fields [5,6] is considered, which includes two steps: determining the current state of the CBMS information entropy based on the AED analysis and developing recommendations for reducing the CBMS information entropy.

The CBMS includes wells that are separate systems. The contours of the influence of wells CBMS can intersect, and during the transportation of products along pipes in some cases there can be a flow squeezing, leading to a decrease in the efficiency of operation of the CBMS as a whole. The boundaries of the system under consideration are the gas-bearing formation and the block of input filaments (BIF).

The internal characteristic of the CBMS is the gas pressure, since the CBMS boundaries are characterized by known pressures [7]. Gas reservoir pressure is characterized by the geology of the reservoir. The pressure on the BIF is set by the technological mode of operation.

With the movement of products from the reservoir to the BIF, there are losses of static pressure in the wellbore zones, during gas rise to the wellhead and friction about the borehole wall, as well as in the plume [8]. Correct control of GDM can be realized only in the case when structural components and characteristics of the system are known for CBMS. In the conditions of the Orenburg field, the design parameters of CBMS are not fully known for all systems of product collection. The design parameters of the CBMS relate to the variety of static parameters that characterize the completeness of knowledge about the state of the system.

The CBMS dynamic parameters include the values of all types of well pressure and the values of gas, condensate and water flow rates for all wells connected to the collector system. The resulting hydrodynamic well survey document (HWS) contains factual information, including the values of reservoir and downhole pressures of wells. Information about the flow rates and the value of wellhead pressure comes to the AED from the geological and technological report (GTR).

Under the conditions of the Orenburg field, for the dynamic parameters of the CBMS we will use the following obsolescence function:

$$F_{obs} = \exp\left(-\sqrt{T}\right), \tag{2}$$

where T – is the number of months from the receipt of the parameter.

The choice of exponential dependence as a obsolescence function is due to a single exponential view of the base curves of the fall of the main operational indicators of the Orenburg field. For the dynamic parameter obtained in the current month ($T_{ex} = 0$), the value of the obsolescence function is one ($F_{obs} = 1$). The assessment of the significance of dynamic parameters stored in the AED is performed by multiplying the value of the parameter weight by the value of the obsolescence function.

During the operation of the LTS, as a rule, the number of system parameter values stored in the AED is less than the number of the specified GDM.

For example, a HWS in the conditions of the Orenburg field should be performed for each of the production wells once a quarter, but the difficulty of ensuring the planned production volumes at the field does not allow for stopping the wells for carrying out HWS four times a year, since the well stops producing at the time of the research. It is worth noting that the design parameters are not fully known for all the collection systems of the Orenburg field.

Thus, when making management decisions of GDM, it is necessary to take into account the amount of informational entropy for the CBMS.

We use expression (1) taking into account the obsolescence function (2) of estimating the value of the information entropy CBMS. Taking into account the fact that for any CBMS, the AED contains the minimum amount of factual information, its informational entropy value will be equal to the "start entropy", and for the CBMS which have all the factual information specified by the GDM rules in the AED, the informational entropy value will be zero. The value of the "starting entropy" value will depend on the number of LTS description parameters specified by the GDM, taking into account their weight and the obsolescence function.

For CBMS, we determine the values of weighting coefficients for static and dynamic parameters, taking into account their significance in simulating the CBMS and the complexity of their difficulty getting (Table 1).

Table 1. The values of the weighting factors for the parameters of the CBMS

Type	Parameter	Weight
Static	Technical parameters of wells CBMS and pipeline (except for its relief)	30
	Relief of the pipeline	20
Dynamic	The flow rates of gas, condensate, water of the well and wellhead pressure	1
	Reservoir pressure	10
	Down-hole formation pressure	3

The relief of the CBMS pipeline at the Orenburg field is not known for all collection systems, and the remaining design parameters are usually known for the vast majority of CBMS and are stored in the AED. The lack of information on the relief of the pipeline is a serious obstacle in its modeling, which is especially important for the CBMS in which water is present, therefore the design parameters of the CBMS, except for the relief, are separated into a separate static parameter, which has the greatest weight value, and the relief of the is represented by a separate parameter.

Since the GTR for wells that are part of the CBMS, contains information about the flow rates of gas, condensate, water and the value of wellhead pressure can be characterized like by one dynamic parameter.

The values of reservoir and down-hole formation pressures at wells can be characterized by two dynamic parameters. Due to the greater complexity of obtaining reservoir pressure, the weight of its parameter is higher than the weight of the parameter for down-hole formation pressure. The latter is due to the peculiarity of reservoir pressure recovery under the conditions of the Orenburg field.

Using the AED factual information, we will evaluate the informational entropy for similar characteristics of CBMS 1 and CBMS 2. The analyzed collection systems functioned for the entire period of operation in the steady state, their technical parameters did not modify, and the wellhead valves were in the fully open position for the entire period of operation. For CBMS 2 in the AED there is no factual information characterizing the relief of the pipeline.

Thus, applying the CBMS informational entropy estimation method described in the paper, the following results were obtained for the CBMS (Table 2).

Table 2. The results of the calculation of the information entropy

Information entropy	CBMS 1	CBMS 2
Excluding static parameters	0.624	0.490
Including static parameters	0.164	0.711

Assessing the results of the CBMS 1 and CBMS 2 information entropy values, taking into account the factual data stored in the AED, the following conclusion can be formulated: CBMS 2 has a smaller value of informational entropy without taking into account static parameters, which is due to regular HWS for all its wells, but when static parameters are taken into account, the value of its informational entropy is higher than that of CBMS 1. The latter is due to the lack of factual information in the AED about the relief of the pipeline.

The quantitative value of the information entropy allows the GDM to estimate its level for the considered LTS, which can be used to determine the admissible value of entropy, compliance with which will prevent the adoption of incorrect control actions.

For example, as a valid value of the information entropy, one can choose the average value of the entropy for the all analyzed LTS.

For correct management of the system, it is necessary to comply with the condition on the minimum allowable value of the information entropy. In cases when the threshold value of the information entropy of the LTS is exceeded, the development of recommendations for its reduction is implemented. Consider the algorithm for reducing the informational entropy for the CBMS include of five wells.

Suppose that there are three ways available to reduce informational entropy, which implies well surveys conducted by several contracting companies. In this case, for each method, there are limitations on the number of simultaneous studies. The latter is associated with a limitation in the available equipment and personnel of the contracting companies. For each method of reducing the information entropy, the magnitude of its reduction and the complexity of the study for each of the wells are determined, which will be individual, as they depend on the particular design of the well, its location, etc. For each of the wells only one of the ways to reduce the information entropy can be applied, since the simultaneous carrying out of two studies on one well is impossible due to their laboriousness and complexity.

4 Methods

The task of reducing entropy by a certain amount in five wells with minimal labor intensity is similar to the traveling salesman problem, which can be solved by the simplex method in MS Excel using "Finding solution" tool [9,10] (Fig. 2).

Entropy reduction method	Well 1	Well 2	Well 3	Well 4	Well 5	Count of methods	Available methods	Entropy reduction for the method
Method 1	1	0	1	1	0	**3**	4	0,1
Method 2	0	0	0	1	1	**2**	8	0,05
Method 3	1	1	0	0	0	**2**	5	0,25
Total decrease in entropy	**0,35**	**0,25**	**0,1**	**0,15**	**0,05**			

Requirements for reducing the entropy of wells	0,3	0,25	0,1	0,15	0,05

The laboriousness of the work

Entropy reduction method	Well 1	Well 2	Well 3	Well 4	Well 5	Total laborintensity
Method 1	4	8	6	5	6	
Method 2	6	5	8	3	4	
Method 3	20	24	26	30	29	
The laboriousness of reducing the entropy of the well	24	24	6	8	4	**66**

Fig. 2. An example of solving the problem of reducing information entropy in five wells include in one CBMS

When developing recommendations, a decision with a decrease in the information entropy is more than required in cases where it minimizes the overall complexity of the work. For example, for "Well 1", the need for a decrease in the information entropy is 0.3, while the proposed solution has achieved a decrease of 0.35. This is explained by the fact that "Method 1", which reduces entropy by a value of 0.1, is less labor intensive than "Method 2".

5 Conclusion

The described method of automated control of information entropy CBMS increases the efficiency of management decision-making through the automated assessment of the completeness of knowledge about the system and the development of optimal recommendations for reducing it to a permissible threshold value. Also, the method under consideration implements the effective expenditure of funds allocated for the study of wells and CBMS, as it allows to carry out these works on objects with a maximum discrepancy between the actual operating parameters and their values in the accompanying operational content.

References

1. Shafiee, M., Animah, I., Alkali, B., Baglee, D.: Decision support methods and applications in the upstream oil and gas sector. J. Petrol. Sci. Eng. pp. 1173–1186 (2019) https://doi.org/10.1016/j.petrol.2018.10.050
2. Petrov, B.: The Theory of Automatic Control, pp. 1–432. Science, Moscow (1983)
3. Bratvold, R.B., Bickel, J.E., Lohne, H.P.: Value of information in the oil and gas industry: past, present and future. SPE Reservoir Eval. Eng. **12**(4), 630–638 (2009). https://doi.org/10.2118/110378-PA
4. Polishchuk, Y.: Qualimetric identification of large and complex technical systems by related operational content. Bull. Comput. Inf. Technol. **6**, 34–38 (2014)
5. Volkov, N., Kolev, K.: Reference Book Gas Industry Worker, pp. 1–296. Subsoil, Moscow (1989)
6. Makhlouf, A.S.H., Aliofkhazraei, M.: Handbook of materials failure analysis with case studies from the oil and gas industry. In: Handbook of Materials Failure Analysis with Case Studies from the Oil and Gas Industry, pp. 1-430 (2015). https://doi.org/10.1016/C2014-0-01712-1. www.scopus.com
7. Polishchuk, Y.: Modeling of a collector-ray collection system in the conditions of the Orenburg field. Oilfield Bus. **6**, 60–63 (2007)
8. Ermilov, O., Remizov, V., Shirkovsky, A., Chugunov, L.: Reservoir Physics, Mining and Underground Gas Storage, pp. 1–541. Science, Moscow (1996)
9. Gaidyshev, I.: Solving scientific and engineering problems using Excel, VBA and C/C++, pp. 1–512. BHV-Petersburg, Saint-Petersburg (2004)
10. Kirkup, L.: Data analysis for physical scientists: featuring excel. Data analysis for physical scientists: featuring excel, pp. 1–510 (2012). https://doi.org/10.1017/CBO9781139005258. www.scopus.com

Analytical Decision of Adaptive Estimation Task for Measurement Noise Covariance Matrix Based on Irregular Certain Observations

Sergey V. Sokolov[1], Andrey V. Sukhanov[1,2(✉)], Elena G. Chub[1], and Alexander A. Manin[3]

[1] Rostov State Transport University, Rostov-on-Don, Russia
s.v.s.888@yandex.ru, iiti16@rgups.ru
[2] JSC NIIAS, Rostov Branch, Rostov-on-Don, Russia
[3] Moscow Technical University of Communications and Informatics, Rostov-on-Don, Russia

Abstract. The problem of adaptive estimation for measurement noise covariance matrix in Kalman filter is analytically solved based on accurate observations obtained irregularly. The results of numerical modeling are provided. These results illustrate the key advantages of state vector stochastic estimation algorithm based on proposed approach in comparison to conventional one.

Keywords: Irregular accurate observations · Measurement noise covariance matrix · Kalman filter · Adaptive estimation

1 Introduction

Stochastic estimation of dynamical systems' state based on filtration theory techniques and, in particular, Kalman filter, considers accurate prior parameter settings for dynamical system equation and its noises. As well, it requires data about probabilities of measurement noises [1]. However, in practice, parameters of system and measurement noises are very inexact or changed in stochastic manner. From the point of accuracy, the most critical estimation here is covariance matrix, which directly defines the value of filter gain and, consequently, defines convergence rate of estimation process. Nowadays, three approaches are used to provide stability of Kalman filtering in conditions of prior uncertainty of noise covariance matrix: introduction of empirical scale coefficients for posteriori covariance matrix and noise covariance matrix of measurements [2–4], noise covariance matrix scaling [5,6], estimation of noise covariance matrices using condition of minimum covariance of innovation sequence [7,8].

The disadvantage of first two approaches is a lack of strict criteria for choosing scaling coefficients and, consequently, lack of criteria computing. The disadvantage of the last one is impossibility of adaptive estimation in real time because of necessity to prior computing of innovation sequence covariance.

S. Kovalev et al. (Eds.): IITI 2019, AISC 1156, pp. 589–596, 2020.
https://doi.org/10.1007/978-3-030-50097-9_60

Because of this, it is required to develop an approach, which can provide stability for Kalman filter using adaptive computation of elements at noise covariance matrix during current estimation of state vector. To solve this problem, it is considered that the key idea of construction for many information measurement systems is to correct prior measurements from sensors, which have unstable errors, using measurements from another ones, which are considered as references. Usually, correction is performed at certain time intervals that exceed the tact of the primary measurements and are not always the same (often random). As an example, the following systems can be described: inertial satellite navigation systems (NS), where correction is performed for those measurements of inertial NS, errors of which rise during time, by measures of satellite NS, which act as references [9,10]; NS of robots, where correction is realized according to zero speed of its foot (bottom point of wheel) at the time of touching the surface of ground [11]; information-measurement systems of several transportation systems (marine, railway, etc.), where correction is performed at the time of pass through the basis points with known coordinates [12]; combined NS providing navigation inside closed spaces based on inertial sensors [13], etc.

Unfortunately, today the similar problem is decided by utilization of existing methods for the change of real estimations without measures of parameters of estimation algorithm. Obviously, such approach do not decrease errors of estimations at time interval defined by the time of next accurate estimation [11,13–15].

Based on above mentioned, the following opportunity of use obtained accurate observations is considered for construction of adaptive algorithm for formation of measurement noise covariance matrix, which allows to significantly increase the accuracy and stability of estimation process for state vector at all.

2 Problem Statement

Since accurate observations are performed in discrete time points, then discrete Kalman filter is considered for the solution of the problem. The estimation of state vector X_{k+1} at $(k+1)$th time point is defined as follows [1–5]:

$$\hat{X}_{k+1} = \Phi_k \cdot \hat{X}_k + K_k \cdot (Z_k - H_k \cdot \hat{X}_k), \tag{1}$$

where $\hat{X}_k$ is the estimation of state vector at kth time point, Φ_k is the transition matrix, Z_k is the measurement vector:

$$Z_k = H_k \cdot X_k + V_k \tag{2}$$

H_k is the measurement matrix mapping state vector space int measurement vector space; V_k is the centered Gaussian sequence with desired diagonal covariance matrix R_k, which is estimated by accurate observations, K_k is the coefficient of filter gain.

K_k is defined as follows:

$$K_k = P_{k/k-1} \cdot H_k^T \cdot (H_k \cdot P_{k/k-1} \cdot H_k^T + R_k)^{-1}, \tag{3}$$

$$P_{k/k-1} = \varPhi_k \cdot P_{k-1} \cdot \varPhi_k^T + Q_k, \tag{4}$$

where $P_{k/k-1}$ is extrapolated covariance matrix, Q_k is noise covariance matrix.

Based on presented form of K_k, the problem of adaptive estimation of noise covariance matrix by accurate observation can be formulated as problem of computing of R_k based on condition of coincidence of $\hat{X}_{k+1}$ and X_{k+1}.

3 Problem Solution

To solve the problem, the full expression of Kalman gain, which is observed as substitution of (4) in (3), is used:

$$\begin{gathered} K_k = (\varPhi_k \cdot P_{k-1} \cdot \varPhi_k^T + Q_k) \cdot H_k^T \\ \times \left[H_k \cdot (\varPhi_k \cdot P_{k-1} \cdot \varPhi_k^T + Q_k) \cdot H_k^T + R_k\right]^{-1}. \end{gathered} \tag{5}$$

For the convenience of the following description, let $(\varPhi_k \cdot P_{k-1} \cdot \varPhi_k^T + Q_k) \cdot H_k^T = \gamma_k$. Hence, (5) can be presented in the following form:

$$K_k = \gamma_k \cdot (H_k \cdot \gamma_k + R_k)^{-1}. \tag{6}$$

Therefore, Eq. (1) is presented as:

$$\hat{X}_{k+1} - \varPhi_k \cdot \hat{X}_k = \gamma_k \cdot (H_k \cdot \gamma_k + R_k)^{-1} \cdot (Z_k - H_k \cdot \hat{X}_k). \tag{7}$$

Equation (1) is nonlinear vector equation reflecting to R_k, which require the expensive multiple matrix inversion. To analytically solve the problem, the following construction is proposed.

Since problem conditions state that $\hat{X}_{k+1} = X_{k+1}$, then letting $X_{k+1} - \varPhi_k \cdot \hat{X}_k = X_*$ and $Z_k - H_k \cdot \hat{X}_k = Z_*$, the Eq. (7) is presented in the following form:

$$X_* = \gamma_k \cdot (H_k \cdot \gamma_k + R_k)^{-1} \cdot Z_*. \tag{8}$$

The following equation is obtained by multiplying both sides of (8) by inverse matrix $[\gamma_k(H_k\gamma_k + R_k)]^{-1}$:

$$(H_k \cdot \gamma_k + R_k) \cdot {\gamma_k}^{-1} \cdot X_* = Z_*, \tag{9}$$

or

$$R_k \cdot {\gamma_k}^{-1} \cdot X_* = Z_* - H_k \cdot X_*. \tag{10}$$

Let ${\gamma_k}^{-1} \cdot X_*$ as $\varDelta_*$ to simplify the following solution. Then, Eq. (10) easily admits an analytical solution for all elements of R_k if $R_k \varDelta_*$ is possibly represented in form $\varDelta_{*diag} R_{k\ vect}$:

$$\varDelta_{*diag} = \begin{vmatrix} \varDelta_{*1} & 0 & 0 & dots & 0 \\ 0 & \varDelta_{*2} & 0 & \dots & 0 \\ \dots & \dots & \dots & \dots & \dots \\ 0 & 0 & 0 & \dots & \varDelta_{*n} \end{vmatrix}, R_{k\ vect} = \begin{vmatrix} R_{k\ 1} \\ R_{k\ 2} \\ \vdots \\ R_{k\ n} \end{vmatrix}, \tag{11}$$

In this case, the desired expression for vector of elements from covariance matrix is following:

$$\Delta^{-1}{}_{*diag}(Z_* - H_k X_*) = R_{k\ vect} \tag{12}$$

where $\Delta^{-1}{}_{*diag}$ is inverted matrix, which is simply computed because it is diagonal:

$$\Delta^{-1}{}_{*diag} = \begin{vmatrix} \frac{1}{\Delta_{*1}} & 0 & 0 & \dots & 0 \\ 0 & \frac{1}{\Delta_{*2}} & 0 & \dots & 0 \\ \dots & \dots & \dots & \dots & \dots \\ 0 & 0 & 0 & \dots & \frac{1}{\Delta_{*n}} \end{vmatrix}.$$

Therefore, found decision of nonlinear vector equation (7) in form (12) allows to analytically solve the problem of adaptive estimation of noise covariance matrix using accurate measurements.

4 Computational Example

The equations for mass center motion for moving object on spherical Earth in geographical coordinate system in common case of several navigation tasks decisions are in the following form:

$$\begin{aligned} \dot{\phi} &= \frac{V_y}{(r+h)}, \\ \dot{\lambda} &= \frac{V_x}{\cos\phi(r+h)}, \end{aligned} \tag{13}$$

where ϕ and λ are geographical latitude and longitude of an object, r is the Earth radius, h is the object height, V_y and V_x are projections of object speed into the corresponding axes of geographical coordinate system.

The example considers that an object moves from point with coordinates $\phi_0 = 0.8$ rad, $\lambda_0 = 0.3$ rad during time interval [0; 1000 s] with constant speed $V = 20$ m/s along the loxodromic trajectory with azimuth angle $A = 0.2$ rad on surface of Earth, relief of which leads to random changes of object height h with zero mean and variance $Q = (0.15\ \text{m})^2$. In such case, its speeds on geographical axes are the following:

$$\begin{aligned} V_x &= V \cdot \sin A, \\ V_y &= V \cdot \cos A. \end{aligned}$$

NS of an object is a complex system based on integration of inertial and satellite NS. Satellite measures are obtained at intervals of 20, 15 and 30 s. These measures are considered as accurate measurements. Other time points with tact $\tau = 0.1$ s use measures of inertial NS via λ and ϕ channels, where noise measurement matrix is $R_k = \begin{vmatrix} 4.2 \cdot 10^{-10} & 0 \\ 0 & 9.5 \cdot 10^{-10} \end{vmatrix}$.

Because of motion character of an object, the measurement of its coordinate for the certain interval is insignificant, which allow to use linearized navigation equations (13):

$$\dot{\varphi}(t) = \frac{V_y \cdot (r + 2h_0)}{(r + h_0)^2} - \frac{V_y}{(r + h_0)^2} \cdot h\,,$$

$$\dot{\lambda}(t) = \frac{V_x}{\cos\varphi_0 \cdot (r+h_0)} \cdot \left[\frac{r+2h_0}{r+h_0} - \frac{sin\varphi_0 \cdot \varphi_0}{\cos\varphi_0}\right] - \frac{V_x}{\cos\varphi_0 \cdot (r+h_0)^2} \cdot h \tag{14}$$

$$+ \frac{V_x \cdot sin\varphi_0}{\cos^2\varphi_0 \cdot (r+h_0)} \cdot \varphi.$$

Discrete filter constructed using (14) can be presented in form:

$$\hat{X}_{k+1} = \Phi_k \hat{X}_k + \Omega_k + K_k(Z_k - H_k \hat{X}_k) \tag{15}$$

where

$$\hat{X}_k = \begin{vmatrix} \hat{\phi} \\ \hat{\lambda} \end{vmatrix},$$

$$\Phi_k = \begin{vmatrix} 1 & 0 \\ \tau \frac{V_x \tan\phi_0}{r\cos\phi_0} & 1 \end{vmatrix},$$

$$\Omega_k = \begin{vmatrix} \tau \frac{V_y}{r} \\ \tau \frac{V_x}{r\cos\phi_0}(1 - \phi_0 \tan\phi_0) \end{vmatrix},$$

$$Q_k = \begin{vmatrix} (\tau \frac{V_y}{r^2} \cdot 0.15)^2 & \frac{V_y}{r^2} \cdot \frac{V_x}{r^2\cos\phi_0} \cdot (\tau \cdot 0.15)^2 \\ \frac{V_y}{r^2} \cdot \frac{V_x}{r^2\cos\phi_0} \cdot (\tau \cdot 0.15)^2 & (\tau \frac{V_x}{r^2\cos\phi_0} \cdot 0.15)^2 \end{vmatrix},$$

$$H_k = \begin{vmatrix} 1 & 0 \\ 0 & 1 \end{vmatrix}.$$

Since covariance matrix R_k is unknown according to the task conditions, estimation of navigation variables is performed for 2 cases: when value of R_k is computed with error for all interval $\left(R_k = \begin{vmatrix} 1.1 \cdot 10^{-10} & 0 \\ 0 & 2.5 \cdot 10^{-10} \end{vmatrix}\right)$ and when R_k is estimated according to the proposed scheme.

Estimation errors' graph obtained for (15) in the first case is presented in Fig. 1 and Fig. 2, where obvious error is from 100 to 350 m for latitude and from 50 to 200 m for longitude.

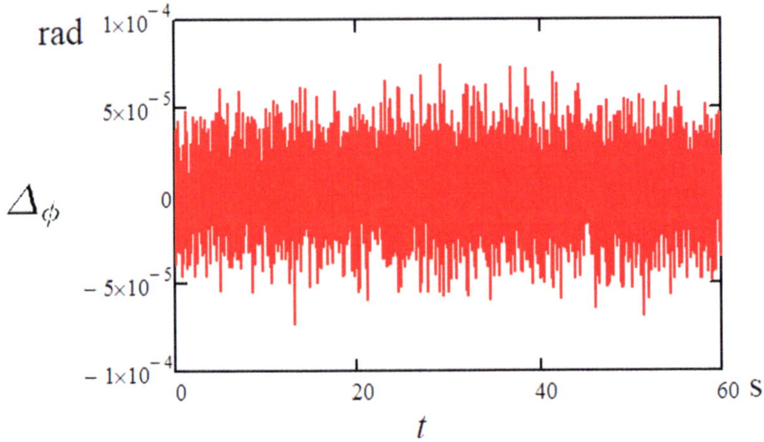

Fig. 1. Estimation errors' graph for ϕ obtained for conventional Kalman filter

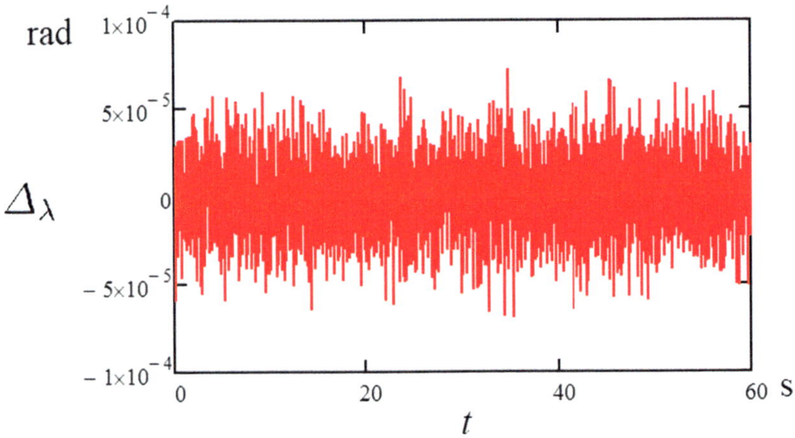

Fig. 2. Estimation errors' graph for λ obtained for conventional Kalman filter

Estimation errors' graph for the proposed case is presented in Fig. 3 and Fig. 4.

It can be seen that proposed algorithm reduce errors with comparison to conventional one (from 2 to 7 m by latitude and from 1.5 to 4 m by longitude).

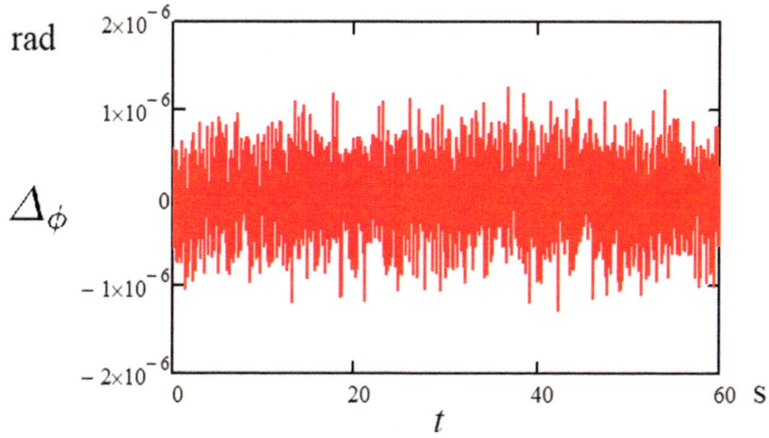

Fig. 3. Estimation errors' graph for ϕ obtained for adaptive Kalman filter

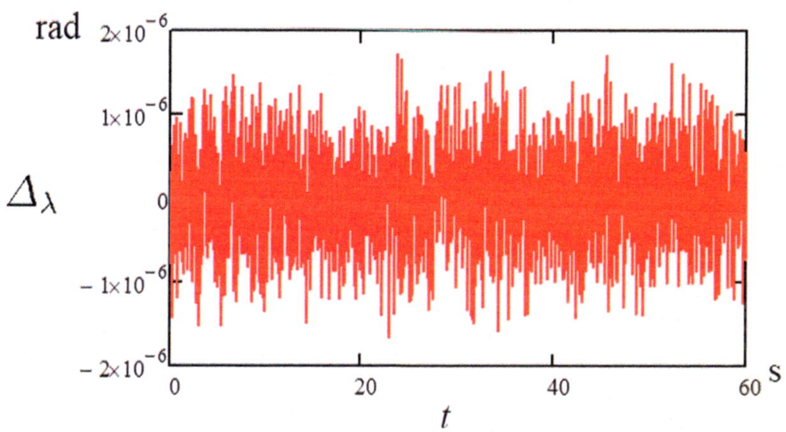

Fig. 4. Estimation errors' graph for λ obtained for adaptive Kalman filter

5 Conclusion

According to comparison of traditional and proposed adaptive filtration, the advantages of the second one are obvious in spite of the fact of more computational costs (these costs are very insignificant in comparison with general filter costs).

Simplicity and accuracy of the proposed algorithm provide the possibility of its effective implementation in wide range of measurement information systems.

Acknowledgement. This work was supported by RFBR (Grants No. 17-20-01040 ofi_m_RZD, No. 18-07-00126).

References

1. Welch, G., Bishop, G., et al.: An introduction to the kalman filter (1995)
2. Anderson, B.D.: Exponential data weighting in the kalman-bucy filter. Inf. Sci. **5**, 217–230 (1973)
3. Sasiadek, J., Wang, Q.: Low cost automation using INS/GPS data fusion for accurate positioning. Robotica **21**(3), 255–260 (2003)
4. Herrera, E.P., Kaufmann, H.: Adaptive methods of Kalman filtering for personal positioning systems. In: Proceedings of the 23rd International Technical Meeting of the Satellite Division of the Institute of Navigation, Portland, OR, USA, pp. 21–24 (2010)
5. Hide, C., Moore, T., Smith, M.: Adaptive Kalman filtering algorithms for integrating GPS and low cost ins. In: PLANS 2004. Position Location and Navigation Symposium (IEEE Cat. No. 04CH37556), pp. 227–233. IEEE (2004)
6. Hu, C., Chen, W., Chen, Y., Liu, D., et al.: Adaptive kalman filtering for vehicle navigation. J. Glob. Positioning Syst. **2**(1), 42–47 (2003)
7. Mehra, R.: On the identification of variances and adaptive kalman filtering. IEEE Trans. Autom. Control **15**(2), 175–184 (1970)
8. Mohamed, A., Schwarz, K.: Adaptive kalman filtering for INS/GPS. J. Geodesy **73**(4), 193–203 (1999)
9. Litvin, M., Malyugina, A., Miller, A., Stepanov, A., Chirkin, D.: Types of errors in inertial navigation systems and methods of their approximation (tipy oshibok v inertsial'nykh navigatsionnykh sistemakh i metody ikh approksimatsii). Informatsionnye processy **14**(4), 326–339 (2014). (in Russian)
10. Reznichenko, V., Maleev, P., Smirnov, M.: Types of errors in inertial navigation systems and methods of their approximation (tipy oshibok v inertsial'nykh navigatsionnykh sistemakh i metody ikh approksimatsii). Navigatsiya i gidrografiya **27**, 25–32 (2008). (in Russian)
11. Looney, M.: Inertial sensors facilitate autonomous operation in mobile robots. Analog Dialogue **44**, 1–4 (2010)
12. Tsyplakov, A.: Introduction to state space modeling (vvedeniye v modelirovaniye v prostranstve sostoyaniy). Kvantyl **9**, 1–24 (2011). (in Russian)
13. Shilina, V.: Inertial sensor system for indoor navigation (sistema inertsial'nykh datchikov dlya navigatsii vnutri pomeshcheniy). Molodezhnyy nauchno-tekhnicheskiy vestnik **4**, 39 (2015). (in Russian)
14. Polyakova, M., Sokolova, O.: Improving the accuracy of adaptive filtering based on the use of non-periodic accurate observations (povysheniye tochnosti adaptivnoy fil'tratsii na osnove ispol'zovaniya neperiodicheskikh tochnykh nablyudeniy). Tekhnologii razrabotki informatsionnykh sistem TRIS, pp. 61–64 (2016). (in Russian)
15. Velikanova, E., Voroshilin, E.: Adaptive filtering of the coordinates of a maneuvering object when the transmission conditions in the radar channel change (adaptivnaya fil'tratsiya koordinat manevriruyushchego ob"yekta pri izmeneniyakh usloviy peredachi v radiolokatsionnom kanale). Doklady Tomskogo gosudarstvennogo universiteta sistem upravleniya i radioelektroniki **2–1** (26), 29–35 (2012). (in Russian)

Analytical Analogies Calculus

Nikolay N. Lyabakh[1,2] and Yakov M. Gibner[1,2(✉)]

[1] Rostov Branch, JSC «NIIAS», Rostov-on-Don, Russia
gibner88@gmail.com
[2] Rostov State Transport University, Rostov-on-Don, Russia

Abstract. Convergence of machine intelligence to nature (human) intelligence is one of the most important tasks of artificial intelligence. Decision making using analogies is one of such gears. The paper improves the mathematical model for describing analogies and proposes a computational mechanism for calculating the parameters of this model. The first and second kind analogies together with the mixed analogies are introduced and mathematical methods for their description are considered.

Keywords: Psycho-emotional basis of artificial intelligence · First and second kind analogies · Mathematical modeling of analogies

1 Introduction

Let the further presentation of the material be introduced with several important remarks defining the essence of the proposed research, which is consideration of the psycho-emotional basis of artificial intelligence (AI):

1. In the study, "artificial intelligence is not identical to machine intelligence" meaning machine intelligence (MI) is intelligence produced by a machine. AI can be formed in the environment of natural intelligence. For example, in the work of expert communities [2], multi-agent systems creation [1] form AI from natural intelligence, if their agents are individuals. AI also arises in the synthesis of systems characterized as the "Internet of Things" [3]. This is the so-called emergent intelligence, and it depends not only on the properties of individuals, but also on the scheme and logic of their interaction. It is well known that the intelligence of the crowd is lower than the intelligence of its individual members. This type of intelligence is an artificial creation (refers to AI), and its value essentially depends on the social environment, its intelligent potential, and the "contamination" of individuals with national and other socially significant ideas and attitudes.
2. Understanding of AI is deformed developing over time. If someone had declared a spell-checking system that automatically corrects grammatical and stylistic mistakes made by man in the middle of the last century, then it was exactly attributed to the area of intelligent activity. Nowadays, this fact is not considered as form of AI.

The work was supported by Russian Fundamental Research Fund, project No. 17-20-01040 and 19-07-00263.

S. Kovalev et al. (Eds.): IITI 2019, AISC 1156, pp. 597–606, 2020.
https://doi.org/10.1007/978-3-030-50097-9_61

Comprehending new and new intelligent functions of man (not always identical to him), we move to the core of natural intelligence (NI). The awareness of the fact that the NI is inextricably linked with the psycho-emotional activity of man is coming. And there is nothing can be done without modeling this human activity when creating AI systems. Computer modeling of insight, reflection, analogies and other attributes of human mental activity is the most important aspect of creating AI.

There are some works in this area. This is, for example, the monograph [4] devoted to the mathematical models of reflection. Following [5, 6], this study is devoted to mathematical modeling of analogies. This can expand the capabilities of MI through the formalization of the algorithms of the NI and AI.

To increase exemplariness of the proposed theoretical material, results of the research are commented on examples, which are obtained during solution of railway sorting tasks, because railway freight sorting station is one of common example of integration of AI and NI [6, 8, 10].

Moreover, to the numerical data, which illustrates analyzed effects of analogues formalization, was generated to perform the computational experiments.

2 Definitions

According to [7], the logical-linguistic definition of analogy is following: "Analogy is the establishment of the similarity of phenomena, objects, processes by any features using association, comparison, reflection; common method of scientific and philosophical research". That is, the analogy is: the decision-making procedure and the method of scientific research. This general definition well emphasizes the mental (intelligent) component of the category under study. An analytical study of this category requires its formalized definition. In [5], two definitions are given:

- *Definition 1.* If objects O_1, O_2, ..., O_n of a certain class have the property S, then it is likely that an object of the same class O_{n+1} has the property S.
- *Definition 2.* If the object O_1 has the properties S_1, S_2, ..., S_n, S_{n+1}, and the object O_2 has the properties S_1, S_2, ..., S_n, then the object O_2 also has the property S_{n+1} with some probability.

Some class of objects can act as S_{n+1}. That is, if the object O_1 belongs to a given class, then the object O_2 belongs to this class if there is an analogy between them.

Next, we consider three types of analogies: the first kind, the second kind, and mixed analogies. In the first case, the analogy "works" with one property of the considered class of objects. In the second case, analogies are derived from a comparison of various properties of objects. Comparative analysis of these concepts is performed on the following examples. In the third case, the effects under consideration act simultaneously.

The following examples describe different types of analogies.

- *Example 1.* An observer deals with objects that are characterized by only one property: grass is characterized by shades of green, animals (including human) are characterized by one of the possible properties (height, weight, knowledge, etc.), cuts, which rolling down from railway hump [8], are characterized by mass or speed, etc. All these objects are presented to the observer in turn, and he concludes about the possible value of the property of the next object according the previous observations. For example, a large-sized hare ran at the hunter twice: there is an increased probability that the third case it would be of the same size (assuming that conditions are obviously favorable for the growth of hares in nature).
- *Example 2.* The observer deals with objects that are characterized by at least two properties. In this case, when the object is presented, in addition to the first effect (property S_1), the second (property S_2) one also appears: the "value" of the property of a new object is largely determined by the value of the property of another. So, during the trains humping, a high speed of rolling corresponds to a cut with a larger mass.

It is obvious that the above formalized definition 1 corresponds to analogies of the first kind, and definition 2 to analogies of the second kind.

3 Mathematical Tools for First Kind Analogies Description

The studied situation is the following: objects are characterized by only one property, the measured quantity of which is denoted further by x_0. Examples are:

1. The broker analyzes the stock market (securities, currencies, prices for key production resources, etc.). The stock price at time i (object O_i) is x_i. Over a period of observations, he developed the idea of their volatility (price spread). If the price is risen sharply, then there is a general recommendation not to buy these stocks, since the price "correction" follows inevitably adjusting it in the opposite direction, and those who bought the stock at an inflated price would lose. This is the reasoning of the intuitive level. It is based on the fact that the observer formed an opinion on the average price of a stock, the magnitude of the spread, and he decides "by analogy" using this past experience. At the same time, the broker does not have any numerical estimates.
2. In the same situation, there may be a different development scenario: the observer fixes prices equal to x_{i0} (he observes objects O_i), where $i = 1, 2, 3, \ldots, n$. These values have some tendency, for example, each following value is greater than the previous one. The broker has a different (than in the first example) analogy: "at time $n + 1$, as it was before, a price increase is expected". Using this analogy, the decision is different: there is a desire to take advantage of the detected trend of growth in the stock price. If he buys some resource, then it makes sense to buy it in reserve, because the price would be increase in future and the cost of products manufactured from this resource would be increased. But, as in the first case, the observer can only estimate the price increment "by eye".

For the considered cases, it is necessary to use various mathematical tools. For the first case, two forms of calculation are proposed:

1. After checking the stationarity of the observations, the average x_c is calculated. The value of $x_{n+1,0}$ is calculated as follows:

$$x_{n+1,0} = x_c - \left(x_{n,0} - x_c\right) = 2x_c - x_{n,0} \tag{1}$$

In other words, the symmetric value in relation to x_c is proposed as solution. It preserves the found tendency.

Example. Let data be established by Table 1 (rows 1 and 2) (Fig. 1):

Table 1. Initial data of illustrative calculations.

i	1	2	3	4	5	6	7	8	9	10
x_{i0}	3	5	3	4	6	2	4	5	3	5
x_{i0}	1	1	2	3	4	3	4	5	5	6

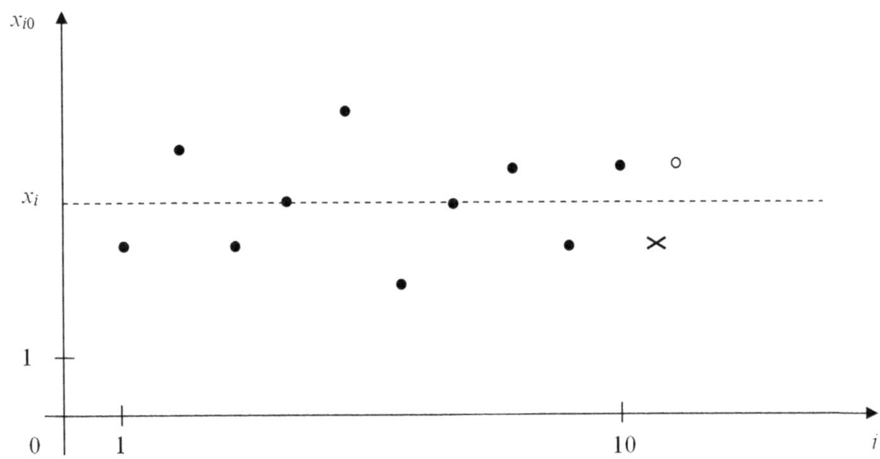

Fig. 1. Graphical illustration of decision making by mean measure

It follows from table that $x_c = 4$. If $x_{n,0} = 3$ (crosses in Fig. 1), then $x_{n+1,0} = 5$ (circles in Fig. 1). It follows from (1).

2. The second form of the calculation is more complicated, but it carries more information about the making decision. The following algorithm can be implemented:
   a) Build the distribution histogram for the observed x_{i0}, calculate the characteristics of the appropriate distribution laws, check the hypothesis about the compliance of the obtained law with the empirical data.

b) Calculate various numerical characteristics of the distribution: mean value, volatility, traditional indicators of the variance of the studied random variable (variance, standard deviation), mode, median, etc.
c) Select the value x_{n+1}, 0 (object O_{n+1}) and evaluate its parameters (reliability, the boundaries of the choice).

Now x is not chosen just "by analogy", but the choice is confirmed with numerical characteristics, i.e. the analogy is described clearly and statistically substantiated.

In the second case, the time series (TS) tool is well suited to describe the analogy. After checking the stability of the observed trend (trend assessment) for the training sample x_i, where $i = 1, 2, 3, \ldots, n$, TS autocorrelation model is built as follows:

$$x_{n+1,0} = a_1 x_1 + a_2 x_2 + \ldots a_n x_{n,0}, \tag{2}$$

This model considers the observed prehistory (previous values of x). Theory of TS is well considered in [9], which allows to continue without describing this question.

The following particular approach is considered for TS modeling.

Example (data is given in Table 1, rows 1 and 2, and Fig. 2).

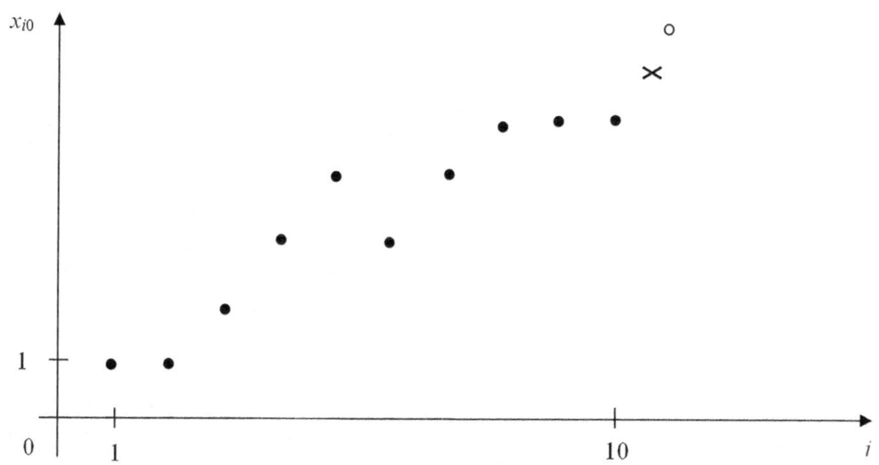

Fig. 2. Graphical illustration of decision making based on TS prediction

The simplified model (2) can be constructed using only one previous value:

$$x_{n+1,0} = a x_{n,0}, \tag{3}$$

Unknown coefficient a is calculated using certain values for every pair $x_{i+1,0}$ and $x_{i,0}$ from Table 1 ($a = \frac{x_{n+1,0}}{x_{n,0}}$ follows from (2)) and averaged. As a result, a = 1.263.

Thus, if $x_{n,0}$ = 6 (cross in Fig. 2), then $x_{n+1,0}$ = 7.58 (circle in Fig. 2) using Eq. (3).

4 Mathematical Tools for Second Kind Analogies Description

Example. Rolling cut O_i approaches to the brake position (BP). A human assessing the situation should make a decision about applying the various control actions on the BP (stage of braking, moment of its switching on and off, duration of braking). The analyzed situation is characterized by a number of properties: S_1 is the mass of the cut (x_1), S_2 is the unrolling velocity of the cut (x_2), S_3 is the distance to the foregoing cut (x_3), S_4 is the distance to the following cut (x_4), S_5 is the number of cars in the cut (x_5), etc.

Human operator applies one or another control action, then he can see the result and accumulate the experience. The experience (or its positive points) are stored in a training set (Table 2).

Table 2. Initial data of model for predicting the object's properties by analogy.

O_i	S_1 (x_1)	S_2 (x_2)	...	S_m (x_m)
O_1	x_{11}	x_{12}	...	x_{1m}
O_2	x_{21}	x_{22}	...	x_{2m}
...	...	...	...	...
O_n	x_{n1}	x_{n2}	...	x_{nm}
O_{n+1}	?	$x_{n+1,2}$	...	$x_{n+1,m}$

The first row shows the semantic content of the columns: O_i is the column of the researched objects (i = 1, 2, ... n), S_j is the j-th property of the object, which is characterized by x_j (j = 1, 2,..., m).

The most common case considers x be the membership grade of j-th property to i-th object. This representation assumes applying the theory of fuzzy sets [10]. So, the columns of the table are the membership functions of a given property to objects, i.e. $\mu S_j(O_i)$, and the rows of the table are the membership functions of a specific object to properties, i.e. μO_i (S_i).

Rows of the table from the second to the n-th are the training data set: the relationships of all objects with all properties are known. According to it, a model should be built to decide whether the property S_1 belongs to O_{n+1} object, if the measured values of the attributes are specified (the membership degrees of the properties $S_2 - S_m$ to this object), i.e. the values $x_{n+1,2}, \ldots, x_{n+1,m}$, respectively.

Similarly, it is possible to establish a problem regarding the membership of other properties to the object O_{n+1} (with certain membership degrees of all others).

The problem is solved analytically if it is possible to construct the dependencies according to the training sample:

$$x_1 = f_i(x_2, x_3, \ldots, x_m), \tag{4}$$

$$x_2 = f_i(x_1, x_3, \ldots, x_m), \tag{5}$$

...

$$x_m = f_i(x_1, x_2, \ldots, x_{m-1}), \tag{6}$$

5 Mathematical Tools for Second Kind Analogies Description (Mutual Analogies of First and Second Kind)

In this case, properties of TS models (2) and dependencies (4)–(6) are combined. Otherwise, the terms describing the prehistory of the process are introduced into the right-hand sides of relations (4)–(6). Let this case be analytically written for two variables and minimum prehistory as follows:

$$x_{n+1,1} = f\left(x_{n,1}, x_{n,2}\right), \tag{7}$$

$$x_{n+1,2} = g\left(x_{n,1}, x_{n,2}\right), \tag{8}$$

This problem statement leads to the flow theory [11, 12]. The dependencies (7) and (8) can be constructed on the basis of training set.

For the beginning, let the case on linear dependencies (7) and (9) be as follows:

$$x_{n+1,1} = ax_{n,1} + bx_{n,2}, \tag{9}$$

$$x_{n+1,2} = cx_{n,1} + dx_{n,2}, \tag{10}$$

For the convenience, the particular case is used:

$$x_{n+1,1} = x_{n,1} - x_{n,2}, \tag{11}$$

$$x_{n+1,2} = 2x_{n,1} - x_{n,2}, \tag{12}$$

Let the initial conditions ($x_{0,1}$ and $x_{0,2}$) be defined. Then, the dynamics of $x_{n,1}$ and $x_{n,2}$ is watched (Table 3).

Table 3. Results of the flow modeling.

$x_{n,1}$	−1	−1	1	1	−1	0	1	0	1	−2	−5	2	5	−2
$x_{n,2}$	0	−2	0	2	−1	−1	1	1	−1	3	−7	−3	7	3

After some transition period the cycle is observed (see Fig. 3). It begins in Table 2 from column 10. Column 14 is identical to it, i.e. the calculations show the identical results. It is easy to show that the position of the cycle does not depend on the choice of the starting point.

In Fig. 3, point (1, 1) is chosen as the initial point of calculations.

The cycle presented in Fig. 3 has its center of origin. Such cycle is called a cycle with zero center. The system of dependencies (11), (12) does not take into account the damping forces. If they are available, the process converges to the origin.

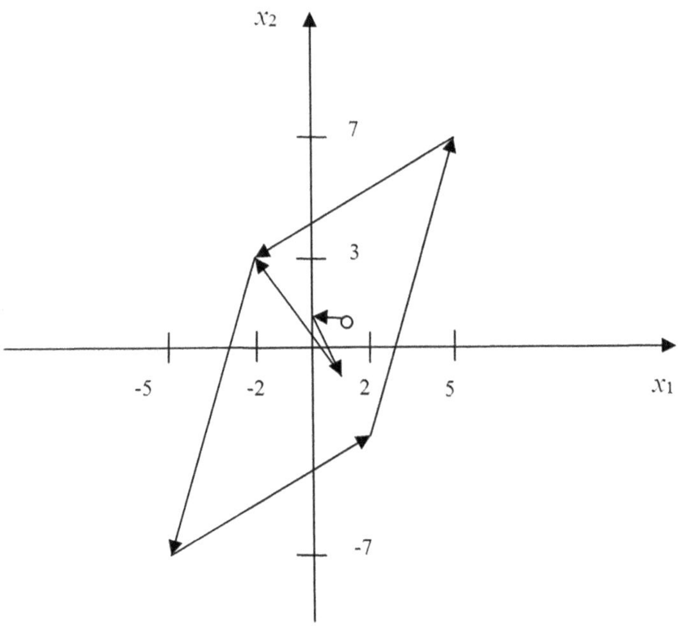

Fig. 3. The transition and the stationary processes in task of decision making by analogies for (11) and (12)

One more example:

$$x_{n+1,1} = 0.5x_{n,1}, \tag{13}$$

$$x_{n+1,2} = -x_{n,1} + x_{n,2}, \tag{14}$$

It is easy to calculate that motion of the corresponding point in coordinate system $x_1 0 x_2$ is parallel with each other to axis $0x_2$ from both sides for different initial conditions (initial values of x_1 and x_2) (Fig. 4).

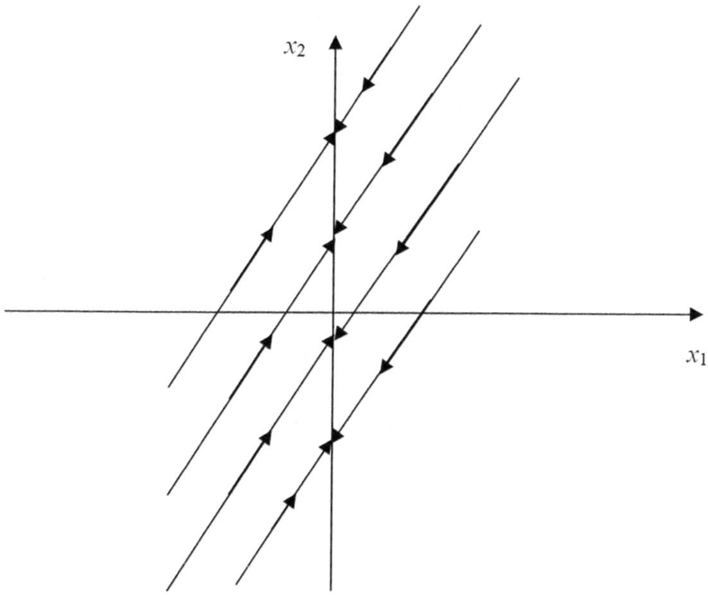

Fig. 4. Transition and stationary processes in task of decision making by analogies for (13) and (14)

The described above model constructions allow to numerically estimate the structure and dynamics of machine to intelligent decision making by analogies, i.e. it helps to MI development.

6 Conclusions

The paper contributes the following results:

1. The necessity of psycho-emotional human activity is proved for AI systems' creation.
2. The categorical analogues engine is made: definitions are given, classification is performed.
3. The mathematical description tools for first-, second- and mixed-type analogies are proposed and researched.

References

1. Chastikov, A., Nosova, Yu.: The architecture of intellectual agents by Barbucchean-Fox. In: Proceedings of the Kuban State Technological University, vol. 25, no. 3, pp. 84–85 (2005). (in Russian)

2. Liabakh, N., Kiba, M.: Improving the system of reviewing articles in scientific journals. Bull. GUU (7), 23–26 (2015). (in Russian)
3. Flavio, B., Rodolfo, M., Jiang, Z., Sateesh, A.: Fog computing and its role in the Internet of Things. In: SIGCOMM (2012)
4. Novikov, D., Chkhartishvili, A.: Reflexive Games. SINTEG, Moscow (2003). (in Russian)
5. Shabelnikov, A., Liabakh, N.: Theory of sorting processes. Intell. Inf. Technol. Ind. **1**, 138–145 (2017)
6. Liabakh, N., Pushkarev, E.: Simulation of decision-making procedures in man-machine complexes based on analogies. In: Proceedings of the VI Scientific and Technical Conference of Intelligent Management Systems in Railway Transport (ISUZHT 2018), pp. 59–61 (2018). (in Russian)
7. Uemov, A.: Analogy in the Practice of Scientific Research. Science, Moscow (1970). (in Russian)
8. Rosenberg, I., Shabelnikov, A., Liabakh, N.: Control Systems for Sorting Processes Within the Framework of the Ideology of the Digital Railway. VINITI RAS, Moscow (2019). (in Russian)
9. Mishulina, O.: Statistical Analysis and Processing of Time Series. MEPHI, Moscow (2004). (in Russian)
10. Adadurov, S., Gapanovich, V., Liabakh, N., Shabelnikov, A.: Rail Transport: Towards Intellectual Management. SSC RAS, Rostov-on-Don (2010). (in Russian)
11. Kuznetsov, A., Kuznetsov, S., Pozdnyakov, M., Sedova, Yu.: Universal two-dimensional mapping and its radiophysical realization. Nonlinear Dyn. **8**(3), 461–471 (2012). (in Russian)
12. Kuznetsov, S.: Dynamic Chaos. Fizmatlit, Moscow (2006). (in Russian)

Case-Based Reasoning Tools for Identification of Acoustic-Emission Monitoring Signals of Complex Technical Objects

Alexander Eremeev, Pavel Varshavskiy, and Anton Kozhevnikov(✉)

National Research University "MPEI", Krasnokazarmennaya St., 14, Moscow 111250, Russia
eremeev@appmat.ru, VarshavskyPR@mpei.ru, antoko@yandex.ru

Abstract. The article discusses actual issues of developing a tool for identifying signals received during acoustic emission (AE) monitoring of complex technical objects using case-based reasoning (CBR – Case-Based Reasoning). The main task of AE-monitoring is to detect AE-sources that are associated with defects at the monitored objects, information about them is contained in the parameters of AE-signals.

Unfortunately, at each individual object, the difference between AE-signals cannot be quantitatively described, but can be established empirically with the help of an expert operator. For a qualitative analysis of AE-monitoring data, it is necessary to accumulate a sufficient database for each individual object, which is a difficult rather time-consuming task. To improve the efficiency of the operator in accumulating the database for identifying AE-signals, it is proposed to use case-based reasoning methods and systems (CBR-systems) that can work with relatively small training samples and at the same time produce high-quality results.

Developed in MS Visual Studio in C# language, the CBR-tools for identifying acoustic emission signals allows not only to accumulate a sufficient database for identifying AE-signals, but also to use it later as an independent tool for identification. The work of the proposed tool was tested on valid expert data obtained during AE-monitoring of metal structures.

Keywords: Case-based approach · Monitoring · Data analysis · Acoustic emission

1 Introduction

Nowadays, to assess the state of complex industrial facilities are increasingly using monitoring – constant surveillance of the technical condition of the structure or unit. The development and distribution of monitoring systems is associated with the aging of equipment and the need to extend its service life, and their successful operation is ensured by a high technical level of non-destructive testing, complication of their

This work was supported by RFBR projects №. 18-01-00459, №. 18-29-03088, №. 18-51-00007.

S. Kovalev et al. (Eds.): IITI 2019, AISC 1156, pp. 607–615, 2020.
https://doi.org/10.1007/978-3-030-50097-9_62

algorithmic and software [1]. The ability to track the occurrence and development of cracks, fractures and other defects allows you to plan repair work or preventive maintenance, as well as prevent accidents.

The most effective method for monitoring especially dangerous objects is the acoustic emission method (AE), which refers to the phenomenon of the appearance and propagation of elastic vibrations (acoustic waves) in various processes, such as deformation of a stressed material, the outflow of gases, liquids, combustion, explosion, etc. [1, 2]. AE is a quantitative criterion for the integrity of a material, which is determined by the sound radiation of a material at its control load. The effect of AE can be used to determine the formation of defects at the initial stage of decay of construction and to monitor the nature of the formation and development of defects in the material of the entire object as a whole.

The basis of AE-monitoring is the detection and transformation of elastic waves into an electrical signal. Analysis of these signals provides valuable information on the presence and origin of defects in the material [3]. One of the problems arising during AE-monitoring in the task of diagnosing the state of a monitored object is a large volume of heterogeneous diagnostic information – AE-signals, their parameters, readings of temperature, pressure, humidity sensors, etc. [4]. In addition, during monitoring, the monitored object operates normally, but equipment vibration, product flow (for example, liquid, oil, etc.) in the pipeline, mechanical effects (for example, loosened bolts or loose connectors that move when exposed to wind, rain, flying objects, dust) and other influences create acoustic disturbances that carry deliberately false information, which makes monitoring difficult. To eliminate noise signals, it is necessary to solve the problems of signal identification and data filtering.

Some realistic approaches to compensation for background noise include the manufacture of special sensors with electronic filters to block noises [5], taking into account the placement of sensors as far as possible from the noise source, and electronic filtering (by either taking into account the time of appearance of the signal or differences in the spectral composition of true AE-signals and background noise) [4]. However, even the application of these approaches does not always and not completely exclude the appearance of noise signals among the data provided to the expert for further diagnostics. In addition, the expert often needs to choose from a large number of oscillograms certain types of signals for analyzing a defect on the object under consideration.

To assist an expert (decision making person, operator) in solving the problem of identifying signals, artificial intelligence methods and, in particular, a CBR-tools can be applied. Each signal can be considered as a separate precedent, and then, having information about previously detected signals, it is possible to identify certain signals for subsequent analysis by the expert without the active involvement of the expert [2].

2 Presentation of Information on AE-Signals as Precedent

Previously, the authors have already published a solution to this problem [2], which processed the secondary calculated parameters of the AE-signal for its identification. In this paper, an identification algorithm for processing the primary data of the AE-signal (its entire waveform) is proposed.

The case is proposed to include a description of the situation, a solution for the situation and information on the result of the application of the decision [6]:

$$CASE = (Situation, Solution, Result),$$

where *Situation* – description of the situation, *Solution* – solution, for example diagnosis and recommendations for user or decision-maker, *Result* – result of applying the solution, which may include list on actions taken, additional comments and references to other cases, as well as the rationale for the choice of this solution and possible alternatives.

The received signal contains a lot of data that are intermediate and do not carry important information. Therefore, the signal data can be simplified as a broken line of extrema of waveform. The extremum will be the point with the maximum/minimum amplitude in the ε-neighborhood, which is determined by the length of the signal.

A simplified version of the signal is used to build a sequence of rises and falls in the signal with fixing the values of their time ranges and relative amplitudes:

$$\delta A_i = \frac{A_i}{A_{MAX}},$$

where A_{MAX} is the maximum absolute value of the oscillogram amplitude, A_i is the absolute value of the amplitude for respective rise or fall. In other words, the sequence $r_1, r_2, \ldots, r_K$ is constructed, in which each member r_i is represented by a quadruple $\{T, \delta A, t_s, t_f\}$, where T is a type (rise or fall), $\delta A \in [-1; 1]$ – relative amplitude in this area, $t_s, t_f \in [0; 1]$ – relative time limits of the time range.

Next, the resulting sequence goes through the procedure of grouping according to the following rules:

- the pair of rise and fall of the signal acting within the limits $\leq \varepsilon$ is considered a peak;
- the pair of rise and fall of the signal, acting within the limits $> \varepsilon$ is considered oscillation;
- consecutive oscillations with similar values of the relative amplitude are grouped (combined into one with the total time range).

After grouping, a similar sequence $S_1, S_2, \ldots, S_L$ is formed, where S_i is represented by the same quadruple $\{T, \delta A, t_s, t_f\}$, but now $T \in$ {peak; oscillations}.

The formation of this sequence can significantly reduce the volume of signal data, and that leads to a significant reduction in time spent on the search for similar signals in the knowledge base (KB).

As an illustration, we give the following example: for a signal (Fig. 1a) with a length of 2000 points, the following sequence of 30 quadruples was obtained (Fig. 1b). In the graphical display of the formed sequence (Fig. 1c) the oscillations are marked in gray, the peaks are marked in black.

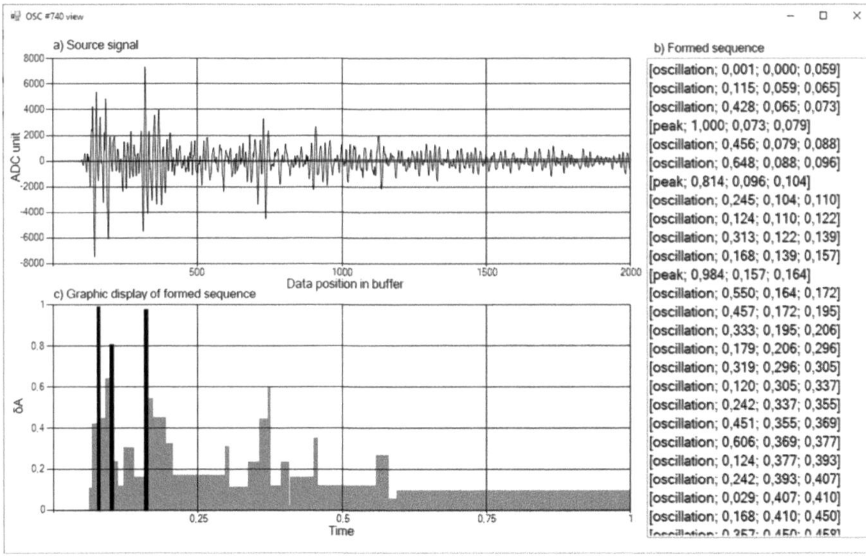

Fig. 1. An example of the transformation of waveform data into a precedent

The resulting sequence can be saved in the KB as a case for further comparison with the existing cases in the KB, which can be stored as cases with noise and actual signals.

Note that the case does not necessarily contain complete information about the signal (for example, there may be no specific waveform data), but it must contain information necessary for expert assessment of the relevance of the signal [2].

3 A Case-Based Reasoning Approach for Identification of AE-Signals

CBR-methods for finding solutions include four main stages that form the so-called CBR-cycle: retrieval, reuse, adaptation and retaining of the case [7]. There are a number of methods for extracting cases, for example, the Nearest Neighbor (NN) method, the method based on decision trees, the method taking into account the applicability of the precedent, etc. [6].

To determine the similarity of the current signal with cases from the KB, it is proposed to use the NN-method [8] – the most used method of comparison and extraction of cases. NN-method allows you to simply calculate the degree of similarity of the current problem situation and cases from the KB. To determine the degree of similarity on the set of parameters used to describe cases and the current situation, a certain metric is introduced. Further, in accordance with the selected metric, the distance from the target point corresponding to the current problem situation to the points representing the cases from the KB is determined, and the nearest point to the target is selected.

The effectiveness of the NN-method depends largely on the choice of the metric (measure of similarity). The two sequences can differ both by the members themselves and by the number of members, so for comparing the sequences of oscillations, a metric was selected based on the Levenshtein distance [8].

The Levenshtein distance for the sequences S_1 and S_2 (of length M and N, respectively) can be calculated by the formula $d(S_1, S_2) = D(M, N)$, where

$$D(i,j) = \begin{cases} 0, & i=0, j=0 \\ i, & j=0, i>0 \\ j, & i=0, j>0 \\ \min\begin{pmatrix} D(i,j-1)+1 \\ D(i-1,j)+1 \\ D(i-1,j-1)+m(S_1[i],S_2[j]) \end{pmatrix}, & j>0, i>0 \end{cases},$$

where S_1, S_2 are input sequences, and $m(a, b) = 0$ if $a = b$, otherwise $m(a, b) = 1$.

The distance d_{CT} between the new case T and the cases C stored in the database in the selected metric are determined. To determine the value of the degree of similarity $Sim(C, T)$, it is necessary to find the maximum distance d_{MAX} in the selected metric using the range boundaries of the corresponding parameters [2].

NN-method estimate:

$$Sim(C,T) = 1 - \frac{d_{CT}}{d_{MAX}},$$

where d_{CT} is the distance between the current signal and the case from the KB, d_{MAX} is the maximum distance in the selected metric.

When determining the class of a signal, the expert compares the images of the signals, in particular, the presence of oscillations and their amplitude, instead of concrete numbers. The proposed approach to constructing a sequence of oscillations for a new case and the search for a similar case in the KB allow us to speed up and simplify the work of an expert in identifying signals while analyzing the waveform data file.

4 The Architecture of CBR-Tools for Identifying AE-Signals

The developed CBR-tools (Fig. 2) are an add-on that can be connected as a separate library to software systems of the appropriate purpose implemented in various programming languages. The CBR-tools can receive from the main program a buffer with waveform data, which then goes to the waveform extraction unit and goes through the stages of the CBR-cycle.

A case is formed from the waveform data then the resulting case goes through the stages of retrieval and retaining in the KB. Saved cases go through a marking procedure depending on user-defined classes. However, with an increase in the number of stored cases in the KB, there is a problem of reducing system performance [9] and a block reducing the number of precedents is used to solve it.

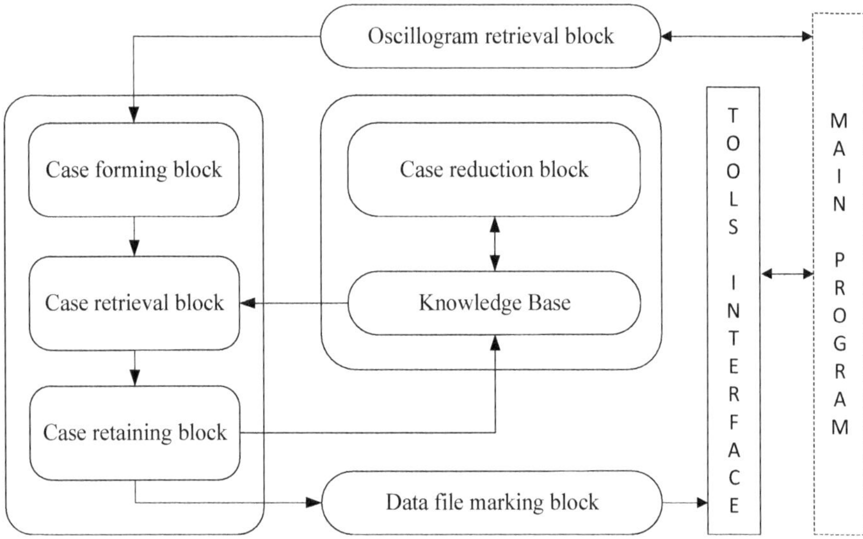

Fig. 2. Architecture of the developed CBR-tools

The CBR-tools solve the problem of marking and filtering the waveform data file to facilitate and speed up the work of the expert. After analyzing the contents of the data file, the tools transmit (using the message interface) the numbers of the waveforms of interest (positively marked) to the main program, which allows the decision making person (the operator) to quickly move around the file with AE-monitoring data.

5 The Interaction Between the Developed CBR-Tools and the Main Program

A set of messages has been developed for communication between the main program of AE-monitoring and the implemented CBR-tools. To start the work of the CBR-tools, it is necessary to load the configuration with oscillograms selected and marked by an expert. After that, the tools are ready to receive messages with oscillograms for identification. Messages must contain the following information:

- buffer size (in bytes);
- size of data unit (in bytes);
- waveform number;
- the beginning of the block of waveform points (optional);
- number of waveform points (optional).

If the last two points contain a zero value, the entire waveform is extracted from the buffer, otherwise the specified set of points is extracted. The extracted set is saved as an array of real numbers (double) and is processed by the CBR-cycle. The corresponding algorithm is presented in Fig. 3. After processing in the KB of the CBR-tools, the generated sequence, the result of the marking and the resulting oscillogram number are saved.

The CBR-tools can receive messages that indicate which oscillogram classes need to be viewed and studied by the operator. In this case, the selection of the priority for displaying the classes of marked waveforms is configured.

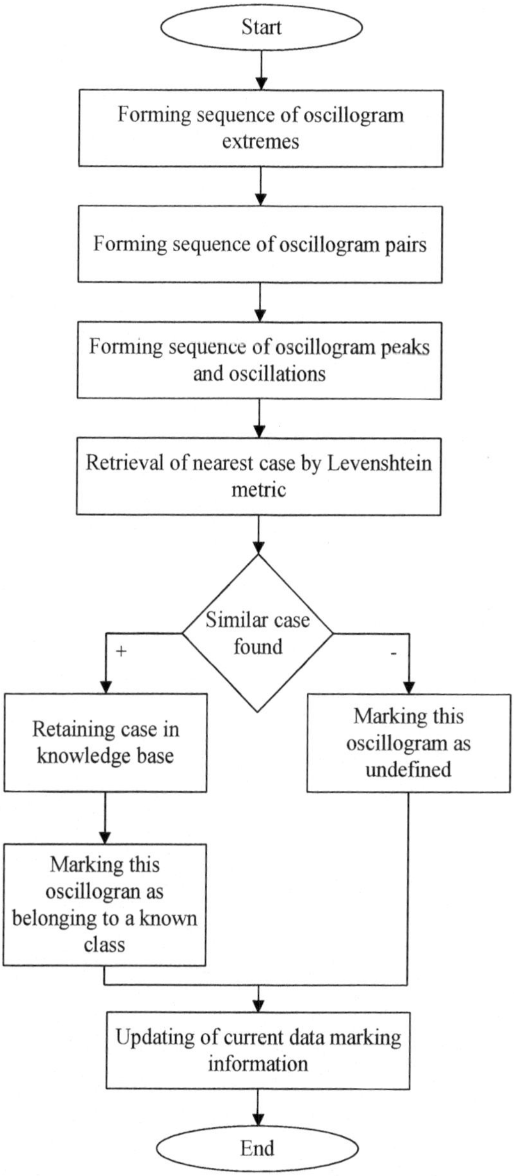

Fig. 3. The flowchart of the processing of the received oscillogram

Oscillogram numbers are used to quickly navigate through the data in the main program. You need to send a message that will contain the current waveform number of the main project, and the direction of the search for the nearest matching waveform in the label. Thus, the user is not restricted to navigating through the data file.

In the case when the user is not satisfied with the result obtained by the CBR-tools, you can send a message about the change of the marking class.

6 Software Implementation of a CBR-Tools for Identifying AE-Signals

The proposed CBR-approach for identifying AE-signals based on the NN-method using the Levenshtein metric was implemented in the software tools for analyzing files containing AE-monitoring data (Fig. 4) in C# (.NET 4.5) [10] in the environment Microsoft Visual Studio 2012 for MS Windows.

The user sets the CBR-tools operation settings: sets the types of signals already known (classified by an expert) and selects the classes of signals necessary for display. After that, the CBR-tools perform the procedure of marking the data file and provides the user with a simplified navigation through the file contents, which allows viewing only the user-defined signal classes. But if desired, the user can combine the usual and simplified views for detailed analysis of the data file.

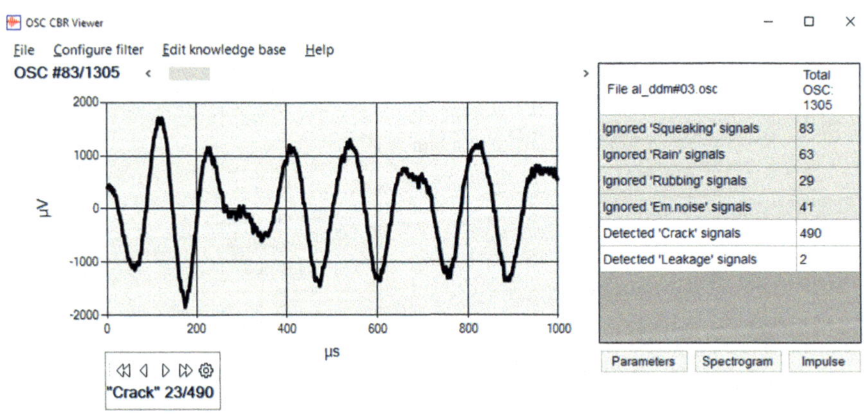

Fig. 4. An example of the work of the developed CBR-tools

The work of the CBR-tools was verified by experts on test data obtained during AE-monitoring of metal structures (more than 20 measurement files), which confirmed the effectiveness of the use of a precedent approach for identifying signals received during AE-monitoring of complex technical objects. The use of the implemented CBR-tools can significantly reduce the time of data analysis due to the marking of the file content (more than 106 oscillograms) obtained based on precedents.

7 Conclusion

This paper investigates the possibility of using decision search methods based on CBR for identifying AE-monitoring signals of complex technical objects. To extract precedents from the knowledge base (case base), it was proposed to use the NN-method and the Levenshtein distance as a measure of similarity. A method for presenting AE-monitoring signals as a precedent has been developed. Based on the proposed approach, the CBR-tools for identifying AE-monitoring signals were implemented. The tools were tested on test data obtained during AE-monitoring of metal structures. The results showed the effectiveness of the proposed approach and its implementation in the form of the CBR-tools.

References

1. Barat, V.A., Alyakritski, A.L.: Statistical processing of data from acoustic emission monitoring: Mozyrski NPZ hydrotreater case study. NDT World **4**, 52–55 (2008)
2. Alekhin, R.V., Varshavsky, P.R., Eremeev, A.P., Kozhevnikov, A.V.: Application of the case-based reasoning approach for identification of acoustic-emission control signals of complex technical objects. In: 3rd Russian-Pacific Conference on Computer Technology and Applications (RPC), pp. 28–31 (2018)
3. Terentyev, D.A.: Time frequency analysis for identification of acoustic emission signals. NDT World **2**, 55 (2013)
4. Kwon, O.-Y., Ono, K.: Acoustic emission characterization of the deformation and fracture of an SiC-reinforced aluminum matrix composite. J. Acoust. Emission **9**, 127 (1990)
5. Kuz'min, A.N., Zhuravlev, D.B., Filippov, S.Yu.: Corrosion – sentence or diagnosis? On the issue of technical diagnostics of heat networks. Tehnadzor **3**, 32–35 (2009)
6. Eremeev, A., Varshavskiy, P., Alekhin, R.: Case-based reasoning module for intelligent decision support systems. In: Proceedings of the First International Scientific Conference – Intelligent Information Technologies for Industry (IITI 2016), vol. 1, pp. 207–216. Part III. Springer International Publishing (2016)
7. Aamodt, E.P.: Case-Based Reasoning: Foundational Issues, Methodological Variations, and System Approaches. Artificial Intelligence Communications, vol. 7, no. 1, pp. 39–59. IOS Press (1994)
8. Russell, S., Norvig, P.: Artificial Intelligence: A Modern Approach, 2nd edn edn, p. 1132. Prentice Hall, US (2002)
9. Eremeev, A., Varshavskiy, P., Alekhin, R.: Improving the efficiency of solution search systems based on precedents. In: Kacprzyk, J. (ed.) Proceedings of the Third International Scientific Conference "Intelligent Information Technologies for Industry" (IITI 2018). Advances in Intelligent Systems and Computing, vol. 874, pp. 313–320. Series editor. Springer Nature Switzerland AG (2019)
10. Albahari, J., Albahari, B.: C# 4.0 in a Nutshell. The Definitive Reference, p. 1056. O'Reilly Media, Inc. (2010)

Railway Sorting Robotics

A. N. Shabelnikov[1,2(✉)]

[1] Russia JSC «NIIAS», Rostov Branch, Rostov-on-Don, Russia
shabelnikov@rfniias.ru
[2] Rostov State Transport University, Rostov-on-Don, Russia

Abstract. New paradigm of the transport automation and transport artificial intelligence can be formulated as railway transport digitalization or, in particular, railway station digitalization. The digitalization provides development and implementation of fundamentally new forms of intelligent functioning and the methods of intelligent complex systems. The paper reveals the possibilities of railway sorting robotics. The cyber-physical system, which uses internet of things and multi-agent ideology, is formed. The algorithm of control over robots-uncouplers and mathematical model of their interaction are provided.

Keywords: Digital marshalling yard · Cyber-physical systems · Internet of things · Multi-agent systems · Robots-uncouplers

1 Introduction

Integrated Automation System for Marshalling Process Control (KSAU SP) was developed for the automated control over humping operations [1]. However, its capabilities are currently limited, since marshalling processes have a number of significant problems, among of which are high scores of:

- Uncertainty and fuzziness of marshalling processes;
- Dimensionality of control and management tasks inaccessible to centralized data processing;
- Noisy of initial data;

Accounting for these properties within the paradigm of automated control is not possible. The trends of recent years in the economical and industrial digitization predict a new quality of railway marshalling control. Digitalization of marshalling yard allows to extend the area of decided tasks and list of its intelligent functions [2–4]. It is the basic essence of digital station (DS) developed in JSC NIIAS (Russian Federation). In this case. KSAU SP is transformed into a cyber-physical system (CPS) [4]. Actually, one of the main key points of CPS is that it uses Internet of Things (IoT). As interacted things, experts, station operators, equipment (sensors, compressor stations, retarders, locomotives, etc.) can act inside of CPS of marshalling yard. Mathematical basis of this interaction is Multi-Agent Systems (MAS) [4].

The work was supported by Russian Fundamental Research Fund, project No. 17-20-01040.

S. Kovalev et al. (Eds.): IITI 2019, AISC 1156, pp. 616–622, 2020.
https://doi.org/10.1007/978-3-030-50097-9_63

Inside DS, a set of complicated functions can be implemented on the basis of MAS:

- shunting work of a station;
- train formation control;
- technical monitoring;
- maintenance;
- repair of station equipment.

KSAU SP successfully automates many functions (control of retarders, compressor stations, interaction with the information field and the top level of train operation control). However, nowadays there are still a number of complex processes that cannot be controlled by traditional methods of the theory of automatic control and regulation. Among them, the separation of cuts on the top of the hump yard can be highlighted. This paper proposes specially created robots for this problem solution and describes the algorithm of their work, as well as mathematical description of the process.

2 Robots in KSAU SP

As robots, the following items can be used:

- Robots-uncouplers on the top of hump. Cut uncoupling is one of the most dangerous procedures, which still does not have any acceptable automated decision.
- Robots-locomotives performing shunting motions on station. This topic is promoted rather good in Shunting Automatic Locomotive Signaling (MALS) [5, 6].
- Retarders providing interval and target regulation over cuts on hump.

In further, the first task is considered. Automatic system of control over robots are classified as follows:

- Program control (robot motion is calculated in advance).
- Adaptive control (robot motion is organized by flexibly changed programs) is a program reconstruction responding to changes of external conditions.

These systems are described in the following papers detaily [7–9].

Robots-uncouplers belong to the second type of automatic control systems, since many external factors that are difficult to take into account affect their work (position of the cut's center of gravity, limited ability to measure and accurately realize the humping velocity, weather conditions affecting the running properties of cuts (wind, temperature, humidity), etc.).

Adaptive control systems contain various tools sensing intelligent control. The movement program of proposed robot is not established, but is synthesized by the control system based on the description of the external environment, a set of rules of possible behavior in this environment in order to solve the considered task. The intelligent robotic system (IRS) consists of three subsystems: the perception of the internal and external environment (moving cuts, identification of the place of uncoupling), the problem solving block (matching the velocities of the train and the robot, the calculation of the uncoupling time), control subsystem.

The perception subsystem is associated with two sources of information: the external world and the human operator. With the first one, IRS communicates through sensors and input systems. With the operator, IRS communicates through standard terminals of interactive systems. The robot control subsystem is a set of actuators (a manipulator for gripping and rotating a couple) that are included according to the "indications" of the decision block.

The scheme of the installation of robots-uncouplers (R_1, R_2 and R_3) on hump is presented in Fig. 1. Zones of their effect ($0 - S_2$, $S_1 - S_4$ and $S_3 - S_5$, respectively) are intersected (intersected zones are $S_1 - S_2$ and $S_3 - S_4$).

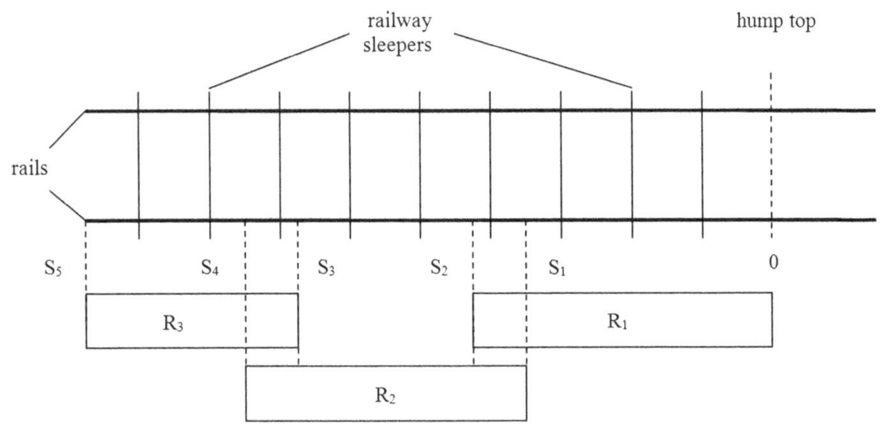

Fig. 1. The block diagram of the installation of robots-uncouplers on hump yard

3 Membership Functions of Robots

Favorable time for uncoupling is the time of intersection of the hump top by the cut's center of gravity. At this point, the pressure is relieved from the locomotive pulling the train to the couple, but the tension forces from the accelerating are not emerged yet. During this time interval, the speeds of movement of the cut and the robot must be synchronized with each other, and the manipulator must have time to uncouple a train.

If the uncoupling point falls into intersecting zones, then uncoupling can be carried out by each robot, but there are also preferences that are established by membership functions $\mu R_i\ (s)$, see Fig. 2.

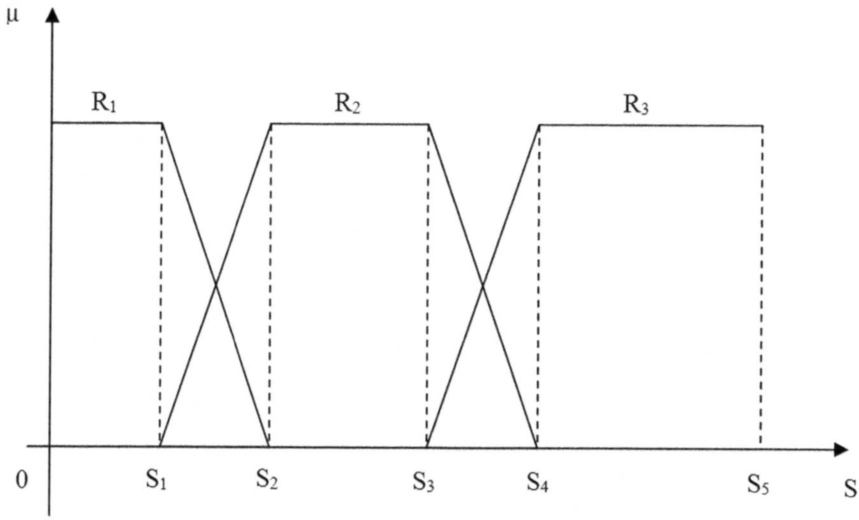

Fig. 2. Membership functions of robots-uncouplers

The engine of decision making is described as follows. If the uncoupling point is at $[0, S_1]$, then uncoupling is made by the first robot ($\mu_{R1}(s) = 1$). If this point belongs to $[S_2, S_3]$, then uncoupling is made by the second one ($\mu_{R2}(s) = 1$). If it belongs to $[S_4, S_5]$, then uncoupling is made by the third one ($\mu_{R3}(s) = 1$). If the point belongs to intersections, then uncoupling can be made by random one, but the preferences can be defined by the membership functions.

Critical points for installation and operation of robot Ri are determined at an extra-systemic level, even at the stage of analyzing the flow of cuts rolling down from the hump (masses, lengths, speeds). They form the basis of the design of the robotic complex.

The number of robots installed is influenced by several factors:

- Curvature of the hill. After all, in fact, with the help of straight sections, on which robots move, the curvature of the hump is approximated. Increasing the accuracy of the approximation requires an increase of the number of approximating segments.
- The length of a cut. The longer it is, the farther separation of cut from the top will be.

It is easy to calculate (see Fig. 2) that analytically the membership functions of robots are given by the relations:

$$\mu_{R1}(s) = \begin{cases} 1, & \text{if } S \in [0; S_1] \\ \frac{S}{S_1 - S_2} + \frac{S_1}{S_2 - S_1}, & \text{if } S \in [S_1; S_2], \\ 0, & \text{else.} \end{cases} \quad (1)$$

$$\mu_{R2}(s) = \begin{cases} 1, & \text{if } S \in [S_2; S_3] \\ \frac{S}{S_2 - S_1} + \frac{S_1}{S_1 - S_2}, & \text{if } S \in [S_1; S_2] \\ \frac{S}{S_3 - S_4} + \frac{S_4}{S_4 - S_3}, & \text{if } S \in [S_3; S_4] \\ 0, & \text{else.} \end{cases}, \tag{2}$$

$$\mu_{R3}(s) = \begin{cases} \frac{S}{S_4 - S_3} + \frac{S_3}{S_3 - S_4}, & \text{if } S \in [S_3; S_4] \\ 1, & \text{if } S \in [S_4; S_5] \\ 0, & \text{else.} \end{cases}. \tag{3}$$

4 Internet of Things and Internet of Services for Marshalling Stations

Block diagram of robot-uncoupler functioning is presented in Fig. 3.

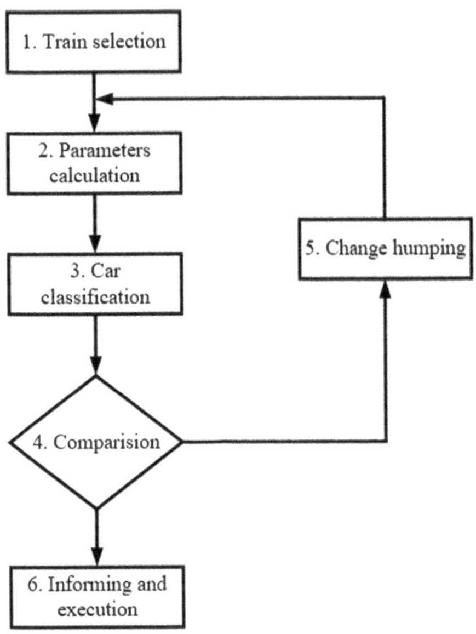

Fig. 3. Block diagram of robot-uncoupler functioning

The algorithm is the following:

1. At the first stage, information about the following humping train is obtained at system of control over robots-uncouplers: structure of the train (splitting by cuts), characteristics of cuts (car mass, length) (unit 1).
2. On the basis of this information, the following items are determined for each cut in block “Parameter calculation”: the center of gravity, the estimated place of uncoupling of the cut from the train.

3. In block 3, all cuts are classified into classes according to belonging to one or another robot (according to the place of uncoupling: each robot has its own zone defined by the corresponding membership function, see formulas (1)–(3)).
4. Comparison block 4 verifies the fact: all uncouples are "tied" to the robots-uncouplers. It is possible that the uncoupling point goes beyond the capabilities of the robots.
5. In this case, a recommendation is issued (block 5) to change the program of humping. Problem cut is divided into several smaller sizes, which will fall within the area of responsibility of the robots.
6. With a favorable outcome (all cuts are distributed between the robots), the calculated program is issued to the robots for execution and to the operator for informing (block 6).

5 IoT and Multi-agent Technologies of DS

In the considered technology of train cut control, IoT can be shown as follows. There are several active sides of the process of "things": operators of the hump yard, robots-uncouplers, program-mathematical blocks of decision-making (calculation of system parameters, classification of cuts, see Fig. 3), which informationally interact with each other. As a result of this interaction, a new quality of transport service is formed. It is characterized by:

- increasing the safety of the train humping;
- increasing the degree of automation and intelligence integration of the humping process;
- facilitate the work of hump staff.

The regulation of interactions is carried out by means of MAS. An example of one such interaction is the following. The cuts have different lengths: from one car (the place of uncoupling in this case almost coincides with the top of a hump) and up to twenty or more cars (the place of uncoupling is shifted from the hump to 150–200 m). In this case, one robot-uncoupler does not solve the considered problem (high intensity of work, difficult relief of movement). Each robot evaluates its capabilities and reports them to a common control unit for a set of robots, which analyzes the received information and transfers control to the most suitable robot.

Developed technical and technological procedures of railway sorting processes and methods of their formalization allow to increase the degree of intelligence in these processes [10].

6 Conclusions

The following contributions are made by the current work:

1. The new approach to automation of uncoupling process on hump is proposed. The approach is based on robots-uncouplers implementation.
2. The scheme of interaction between robots is developed.
3. The models of membership functions together with gears of their interaction are proposed for robots.

References

1. Shabelnikov, A.: Integrated System for Automated Control of Sorting Processes - An Innovative Project of Russian Railways. VINITI RAS, Moscow (2016)
2. Kupriyanovskiy, V., Sukonnikov, G., Bubnov, P., Sinyagov, S., Namiot, D.: Digital railway - forecasts, innovations, projects. Int. J. Open Inf. Technol. **4**(9) (2016)
3. Lyovin, B., Tsvetkov, V.: Digital railway: principles and technologies. World Transp. **16**(3), 50–61 (2018)
4. Rosenberg, I., Shabelnikov, A., Liabakh, N.: Control Systems for Sorting Processes within the Framework of the Ideology of the Digital Railway. VINITI RAS, Moscow (2019)
5. MALS, Shunting automatic locomotive signaling system. http://mals.su/en.html. Accessed 25 Apr 2019
6. Ventsevich, L.: Locomotive Devices for Ensuring the Safety of Train Traffic and Decoding the Information Data of Their Work. Route, Moscow (2006)
7. Unbehauen, H.: Control Systems, Robotics, and Automation, vol. I (2009)
8. Ghosh, B., Bijoy, K., Tarn, T.: Control in Robotics and Automation: Sensor-based Integration. Academic Press Series in Engineering (1999)
9. Kozlowski, K.: Robot motion and control. LNCIS **396**, 123–132 (2009)
10. Adadurov, S., Gapanovich, V., Liabakh, N., Shabelnikov, A.: Rail Transport: Towards Intellectual Management. SSC RAS, Rostov-on-Don (2010)

Procedural Generation of Virtual Space

Vladimir Polyakov(✉) and Aleksandr Mezhenin

ITMO University, Kronverksky Avenue 49, St. Petersburg, Russia
v_i_polyakov@mail.ru, mejenin@mail.ru

Abstract. At present, various approaches are used to synthesize the virtual environment, among which the most common are computer graphics and photogrammetric modeling. In addition, neural network information processing systems are being developed to create fully artificial interactive worlds based on real-world video recordings. This paper addresses the procedural generation of virtual space. It is a continuation of a series of articles devoted to the research of authors in this field. The generated content is represented as a cloud of points of different density. The construction of a cloud of points of three-dimensional space takes place on the basis of video data obtained as a result of shooting with a single camera moving along an arbitrary trajectory. Particularly addressed the problem of incomplete data and issues of preserving meaningful information. The result of the study is a conceptual description of methods for modeling a virtual environment. As methods of modeling a virtual environment, it is proposed to use the mathematical apparatus of parametric and non-parametric restoration of the density of the distribution of point objects in space according to the available sample. The results of testing the considered methods implemented in the MATLAB environment are presented.

Keywords: Procedural content generation · Cyber-visualization technologies · Modular objects · Real-time algorithm · Designing · Virtual modeling · Point clouds · Thermal maps

1 Introduction

Comprehensive information, the use of virtual presence and cyber-visualization technologies will raise to a new level the construction of intelligent security systems, control and prevention of emergency situations [1, 2].

Currently, various approaches are used to synthesize the virtual environment, among them the most common are computer graphics and photogrammetric modeling [3, 4]. In addition, neural network information processing systems are being developed to create fully artificial interactive worlds based on real-world video recordings [5]. In these methods, in most cases, polygonal models are used, to work with which many algorithms have been developed and there is hardware support for geometric calculations for 3D visualization. The disadvantages of their use include a large amount of source data necessary for complex 3D scenes, the difficulty of representing surfaces of complex shape and getting the proper sense of the effect of depth. These shortcomings become significant with the implementation of real-time systems [6].

S. Kovalev et al. (Eds.): IITI 2019, AISC 1156, pp. 623–632, 2020.
https://doi.org/10.1007/978-3-030-50097-9_64

More progressive can be considered the method of representing space objects in the form of clouds of points of different densities. A point cloud is a collection of vertices in three-dimensional space that represent the surface of simulated objects. Data for constructing a point cloud can be obtained by scanning 3D objects with special devices, or by processing optical scanning data [2]. To increase the perception efficiency, a point cloud is represented in the form of heat maps.

2 The Task of Building a Cloud of Points of Three-Dimensional Space

Building a point cloud is not a trivial task. In this paper, we consider the problem of building a cloud of points of three-dimensional space, based on video data obtained as a result of shooting with a single camera moving along an arbitrary trajectory [6, 7]. Particularly addressed the problem of incomplete data and issues of preserving meaningful information [8, 9].

Despite the large number of different methods for constructing a cloud of points in three-dimensional space, the typical mathematical approaches used in them turn out to be ineffective in problems, the initial data of which are a set of discrete vertex points in the space of point objects [10]. Representation of data in this form follows either from the specifics of the distribution in the space of point objects, or, for example, from the insufficiency of an array of measurement data at points in space [11, 12]. In such cases, it is usually used to refer to the initial set of point objects as a special term - point cloud. In the task of analyzing the distribution of such a cloud of points, first of all, the probability of finding points in a particular area, or rather the density of their distribution, is of interest. Such a problem is solved in using voxels [13]. However, the practical implementation of this method requires colossal computational resources, and in cases of strong sparseness of point data in space, the methods of reconstruction using voxels do not allow revealing information that is significant for analysis. A direct approach to solving the problem of reconstructing the spatial density of distribution of points is an approach in which the set of points $X = \{x(1), \ldots, x(m)\}, x \in Rr$ is considered as the implementation of a sample from one unknown distribution with density $\rho(x)$, and some approximation of the density is $\hat{\rho}(x) \approx \rho(x)$. There are three main types of distribution density search algorithms: non-parametric, para-metric, and recovery of mixtures of distributions.

Nonparametric recovery of distribution density. The basic non-parametric method for restoring the density distribution is the Parzen-Rosenblatt method (kernel density estimation), an algorithm for Bayesian classification based on non-parametric density recovery from an existing sample. The approach is based on the idea that the density is higher at those points next to which there is a large number of sample objects. Parsen density estimate is:

$$\rho_h(x) = \frac{1}{m}\sum_{i=1}^{m}\prod_{j=1}^{r}\frac{1}{h_j}K\left(\frac{x_j - x_j^{(i)}}{h_j}\right) \quad (1)$$

where $h \in Rr$ is the width of the window, and $K(u)$ is the kernel (an arbitrary even, normalized function) specifying the degree of smoothness of the distribution function. The term "window" comes from the classic form of the function:

$$K(u) = \frac{1}{2} I_{\{|u| \leq 1\}} \tag{2}$$

where $I\{\ldots\}$ is an indicator function, but in practice, smoother functions are usually used, for example, the Gaussian kernel function

$$K(u) = \frac{1}{\sqrt{2\pi}} e^{-\frac{1}{2}u^2} \tag{3}$$

The window width h and the type of the kernel $K(u)$ are the structural parameters of the method on which the quality of the restoration depends. At the same time, the width of the window has the main influence on the quality of restoration of the density distribution, whereas the type of core function does not affect the quality in a decisive way.

This method is widely used in machine learning for classification problems in cases where the general form of the distribution function is unknown, and only certain properties are known, for example, smoothness and continuity. To find the optimal width of the window, the maximum likelihood principle is usually used with the exception of objects one by one -leave-one-out (LOO).

Parametric recovery of distribution density. Parametric estimation relies on families of density functions, which are specified using one or several numerical parameters: $\rho(x) = \phi(x; \theta), x \in Rr, \theta \in \Theta$. One of the ways to choose the density function from this family that best approximates the original one is the maximum likelihood method.

$$\theta^* = \arg\max_{\theta \in \Theta} \prod_{i=1}^{m} \varphi(x^{(f)}; \theta), \hat{\rho}(x) = \varphi(x; \theta^*), \tag{4}$$

For example, for the multivariate normal distribution function:

$$\varphi(x; \theta) = N(x; \mu, \sum) = \frac{\exp\left(-\frac{1}{2}(x-\mu)^T \sum^{-1}(x-\mu)\right)}{\sqrt{(2\pi)^R \sum}}, x, \mu \in R^R, \sum \in R^{r \times r}, \tag{5}$$

maximum likelihood estimates are written explicitly:

$$\mu^* = \frac{1}{m}\sum_{i=1}^{m} x^i, \overset{*}{\sum} = \frac{1}{m}\sum_{i=1}^{m} (x^i - \mu^*)(x^{(i)} - \mu^*)^T, \tag{6}$$

This approach can be considered a complication of the parametric for cases where the distribution has a complex form, which is not accurately described by a single distribution. The distribution density $\rho(x)$ in the framework of this approach is represented as a mixture, i.e. the sum of distributions with certain coefficients:

$$\rho(x) = \sum_{j=1}^{k} w_j \rho_j(x) = \sum_{j=1}^{k} w_j \varphi(x;\theta_j) = \rho(x|\{w_j,\theta_j\}), x \in R^n, \sum_{j=1}^{k} w_j = 1, w_j \geq 0 \quad (7)$$

In (7), $\rho_j(x)$ is the density of the distribution of the components of the mixture, belonging to one parametric family $\phi(x;\theta_j)$, w_j is its a priori probability (weight), k is the number of components in the mixture. The function $\rho(x \mid \{w_j, \theta_j\})$ is called the likelihood function.

Each of the above methods for determining the density of distributions (non-parametric, parametric, and recovery of mixtures of distributions) is applied with certain a priori knowledge of the density of distribution (of the form or properties of a function). Despite the fact that these approaches seem to have different areas of applicability, it is possible to identify similarities between them. So the non-parametric method can be considered as a limiting special case of a mixture of distributions, in which each $x(i)$ corresponds to exactly one component with a priori probability $w_i = 1/m$ and the selected density function (core) with center at point $x(i)$. The parametric approach is another extreme case of a mixture consisting of one component.

Thus, the three approaches described differ, first of all, in the number of additive components in the distribution model: $1 \leq k \leq m$. Therefore, restoring a mixture from an arbitrary number of components k is in some sense a more general case of restoring a continuous distribution density over a discrete sample.

3 Analysis of the Results of Point Cloud Synthesis by the Hausdorff Metric

To analyze the quality of synthesis, it is proposed to use modified Hausdorff metric [8]. For the sake of simplicity, let us imagine discrete 3D models represented by point cloud, since this is the most general way of representation of such data. The point cloud M will be a representation of the ensemble of points P in R^3 (apices) and the ensemble of points T that describe the connection between the apices of P.

Let us denote the two continuous surfaces S and S', and

$$d(p, S') = \min_{p' \in S'} \|p - p'\|_2, \quad (7)$$

where $\|\cdot\|_2$ – is the Euclidian norm.

Therefore Hausdorff metrics between S and S$'$: $d(S, S') = \max_{p' \in S'} d(S', S)$. It is important to understand the fact that the metrics is not symmetrical, h.e. $d(S, S') \neq d(S', S)$. Let us denote $d(S, S')$ as the direct distance, $d(S', S)$ as inverse distance. Then the symmetrical metrics:

$$d_2(S, S') = \max\left[d(S, S'), d(S', S)\right]. \quad (9)$$

Symmetric metrics ensures a more precise measurement of an error between two surfaces, since the calculation of a "one-way" error can lead to significantly underestimated distance. One can see that $d(S, S')$ is smaller than $d(S', S)$, since $d(A, S') < < d(S, B)$. Thus, a not very large one-way distance does not mean a small presentation. The calculation of the Hausdorff distance between two discrete surfaces $M(P, T)$ and $M'(P', T')$ is related to the preceding definitions. Let us focus on calculation of the Hausdorff direct distance, h.e. $d(M, M')$, since the symmetric distance can be calculated from the direct and inverse distances. The distance between any point *p* from *M* (*p* is assumed not to be from *P*) and M' can be calculated from the estimation of the distance minimum between *p* and all triangles $T \in T'$.

When the orthographical projection p$'$ of p on the plane T$'$ is inside the triangle, the distance between the point and the triangle is simply a distance from the point to the plane. When the projection remains outside T$'$, the distance between the point and the plane is the distance between p and the closest p$''$ from T$'$, which should lie on one of the sides of T$'$.

MESHLAB - open source solution allowing to compare polygon meshes. The authors modified the Hausdorff distance filter, which calculates the distance from the grid X to Y. First and foremost, the Hausdorff metric between two meshes is the maximum between two so-called one-sided Hausdorff distances (technically speaking, it is not distance). These two measures are not symmetrical (for example, the results depend on which mesh is given as X). The Hausdorff MeshLab filters enable the user to calculate only a one-sided version. A sample based on the ensemble of X mesh points is used for calculation while for each x the nearest point y on the grid Y is recognized. This means that the result depends heavily on how many points on X are taken. A general approach is to use the mesh apex with the highest density as sample points for this purpose we select the "Apex Sampling" option in the dialog box. It is important to become certain that the number of samples is greater than or equal to the number of apices. The gathered information is recorded in the layers log window.

4 Experimental Results

The iterative closest point (ICP), algorithm is the most fitting one for comparing two different point clouds of the same given object [14, 15]. This algorithm is based on the minimization of the average distance between two point clouds. A special procedure is used to minimize the distance between the two point clouds in question. The algorithm can be used to compare point clouds based on the same object, which has had some changes made to it. Also it works if the object was scanned from a different angle. In either case, the point clouds must have similar areas that cover each other. This algorithm moves and adjusts these corresponding areas until they overlap to the extent denoted by E – the margin of error.

Below is the practical implementation of the algorithm for modeling a cloud of points in three-dimensional space based on a video sequence obtained by a camera moving along an arbitrary trajectory.

1. Decomposition of a video into a set of sequential images. The files needed for the reconstruction are obtained from the frame-by-frame decomposition of the video sequence into a sequence;
2. Selection on the images of key points and their descriptors;
3. By comparing descriptors, key points corresponding to each other in different images are found;
4. Based on the set of matched key points, an image transformation model is built, with which you can get another from one image;
5. Knowing the model of camera transformation and correspondence of points on different frames, three-dimensional coordinates are calculated and a cloud of points is built.

To test the modified Hausdorff filter, 3D scene frames obtained in a computer simulation system were taken. In Fig. 2. shows the frames of source images (Fig. 1).

Fig. 1. Source frames.

To obtain stabilization and determine key points, software was used based on the considered algorithms and the Point Cloud Library [19, 20] library. The point cloud obtained as a result of synthesis is presented in Fig. 2. Visualization done in MeshLabJS [21].

Fig. 2. The result of the analysis of the density of the constructed point cloud obtained.

The results of the analysis of the synthesized point cloud with a modified Hausdorff filter are presented in Fig. 3.

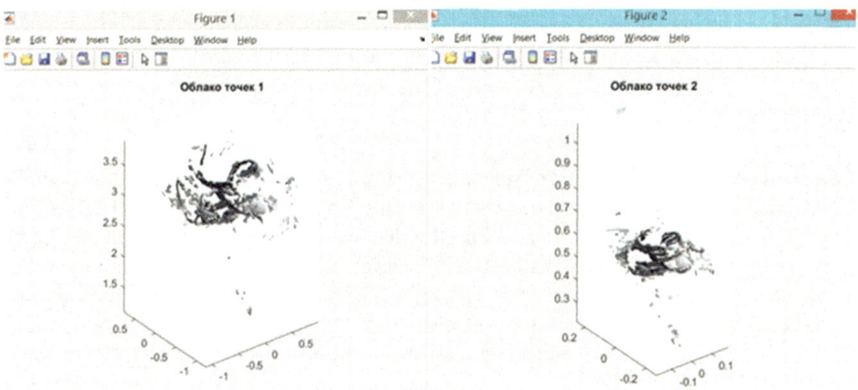

Fig. 3. The result of the analysis of the density of the constructed point cloud obtained.

5 Conclusion

The article considers the modeling of a virtual environment and its presentation in the form of a point cloud for various information applications. The terms cyber visualization and virtual presence are discussed. The mathematical apparatus used to build a cloud of points of three-dimensional space is considered. Examples of testing the developed algorithms are given: visualization and analysis of the distribution density of a point cloud; building a heatmap with an adaptive scale; modeling of a point cloud of three-dimensional space based on a video sequence; and results of an analysis of the density of distribution of points.

The proposed approach to the modeling of coverage areas and their visualization will allow obtaining more effective design solutions for video surveillance systems. The possibility of using universal 3D modeling and data processing systems for solving a specific applied problem is shown. The results of the comparison, presented in the form of a heat map, make it possible to estimate the degree of coverage of the observation zone and reveal the presence of dead zones. These provisions are the basis for further ongoing research in this area.

The result of the study is a conceptual description of methods for modeling a virtual environment. As methods of modeling a virtual environment, it is proposed to use the mathematical apparatus of parametric and non-parametric restoration of the density of the distribution of point objects in space according to the available sample. Visually, the simulation results are presented in the form of a cloud of points of uneven density. The proposed approach will improve the accuracy and visibility of the virtual environment for the subsequent visualization and detailed analysis of the simulated space. Based on the materials presented above, we can conclude that the proposed concepts are robust (relevant) and are subject to further elaboration and elaboration.

Acknowledgments. The research has been supported by the RFBR, according to the research projects No. 17-07-00700 A.

References

1. Afanas'yev, V.O.: Sistemy 3D-vizualizatsii indutsirovannoy virtual'noy sredy. Avtoreferat dissertatsii doktora fiziko-matematicheskikh nauk. Korolev-Moskva (2007). (Afanasyev V. O. 3D visualization systems of an induced virtual environment. Abstract of the dissertation of the doctor of physical and mathematical sciences. Korolev-Moscow) (2007). (in Russian)
2. Mezhenin, A., Polyakov, V., Izvozchikova, V., Burlov, D., Zykov, A.: The synthesis of virtual space in the context of insufficient data. In: The International Symposium on Computer Science, Digital Economy and Intelligent Systems, CSDEIS 2019. AISC, vol. 1127, pp. 39–46 (2019)
3. Bolodurina, I.P., Shardakov, V.M., Zaporozhko, V.V., Parfenov, D.I., Izvozchikova, V.V.: Development of prototype of visualization module for virtual reality using modern digital technologies. In: Proceedings - 2018 Global Smart Industry Conference, GloSIC (2018)
4. Mezhenin, A., Izvozchikova, V., Shardakov, V.: Reconstruction of spatial environment in three-dimensional scenes. In: The International Symposium on Computer Science, Digital Economy and Intelligent Systems. AISC, vol. 1127, pp. 47–55 (2019)
5. New NVIDIA Research Creates Interactive Worlds with AI (2018). https://nvidianews.nvidia.com/news/new-nvidia-research-creates-interactive-worlds-with-ai?utm_source=ixbtcom
6. Paramonov, P.P.: Metody predstavleniya slozhnykh poligonal'nykh modeley v graficheskikh sistemakh, rabotayushchikh v rezhime real'nogo vremeni/Paramonov, P.P., Vidin, B.V., Mezhenin, A.V., Tozik, V.T.: Izvestiya vysshikh uchebnykh zavedeniy. Priborostroyeniye.- 2006. -T.49. -№. 6, pp. 17–19. (Paramonov, P.P.: Methods for representing complex polygonal models in graphic systems operating in real time. In: Paramonov P.P., Vidin, B.V., Mezhenin, A.V., Tozik, V.T. (eds.): News of higher educational institutions. Instrumentation -T. 49. No. 6, pp. 17–19) (2006). (in Russian)

7. Izvozchikova, V.V., Mezhenin, A.V.: Razmernost' Khausdorfa v zadachakh analiza podobiya poligonal'nykh ob"yektov INTELLEKT. INNOVATSII. INVESTITSII - 2016 №. 2, pp. 109–112. (Izvozchikova, V.V., Mezhenin, A.V.: Hausdorff dimension in problems of analysis of similarity of polygonal objects intelligence. Innovation. Investments No. 2, pp. 109–112 (2006). (in Russian)
8. Mezhenin, A., Zhigalova, A.: Similarity analysis using Hausdorff metrics. In: CEUR Workshop Proceedings, vol. 2344 (2019)
9. Mezhenin, A.V.: Metody postroyeniya vektorov normaley v zadachakh identifikatsii ob"yektov/ Mezhenin, A.V., Izvozchikova, V.V.: Kibernetika i programmirovaniye, pp. 51–58, №. 4, (2013). (Mezhenin, A.V.: Methods of constructing normal vectors in problems of identification of objects/Mezhenin, A.V., Izvozchikova, V.V.: Cybernetics and programming No. 4, pp. 51–58 (2013). (in Russian)
10. Mezhenin, A.V.: Rekonstruktsiya trekhmernykh modeley po rastrovym izobrazheniyam Mezhenin, A.V.. Tozik, V.T.: Nauchno-tekhnicheskiy vestnik informatsionnykh tekhnologiy, mekhaniki i optiki.- 2007 №. 45, pp. 203 – 207. (Mezhenin, A.V.: Reconstruction of three-dimensional models from raster images/ Mezhenin, A.V., Tozik, V.T.: Scientific and Technical Journal of Information Technologies, Mechanics and Optics. №. 45, pp. 203–207) (2007). (in Russian)
11. Eisert, P.: Multi-hypothesis, volumetric reconstruction of 3-D objects from multiple calibrated camera views. In: Eisert, P., Steinbach, E., Girod, B. (eds.) Proceedings of the International Conference on Acoustics, Speech, and Signal Processing, pp. 3509–3512 (1999)
12. Huang, J.: Automatic data segmentation for geometric feature extraction from unorganized 3-D coordinate points. In: Huang, J., Menq, C.-H. (eds.) IEEE Conference on Robotics, Automation, vol. 17, pp. 268–279 (2001)
13. Chmielewski, S., Tompalski, P.: Estimating outdoor advertising media visibility with voxel-based approach. Appl. Geogr. **87**, 1–13 (2017)
14. Izvozchikova, V.V., Mezhenin, A.V.: 3D-modelirovaniye metodov s"yemki mobil'nymi videosistemami. Programmnyye produkty i sistemy № 3, pp. 163–167 (2016). (Izvozchikova, V.V., Mezhenin, A.V.: 3D modeling of shooting methods by mobile video systems. Software products and systems, No. 3, pp. 163–167 (2016). (in Russian)
15. Mezhenin, A., Izvozchikova, V., Ivanova, V.: Use of point clouds for video surveillance system cover zone imitation. In: CEUR Workshop Proceedings, vol. 2344 (2019)
16. Burlov, D.I., Mezhenin, A.V., Nemolochnov, O.F., Polyakov, V.I.: Avtomatizatsiya vybora metoda szhatiya tsifrovogo video v intellektual'nykh sistemakh zheleznodorozhnogo transporta, Pechatnyy, Vestnik RGUPS. - 2014. –vyp.54, ISSN: 0201-727KH, c. 5. (Burlov D.I., Mezhenin A.V., Nemolochnov O.F., Polyakov V.I. Automation of the choice of digital video compression method in intelligent railway systems, Pechatny, Vestnik RGUPS Iss. 54, p. 5) (2014). ISSN: 0201-727X. (in Russian)
17. Sizikov, V.S., Stepanov, A.V., Mezhenin, A.V., Burlov, D.I., Ekzemplyarov, R.A.: Opredeleniye parametrov iskazheniy izobrazheniy spektral'nym sposobom v zadache obrabotki snimkov poverkhnosti Zemli, poluchennykh so sputnikov i samolotov//Opticheskiy zhurnal - 2018. - T. 85. - №. 4, pp. 19–27. (Sizikov, V.S., Stepanov, A.V., Mezhenin, A.V., Burlov, D.I., Ekzemplyarov, R.A.: Determining the parameters of image distortion by a spectral method in the task of processing images of the Earth's surface obtained from satellites and aircraft//Optical magazine - 2018. - T. 85. - No. 4, pp. 19–27.) (2018) (in Russian)

18. Sizikov, V.S., Stepanov, A.V., Mezhenin, A.V., Burlov, D.I., Eksemplyarov, R.A.: Determining image-distortion parameters by spectral means when processing pictures of the earth's surface obtained from satellites and aircraft. J. Opt. Technol. **85**(4), 203–210 (2018)
19. PCL - Point Cloud Library (PCL). http://www.pointclouds.org/
20. CloudCompare - Open Source project. https://www.danielgm.net/cc/
21. MeshLabJs. http://www.meshlabjs.net/

High-Speed Induction Motor State Observer Based on an Extended Kalman Filter

Pavel G. Kolpakhchyan(✉), Alexander E. Kochin, Boris N. Lobov, Margarita S. Podbereznaya, and Alexey R. Shaikhiev

Rostov State Transport University, Rostov-on-Don, Russian Federation
kolpahchyan@mail.ru

Abstract. This article deal with the developing of a high-speed electric generator based on the induction motor with massive rotor. The displacement current in a massive magnetic core has a significant impact on the characteristics of this electric machine that is why the state observer must adapt to the rapidly changing of the control object parameters. For this purpose we proposed to use the state observer based on the extended Kalman filter developed on the basis of the induction electric machine equations. The processes modeling in the system of a high-speed electric generator and a state observer were made to verify the taken decisions. The mathematical model of an induction electric machine with a massive rotor and three circuits on the rotor was developed. The simulation results confirmed the correctness of chosen approach.

Keywords: Induction motor · Hi-speed electrical machine · Flux observer · Kalman filter

1 Introduction

The use of electric grid with distributed generation and diverse energy sources created on the base of "smart-grid" technology is one of the future-oriented area of energy systems development [1,2]. In this case, the power generation takes place in the vicinity of the consumer. The use of power plants of low and medium power that works on fossil fuels is rational is it is a main source of electrical and thermal energy [3]. The use of solid or liquid fuel is ecologically unacceptable, that is why the use of gaseous fuel, natural gas or biogas is more suitable [4].

The power plants with capacity of 60–300 kW are the most requested in systems with distributed generation [3,5]. In the specified power range, the use

S. Kovalev et al. (Eds.): IITI 2019, AISC 1156, pp. 633–644, 2020.
https://doi.org/10.1007/978-3-030-50097-9_65

of micro gas-turbine system is more acceptable. Efficient gas turbine operation is possible in this case with rotation speed more than 50 000 rpm [4,6–8]. The development of high-speed electrical generator located with a gas turbine on a common shaft is the actual problem.

In most cases, the high-speed electrical generator is made as a brushless electrical machine working together with semiconductor power converter. The control methods largely determine the characteristics and performance of such a system. Therefore the technical solutions development is one of the most important tasks of creating a high speed generator.

2 Problem Formulation

In the process of work under an agreement with the Ministry of Education and Science of Russia (unique identity code is RFMEFI60417X0174), the team of authors developed the high-speed electric generator with 100 kW of power and 100 000 rpm of rotating speed [5,7,9]. This power and speed combination when using a synchronous electric machine with permanent magnets on the rotor cannot provide durable work. Therefore, we decided to use the induction electrical machine with massive rotor [7].

The use of the rotor in the form of a cylinder of solid material is the design feature of the electrical machine under consideration. The current displacement in a massive rotor greatly affects the electromagnetic processes in such a machine. Therefore, the traditionally used methods of modeling and control of an induction electrical machine cannot be used without changes, taking into account this phenomenon. The electric generator regulation must be realized with high dynamic performance over the entire speed range. For this purpose, we propose to use the field-oriented control method. The flow control rotor provides the best regulatory performance. The efficiency of such a control system depends on the accuracy of estimation of the magnitude and position of the flux vector. So, induction electric machine state observer is one of the most important elements of this system.

While in operation, the parameters of the induction electric machine are changed under the influence of various factors. The state observer must adapt to these changes for efficient operation of the control system. For electric machines with "squirrel-cage", winding the change of parameters occurs mainly due to oscillations of the stator and the rotor active resistance under the influence of the windings temperature changing. Such changes are relatively slow in comparison with electromagnetic processes. Most known types of state observers, such as the Luenberger observer [10], the Kalman Filter [11], the model reference adaptive system observer [12] can adapt to this slow change of induction electrical machine parameters. The rotor contour parameters of induction electric machine with massive rotor (active resistance and leakage inductance) depend on sliding motion and significantly change with saturation of the magnetic circuit. Therefore, the creation of the state observer of induction electrical machine with massive rotor that would take this feature into account is necessary.

Consider next the choice of type and implementation of the state observer of high-speed induction electric generator with massive rotor, the research results of the state observer in different moods using a model of electromagnetic processes.

3 The Choice of State Observer Type of Induction Electric Generator with Massive Rotor

Luenberger observer [10,12] and different varieties of Kalman filter [11,13] are suitable as the induction electrical machines state observer with adaptation to parameters change. In addition, it is possible to use the method of least squares. In most cases, their use allows you to control the induction electrical machine without rotating sensor. The use of such observers allows you to know a number of hidden variables, such as the rotor flux linkage vector, electromagnetic moment, rotor angular velocity, moment of inertia of rotating parts, moment of resistance [14–16]. In some cases, the observer adapts to changing of the induction electrical machine parameters, such as the rotor active resistance or less often stator' one. The inductive parameters vary slightly that is why they are identified before starting the electric drive operation and they are not controlled during the observer operation.

The use of sensor changes the rotor speed. It is one of the feature of the high-speed electric generator state observer working together with a gas micro-turbine. It allows you to control the micro-turbine operation regardless of the generator state and its control system. Therefore, there is not any need to define the rotor angular velocity with the use of state observer. The second feature of the high-speed electric generator state observer is the need to adapt to quick changes of resistance and leakage inductance. According to various estimates, in this case the best results can be obtained using the Kalman filter [11,14,15].

The measured values are phase voltages and phase currents of the high-speed electric generator. Using these parameters, the state observer restores variables, which are necessary for the field-oriented control system operation, such as the projection of the stator current and the rotor flux linkage on the coordinate system axes, electromagnetic moment.

In the light of specified features and requirements for the state observer high-speed electric generator, the use of a linear Kalman filter is not possible as when adapting to rotor circuit changing parameters the task becomes nonlinear and the use of an advanced Kalman filter is required.

To make a state observer based on an advanced Kalman filter it is necessary to create the mathematical model of the object of observation. In our case, it is an induction high-speed electric generator with massive rotor.

4 Mathematic Model of High-Speed Induction Electric Generator with Massive Rotor

The high-speed electric generator under considering is the induction electrical machine with a massive rotor. Five stator winding phases reduce the load on the power semiconductor devices of the power converter and reduce the influence of magnetic field higher harmonics in the gap. The stator construction is traditional for electric machines and has two-layer winding. Magnetic wedges close the grooves from the air gap. It allows reducing the leakage inductance of the stator winding and improving the shape of the magnetic field in the air gap. The wedges isolate the winding from the environment and form a cylindrical shape of the gap on the side of the stator that is necessary to reduce the air friction losses at high rotating speed. Figure 1 shows the drawing of the high-speed electric generator active part.

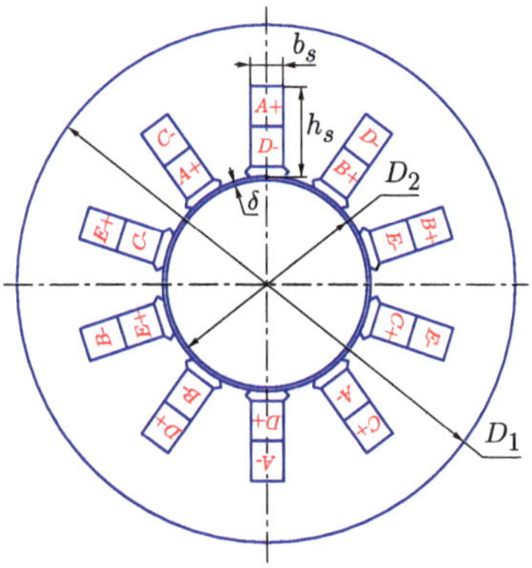

D_1 = 110 mm – the stator outer diameter;
D_2 = 45 mm – rotor diameter;
δ = 0.35 mm – air gap;
b_s = 7.2 mm – stator slot width;
h_s = 19.2 mm – stator slot height

Fig. 1. Drawing of active part of the induction high-speed electric generator with massive rotor

Mathematical model of the high-speed electric generator is developed on the base of the model of generalized electrical machine. It was admitted that:

- only the main harmonic of magnetic induction in the air gap is considered;
- system of two orthogonal, diametral concentrated windings replace the distributed five-phase stator winding;
- massive rotor is the presents as a system of several pairs of orthogonal diametrically concentrated windings.

The use of a single-loop rotor model does not provide the necessary accuracy of process representation because of the strong effect of current displacement in a massive rotor [17,18].

We determined the number of loops on the rotor and the equivalent circuit parameters of the induction electric machine with massive rotor using the method that is described in [19,20]. For this purpose, we calculated the distribution of the magnetic field in the active area of the electric machine in short circuit mode (when the rotor is motionless) and stator winding power supply by currents of different frequency.

To approximate the resulting dependency of imaginary part of phase inductance on current frequency we used the least squares method. The calculations were performed using FEMM v.4.2 (© David Meeker) software package [21].

Figure 2, a shows the results of magnetic field distribution calculation when the magnitude of the stator phase current is 150 A (amplitude value) and the frequency is 12 Hz. The calculation analysis showed that it is necessary to use the induction electrical machine with three-loop rotor to account for the effect of current displacement in the massive rotor in the working range of variation of the rotor slip frequency. Figure 2, b shows the equivalent circuit of a high-speed induction electrical generator with a massive rotor.

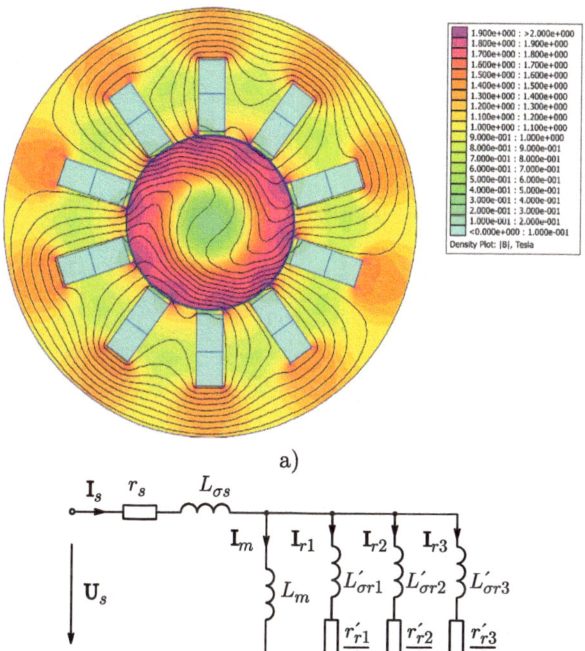

b)

Fig. 2. The equivalent circuit of a high-speed induction electrical generator with a massive rotor (a); distribution of magnetic field induction in the high-speed electric generator (b)

We received the following equivalent circuit parameters (see Fig. 2, b):
$m_f = 5$ is the number of stator phases; $p = 1$ is the number of pairs of poles;
$L_m = 0{,}001407$ Hn; $L_{\sigma s} = 0.000038525$ Hn; $r_s = 0.075$ Ohm;
$L'_{\sigma r1} = 0.000537$ Hn; $L'_{\sigma r2} = 0.000057375$ Hn; $L'_{\sigma r3} = 0.000154675$ Hn;
$r'_{r1} = 0.02201$ Ohm; $r'_{r2} = 0.10385$ Ohm; $r'_{r3} = 1.5651425$ Ohm.

The system of equations describes electromagnetic processes in the induction electrical machine with massive rotor and is presented in coordinate system stationary about the stator:

$$\begin{cases} \mathbf{U}_s = r_s \mathbf{I}_s + \dfrac{d}{dt}\boldsymbol{\Psi}_s \\ 0 = r'_{r1}\mathbf{I}_{r1} + \dfrac{d}{dt}\boldsymbol{\Psi}_{r1} + jp\omega\boldsymbol{\Psi}_{r1}; \\ 0 = r'_{r2}\mathbf{I}_{r2} + \dfrac{d}{dt}\boldsymbol{\Psi}_{r2} + jp\omega\boldsymbol{\Psi}_{r2}; \\ 0 = r'_{r3}\mathbf{I}_{r3} + \dfrac{d}{dt}\boldsymbol{\Psi}_{r3} + jp\omega\boldsymbol{\Psi}_{r3}, \end{cases} \tag{1}$$

where $\mathbf{U}_s$— stator voltage vector;

$\boldsymbol{\Psi}_s$, $\boldsymbol{\Psi}_{r1}$,$\boldsymbol{\Psi}_{r2}$,$\boldsymbol{\Psi}_{r3}$ is the stator and rotor contours flux linkage vectors;
$\mathbf{I}_s$, $\mathbf{I}_{r1}$, $\mathbf{I}_{r2}$, $\mathbf{I}_{r3}$ is the stator and rotor contours current vectors;
ω is the rotation speed of the rotor.

Flux coupling and currents are related as follows:

$$\begin{cases} \boldsymbol{\Psi}_s = (L_m + L_s)\,\mathbf{I}_s + L_m\mathbf{I}_{r1} + L_m\mathbf{I}_{r2} + L_m\mathbf{I}_{r3}; \\ \boldsymbol{\Psi}_{r1} = L_m\mathbf{I}_s + (L_m + L'_{r1})\,\mathbf{I}_{r1} + L_m\mathbf{I}_{r2} + L_m\mathbf{I}_{r3}; \\ \boldsymbol{\Psi}_{r2} = L_m\mathbf{I}_s + L'_{r1}\mathbf{I}_{r1} + (L_m + L'_{r2})\,\mathbf{I}_{r2} + L_m\mathbf{I}_{r3}; \\ \boldsymbol{\Psi}_{r3} = L_m\mathbf{I}_s + L'_{r1}\mathbf{I}_{r1} + L_m\mathbf{I}_{r2} + (L_m + L'_{r3})\,\mathbf{I}_{r3}. \end{cases} \tag{2}$$

The expression below defines the electromagnetic moment:

$$M_{em} = \frac{m_f}{2}\left(\boldsymbol{\Psi}_s \times \mathbf{I}_s\right)$$

The Eqs. (1) and (2) are a system of differential algebraic equations. Their solving is difficult. For further use, we wrote down these equations in projections on the axes of the $\alpha - \beta$ coordinate system and reduced them to the form of Cauchy. The equations are as follows in the matrix form:

$$[\boldsymbol{\Psi}] = [\mathbf{L}]\,[\mathbf{I}]\,,$$

where $[\boldsymbol{\Psi}]$, $[\mathbf{I}]$ is the vectors of flux links and stator currents projections and rotor contours on the axes of the a - b coordinate system;

$[\mathbf{L}]$ is the matrix of self and mutual inductances.

The differential equations in the form Cauchy are as follows:

$$\frac{d}{dt}[\boldsymbol{\Psi}] = \left(p\omega\,[\mathbf{D}] - [\mathbf{r}]\,[\mathbf{L}]^{-1}\right)[\boldsymbol{\Psi}] + [\mathbf{U}] \tag{3}$$

where $[\mathbf{r}]$ is the active resistance winding diagonal matrix;

$[\mathbf{U}]$ is the vector of voltages applied to the windings. Electromotive force (E.M.F.) of rotation is taken into account by the matrix

$$[\mathbf{D}] = \begin{bmatrix} [\mathbf{D}_s] & & & 0 \\ & [\mathbf{D}_r] & & \\ & & [\mathbf{D}_r] & \\ 0 & & & [\mathbf{D}_r] \end{bmatrix},$$

where $[\mathbf{D}_s] = \begin{bmatrix} 0\,0 \\ 0\,0 \end{bmatrix}$ and $[\mathbf{D}_r] = \begin{bmatrix} 0 & 1 \\ -1 & 0 \end{bmatrix}$ are matrices taking into account the stator and rotor windings rotation.

5 State Observer Based on the Kalman Filter

The use of high-speed electric generator with three circuits on the rotor is impossible for the Kalman filter design because of high difficulty. The current displacement effect in a massive rotor significantly influences on the electromagnetic processes when the slip frequency is much higher than the operating frequency. That is why the state observer can be can be created on the basis of a simpler model, with one circuit on the rotor.

A state observer based on an extended Kalman filter is based on the dynamic description of the control object obtained with the Eq. (3). The state variables of the object are the projections of the stator and rotor flux link vectors on the axes of the $\alpha - \beta$ coordinate system. The phase voltages applied to the stator winding represent the control action, the phase currents are measured.

The equations of state are as follows:

$$\begin{cases} \dfrac{d}{dt}[\mathbf{\Psi}] = \left(p\omega[\mathbf{D}] - [\mathbf{r}][\mathbf{L}]^{-1}\right)[\mathbf{\Psi}] + [\mathbf{B}][\mathbf{u}_{f\,mean}] \\ [\mathbf{I}_f] = [\mathbf{C}][\mathbf{\Psi}], \end{cases} \tag{4}$$

where $[\mathbf{B}]$, $[\mathbf{C}]$ are control and measurement matrices. The control matrix converts the phase voltages in the projections on the axes of the $\alpha - \beta$ coordinate system (direct Clarke transform):

$$[\mathbf{C}] = [\mathbf{T}_{inv}]\left[\mathbf{L}^{-1}\right],$$

where $[\mathbf{T}_{inv}]$ is the inverse Clarke transform matrix. Two variables (active resistance and rotor leakage inductance) complement the state variable vector.

The Kalman filter is considered in discrete position. The filter operation at each time step consists of two stages such as a forecasting and correction. The system state forecast is carried out using numerical differentiation at the $k+1$ time step:

$$[\mathbf{\Psi}]^*_{k+1} = f\left(t_k, [\mathbf{\Psi}]_k\right) = [\mathbf{\Psi}]_k + \Delta T \cdot \frac{d}{dt}[\mathbf{\Psi}]_k, \tag{5}$$

where ΔT is the filter sample spacing.

We used the Euler's explicit method as the simplest method of numerical differentiation. It is enough for stable filter operation at a sufficiently small sample spacing. To improve the forecast accuracy, one of the one-step numerical integration methods, such as "predict – correct" can be used (i.e. Adams-Moulton method).

Also, at the forecast stage, predictable state vector error covariance matrix can be calculated:

$$[P]_{k+1}^{*} = [F]_k \, [P]_k \, [F]_k^T + [Q], \tag{6}$$

where $[F]_k = \dfrac{\partial f\left(t_k, [\mathbf{\Psi}]_k\right)}{\partial [\mathbf{\Psi}]_k}$ is a system evolution matrix for a time moment t_{k+1}.

$[\mathbf{Q}]$ is process covariance matrix.

The covariance matrix for deviation vector and optimum Kalman gain matrix are calculated at the stage of correction:

$$[\mathbf{S}]_k = [\mathbf{C}] \, [\mathbf{P}]_{k+1}^{*} \, [\mathbf{C}]^T + [\mathbf{R}], \tag{7}$$

where $[\mathbf{R}]$ is the measurement error covariance matrix. Then, the system state assessment vector and its covariance matrix at the k+1 step are defined:

$$\begin{aligned} [\mathbf{\Psi}]_{k+1} &= [\mathbf{\Psi}]_{k+1}^{*} + [\mathbf{K}]_{k+1} \left([\mathbf{i}_{f\,mean}] - [\mathbf{C}]\right) [\mathbf{\Psi}]_{k+1}^{*}), \\ [\mathbf{P}]_{k+1} &= \left([\mathbf{I}] - [\mathbf{K}]_{k+1} \, [\mathbf{C}] \, [\mathbf{\Psi}]_{k+1}^{*}\right) \end{aligned} \tag{8}$$

where $[\mathbf{I}]$ is the identity matrix;

$[\mathbf{i}_{f\,mean}]$ is the phase currents vector (measured values).

The process covariance matrix is based on a priori information about the equations structure of the electromagnetic processes mathematical model and accuracy of high-speed generator parameters calculation. The measurement errors covariance matrix is calculated by analyzing a test sample of phase current measurements and must be corrected during the calculation.

6 The Mathematical Modeling Results

To assess the correctness of the chosen approach to create a state observer based on the base of Kalman filter (Eq. (4)–(8)) we performed the modeling of its operation together with the model of high-speed induction electric generator. We used the model with three circuits on the rotor (3). The high-speed electric generator operates together with a semiconductor power converter that has functions of an active rectifier. The converter is a bridge five-phase self-commutate voltage inverter. A single-pulse modulation of the output voltage was used in the calculations. DC link voltage is 600 V. The electric generator rotor rotates at a frequency of 100 000 rpm. We considered the transition from idle mode to generation mode. The frequency of the voltage applied to the generator windings changed in step from synchronous frequency to the amount of slip in the nominal mode (12 Hz). The voltage and generator phase's current signals were fed from the output of the electric machine model to the input of the state observer after summation with normally distributed noise with a dispersion of about 10%. It corresponds to the used measurement systems. The observer sampling spacing was taken as 5 us. Figure 3 and 4 show the modeling results. Figure 3 shows the system transition from one steady state to another; Fig. 4 shows the dependencies of currents, voltages, electromagnetic moment, and rotor flux linkage on a larger time scale.

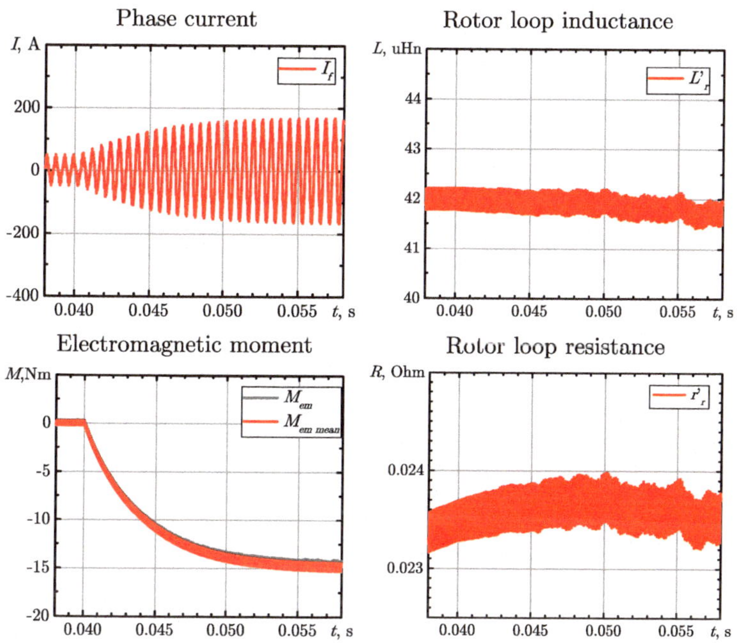

Fig. 3. The modeling results in the system of a high-speed generator and a state observer based on an extended Kalman filter

An analysis of the results shows that the developed state observer provides good filtering of the measured signals (phase currents) and allows us to recover hidden variables (rotor flux linkage and electromagnetic moment) with an accuracy sufficient to implement the field-oriented control methods. The adopted sampling step is valid for the filter stability. The filter sampling is shown in the signals shift along the time axis on the amount corresponding to the sampling step and is easily adjusted in the control system. To improve the accuracy of the state observer, it is necessary to reduce the filter-sampling step. However, allows using a high-speed ADC and significantly increases in computational costs.

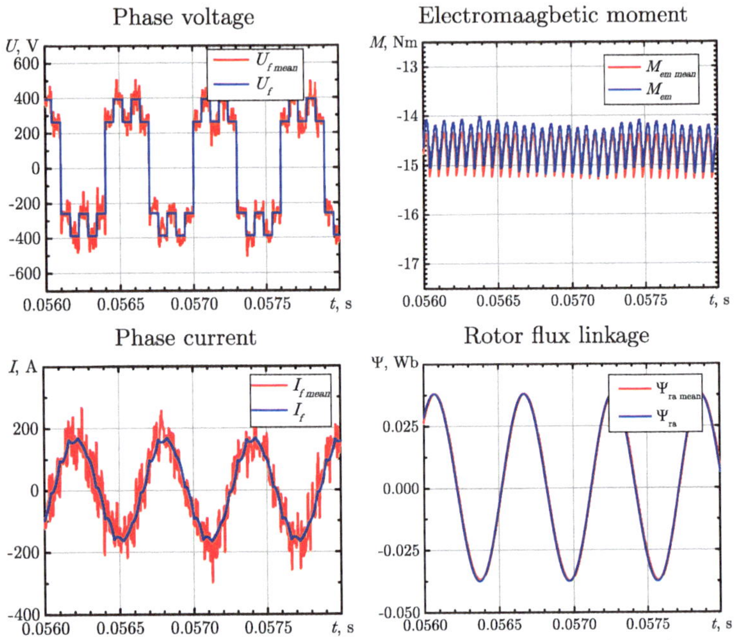

Fig. 4. Determination of flux linkage of a rotor and electromagnetic moment of a high-speed generator using a state observer based on an extended Kalman filter

7 Conclusions

The control of the high-speed induction electric generator with massive rotor operating together with the micro-gas turbine must be realized with a high speed and stringent quality requirements for transient processes. For this purpose it is necessary to use the field-oriented control methods. The development of a high-speed electric generator state observer is the most important problem. The significant effect of a rotor current displacement is the feature of the high-speed electric generator with massive rotor. Therefore, the state observer must adapt to the rapid changes in the active resistance and the leakage inductance of the rotor. The state observes based on the extended Kalman filter is the most appropriate for this purpose.

Acknowledgements. The work is done by the authors as part of the agreement No 14.604.21.0174 about subsidizing dated 26.09.2017. The topic is "Development of scientific and technical solutions for the design of efficient high-speed generator equipment for a gas micro turbine" by order of the Ministry of Education and Science of the Russian Federation, Federal Targeted Programme (FTP) "Researches and developments in accordance with priority areas of Russian science and technology sector evolution for 2014–2020 years". The unique identity code of applied researches (the project) is RFMEFI60417X0174.

References

1. Ahmad, A.: Smart Grid as a Solution for Renewable and Efficient Energy. Advances in Environmental Engineering and Green Technologies. IGI Global (2016). https://books.google.ru/books?id=c6EoDAAAQBAJ
2. Sioshansi, F.: Smart Grid: Integrating Renewable, Distributed & Efficient Energy. Academic Press (2012). https://books.google.ru/books?id=MQMrLNPjZVcC
3. Wang, Q., Fang, F.: Optimal configuration of CCHP system based on energy, economical, and environmental considerations. In: 2011 2nd International Conference Intelligent Control and Information Processing (ICICIP), vol. 1, pp. 489–494 (2011)
4. Soares, C.: Gas Turbines: A Handbook of Air, Land and Sea Applications. Elsevier Science (2014). https://books.google.ru/books?id=ubJZAwAAQBAJ
5. Kolpakhchyan, P., Shaikhiev, A., Goltsman, B., Oshchepkov, A.: Local intelligent energy systems based on gas microturbines with high-speed electric generators. Int. J. Mech. Eng. Technol. **9**(51–57), 51–57 (2018)
6. Giampaolo, T.: Gas Turbine Handbook: Principles and Practice, 5th edn. Fairmont Press, 15 November 2013. https://www.amazon.com/Botanicals-Phytocosmetic-Frank-DAmelio-Sr/dp/0849321182?SubscriptionId=0JYN1NVW651KCA56C102&tag=techkie-20&linkCode=xm2&camp=2025&creative=165953&creativeASIN=0849321182
7. Kolpakhchyan, P., Parshukov, V., Shaikhiev, A., Kochin, A., Podbereznaya, M.: High speed generator for gas microturbine installations. Int. J. Appl. Eng. Res. **12**(23), 13874–13878 (2017)
8. Gerada, D., Mebarki, A., Brown, N.L., Gerada, C., Cavagnino, A., Boglietti, A.: High-speed electrical machines: technologies, trends, and developments. IEEE Trans. Ind. Electron. **61**(6), 2946–2959 (2013)
9. Kolpakhchyan, P., Stýskala, V., Shaikhiev, A., Kochin, A., Podbereznaya, M.: Grid-tie inverter intellectual control for the autonomous energy supply system based on micro-gas turbine. Adv. Intell. Syst. Comput. **875**, 399–408 (2019)
10. Jouili, M., Jarray, K., Koubaa, Y., Boussak, M.: Luenberger state observer for speed sensorless ISFOC induction motor drives. Electr. Power Syst. Res. **89**, 139–147 (2012). https://doi.org/10.1016/j.epsr.2012.02.014. http://www.sciencedirect.com/science/article/pii/S0378779612000648
11. Zhang, Y., Zhao, Z., Lu, T., Yuan, L., Xu, W., Zhu, J.: A comparative study of luenberger observer, sliding mode observer and extended Kalman filter for sensorless vector control of induction motor drives. In: 2009 IEEE Energy Conversion Congress and Exposition, pp. 2466–2473 (2009). https://doi.org/10.1109/ECCE.2009.5316508
12. Messaoudi, M., Lassaad, S., Mouna, B., Kraiem, H.: Mras and luenberger observer based sensorless indirect vector control of induction motors. Asian J. Inf. Technol. **7**, 232–239 (2008)
13. Yin, Z., Li, G., Zhang, Y., Liu, J., Sun, X., Zhong, Y.: A speed and flux observer of induction motor based on extended Kalman filter and Markov chain. IEEE Trans. Power Electron. **32**(9), 7096–7117 (2017). https://doi.org/10.1109/TPEL.2016.2623806
14. Tiwari, V., Das, S., Pal, A.: Sensorless speed control of induction motor drive using extended Kalman filter observer. In: 2017 IEEE PES Asia-Pacific Power and Energy Engineering Conference (APPEEC), pp. 1–6 (2017). https://doi.org/10.1109/APPEEC.2017.8308989

15. Basheer, O., Obaid, M.: Flux and speed estimation of induction motors using extended Kalman filter. Int. J. Comput. Appl. **181**, 27–31 (2018). https://doi.org/10.5120/ijca2018917586
16. Alonge, F., D'Ippolito, F., Sferlazza, A.: Sensorless control of induction-motor drive based on robust Kalman filter and adaptive speed estimation. IEEE Trans. Ind. Electron. **61**, 1444–1453 (2014)
17. Levi, E.: General method of magnetising flux saturation modelling in d-q axis models of double-cage induction machines. IEE Proc. Electr. Power Appl. **144**(2), 101–109 (1997). https://doi.org/10.1049/ip-epa:19970781
18. Pedra, J., Candela, I., Sainz, L.: Modelling of squirrel-cage induction motors for electromagnetic transient programs. IET Electr. Power Appl. **3**(2), 111–122 (2009). https://doi.org/10.1049/iet-epa:20080043
19. Wilow, V.: Electromagnetical model of an induction motor in comsol multiphysics. Master's thesis, KTH, Electrical Energy Conversion (2014)
20. Dolinar, D., De Weerdt, R., Belmans, R., Freeman, E.M.: Calculation of two-axis induction motor model parameters using finite elements. IEEE Trans. Energy Convers. **12**(2), 133–142 (1997). https://doi.org/10.1109/60.629695
21. Meeker, D.: Finite Element Method Magnetics. Version 4.2. User's Manual (2018). http://www.femm.info/wiki/Files/files.xml?action=download&file=manual.pdf

Development of an Industrial Internet of Things Ontologies System

Alena V. Fedotova[1(✉)], Kai Lindow[2], and Alexander Norbach[3]

[1] Bauman Moscow State Technical University, 2-Ya Baumanskaya, 5, 105005 Moscow, Russia
afedotova.bmstu@gmail.com

[2] Fraunhofer Institute for Production Systems and Design Technology IPK, Pascalstr. 8-9, 10587 Berlin, Germany
kai.lindow@ipk.fraunhofer.de

[3] Bremen University, Otto-Hahn-Allee, 1, 28359 Bremen, Germany
anorbach@gmx.de

Abstract. The work is aimed at researching and developing an innovative subject field Industrial Internet of Things. The article deals with the system of the structure of the Industrial Internet of Things, the relationship between the objects of structure and the method of knowledge visualization with the help of ontological modeling of systems. Using the ontological approach allows us to provide support for design management, as well as further improvement of the complex technical system under consideration. The paper presents the ontology of the Industrial Internet of Things.

Keywords: Industrial Internet of Things · Ontologies system · Ontological modeling

1 Introduction

The industrial sector is the basis of economic growth and productivity growth in any country. The development of industry is accompanied by the growth of scientific and research activities, contributes to the formation of a base of new knowledge and new industries, the emergence of innovations and inventions within the country [1].

The global industry today is on the threshold of the fourth technological revolution, with which they associate the possibilities of a radical modernization of production and the economy, as well as the emergence of phenomena like: digital production and services, the economy of "shared use" (shared economy), collective consumption, "uberization" of the economy, cloud computing model, distributed networks, decentralized management, etc. The technological basis for the transition to the new economic paradigm is the Internet of Things (IoT) [1]. While the IoT permeates society as a whole, there is a version of this physical-digital link in industrial manufacturing: Industrial Internet of Things (IIoT). IIoT is mostly used for monitoring and controlling production processes with the aim of controlling processes more granularly and differentiated. The use of IIoT implies the creation of an integrated solution combining information processes with production processes. This is a fairly new task for many

S. Kovalev et al. (Eds.): IITI 2019, AISC 1156, pp. 645–654, 2020.
https://doi.org/10.1007/978-3-030-50097-9_66

companies, and when solving it, it is necessary to take into account many factors, including industry standards and processes, technological safety and regulatory framework [1]. The research problem of the paper is to find a way to organize IoT knowledge with using ontologies.

2 Internet of Things

Currently, an increasing number of components (controllers, sensors) are being created to build solutions in the field of IoT. According to CISCO, the number of "things" in the world of IoT will exceed 50 billion pieces by 2020, and the traffic generated by such devices will exceed the current one by an order of magnitude, which is connect-ed both with the frequency and volume of data generated and with additional interactions, independent data exchange between devices and data centers [2].

IoT is the scientific view that every object in the world has the ability to be connected to the Internet and transmit information about itself by performing its operations or through other objects interacting with it [3].

IoT is a network of networks consisting of uniquely identifiable objects (things) that can interact with each other without human intervention via an IP connection [1]. The key to this definition is the autonomy of devices and their ability to transfer data independently, without human intervention.

The following main segments of IoT are distinguished:

1. The production segment, which includes implementation in various industries and most closely fits the definition of the industrial Internet of things within this document.
2. The state segment, which includes solutions to improve the efficiency of the federal and municipal authorities and ensure the safety of the population.
3. The consumer segment, covering solutions for home users and solutions for smart homes.
4. Cross-industrial segment covering IoT solutions applicable in all industries.

The Industrial Internet of Things (IIoT) is the Internet of Things (IoT) for corporate/industry use – a system of integrated computer networks and connected indus-trial (production) facilities with built-in sensors and software for collecting and ex-changing data, with remote control and control in an automated mode, without human intervention [4].

3 Ontology

An ontology is an explicit formal specification of conceptualization shared by a group of agents [5].

Here the term "conceptualization" means the construction of a conceptual model of the phenomena of the external world by identifying the key concepts associated with these phenomena, and the relations between them. The word "formal" means the conceptualization of the subject area in a machine-readable format that is understandable for computer systems. The word "explicit" means that the concepts of ontology and restrictions on their use are explicitly defined.

According to N. Guarino, it is a logical theory that consists of a dictionary of terms that form a taxonomy, their definitions and attributes, as well as the axioms associated with them and the rules of inference [6]. In essence, ontologies reflect agreements on common ways of building and using conceptual models. They act as a convenient method of presenting and reusing knowledge, a means of knowledge management, and a way of learning.

Due to the great complexity of the concept of IIoT, the construction of a single understandable and consistent subject ontology is often impossible, therefore, ontologies of tasks and applications are built on the lower level, and ontologies of basic categories occurring in different subject areas are built separately. They also distinguish meta-ontology ("ontology of ontologies"), which includes methods and forms of presentation, integration, and merging of various ontologies [7, 8].

In this paper, we use the top-down approach of designing the ontology IIoT.

The top-down approach (Fig. 1) to the development of ontologies begins with the definition of the most general concepts that are invariant with respect to the subject domain, and then their sequential detailing is carried out. Here, in the first place, defines a metaontology that defines the properties of both the ontological system as a whole and its individual ontologies in particular. Namely, metaontology determines the choice of ontologies of the upper and lower levels.

The hierarchical system of ontologies in this case takes the following form, Fig. 1.

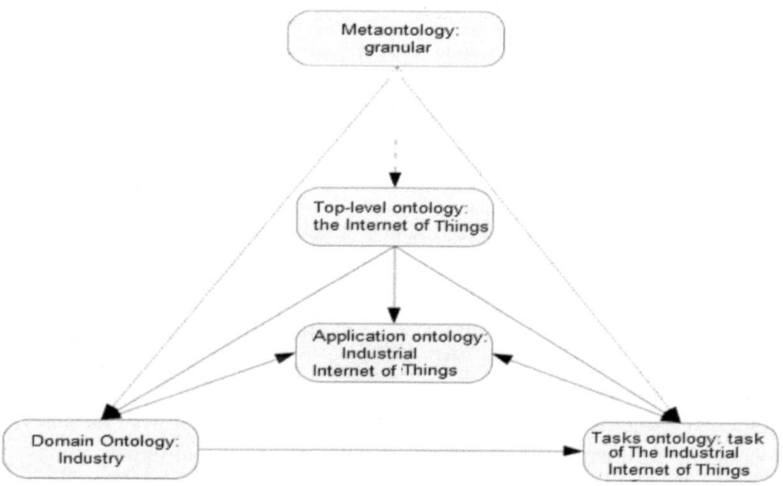

Fig. 1. The top-down hierarchical system of ontologies of the Industrial Internet of Things

The ontology of the subject area (domain ontology): industry, tasks ontology: tasks of IIoT and application ontology: IIoT are built on the lower level, and ontology of the basic categories related to IIoT, for example, the ontology of the Internet of Things, are built on the top level. According to J. Sowa, top-level ontologies describe the most common, paradigmatic conceptualizations, independent of the subject area and its tasks, which characterize the state of some professional community [9]. In contrast, low-level ontologies are local, specific, and directly depend on the type and roles of the Industrial Internet of Things agents for which they are used [10].

In the domain ontology, the concepts of the Industrial Internet and their interrelations are considered (Communication Means: Communication Means for Data Collection and Communication Means for Data Transmission), i.e. coarse-grained and fine-grained granules.

The ontology of the tasks of IIoT contains, for example, such tasks as tracking industy data, data mining and decision-making tasks [11].

The term "meta-ontology" (i.e., ontology over ontologies) is understood as the basis of the downward ontological design. Meta-ontology provides both an accurate mathematical specification of ontologies and a formal analysis of their properties [9]. With its help, a correspondence is established between the type of information available (the level of uncertainty) and the selected description language. From Fig. 1 it is clear that the choice of a particular meta-ontology directly determines the composition of ontologies, the relationships between them, the choice of formal models and languages for the representation of ontologies of both the upper and lower levels. Granular meta-ontologies suggest consideration of the basic concepts of the underlying ontology at various levels of abstractness.

Considering the system of ontologies in the context of ontological engineering we can conclude that this approach allows integrating information and knowledge from different subject areas, as well as linking it logically and dynamically [12, 13].

4 The Structure of the Industrial Internet of Things

In terms of technology, IIoT includes the following components:

- Devices and sensors that can capture events, collect, analyze data and transmit them over the network;
- Communication facilities are heterogeneous network infrastructure uniting heterogeneous communication channels – mobile, satellite, wireless (Wi-Fi) and fixed;
- Platforms for the IIoT from various IT suppliers and industrial companies designed to manage devices and communications, applications and analytics. The IIoT platform, among other things, also provides a development environment and IT security solutions;
- Applications and analytic software – a software layer responsible for analytic data processing, the creation of predictive models and intelligent device management;
- Data storage systems and servers capable of storing and processing large volumes of various information, including edge processing. IT will allow developers to orchestrate data flows between existing IoT devices at the edge, as well as granting the capability to forward the data flows to the cloud;
- IT services for the creation of solutions in the field of industrial Internet, requiring industry knowledge and business specifics;

– Security solutions, that are responsible not only for the information security of all components of the solution, but also for the security of the operational process. Due to the fact that the IIoT implies a tight integration of IT and production processes, the task of security goes beyond ensuring the smooth operation of the IT infrastructure [14].

5 Designing Industrial Internet of Things Ontology

In the course of work, the ontology system of the IIoT is presented in the paper. The general ontology of IIoT is constructed in the format .owl in Protége.

First of all simple ontologies should be constructed (Figs. 2, 3, 4, 5, 6, 7, 8, 9, 10 and 11). Then should be determine variants of axioms, used language of descriptive logic and creating system of ontologies in Protége (Simple ontologies - > Variants of axioms - > Language of descriptive logic - > Protégé).

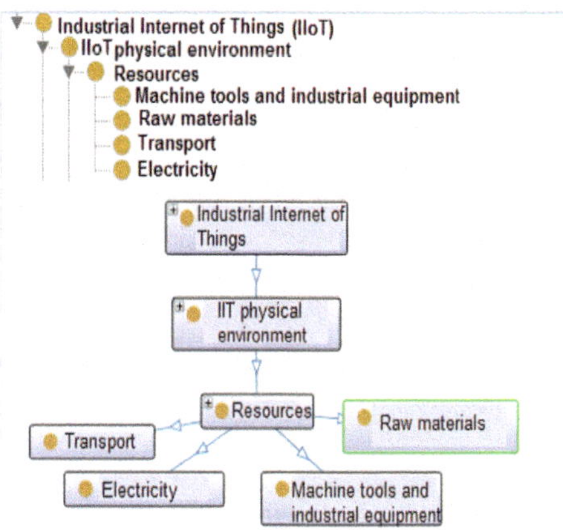

Fig. 2. Resources

This ontology is intended to ensure the publication of information about the organizational structures of IIoT. The basic concepts are: the physical environment of IIoT, the intelligent environment of IIoT, resources, devices and sensors, data transmission environment, data collection environment, data processing center, platforms, applications and software, staff, security, processes. Each of the classes is divided into subclasses, which you can see in the Figs. 2, 3, 4, 5, 6, 7, 8, 9, 10 and 11).

The ontologies of the Industrial Internet were developed and then merged with the help of rules and restrictions in Co-Protége. The developed ontologies system is presented below (Fig. 12).

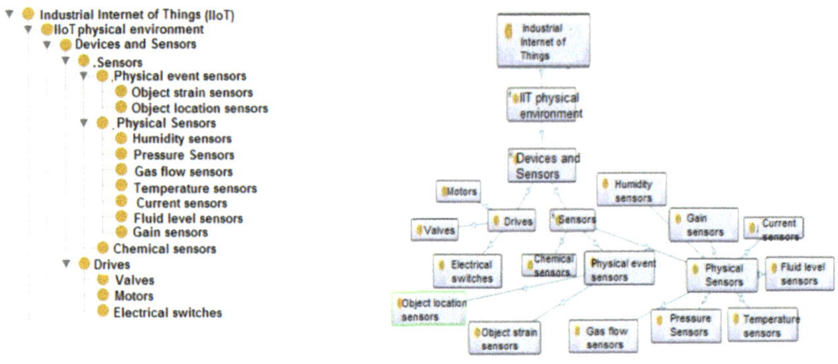

Fig. 3. Devices and sensors

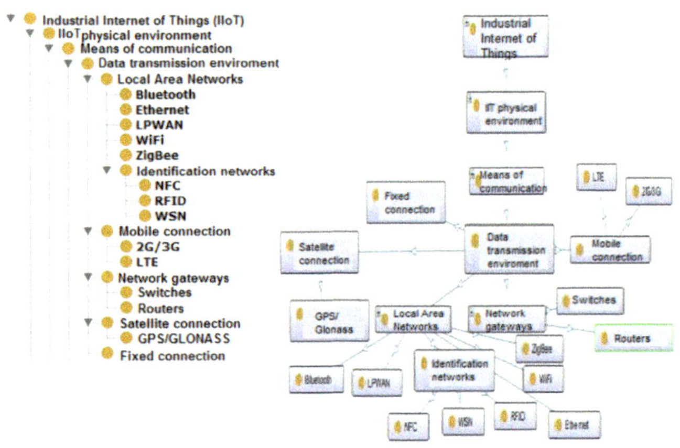

Fig. 4. Data transmission environment

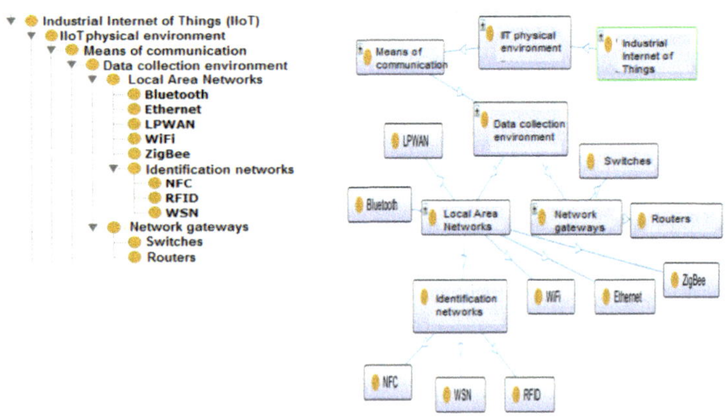

Fig. 5. Data collection environment

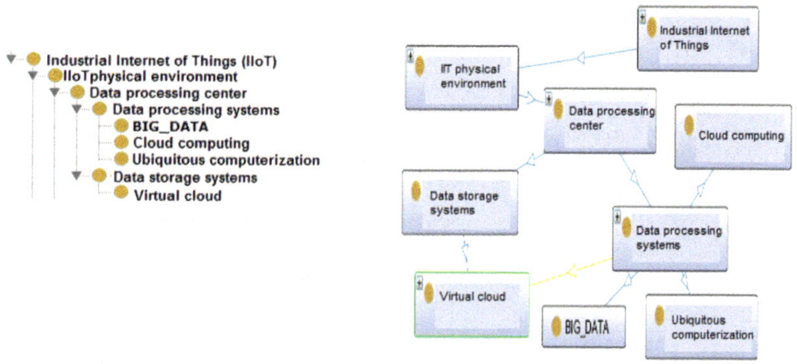

Fig. 6. Data processing center

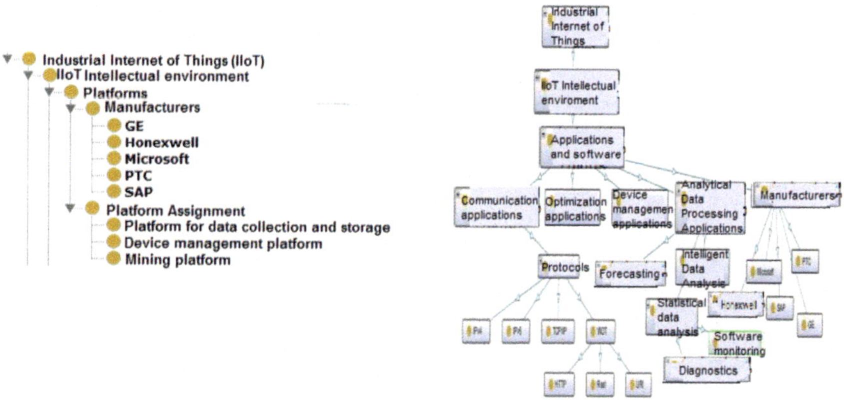

Fig. 7. Platforms

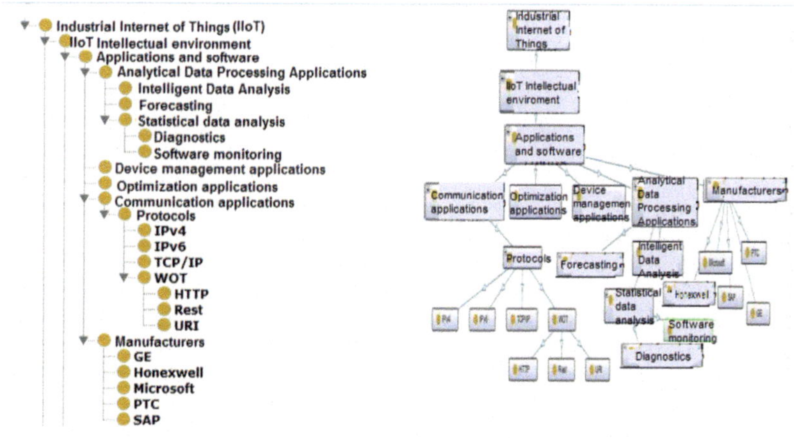

Fig. 8. Applications and software

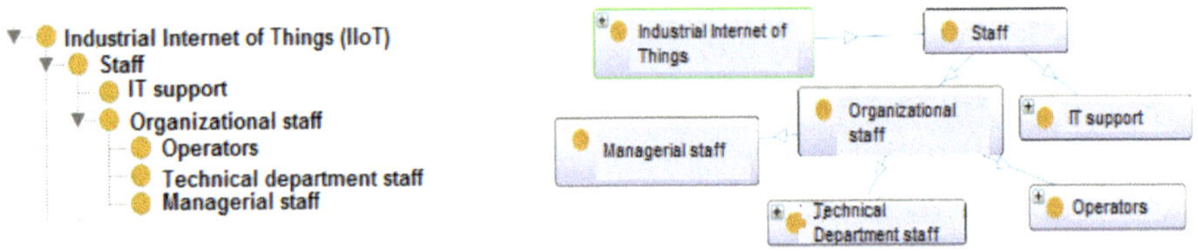

Fig. 9. Staff

Fig. 10. Security

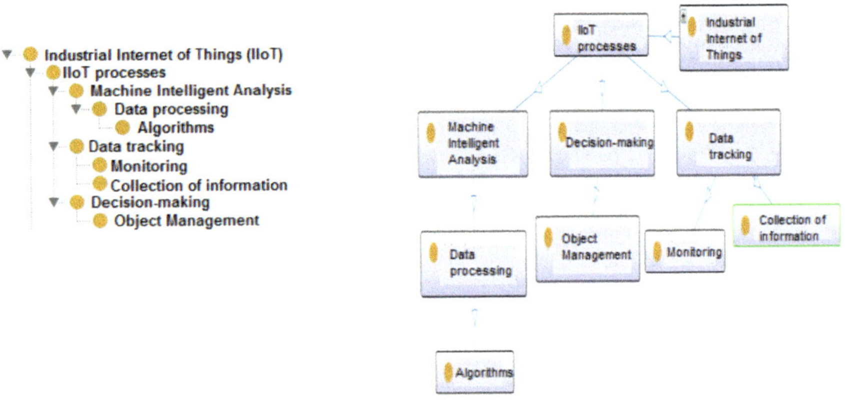

Fig. 11. Processes

After displaying, comparing and integrating the compiled ontologies, we obtain the ontology of IIoT (Fig. 12).

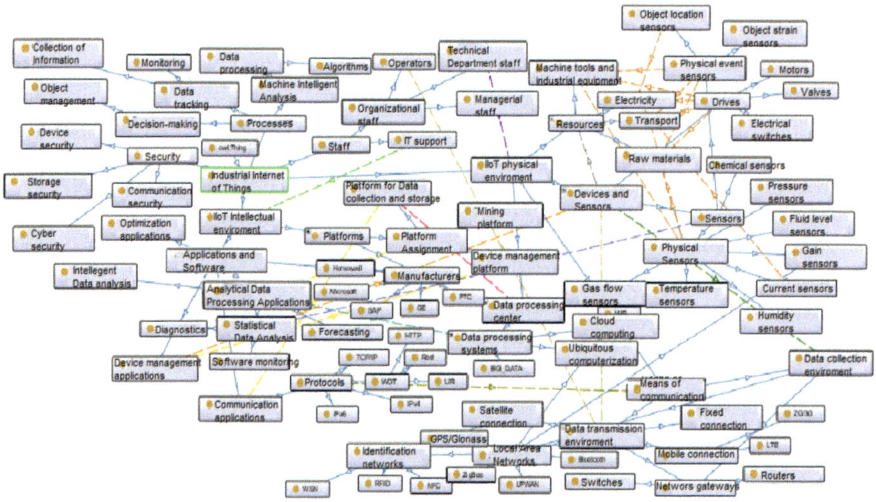

Fig. 12. Ontology of IIoT

6 Conclusion

The article describes an example of ontological modeling of the innovative area IIoT. The article describes an ontological approach for building and visualizing the IIoT system. The top-down approach for designing of ontologies are discussed. The main classes and subclasses of the ontology of IIoT are presented. Considering the system of ontologies in the context of ontological engineering we can conclude that this approach allows integrating information and knowledge from different subject areas, as well as linking it logically and dynamically. The ontologies system allows us to provide support for design management, as well as further improvement of a complex technical system. IIoT allows us to solve problems at a radically new level. The introduction of the concept requires the management of the company to make radical decisions on setting the course, but as a result, it pays off in full, both in terms of real saving and increasing profits, and in terms of improving the reliability and safety of the enterprise.

This work has been supported by DAAD and Russian Ministry of Science and Higher Education according to the research project 2.13444.2019/13.2, by RFBR according to the research project № 19-07-01208.

References

1. Roslyakov, A.V.: Internet of Things. news of the Russian academy of sciences. Theory Control Syst. **5**, 75–88 (2015)
2. IoT Reference Model Whitepaper. http://www.iotwf.com/resources/71
3. Sundmaeker, H., Guillemin, P., Friess, P., Woelffle, S. (eds.): Vision and Challenges for realizing the Internet of Things. EU, CERP IoT, Luxembourg (2010)

4. Industrial Internet of Things (electronic resource). http://json.tv/ict_telecom_analytics_view.ru,free. Title from the screen
5. Gruber, T.R.: A translation approach to portable ontologies. Knowl. Acquis. **5**(2), 199–220 (1993)
6. Guarino, N.: Formal ontology, conceptual analysis and knowledge representation. Int. J. Hum. Comput. Stud. **43**(5–6), 625–640 (1995)
7. Kovalev, S.M., Tarassov, V.B., Koroleva, M.N., et al.: Towards intelligent measurement in railcar on-line diagnostics: from measurement ontologies to hybrid information granulation system. In: Abraham, A., Kovalev, S., Tarassov, V., et al. (eds.) (IITI 2017). AISC, vol. 679, pp. 169–181. Springer, Varna (2017)
8. Petrochenkov, A.B., Bochkarev, S.V., Ovsyannikov, M.V., Bukhanov, S.A.: Construction of an ontological model of the life cycle of electrotechnical equipment. Russ. Electr. Eng. **86** (6), 320–325 (2015)
9. Sowa, J.F.: Top-Level ontological categories. Int. J. Hum. Comput. Stud. **43**(5–6), 669–685 (1995)
10. Fedotova, A.V., Davydenko, I.T., Pförtner, A.: Design intelligent lifecycle management systems based on applying of semantic technologies. In: Proceedings of the First International Scientific Conference Intelligent Information Technologies for Industry (IITI 2016), vol. 1, pp. 251–260. Springer Sochi (2016)
11. Fedotova, A.V., Tabakov, V.V., Ovsyannikov, M.V.: Ontological modeling for industrial enterprise engineering. In: Abraham, A., et al. (eds.) IITI 2018. AISC, vol. 874, pp. 182–189. Springer, Sochi (2019)
12. Sokolov, A.P., Pershin, A.Yu.: Computer-aided design of composite materials using reversible multiscale homogenization and graph-based software engineering. Key Eng. Mater. **779** (2018), 11–18 (2018)
13. Kuzenov, V.V., Ryzhkov, S.V.: Numerical modeling of laser target compression in an external magnetic field. Math. Models Comput. Simul. **10**, 255–264 (2018)
14. Tarasov V.B.: Enterprise engineering and organizational ontologies. In: Enterprise Engineering and Knowledge Management. Collection of Scientific Papers of the XVIIIth Scientific-Practical Conference (IP & UZ). pp. 25–41. MESI, Moscow (2015)

Segmentation of CAPTCHA Using Corner Detection and Clustering

Yujia Sun[1,2] and Jan Platoš[1](✉)

[1] Technical University of Ostrava, 17. listopadu 2172/15, 70800 Ostrava-Poruba, Czech Republic
{yujia.sun.st, jan.platos}@vsb.cz
[2] Hebei GEO University, No. 136 East Huai'an Road, Shijiazhuang, Hebei, China

Abstract. Character segmentation is the key to CAPTCHA recognition. In order to solve the problem of low success rate of CAPTCHA segmentation caused by adhesive characters, an adhesion character segmentation algorithm based on corner detection and K-Means clustering was proposed. The algorithm performs corner detection on the CAPTCHA image of the adhesive characters, then uses K-Means clustering method to cluster the corner points of ROI, and determines the adhesion character segmentation line from the clustering results. The experimental results are compared with the drop-fall and skeletonization, in which the recognition accuracy of the image with serious adhesion is 92%. The result shows the superiority of the segmentation algorithm and provides a new method for the segmentation of adhesive characters.

Keywords: Corner detection · Clustering · ROI

1 Introduction

In the process of CAPTCHA recognition, segmentation is a more difficult problem to solve than recognition. As long as the characters are successfully segmented, the algorithm can achieve better recognition results. Developments of adhesion characters in the CAPTCHA constitute the most difficult part of the process. In order to improve the recognition rate of the CAPTCHA, the adhesion characters must be separated accurately [1, 2]. In view of the difficulty of CAPTCHA segmentation formed by adhesion characters, we proposes an adhesion character segmentation algorithm based on the combination of corner detection and K-Means clustering [3]. Firstly, the corner points of the image were obtained by using Shi-Tomasi [4] corner point detection; then the ROI (region of interest) was established in order to extract the corner points according to the adhesion position; and, then the K-means clustering method was used to classify the corner points in order to determine the dividing line. Finally, the feasibility of the algorithm was verified by testing a CAPTCHA image set on behalf of an online bank in China.

S. Kovalev et al. (Eds.): IITI 2019, AISC 1156, pp. 655–666, 2020.
https://doi.org/10.1007/978-3-030-50097-9_67

2 Methodology

2.1 Image Preprocessing

The CAPTCHA used in this study has the following features: numbers and characters (some of which are merged). In order to highlight the character-related information in the image, so as to weaken or eliminate interfering information, the CAPTCHA must be preprocessed prior to segmentation. The preprocessing of CAPTCHA includes binarization and denoising.

2.2 Determination of Adhesion Area

Our approach was to use a binary image vertical projection to locate the adhesion characters. The projection technique is based upon the idea of projection the image data onto the X-axis [5]. In practice, this is implemented by summing the number of black pixels in each column of the image parallel to the Y-axis. We performed vertical projection on the preprocessed CAPTCHA image, the width of the projection area was calculated, the empirical threshold w was defined, and the adhesion character area was found. If the projection width N is greater than the adaptive threshold w, it is considered to be an adhesion character; otherwise, it is considered to be a single character that is successfully segmented. As shown in Fig. 1.

$$\begin{cases} N < w, & N \text{ is single character.} \\ N > w, & N \text{ is adhesion character.} \end{cases} \tag{1}$$

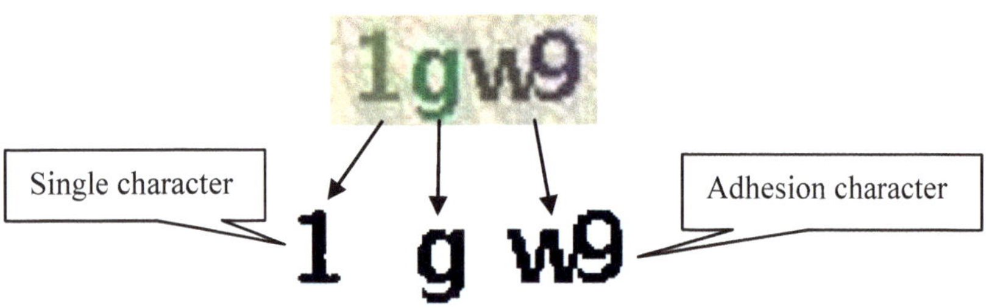

Fig. 1. Determination of adhesion area.

2.3 Feature Extraction

In this study, when the CAPTCHA image is segmented using the projection method, a single character was normalized to an image of 40 * 40 pixels. Then, the feature extraction method based on grid pixel statistical is employed. As shown in Fig. 2, the image was first segmented into 16 grids regions of equal size. Calculated the percentage of black pixels in each grid area and combined the statistics into a 16-dimensional feature vector.

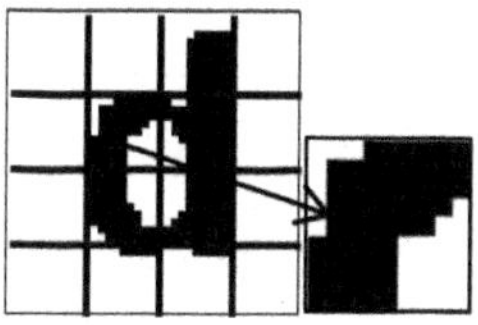

Fig. 2. Feature extraction.

2.4 Kohonen Maps

Kohonen maps is a machine learning techniques, which is special types of neural network algorithms that learn on their own through unsupervised competitive learning [6]. So work "mapping" is used because they attempt to update the winner neuron weights and adjust the weights of the neighbors according to the neighboring function to match the given input data.

The data set has a total of 3,868 images, the character image 15472 samples by vertical projection segmentation, including single character images and adhesion character images. 20 samples each character were selected and imputed into the Kohonen maps for training. Since the feature vector extracted by the feature extraction method of the previous step of each character type has 16 dimensions, the number of neurons in the input layer is 16. This is a neural network that recognizes 0–9, a–z, and maps the input so that it can be identified in the output, indicating which type it belongs to. There are 36 neural network output layers. In this experiment, the learning rate of the Kohonen maps training character images was set to 0.2, sigma was set to 8, and the training result of 5000 training times is shown in Fig. 3.

2.5 Adhesive Character Image Corner Detection

After vertical projection segmentation, the CAPTCHA data set will have some images with severely adhered characters, as shown in Fig. 1. The corner point refers to the point on the boundary curve of the image that has a maximum value of curvature. The gray value near the point will change drastically, which reflects the salient features of an image. The Shi-Tomasi corner detection algorithm is an improvement of Harris [7]. For the image I (u, v) is recorded as:

$$I_u = \frac{\partial}{\partial_u} I(u, v) \; I_v = \frac{\partial}{\partial_v} I(u, v)$$

$$A = I_\upsilon^2 \; B = I_v^2 \; C = I_\upsilon I_v$$

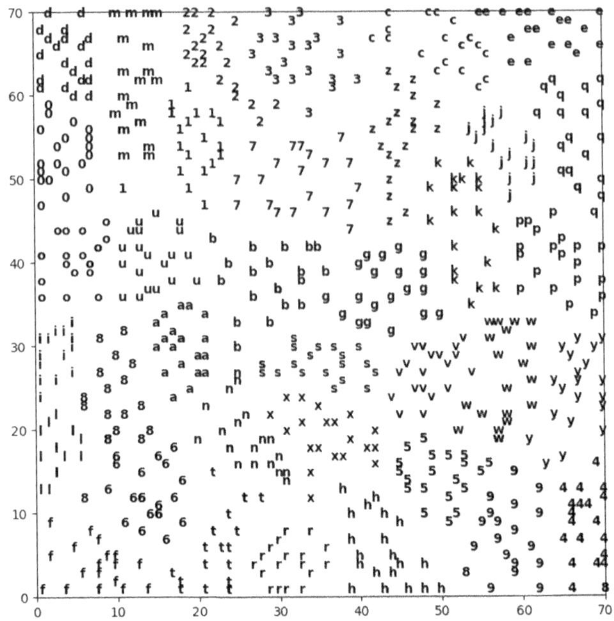

Fig. 3. Kohenen maps train single character image results.

Construct a local structure matrix:

$$M = \begin{pmatrix} A & C \\ C & B \end{pmatrix}. \tag{2}$$

Smooth with a Gaussian filter HG yields:

$$\overline{M} = \begin{bmatrix} A \times H_G & C \times H_G \\ C \times H_G & B \times H_G \end{bmatrix}. \tag{3}$$

Diagonalize matrix $\overline{M}$ to:

$$\overline{M} = \begin{bmatrix} \lambda_1 & 0 \\ 0 & \lambda_2 \end{bmatrix}. \tag{4}$$

Where λ_1 and λ_2 are the two eigenvalues of the matrix $\overline{M}$.

In the flat image area, $\overline{M} = 0$, so $\lambda_1 = \lambda_2 = 0$. If the smaller of the two eigenvalues is greater than the specific threshold, a corner point is obtained. Figure 4 shows the corner detection of some adhesion characters.

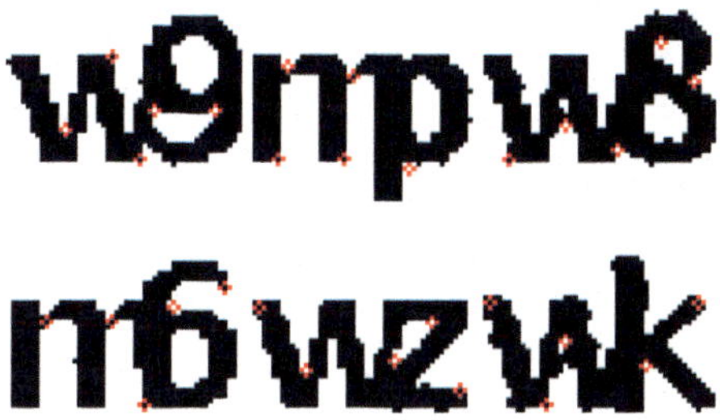

Fig. 4. Corner detection of adhesion area.

2.6 Determination Candidate Segmentation Regions

We considered that the adherent portion of the adherent character usually occurs in the middle of the region of the adherent character. According to the feature, a candidate segmentation point region ROI was established in the congruent character connected domain, and the corner points located in the ROI were used as the segmentation feature points [8]. We defined a window factor to determine the left and right boundaries of the ROI:

$$w = \frac{m \times log_2 \tau}{4}. \tag{5}$$

Whereby w is the window factor; m is the central axis of the character image; and t is the stroke width.

From Eq. 5, it can be concluded that the size of the ROI area is completely dependent on the character stroke width and character image width. The larger the stroke width and adhesion character width, the larger the ROI area accordingly. Based on this, the logarithm of the stroke width is used to make the window factor change smoother. The ROI size of the *roi_range* can be defined as:

$$roi\_range = m \pm w. \tag{6}$$

Whereby $m - w$ is the left boundary of the ROI, and $m + w$ is the right boundary of the ROI. Figure 5 shows the corners in the ROI area.

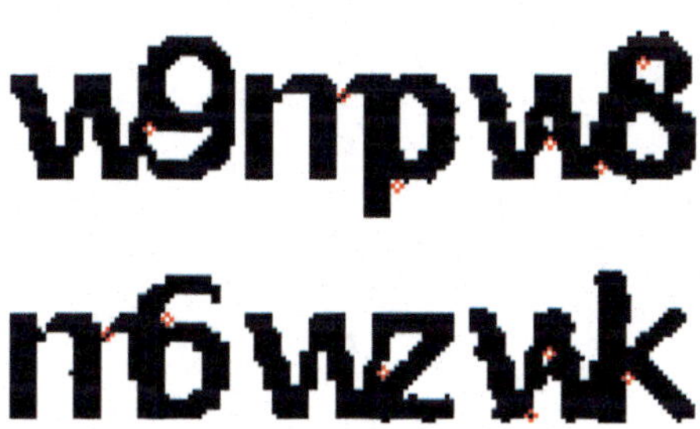

Fig. 5. The corners in the ROI area.

2.7 Corner Clustering

Our algorithm uses the K-Means clustering algorithm for corner clustering. K-Means method steps as follows:

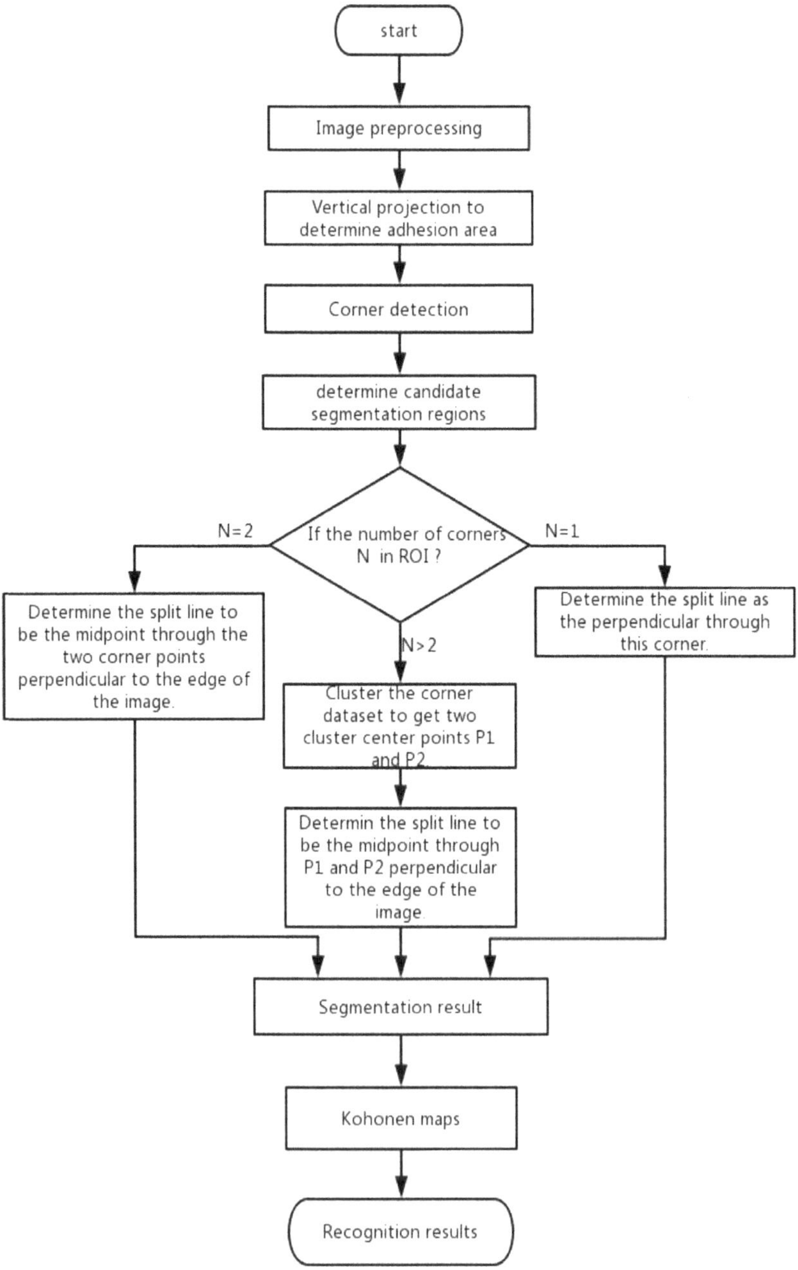

Fig. 6. The flowchart of the proposed method.

Step 1: Randomly select two initial cluster centers.
Step 2: Calculate the distance between each corner point and two cluster centers, assigning each object to the nearest cluster center.
Step 3: The clustering center is re-adjusted by the principle that the clustering object and the cluster center have the smallest distance difference.
Step 4: After the first two iterations are updated, if the cluster centers remain unchanged, the iteration ends, otherwise, continue iteration.

The vertical bisector of the two types of cluster center lines is used as the dividing line. The flowchart of the proposed method is shown in Fig. 6. Figure 7 shows the process from corner detection to candidate segmentation area determination to segmentation line determination.

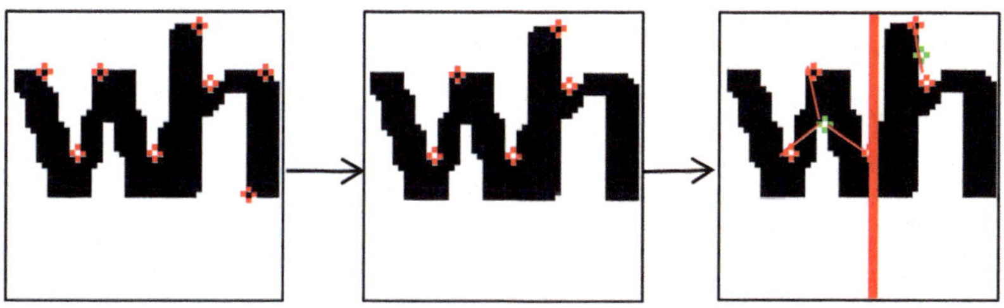

Fig. 7. The process of split line determination.

2.8 Skeletonization

The skeleton removes the extra pixels on the premise of maintaining the original image frame structure information and obtains a single pixel width skeleton structure for analyzing the image [9]. The skeleton maintains the topological and geometric invariance of the graphic. The target image point pixel is set to 1 and the background point pixel is 0, and the determination condition is set according to the refinement requirement to eliminate the point on the non-frame. The basic idea is to peel off the layers, that is peeling from the boundary layer by layer until the last skeleton. Mark the center point of the boundary point p_0, and its 8 neighborhood pixels are denoted clockwise as p_1, p_2, p_3, p_4, p_5, p_6, p_7, p_8, where p_1 is above p_0, as shown in the Fig. 8.

p8	p1	p2
p7	p0	p3
p6	p5	p4

Fig. 8. Boundary point and its neighborhood distribution.

$N(p_0)$ represents the number of non-zero points in the neighborhood of P_0 pixel 8, and $S(p_0)$ represents the number of times the pixel value of the pixel in the neighborhood of p_0 pixel 8 changes from 0 to 1 in a clockwise turn. The specific algorithm steps are as follows:

Sub-iteration A: If the 8 neighborhood of p_0 point satisfies the following conditions:

(1) $2 \leq N(p_0) \leq 6$
(2) $S(p_0) = 1$
(3) $p_1 \times p_3 \times p_5 = 0$
(4) $p_3 \times p_5 \times p_7 = 0$

The flag p_0 is to be deleted. After the end of the scan, the marked pixels are deleted, that is, $p_0 = 0$.

Sub-iteration B: If p_0 point satisfies both (1) and (2) in sub-iteration A and the following conditions (5) and (6), the flag p_0 is to be deleted.

(5) $p_1 \times p_3 \times p_7 = 0$
(6) $p_1 \times p_5 \times p_7 = 0$

Repeat the above operations until there are no pixels that can be deleted in the target image, and the skeletonization process ends. Figure 9(b) shows the result of the above-described skeletonization algorithm acting on Fig. 9(a).

Fig. 9. Example of the skeletonization algorithm.

Similar to single character feature extraction, the character skeleton image was first normalized to an image of 40 * 40 pixels. Then, the image was segmented into 16 grids regions of equal size. Calculated the percentage of black pixels in each grid area and combined the statistics into a 16-dimensional feature vector. Figure 10 shows the feature extraction of character skeleton image.

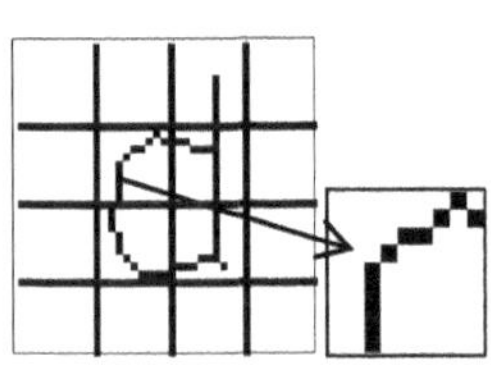

Fig. 10. Feature extraction of character skeleton image.

Selected 20 samples for each skeleton character and imputed into the Kohonen maps for training. In this experiment, the learning rate of the Kohonen maps training skeleton character images was set to 0.2, sigma was set to 8, and the training result of 5000 training times is shown in Fig. 11.

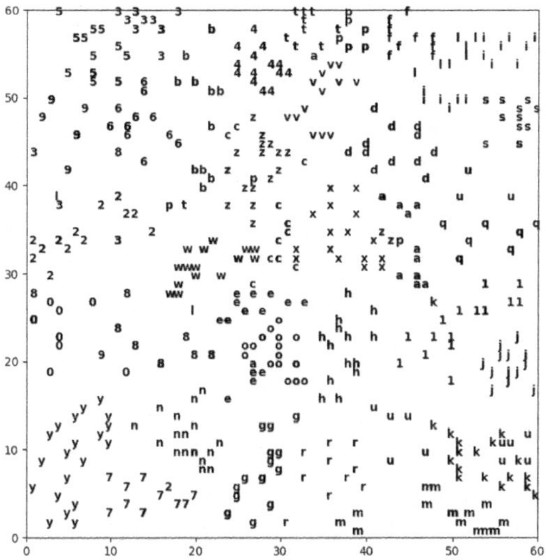

Fig. 11. Kohenen maps train skeleton single character image results.

The segmentation line of the adhesion character skeleton image is determined in the same manner as the flowchart of Fig. 6, and is also subjected to corner detection, determination candidate segmentation regions, corner clustering, and determining the split line.

3 Experimental Results and Analysis

The test consisted of 30CAPTCHA pictures with obvious character adhesion (most of which are complex adhesion cases, and a small number of which contained two characters sharing close proximity. These adhesion characters images were extracted using the projection method). Figure 12 shows adhesion character image test set. Figure 13 shows adhesion character skeleton image test set. First, the preprocessing operation (including binarization of the CAPTCHA image) was performed; then the CAPTCHA character sequence images were extracted, and then character segmentation using the predetermined algorithm was performed.

Fig. 12. Adhesion character image test set.

Fig. 13. Adhesion character image skeleton test set.

Table 1. Comparison of the segmentation results

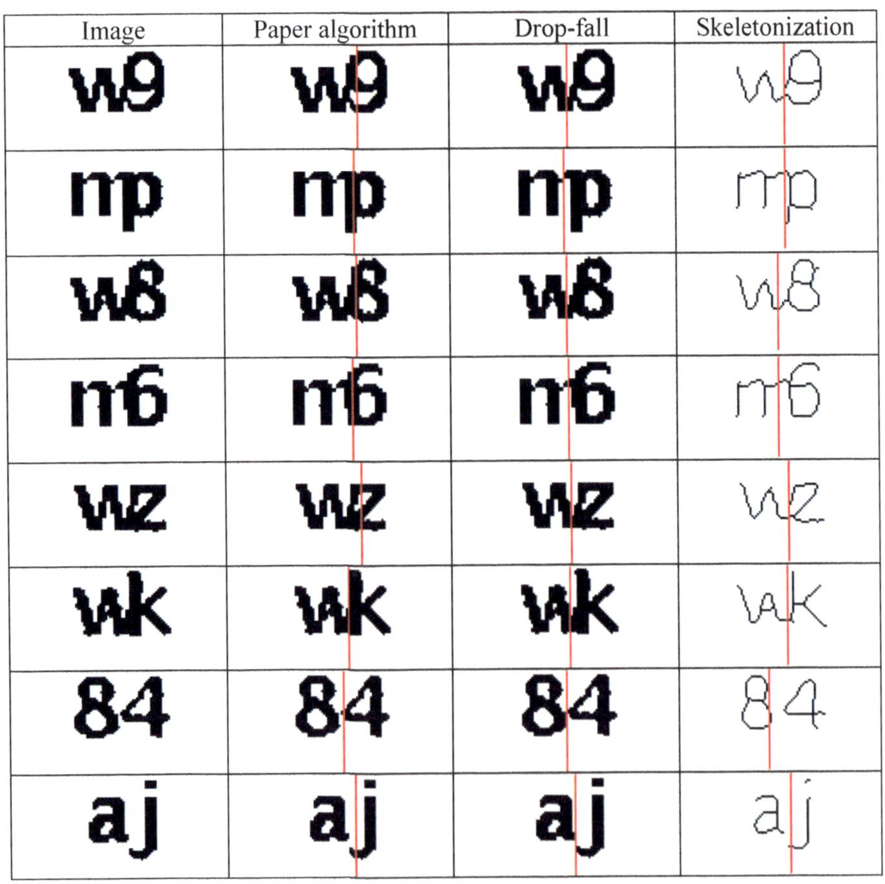

The classical drop-fall algorithm and the skeletonization were selected and compared with the algorithm described in this paper. Table 1 gives a comparison of the segmentation results of the three methods. Each method went through the same pre-processing process. The test image to be recognized was segmented according to the paper algorithm, drop-fall algorithm and skeleton algorithm, and then input into the Kohonen maps for recognition, with the output being the recognition result. The statistics results are shown in Table 2.

In summary, in the case of processing adhesion characters, the algorithm proposed in this paper leads to significantly improvement in the accuracy and universality of segmentation. During analysis of character images which failed segmentation, it was found that the segmentation error is ascribed mainly to inaccuracy of corner detection, which causes wrong location of the segmentation point and which affects the segmentation result.

Table 2. Experimental testing result statistics

Character adhesion type	Total	Paper algorithm	Drop-fall	Skeletonization
		Correct segmentation/ accuracy (%)	Correct segmentation/ accuracy (%)	Correct segmentation/ accuracy (%)
Complex adhesion	50	46/92.0%	29/58.0%	18/36.0%
Two characters close proximity	10	10/100%	10/100%	7/70.0%

4 Conclusion

In view of the difficulty of segmentation of a CAPTCHA composed of adhesive characters, this paper introduces a new algorithm of adhesive characters. Taking into according the corner points of the image, the algorithm finds the segmentation points through K-Means clustering, and draws the segmentation line from the segmentation points, and segments the image of adhesive characters. The experimental analysis has shown the effectiveness of the algorithm, although there is a need for further improvement. In future research, the method can be improved by enhancing the accuracy of corner detection and segmentation line positioning.

References

1. Platoš, J., Nowaková, J., Krömer, P., Snášel, V.: Space-filling curves based on residue number system. In: Barolli, L., Woungang, I., Hussain, O. (eds.) Advances in Intelligent Networking and Collaborative Systems, INCoS 2017. Lecture Notes on Data Engineering and Communications Technologies, vol. 8. Springer, Cham (2018)

2. Snášel, V., Drazdilova, P., Platoš, J.: Closed trail distance in a biconnected graph. PLoS ONE **13**(8), e0202181 (2018)
3. Platoš, J., Krömer, P., Snášel, V., Abraham, A.: Searching similar images—vector quantization with S-tree. In: 2012 Fourth International Conference on Computational Aspects of Social Networks (CASoN), Sao Carlos, pp. 384–388 (2012)
4. Tommasini, T., Fusiello, A., Trucco, E., Roberto, V.: Making good features track better. In: Proceedings 1998 IEEE Computer Society Conference on Computer Vision and Pattern Recognition (Cat. No. 98CB36231), Santa Barbara, CA, pp. 178–183 (1998)
5. Sun, Y., Platoš, J.: CAPTCHA recognition based on Kohonen maps. In: Barolli, L., Nishino, H., Miwa, H. (eds.) Advances in Intelligent Networking and Collaborative Systems, INCoS 2019. Advances in Intelligent Systems and Computing, vol. 1035. Springer, Cham (2020)
6. Kohonen, T.: Self-Organizing Maps, 3rd edn. Springer, New York (2001)
7. Harris, C., Stephens, M.: A combined corner and edge detector. In: Alvey Vision Conference, pp. 147–152 (1988)
8. Everton, B.L., Carlos, B.M.: Segmentation of connected handwritten digits using self-organizing maps. Expert Syst. Appl. **40**, 5867–5877 (2013)
9. Ogniewicz, R.L., Kubler, O.: Hierarchic Voronoi skeletons. Pattern Recogn. **28**(3), 343–359 (1995)

Using the Linked Data for Building of the Production Capacity Planning System of the Aircraft Factory

Nadezhda Yarushkina, Anton Romanov(✉), and Aleksey Filippov

Ulyanovsk State Technical University, Street Severny Venets 32, 432027 Ulyanovsk, Russian Federation
jng@ulstu.ru,
romanov73@gmail.com, al.filippov@ulstu.ru
http://ulstu.ru

Abstract. The basic principles of data consolidation of the production capacities planning system of the large industrial enterprise are formulated in this article. The article describes an example of data consolidation process of two relational databases (RDBs). The proposed approach involves using of ontological engineering methods for extracting metadata (ontologies) from RDB schemas. The research contains an analysis of approaches to the consolidation of RDBs at different levels. The merging of extracted metadata is used to organize the data consolidation process of several RDBs. The difference between the traditional and the proposed data consolidation algorithms is shown, their advantages and disadvantages are considered. The formalization of the integrating data model as system of extracted metadata of RDB schemas is described. Steps for integrating data model building in the process of ontology merging is presented. An example of the integrating data model building as settings for data consolidation process confirms the possibility of practical use of the proposed approach in the data consolidation process.

Keywords: Relational databases · Data model schema · Metadata · Ontology · Ontology merging · Data consolidation · Integrating data model · Production capacity planning system

1 Introduction

Modern airplane is a system with quantitative and qualitative complexity. Quantitative complexity is defined by the number of components. Qualitative complexity is defined by the complexity of production processes and a high level of uncertainty in consequence of the many external and internal factors.

The modern aircraft factory produces a set of product lines and their modifications. This product diversity determines the dynamic nature of production processes and the need for their adaptation to the changing nature of the problem area. The complex and dynamically changing production processes delay the decision-making, increases production time, and involve huge financial losses.

S. Kovalev et al. (Eds.): IITI 2019, AISC 1156, pp. 667–677, 2020.
https://doi.org/10.1007/978-3-030-50097-9_68

The modern aircraft factory has a large number of disparate information systems that store a wide massive of data about real state of production and production processes. Consolidation with the internal information systems (IS) of the aircraft factory is required in the process of creation the production capacity planning system. The main objectives of the capacity planning system of an aircraft factory are:

1. Operational formation of objective information about the available capacity of the aircraft factory.
2. Identification of the shortage of production capacities.
3. Redistribution of production volumes between productions and workshops for similar types of work.
4. Identification of the shortage of production capacity for productions, workshops, and types of work.
5. Definition of needs for commissioning additional capacity in the context of equipment, area, and personnel.

The large factories usually have many ISs that using different approaches and technologies for processing with different data models. Data consolidation in the process of information interaction involves combining data from different data sources and providing data to users in a unified form. Some of the methodological problems arise in the process of data consolidation [1–6]:

1. Fragmented automation and the impossibility of implementing a complete data processing cycle.
2. The different versions of the same information systems are used to work with different product lines.
3. The use of special or unique information systems.
4. The use of various and/or deprecated technologies, approaches, and software systems.
5. The inability to change anything in the architecture of the information environment of the aircraft factory.

Thus, the development of data-integration methods for interaction with the existing information systems of the aircraft factory is needed.

2 State of the Art

The following possible methods of organizing the data consolidation of existing on the aircraft factory ISs and the system for balancing the production capacity are [1–6]:

1. Direct exchange.
2. File sharing.
3. Service-oriented approach (SOA).

The easiest way to organize the data consolidation of several data sources is the development of the metasystem that will execute SQL queries to source RDBs on a schedule and will save data to RDB of that metasystem. The main disadvantage of this approach is the need to maintain correspondence between SQL queries for data extraction and RDB schemas of a large number of data sources. In our case, data sources are RDBs of the aircraft factory ISs. The RBDBs schemas may change in the process of their operation and development. This problem can be solved with the integrating data model.

Linked Data methods are commonly used to solve the methodological problem of building an integrating data model in the process of data consolidation. Tim Berners-Lee introduced the term Linked Data [7]:

1. Uniform resource identifiers.
2. The HTTP protocol is used for accessing the list of resources.
3. Standard Semantic Web technologies: RDF, OWL, SWRL, SPARQL.
4. Hyperlinks are used to identify web documents and domain entities.

The OWL knowledge representation language is used in the integrating data model as a single unifying data metamodel [8]. The integrating data model based on OWL uses common dictionaries containing terms from various dictionaries of external data sources.

The proposed data consolidation algorithm consists of the following steps:

1. Extracting metadata from the RDB schemas for automatic generation of ontologies for the source and target RBDs.
2. Creation of integrating data model. Ontology merging to configure correspondence between objects, attributes, and relationships of integrated ISs.
3. Using the integrating data model to perform the data consolidation procedure on a schedule.

The integrating data model is the settings contains correspondences between RDB schemas (tables and columns) of integrated ISs.

At the moment, a lot of researchers use the ontological approach for extracting metadata from the RDB schema:

1. Relational.OWL [9] currently supporting only MySQL and DB2 database management systems (DBMS). The main disadvantage of ontology generated by Relational.OWL is the presence of limited coverage of the domain, not considering, for instance, data type, foreign keys, and constraints.
2. OWL-RDBO [10,11] currently supporting only MySQL, PostegreSQL and DB2 DBMSs. The main disadvantage of ontology generated by OWL-RDBO is the presence of concepts external to the domain, such as RelationList to group a set of Relation, and AttributeList to group a set of attributes.
3. Ontop [14] build a conceptual layer over data sources in the form of an ontology that hides the structure of that data sources. Ontop allow to execute unified semantic queries in SPARQL to this conceptual view, and the users no longer need an understanding of the data sources models (relations between RDB entities, differences between data types and data encodings). Ontop

translated semantic user queries into queries over various integrated data sources. The ontology is connected to the data sources through a declarative specification given in metaontology that relate the ontology classes and properties to SQL views over data. However, the primary purpose of Ontop is querying data from several data sources with semantic user queries, and this is not data consolidation.
4. Other approaches, such as [12,13] extract the real world relations from the RDB structure, and unable to reconstruct the original schema of the RDB.

Thus, it is necessary to develop a method for data consolidation based on ontology merging of metadata extracted from the RDB schemas of several ISs of the aircraft factory.

3 Metadata of the RDB Schema

The logical representation of the integrating data model is based on the $\mathcal{ALCHIF}(\mathcal{D})$ description logic.

Let's describe the ontological representation of the integrating data model in $\mathcal{ALCHIF}(\mathcal{D})$ description notation as follows:

$$O_i^{IS} = \langle TBox, ABox \rangle,$$

where $TBox$ is the terminological box;

$ABox$ is the assertional box.

The $TBox$ of the ontological representation of the integrating data model can be represented as:

$$RDBTable \sqsubseteq E \qquad RDBAttribute \sqsubseteq A$$
$$RDBDataType \sqsubseteq C \qquad RDBConstraint \sqsubseteq C$$

$$\begin{aligned} C \equiv\ & \top \sqcap \exists hasType.String \sqcap \forall hasType.String \sqcap \\ & \sqcap \forall hasValue.\,(Integer \sqcup String \sqcup Double \sqcup E) \\ A \equiv\ & \top \sqcap \exists hasName.String \sqcap \forall hasName.String \sqcap \\ & \sqcap \forall hasConstraint.C \\ E \equiv\ & \top \sqcap \exists hasName.String \sqcap \forall hasName.String \sqcap \\ & \sqcap \forall hasAttribute.A \sqcap \forall hasParent.E \end{aligned}$$

where E is a entity;

A is a attribute;
C is a constraint;
$hasName$ is a "has the name" role;
$hasValue$ is a "has the value" role;
$hasAttribute$ is a "entity has the attribute" role;
$hasConstraint$ is a "attribute has the constraint" role;
$hasParent$ is a "child has the parent" role.

The process of extraction of the metadata form the RDB schema presents in [15,16] in more detail.

4 Integrating Data Model

It is necessary to form an integrating data model based on metadata extracted from RDB schemas of each integrated information systems. The definition of an ontological system is used as a formal representation of an integrating data model:

$$\sum^{O} = \langle O^{META}, O^{IS}, M \rangle,$$

where O^{Meta} is a result of merging the set of ontologies O^{IS};

$O^{IS} = \{O_1^{IS}, O_2^{IS}, \ldots, O_p^{IS}\}$ is a set of metamodels extracted from various data sources;

M is the consolidation reasoner that realize several functionality:

1. Data extraction.
2. Data types compatibility validation.
3. Constraints checking.
4. Data types conversion.
5. Saving data to the target vault.

The process of an integrating data model formation based on the set of metadata extracted from information systems presents in [15,16] in more detail.

5 Example of Integrating Data Model Creation

Let see the following example of the integrating data model formation. The RDB of an aircraft factory IS is the source, and the RDB of the production capacity planning system is the target.

Table 1 shows the structure of the "Equipment and Tools" table of the aircraft factory IS.

Thus, the metadata of the "Equipment and Tools" entity (Table 1) of the source RDB can be represented as:

$$EquipmentAndTools(ET)\colon RDBTable$$

$$\begin{array}{rr}
ET.t2_ob\colon RDBAttribute & ET.t2_ng\colon RDBAttribute \\
ET.t2_nn\colon RDBAttribute & ET.t2_r1\colon RDBAttribute \\
ET.t2_r2\colon RDBAttribute & ET.t2_r3\colon RDBAttribute \\
ET.t2_gm\colon RDBAttribute & ET.up_dt\colon RDBAttribute \\
ET.up_us\colon RDBAttribute & ET.t2_dc\colon RDBAttribute \\
ET.t2_vid\colon RDBAttribute & ET.t2_doc\colon RDBAttribute \\
ET.t2_prim\colon RDBAttribute & ET.t2_yyy\colon RDBAttribute
\end{array}$$

$$\begin{array}{rr} CHAR\colon RDBDataType & NUMBER\colon RDBDataType \\ BLOB\colon RDBDataType & DATE\colon RDBDataType \\ nullable\colon RDBConstraint & LENGTH\_200\colon RDBConstraint \\ \ldots & \ldots \\ LENGTH\_4\colon RDBConstraint & PK\colon RDBConstraint \end{array}$$

$$\begin{array}{l} (ET, ET.t2\_ob) : hasAttribute \\ \ldots \\ (ET, ET.t2\_yyy) : hasAttribute \\ (LENGTH\_200, 200\colon Integer) : hasValue \end{array}$$

$$\begin{array}{l} (ET.t2\_ob, CHAR) : hasConstraint \\ (ET.t2\_ob, LENGTH\_200) : hasConstraint \\ \ldots \\ (ET.t2\_nn, PK) : hasConstraint \\ \ldots \\ (ET.t2\_yyyy, CHAR) : hasConstraint \\ (ET.t2\_yyyy, LENGTH\_4) : hasConstraint \end{array}$$

Table 1. The "Equipment and Tools" table of the aircraft factory IS

Column	Data type	Constraint	Description
t2_ob	CHAR(200)	Nullable	Name
t2_ng	NUMBER(5)		Group
t2_nn	NUMBER(6)		Position
t2_r1	CHAR		Type #1
t2_r2	CHAR		Type #2
t2_r3	CHAR		Type #3
t2_gm	BLOB		Geometric model
up_dt	DATE		Date of last update
up_us	CHAR(32)		User
t2_dc	BLOB		Attachment
t2_vid	CHAR(4)		Tooling type
t2_doc	CHAR(100)		Document name
t2_prim	CHAR(100)		Notes
t2_yyyy	CHAR(4)		Production date

The "Equipment and Tools" entity of the target RDB consists from several tables:

1. Tools (Table 2).
2. Tool types (Table 3).

Table 2. The "Tools" table of the production capacity planning system

Column	Data type	Constraint	Description
id	INTEGER	Nullable	Primary key
name	VARCHAR(255)		Name
serial_number	VARCHAR(255)		Serial number
inventory_number	VARCHAR(255)		Inventory number
production_date	DATE		Production date
tool_types_id	INTEGER		Link to "Tool type" table (foreign key)

Table 3. The "Tool types" table of the production capacity planning system

Column	Data type	Constraint	Description
id	INTEGER		Primary key
name	VARCHAR(255)		Name

The metadata of the "Equipment and Tools" entity (Table 2, 3) of the target RDB can be represented as:

$$
\begin{aligned}
Tools(T)&\colon RDBTable & Tool\_types(TT)&\colon RDBTable\\
T.id&\colon RDBAttribute & T.name&\colon RDBAttribute\\
T.serial\_number&\colon RDBAttribute & T.inventory\_number&\colon RDBAttribute\\
T.production\_date&\colon RDBAttribute & T.tool\_types\_id&\colon RDBAttribute\\
TT.id&\colon RDBAttribute & TT.name&\colon RDBAttribute\\
INTEGER&\colon RDBDataType & VARCHAR&\colon RDBDataType\\
DATE&\colon RDBDataType & TT\_FK&\colon RDBConstraint\\
nullable&\colon RDBConstraint & LENGTH\_255&\colon RDBConstraint\\
PK&\colon RDBConstraint & FK&\colon RDBConstraint
\end{aligned}
$$

$$(T, T.id) : hasAttribute$$
$$\ldots$$
$$(T, T.tool\_types\_id) : hasAttribute$$

$$(TT, TT.id) : hasAttribute$$
$$(TT, TT.name) : hasAttribute$$

$$(LENGTH\_255, 255 \colon Integer) : hasValue$$
$$(TT\_FK, TT) : hasValue$$

$$(T.id, INTEGER) : hasConstraint$$
$$(T.id, PK) : hasConstraint$$
$$(T.name, VARCHAR) : hasConstraint$$
$$(T.name, LENGTH\_255) : hasConstraint$$
$$\ldots$$
$$(T.tool\_types\_id, FK) : hasConstraint$$
$$(T.tool\_types\_id, TT\_FK) : hasConstraint$$

$$(TT.id, INTEGER) : hasConstraint$$
$$(TT.id, PK) : hasConstraint$$
$$(TT.name, VARCHAR) : hasConstraint$$
$$(TT.name, LENGTH\_255) : hasConstraint$$

The proposed method allows organizing data consolidation without the participation of developers in contrast to the traditional approach, based on the method of direct data exchange. In our case, an expert can organize the data consolidation process by creation of the integrating data model in the process of ontology merging in Protege [18].

Let see the example of the integrating data model O^{Meta} formation for metadata O^S and O^T:

Step 1. Formation of the universal concept dictionary for the current domain.

Expert defined that:

- Term "Position" (Table 1) of source is the same as term "Primary key" (Table 2) of target.
- Term "Notes" (Table 1) of source is the same as term "Inventory number" (Table 2) of target.

Step 2. Setting data source properties.

$$(Source1, Oracle) : hasDBType$$
$$(Target1, PostgreSQL) : hasDBType$$

Step 3. Entity merging (manually).

$$ET = Source1 \quad T = Target1 \quad TT = Target1$$

Step 4. Attribute merging (manually).

$$ET.t2\_nn = T.id \quad \ldots \quad ET.r3 = TT.name$$

Step 5. Constraint and data type matching (automatically based on previous steps). The reasoner executes this step in the process of consolidation. Reasoner allows to check:

- data types compatibility. For example, it is impossible to convert data value to boolean;
- constraints compatibility. For example, it is impossible to write nullable value to non nullable attribute.

The resulting integrating data model is passed to the data exchange component.

6 Conclusion

We have developed an ontology-based approach to consolidate heterogeneous relational databases. Consolidation with several ISs of the aircraft factory is required in the process of creation the production capacity planning system. The results of searching for similar software implementations for organizing of data consolidation process with ISs allows concluding about the relevance of the proposed approach. The proposed method allows organizing information interaction between ISs without the participation of developers, which helps to increase the flexibility and efficiency of data consolidation.

Our approach allows organizing data consolidation without the participation of developers in contrast to the traditional approach, based on the method of direct data exchange. However, more work needs to be done for implementation of the data type casting algorithms in case of their mismatch for each DBMS, and adapting the proposed method implementation to the SQL dialect of DBMS involved in the exchange process. Random DBMS cannot be supported by this implementation.

An interesting future direction might be to integrate our approach with some database migration tool to reduce labor costs for work with different DBMS [17]. Database migration tools allow to literally move data from one type of database to another or to another destination like a data warehouse or data lake. But migrating databases requires careful planning to ensure that all your data is properly accounted for, transported, and protected.

Acknowledgments. The study was supported by:

– the Ministry of Science and Higher Education of the Russian Federation in framework of projects 2.4760.2017/8.9 and 2.1182.2017/4.6;

– the Russian Foundation for Basic Research (Projects No. 18-47-732016, 18-47-730022, 17-07-00973, No. 18-47-730019).

References

1. Clark, T., Barn, B.S., Oussena, S.: A method for enterprise architecture alignment. In: Practice-Driven Research on Enterprise Transformation, pp. 48–76. Springer (2012)
2. Rouhani, D.B., et al.: A systematic literature review on enterprise arquitecture implementation methhodologies. Inf. Softw. Technol. **62**, 1–20 (2015)
3. Medini, K., Bourey, J.P.: SCOR-based enterprise architecture methodology. Int. J. Comput. Integr. Manuf. **25**, 594–607 (2012)
4. Poduval, A., et al.: Do More with SOA Integration: Best of Packt. Packt Publishing Ltd, Birmingham, UK (2011)
5. Caselli, V., Binildas, C., Barai, M.: The Mantra of SOA. Service Oriented Architecture with Java, Birmingham, UK (2008)
6. Berna-Martinez, V.J., Zamora, C., Ivette, C., Pérez, M., Paz, F., Paz, L., Ramón, C.: Method for the integration of applications based on enterprise service bus technologies. https://www.wseas.org/multimedia/journals/computers/2018/a445905-1342.pdf. Accessed 20 July 2019 (2018)
7. Bizer, C., Heath, T., Berners-Lee, T.: Linked Data - The Story So Far. http://tomheath.com/papers/bizer-heath-berners-lee-ijswis-linked-data.pdf. Accessed 20 July 2019
8. Gruber, T.: Ontology. http://tomgruber.org/writing/ontology-in-encyclopedia-of-dbs.pdf. Accessed 20 July 2019
9. de Laborda, C.P., Conrad, S.: Relational.OWL: a data and schema representation format based on owl. In: Proceedings of the 2nd Asia-Pacific Conference on Conceptual Modelling, vol. 43, pp. 89–96. Australian Computer Society, Inc (2005)
10. Trinh, Q., Barker, K., Alhajj, R.: RDB2ONT: a tool for generating OWL ontologies from relational database systems. In: International Conference on Internet and Web Applications and Services/Advanced International Conference on Telecommunications 2006, AICT-ICIW 2006, p. 170. IEEE (2006)
11. Trinh, Q., Barker, K., Alhajj, R.: Semantic interoperability between relational database systems. In: 11th International Database Engineering and Applications Symposium, IDEAS 2007, pp. 208–215. IEEE (2007)
12. Barrett, T., Jones, D., Yuan, J., Sawaya, J., Uschold, M., Adams, T., Folger, D.: RDF representation of metadata for semantic integration of corporate information resources. In: International Workshop Real World and Semantic Web Applications, vol. 2002. Citeseer (2002)
13. Bizer, C.: D2R MAP – a database to RDF mapping language. In: Proceedings of the 12th International World Wide Web Conference – Posters (2003)
14. Calvanese, D., Cogrel, B., Komla-Ebri, S., Kontchakov, R., Lanti, D., Rezk, M., Rodriguez-Muro, M., Xiao, G.: Ontop: answering SPARQL queries over relational databases. Semant. Web J. **8**, 471–487 (2017)
15. Yarushkina, N., Romanov, A., Filippov, A., Guskov, G., Grigoricheva, M., Dolganovskaya, A.: The building of the production capacity planning system for the aircraft factory. In: Research Papers Collection Open Semantic Technologies for Intelligent Systems, issue 3, pp. 123–128 (2019)
16. Yarushkina, N., Romanov, A., Filippov, A.: Using ontology engineering methods for organizing the information interaction between relational databases. In: Kuznetsov, S., Panov, A. (eds.) Artificial Intelligence, RCAI. Communications in Computer and Information Science, vol. 1093. Springer, Cham (2019)

17. Alley, G.: Database Migration Tools. In Database Zone (2019). https://dzone.com/articles/database-migration-tools-1. Accessed 20 July 2019
18. Protege — Free open-source ontology editor. https://protege.stanford.edu/. Accessed 20 July 2019

Method for Recognizing Anthropogenic Particles in Complex Background Environments

B. V. Paliukh and I. I. Zykov(✉)

Tver State Technical University, Tver, Russia
pboris@tvstu.ru, kemer0802@mail.ru

Abstract. The technogenic pollution of the near-Earth space is a significant negative consequence of its development. Further exploration of the near-Earth space is impossible without data mining of the current state of pollution. One of the solutions to this problem is to develop some methods for recognizing anthropogenic particles whose work is complicated due to the complex background conditions in the near-Earth space. Object recognition technologies have delivered good but not sufficient results to solve the tasks of space exploration with human presence being unavailable. Thereby, recognition methods are to be improved continuously with modern methods and algorithms. This article describes the method for recognizing anthropogenic particles with wavelet, fractal, and correlation analyses of images obtained with an optical-electronic device located on an orbiter.

Keywords: A wavelet analysis · A fractal analysis · A correlation analysis · Anthropogenic particles · Recognition

1 Introduction

The data mining of the current state of pollution is particularly acute for the orbits up to 2000 km, which are called low, as well as for geostationary orbits containing a large mass of man-made particles and objects and, hence, representing much probability of dangerous catastrophic collisions of anthropogenic particles with space objects. With the number of satellites on orbit and their evident obsolescence growing, the risk of the avalanche-like development of the Kessler syndrome has been steadily increasing. The main feature of the Kessler syndrome is the domino effect [1]. The collision of two sizable man-made objects will lead to a larger number of new anthropogenic particles. Each of these particles is capable of colliding with other objects or operating spacecraft which could cause a cascade of anthropogenic particle generation.

Anthropogenic particle data mining can determine the actual and predicted danger to space operations. Unfortunately, near-Earth debris is difficult to characterize accurately. Only the largest anthropogenic particles are constantly monitored by ground sensors, there being much more difficulty in producing new sensors to detect numerous small anthropogenic particles [2].

S. Kovalev et al. (Eds.): IITI 2019, AISC 1156, pp. 678–687, 2020.
https://doi.org/10.1007/978-3-030-50097-9_69

The Scientific and Technical Subcommittee of the UN Committee on the Peaceful Uses of Outer Space approved the following notion of space debris, “Space debris is defined as all man-made objects, including fragments and elements thereof, in Earth orbit or re-entering the atmosphere, that are non-functional.” [3].

Considering the mass-geometric and orbital characteristics of space debris the US Space Command catalog currently consists of more than 16,200 observable space objects of man-made origin, generated during the entire history of space activities (4,765 carrier rocket launches) including 251 space object destructions occurring randomly or initiated deliberately [4].

Currently, over 110 thousand objects of space debris larger than 1 cm in diameter and 40 million objects with a size greater than 1 mm have accumulated around the Earth.

The number of objects which size is between 1 and 10 cm can only be estimated statistically as about 70–150 thousand objects since they can be observed with neither telescopes nor radars and, therefore, they cannot be recorded in any catalogs. The collision of any space debris of more than 1 cm in size with an operational satellite is dangerous for the latter due to the large kinetic energy of the former and may cause its functional activities to cease which is not the worst consequence if we consider that the satellite may have a nuclear reactor.

The scientific and technical Subcommittee of the United Nations Committee on the peaceful uses of outer space approved the following concept of space debris: “Space debris – all man-made objects in earth orbit or returning to the atmosphere, including their fragments and elements that are non-functional”.

2 Analysis of Anthropogenic Particle Recognition Methods

In paper proposes to develop a Kapton film net in order to detect small space debris and micrometeorites [5]. Another similar method based on a shock detector uses solar panels for impact detection. The concept of the method consists in modifying the insulating layer of solar cells for common solar batteries [6]. The shock detector methods have a significant drawback of damage beyond repair which appears on the detector surfaces after their colliding with anthropogenic particles and objects. Besides, new anthropogenic particles are likely to generate in this case.

Currently, the Japan Aerospace Exploration Agency, JAXA, is developing an on-board detector for the in-situ measurement system of space micro-debris (less than 1 mm in diameter) which cannot be detected from the Earth. The detector is called an indicator of space debris particles. The principle of operation of its sensor is based on the first use of conductive (resistive) lines [7]. But the problem remains in covering the entire near-Earth space with the resistive lines.

In paper proposes to use gamma spectrometers based on scintillation detectors and xenon gamma detectors to solve the problems of detecting and identifying radioactive space debris. A scintillation detector serves mainly to detect radioactive space debris objects [8].

The model of using laser for space debris location is based on the classical location equation [9] and used for monitoring separate space debris objects at considerable distances. It represents a solution to a classical laser location problem of a diffuse reflector of an arbitrary shape, the angular dimensions of which are less than the angular divergence of a laser locator radiation.

The first NEOSSat orbiter is unable to detect an object of more than 500 m in diameter. Nevertheless, NEOSSat efficiency is expected to be higher than that of ground observatories since orbital telescopes will not be affected by intense sunlight [10].

The proposed approach eliminates the disadvantages described above since it does not create additional man-made particles and its main task is to detect and recognize objects smaller than 10 cm, being able to detect larger anthropogenic particles at the same time.

3 Problem Statement and Proposed Method Description

Intelligent Information Technologies can help solve the problem of recognizing anthropogenic particles in complex background environments. The method of recognizing and calculating the coordinates of anthropogenic particles in the near-Earth space is based on the initial detection of anthropogenic particle regions with wavelet transforms. Then the fractal dimension and the maximum eigenvalues of an autocorrelation matrix are calculated for the detected regions to confirm the hypothesis that an anthropogenic particle is present in this region.

The first stage of the method consists in the following. The images from an opto-electronic device are converted into halftone ones to reduce the core memory capacity. Further, all the pixel brightness values of the resulting image are converted into an array of double-precision real numbers.

Applying the Cauchy–Bunyakovsky–Schwarz inequality to the correlation in view of the image function, we obtain

$$\begin{aligned}
&\sum_{x\in X}\sum_{y\in Y} f(x,y)\psi_{mk}(x,y) = \sum_{x\in X}\sum_{y\in Y}\phi(x,y)S(x,y)\psi_{mk}(x,y)\\
&+\sum_{x\in X}\sum_{y\in Y} B\phi(x,y)\psi_{mk}(x,y) + \sum_{x\in X}\sum_{y\in Y} n(x,y)\psi_{mk}(x,y)\\
&\leq \leq \left[\sum_{x\in X}\sum_{y\in Y}\phi^2(x,y)S^2(x,y)\right]^{\frac{1}{2}}\left[\sum_{x\in X}\sum_{y\in Y}\psi^2_{km}(x,y)\right]^{\frac{1}{2}}\\
&+\left[\sum_{x\in X}\sum_{y\in Y}B^2_{\phi}(x,y)\right]^{\frac{1}{2}}\left[\sum_{x\in X}\sum_{y\in Y}\psi^2_{km}(x,y)\right]^{\frac{1}{2}}\\
&+\left[\sum_{x\in X}\sum_{y\in Y}n^2(x,y)\right]^{\frac{1}{2}}\left[\sum_{x\in X}\sum_{y\in Y}\psi^2_{km}(x,y)\right]^{\frac{1}{2}}
\end{aligned} \tag{1}$$

Where $\phi(x,y)$ – the function of changing the characteristics of the medium of radiation propagation, $S(x,y)$ – the function of changing the brightness of anthropogenic particles and objects, $B_{\phi}(x,y)$ – the function of changing the background brightness, $n(x,y)$ – the function of the resulting image noise, $(x,y) \in S$, $S = X \times Y$ – integer coordinates in the picture plane.

When solving any of the equalities in processing the image at the output of the discussed wavelet correlator as a matched filter we will have the maximum ratio of anthropogenic particles to background noise.

The second stage is the calculation of the wavelet coefficients of columns WTv(x, y), rows WTg(x, y), and one of diagonals WTd(x, y). This is required for greater refinement since the detected area in the image can shift from the actual position of the anthropogenic particle at different angles of the optoelectronic device inclination. To obtain the resulting wavelet array, coefficients WT(x, y), without losing the structural features of the rows, columns, and diagonals we use the following equation:

$$\mathrm{WT}(x, y) = \mathrm{WTg}(x, y) \vee \mathrm{WTv}(x, y) \vee \mathrm{WTd}(x, y) \quad (2)$$

At the next stage the image is segmented by a threshold value. All the values being less than the specified threshold take the values equal to 0, and those being above the threshold take the value of 1, with the iteration results being stored in the segmentation array. The structural features are localized in the coordinate system of images obtained from an optoelectronic device and have their own fractal dimensions and maximum eigenvalues of autocorrelation matrices. The segmentation consists in implementing two properties: high spatial and frequency resolutions and the maximum ratio of pixel brightness values to those of background noise brightness.

When the segmentation is completed, the binarization of the array and the filling of the holes are performed sequentially. The hole filling operation uses an algorithm based on morphological reconstruction which allows the holes with 0 values and surrounded with values of 1 for 8-connected background neighborhoods to be filled. Morphological reconstruction [11] is a non-linear filter based on mathematical morphology. It uses expansion and erosion functions which are the simplest operations of mathematical morphology [12]. The reconstruction procedures help restore the main contours of the objects. The original image is simplified, but, at the same time, the information about the basic contours is saved.

The resulting regions of the localized anthropogenic particles are of an arbitrary shape, but they are required to be rectangular in order to build their autocorrelation matrix. To provide the regions with the required shape we apply the following rule:

$$\begin{aligned}&\mathrm{IF}((\mathrm{W}(i,j) = 1)\,\mathrm{AND}\,(\mathrm{W}(i+1,j+1) = 1))\,\mathrm{OR}\,((\mathrm{W}(i+1,j) = 1)\,\mathrm{AND}\,(\mathrm{W}(i,j+1) = 1))\\&\mathrm{THEN}\,\mathrm{W}(i,j) = 1, \mathrm{W}(i+1,j) = 1, \mathrm{W}(i,j+1) = 1, \mathrm{W}(i+1,j+1) = 1\,\mathrm{ELSE}\,\mathrm{W}(i,j) = 0,\\&\mathrm{W}(i+1,j) = 0, \mathrm{W}(i,j+1) = 0, \mathrm{W}(i+1,j+1) = 0\end{aligned} \quad (3)$$

The fourth stage consists in finding the vertex coordinates of the formed regions and results in generating a corresponding two-dimensional array of coordinates.

Further, the fractal dimension is estimated with the covering method [13] and the maximum eigenvalue of the autocorrelation matrix is calculated for each detected rectangular area covering an anthropogenic particle in the image. In the general case, the use of fractal characteristics is aimed at improving the information content of the feature space and distinguishing the classes of natural and artificial objects [14]. The maximum eigenvalues substantially dominate all other indicators and contain practically complete information about the actual situation in the corresponding image area being analyzed.

The simulation results provide the frequency diagrams of the fractal dimension (d_{vi}) and maximum eigenvalues of the autocorrelation matrix ($\lambda_{max\,vi}$) for the situations of an anthropogenic particle, a false target, a star or a planet.

The frequency diagram of the fractal image dimensions for a situation of an anthropogenic particle ($\beta_d = 5.2$, $\alpha_d = 3.4$) is left-handed relative to the situation of a star or a planet ($\beta_d = 7.7$, $\alpha_d = 2.6$) and right-handed relative to the situation of a false target ($\beta_d = 2$, $\alpha_d = 3.6$) (Fig. 1). Figure 2 shows that beta probabilities overlap significantly, but, at the same time, the situations can be solved among themselves.

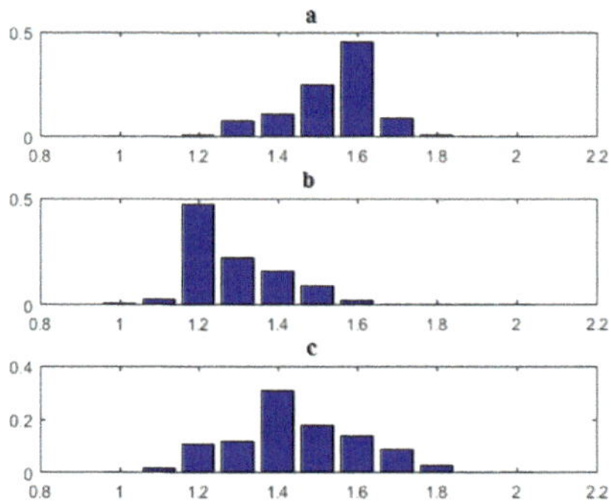

Fig. 1. Frequency diagrams of fractal image dimensions. a is for a false target situation, b is for a star or a planet situation, and c is for an anthropogenic particle situation.

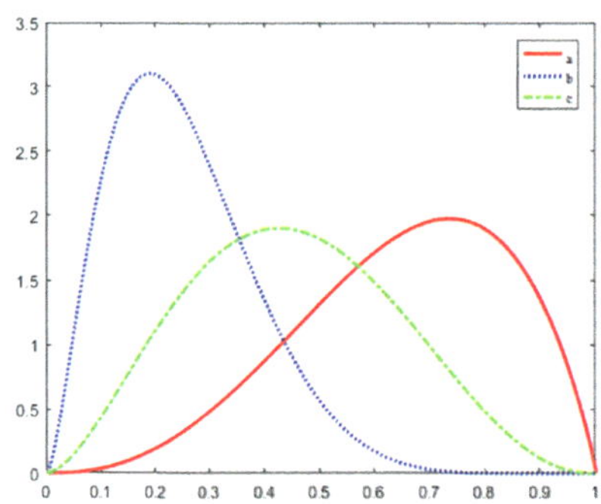

Fig. 2. Beta distributions of the fractal image dimensions. a is for a false target situation, b is for a star or a planet situation, and c is for an anthropogenic particle situation.

The frequency diagram of the maximum eigenvalues of the autocorrelation matrix of images for an anthropogenic particle situation ($\beta_\lambda = 0.7$, $\alpha_\lambda = 1.5$) is right-handed relative to the situations of a false target ($\beta_\lambda = 3.3$, $\alpha_\lambda = 0.7$) and a star or a planet

($\beta_\lambda = 9.4$, $\alpha_\lambda = 3.1$) (Fig. 3). Figure 4 shows that the beta probabilities of the maximum eigenvalues have the same properties as the beta probabilities of the fractal dimension.

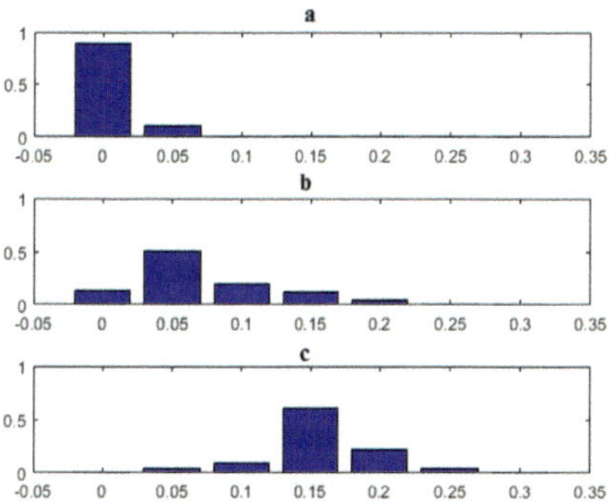

Fig. 3. Frequency diagrams of the maximum image eigenvalues. a is for a false target situation, b is for a star or a planet situation, and c is for an anthropogenic particle situation.

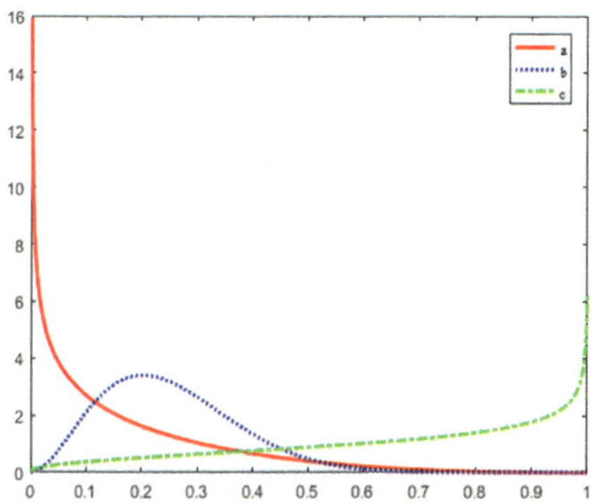

Fig. 4. Beta distributions of the maximum image eigenvalues. a is for a false target situation, b is for a star or a planet situation, and c is for an anthropogenic particle situation.

To calculate the threshold values between the pairs of fractal dimension situations we should use the following expressions:

$$\frac{\Gamma(\beta_1+\alpha_1)}{\Gamma(\beta_1)\Gamma(\alpha_1)}\int_{\Gamma(\alpha_{\mathrm{add},d})}^{\infty}\exp\{-y\beta_1\}(1-\exp\{-y\})^{\alpha_1-1}dy \le \alpha_{\mathrm{add},d} \tag{4}$$

$$\Pi(\alpha_{\partial on}; d_{vi}) = \frac{\partial}{\partial\alpha_1}\ln\Gamma(\beta_1+\alpha_1) - \frac{\partial}{\partial\alpha_1}\ln\Gamma(\alpha_1) + \ln(\mu_{11}-\mu_{10}) + \frac{\rho_d}{d\alpha_1/d\beta_1} \tag{5}$$

For $\alpha_{add,d} = 0.2$, $\beta_d = 5.2$, $\alpha_d = 3.4$, we obtain $P(\alpha_{add}; d_{vi}) = 0.55$ for the priority limit and $P(\alpha_{add}; d_{vi}) = 0.36$ for the lower limit. Using $P(\alpha_{add}; d_{vi})$ and expression for them we can calculate the value of the threshold constant $P_d = 1.55$ for the priority limit and $P_d = 1.36$ for the lower limit.

Similarly (taking into account that there is only one border) we define a decision criterion on the maximum eigenvalue found by statistics $P(\alpha_{add}; \lambda_{maxvi}) = 0.48$ $p_\lambda = 0.144$, with $\alpha_{add,\lambda} = 0.2$, $\beta_\lambda = 0.7$, $\alpha_\lambda = 1.5$.

At the sixth stage of the method the Neumann-Pearson criterion is used to confirm the detection of an anthropogenic particle. According to this criterion we choose the detection rule that ensures the minimum value of a signal skip probability (the maximum probability of a correct detection), provided that a false alarm probability does not exceed a desired value [15]. The optimal decision threshold is obtained for each criterion experimentally.

Based on the results obtained a criterion structure for detecting man-made particles and objects should be defined. According to the criteria of maximum eigenvalues and fractal dimensions the following combinations are possible: (true, false), (false, true), (false, false), and (true, true). In connection with a prior uncertainty about the situation of combination (true, false) and (false, true) they should be transformed into a combination (false, false). Therefore, the criterion for detecting man-made particles and objects should be based on a combination (true, true), and the desired criterion for making a decision on detecting anthropogenic particles should be defined on the logic of a binary accumulator, i.e. on the logic of a binary serial procedure.

4 Practical Results of Developed Method

We chose and tested several situations obtained by the Canadian NEOSsat Space Telescope designed to track potentially dangerous asteroids and space debris.

Fig. 5. The image obtained by the Canadian space telescope NEOSsat.

Figure 5 shows one anthropogenic object and about 12 stars. The program processed and selected one object with the fractal dimension of 1.43 and the maximum eigenvalue of the autocorrelation matrix of 0.21. The brightest stars in this image that were not rejected after using the wavelet coefficient thresholds have the fractal dimension greater than 1.58 and the maximum eigenvalue of the autocorrelation matrix less than 0.11.

Fig. 6. The image obtained by the Canadian space telescope NEOSsat.

Figure 6 shows one anthropogenic object and about 7 stars. The program processed and selected one object with the fractal dimension of 1.45 and the maximum eigenvalue of the autocorrelation matrix of 0.17.

The method provides high efficiency indices in any target environment without any prior information including that on the law of the probability distribution of optical-electronic device measurement errors.

5 Conclusion

The authors have revealed that the operation of an optoelectronic device can cause the situations when some distortions associated with the noise from the orbiter power facility appear which lead to the generation of a complex background environment. In this regard, an approach to recognize anthropogenic particles with the above mentioned methods and algorithms has been developed.

The proposed approach shows the highly efficient recognition of anthropogenic particles in complex background environments. The image processing based on the wavelet transform allows us to solve a number of problems related to noise reduction, local spatial inhomogeneity detection, image compression, and texture analysis [16]. One of the main advantages that wavelet provides is the ability to present a local analysis, i.e. to analyze the localized area in a large signal [17]. The information content is not violated when the background conditions change; the properties of fractal dimension as the minimum sufficient statistics [18] and the maximum invariant to confirm the hypothesis of anthropogenic particle detection are confirmed. By definition, the maximum eigenvalues are the maximum invariants and the minimum sufficient statistics [19] to confirm the hypothesis of anthropogenic particle detection. The final decision on detection is made with Neumann-Pearson criterion along with the threshold values calculated, the beta probability distribution over fractal dimensions and the maximum eigenvalues of the autocorrelation matrices.

The research was done within the government task of the Ministry of Education and Science of the Russian Federation. The number of the publication is 2.1777.2017/4.6.

References

1. Adushkin, V.V., Veniaminov, S.S., Kozlov, S.I., Silnikov, M.V.: Counter-terrorism technical devices, on technogenious contamination of space and some special consequences. Mil. Eng. (7–8), 16–21 (2015)
2. Stansbery, G.: Orbital Debris Research at NASA. Johnson Space Center. http://aero.tamu.edu/sites/default/files/faculty/alfriend/S2.2%20Stansbery.pdf. Accessed 21 Nov 2018
3. Space Debris Mitigation Guidelines of the Committee on the Peaceful Uses of Outer Space (2007)
4. Veniaminov, S.S., Chervonov, A.M.: Space Debris - a Threat to Mankind. Space Research Institute of Russian Academy of Sciences, Moscow (2012). 192 p.
5. Bauer, W., Romberg, O., Krag, H.: Debris in-situ impact detection by utilization of cube-sat solar panels. In: The 4S Symposium (2016)
6. New, J.S., Price, M.C., Cole, M.: ODIN: a concept for an orbital debris impact detection network. In: 47th Lunar and Planetary Science Conference (2016)
7. The United Nations Organization: The Committee on the Peaceful Uses of Outer Space. The Scientific and Technical Subcommittee. National Research on Space Debris, Safety of Space Objects with Nuclear Power Sources on Board and Problems of their Collision with Space Debris, Vienna (2015)
8. Anikeeva, M.A., Boyarchuk, K.A., Ulin, S.E.: Detecting radioactive space debris by spacecraft. Space electro-mechanics, spacecraft. Electromech. Matters **126** (2012)
9. Garnov, S.V., Moiseeva, A.V., Nosatenko, P.Ya., Fomin, V.N., Tserevitinov, A.B.: Evaluation of perspective orbital laser locator characteristics for space debris monitoring. In: Proceedings of Prokhorov General Physics Institute of the Russian Academy of Sciences, vol. 70 (2014)
10. NEOSSat's Dual Mission – HEOSS, NESS - Near Earth Space Surveillance. http://neossat.ca/?page_id=99. Accessed 22 May 2018
11. Vincent, L.: Morphological gray scale reconstruction in image analysis: applications and efficient algorithms. IEEE Trans. Image Process. **2**(2), 176–201 (1993)
12. Soille, P.: Morphological Image Analysis: Principles and Applications, pp. 173–174. Springer, New York (1999)
13. Petukhov, N.Yu.: Covering method for calculating fractal dimensions of landscape images. digital signal processing and its application. In: Proceedings of Abstracts of 11th International Conference and Exhibition, Moscow, vol. 2, pp. 393–396 (2009)
14. Avramenko, D.V.: Detection and allocation of space objects artificial and natural origin based on the wavelet transformation. Science of yesterday, today and tomorrow. In: Collection of Essays on 35th International Scientific and Practical Conference, Novosibirsk, vol. 28, pp. 77–83 (2016)
15. Katulev, A.N., Kolonskov, A.A., Khramichev, A.A., Yagol'nikov, S.V.: Adaptive method and algorithm for detecting low-contrast objects with an optoelectronic device. J. Opt. Technol. **81**, 29–39 (2014)
16. Ercelebi, E.: Electrocardiogram signals de-noising using lifting-based discrete wavelet transform. Comput. Biol. Med. **34**(6), 479–493 (2004)

17. Singh, B.N., Tiwari, A.K.: Optimal selection of wavelet basis function applied to ECG signal denoising. Digit. Signal Process. **16**(3), 275–287 (2006)
18. Potapov, A.A.: Fractals in Radiophysics and Radiolocation. Logos, Moscow (2005). 848 p.
19. Kolmogorov, A.N., Fomin, S.V.: Elements of the Theory of Functions and Functional Analysis, Chicago (1968). 496 p.

Technology and Mathematical Basis of Digital Twin Creation in Railway Infrastructure

Alexander. N. Shabelnikov and Ivan. A. Olgeyzer(✉)

JSC "NIIAS" Rostov Branch, Rostov-on-Don, Russia
olgeyzer@rfniias.ru

Abstract. The paper considers the problems of development and implementation of industrial control systems in conditions of digitalization of railway transport. The actuality of digital twin creation is shown for railway infrastructure and rolling stocks. The meaning of "Digital twin of railway freight station" is revealed. The advantages of digital twin usage in conjunction with big data, industrial internet of things are presented for development, design and implementation of industrial control systems.

Keywords: Digital station · Automation systems · Industrial control system of railway transport · Digital twin

1 Introduction

Nowadays, Industrial control systems (ICSs) work in the conditions of continuous specification and correction of technological and economic aspects of their functioning.

Rapid updating and modernization of the existing nomenclature of trackside devices, indoor computing and control equipment and common microelectronic element base leads to situation when components of developed ICS become outdated either right after, or even before implementation process is completed. Here, "outdating" is understood as moral obsolescence, impossibility of spare parts acquisition and rather low industrial guaranteed terms of maintenance and support of the equipment by manufacturers.

Changeable operating conditions, emergence of new samples of the equipment in coupe with specifications of technological process (for example, impossibility to stop humping in decisive freight stations) leads to necessity of new researches in area of railway ICS.

One of the decision is to put the following things into development of ICS [1]:

1. Possibility of modularity;
2. Universality in use of the outdoor and indoor equipment;
3. Smaller connectivity of the used components;
4. Possibility of technical and technological maneuver for synthesis of ICS.

However, long terms of development and implementation also negatively affect project economy especially taking into account specifics of implementation of large complex projects for development of railway infrastructure.

S. Kovalev et al. (Eds.): IITI 2019, AISC 1156, pp. 688–695, 2020.
https://doi.org/10.1007/978-3-030-50097-9_70

The use of digital twins in railways including sustainability assessment for the infrastructure is the rapidly growing task [2].

The present work proposes the use of digital twins at all stages of life cycle of any infrastructure project to accelerate the design of ICS and commissioning works during its implementation together with cost minimization for choosing optimum set of technical equipment.

2 Problem Statement

Digital twin is the counterpart of the physical device, process or a system transferred into digital environment. The digital twin is a mathematical model of high level of adequacy, which allows to describe behavior of an object in all situations, at all stages of life cycle including emergencies with a big accuracy. Use of digital twins allows to quickly simulate evolution of events depending on one or another factor, define potential risks, find the most effective operating modes and build steps on safety [3].

The digital twin is created as the copy of a physical object using information from all available devices and sensors taking into account the principles of isomorphism. Here, homomorphism is mapping of area from the state space of a real object to the certain point of the state space of a digital twin. Graphically, it is presented in Fig. 1.

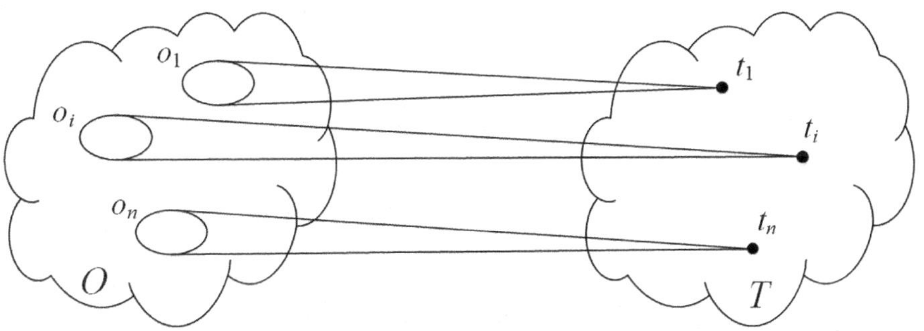

Fig. 1. Mapping of real object's and digital twin's states

In Fig. 1, O is the state space of a real object, T is the state space of the digital twin, $o_i (i = 1, \ldots, n)$ is the i-th state subspace of a real object, t_i $(i = 1, \ldots, n)$ is the i-th state of a digital twin.

In algebraic form, it means

$$\forall o_i \subseteq O \; \exists t_i \in T \quad f(x \in o_i) = t_i \tag{1}$$

In fact, (1) means that digital twin T is capable to consider the most significant factors affecting the real object. Thus, calculated $f(\cdot)$ gives an opportunity not only to provide full mapping of the digital twin into a real object, but also to model different options of evolution of technological process on the digital twin, imitating their occurrence on a real object with high degree of reliability. It gives an opportunity to analyze, test and optimize the digital copy of a real object, minimizing at the same time risks and losses if cancellation of the made changes is necessary. Real object uses the optimization model only after its debugging on the digital twin.

3 Proposed Approach

On a real object of railway infrastructure (e.g. hump yard), it is not often possible to obtain $f(\cdot)$ in analytical form with required accuracy because of a large number of affecting factors, which measures have a lack necessary points or their time is exceeds admissible limits for the real-time decision making.

It is proposed to divide the task into several iterations. Here, on the first step, the known (easily measured) parameters for creation of analytical dependence are used. For the following iterations, it is proposed to specify analytical dependency obtained at the first step using similarity theory [4], stochastic approach for approximation unknown (unavailable to measure) parameters and accumulated statistics on the real object operation. The coefficients are proposed to be used as the specifiers for analytical dependency.

Railway cut's (group of cars) velocity is changed during rolling down due to the transition from potential energy to kinetic one taking into account the rolling resistance. The transition is calculated as follows:

$$m\frac{(V^2 - V_0^2)}{2} = mgh, \tag{2}$$

where m is the cut's weight, V_0 and V are the input and output velocities at a track, respectively, $g = 9.6$ is the acceleration of gravity including rotational inertia and h is the height change between start and end point of a track.

Height change can be computed via length and incline of track taking into account specific rolling resistance and specific resistance of environment:

$$h = l(i - w_0 + w_{av}), \tag{3}$$

where l is the length of explored track, i is the incline of longitudinal profile of track, w_0 is the specific rolling resistance, w_{av} is the specific resistance of environment. It should be noted, that i, w and w_{av} are used in thousandths. Environmental resistance sign show whether the environment give additional resistance (headwind, snowed rails, etc.) or help to cut's motion (tailwind, high humidity, etc.) compared to standard conditions. Standard conditions mean the absence of wind load and precipitation, temperature is about 20° centigrade.

Thus, simplified function of velocity $f_p(V)$ for a rolling cut constructed on the first step by the known parameters in a general form is the following

$$f_p(V) = \sqrt{V_0^2 + 2gl(i - w_0 + w_{av})}. \quad (4)$$

In Eq. (4), coefficients of w_O and w_{av} represent coefficients, which depend on a number of poor formalized factors and demand specification during iterations on specification of models of the studied processes. They, in fact, can be used as criteria of similarity [5].

It is obvious that w_o depends on weight load on an axis of the rolling stock and characteristics of bushed sheaves. It is not difficult to define analytical dependency on w_O, because this information can be found in the documents of rolling stock.

w_{av} depends on the current combination of weather and climatic factors (speed and the direction of wind, air temperature, humidity, quantity and a type of rainfall, etc.). Problems, solutions and calculations of weather and climatic factors affects are presented in [6].

Therefore, it is necessary to compute the meteorological affect for statistical data if the data available from the meteorological station located on a real object. The statistical observations of weather factors affecting on the velocity of the rolling stock are presented in the form of Table 1 as an abstract example.

Table 1. Statistical data of the current weather conditions and velocity of the rolling-down cars collected from Vkhodnaya Station

j	1	2	...	m
x_1	1	5	...	1
x_2	2	4	...	3
x_3	6	3	...	2
...	...	...	...	...
x_n	4	2	...	5
V	1	3	...	1

In Table 1, x_1 is the direction and force of wind, x_2 is temperature of the environment, x_3 is the character and size of rainfall, etc., V is the velocity at the end of the track (initial velocity is identical); j = 1, 2. m are the observation numbers.

Two alternative approaches to modeling for the studied process are proposed:

1. According to Table 1, the regression dependence is constructed

$$V = f(x_1, x_2, \ldots, x_n) \quad (5)$$

2. The feature space $X = \{x_k, k = 1, 2, \ldots, x_n)$ is divided into homogeneous values areas. For each of them, the coefficient of w_{av} is assigned.

The conditions for the first one at construction of (5) are the following [7]:

- x_k should be set without errors;
- x_k should be independent from other factors;
- the observed values of output variable V should be uncorrelated among themselves;

- correct form of the regression equation should be known in advance;
- errors of observations ε should be normally distributed with the mathematical expectation equal to zero and continuous dispersion;
- the volume of selection of observations should be sufficient for receiving statistically valid conclusions.

It is obvious that these conditions are not satisfied for real automation objects of sorting processes and therefore this method has limited application.

The second approach leads to the problems of classification [8]:

1. Automatic classification allows to carry out the objective choice of homogeneous areas (for example, k-means and potential functions methods using non-stationarity of weather conditions and status of engineering devices).
2. Supervised classification demands the choice and development of methods of the decisive rule identification.

The performed analysis shows that adaptive methods of finding solutions by results of statistical observations are required for this purpose since processes of changes of weather conditions and state of engineering devices are reflected in statistics of real automation object work.

The special algorithm was developed for modeling of climatic factors affects rolling stock by statistical data [9].

4 Experimental Results

On the basis of accumulated experience of railway humping automation [10], specialists of JSC NIIAS perform the development of the digital twin of hump yard. Hump yard is the most complex subsystem of freight railway station from the point of automation and security views. At the present day, analytical dependencies are implemented in simplified form. The statistical accumulation and specifications are being provided.

The example of 3D visualization of the digital twin of the automated railway hump yard of Kinel freight station is presented in Fig. 2.

Fig. 2. 3D visualization of the digital twin of the hump yard of Kinel station

It is planned to use obtained digital twin for the design of automated marshalling yards.

The main value of the digital twin is an exact compliance of dynamic processes and life cycle of the original hump [11].

The digital twin of an automation object of industrial control system at different stages of life cycle allows to:

- *Design stage*: check correctness of technical solutions on coordination and connection of the equipment reducing time for elimination of errors;
- *Implementation stage*: debug mounting errors on model before complex approbation on a real automation object;
- *Commissioning stage*: model required tests on the digital twin to increase safety in advance;
- *Operational stage*: specify models of devices and provide training of the operating staff in use, management and system maintenance at the digital twin of an object in parallel of its functioning with the maximum approach to reality and full safety. For example, a station is an automation object. Simulating the existing plan and a profile of its hump yard, train flows coming to the station (intensity of train approaches, laws of their distribution on length, weight), humping cut flows (their laws of distribution on length, weight) digital twin allows to determine, on the one hand, the optimum parameters of humping, which provide the parameters set of its safety. On the other hand, it is possible to calculate hump parameters (height, inclines) ensuring the required intensity and safety of humping.

Taking into account 3D visualization of technological processes the real possibility of distance learning, retraining, professional development and management of technological processes from as much as remote working places arises.

Besides, the digital twin allows to use technologies of augmented reality: to show the predicted production results of the station work (i.e. to reflect to service staff information about various horizons of scheduling), to reflect current and give the forecast of a state of infrastructure facilities on the station (i.e. to reflect residual resource of equipment and show time of the following maintenance). The digital twin working simultaneously with real process is augmented reality itself and expands possibilities of management at all levels.

After creation of digital twins of an object (hump yard or station in general), the following advantages are obtained:

- Digital dynamic image of an object, where each involved specialist can get acquainted via the screen of the monitor or by means of augmented reality;
- Optimization of number of staff servicing and managing an object;
- Modeling of the most optimum operation modes of an object depending on a specific situation and the selected criterion of optimization simply clicking computer mouse;
- A possibility of the maximum replication of the digital twin at start of a new object;
- A possibility of warning of unforeseen situations due to intelligent processing of Big Data reflecting the work of an object.

Development of the digital twin of railway freight station produces a number of problems, which are nonconventional for the creators of ICS [12]:

1. Processing considerable on volume, separated in the position, diverse in a form of representation, noisy and non-stationary data. The solution of this task seems in use of technologies of Big Data processing, which provides systematization, structuration and convenient interface of data usage.
2. The organization of convenient access to work with the digital twin and "friendliness" of the interface to users with different qualification and different areas of interest (to economists for assessment on models of operational and financial performance, to mathematicians for adaptation of models of the studied technological processes, to managers for assessment of administrative parameters of a system, etc.). The solution of this problem lies in development of the Internet of services (along with the Internet of knowledge, people, things).
3. Providing digital twin with technologies of augmented reality. For example:
   - during the operation of devices of automation, infrastructure facilities of railway transport, the forecast of their functioning until necessary technical maintenance and/or repair is displayed to user;
   - during the humping, boundary admissible parameters of control over this process are specified (humping speed on sections; brake efforts on brake positions; the mass of a cut for the different tracks etc.)
4. Interaction organization of objects and subjects of the different nature. So, at technical maintenance of the station equipment, electricians "communicate" among themselves, with the inspected devices (check data, receive signals about emergency operations, issue instructions for change of a functional state, etc.), with the documents regulating technical maintenance. The solution of problems of such interaction seems in use of the special digital platforms implementing Internet of things, Internet of services and theory of multi-agent systems.

5 Conclusion

As a result, implementation of digital twins improves economic indicators of work of an object, reduces Total Cost of Ownership, increases safety, provides interactivity of warning of fault situations.

Thus, digital twin becomes carrier of artificial intelligence during process of accumulation in its structure of elements, which are digital copies of infrastructure devices, statistics of their functioning in different seasons and with different loading.

Using the developed digital twin on the basis of JSC NIIAS it is going to carry out debugging of the developed software, simulation of different unusual situations and determination of boundary parameters of functioning of one or another infrastructure element.

References

1. Chertkov, A.A.: Industrial enterprise of the future. Industrial Production, In: Innovations and Nanotechnologies, Moscow, no. 2 (2018)
2. Kaewunruen, S., Ningfang, X.: Digital twin for sustainability evaluation of railway station buildings. Front. Built Environ. **4**, 77 (2018)
3. Rozenberg, I.N., Shabelnikov, A.N., Olgeyzer, I.A.: Platform of creation of digital twins of infrastructure objects. In: Railway Transport, no. 8, Moscow (2019)
4. Guhman, A.A.: Introduction to the theory of similarity. In: Mechanical engineering, Moscow (1985)
5. Sedov, L.I.: Methods of similarity and dimension in mechanics. 7 prod, Moscow (1972)
6. Olgeyzer, I.A.: Development of means and methods of taking note of climatic conditions in management of sorting processes. In: The Thesis for a Degree of Candidate of Technical Sciences, Rostov-on-Don (2010)
7. Lyabah, N.N., Shabelnikov, A.N.: Technical cybernetics on railway transport. In: Textbook, Rostov-on-Don (2002)
8. Lyabah, N.N., Pirogov, A.E.: Automation of technological processes on railway transport with application of methods of recognition. In: RIIGT: Textbook, Rostov-on-Don (1984)
9. Olgeyzer, I.A.: Methods of assessment and compensation of influence of weather conditions on dissolution of structures on a hump yard. In: The Collection of Works of Young Scientists. RGUPS, Rostov-on-Don (2008)
10. Shabelnikov, A.N., Olgeyzer, I.A., Rogov, S.A.: The modern switchyard. From mechanization to digitalization. In: Automatic Equipment, Communication and Informatics, no. 1, Moscow (2018)
11. Komrakov, A.V., Suhorukov, A.I.: The concept of the digital twins in management of life cycle of industrial objects, no. 3. Scientific idea, Moscow (2017)
12. Shabelnikov, A.N.: The complex system of automated management of sorting processes – the innovative project of Russian Railways: Monograph. RAS, Moscow, p. 242 (2016)
13. Rozenberg, I.N., Shabelnikov, A.N., Lyabah, N.N.: Control systems of sorting processes within ideology of the Digital railroad: Monograph. RAS, Moscow, p. 244 (2019)

Early Age Education on Artificial Intelligence: Methods and Tools

Feng Liu[1,2] and Pavel Kromer[1(✉)]

[1] Department of Computer Science, VŠB-Technical University of Ostrava, Ostrava, Czech Republic
{feng.liu.st,pavel.kromer}@vsb.cz

[2] Department of Information and Engineering, Hebei GEO University, Shijiazhuang, China

Abstract. The development of artificial intelligence has become one of the key drivers of modern society. An attention has to be paid to the methods and tools available for training and education of users of intelligent systems. The youngest generation is exposed to artificial intelligence from an early age and its education is therefore of utmost importance. With respect to the early childhood development, the concept of artificial intelligence should be introduced into education as soon as possible. It can greatly contribute to the development of children's creativity, collaboration, comprehension and other abilities. There are many tools that can be used to aid the early age education on artificial intelligence. In this article, a brief survey of tools for the education of children in different age groups on artificial intelligence is provided.

Keywords: Artificial intelligence · Early age education · Tools · Programming

1 Introduction

Nowadays, children are born into the digital information age. From the very start of their lives, children are exposed to various products with artificial intelligence. The typical ones include smart homes, early education robots, and there will be intelligent education when entering kindergarten. Artificial intelligence (AI) already affects the life, play, and study of the current generation of children and will have a huge impact on the future of the whole society. A recent UNESCO working paper predicts the impact of AI on education and points out that AI can promote individualization and improve learning results, improve the fairness and quality of education in developing countries, and make the learners more competitive in the future [1]. At present, there is already a big gap between the demand for AI talents and the available number of talents [29]. Looking towards the near future, jobs will largely be related to AI. In this context, AI and computer science literacy are becoming as important as the traditional (reading/writing) literacy. The education on AI should not be delayed but introduced to the curricula as early as possible in the education process. An early education on AI can empower children to understand and use intelligent devices. By learning the concept of appropriate framework content, preschoolers can interact with, e.g., smart toys and smart speakers at home and young children can use artificial intelligence to explore and

S. Kovalev et al. (Eds.): IITI 2019, AISC 1156, pp. 696–706, 2020.
https://doi.org/10.1007/978-3-030-50097-9_71

create. These activities will affect children's perception of AI and their attitude towards themselves as engineers [2].

There is an increasing amount of research showing the benefits of learning AI at an early age. AI, and especially robotics, can help young children in developing of a variety of cognitive skills, including the sense for numbers, language skills, and visual memory. It helps the children to better understand mathematical concepts such as math, size and shape. Robotic manipulation allows them to develop fine motor skills and improve the hand-eye coordination. It also encourages the children to participate in cooperation and teamwork. Robotics and computer programming courses also promote the *computational thinking*, i.e. the skills, habits, and methods used in computer science, of young children. The education on AI, the Internet, big data, and other relevant technologies will also become an important element of the training of the next generation of workers and staff in the factories of the future, smart and cognitive cities, smart buildings (e.g., hotels), various industries, etc.

At present, many countries have included computer and AI skills in compulsory courses in primary schools, such as Spain [3], the United Kingdom [4], Australia [5], and many other countries are ready to start with an early training of computer and AI skills, such as Sweden [6], Mexico [7], Japan [8]. In this work, a brief survey of existing tools that can aid in the early education on AI and computer skills is provided. The remainder of this paper is structured in the following way: the problem of early education on AI is summarized in Sect. 2. Several popular tools for early age education on AI and computer skills are presented in Sect. 3. Section 4 outlines the features and properties important for efficient transfer of AI and computer skills to children and Sect. 5 concludes the work.

2 Problem Statement

Nowadays, many tools that can help with the learning of AI for people who are not computer experts and do not have programming skills are appearing. In the near future, practical and user-friendly tools for AI applications will become more frequent and easier to use. The development in AI follows the usual path where end-users (e.g., data analysts, expert system clients, public authorities) are increasingly isolated from underlying the technological complexity of the tools they use. This is a direct analogy to the past and present development in other fields, for example, automotive transportation, in which the end-user (driver) does not need to have detailed knowledge about the working of the components such as the internal combustion engines, electrical powertrain, and, e.g., advanced driver-assistive systems. In contrast, they need to be properly trained to use the car as a method of transportation. However, the development in automotive also shows that the requirements on end-users' skills change in time and may be relaxed in the future thanks to more advanced technology, just like in the case of self-driving cars.

The rapid development of AI is the key driver of contemporary society and has the potential to radically change the technology, society, and the world in a short period of time. It is already used in many aspects of various industries globally. As a consequence, the inclusion of the education on the Internet, Big Data, and AI in the curricula on all levels of education is inevitable and will come hand in hand with the digitization of education. In the same way AI is perceived as a major strategic opportunity, an efficient, effective, and sustainable education on AI and computer skills is becoming a strategic agenda. Several studies have shown that an early age education on AI can help the children grasp the basic concepts of the technology and help them to adopt future new developments in the field more easily. The training on AI in early age can not only deliver the needed skills but also stimulate children's creativity, imagination, and collaboration. It can exercise children's ability to think logically and to solve problems, which can, in turn, cultivate children's self-confidence. At the same time, parents are realizing the importance of AI and computer skills and do invest in a quality education of the children in these areas. This is seen as a clear business opportunity and a variety of tools is being developed to aid in this process.

3 Early Age Education on AI

The early age education on AI has been the subject of several recent studies. In 2016, Burgsteiner et al. [9] conducted a trial run of teaching AI and basic computer science in a high school (grades 9–11, 9 students) for a total of seven weeks. The time of instruction was two hours per week. The results of the experiment showed that the teaching of AI in high school is very effective. The children became familiar with the concepts taught and benefited from the obtained knowledge in future study life. As a matter of fact, children of different ages have different characteristics. As long as the methods are appropriate, it is not a mistake to start with the learning of AI early. In the same year, Sullivan et al. [10] conducted an 8-week robot programming training for 60 children from kindergarten to second grade. The training took place once a week and lasted for about 1 h every time. The evaluation of the trial showed that the concept of robots can be well mastered, no matter what grade are the children at. The average programming score shows that the children's performance is good at all grades. The results suggest that kindergarten, first, and second grade students can adopt the skills relevant for AI and computer use. Needless to say, the pace and method of the training need to be adapted to the age and capabilities of the children, set with respect to the stage of their development.

The learning of AI skills is a gradual process. It ought to start from kindergarten and progress to the university level. There are different learning methods and sets of skills for different stages. For children aged 4 to 8 years, computer science knowledge should be decomposed in an interesting way, encourage them to participate and to explore computer science knowledge [11]. For example, discovery and inquiry learning, the storytelling technique, as well as the use of educational robots and smart productions are efficient.

The results of studies focused on the use of AI in early childhood education show that this approach is an effective way to provide AI skills. Children are interested in AI focused courses and consider the use of computers and programming useful and interesting. They are willing to learn for a long time and some that have special needs benefited from an early introduction of AI to the curricula [12–14].

Experiments have shown that, when the right methods and tools are used, 4-year-old children in kindergartens are interested in robotics, programming, and computational thinking, and have the ability to learn the associated knowledge [15]. The use of educational robots has been successful, too [16]. If humanoid robots are placed together with children before the age of 2, the children accept them and consider them companions rather than machines. It was shown that they can remain interested in interaction with the robot for more than 10 h and enjoy the interaction [17]. This is in line with the basic principle of constructivist education which is that learning occurs when the learner is actively involved in a process of knowledge construction [18]. The use of toys, such as LEGO, motivates the children for an active involvement throughout the learning process. The children are happy and receive skills useful for the development of logical thinking and problem-solving skills in different fields in the future [19–23].

4 Methods and Tools for AI and Computer Skills Training

The use of tangible tools makes children feel more comfortable and enjoyable during the programming training process [24, 25]. The absence of a computer makes it easier for the children to focus on the merit and obtain the skills [26, 27] and physical programming makes people more willing to accept this educational agenda [28]. An important point for early age education is the curiosity of the students. It motivates them for a long and in-depth study of the subject, including AI and computer skills. The topics from these fields ought to be presented to the children in an attractive and fun way and motivate them for self-learning through play and interactive behavior. Many tools for computer skills' education take advantage of this principle. Some of them are shown in Table 1.

Table 1. Popular programming tools for children.

Number	Name	Features	Suitable age
1	Lego Mindstorms	Block, robot	6+
2	App Inventor	Block, for android app	10+
3	Alice	3D animation, story	10+
4	Codea	Code	10+
5	Etoys	Free, simple	7+
6	Hopscotch	For ipad, block	7+
7	Kadable	For ipad, free, game	5+
8	Stencil	Complex, game, block	10+
9	Waterbear	Complex, code	10+
10	Robomind	Game & AI, pay	10+
11	Hackety hack	Code	10+
12	Tynker	Pay, can use to reality life	7+

(*continued*)

Table 1. (*continued*)

Number	Name	Features	Suitable age
13	Crunchzilla (code monster)	Code	6+
14	Khan academy	Drawing	6+
15	Kargobot	Ipad, game	7+
16	Robologic	Ipad, iphone, game	7+
17	Movetheturtle	Game, ipad	6+
18	Swift playgrounds	Apple	10+
19	Scratch JR	Scratch for baby, only tablet	4+
20	Lightbot	Game	5+
21	Osmo Coding	Virtual & reality	4+
22	Code spark academy with foos	Game, pay, only for USA area	4+
23	Code Monkey	Game, block	7+
24	Code combat	Game, block	10+
25	Daisy the dinosaur	Game, pad	4+
26	Code academy	Simply UI, for older kids	10+
27	Scratch	Block	7+
28	Mblock	Block & AI, based on scratch, support python	10+
29	Blockly	Block, based on scratch	7+
30	Codemao	Block	7+

As can be seen from Table 1, the programming tools for children below the age of 12 can be roughly divided into two categories. The first one allows programming based on *visual building blocks* and the second one is based on the *extensions of stub games*. Both methods are attractive and make the children feel involved in the design and development of an original, novel product. On the other hand, middle and high schoolers older than 12 are more prone to efficient training through a proper programming language such as Python, C/C++, Java, Javascript, etc.

5 Features of Tools for AI and Computer Skills Training

The training of children in computer and programming skills mainly follows the three-stage system of *graphic-code-algorithm*. The first stage of *graphic learning* is very important and can have a large impact on the training procedure. If it is too difficult, it may cause the loss of interest in the learning for children who lack patience, but if it is too simple, it makes only a little contribution to the adoption of new skills in the future and the true value of the process is not high. Therefore, an efficient tool for training of children in AI, computer, and programming skills should be selected on the basis of the following features: attractiveness (fun), ease of use (simplicity), added value for future learning (transfer of knowledge and skills).

6 Attractiveness

Attractiveness is a key aspect of the training process that helps to keep the children focused on the activity. The concepts that contribute to attractiveness are: visual expression, animations, gameplay, visual coding, storytelling, competition, etc. For an attractive tool, it is easier to keep the children engaged in the learning through a play. Figure 1 shows a child using a physical module to control the on-screen animation in an interactive way. Different modules represent different actions and commands and the process of algorithmization and abstraction can be learned by the children through a combination of tangible toys and digital animations. The children then have a very concrete illustration of what a sequence of instruction is and how to combine them into more complex computer programs to solve problems.

Fig. 1. Osmo coding programming.

7 Ease of Use

The ease of use is another important feature of tools aiding in education on AI and computer skills. The simplicity of the tool ought to be adjusted to the age and social and cultural context of the children. Some children may not recognize the 26 letters of the Latin (English) alphabet but can easily see symbols and manipulate (drag and drop) visual building blocks. The commands and parameters that make up a program are then composed by the child by arranging the modules into a program edit bar. An example of building block programming is shown in Fig. 2. When the user of the Scratch application selects appropriate building blocks they can be combined in the code interface and the running result of the code can be seen in the visualization window, which is easy to operate and convenient to use.

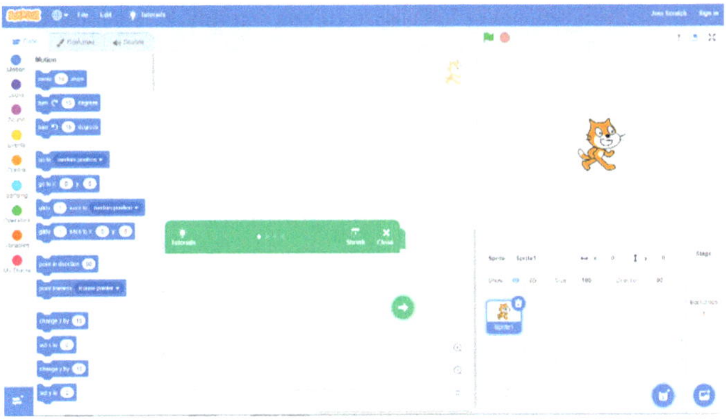

Fig. 2. Scratch interface

The user interface of another application, Lightbot, is shown in Fig. 3. The training process is similar to the process of writing a program but in the form of a game. Several basic instructions can be given to the robot, including those implementing the sequence, conditional branching, and loop control structures in programming languages. In order to accomplish the goal within a limited number of instructions, it is necessary to write functions and call the functions. Through the process of playing the game, the children are educated on analytical and *programming* thinking.

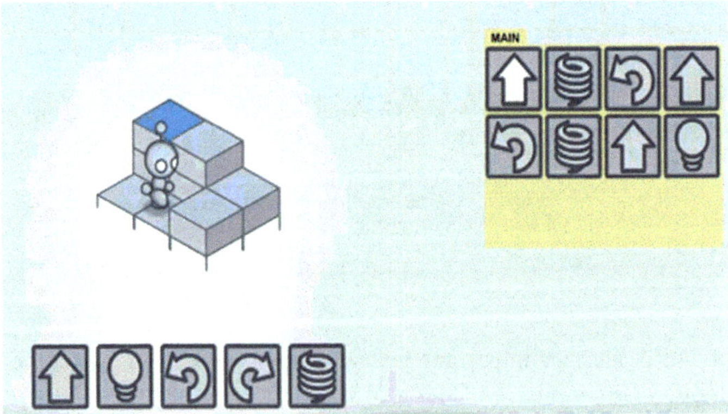

Fig. 3. Lightbot interface

8 Transfer of Skills

The main purpose of tools for the aid in AI and computer skills education at an early age is to provide a short-term transfer of elementary knowledge and skills and to facilitate a life-long habit of engagement with technology, curiosity, and continuous self-education. The tools ought to help with the adoption of analytical, programming thinking and understanding of high-level concepts such as sequences of instructions, loop and branching structures, as well as as many common programming structures and concepts as possible. They should help the children to see the problems in a structured way, suitable for algorithmization of the solutions and expression of the algorithms in a way understandable to computers. It is therefore apparent that children of different age groups are prone to efficient and useful training on AI and computer skills with the help of different tools.

In the course of this survey, the tools outlined in Table 1 were thoroughly analyzed from the attractiveness, ease of use, and skill transfer (added value) perspectives. Figure 4 shows their proposed classification with regard to the age groups for which they appear to be most beneficial.

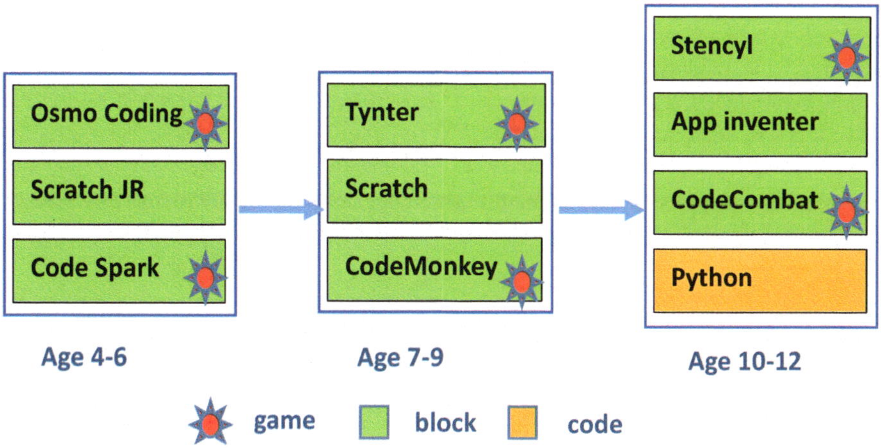

Fig. 4. Children's programming road

The first age group (Age 4–6) corresponds to the kindergarten level. This is an enlightenment period in which children learn through games, through the building blocks, to specify instructions (commands) for the computer. To serve this purpose efficiently, the tools aiding in this education should be simple and the functional modules (building blocks) should be as small as possible. This is well accomplished by, e.g., Osmo Coding, Scratch JR, and Code Spark. The second age group (Age 7–9) corresponds to the lower grades of elementary school. The children, already familiarized with the control of computer systems and the effects different instructions (commands) can have, can compose the instructions into more complex programs solving bigger problems. They can also try interaction with slightly more difficult programming tools. The suggested tools for this period include Scratch, Tynker, CodeMonkey, etc. Finally the last considered age group (Age 10–12) corresponds to the upper grades of elementary school. The children are at this stage ready for the use of more complex tools including simple programming software that allows coding in simple and accessible programming languages. Needless to say, it is important to let the children highlight their own interests and introduce the training in a lightweight and properly paced way. In this period, CodeCombat, App Inventor, Stencyl, and even Python can be recommended.

9 Conclusions

Kindergarten and elementary school studies represent important periods in the life of children. They have a strong curiosity and a desire to learn. They want to explore the world in which they live, including the real physical world and the artificial virtual world that is more and more present in our daily lives. Children need to learn the skills needed for their future life and profession in the technically highly advanced environments early and be prepared to adapt to changes their whole life. Properly designed and at the right moment used support tools can help with this goal significantly. It is not possible to train everyone into a computer expert and engineer. However, almost every person in the modern world would benefit from being an efficient user of modern technologies including AI and computers. This is even more underpinned by the fact that more and more professions include the production of digital content as their part. It can also help with an early identification of talents and proper, personalized education to maximize the satisfaction of children who can do what they desire to and benefits for the society which takes advantage of the good job done by satisfied and well-motivated people.

Acknowledgement. This work was supported by the projects SP2019/135 and SP2019/141 of the Student Grant System, VSB - Technical University of Ostrava.

References

1. Pedro, F., Subosa, M., Rivas, A., Valverde, P.: Artificial intelligence in education: challenges and opportunities for sustainable development. UNESCO Working Papers on Education Policy (07) (2019)
2. Williams, R., Park, H.W., Breazeal, C.: A is for artificial intelligence: the impact of artificial intelligence activities on young children's perceptions of robots. In: Proceedings of the 2019 CHI Conference on Human Factors in Computing Systems (2019). 447
3. Muñoz-Repiso, A.G.V., Gómez-Pablos, V.B., Garcá, C.L.: ICT in collaborative learning in the classrooms of primary and secondary education. Comunicar Revista Cientfica de Comunicación y Educación **21**(42), 65–74 (2014)
4. Brown, N.C., Sentance, S., Crick, T., Humphreys, S.: Restart: the resurgence of computer science in UK schools. ACM Transactions on Computing Education (TOCE) **14**(2), 9 (2014)
5. Falkner, K., Vivian, R., Falkner, N.: The Australian digital technologies curriculum: challenge and opportunity. In: Proceedings of the Sixteenth Australasian Computing Education Conference, vol. 148, pp. 3–12 (2014)
6. Otterborn, A., Schönborn, K., Hultén, M.: Surveying preschool teachers' use of digital tablets: general and technology education related findings. Int. J. Technol. Des. Educ. **29**, 717–737 (2019)

7. Ponce, P., Molina, A., Mata, O., Baltazar, G.: Lego EV3 platform for stem education in elementary school. In: Proceedings of the 2019 8th International Conference on Educational and Information Technology, pp. 177–184 (2019)
8. Ohashi, Y., Kumeno, F., Yamachi, H., Tsujimura, Y.: Readiness of Japanese elementary school teachers to begin computer-programming education. **12**, 807–810 (2018)
9. Burgsteiner, H., Kandlhofer, M., Steinbauer, G.: IRobot: teaching the basics of artificial intelligence in high schools. In: Thirtieth AAAI Conference on Artificial Intelligence, pp. 4126–4127 (2016)
10. Sullivan, A., Bers, M.U.: Robotics in the early childhood classroom: learning outcomes from an 8-week robotics curriculum in pre-kindergarten through second grade. Int. J. Technol. Des. Educ. **26**(1), 3–20 (2016)
11. Kandlhofer, M., Steinbauer, G., Hirschmugl-Gaisch, S., Huber, P.: Artificial intelligence and computer science in education: from kindergarten to university. In: 2016 IEEE Frontiers in Education Conference (FIE), pp. 1–9 (2016)
12. Sáez-López, J.M., Román-González, M., Vázquez-Cano, E.: Visual programming languages integrated across the curriculum in elementary school: a two year case study using â €œscratchâ€ in five schools. Comput. Educ. **97**, 129–141 (2016)
13. Prentzas, J.: Artificial intelligence methods in early childhood education. In: Artificial Intelligence, Evolutionary Computing and Metaheuristics, pp. 169–199. Springer (2013)
14. Jung, J.H., Bang, Y.S.: A study of the use of R-learning content in kindergartens. In: 2011 8th International Conference on Ubiquitous Robots and Ambient Intelligence (URAI), pp. 708–710. (2011)
15. Bers, M.U., Flannery, L., Kazakoff, E.R., Sullivan, A.: Computational thinking and tinkering: exploration of an early childhood robotics curriculum. Comput. Educ. **72**, 145–157 (2014)
16. Tanaka, F., Isshiki, K., Takahashi, F., Uekusa, M., Sei, R., Hayashi, K.: Pepper learns together with children: development of an educational application. In: 2015 IEEE-RAS 15th International Conference on Humanoid Robots (Humanoids), pp. 270–275 (2015)
17. Tanaka, F., Cicourel, A., Movellan, J.R.: Socialization between toddlers and robots at an early childhood education center. Proc. Natl. Acad. Sci. **104**(46), 17954–17958 (2007)
18. Fridin, M.: Storytelling by a kindergarten social assistive robot: a tool for constructive learning in preschool education. Comput. Educ. **70**, 53–64 (2014)
19. Shih, B.Y., Shih, C.H., Li, C.C., Chen, T.H., Chen, Y.H., Chen, C.Y.: Elementary school students acceptance of lego nxt: the technology acceptance model, a preliminary investigation. Int. J. Phys. Sci. **6**(22), 5054–5063 (2011)
20. Zaharija, G., Mladenovic, S., Boljat, I.: Íntroducing basic programming concepts to elementary school children. Procedia Soc. Behav. Sci. **106**, 1576 (2013)
21. Wang, E., Wang, R.: Using Legos and RoboLab (labview) with elementary school children. In: 31st Annual Frontiers in Education Conference. Impact on Engineering and Science Education. Conference Proceedings (Cat. No. 01CH37193), vol. 1, pp. T2E–T11 (2001)
22. Ohnishi, Y., Mori, S.: A practical report of programing experiment class for elementary school children. In: Proceedings of the 2014 International Conference on Advanced Mechatronic Systems, pp. 291–294 (2014)
23. Ramrez-Benavides, K., Guerrero, L.A.: MODEBOTS: environment for programming robots for children between the ages of 4 and 6. ÍEEE Revista Iberoamericana De Tecnologias Del Aprendizaje **10**(3), 152–159 (2015)
24. Noma, H., Sasamoto, H., Itoh, Y., Kitamura, Y., Kishino, F., Tetsutani, N.: Computer learning system for pre-school-age children based on a haptized model railway. In: 2003 First Conference on Creating, Connecting and Collaborating Through Computing. C5 2003. Proceedings, pp. 118–119 (2003)

25. Sapounidis, T., Demetriadis, S.N.: Exploring children preferences regarding tangible and graphical tools for introductory programming: evaluating the proteas kit. In: 2012 IEEE 12th International Conference on Advanced Learning Technologies, pp. 316–320 (2012)
26. Wyeth, P., Purchase, H.C.: Programming without a computer: a new interface for children under eight. In: Proceedings First Australasian User Interface Conference, AUIC 2000 (Cat. No. PR00515), pp. 141–148, January 2000
27. Meadthaisong, S., Meadthaisong, T.: Tangible programming for developmenting of computer scientist thinking skill new frameworks to smart farm in elementary school student. In: 2018 15th International Conference on Electrical Engineering/Electronics, Computer, Telecommunications and Information Technology (ECTI-CON), pp. 640–643 (2018)
28. Smith, A.C.: Rock garden programming: programming in the physical world. In: 2014 Fourth International Conference on Digital Information and Communication Technology and its Applications (DICTAP), pp. 430–434 (2014)
29. Tencent, 2017 Global Artificial Intelligence White paper. http://www.tisi.org/Public/Uploads/fie/20171201/20171201151555_24517.pdf. Accessed 01 Oct 2019

Author Index

S. Kovalev et al. (Eds.): IITI 2019, AISC 1156, pp. 707–709, 2020.
https://doi.org/10.1007/978-3-030-50097-9

Printed by Printforce, the Netherlands